ADVANCES IN ENERGY UTILIZATION TECHNOLOGY

ADVANCES IN

ENERGY UTILIZATION TECHNOLOGY

ADVANCES IN ENERGY UTILIZATION TECHNOLOGY

ISBN: 0-915586-62-2 *Hardcover*
ISBN: 0-915586-61-4 *Softcover*

Published by
THE FAIRMONT PRESS, INC.
P.O. Box 14227
Atlanta, Georgia 30324

PREFACE — CONFERENCE SPONSOR'S STATEMENT

The papers which are compiled herein are based upon presentations made at the 4th World Energy Engineering Congress, October 12-15, 1981. The Congress is sponsored by the Association of Energy Engineers and supported by the United States Department of Energy, Office of Industrial Programs.

The sharing of information is essential to the continued growth of the energy profession. AEE is proud to have played a major role in sponsoring this important conference.

Albert Thumann, PE
Executive Director
Association of Energy Engineers

CONTENTS

CONTRIBUTORS

Adams, John, Jr., *Georgia Institute of Technology*
Armwood, Levin G., *The Mogul Corporation*
Ball, Joe Dale, P.E., *The University of Texas*
Bass, Robert W., Ph.D., *United Technologies Research Center*
Baumgartel, John, P.E., *Energy Management Services, Inc.*
Bona, Mario, *Tecnotessile Research Center*
Bonne, Ulrich, Sr., Ph.D., *Honeywell Corporate Technology Center*
Brewer, David C., *Armco, Inc.*
Brown, Michael L., P.E. *Georgia Institute of Technology*
Brown, Thomas, S., *Engelhard Minerals & Chemicals Corporation*
Bullock, Cary G., P.E., *Xenergy, Inc.*
Carnevale, Ennio, *University of Calabria*
Cox, Norman D., *Jefco Laboratories, Inc.*
Crane, Richard, *Synergic Resources Corp.*
Crise, Daniel J., AIA, *Daniel International Corporation*
Crow, Stuart J., *U.S. Army Materiel Development and Readiness Command*
Cummins, Thomas P., *Energy Management Services, Inc.*
Cutler, Robert R., P.E., *Synergo, Inc.*
Darby, Douglas F., *Deere & Company*
Darcy, Jon, *Thermo Engineering, Inc.*
Desmon, Leland G., P.E., *Peabody Gordon-Piatt, Inc.*
DiPaolo, Donna M., *Kaye, Scholer, Fierman, Hays & Handler*
Dubey, Michael B., P.E., *D.B.S. Pacific, Inc.*
Evans, Deane, *Steven Winter Associates, Inc.*
Feledy, Charles F., *United Technologies Corporation*
Foster, John W., *HDR Energy Corporation*
Fowler, Delbert M., *Fowler/Blum Energy Consultants, Inc.*
Gardstein, Cynthia, *Steven Winter Associates, Inc.*
Gengler, Mark S., *Water Services of America, Inc.*
Gilbert, Joel S., P.E., *Mechanical Technology Incorporated*
Gossett, J. Stephen, *Delta Energetics Corporation*
Greene, Deni, *California — Office of Planning and Research*
Gupta, Vijay K., Ph.D., *Central State University*
Gustinis, John P., *The Bickle Group*
Hall, David C., *Mechanical Technology, Inc.*
Hansen, Wayne, P.E., *Giampaolo & Associates, Inc.*
Hardison, L. C., *Air Resources, Inc.*
Hassett, James E., P.E., *A.C. Manufacturing Company*
Hill, Douglas, Eng. Sc.D., P.E., *Brookhaven National Laboratory*
Hill, Gerald R., Ph.D., *Southern States Energy Board*
Hirst, Eric, Ph.D., *Oak Ridge National Laboratory*
Howe, S. O., Ph.D., *Brookhaven National Laboratory*
Johnson, Charles H., *Hercules Incorporated*
Landerman, Abraham M., *United Technologies Research Center*
LeBlanc, Richard J., P.E., *MCC Powers*
Leibowitz, L. P., *California Institute of Technology*
Leigh, R. W., Ph.D., *Brookhaven National Laboratory*
Levy, Emanuel, *Steven Winter Associates, Inc.*
Liu, T. M., Ph.D., *California Institute of Technology*
Lonsdale, Joseph T., *Raychem Corporation*
MacCracken, Calvin D., *Calmac Manufacturing Company*
Macdonald, Copthorne, P.E., *Enersave for Industry & Commerce*
MacLeod, J. Lee, P.E., *Enersave for Industry & Commerce*
McKay, Michael E., P.E., *Princeton University*
McKibben, Francis, R.S., *The Vitetta Group*
Mehta, D. Paul, Ph.D., *Bradley University*
Mesko, John E., *Pope, Evans, and Robbins Inc.*
Michalek, Romuald, P.E., *Engelhard Minerals & Chemicals Corporation*
Molloy, Robert C., *Dynatron Inc.*
Mozzo, Martin A., Jr., P.E., *American Standard Inc.*
Mueller, James H., *REECO*
Mytels, Michael, P.E., *Stanford University*
Nagl, G. J., *Air Resources, Inc.*
Nelson, John S., P.E., *Affiliated Engineers, Inc.*
Newman, H. Michael, *Cornell University*
Niess, Richard C., *Westinghouse Electric Corporation*
O'Connor, Paul J., *Nutmeg Chemical Company*
Omurtag, Yildirim, Ph.D., P.E., *University of Missouri-Rolla*
Paganelli, Marcello, *Tecnotessile Research Center*
Patani, Amin, Ph.D., *Honeywell Corporate Technology Center*
Patterson, G. V., *Fanjet Solar Roof Cooling Company*
Pham, H. Q., *California Institute of Technology*
Pierce, Barbara L., *Brookhaven National Laboratory*
Pincelli, Richard D., *Energy Systems, Inc.*
Pittas, J. J., *Air Resources, Inc.*
Polisner, Lee, *Omnia Engineering, Inc.*
Qureshi, A. Salim, P.E., *Michael Baker, Jr. of New York, Inc.*
Rabiah, Talal M., *University of Michigan*
Reaves, Frank M., *R.M. Thornton, Inc.*
Reaves, John, *R.M. Thornton, Inc.*
Reiter, Sydney H., P.E., *Englehard Minerals & Chemicals Corporation*
Roberts, Michael M., *Tempmaster Corporation*
Root, T. E., *Detroit Edison Company*
Rossi, Sharon M., *Vitro Laboratories*
Ryan, William A., *Affiliated Engineers, Inc.*
Sallman, Charles L., P.E., *Jefco Laboratories, Inc.*
Scheuch, Karl E., *The Bickle Group*
Schlom, Leslie A., *DBS Pacific, Inc.*
Seale, Robert L., Ph.D., P.E., *University of Arizona*
Simon, Gary D., *Energy Business Consultant*
Smith, Donald O., P.E., *Viron Corporation*
Smith, Gregory C., *Wolverine Aluminium Corporation*
Smith, Hugh W., *Princeton University*
Smith, Walter P., Jr., *Badische Corp.*
Snell, L. Robert, *Cornell University*
Sobczak, Nancy L., *Johnson Controls, Inc.*
Spollett, Fred, *H. P. Hood Inc.*
Stecco, Sergio S., *University of Florence*
Steigelmann, William, P.E., *Synergic Resources Corporation*
Thomas, Malcolm H., P.E., *Kraft, Inc.*
Van Schelt, Jules E., Jr., *United Technologies Corporation*
Wieczorek, Catherine A., *JRB Associates, Inc.*
Zeigler, E., *Hercules, Incorporated*

ACKNOWLEDGMENTS

Appreciation is expressed to all those who have contributed their expertise to this volume, to the conference chairmen for their outstanding contribution to the 4th World Energy Engineering Congress, and to the officers of the Association of Energy Engineers for their help in bringing about this important conference.

INTRODUCTION

The energy field is in a state of transition. The quick fix approaches of the 1970's and the government subsidies of energy utilization expenditures are rapidly dimishing. The energy engineer and manager is now faced with the challenge of applying the latest technology for facilities improvement which can be justified on its own economic merits. The energy engineer and manager must broaden his or her background to take into account a host of new parameters.

This text represents the latest thinking with regard to the application of energy engineering and management technology. It covers a host of specific technical subjects including: low and medium temperature heat recovery, fluidized bed combustion and combustion efficiency. Sections deal with energy management principles and the role of the energy manager as well.

It is hoped that this comprehensive collection of state of the art applications will serve as a valuable tool for the dissemination of information to the professional community. The energy experts who have contributed to this important work share one goal in common: to apply the latest technology to reduce energy consumption in a cost-effective manner.

SECTION 1
LOW AND MEDIUM TEMPERATURE HEAT RECOVERY

Chapter 1

AN OVERVIEW OF LOW-MEDIUM TEMPERATURE HEAT RECOVERY

S. H. Reiter

1. INTRODUCTION

Heat recovery systems are an important part of energy conservation. They recover heat being exhausted and convert it into a usable form, reducing the amount of purchased energy and the energy cost. Reduction of amount of purchased energy results in improved ability to deal with fuel shortages. Energy cost reduction provides both immediate and long term benefits. Prices of energy supplies are increasing every year with no abatement in sight. As an example, consider a heat recovery system saving $10,000 in energy costs during its first year of installation. If the cost of energy increases 15% annually, at the tenth year it is saving $35,178 annually. Over a ten year life, the system would save $203,037. As a safeguard against inflation and rising energy costs, heat recovery is equivalent to having an oil well on your property.

This paper is presented as a brief and concise introduction to heat recovery systems. The steps necessary to design the system, calculate its performance, and make a financial evaluation are included. Descriptions of the major types of heat recovery equipment and several system examples are provided. The reader will be able to follow the steps outlined in the paper when investigating heat recovery for his plant, building orfacility.

2. PRINCIPLES OF HEAT RECOVERY

Heat Recovery refers to the capture of waste heat, its conversion into a useful form of heat, and the re-use of the heat in a process or building. The amount of waste heat available for recovery is determined by the temperature and flow of the heat source. The amount of heat converted into a useful form is determined by the design of the heat recovery equipment. Finally, the amount of heat re-used is fixed by the capacity of the use to absorb the recovered heat. Figure 1 illustrates the basic elements of the heat recovery system.

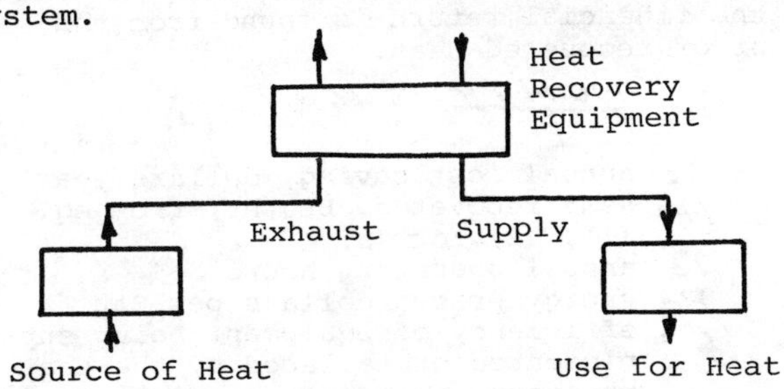

FIGURE 1

The sources of recoverable heat are heated exhausts from the facility. A partial list of these is shown in Figure 2.

Sources of Recoverable Heat

Air

Roof ventilation exhaust
Laboratory exhaust
Oven/dryer exhaust
Transformer vault exhaust
Dust collector exhaust
Kitchen hood exhaust
Cooling air exhaust
Boiler room exhaust

Combustion Products

Boilers	Air Heaters
Bakery Ovens	Engines/Turbines
Furnaces	Space Heaters
Textile Dryers	Kilns
Ovens	Paper Dryer
Incinerators	Roasters

Hot Water

Textile Dye Works	Boiler Blow Down
Compressor Coolant	Paper Mills
Boiler Condensate	Refrigeration Compressor
Food Preparation	Process Cooling Water
Cooling Towers	

FIGURE 2

The uses for recovered heat include any consumer of heat in the facility. A partial list of these is shown in Figure 3.

Uses of Recoverable Heat

Building heat
Pre-heating dryer air
Pre-heating combustion air
Pre-heating heat transfer fluid
Heating boiler feedwater
Heating chemical feed stock
Heating laundry water
Absorption air conditioning

FIGURE 3

2.1 Heat Recovery Survey

The identification of sources and uses for recovered heat is carried out in a preliminary survey of the building, plant, or facility. The survey covers all the sources and uses for heat in the area covered. The information should consist of the information shown in Figure 4.

Heat Recovery Survey Information

Flow	Composition
Temperature	Contaminants
Operating Schedule	Operation
	Location

FIGURE 4

The flow can be measured, estimated or taken from equipment specifications. Air and gas flow measurements are made by a traverse of a stack or duct. The traverse is made with a velocity-measuring instrument, such as a Pitot-tube or Velometer. The average velocity is multiplied by the flow area to find the actual volume flow. The actual flow is reduced to standard conditions of 70°F and one atmosphere for use in the heat calculations of the next section. Further information may be found in reference 1. Liquid volume flow is measured by a flow meter, displacement from a tank, or using a catch container and a timer. Temperature is measured using a thermometer or pyrometer. Thermometers take several minutes to reach a steady reading. Cycling processes should be measured at high and low points for both flow and temperature and an average estimated. If measurements cannot be made, process conditions, fuel consumption, fan rating, or pressure drop will help estimate flow and temperatures.

Operating schedule is important for two reasons. The economic benefit from a heat recovery system depends on the annual hours of operation. Daily, weekly and monthly schedules are necessary to calculate hours both at operating condition or at a part load condition. The source of recoverable heat and the use of recoverable heat must have similar operating schedules for them to be joined in a sucessful heat recovery system. Otherwise, system annual hours of operation are reduced by the hours of mis-matching.

Composition information is required for heat recovery and system design calculations. Data on specific heat, specific gravity and viscosity are required.

Contamination is a very important part of the survey. Heat exchangers can foul, plug or be destroyed by contaminants. Particular care should be taken to note the presence of: dust or particulates, moisture, organic solvents, paint carryover, combustion products, oil and grease, or any other material carried by the source. The condition of the stack, duct, or piping will provide a valuable insight into problem areas likely to occur inside heat recovery equipment. This information should be part of any performance evaulation by both the surveyor and the heat recovery equipment manufacturer.

Operation refers to transient or unsteady operating conditions. Burners operating on high and low fire, furnaces on cyclic schedules, dryers starting at moist, low temperatures and ending their cycles at dry, high temperatures are examples of areas to evaluate for contamination, extreme temperature and flow, and average operating conditions.

Lastly, location refers to the physical parameters of the installation. Available space for the installation of equipment, size and location of exhausts, drains, inlets, obstructions, supporting structure and access for cranes or construction equipment are all important. Particular attention should be paid to sources and uses that are reasonably close to-gether or particularly accessible. The closer the source and use, the less complicated and less costly will be the installation.

2.2 Heat Recovery Calculations

Following the survey, the amount of heat recovered from the source or delivered to the use is calculated from the basic heat equation:

$$H = V d C_p \Delta T \qquad (1)$$

Where: H = heat recovered, Btu/hr
V = volume flow, standard conditions, units of measurement
d = density, lbs/unit of measurement
C_p = specific heat, Btu/lb°F
ΔT = temperature change of gas or liquid during heat recovery, °F

FIGURE 5

All of the data, except ΔT, is available from the survey. The temperature change, ΔT, must be obtained from heat recovery equipment manufacturers; their literature, or from process requirements. A conservative estimate is to split the temperature difference between the source and use for ΔT.

If standard cfm is used for air, equation (1) becomes:

$$H = 1.085 V_{STD} \Delta T \qquad (2)$$

Where: H = heat recovered, Btu/hr
V_{STD} = volume flow in standard cfm
ΔT = temperature change during heat recovery, °F

FIGURE 6

Equation (1) for water becomes:

$$H = 500 V \Delta T \qquad (3)$$

Where: $H, \Delta T$ = same as (1), (2)
V = volume flow in gpm

FIGURE 7

2.3 Financial Calculations

Now the financial return is found from the amount of recovered heat:

$$C = \frac{100 HTP}{E} \qquad (4)$$

Where: C = annual cost saving, dollars/year
H = heat recovered, Btu/hr, from eq. (1), (2), or (3)
T = annual operating hours
P = energy price, dollars per Btu
E = efficiency of equipment being supplemented or replaced by the heat recovery equipment, percent

FIGURE 8

The energy price can be found from the fuel price divided by the heat content of the fuel. The efficiency in the denominator credits the heat recovery system for the wasted fuel or energy of the process it supplements or replaces.

The evaluation of the heat recovery system is carried out by calculating certain financial ratios. The simplest ratio is the payback:

$$PB = \frac{I}{C} \qquad (5)$$

Where: PB = payback time, years
I = investment, dollars
C = annual cost saving, dollars/year, from (4)

FIGURE 9

The investment, I, is the total heat recovery system cost, including equipment, installation, contingency and taxes. It is obtained from manufacturers' prices, contractors' bids and estimates by consultants or plant engineers. No tax benefits or income taxes are included. Values of one to four years span the usual acceptable range. If payback is calculated including investment tax credit, then the cost saving must be reduced by the applicable income taxes. The proper method for after-tax financial evaluation is to make a cash flow calculation, then calculate Internal-Rate-of Return or Net Present Value. Return-on-Investment is usually the inverse of Payback. More information can be found in reference 2.

3. HEAT RECOVERY EQUIPMENT

Heat recovery equipment is available in four basic forms. Each form has a number of variations, as shown below:

Heat Recovery Equipment

Gas/Air Heat Exchangers

Counter-flow thin plate	Heat pipe
Cross-flow thin plane	Heat wheel
Shell-tube	Packed bed

Gas/Liquid Heat Exchangers

Finned tube

Liquid/Water Heat Exchangers

Shell-tube
Plate

Waste Heat Boilers

Fire Tube
Water Tube

FIGURE 10

The following sections briefly describe the above equipment.

3.1 Gas/Air Heat Exchangers

Counter-flow and cross-flow types use their plates to separate hot and cold flows. Heat is transferred across the plates. Plate separation varies with design. Some units use fins between the plates to aid heat transfer. In the counter-flow type, hot and cold flows move at 180° to each other, or in opposite directions. In the cross-flow unit, the flows are 90°, or at right angles to each other. The shell-tube exchanger contains numerous tubes sealed into end plates and surrounded by a casing. One flow passes through the tubes and the other through the casing around the tubes. A heat pipe is a tube with fins on the outside, filled with a volatile liquid and a wicking. Hot gas passing over a portion of the tube vaporizes the liquid. The vapor moves to the cold end, condenses, and gives up heat to the cold gas. Condensate is moved by the wicking back to the hot end, repeating the process. The heat wheel is a slowly revolving porous metal wheel. As the wheel rotates, it moves between hot and cold streams. The hot stream heats the porous metal wheel packing and is cooled. The opposite process occurs in the cold stream. The choice between the various types of gas/air heat exchangers will depend on the contaminants present, operating temperature and pressure, flow, performance, cost and size. In heat recovery system design the various types are explored, a choice made, and the manufacturers' ΔT used in equations (2), (3) and (4) to calculate the energy and cost savings.

3.2 Gas/Liquid Heat Exchangers

Most gas/liquid heat exchangers are the finned tube type. Tubes with external fins span the gas flow area, often several rows deep in the gas flow direction. The tube ends are manifolded to-gether to form a continuous liquid path. The path may be arranged to provide parallel or series traverses across the gas flow. Heat is transferred by the fins and tube walls between the gas and liquid. Manufacturers are able to provide a wide latitude of ΔT, flow volume, and pressure drop for both gas and liquids depending on the liquid flow path and the number of rows of finned tubes in the gas flow direction. Cost will determine the most suitable design. The manufacturers recommended ΔT's for gas and liquid are used in equations (2) and (3).

3.3 Liquid/Liquid Heat Exchangers

The shell-tube heat exchanger has traditionally been the most popular liquid/liquid heat exchanger. It consists of many small tubes fastened between tube sheets. A surrounding jacket contains plate baffles. One flow passes through the tubes. The other flow passes over the tubes inside the jacket. The baffles provide good distribution of the jacket flow around the tubes. Baffles in the end chambers allow several passes through the tubes. The plate type consists of many thin plates clamped to-gether in a frame. The plates are corrugated to provide well-defined flow passages. Gasketed feed and drain connections separate the flows on either side of the plates. Both types are in common use. Their cost and performance is determined from manufacturers' data and ΔT used for equation

(3) to calculate the amount of heat recovered.

3.4 Waste Heat Boilers

Waste heat boilers are normally used at high temperatures, but are sufficiently important to be included in this section. They transfer heat from hot gas to water, making steam for heating and process use. The gas is cooled at the same time. There are both fire tube and water tube waste heat boilers. The hot gas passes through the tubes of the fire-tube boiler, which is a large shell-tube exchanger, with baffles in the end chamber to allow for multi-pass operation. The hot gas passes through a chamber, and the water passes through tubes in the chamber of a water tube boiler. Various circuiting arrangements are possible to produce the desired flows and temperatures. The choice between waste tube and fire tube will depend on size, contaminants and steam pressure versus cost. As in the other heat recovery equipment, several proposals will identify the most suitable choice.

4. HEAT RECOVERY SYSTEMS

The preceding sections have outlined the steps leading up to the design of a heat recovery system. The temperatures, flows, heat recovery equipment, structural limitations and special material requirements have been determined. Design of the actual heat recovery system is like any other engineering project or installation. Very often, the heat recovery equipment manufacturer will help layout the system and its components. Two examples of heat recovery systems follow, to illustrate typical arrangements.

4.1 Hospital Ventilation

Hospitals provide a unique opportunity for heat recovery. They operate continuously with no night, weekend, or holiday shut-down. Fresh air requirements are very high, since in many cases, no re-circulation is allowed. A typical system schematic is shown in Figure 11 for winter conditions.

VENTILATION HEAT RECOVERY

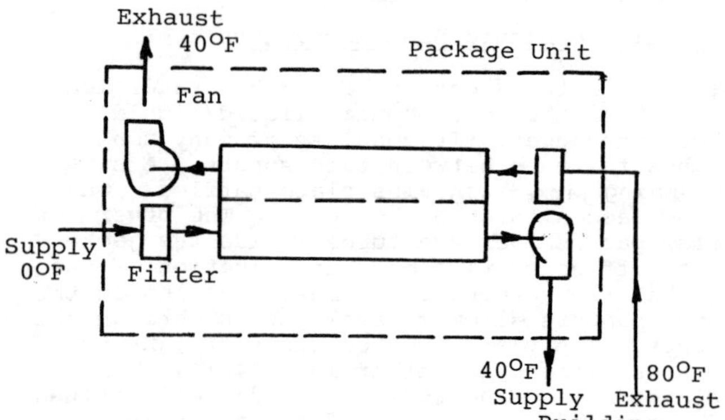

FIGURE 11

The air/air heat exchanger is supplied as part of a packaged unit. The unit contains fans, filters, control dampers, and heat exchangers. It may also contain heating and cooling coils. The use of a packaged unit requires that the building supply and exhaust connections are relatively close to each other. The same type of installation is used on many commercial buildings. It makes an attractive proposition because all components are self-contained, easily installed, with little or no system design. Where the supply and exhaust connections are far apart, two air handlers are connected with pipes. A glycol-water solution in the loop transfers heat between the two air handlers.

4.2 Plastics Manufacture

Continuous ovens are used in the manufacture of coated products to evaporate solvents or cure the coating. Large amounts of heated air flow along with the coated material, supplying heat and carrying off solvent or organic vapors. The heated air must be exhausted to avoid any build-up of explosive vapors in the oven. Because of pollution laws, the exhaust air is passed through an incinerator where the vapors are burned. Very often, almost enough fuel value is contained in the exhaust to supply the incineration requirements. The additional heat can be supplied by an air/air heat exchanger in the incinerator exhaust. A schematic of the installation is shown in Figure 12. A second heat exchanger pre-heats the oven supply air, using the first heat exchanger's exhaust.

PLASTICS HEAT RECOVERY

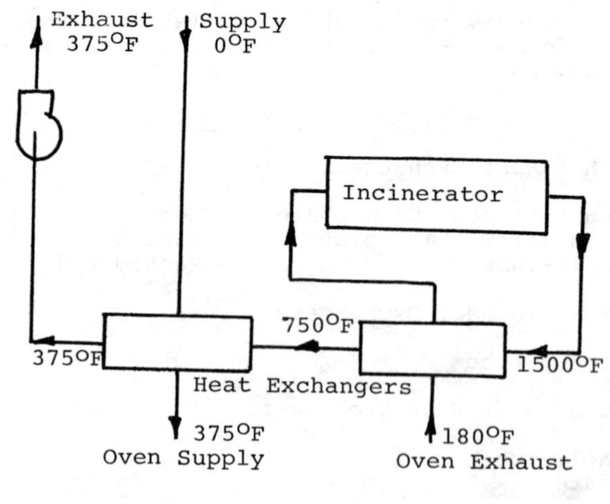

FIGURE 12

4.3 Closure

This paper has presented heat recovery principles and system design in a very brief and condensed form. The reader is encouraged to use the information as a basis for further work at his own building, plant or facility.

References:

1. Industrial Ventilation
Committee on Industrial Ventilation
P.O. Box 453, Lansing, MI 48902, 1972

2. Brown, R.J. and Yanuck, R.R.
Life Cycle Costing
Fairmont Press, Atlanta, GA., 1980

Chapter 2
HEAT RECOVERY FROM BOILER ECONOMIZERS AND WASTE INCINERATORS

M. H. Thomas

The Kraft Plant in Decatur, Georgia is a manufacturing and distribution center for Kraft products for the Southeastern United States. We produce processed cheese, margarine, salad dressings and BBQ Sauce as well as distribute Kraft products. The facility is 750,000 square feet under roof.

Our central power plant includes one 40,000#/hour Murray Boiler and two 20,000#/hour Wickes Boilers all tied to a common steam header. This steam is used for both building heat and process heat. Fuel is Natural Gas on an interruptable rate with #2 fuel oil as a back-up fuel. Because of our intensive energy program, we have eliminated the Murray Boiler from service and we rely on one or two Wickes Boilers for steam.

HEAT RECOVERY FROM BOILER ECONOMIZERS

The first project is the installation of waste heat economizers on both Wickes Boilers. In 1978, we initiated a project to install one economizer on the larger Murray Boiler. Discussions with a vendor led to a budget price. Between the time of this activity and the actual letting of contracts, we had done some extensive work on increasing efficiency of our small boilers and steam distribution system and could now get by with running only the two small boilers even in the winter.

We recalculated our money requirements based on economizers for two small boilers and scaled down the justification from original information on the large economizer. We were anticipating a two year payout. Bids were let for the economizers. The contractor who had done much of our boiler repair work was selected. The unit purchased has a rated performance at a set steam flow. Our steam flow is erratic. Our steam flow for processing is used to pasteurize or cook products. Some of our operations are batch operations with short cook cycles. Some operations are continuous cooking but shut themselves off when a hold tank is full and may run four or five cycles in a shift. Some steam is used to make hot water. This load varies during the day and is very large during sanitation clean-up. Our steam flow for space heating is used only in the winter and varies as to time of day. Taking this fluctuating steam flow into account, we were still estimating a 2½ year payback.

The economizer is a helical wound finned tube gas to liquid heat exchanger. This unit transfers heat from the hot exhaust gas to the boiler feedwater. We discussed installing an automatic bypass system to prevent overcooling of the exhaust stack below the dew point of sulphurous acid and sulphuric acid. We use natural gas almost all of the time. Natural gas has such a low amount of sulphur it does not present a problem. We decided the much simpler three valve by-pass would suffice.

The economizers were too tall to install in the building and we had to design a support structure to support the economizers from the roof. We took the load to main beams of the building. The unit came with a soot blower and safety relief valve on the water line. We insulated the economizer,piping, and the exhaust stack to the economizer. We also have provisions to drain the economizer in the event we do not use the boiler during the winter time; this prevents freezing.

We were particularly interested in determining the savings. Data was taken periodically as the boiler feedwater was flowing into the boiler. We have a 40° F rise in boiler feedwater. This represents a three year simple payback which is longer than the anticipated two and one-half year payback but is still acceptable.

TABLE I

CALCULATIONS FOR BOILER ECONOMIZERS

	PRESS. PSI	TEMP. °F	ENTHAL. BTU/#
FEEDWATER	5	228	169.19
FEEDWATER AFTER ECONOMIZER		268	236.79
STEAM	125	353	1193.00

ECONOMIZER ADDS
236.79 - 196.19 = 40.6 BTU/#

TOTAL HEAT ADDED
1193.0 - 196.19 = 996.81 BTU/#

% ADDED BY ECONOMIZER
40.6/996.81 × 100 = 4 %

4% SAVINGS IN FUEL EQUALS
3 YEAR PAYBACK

The experience we have had with the heat recovery system has been very good. There has not been any maintenance work necessary to keep them running for the seven months they have been in operation. We run the soot blower about once every month to keep the soot from accumulating.

HEAT RECOVERY FROM WASTE INCINERATOR

The second project is an incinerator and waste heat boiler where wood, cardboard and plastic trash is burned, the heat converted into steam, and fed to the main steam header.

A study was made in January 1976 which indicated that the project was not justified. This was before the big rise in fuel prices. A second preliminary engineering study was made in December 1978. This study showed that a ROCE of 13.4% could be achieved after four years. The study recommended a program with three phases:

Phase I: Verify waste at Decatur and prepare specifications and solicit bids for incinerator. Recalculate the return of capital employed and submit a proposal for funds.

Phase II: Purchase and install an incinerator and waste heat boiler, then evaluate performance.

Phase III: Purchase and install incinerators at other locations.

Phase I was approved. A detailed study of the combustible materials generated by Decatur was made and included in the specifications for an incinerator and waste heat boiler. Five vendors were selected to bid. Three bids were analyzed to determine which was the best value for Kraft. The best bid projected a 26% ROCE after five years with a simple payback of three years and was accepted.

Phase II was approved. A local engineering firm was hired to design the foundation pit for the incinerator and the electrical and mechanical tie-ins to plant utilities. A local contractor was selected to install the incinerator and boiler.

The heat recovery system is basically two parts; a semi-automatic incinerator and waste heat boiler. The incinerator is semi-automatic as the material to be burned has to be put in the loading hopper by the operator when the unit indicates it is ready to be loaded. The incinerator automatically moves this material into the main burning chamber and controls the burners, carburetors and moving of ashes through the chamber to the ash cart. The primary combustion chamber has a minimum temperature of 650° F. The exhaust goes to the secondary chamber which has a temperature of 1800-2000° F for complete combustion of smoke and any odors. The exhaust gases then can be vented to the atmosphere or sent to the boiler.

The waste heat boiler is a fire tube boiler which converts the heat in the exhaust gas to steam. If low boiler feedwater, high steam pressure, or no exhaust fan occurs, the incinerator goes into a shutdown sequence and the hot gas is exhausted to the atmosphere without going to the boiler. The exhaust gas leaving the boiler is about 450° F.

We have been able to get 7500#/hour steam flow for only short periods of time. Results of four months of operation show we have averaged 4300#/hour. We were given an incinerator which would handle the quantity of waste we generate and the boiler is capable of generating 7500#/hour.

The steam that is generated in the waste boiler reduces our natural gas usage. Natural gas is used to start the primary burner and to keep the secondary burner at the appropriate temperature. Steam produced by the two conventional boilers uses 6.4 times as much natural gas as the incinerator/waste heat boiler. This project has done much to reduce the amount of natural gas usage at this plant.

TABLE II

CALCULATIONS FOR INCINERATOR AND WASTE HEAT BOILER

FT^3 GAS USED	1,321,500
# STEAM GENERATED	6,483,830
HRS. RUN	1,505
DAYS RUN	83

(BASED ON 17 WK. TOT'L)

AVG. #/HR.

$$\frac{6,483,830}{1,505} = 4,308 \text{ \#/HR.}$$

AVG. HOURS/WK.

$$\frac{1,505}{17} = 88.5 \text{ HRS./WK.}$$

AVG. DAYS/WK.

$$\frac{83}{17} = 4.88$$

RATIO OF COST OF STEAM (WICKES BOILER VS. WASTE HEAT BOILER) = 6.4

PAYBACK = 4 YRS.

We were expecting a three year payback. In reality, we are experiencing a four year payback. There are two reasons for this change. The installation has been slightly more expensive as we had to house the loading hopper to keep flies from coming into the building during loading. This, coupled with the decreased steam flow, results in a longer payback.

The original intent was to burn waste during two shifts. We currently are spreading this load over three shifts as we have a significant steam load on third shift heating water for sanitation. This has contributed to our reduction in steam flow.

We have had some bad experiences with the incinerator. We have had a build-up of clinker material in the bottom of the incinerator that clogs the ash pusher system. Ashes then cannot be moved through the incinerator. We seem to have more maintenance in general with this unit, however, we have had this equipment running for about 6 months and are still learning how to maintain it.

We will continue to monitor steam flow, gas usage, and other pertinent data for one year. We will recalculate a ROCE (Return On Capital Employed). This will complete Phase II.

Phase III will be implemented if the ROCE is acceptable.

Chapter 3

THE PRACTICAL APPLICATION OF HEAT RECOVERY CONCEPTS

J. S. Nelson, W. A. Ryan

INTRODUCTION

This paper describes an energy engineering project in which a business engaged and worked with a consultant on an energy conservation equipment retrofit project. Described herein are the detailed steps involved in successfully completing such a project from inception through startup, including the time required and the responsibilities of both the business and the consultant.

In this project the business was Western Products in Milwaukee, Wisconsin and the consultant was Affiliated Engineers, Inc. of Madison, Wisconsin. The project was an HVAC expansion featuring heat reclamation.

PROBLEM DEFINITION

Facility Description

Western Products' facility is a 120,000 sq ft, two floor building in Milwaukee, Wisconsin. The building, originally 34,000 sq ft in 1965, was expanded in 1968 and in 1974. Remodeling was done in 1976.

The facility houses offices, a computer room, warehouse space and production facilities, including welding, painting and general machining. The product manufactured is a steel vehicle-mounted snowplow.

Problem Identification

By 1977, the HVAC system in the facility had become inadequate. Ventilation in offices, welding, and machine areas was insufficient. The building had a significant negative pressure, due to the high exhaust rates necessary for welding areas. The HVAC system was undersized, leading to noticeable loss of comfort during cold winter periods.

Normally, such difficulties would be handled by expansion of the existing HVAC system to provide more heated ventilation makeup air. However, the winter heating costs were already up to $5,000 per month in the winter of 1976-77. More importantly, the local natural gas supplier, facing shortages, refused to provide expanded service to industrial customers including Western Products. These concerns, together with large heat losses resulting from high exhaust rates, suggested a potential heat recovery system application.

IDENTIFYING AND EVALUATING OPTIONS

Selection of a Consultant

Western Products had limited experience in HVAC modifications or in energy conservation and reclamation equipment. They therefore decided to obtain professional help from an outside consultant. In selecting the consultant, Western Products based their evaluations primarily on the professional qualifications of the firms under consideration. Once the firm best suited to execute the project was identified, negotiations were conducted to determine compensation.

Relevant qualifications considered included: 1) the firm's size, relative to the size of the proposed project, 2) the firm's experience, both overall and in similar projects, 3) the qualifications and professional registration of the firm's staff, and 4) references from the firm's current and/or former clients.

The firm's proximity to the client was not overemphasized. An experienced firm at a remote location from the client is usually cost competitive with a less experienced local firm.

As a result of their consultant search and evaluation, Western Products contracted with Affiliated Engineers in January of 1977.

Site Survey

As a first step in identifying and evaluating options, the consultant carried out a site survey. The purpose of the survey was twofold; it allowed members of the consultant's staff to become familiar with the facility, and it provided an opportunity for the consultant to begin to identify potential solutions to the problem. The survey was conducted by a team of engineers who had experience with similar applications.

Physical data, such as dimensions, temperatures, and air flows were obtained. The condition of the existing equipment was assessed. Discussions were held with operating personnel, to identify deficiencies and desired additional features. Potential options were identified, and other data necessary for their evaluation was obtained.

Evaluation of Options

After the site survey and discussions with the owner, the next step was to evaluate the possible options in depth. The four options identified for evaluation were:

1. Add new ductwork and additional cooling for offices and computer room.

2. Add a new makeup air unit, with heat reclamation in the machine shop area.

3. Improve the exhaust of smoke in the welding shop by increasing air flow. This option also includes heat reclamation.

4. Add a machine shop ventilating unit and rebalance the supply air to correct building pressurization.

The results of the evaluations were presented in a report, in which the estimated construction cost of each alternative was presented within ± 20%. A generic description of each option was also presented. The report then summarized the results and the consultant's recommendations. The report was dated April 1977.

Recommendations

After examining the benefits and costs of the various options, the owner then determined the most appropriate course to follow. These actions were as follows:

1. Add ductwork and fans to improve air distribution in machine shop, welding shop, and office space.

2. Add heat reclamation on the increased air flow supply in the welding shop.

3. Balance the air flow in the facility to set the building at a 1% positive air supply.

These solutions represent a combination of elements of each of the consultant's recommendations. They were arrived at by the owner after consideration of both desired system improvements and cost limitations.

DESIGN OF PROBLEM SOLUTION

Once the solution of the problem was determined qualitatively, the client then instructed the consultant to prepare construction documents. These documents were to be used to obtain competitive bids for construction of the solution.

Design Criteria

The first step in this process was establishing design criteria. Design criteria form the basis upon which the design is executed, and include both subjective and objective elements, such as relative quality of construction, design life, type and supplier of equipment to be used, and future space functional changes. Input from the owner was critical in establishing these criteria, and included specific requirements of the system and its components, as well as any future changes in facility layout.

In establishing the design criteria for the type of heat reclaim equipment to be used, the consultant evaluated three alternative types of equipment. They were: 1) heat recovery wheels, 2) heat pipe heat exchangers, and 3) air-to-air heat exchangers.

The owner indicated that the exhaust air in the welding shop was extremely dirty. This precluded the use of a heat recovery wheel, because the dirty exhaust airstream would plug the media in the wheel, which would result in low effectiveness and high maintenance costs. An economic comparison of the air-to-air unit with the heat pipe unit showed that the heat pipe unit was more cost effective, becuase of its higher effectiveness and smaller space requirement. As a result, the heat pipe unit was selected as the basis for the design.

The owner also indicated that the welding shop would not be cooled in the summer, and that fresh air ventilation was highly desirable. Therefore in the summer, a necessary design criteria was that a bypass around the heat pipe unit be provided to prevent overheating of the space. The heat pipe unit is unused during the summer. Therefore, variable tilt devices, which allow the heat pipes to reverse heat flow direction and precool outside air for air conditioning, were unnecessary.

Construction Documents

Based on the design criteria, the consultant then developed the construction documents for the project. These included six drawings and a thirty page specification, which described the required equipment and installation in detail.

The specification wad divided into various areas of concern, for each piece or type of equipment or for certain procedures. The equipment specification sections were further divided in three subsections: General, Products, and Execution.

The general section covers testing, shop drawing submittals and details, and related comments. The Products section covers the specifics of the manufacture and operation of the equipment. The Execution section details the equipment installation.

The wording of the specification was as precise and as explicit as possible. To serve as an example, the following paragraphs are excerpted from the Products subsection of the heat pipe unit specification.

FABRICATION

"Casings on coil shall be fabricated from No. 16 gauge galvanized steel, of a design proven to mount and support such coils in ductwork or in a piece of apparatus. All casings shall be furnished with a minimum of 1-1/2" wide flange on each of the four sides, both front and back for mounting. Bolt slots for mounting shall be punched on 3" centers throughout. Intermediate core supports shall be furnished to prevent sagging or bowing when mounted in the recommended manner. Coil to have aluminum noncorrugated plate fin with 1" tube.

Coil shall be complete with drip pan.

Working fluid shall be commercially pure group 1 refrigerants as classified in the American National Standard Safety Code for Mechanical Refrigeration. Prior to charging at the factory, each tube shall be cleaned, dried and evacuated.

Protective end covers shall be provided to protect the tube ends which extend through the casing. They are to be constructed of No. 20 gauge galvanized steel and will be factory sealed.

Total coil size shall be 144 inches long by 108 inches high. The coil bank shall be comprised of two 144 inch by 54 inch coils. Fresh air side shall be 6 rows, 14 fins per inch. The exhaust air side shall be 6 rows, 11 fins per inch. The coil bank shall have a total free area of 108 sq ft. The coils shall be capable of handling 26000 cfm of exhaust air at 60°F and 50% RH, and 24000 cfm of fresh air at -15°F.

The coil shall be capable of transferring sensible heat between outgoing-incoming counterflow air streams at approximately 61% effectiveness. Thermal recovery unit shall be capable of operating at temperatures from a minimum of -40°F up to a maximum of 140°F.

Entire unit must be capable of withstanding entrained moisture from outside air. If required, unit must withstand washdown of cleaning without harm to operation."

Bids

Once the construction documents were completed, the consultant assisted the owner in identifying qualified and experienced contractors. A complete set of construction documents was then sent to each of the four approved contractors, as well as an invitation for bids on July 22, 1977. Bids were received and tabulated by August 4, 1977. Bids were for the total construction of the project and differed by approximately 25% from high to low. In this case, after evaluation by the owner and the consultant, the low bid was accepted.

Construction

Once the contractor was selected, a construction schedule was established by the consultant, owner, and contractor. Construction was carried out without interruption of the facility's operation and was completed with a final air balancing on February 27, 1978, slightly later than originally anticipated.

RESULTS

Economics of Heat Reclamation

In this section, the owner's additional investment in energy conservation equipment included only the heat pipe unit, and some additional ductwork (Figure 1). The remainder of the system components were necessary to provide improved ventilation. Therefore, the economic practicality of the investment in energy savings equipment depended only on the cost of installing the heat pipe unit.

The design conditions specified for the heat pipe unit are -15°F outside ambient and 60°F indoor temperature. Under these extreme conditions, the 24000 cfm of make-up air would be preheated from -15°F to 31°F, representing a heat reclamation rate of 1.192 MMBtus/hr. However, under more typical outdoor conditions, heat reclamation, as well as overall building heating requirements, will be lower.

The average ambient temperature during the 30 week heating season in Milwaukee is 32.9°F. Assuming the ambient to be a constant of 32.9°F with 60°F exhaust air, the heat pipe unit will save 2,124 MMBtus per heating season.

A simplified economic analysis of the investment is as follows:

Installed Costs (1978)	
Heat Pipe Unit	$18,130
Extra Ductwork	469
Cost for Larger Fan	754
Total First Cost (Installed)	$19,353

These values include labor, supervision, contractor's overhead and profit, and taxes.

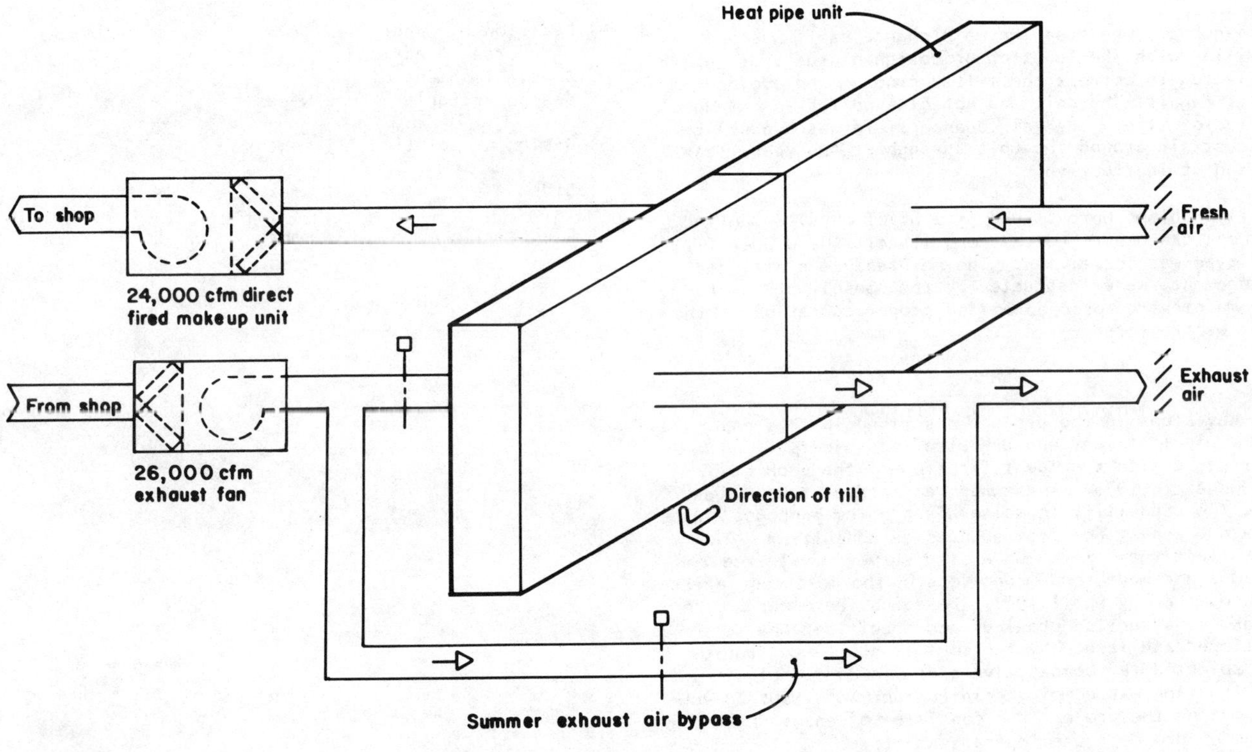

Fig. I - HEAT RECLAIM SYSTEM SCHEMATIC

The yearly savings are calculated as:

Energy Savings	2,124 MMBtu/yr
Equivalent Dollar Savings (Natural Gas @ $0.22/therm)	$4,673
Maintenance of Reclaim Coil (12 hr/yr @ $14.28/hr)	($171/yr)
Net Savings	$4,502/yr

The simple payback period is therefore calculated as:

$$\frac{\$19,353}{\$4,502} = 4.3 \text{ yrs}$$

Operational Experience

The system has been in operation for 3 years and modifications have been made. In the original design, warm and dirty exhaust air passed through 20% filters before entering the heat pipe unit. As clogging of the coil was occurring, the general contractor changed to an 80% filter. A filter of such density is generally only recommended for hospitals. In this industrial situation, these filters clogged within days. Also, this large increase in filter density so restricted exhaust air flow that frosting occurred on the exhaust coil.

At about the same time as these filter changes were occurring, the plant engineer who supervised the installation of the heat pipe unit for the owner left his job. His replacement had no previous experience with heat reclamation systems.

Subsequently, because the maintenance staff was not familiar with the function or design of the unit, they neglected to correct the filter problem and avoided servicing it. The unit was not cleaned for an extended period of time. The maintenance staff was bypassing exhaust air around the unit throughout the year, rendering it ineffective.

These problems were solved by a brief on-site meeting between the owner and the consultant. The proper filter type was identified. The maintenance staff, and management, were instructed by the consultant on proper maintenance procedures, and proper operation of the unit was restored.

SUMMARY

The key steps in the project described in this paper are typical of many energy retrofit projects. The project began with the identification of the problem by the owner in 1976. The owner realized that he did not have the capability to solve his problem in-house, and obtained assistance from an outside consultant. The consultant then performed a site survey, evaluated the options and made recommendations on the most cost effective option, by April 1977. Based on the owner's directions, construction drawings and specifications were developed and issued by the consultant. A contractor was selected, by competitive bids, in August of 1977. Construction was completed in February of 1978. Follow-up work on the project has consisted of answering questions on proper system operation.

Several of the lessons learned in this project apply to similar projects. For example, in projects where construction is to be carried out without interruption of normal space functions, construction schedulès should contain some allowances for unexpected delays. Also, the operational difficulties evident in this project point up the need to both communicate maintenance tasks to the owner's maintenance staff and to document maintenance needs in writing. The owner should maintain up-to-date maintenance manuals for reference and for new employees, and should conduct a preventative maintenance program.

The significance of this project is that it demonstrates the ability of a heat reclamation unit to save money, and more importantly to solve a utility source problem. Without heat reclamation, the owner could not have improved ventilation in his facility. The local natural gas supplier was not allowing any expansion of service, and other sources were far less economical. Heat reclamation served as a new energy source to heat larger quantities of makeup air. In addition, the heat pipe unit recovered its own installation cost in 4.3 years of energy cost savings.

The following table summarizes the major steps in this project, the elapsed time required for each, and the areas of responsibility for both the consultant and the owner.

Activity	Time Required To Complete	Owners Activities	Consultants Activities
1) Definition of problem		X	
2) Site survey	1 month	X	X
3) Evaluation of options & recommendations	2 months		X
4) Decision by owner on course of action & establishment of design criteria	2 weeks	X	X
5) Design of solution	3 months		X
6) Bids	1 week	X	X
7) Construction	7 months	X	X

Chapter 4
LOW TEMPERATURE HEAT RECOVERY

A. S. Qureshi

As Energy Secretary James B. Edwards stated recently, "the cheapest way to get energy is to conserve." To enlarge upon the secretary's words, one of the best and surest means of achieving energy conservation is waste heat recovery and specifically, low level heat recovery.

Low level heat recovery (LLHR) is the reclamation of heat energy which prior process either generated or rejected as a by-product. The primary objective of a heat recovery system is to replace other forms of heating and thereby reduce the overall energy consumption of a facility. In a new facility this would result in the reduction of primary fuels (fuel oil, natural gas, electricity, etc.) consumption, reduced thermal pollution and savings in first cost investment due to reduced sizes in heating equipment such as boilers, furnaces, etc. In the case of an already existing facility, the application of LLHR would require an extra investment in the heat recovery system and modification of the existing system to accomodate heat recovery devices, additional piping or ductwork, controls and maintenance costs, etc. The justification to replace or supplement the existing heat source with LLHR would depend on the amount of savings in energy costs.

For a successful application of LLHR as a substitute heat source, it should provide the same level of service and it must also be environmentally and financially superior to alternative methods of heating.

Large amounts of energy are wasted whenever the chemical energy in fuels is transformed into thermal energy for heat or power. Recapturing this heat and beneficially utilizing it is always desirable. However, there are a number of technical and economic constraints that need to be analyzed before a decision of implementing LLHR is made. For an optimum and economical utilization of recovered waste heat, it is important that the temperature level of the LLHR source is compatible to the temperature desired at the consumption point. A good match of temperatures would enhance the economics of LLHR systems. On the other hand, too low a level of temperature of heat source may reduce the technical and economic advantage of the LLHR system. For a desirable application, the quantity of LLHR should be sufficient to satisfy specific energy needs. Availability of large quantities of waste heat with a greater factor of utilization would always improve the economics of an LLHR system.

Similarly, in the application of an LLHR system, the physical distance between heat source and the point of consumption is of great importance. It is desirable to have a minimum distance between the heat source and point of consumption, since it would minimize transmission losses and would require a smaller investment in distribution system, thereby improving the economics of an LLHR system.

In short, it can be safely stated that if a waste heat source has a lower temperature than the consumption point requires, the heat available is of small quantity and there is too great a distance between heat source and consumption, then, the LLHR system does not warrant consideration.

Sources and Methods of LLHR

Sources of LLHR can be found in industrial, commercial, instutional and residential facilities' comfort and production processes such as:

Media	Sources
Exhaust gases	Boilers, furnaces, ovens process equipment, internal combustion engines and gas turbines.
Exhaust air	Heating, ventilating, air conditioning and industrial exhaust.
Process water	Laundries, restaurants, industrial processes, engine jacket water and lube oil cooling water.
Condenser discharge	Power plants, steam turbines, refrigeration condenser and other processes.
Nature	Air and ground heat as a heat source for heat pump, solar energy and low level geothermal energy.

The above sources of low level heat can be transferred by three methods. One method is process-to-process, where heat is captured from the exhaust stream of one process and utilized in another process. The heat transfer in most cases is sensible. Another method is process-to-comfort, where heat rejected from a process is captured and utilized for heating or cooling a facility for comfort purposes. Heat transfer is usually sensible only. The last method is comfort-to-comfort, where heat is recovered from an exhaust air of comfort conditioning system and transferred to heat or cool supply air system. Heat transfer could be total or sensible only.

Working Temperature of LLHR Systems

There is no established upper limit of temperature for low level heat recovery. Depending upon the source and application, temperature may range from 50° to 400°F. For example, in an organic Rankine Cycle application a temperature of 250° to 400° F may be considered as a low level heat. Whereas this temperature may be too high to achieve from a comfort exhaust system. For the purposes of this paper, the temperature of 400°F and below is the assumed area of LLHR. The following is a list of some of the most common heat sources and the range of their temperatures:

Heat Source	Range of Temperature
Ground Heat	50°- 60°F
Condenser heat (Refrigeration)	95°-125°F
Exhaust air (Comfort Conditioning)	75°-115°F
Process water	80°-180°F
Condensate	80°-200°F
Process (moderate) exhaust air	100°-300°F
Lube oil cooling water	140°-180°F
Engine jacket water	160°-220°F
Boiler exhaust	250°-400°F

Balanced LLHR System

To assure economic viability of LLHR, the heat recovery system should operate in harmony with other forms of heating. For optimum utilization the LLHR should satisfy all thermal needs first, and only in case of deficiency should other forms of heating supplement the additional heating requirements. A balanced LLHR system would always vary the amount of heat recovery according to the facility's requirements. In process-to-process LLHR system, heat requirements may not vary appreciably throughout the year and the outdoor temperature may not have any impact on heat requirements. However, in cases of process-to-comfort and comfort-to-comfort the heat energy requirements will vary according to the changes in outdoor temperatures. Therefore, the LLHR system should be capable of responding to varying outdoor conditions. Otherwise, it would result in an unnecessary heat recovery and a possible penalty to energy conservation.

Equipment and Devices of LLHR System

The low level heat can be captured by means of heat exchange devices such as rotary heat wheels, heat pipes, fixed plate heat exchanger, coil energy recovery loop and systems such as heat pump and Organic Rankine Cycle. The effectiveness of these devices depends upon proper application, overall system design and the quality of maintenance.

a. Heat Wheel: Heat Wheel or rotary air-to-air heat exchanger is most commonly used for comfort-to-comfort LLHR. It can recover sensible or total heat depending on the type of application and the type of media which are used. The effectiveness of a heat wheel is dependent upon the amount of pressure drop in the system. Average effectiveness value is in the 70 to 85% range. In a heat wheel, cross contamination or mixing of air between supply and exhaust air stream can occur due to carry over and leakage. The heat wheel is most cost effective when supply and exhaust air streams are in close proximity. The maintenance requirements for a heat wheel are minimal.

b. Coil Energy Recovery Loop: The coil energy recovery loop is one of the simplest devices of LLHR. It consists of two or more extended surface finned tube water coils connected in a closed loop piping containing water or a freeze preventive solution. The coils are placed in the supply and exhaust air streams of a process or building. It is used for sensible heat recovery and the system is seasonably reversible. It is free of cross contamination and requires minimal maintenance. The coil energy recovery loop is well suited for retrofit and industrial applications. The effectiveness of the coil energy recovery loop lies in the 55 to 70% range.

c. Heat Pipe Heat Exchanger: The heat pipe heat exchanger is a passive energy recovery device with no moving parts. It works on a condensation-evaporation cycle which is continuous as long as there is a temperature difference to drive the process. The effectiveness of a heat pipe depends upon the heat transfer surface area, the ration of the heat capacities of the two air flows and the rate at which the air pass through the heat exchanger. Manipulation of the above factors can vary the effectiveness between 40 to 80%. It has no cross contamination and a fairly low amount of maintenance. It can be utilized for process and as well as comfort LLHR.

d. Fixed Plate Heat Exchangers: Fixed plate heat exchangers have the broadest temperature range for air-to-air heat recovery devices. The fixed plate heat exchanger could be counter flow design (pure plate heat exchanger) or it could be a cross-flow design (plate and fin heat exchanger). The range of effectiveness for the pure plate type heat exchanger is 50 to 80% and for the plate-fine-heat exchanger 40 to 60%. Fixed plate heat exchangers are built in modular form and they can match any air flow pressure drop requirements. One of the main advantages of the plate heat exchanger is that it can be built with little or no leakage. The maintenance requirements are usually minimal unless the quality of air is very poor. They can be used for comfort and as well as process LLHR.

e. LLHR With Organic Rankine Cycle (ORC): The Organic Rankine Cycle (ORC) converts the low level heat into usable shaft horsepower. It can utilize waste heat in the 250 to 700°F range. In process industry an appreciable amount of heat is rejected around the 200 to 350°F range. The conversion of this rejected heat into useful power can improve the overall thermal efficiency of a process or facility. The ORC systems as an LLHR device are presently marketed either as a packaged 500-600 kilowatts cogeneration unit or as a Binary Rankine Cycle unit to provide shaft or electric power from waste heat. Compared to the simple ORC system the Binary ORC has the advantage of being able to use high and low level heat simultaneously. Higher temperature heat (450°F and above) is used in the Steam Rankine Cycle which could be open or closed, and the low level heat (200 to 250°F) is used only for the closed Organic Rankine Cycle. For LLHR from deisel exhaust, industrial exhaust and steam condensate, the ORC could be an excellent device.

f. LLHR With Heat Pump: The heat pump is one of the most effective devices for LLHR. It can extract low level heat, upgrade it to a suitable temperature and deliver it at the point of consumption in a very efficient manner. A heat pump can extract heat from air, water, earth, or any process fluids. The temperature range for the commercial type of heat pump is 95 to 125°F. Some new industrial types of heat pumps can approach temperatures of 250°F and above by utilizing special refrigerants. Heat pump applications are numerous. The following are some of the most common applications for LLHR:

. Heat recovery from ground or well water for comfort heating.
. Heat recovery from building exhaust system (internal source heat pump) for transferring internal heat to the exterior of the building, or where it is most needed.
. Heat recovery from low temperature solar heated water, or fluid for comfort or process heating.
. Heat recovery from waste water such as sewage (effluant water), laundry discharge, commercial and

industrial processes waste water.

Examples of LLHR

a. LLHR with Heat Pump: For the Health and Police Science Building of the Agricultural and Technical College at Farmingdale, New York, our firm recently designed an LLHR system. For the LLHR from building exhaust air, two methods were used to reclaim the heat (see Figures 1 & 2). One is by means of heat wheels, and the other is by means of chilled water coils as part of a heat pump transfer system. When the building is occupied, all of the air exhausted from the building is a temperature of at least 70° to 75°F (much higher with heat of light). To replace this exhausted air, outside air must be brought in at temperatures which could be as low as 9°F. For the perimeter areas of the building, this air must be heated in order to maintain comfort conditions at low outside temperatures. In the original design a heat recovery wheel was utilized to transfer the heat from the exhaust air to the cold outside air so that less additional heat would be required. This method of LLHR delivered an efficiency of 70% so that 9°F outside air could be heated to over 40°F without any external

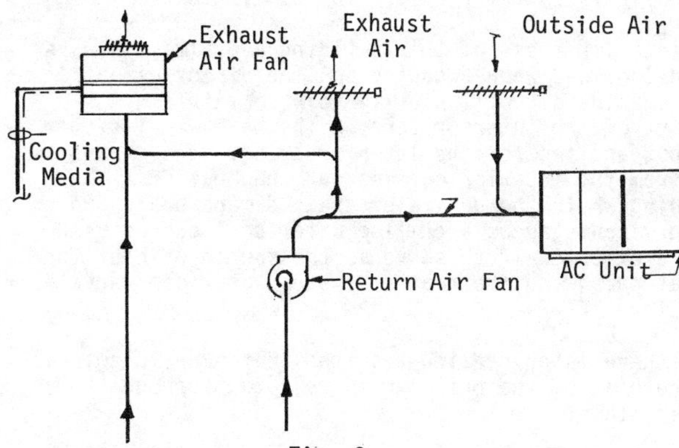

Fig. 2
Interior Zone Outside Air & Exhaust System With Economizer Cycle & Exhaust Heat Pump Recovery System

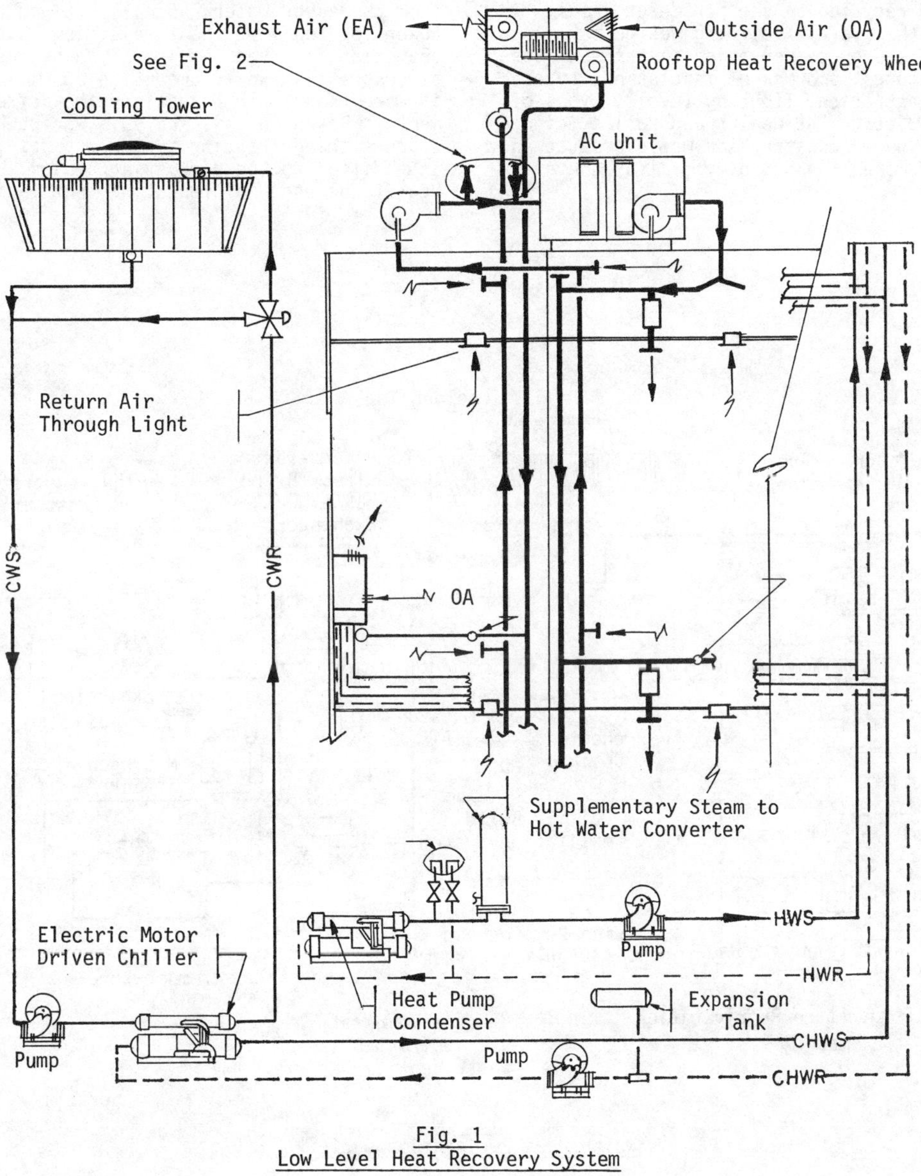

Fig. 1
Low Level Heat Recovery System

energy. Although application of the heat wheel on the exterior zones did save some energy, the poor economics of this system resulted in its elimination.

The interior areas of this building are almost always a cooling load when occupied and the direct use of cool outside air will usually help to maintain the comfort of the interior zones. The heat wheel system was not applied for the interior areas. The exhaust air from the interior carries out the heat from the building while the perimeter areas demand heat. Under these circumstances a chilled water coil can be used in the exhaust air to serve as the source of heat for a heat pump, and the heating system water can serve as the sink.

When there is no heating required in the building, all of the heat of the building is rejected through the cooling tower.

When heating is required, less heat is rejected to the cooling tower and more heat is redirected to the parts of the building that require heating. In this building because of the LLHR, no external heat is required on the coldest day of the year during occupied hours. All of the heat can be extracted from the internal building heat and recycled to the perimeter areas. In addition to LLHR the overall building design takes full advantage of energy conservation in terms of its architectural features, amounts of insulation, types and amount of fenestration, lighting levels, types of lighting fixtures, etc. The Health and Police Science Building which is under construction would produce an energy savings of more than 50% over similar conventional facilities.

Similarly, for LLHR from waste water, our firm recently completed a feasibility study of using effluent water as a heat source for a heat pump at one of the large water pollution control plants in New York City. Our analysis indicated substantial cost and energy savings due to this LLHR system.

b. <u>LLHR From Low Temperature Refrigeration System:</u> Another example of LLHR is the Hunts Point Cooperative Meat and Poultry Market in Bronx, New York, designed by our firm and presently in operation. The Market consists of 6.3 million cubic feet refrigerated warehouse and freezer space. As part of its energy conservation program, the market system recovers heat from the refrigeration process and utilizes it for defrosting its freezer and cooler spaces (see Fig. 3) The plant is designed for a peak capacity of 2400 tons of refrigeration. The refrigerated calcium chloride brine is supplied at -25°F and +19°F in two separate circuits to freezer and cooler spaces throughout the market. The heating requirements to defrost the freezer and cooler coils is approximately 8 million BTU/HR. Ordinarily, this heating requirement would have been satisfied by the boiler plant. It was investigated whether this heating demand could be satisfied by recapturing some of the waste heat from condenser water, normally pumped to the cooling tower for low level heat rejection. It was estimated that heat from the refrigeration process plus the heat of compression would amount to 23 million BTU/HR which is more than ample to satisfy the defrosting requirements. Based on this analysis, an LLHR system to support the defrosting system was designed (see Fig. 3) In addition to the defrosting system LLHR also supplies heat to be used in preheating ventilation air for the

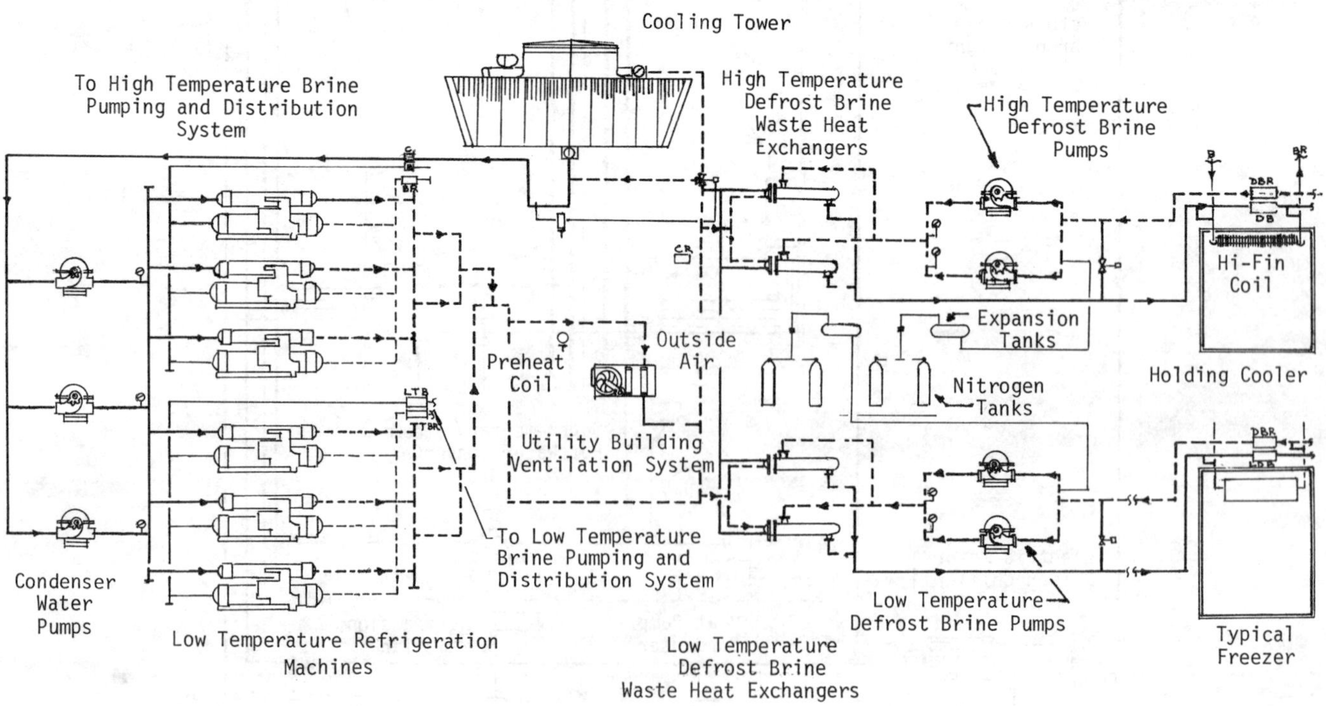

<u>Fig. 3</u>
<u>Waste Heat Reclamation Cycle</u>

utility building during winter. This LLHR was projected to save approximately 220,000 gallons/year or 20% of the peak heating requirements of the market.

From the foregoing, it is apparent that we are throwing away huge quantities of heat at low temperatures. That heat can be recovered and used either at existing temperatures for low level heating needs or at upgraded temperature levels to meet specific requirements. The practicality of implementing a heat recovery system is fundamentally based on the economics of such a system. The initial costs and continued maintenance of equipment must be weighed against the savings in energy costs.

Recycling of low level heat would reduce energy waste, improve economics and also help in maintaing a cleaner environment. High costs of energy are greatly affecting the operating budgets of residential and institutional buildings, as well as the pricing of industrial and commercial products and services. Low level heat recovery provides the best possible option for industry and commerce to remain competitive and to reduce the impact of ever increasing prices of energy.

References:

1. ASHRAE Equipment Handbook-1979, Chapter 34.
2. ASHRAE Systems Handbook-1980, Chapter 7.
3. Mitchel Olszewski: Utilization of Reject Heat (Marcel Dekker Inc., New York, N.Y. 1980).
4. S.W. Brown: Heat Recovery, Energy Conservation and Pollution Control at the Hunts Point Project (ASHRAE Journal Dec. 1972).
5. C.F. Bonilla: Thermal Wheel Rotary Air to Air Heat Exchanger (Archieves of Environmental Health, Vol. 4, March 1962).
6. B.S. Szabo: The Economics of Heat Recovery Systems (Air Conditioning, Heating & Ventilating, August, 1967).
7. F.A. Stanton: The Rotary Air to Air Heat Exchanger (Heating/Piping/Air Conditioning, February 1963).
8. S.E. Gifford: Outside Air System Design Concepts Utilizing Air-to-Air Regenerators (Fourth Annual Technical Meeting, American Association for Contamination Control, Miami Beach, FL, May 1965).
9. M.A. Ruch: Heat Pipe Heat Exchangers as Energy Recovery Devices (ASHRAE Transactions, Vol. 82, Pt. 1, 1976)
10. Richard H. Stamm: The Energy Pump (Heating/Piping/Air Conditioning, March 1981).

Chapter 5

THE DESIGN AND EVALUATION OF A WATER-TO-WATER HEAT PIPE HEAT EXCHANGER

C. H. Johnson, E. Zeigler

INTRODUCTION

The rate of consumption of finite quantities of fossil fuel has initiated far-reaching investigation into methods of energy conservation. One promising approach to energy conservation is the use of heat pipes to recover heat now lost in effluent processing waste water streams. A survey was conducted at Radford Army Ammunition Plant (RAAP) to determine the most appropriate applications for heat pipe technology for energy recovery in unit processing operations. One potential application was the nitrocellulose (NC) purification area. It was decided to design, fabricate, install, and evaluate a prototype water-to-water heat pipe heat exchanger (HPHX) in the boiling tub operation.

In the boiling tub purification processes for NC (figure 1), 171,000 pounds of an NC/water slurry is heated from approximately 50°F to 212°F, using processing steam at 40 psig. Three times during this processing phase, the hot water from this slurry is drained into the waste disposal system, the NC washed, and the tank replenished with 50°F water (figure 2). It was the purpose of this project to extract heat from this waste water and use this recovered heat to preheat the incoming fresh water. This process would save energy by reducing the amount of steam required to bring the slurry to the 212°F operating temperature, because the recovered heat would raise the water, NC, and tank to approximately 95°F, in place of the average annual temperature of 50°F. NC is a material that becomes increasingly sensitive to energy stimuli as the moisture content diminishes. It is, therefore, imperative that the heat exchanger design be capable of thorough cleaning and inspection. For this reason, a heat pipe heat exchanger was designed for heat recovery instead of using a conventional shell and tube heat exchanger.

This project is part of the U.S. Army Manufacturing Methods and Technology program, through the cooperative efforts of ARRADCOM and Radford AAP.

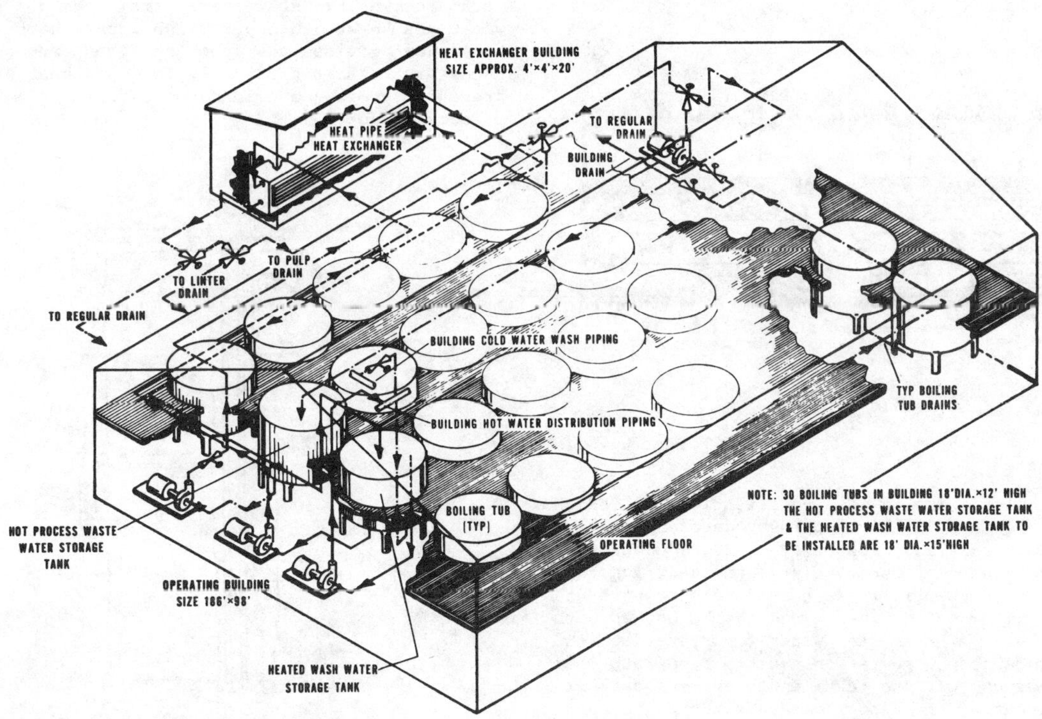

FIGURE 1. FACILITY AND EQUIPMENT LAYOUT

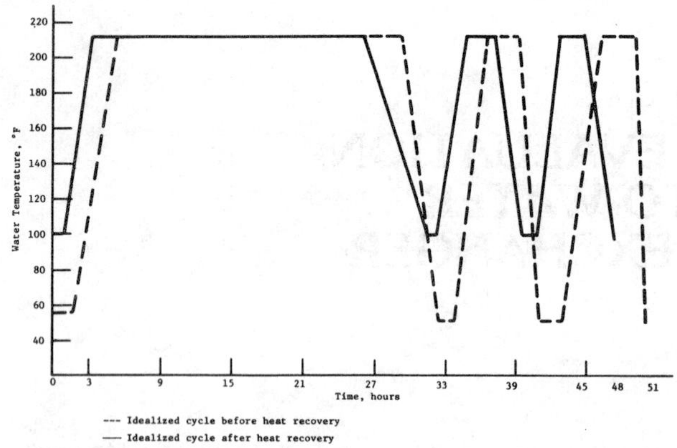

FIGURE 2. IDEALIZED BOILING TUB CYCLE

HEAT PIPE

Heat Pipe Technology is a "state-of-the-arts" device, as the basic prinicples were understood prior to WWII. The escalating costs and diminished availability of energy have renewed interest in this technology in the past few years. The heat pipe presents a novel, highly efficient, and versatile approach to the solution of heat-transfer problems.

The heat pipe is essentially a boiler for heat addition, connected to a condenser for heat removal, and a condensate pump to return the condensate fluid from the condenser to the boiler. This mechanism is housed in a simple casing (figure 3). The heat pipe consists of (1) a sealed enclosure, usually of metal; (2) a wick that acts as a condensate pump through capillary action; and (3) a working fluid that transports the heat. The selection of the optimum combination of the above is the consideration given in the design of a practical, highly efficient heat pipe.

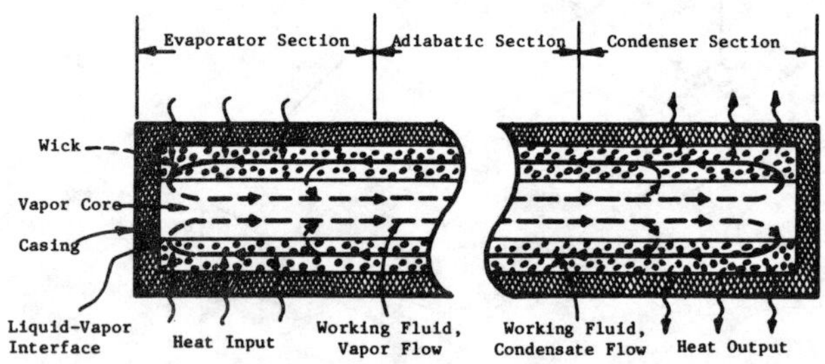

FIGURE 3. HEAT PIPE

Design Objectives

In this particular phase of NC purification, 30 batch tanks (18' diameter x 15' high) per production line are utilized. Each tank has a capacity of 30,000 pounds of NC and 141,000 pounds of water. Three times during this processing cycle, the slurry is heated to a temperature of 212°F, the hot water drained, and the NC washed with cold fresh water. The waste water flow rate from the processing tanks is 600 gpm at an average temperature of 220°F. The purpose of the HPHX is to extract approximately 50 percent of the heat from this hot waste water and preheat fresh water for subsequent use in the process. The flow rate can vary from 100 gpm to 600 gpm. The waste water is acidic (pH = 2, maximum); therefore, all selected construction materials that come into contact with the water must be resistant to this environment, as well as chemically compatible with the nitrocellulose.

Technical Approach

The major factors affecting the design of an HPHX for use in this operation are:

(a) Performance is dependent upon the external heat transfer characteristics between the heat pipe and the water streams, and the internal heat transfer characteristics of the heat pipe.

(b) External heat transfer characteristics are dependent upon the heat pipe lengths and heat pipe spacing in the matrix (tube bundle).

(c) Access to the water water section for cleaning and visual inspection is required.

(d) Provisions for periodic inspection of the functionality of each heat pipe.

(e) Easy removal of a heat pipe(s) from the HPHX, if nonfunctional.

(f) A positive and highly reliable seal between the waste water and fresh water sections is imperative.

Analyses were performed to establish energy recovery efficiency, pressure drop, system conductance, individual heat pipe performance, and total heat recovery.

HPHX Design

The HPHX design, based on the above approach, is shown in figure 4. Each module contains 241 one-inch diameter by 8 feet long heat pipes. The heat pipes within each module transfer heat from the hot lower waste water chamber to an upper fresh water chamber. A metal tube sheet divides the two sections and contains the necessary leak-tight seals for each heat pipe, which prevents cross-contamination between the two water streams (.igure 5).

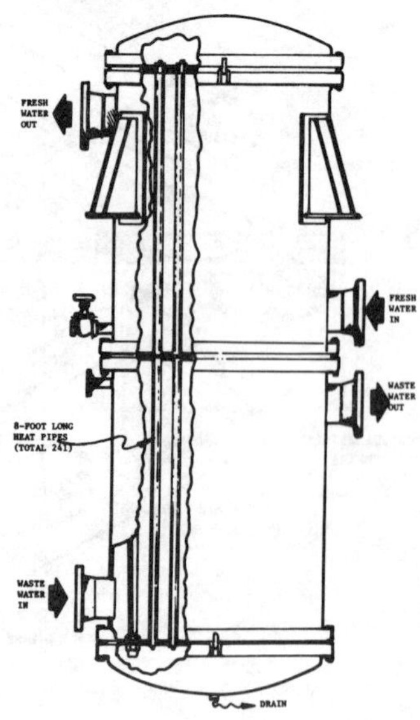

FIGURE 4. HEAT PIPE HEAT EXCHANGER DESIGN

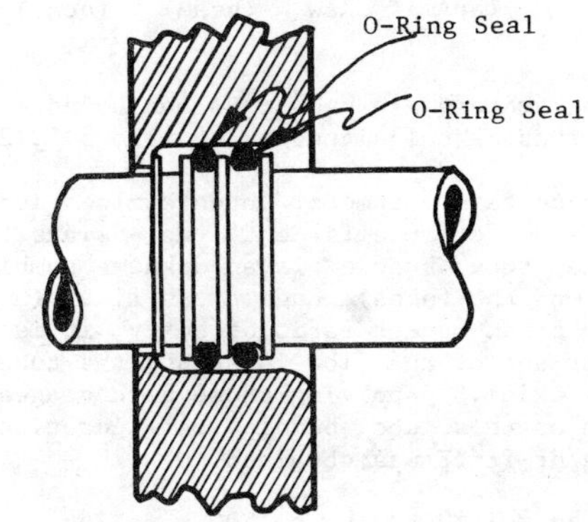

FIGURE 5. HEAT PIPE SEALS

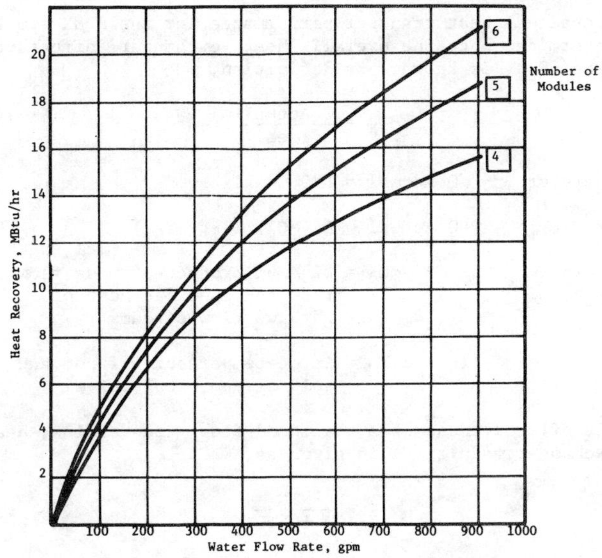

FIGURE 7. HPHX HEAT RECOVERY VERSUS FLOW RATE

To extract 12.5M Btu/hr from the waste water at a flow rate of 600 gpm, four units are required. These units are installed as shown in figure 6. A counterflow arrangement is shown with valving to accommodate the complex process flow in this operation.

Figure 7 shows HPHX heat recovery versus flow rate for 4, 5, and 6 module heat exchanger units.

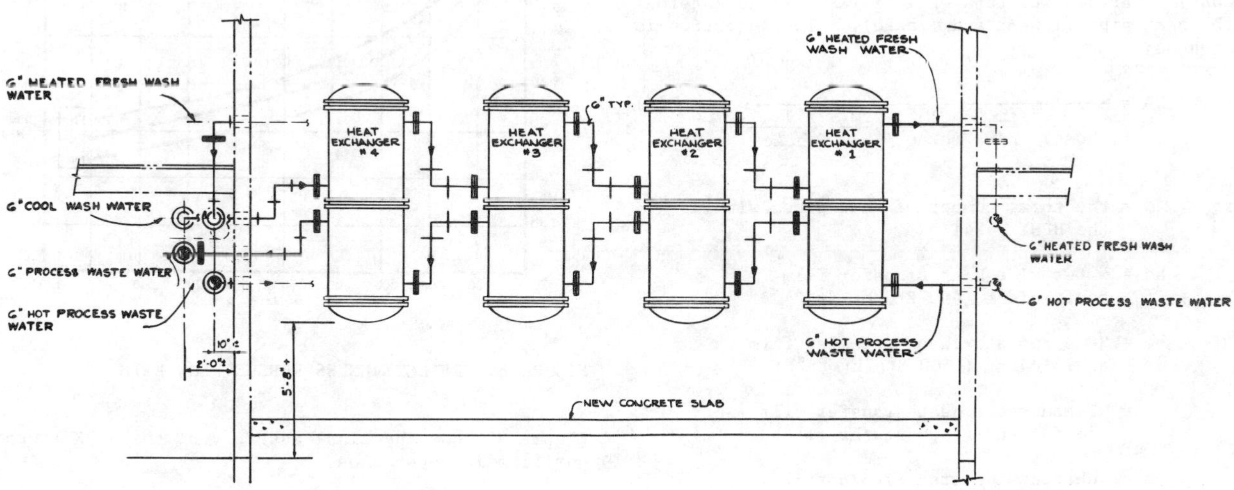

FIGURE 6. INSTALLATION OF HEAT EXCHANGERS

Performance Analysis

The actual heat transfer performance for the HPHX can be determined once the overall heat exchanger effectiveness, E_{HX}, is known. By definition,

$$E_{HX} = \frac{Q\ Actual}{Q\ Ideal} \quad (1)$$

Therefore, $Q\ Actual = E\ Q\ Ideal$

or $$Q\ Actual = E\ (MC_p)(T_{H1} - T_{C1}) \quad (2)$$

where MC_p = minimum product of mass flow rate and specific heat of the two flow streams

T_H and T_C = inlet temperature of hot and cold stream, respectively

The effectiveness of a balanced-flow counterflow heat exchanger module (E) is given as,

$$E = \frac{NTU}{1 + NTU} \quad (3)$$

where $NTU = \frac{UA}{MC_p}$

Note: For 4 heat exchanger modules in series,

$E_{HX} = 4E$

UA = the overall heat conductance between the flow stream. It is the overall heat transfer coefficient times a referenced area.

The product (UA) can be expressed in terms of the individual thermal resistance in series between the hot stream and the cold stream. For each chamber of the unit, these include the resistance on the water side (1/hoAo), through the tube wall (Tw/KAw), and internal to the heat pipe (1/hcAc and 1/heAe). The equation is written as:

$$UA = \frac{N}{\frac{1}{(hoAo)} \frac{tw}{KAw} \frac{1}{(heAe)} \frac{tw}{KAw} \frac{1}{(hcAc)}} \quad (4)$$

where N = the total number of heat pipes within the HPHX = 241

and ho = water side heat transfer film coefficient (400 Btu/hr ft°F)

he = HP evaporator heat transfer film coefficient (1500 Btu/hr ft°F)

hc = HP condenser heat transfer film coefficient (1500 Btu/hr ft°F)

Ao = tube outside area (1.376 ft^2)

Ae, Ac = tube inside area for evaporator and condenser (1.148 ft^2)

Aw = average tube wall area ≈ 1/2 (Ao + Ae), (1.260 ft^2)

tw = tube wall thickness (0.009 ft)

K = tube wall conductivity (10 Btu/hr ft°F)

Assuming equal values for each water side resistance and each wall resistance, the equation can be simplified as,

$$UA = \frac{N}{\frac{2}{(hoAo)} + \frac{2tw}{KAw} + \frac{1}{(heAe)} + \frac{1}{(hcAc)}} \quad (5)$$

Once the parameters are known, UA can be determined and the heat transfer calculated by equations 1, 2, and 3.

The most uncertain parameter to determine is the water side heat transfer coefficient, ho, since the water flow in the tube bundle is an unknown combination of axial (along the tubes) and cross-flow (across the tubes). For an aspect ratio of unity, it is estimated that 40 percent of the flow is across the tubes, and 60 percent is axial. Applying these percentages to estimates made on this tube bundle configuration, a value of 400 Btu/hr-ft^2°F was obtained.

Values of he = 1500 and ho = 1500 were used for heat pipe evaporation and condenser film coefficients, respectively. Heat transfer performance predictions can now be made. Results are presented in Figure 7, as a function of water flow rate in gpm. At 600 gpm, the 4-unit HPHX will recover 12.5 MBtu/hr; whereas, a 6-unit exchanger will extract 17.0 MBtu/hr.

Figure 8 shows the effectiveness versus flow rate for 4, 5, and 6 unit heat exchangers.

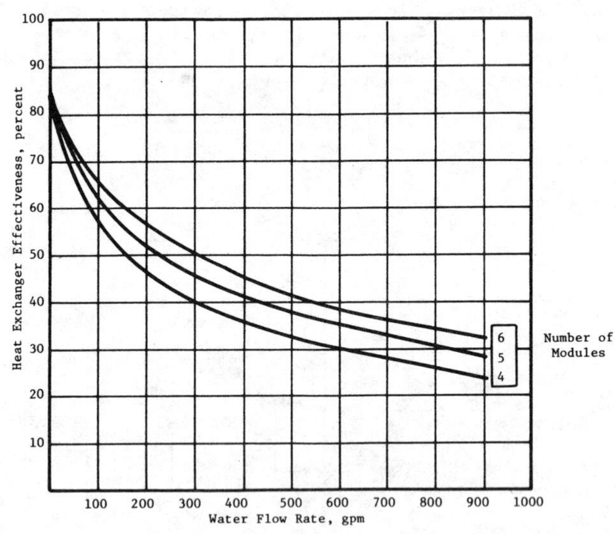

FIGURE 8. EFFECTIVENESS VERSUS FLOW RATE

Figure 9 shows the final assembly of the HPHX at the NC purification operation.

Economic Analysis

Based on mobilization rates at Radford Army Ammunition Plant, a savings of 17.55 x 10^{10} Btu/year will be realized, using the 4-unit HPHX shown in figure 9.

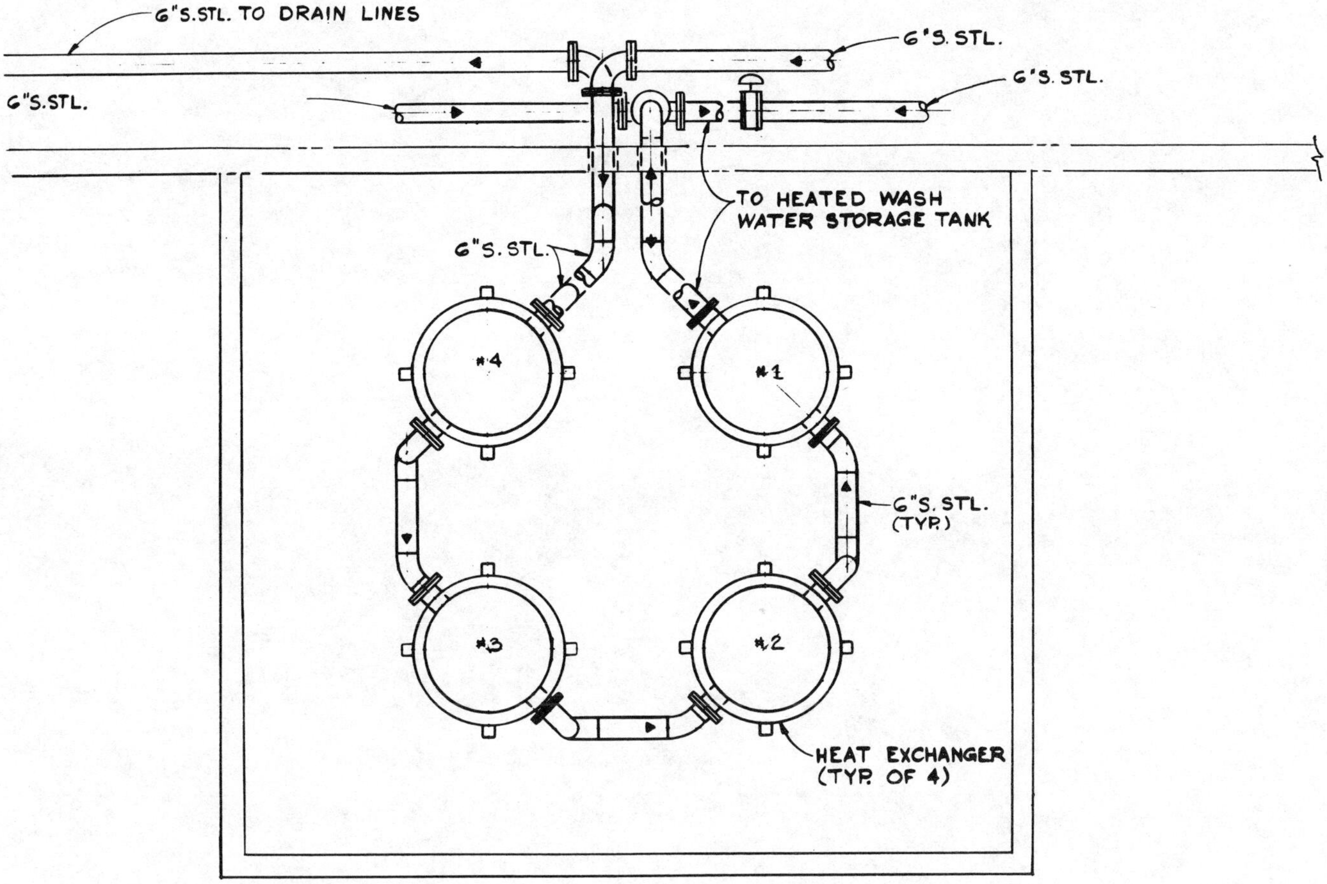

FIGURE 9. ASSEMBLY OF HPHX AT NC PURIFICATION OPERATION

Chapter 6

PRELIMINARY EVALUATION OF LARGE-CAPACITY, HIGH-LIFT HEAT PUMP CONCEPTS

R. W. Bass, A. M. Landerman

SUMMARY

An analytical study was undertaken to identify heat pump system characteristics best suited for recovering low-temperature waste heat from various sources and providing it at temperature levels to match industrial needs. Six distinct waste heat sources and industrial end-uses were identified as representative applications and used for system specifications. Two applications typify large sources of waste heat and co-located industrial park end-uses while the remaining represent in-plant industrial sources and end-uses. Rankine cycle heat pump configurations and working fluid options were assessed and four configurations were selected for further study on the basis of cost and performance. Five different prime movers were examined as potential compressor drivers and gas turbines and diesels were selected on the basis of high overall performance. A brief description is given of an economically attractive system for use at an industrial park near the gaseous diffusion plant at Portsmouth, Ohio. Finally, design considerations for the other applications are discussed.

INTRODUCTION

Low-temperature waste heat (110-150 F) can be converted to useful process heat for industrial needs via a heat pump. In an industrial plant this can be done by converting in-plant waste heat into process hot water or steam. Most process heat requirements are in the form of steam at 250 to 300 F and above. Where large quantities of waste heat exist, for example, at the Federally-owned gaseous diffusion plants, process steam produced by the heat pump can be used to supply the needs of an entire industrial park. Heat pumps with an output capacity in excess of 50 x 10^6 Btu/hr and the temperature capability required to meet these needs are not presently commercially available. Therefore, the objective of this study was to determine the potential cost and performance characteristics of heat pumps designed for selected waste heat applications, evaluate the various cycle, component and working fluid design options and select the systems that could be offered commercially and justified by a favorable return-on-investment (ROI) for the various applications.

APPLICATION CHARACTERISTICS

Use of a heat pump requires both a source of low-temperature heat and an end-use for the higher-temperature output heat. Therefore, a survey was made of published literature and other UTRC data to identify waste heat sources and co-located end-uses. The identified waste heat sources were predominantly cooling water although some process water and nonreturned condensate were also found. In cases where the waste source originates in an industrial plant, the potential in-plant end use requirement was identified. In most of these plants, the end-use was process steam because it represents the most common heat transfer medium and has a wide range of uses from indirect (i.e., jacketed heaters) to direct (i.e., distillation).

In those cases where the waste heat source is cooling water from Federally-owned gaseous diffusion plants or utilities, generally, there was no large co-located process heat need. Therefore, for these cases, all industrial process heat requirement were surveyed to identify steam heat loads and temperatures that are common to a wide variety of industries suitable for an industrial park located near the waste heat source. The resultant data were used to define six different potential applications summarized in Table 1. Two of these applications are for the very large sources of waste heat from Federal facilities and large central power station cooling towers that could be used to provide process heat to a nearby industrial park. The remaining four applications are for waste heat sources and process heat requirements that exist within a given industrial plant.

CONFIGURATION SELECTION

Four types of Rankine cycle configurations were considered: single-fluid closed cycles, semi-open cycles, open-cycles and dual-fluid cycles. Each of these configurations has certain working fluid options. Single-fluid closed-cycles are closed systems and only contact the input and output heat streams across heat transfer surfaces, and, therefore a wide variety of fluids can be employed. A semi-open cycle eliminates the condenser and uses the heat pump fluid directly to satisfy the

TABLE 1. CANDIDATE HEAT PUMP APPLICATIONS.

WASTE HEAT SOURCE			END-USE APPLICATION		
LOCATION	FLOW RATE 10^6 lb/hr	TEMP. F	LOCATION	HEAT REQUIREMENTS 10^6 Btu/hr	TEMP. F
FED. FACILITIES	20	140-165	IND. PARK	400-1000	300
UTILITY COOLING TOWER	20	105-121	IND. PARK	400-1000	300
INDUSTRY	0.2	95-120	IN PLANT	10	300
INDUSTRY	2.0	95-120	IN PLANT	100	300
INDUSTRY	0.3	130-160	IN PLANT	30	300
INDUSTRY	0.5	170-200	IN PLANT	100	300

load. This fluid is usually steam. An open-cycle eliminates not only the condenser but also the evaporator by obtaining the low-temperature steam by flashing a waste water stream. A dual-fluid cycle consists of a closed-cycle heat pump that raises the temperature of the heat to some intermediate value between the waste source and the desired end-use temperatures and a semi-open steam heat pump which raises the temperature from the intermediate value to the end-use temperature. In this section, the analyses used to screen the potential single-fluid closed-cycle and dual-fluid cycle working fluids to a representative few, and select an appropriate compressor driver are discussed.

Preliminary Working Fluid Selection for Single Fluid Closed Cycles

The basic requirement for a Rankine cycle heat pump working fluid is that it undergo a phase change at the heat input temperature. Good performance requires that the critical temperature exceed the condensing temperature by about 50 F. For the applications identified, numerous potential working fluids fit these basic criteria; therefore, several additional criteria which affect system performance and cost must be considered. These include: flammability, toxicity, cost, thermal stability, minimum cycle pressure, and other thermodynamic and heat transfer properties.

The following paragraphs discuss some of the properties of the fluids evaluated in this study. In addition, reasons are given why some commonly used fluids were eliminated.

Steam: Steam is commonly used as a Rankine power cycle working fluid. It is nonflammable, nontoxic, and although requiring chemical treatment, inexpensive. Steam has good temperature properties for a heat pump application. Its drawbacks include low vapor pressure (less than 1 atmosphere for the condenser temperatures required for all of the identified applications) and relatively high isentropic work. These factors will lead to large compressors and higher speed operation.

R113: R113 is the only commonly used refrigerant that has critical temperature (417.4 F) above the 350 F desired for the identified applications and is chemically stable (at 300 to 350 F) under the proper conditions. It is nonflammable and nontoxic. It does, however, have a subatmospheric vapor pressure at temperatures below 117 F and this would increase the potential for noncondensible gas leakage for the lower temperature waste heat sources and during shutdown periods.

Methanol: Methanol is an inexpensive fluid that has the desired temperature properties for use in a 300 F output heat pump. It is flammable, has a subatmospheric vapor pressure below 150 F, and has an OSHA TLV limit of 200 ppm (compared to a R113 value of 1000 ppm).

Propanol: Propanol has the same advantages and disadvantages as methanol. Due to its increased molecular weight (60 compared to 32 for methanol) it may yield smaller, less costly turbomachinery.

Toluene: Toluene, a common aromatic hydrocarbon, is flammable and toxic, having the same OSHA TLV value as methanol (i.e., 200 ppm). It is chemically stable to temperatures in excess of 600 F and has a high critical temperature (609 F) but it has a subatmospheric vapor pressure below 230 F.

Hexafluorobenzene: Hexafluorobenzene is a nontoxic, nonflammable, highly stable working fluid that has been suggested for a number of Rankine cycle applications even though it has a subatmospheric vapor pressure below 180 F and costs about $100 per pound.

Fluorinol-85: Fluorinol-85 is a mixture of trifluoroethanol and water and is considered nontoxic and nonflammable. It is, however, costly ($7.90 per pound) and has a subatmospheric vapor pressure below 170 F.

Rejected Fluids: A number of fluids for which experimentally verified thermodynamic properties are available were eliminated from single-fluid closed cycles for various reasons. Common fluorocarbon refrigerants with the exception of R113 have critical temperatures that are too low or chemical stability limits below the maximum temperature of the heat pump cycle required by the identified application. Ammonia has a critical temperature of 271.4 F and therefore would have to operate supercritically. Benzene and pyridine have OSHA TLV limits of 10 and 5, respectively, which are much too low to insure use of conventional compressor shafts, seals and, therefore, would require more expensive equipment. Furthermore, they have properties very similar to toluene which has a higher OSHA limit.

Preliminary Working Fluid Selection for Dual-Fluid Cycles

Design of a heat pump to recover 150 F waste heat and deliver 300 F heat requires a working fluid that possesses not only a high critical temperature but a low atmospheric boiling temperature and low vapor specific volume. High critical temperature is needed for good performance; the remaining characteristics for prevention of noncondensible gas contamination and to permit use of small components. Since these are somewhat contradictory, the best that can be expected with a single fluid is a compromise. However, the use of a two-fluid heat pump cycle would permit selection of one fluid with both a low atmospheric boiling temperature and a low vapor specific volume for heat pumping from the source temperature to some intermediate temperature, and a second fluid with a high critical temperature for heat pumping up to higher output temperatures. However, this two-fluid configuration does, in general, require one more heat exchanger than a one-fluid configuration and is penalized by the temperature difference across this heat exchanger. This can be eliminated by using steam as the high temperature fluid (see Figure 1). This is a logical use of steam since it eliminates the subatmospheric evaporator, an objection to steam in a single-fluid heat pump. Therefore, three dual-fluid cycles were considered, each using steam as the high temperature fluid and R11, R133a and R216, respectively, as the low temperature fluids. Each of these fluorocarbon refrigerants is nontoxic, nonflammable and stable to temperatures of 275 to 325 F or higher under selected conditions.

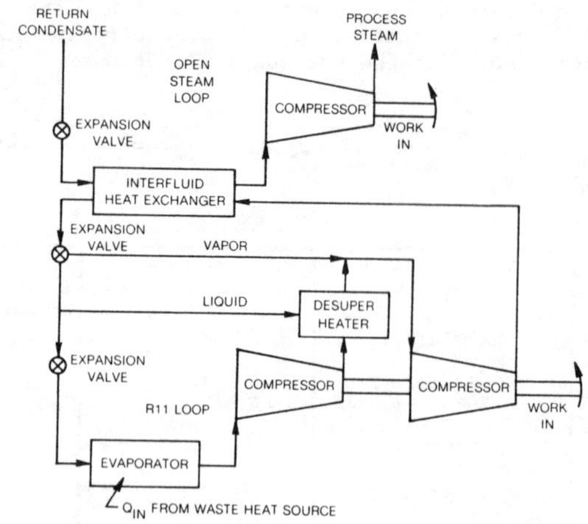

FIGURE 1. DUAL-FLUID RANKINE CYCLE HEAT PUMP.

Working Fluid Comparison

Although final selection of the most appropriate working fluid for any of the defined applications cannot be made until a complete system (i.e., including compressor driver) is defined, the candidate working fluids can be compared in terms of performance and cost, and some eliminated if no advantages are found.

Performance: To assess the relative merits of the candidate working fluids selected for the various cycles, their performance was calculated for the Federal facility application. A 10 F drop in waste heat temperature (equivalent to 200 x 10^6 Btu/hr) and 10 F heat exchanger pinch temperature differences were assumed for the closed-cycle case, yielding evaporator and condenser temperatures of 130 F and 310 F, respectively. The number of compressor stages were estimated based on a maximum tip Mach number of 1.0. The flow out of each compressor stage was desuperheated and added to the portion of expanding fluid that was vaporized at that pressure. The results are summarized in Table 2.

TABLE 2. WORKING FLUID COMPARISON

WASTE HEAT FLUID = WATER AT 150 F
OUTPUT = 300 F STEAM
HEAT INPUT = 200 × 10^6 Btu/hr

OPTION	COP	NO. AND ARRANGEMENT OF COMPRESSORS	TOTAL NO. OF COMPRESSORS	RELATIVE MAJOR EQUIPMENT COST	SIMPLE PAYBACK, YRS
SINGLE-FLUID CLOSED				1.80	9.9
STEAM	3.30	3-2-1-1	7	1.05	∞
R113	2.80	2-1-1	4	1.13	8.2
METHANOL	3.16	2-1-1-1	5	2.93	23.1
PROPANOL	3.16	4-2-1-1	8	4.95	63.0
TOLUENE	3.06	8-5-3-1	17	11.90	∞
HEXAFLUOROBENZENE	2.88	4-3-2	9	2.78	34.7
FLUORINOL-85	3.06	2-2-1-1	6		
SEMI-OPEN				1.56	6.9
STEAM	3.37	3-2-1-1	7		
OPEN				1.37	4.7
STEAM	3.52	3-2-1-1	7		
DUAL-FLUID				1.04	8.1
R11/STEAM	3.14	1-1-1	3	1.00	27.6
R133a/STEAM	2.93	1-1-1	3	3.70	∞
R216/STEAM	2.80	1-1-1	3		

Closed cycle: The COP for the single-fluid closed-cycle heat pump varies from a low of 2.80 for R113 to a high of 3.30 for steam. The low value for R113 is due primarily to the low latent heat at the condenser resulting from the low critical temperature. The low COP for hexafluorobenzene is due to the shape of the fluid saturation curve which yields a large vapor fraction during expansion. Since this vapor must be recompressed without the benefit of absorbing any heat, it acts as a detriment to system performance. The remaining fluids in the single-fluid, closed-cycle heat pump have comparable COP's indicating there is no outstanding working fluid from a performance viewpoint and each fluid must be judged on other merits (i.e., cost, flammability, toxicity, etc.).

Semi-open cycle: Compared with the closed-cycle steam heat pump, the semi-open cycle configuration has a slightly higher COP. Since steam is both the working fluid and the required process fluid, no condenser is needed and the machine internal lift is 10 F less than in the closed-cycle. However, it is assumed that all the condensate returns from the process use and is saturated at 300 F.

Open-cycle: The open-cycle configuration involves two additional considerations. First, heat is recovered by flashing the waste heat stream and recovering the vapor component as saturated steam at 140 F (the heat recovered is equivalent to the same 10 F drop in the waste stream as in the closed-cycle configuration). Second, the return condensate, assumed to be saturated liquid at 300 F, is expanded in steps to permit vapor transfer and desuperheating as before but the portion remaining as liquid at 140 F must be discarded since it is not valuable as a waste heat stream. All other assumptions are the same. The performance (Table 2) shows that the COP of the open-cycle is 4.4 percent higher than the semi-open cycle and 6.7 percent higher than the closed-cycle configuration. However, this concept involves using some of the waste heat stream fluid as the output and if the waste stream is contaminated with noncondensible gases this may be impractical.

Dual-fluid cycle: The performance for the three dual-fluid heat pumps given on Table 2 vary almost as much as the single fluid heat pumps. While the R133a/steam and the R216/steam heat pumps have poorer performance than most of the single fluid heat pumps, the R11/steam working fluid configuration has comparable performance with all but the pure steam single fluid configurations.

Component Sizing: Since the Rankine cycle requires evaporation of the working fluid, the vapor phase thermodynamic properties have a profound effect on the size of the compressors and heat exchangers. This is particularly true when operation is at a low temperature relative to the critical temperature. For large heat pumps with a high specific volume working fluid, the use of a single component (e.g., one compressor) may exceed the economic limits for a commercial application. Since 5 ft diameter compressors are the largest commonly produced, in this study flows are divided into several parallel streams where this dimension is exceeded. The resulting total number of compressors are shown on Table 2 for each fluid. For example, the steam closed-cycle requires 4 stages of compression to achieve the required pressure increase. There must be three units in parallel in the first stage to meet the diameter restriction. There must be two units in parallel in the second stage for the same reason. Only in the third and fourth stages can a single unit under the 5 ft diameter restriction meet the flow requirement. Therefore, 7 individual compressor units are needed. Of the remaining closed-cycles, only R113 and toluene require a significantly different compressor configuration. R113 is more dense than the other fluids and has a high molecular weight. This combination reduces the compressor requirement to 4 units. Toluene has such a high specific volume at the 130 F evaporator condition that 8 5 ft compressors would be required in the first stage and 17 for all four stages. Both the semi-open and the open-cycles have characteristics similar to the closed-cycle steam case. All three of the dual fluid cycles require only three stages (2 organic and 1 steam) since the organic fluids chosen are dense and have high molecular weight.

Preliminary Selection: The cost of the major components of each system assuming a fixed heat input is given relative to the least costly system. Furthermore, a simple payback comparison to an oil-fired boiler. This analysis was made assuming an electric-motor drive, fuel cost of \$3/10^6 Btu and electricity cost of 3.5¢/kWhr. These values can only be used to compare systems with similar configurations. They cannot be used to make a final system selection since the units have different outputs and the driver is not optimized. However, these results provided a simple method of narrowing down the number of potential working fluids for further evaluation. For example, the R113, propanol, toluene, hexafluorobenzene and Fluorinol-85 heat pumps have paybacks at least 2½ times as long as methanol so they can be eliminated from further study. Moreover, the closed-cycle steam heat pump offers no advantage over the semi-open cycle heat pump and can also be eliminated. Finally, the R11/steam dual-fluid cycle shows a 50 percent advantage over its nearest dual-fluid competitor. Therefore, only the methanol, semi-open steam, open steam and R11/steam heat pumps will be studied in more detail.

Heat Pump Driver Selection

Shaft power can be supplied to the heat pump by means of (1) gas turbines, (2) diesels, (3) steam turbines, (4) electric motors or (5) waste heat recovery Rankine systems. Each is characterized by a power conversion efficiency (i.e., the percent of input fuel convertible to useful shaft work) and a heat recovery efficiency (i.e., the percent of input fuel recoverable as useful, 300 F, process heat)(see Table 3). These characteristics are used with typical heat pump COP values to determine the ratio of useful heat output to fuel used as a function of output to fuel used as a function of output temperature for a 150 F waste heat input (Figure 2). Only gas turbines and diesels offer high fuel conversion ratios. Back pressure steam turbines have lower fuel conversion ratios. Condensing steam turbines and electric motors have even lower ratios since no useful prime mover heat recovery is permitted and are eliminated from further study. Waste heat driven-Rankine cycle systems do not use fuel and their economics must be evaluated on a site-specific basis.

TABLE 3. PRIME MOVER CHARACTERISTICS.

	POWER EFFICIENCY	HEAT RECOVERY EFFICIENCY
GAS TURBINE	0.32	0.43
DIESEL	0.42	0.16
STEAM TURBINE		
• BACK PRESSURE	0.17	0.68
• CONDENSING	0.30	0
ELECTRIC MOTOR	0.33	0
WASTE HEAT DRIVEN RANKINE CYCLE	∞	UNDEFINED

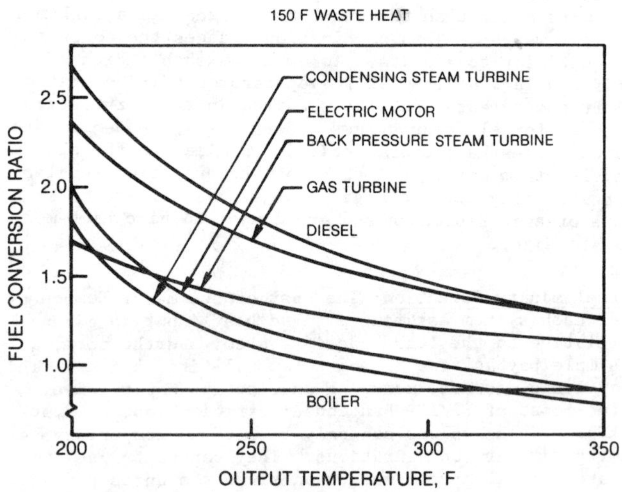

FIGURE 2. HEAT PUMP DRIVER COMPARISON.

SYSTEM SELECTION FOR PORTSMOUTH APPLICATION

The available waste heat resource at Portsmouth is 143,000 gpm in summer and 113,000 gpm in winter at an average temperature of 150 F. A typical process heat requirement for hypothetical industrial park which could effectively utilize this source is 400 x 10^6 Btu/hr of 300 F saturated steam. The heat pump equipment site which is assumed to be located at the industrial park would require a 3-mile pipeline (1½ miles each way) to transport waste water to and from the GDP.

Gas Turbine Drive

Selection of the most attractive heat pump system for a particular application can best be made on the basis of the ROI compared to a conventional gas-fired boiler. ROI values were calculated for the combinations of heat pumps working fluid configuration and drivers previously identified using the assumptions presented in Table 4. Unless otherwise stated, ROI is a before-tax value and is valid for government ownership of the heat pump system. After tax ROI's for privately owned equipment is generally 75 percent of the before-tax ROI. The matrix of configuration and drive options was analyzed by first comparing the ROI of the four configurations with a gas turbine driver. The results are summarized on Table 5 and show that the R11/steam system has the highest ROI although the advantage is not substantial.

TABLE 4. ECONOMIC ANALYSIS ASSUMPTIONS.

SYSTEM LIFE	20 yrs
DEPRECIATION PERIOD*	16 yrs
INVESTMENT TAX CREDIT*	10%
MISC. ANNUAL CHARGES	2%
O&M AS PERCENT OF CAPITAL COST	2.25%
CAPITAL COST YEAR	1980
COST YEAR OF FUEL, ELECTRICITY, ETC.	1980
START-UP YEAR	1982
GENERAL INFLATION RATE	6%
ESCALATION RATE OF FUEL**	4%
ESCALATION RATE OF ELECTRICITY**	2%
COST OF FUEL	$3/10^6 Btu
COST OF ELECTRICITY	3.3¢/kWhr
DEPRECIATION METHOD*	STRAIGHT LINE
TAX RATE*	48%
DISCOUNT RATE	10%

*THESE ITEMS ARE OMITTED FOR BEFORE TAX CALCULATION
**ABOVE GENERAL INFLATION RATE

TABLE 5. COMPARISON OF HEAT PUMP SYSTEMS WITH BOILER FOR PORTSMOUTH APPLICATION.

GAS TURBINE DRIVE
WASTE HEAT TEMPERATURE = 150 F
OUTPUT = 400 x 10^6 Btu/hr OF 300 F STEAM

SYSTEM	COPF	ANNUAL FUEL CONSUMPTION, 10^{12} Btu	TOTAL SYSTEM INSTALLED COST, $\$10^6$ (1980)	BEFORE TAX RETURN ON INVESTMENT, %
BOILER	0.85	3.76	3.2	BASE
R11/STEAM	1.41	2.27	20.5	36.2
METHANOL	1.42	2.26	21.4	34.8
SEMI-OPEN STEAM	1.48	2.16	24.6	32.0
OPEN STEAM	1.52	2.10	22.5	35.7

Diesel Drive

For the total system output of this application, the size of the heat pump needed using a diesel driver is larger than that needed using a gas-turbine driver since significantly less heat can be recovered from the diesel exhaust gases. This larger heat pump plus the higher capital cost for a diesel yield a total R11/steam system installed cost of \$36.3 x 10^6 compared to only \$20.5 x 10^6 for the gas turbine driven system. However, the higher efficiency of the diesel reduces the annual fuel requirement from 2.27 x 10^{12} Btu for the gas turbine to 2.14 x 10^{12} Btu. This yields an ROI of 21.9 percent, considerably below the 36.2 percent for the gas turbine driven system.

Steam Turbine Drive

Using a back pressure steam turbine for the driver of the R11/steam heat pump to produce 400 x 10^6 Btu/hr provides a system that is quite different than the one which uses the diesel. This is due to the large fraction of the steam turbine output which is recoverable as steam and the small fraction of its output as shaft power. For the Portsmouth application, the exhaust heat recovered from the prime mover is 124 x 10^6 Btu/hr, 42 x 10^6 Btu/hr, and 224 x 10^6 Btu/hr for the gas turbine, respectively. Thus, the capacity of the heat pump proper is much less for the steam turbine driver case and the cost for the total system is less. However, the resulting small use of Portsmouth waste heat requires a large fuel consumption. The ROI for such a system is 19.9 percent. Thus, neither the diesel- nor the steam turbine-driven heat pump show a greater ROI than the gas turbine-driven system. Further, a disparity in driver economics is so great that the selection of the R11/steam heat pump over the other fluid configurations for this application would not be altered by use of a diesel or a steam turbine.

Effect of Multiple Units

The Portsmouth gaseous diffusion plant industrial park requirement can, and for high reliability should be provided by a multiple number of smaller units. Using the ROI procedure previously discussed, values of this economic parameter have been calculated for units with smaller outputs both with and without the required 3-mile waste water pipeline. The ROI results are presented in Figure 3 and show the larger units are more economical. However, the use of four 100 x 10^6 Btu units, three 133 x 10^6 Btu/hr units or two 200 x 10^6 Btu/hr units would increase overall system reliability while still providing relatively high values of ROI. However, the use of multiple units each below 100 x 10^6 Btu/hr appears much less acceptable.

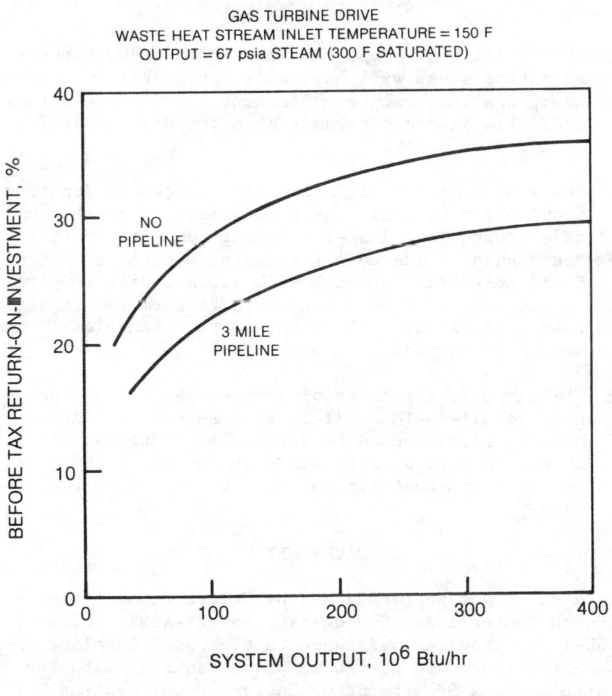

FIGURE 3. HEAT PUMP SYSTEM ECONOMICS FOR POTENTIAL APPLICATIONS.

Coal-Fired Boiler Backup System

Comparing system economics to a gas-fired boiler base case may be misleading since coal-fired boilers are also in use. The installed cost of a 400 x 10^6 Btu/hr coal-fired boiler including blowdown tank, stack gas cleanup, coal storage and handling, solid residue handling, stack, ducts and piping is approximately equal to the installed cost of the 400 x 10^6 Btu/hr gas turbine-driven, duel-fluid heat pump (i.e., \$20 x 10^6). This can be compared to the \$3.2 x 10^6 defined as the installed cost of a more simpler gas-fired boiler.

With a coal-fired boiler having the same capital cost as the heat pump, and assuming yearly maintenance costs are the same, the selection of the preferred process heat producing equipment becomes a function of yearly fuel costs. The data presented in Figure 4 show a comparison of annual fuel costs as a function of process delivery temperature. Coal in the midwest currently costs between \$40 and \$50 per ton. The process heat produced from coal-fired boiler can be provided at the temperature ranges required. Natural gas prices can vary widely depending on location. The cost range shown (\$2.50 to \$3.50/10^6 Btu) represent a typical range. Typically, in Southern Ohio coal can cost \$40/ton while natural gas costs about \$2.75 per 10^6 Btu for large industrial purchasers. At these fuel costs the gas-fired heat pump has an annual fuel cost advantage at delivery temperature of 300 F, but a more detailed evaluation could be required to verify the magnitude of this advantage.

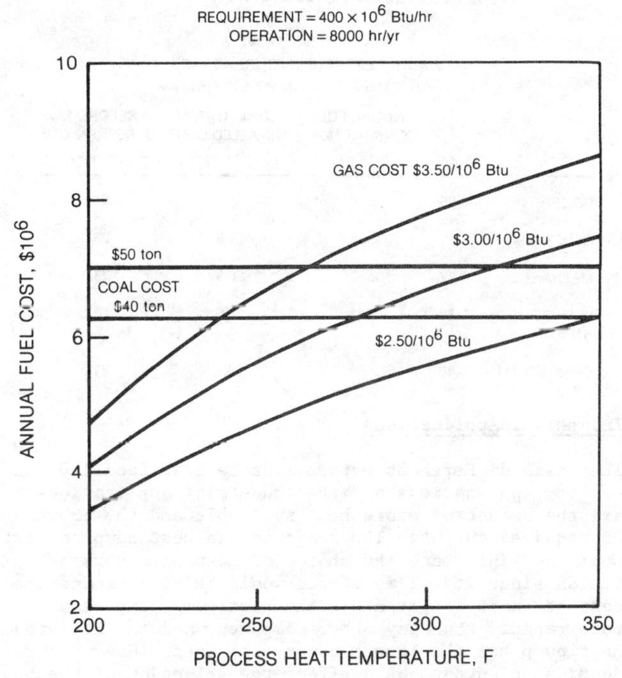

FIGURE 4. ANNUAL FUEL COST FOR PORTSMOUTH INDUSTRIAL PARK PROCESS HEAT.

ALTERNATIVE APPLICATION EVALUATIONS

Utility Applications

For the utility application, 20 x 10^6 lb/hr of waste hot water is available at an average temperature of 110 F and the desired output is saturated steam at 300 F. Thus, this application differs from the Federal facility application primarily in the temperature lift required. Using the same assumptions as those made for the Federal facility application cost estimates for the gas turbine, major components, auxiliaries and installation for the R11/steam, methanol, semi-open steam and open steam heat pump systems were made. There are two reasons why these costs are higher than those for the Federal facility application. First, the increased lift yields a lower heat pump COP (2.58 compared to 3.14 for the R11/steam heat pumps) which requires a larger, more expensive prime mover to produce the same heat output. Secondly, the lower waste heat temperature requires an additional compressor stage, and the use of larger compressors to accommodate the increase in vapor specific volume. This latter reason is quite pronounced in the two steam heat pump configurations where the mass flow had to be split into three streams to permit reasonably sized compressors even for the 150 F Federal facility application.

The return-on-investment values obtained for the four systems (Table 6) show that for this application the R11/steam heat pump is clearly superior to the two steam systems and more than 10 percent better than the methanol system when compared on a before tax basis. The driver selection for the R11/steam heat pump is unaffected by designing to the utility application requirements. Using a diesel will increase the capital cost and not permit any substantial fuel savings. Using a steam turbine drive for this heat pump COP would result in too low a fuel savings compared to a boiler to justify the expense of the heat pump system.

TABLE 6. COMPARISON OF HEAT PUMP SYSTEMS WITH BOILER FOR UTILITY APPLICATION.

GAS TURBINE DRIVE
WASTE HEAT TEMPERATURE = 110 F
OUTPUT = 400 x 10^6 Btu/hr AT 300 F STEAM

SYSTEM	COPF	ANNUAL FUEL CONSUMPTION 10^{12} Btu	TOTAL SYSTEM INSTALLED COST $\$10^6$ (1980)	BEFORE TAX RETURN ON INVESTMENT, %
BOILER	0.85	3.76	3.2	BASE
R11/STEAM	1.26	2.54	23.9	26.0
METHANOL	1.27	2.52	26.9	23.3
SEMI-OPEN STEAM	1.33	2.41	36.6	18.2
OPEN STEAM	1.36	2.35	33.1	21.2

Industrial Applications

The basic differences between the Federal facility and utility applications and the industrial applications are the amount of waste heat available and the amount of required output. The reduction in heat pump capacity would not influence the choice of heat pump working fluids since its prime effect would be to increase the cost per unit of output of the heat pump compressor more rapidly than any other component and the R11/steam heat pump has the lowest compressor cost. However, reduction in output could effect the selection of the heat pump driver. While gas turbines and steam turbines exhibit a fairly continuous variation of cost with size, diesels do not. Large diesels (~10,000 hp) are normally designed for base-load operation and generally operate at low speed. They are quite bulky and are designed for long life with a minimum of downtime and maintenance problems. Industrial diesels in sizes below 10,000 hp have different characteristics. Although there is no sharp dividing line, these diesel engines (commonly referred to as high-speed diesels) operate at higher speeds and tend to weigh only one-third that of the larger units on a per unit output basis. They also are produced in greater quantity. Therefore, a diesel of this design philosophy would cost less than a low-speed model of the same output.

Typical equipment costs for a 4000-10,000 hp diesel of this type is \$125/hp, roughly comparable to gas turbines. Using these driver costs and scaling the heat pump costs to reflect smaller sizes, ROI values were calculated for the two different systems as functions of output for waste heat temperatures of 110 and 150 F. These data show that for the size range where the relatively inexpensive high-speed diesel can be used ROI is insensitive to driver selection for 150 F waste heat. However, the gas turbine has a clear advantage when the waste heat is 110 F (Figure 5).

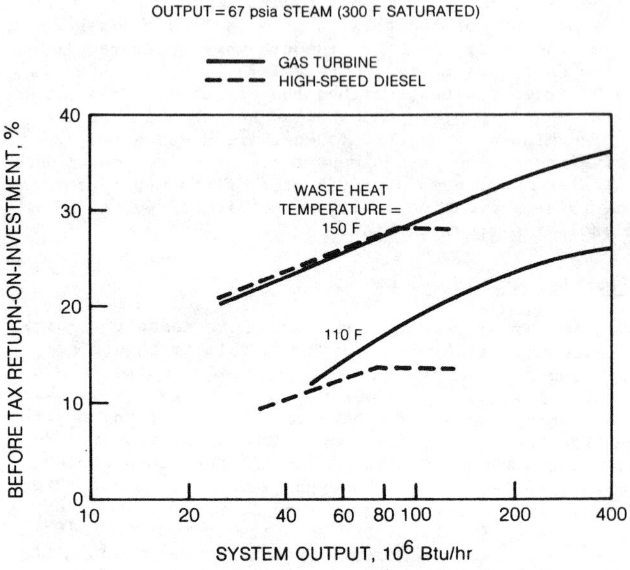

FIGURE 5. THERMALLY DRIVEN HEAT PUMP SYSTEM ECONOMICS.

SIGNIFICANT FINDINGS

- Industrial heat pumps that can provide 300 F steam for process use will have wide applicability. However, low temperature waste heat sources are normally available in quantity only at a temperature of 110 to 150 F or below.
- The most economically attractive system design to provide the process heat requirements from the available source is a dual-fluid heat pump driven by a heat engine. The dual fluid design permits working fluid selection to match both low and high temperatures. Use of a heat engine drive permits exhaust heat recovery at high temperatures to supplement the heat pump output.
- The economic advantage of large-capacity heat pumps over fossil-fueled boilers increases with size. Minimum sizes needed to achieve a before tax ROI of 20 percent are approximately 30 and 100 x 10^6 Btu/hr for waste heat streams of 150 and 110 F, respectively.

ACKNOWLEDGMENT

This study was performed by the United Technologies Research Center under Subcontract No. DE-AC09-77ET12855-SUB-15 to provide assistance to the South Carolina Energy Research Institute for their prime contract with the United States Department of Energy, Power Systems Division. The authors wish to thank Dr. F. Joseph McCrosson and Mr. Paul W. Yngve of the South Carolina Energy Research Institute and Mr. Frank R. Biancardi of United Technologies Research Center for their help during this study.

SECTION 2
ENERGY UTILIZATION IN BUILDINGS

Chapter 7
ENERGY CONSERVATION TECHNIQUES FOR DATA CENTERS

J. E. Hassett

The ever-increasing cost of energy is focusing attention on the design and operation of data centers because of the opportunities to save energy in both the computer room and the office building in which it is located.

The need for a specific environment has been established by the mainframe manufacturer and serves as the basis for the design of the cooling system. It is "process cooling" not "air conditioning" that meets the criteria for satisfactory hardware performance.

A number of questions arise while investigating methods of reducing energy use in data centers. How can the needs of the hardware be met while keeping energy use to a minimum? What are the proper design criteria? What design sources or information are available? What process cooling systems are available? What are the latest techniques for saving overall energy costs?

<u>Design Criteria</u>

Most mainframe manufacturers recommend the following design suggestions and cautions. Room temperature and humidity settings should be maintained to ensure maximum uptime for a center. The recommendation of the site planner or customer engineer should be followed. He knows the optimum conditions for the hardware configuration. The temperature setting is in the range of 72 to 75 F. The relative humidity is normally 50 percent.

The relative humidity of the air under the floor, which enters the room via the cable holes of equipment, must not exceed 80 per cent. Many cooling systems violate this requirement! The designer should verify the underfloor relative humidity for the data center.

The temperature of the air that passes through hardware when it is "down" for service should not be lower than 60 F when the equipment is turned on again. If this is not followed thermal shock will adversely affect system performance for many hours or days.

There are many other criteria, particularly those involving energy conservation, which will be discussed later.

The hardware suppliers do not normally design environmental systems. Instead, they identify optimum design conditions and equipment limits and leave the design and performance of the environment to the user.

However, there are many people available to assist in designing new or expanded facilities and in the diagnosis of problems in existing rooms. The project may warrant putting together a design team. Consider the expertise of the following people:

<u>Installation Planning Representative</u>, site planning specialist, pre-installation planner or whatever name the mainframer calls him. Each knows the conditions for optimum performance and the room should not be designed without his information. Ask for the latest copy of his 'site planning manual.'

<u>In-House Facilities Planners</u>, building engineering and maintenance personnel. They know the services available in unsuccessful systems, etc. They can make recommendations as to suppliers.

<u>Utility Company Personnel</u>, in addition to their knowledge of electric power, are becoming more aware of energy conservation techniques and can also document the magnitude of the increase in the electric rates in a particular area. Their projections will help in preparing operating cost savings figures in reports to management. Some energy conservation techniques may result in an initial expenditure which will have to be justified. Utility company projections will lend credence to energy savings projections.

<u>Qualified Manufacturers' Representative</u>, know the latest techniques for the environmental design, systems and energy conservation. They are professional in their engineering application of systems for data centers. Because customer satisfaction is important to their local future, more attention will be paid to the total requirements of a center. Look for after-sales service capability, too.

"Process cooling" has become synonymous with environemntal control of data centers. It is "process cooling" not "air conditioning" that provides reliable and continuous computer room operation by satisfying the requirements of the hardware. A typical downflow process cooler (see Figure 1) meets the following criteria:

1. Cools, dehumidifies, reheats or humidifies automatically.

2. Provides the high sensible heat capacity to match the room heat dissipation. Air conditioners do not do this.

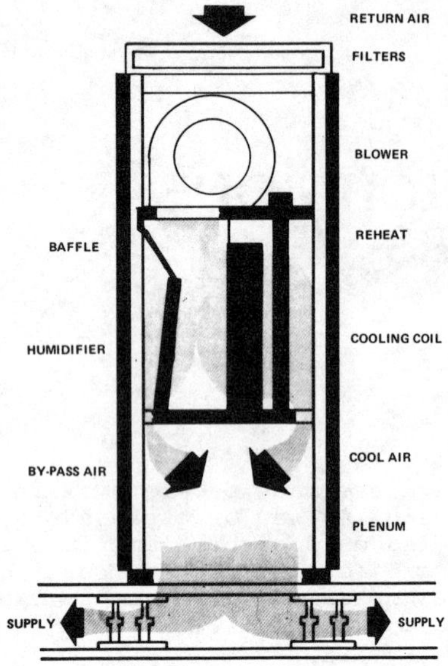

FIGURE 1

3. Limits the discharge air to about 80 per cent relative humidity by using a fan which delivers 20 percent additional air from the room to bypass the cooling coil. The mixture of these air streams (cooled and bypass) leave the process cooler at about 80 percent relative humidity.

4. Built-in redundance with multiple refrigerant circuits.

5. Automatic controls and solid state status and service panels for prompt and accurate display.

6. Static pressure regain plenum for better air delivery to low-height raised floors.

The bypass air stream is the proper place for a humidifier. Since the air passing through the humidifier is at the return air room condition of 50 percent relative humidity, it is easy to add moisture that is only holding 50 percent as much as it can hold at that temperature (72 to 75 F.). If no bypass were available, it would be difficult to add moisture to air before or after the cooling coil.

Factory assembly and thorough testing contribute greatly to the reliability of the installed system. When the system is piped and power-wired, the installer knows that the factory testing should assure him of a quick and satisfactory start-up.

Energy is used in data centers to cool and dehumidify, dehumidify only, or humidify.

In addition, there are many hours of the year when heating is needed for other parts of the building. The heat being rejected from the data center could be used for this purpose, resulting in savings in heating bills.

Two ways in which energy use may be curtailed are by reducing the need for cooling, dehumidifying, dehumidifying only and humidifying; and by reducing the cost of providing the resulting tasks. With this in mind, here are the most common energy conserving techniques being employed in the data center application today.

1. Reduce the ventilation air quantity. Outside air that is brought in the data center must be cooled and dehumidified during warm weather and heated and humidified during cool weather. While cool air would be a benefit in reducing the cooling load, the cost to humidify it outweighs the cooling gain.

Reduce the amount of outside air to about 1 percent of the air in circulation, not the 15 percent usually associated with comfort air conditioning design. Look at the difference: process cooling a data center vs. air conditioning.

In a 5,000 sq. ft. data center, the air in circulation would be about 30,000 cfm. Based on process cooling standards, it would bring in a maximum of 1 percent or 300 cfm. An air conditioning system would call for 4,500 cfm to meet most local ventilation codes!

Process cool the data center. Don't air condition it!

2. Use a vapor barrier to minimize infiltration/exfiltration. Many data centers are constructed within existing buildings, with sheetrock partitions extending from floor to the underside of the hung ceiling. The space surrounding the data center is normally air conditioned offices. In the winter the relative humidity of the office space is comfortable for the occupants at 25 to 30 percent. The data center environmental control tries to maintain 50 percent but cannot do so. Moisture from the 50 percent relative humidity room migrates to the 25 percent side of the wall because no vapor barrier was used in the walls and above the ceiling. Without a vapor barrier, the humidifiers will be on most of the time, and the cost to humidify can be expensive.

3. Minimize system dehumidifying. Almost all cooling systems take out moisture (dehumidify) while they are cooling the air. Very little moisture should be getting into the room at any time of the year, so any dehumidification being done by the cooling apparatus is usually undesirable. If the relative humidity of the data center is being maintained at 50 percent, the moisture removed by the cooling process will have to be put back by the humidifier. Limit the amount of dehumidification whenever possible, depending on the type of cooling system in use.

Avoid the use of comfort air conditioning systems. They were designed to dehumidify an amount equivalent to 30 to 35 percent of their total capability. This is far too much in a data center where so little moisture is present.

The temperature of the chilled water in chilled water systems should be variable so that in the summer a water temperature of about 45 F will be needed for any dehumidifying. In the winter, the temperature should be raised to 50 to 52 F. Leaving the water temperature at 45 F year round will result in excessive dehumidification and very expensive re-humidification.

Almost all compressorized direct expansion systems do some dehumidifying, particularly air-cooled systems operating during outside temperatures below 75-80 F, which is most of the time. Direct expansion systems used in these process cooling applications should be operated at higher condensing temperatures and pressures than normal air conditioning systems to minimize their dehumidifying capability.

It will be shown later that glycol-cooled systems are generally the best choice for data center application.

4. Operate cooling systems at peak efficiency. Simple, normal maintenance will keep the operating costs down and minimize downtime. Filters and humidifiers should be inspected regularly and cleaned or changed as needed.

5. Reduce the cost of humidifying: Humidification is required most of the year to replace moisture that migrates out of the room into surrounding spaces, to add moisture to winter ventilation air and to replace moisture removed by the cooling systems.

Two common ways to humidify air are by flashing water into steam by electricity, and evaporating water from a special dispersing surface.

The electric humidifier requires about 1,000 BTU per pound of water while the evaporative panel type needs only 80 BTU per pound. In addition, the electric humidifier heats the same air stream that needs to be cooled.

On the other hand, the evaporative panel humidifier contributes to the cooling of the system by cooling the air stream.

Also, evaporative humidifiers are 'fail-safe', in that they cannot overhumidify. The electric humidifier will continue to generate steam as long as the humidistat calls for moisture. If the humidistat sticks 'shut' excessive moisture will adversely affect the hardware, cause corrosion, tape stick and card-feed failure.

Evaporative humidifiers are designed to operate efficiently with 140 F water. This water can be heated with waste heat from normal compressor operation, resulting in 'free' humidification.

The savings in energy of evaporative panel humidifiers over electric steam generation humidifiers range from $9,251 for a 300 sq. ft. room to $40,051 for a 1,200 sq. ft. room. The savings are for a ten year period, assuming an electric power rate of 4¢/kwh with an anticipated doubling during the period.

There are additional savings. Since there is no electric service to the humidifier, installation costs are lower, and demand charges and the costs associated with elec= trical maintenance are saved.

Ask for: Evaporative panel humidifiers with 'hot gas' water heaters! For more on humidification and dehumidification see "M/E Update Humidifiers and Dehumidifiers, SE, October 1979, pp. 111-118, and "Humidity control provides subtle savings of energy," SE, November 1979, pp. 108-111.

6. Reduce the cost of "reheating:" There are times during the operation of small data centers when the room temperature is satisfactory but the moisture content is too high. In order to wring the moisture out, the air temperature is reduced. This air must be reheated during this 'dehumidification only' period.

In the past, this reheating was done with an electric coil. However, the waste heat from the compressor that is running to provide the cooling can be used to do all the needed reheating. For more on using compressor waste heat, see "There's energy for use in refrigeration waste heat," SE, June 1979, pp. 96-103; November 1979, pp. 122-126; and January 1980, pp. 74-78.

The savings by using hot gas reheat rather than electric reheat range from $19,505 (300 sq. ft.) to $27,417 (1,200 sq. ft.). Assumptions are the same as in the previous example.

Ask for "Hot Gas Reheat".

7. Reduce the cost of providing cooling: When the outdoor air temperature is 55 F and below, it is possible to cool the data center using cold glycol instead of running the compressors. There are thousands of hours a year when the outdoor temperatures are low enough to permit this type of operation. The savings are impressive.

For instance, in a 5,000 sq. ft. computer room located along a belt which includes New York, Indianapolis and Kansas City, the ten year savings would be approximately $200,000!

Here's how the system works:

When the outdoor temperature is more than 55 F, glycol from each of the in-room conditioners is warm, due to the heat rejection from the refrigeration compressors and condensers. The warm glycol must be cooled through the coil of the dry cooler located outside the building and returned to the process coolers in the room.

However, when the outdoor temperature drops to 55 F and below, the glycol, which is now very cool, is automatically diverted to a cooling coil mounted on top of the process cooler. At this point the refrigeration cycle is shut off. This cycle is shown in Figure 2.

One might think that outside air can be used directly in the data center rather than to cool glycol. While it is true that cool outdoor air could cool the room, its moisture content would be so low that it would cost more to humidify the air stream than the savings in compressor operating cost.

By cooling glycol, which cools the air, outdoor air is not introduced directly into the space. This process avoids a large humidification cost.

Ask for: "Glycol cooling at low outdoor temperatures."

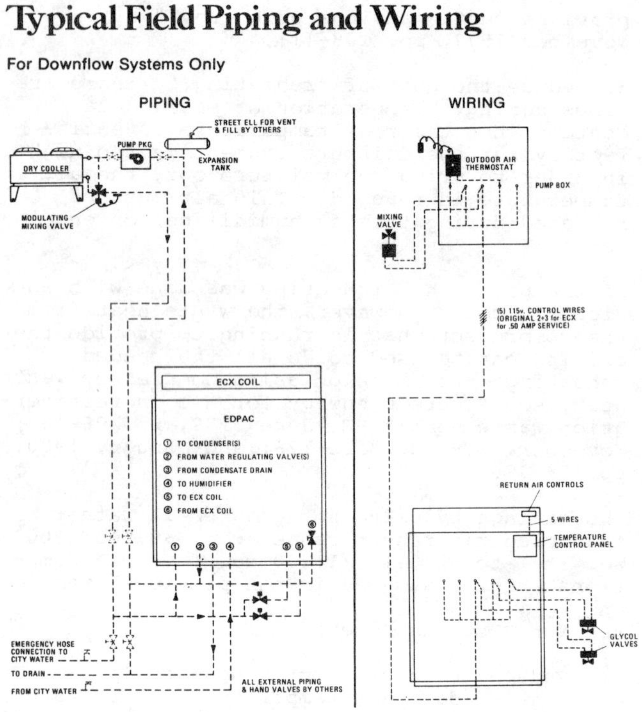

FIGURE 2

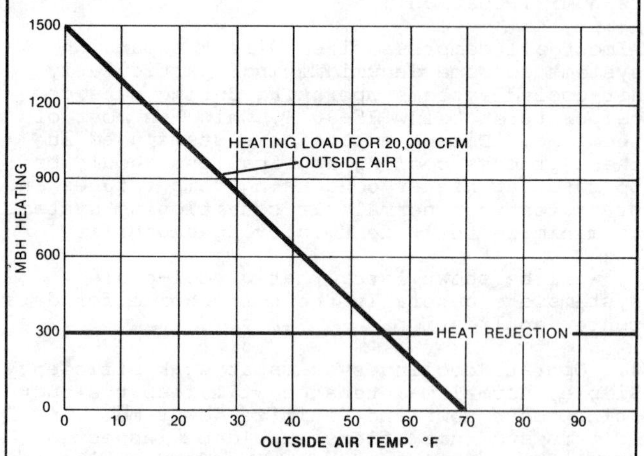

	OUTSIDE AIR TEMP. RANGE °F							
	0-10	10-20	20-30	30-40	40-50	50-60	60-70	ANNUAL:
HOURS	32	238	683	1508	1391	1391	1581	HOURS IN RANGE
MBH	300	300	300	300	300	284	87	RECOVERED HEAT
MBTU	9,600	71,400	204,900	452,400	417,300	395,044	137,547	1,688,191 MBTU TOTAL
								= 494,640 kwh

FIGURE 3

8. Use rejected heat for the building heating needs: In the previous example, the heat of rejection from the compressors and condensers in the process coolers heated the glycol to about 110 to 115 F. This warm glycol would normally be pumped to the dry coolers where it would be reduced in temperature and sent back to the process coolers.

Warm glycol could be used to heat other parts of the building and save large amounts of money and energy. Or, it could be used for other heating needs such as : preheating city water for the domestic hot water system, preheating or reheating air for the building air conditioning system, saving the heat in large storage facilities, or for snow-ice-melting systems. Figure 2 shows the simple piping system.

For each heat-using device or system, (air heater, water heater, etc), there is a secondary pump and three-way modulating valve. When the return air or water temperature drops below the set point of the control or thermostat, the secondary pump starts and the valves modulate the flow of glycol through/around the coil or heater.

Any competent piping designer knows how to specify the piping requirements for this type application.

The savings associated with recovering the heat of rejection from data centers represents one of the designer's greatest opportunities to reduce operating costs. While the actual saving depends on how the heat of rejection can be used and the electric power rate, the designer should look at all the possibilities.

For instance, consider an all-electric facility in New Jersey which has a 2,000 sq. ft. data center with a heat rejection of 300,000 Btuh around the clock.

Another part of the building has a need to heat 20,000 cfm of outside air from varying temperatures up to 70 F; the heating load is expressed as the line from 70 F (zero heating need) down to 1,500 MBH at 0 F. The data center heat rejection provides the first 300 MBH, leaving the rest of the heating to be paid to the electric utility company. But what is the value of the free 300 MBH?

Figure 3 shows that all 300 MBH can be used up to about 57 F, where the load is less than the heat rejection. Given the number of hours in each ten degree outdoor temperature range and multiplying by the MBH energy saving, it yields an annual saving of 1,688,191,000 BTU (equivalent to 494,640 kilowatt=hours). At an average electric rate of 6¢/kwh, the total saving over a ten year period is $296,784.

Summary

The designer has many opportunities for energy conservation at his disposal, namely: reducing ventilation air quantity, using a vapor barrier, minimizing dehumidifying, maintaining performance at peak efficiency, using evaporative panel humidifiers with hot gas water heating, using hot gas reheat coils, using glycol cooling at low outdoor temperatures and using heat of rejection for other building heating needs.

Leading suppliers of process cooling equipment will prepare an energy analysis of various types of systems (air-cooled, water-cooled, glycol-cooled, chilled water). The computer analysis gives a fair evaluation of all the costs of operation: compressor, fans, pumps, humidifiers, etc. Do not make the mistake of comparing one item alone for best performance. What you pay for is the total composite system.

Chapter 8

ENERGY SAVING DESIGN IN THE OPERATION OF THE HEATING AND VENTILATING SYSTEM FOR THE WOLVERINE'S NEW FIELD HOUSE

T. M. Rabiah

ABSTRACT

The University of Michigan Wolverines new Field House that opened in December 1980, before The University of Michigan football team won the Rose Bowl game, is 380 ft (115 m) long by 190 ft (58 m) clearspan wide by 56 ft (17 m) high.

The building has a steel rigid frame with 4 inch (100 mm) thick vinyl faced fiberglass insulation on walls and in ceiling. Floor is concrete with 1/2 inch (13 mm) thick foam covered with grass-like carpeting.

The heating system consists of two direct gas fired make-up air units, 50,000 CFM (85,000 m^3/hr), 1,500,000 BTU/HR (440 kw) each. They are used to provide winter heating with 50% outside air and 50% return air. They also provide summer ventilation by supplying 100% outside air. Even though there are no distribution ducts, temperature variation throughout the building is less than 2°F.

The building exhaust system is designed to operate automatically according to building pressurization through gravity type weight counterbalanced dampers. They open and close according to building pressure while maintaining the design pressure. Therefore, electrically driven exhaust fans were eliminated which resulted in electric power saving of 40 hp (30 kw).

Air exhausted through the counterbalanced dampers is directed downward into the perimeter glass windows. Thereby this eliminates frost accumulation or condensation on the glass and at the same time reducing heat loss through it.

BUILDING LOCATION

The building is located in the south side of the campus adjacent to the existing sports services building and the outdoor football practice field.

BUILDING USAGE

The building was mainly designed on the idea that it will be used as an indoor football practice field. One goal post at each end of the building is hung from the ceiling where full place-kicking drills can be held. Punt-blocking and punt returns can be practiced as well. The building may also be used for other various athletic activities.

HEATING SYSTEM

The original thought for heating the building was to construct a new steam supply and condensate return system from an existing power house in that part of the campus. But after considerable thought, study and many conferences it was decided that the pay-back for such an investment was too long. Finally a decision was made to use two gas fired heating units.

A careful consideration of characteristics was made for both, the direct fired and the indirect fired air heaters. The direct fired was selected due to many advantages, such as:

1. Higher efficiency (90% or higher) which leads to lower operating cost, hence, higher energy savings.
2. No heat exchanger to corrode or leak.
3. Can be easily adapted to take all combustion air from outdoors, or provide 50% outdoor air and 50% recirculated air (after burner) through an automatic recirculation motorized damper.
4. Can heat over a wide temperature range.
5. Better turn-down ratio, 25:1 which leads to better control and lower operating cost.
6. First cost is lower in larger sizes. (Ref. 1)

The two units were installed as completely housed, wired and internally piped. One unit was installed on the roof of an existing, adjacent building on the east side of the field house. The other unit was installed on the west end of the field, on the roof of the electric trans former room.

The discharged air from each unit is blown into the building through wall mounted supply air grilles, with the return air grilles adjacent to them.

EXHAUST SYSTEM

There are twelve counterweighted horizontal relief dampers with louvers located in the overhang along the periphery of the building. These dampers were set to open and close automatically in accordance with building pressure. Dampers will be in full open position during summer operation and half open position during winter operation, thereby, maintaining the same building design pressure all year around.

The exhausted air through these relief dampers is directed toward the slanted glass along the periphery of the building. During winter heating, the warm exhausted air through the louver-damper combination would tend to create a warm air circulation around the glass and there by reducing building heat loss through it and keeping moisture or snow from accumulating on it.

The following formula, which was obtained from the counterweighted dampers manufacturer, was used to determine the dampers sizes (in English units). See page (3) for units conversion factors.

For 15 M.P.H. wind

$$X = \frac{A(3100)}{700} + \frac{\text{Ventilation CFM}}{1500}$$

Where
X = Damper face area in square feet

A = Estimated total area of building air leakage space (cracks), ignoring the damper leakage since it was already accounted for in the formula

Ventilation CFM - Amount of air exhausted through the damper in cubic feet per minute

The numbers 3100, 700 and 1500 are constants. (Ref. 2)

BUILDING PRESSURIZATION

The heating and ventilating system design attached alot of importance to pressurizing the building which would result in exfiltration of air out of the building, rather than infiltration of air into the building. This internal pressure, which is in the range of 0.1 inch H_2O (25 Pascal), would eliminate infiltration through walls, doors or roof and, thereby, create an even temperature throughout the building. The internal pressure would eliminate freezing cold air from entering the building through the doors even when they are open.

Exfiltration in large, open buildings is normally spread quite uniformly throughout the building. Similarly, the heated air distribution is equally uniform. This agrees with Bernoulli's Principle which states that pressure enclosed in a vessel exerts an equal pressure in all directions. In other words, it is impossible to have a pressure dead spot in the building.

Building pressurization actually falls into two categories, these are: pressurization without ventilation and pressurization with ventilation.

Pressurization Without Ventilation

In this situation, the equipment used for pressurization is selected to maintain a predetermined building pressure only and without the need for replacing the air inside the building. Therefore, the supply air fan capacity needs to be only that required to overcome the velocity head pressure of the prevailing winds in the area plus the pressure drop of the system. In this case the counterweighted dampers will remain closed most of the time. Normally the system will be operating at the design pressure and the counterweighted dampers will not be open as there is no external resistance to the air leaving through the cracks. When the wind picks up, it produces resistance to this air flow and the pressure in the room begins to rise. As the design pressure is exceeded, the counterweighted dampers will begin to "bleed off" by oscillating a closed and partially open position, thereby allowing the building to return to the design pressure.

Pressurization With Ventilation

This is the type of application we are dealing with, which provides both, ventilation in the building in addition to maintaining the design building pressure. Therefore, the supply fans on the make-up air units were sized with pressure (equal to velocity head pressure of prevailing winds plus the system pressure drop) and amount of ventilation air both in mind.

The following formula, which was obtained from the counterweighted dampers manufacturer, was used to determine the capacity of the supply fan (in English units). See page (3) for units conversion factors.

For 15 M.P.H. wind and 0.10 inch H2O:

Fan C.F.M. = A (3100) + VENTILATION C.F.M.

Where:

A - Estimated total area of building cracks

VENTILATION C.F.M. = Amount of air exhausted

The number 3100 is a constant. (Ref. 3)

Self-Modulating Air Delivery

Another energy saving feature is accomplished from build ing pressurization is the self-modulation of the supply fans, which leads to self-modulation of the power required to run these fans. It is evident that if the building is not pressurized and open, the air delivery from the supply fans is high. But as the building becomes pressurized and closed-up, the air delivery and horsepower go down since any increase in the system external static pressure leads to decline in fan air delivery and horsepower requirements. Therefore, the system can be self-regulating without the need for expensive multi-speed motors and controllers. (Ref. 4)

UNIT CONTROLS AND OPERATION

Each of the two units is provided with a large control panel attached to the unit and contains all electrical components such as motor starter, relays, fused disconnect switch, pilot lights, etc. Each unit is also provided with a separate remote control panel. This panel contains a summer/winter switch and an adjustable knob to accurately control the discharge temperature across the full range of burner firing.

Winter Operation

Summer/winter switch is manually set on "WINTER" position. When the supply fan is energized, a minimum position switch inside the unit allows return air damper on maximum position which is equal to 50% of fully open position of outside air. Space temperature stat modulates gas fired heating section through a controller inside the unit panel. Flame modulation can be made through 25 stages, (turn down ration 25:1). Gas fired heating is off, O.A. damper closed when fan is deenergized.

Summer Operation

Summer/winter switch is manually set on "SUMMER" position. When supply fan is energized, outside air damper opens to maximum 100% position and return air damper is closed. The unit operates on ventilation cycle only.

CONCLUSION

The building heating and ventilating system, as described, required little space for installation outside the building and none to speak of inside the building. Installation cost was greatly reduced as no boiler room, pumps, traps, valves, etc. were needed. Sheet metal ductwork was not necessary for air distribution throughout the building. Installation of roof exhaust fans was eliminated and probably would not accomplish satisfactory results anyway. In fact, excessive heat would be exhausted from the building and be replaced with cold, damp air pulled in through the doors and windows, etc.

Maintenance was reduced as there is less equipment operating. The only maintenance required would be burner controls adjustment and replacement of fan belts.

"REFERENCES"

1. Industrial Ventilation Manual - 16th Edition American Conference of Governmental Industrial Hygienists, 1980, PF-12

2. American Warming and Ventilating, Inc. Engineering Bulletin, April, 1973, P-3

3. Ibid., P-2

4. Rapid Engineering, Inc. Information Bulletin #7, P-2

Conversion Factors:

M.P.H. = Kilometer/Hr x 0.6214
C.F.M. = $(Meter)^3$/Hr x 0.588
FT^2 = $(Meter)^2$ x 10.75
1 inch H_2O = Pascal x 0.004

OR:

Kilometer/Hr = M.P.H. x 1.61
$(Meter)^3$/Hr = C.F.M. x 1.7
$(Meter)^2$ = FT^2 x 0.093
Pascal = Inch H_2O x 250

FIGURE-1

EXTERIOR VIEW OF ATHLETIC FIELD HOUSE

FIGURE-2

INTERIOR VIEW OF ATHLETIC FIELD HOUSE

FIGURE-3

CLOSE-UP VIEW OF SUPPLY AND RETURN AIR GRILLES

FIGURE-4

ROOF MOUNTED MAKE-UP AIR UNIT ON EAST SIDE OF FIELD HOUSE

FIGURE-5

WEST SIDE MAKE-UP AIR UNIT, ABOVE ELECTRIC TRANSFORMER ROOM. AIR INTAKE LOUVER SHOWN ONLY

FIGURE-6

EXTERIOR VIEW OF TYPICAL EXHAUST LOUVER WITH COUNTERWEIGHTED DAMPER

Chapter 9
CONSERVATION ATTAINED AND MAINTAINED - A CASE STUDY

L. R. Snell, H. Michael Newman

Ideas for conserving energy through retrofit and improved control have become fairly standard. Actually implementing these ideas in a research facility is, however, a delicate matter.

OVERVIEW

Cornell's Center for Radiophysics and Space Research (CRSR) is a multi-use research facility built in the 1960's. This facility contains laboratories, offices, classrooms, the computer operations, and is a national respository for the photographic records from NASA's space program. A total energy conservation program was provided for CRSR in three phases: Phase I - Design Optimization; Phase II - Computer Monitoring and Control; and Phase III - Preventive Maintenance. Control of all three phases has resulted in both attaining and maintaining a 26% reduction in energy consumption (see Figs. 1 and 2) and improved environmental conditions. The payback on this investment was one year, three months.

INTRODUCTION

A key element of any successful energy conservation project in such a facility must include the balancing of concerns for air quality, research requirements, and the dynamic nature of building functions. Overcoming the challenges that are posed by these constraints requires both communication and compromise.

It is fundamental that any energy program be supported by top management in all areas that are affected. In this project support had to come from both the people responsible for physical plant operations as well as people in the research facility. Engineering personnel and building occupants have worked together to arrive at an optimized operating condition.

PHASE I - DESIGN OPTIMIZATION

Redesign, repair, and calibration of the existing mechanical systems is generally required for buildings constructed in the 1960's because of the lack of energy considerations during that period. In extreme cases, it may be necessary to actually replace major portions of the system. At CRSR the original mechanical systems themselves were not replaced or even substantially modified as part of this project. Instead the retrofit involved extensive control changes and connection to the University's central energy monitoring and control system computer.

The Building

CRSR is a four story brick and glass building of approximately 40,000 square feet. Located in the center of Cornell's campus, it supports research that involves both physical and biological chemistry as well as radiophysics and planetary studies.

Heating and cooling are provided by three terminal reheat, mixed air systems. A supply of cooled air is delivered by the central systems and reheated in terminal units when the space cooling load is less than maximum. The building is supplied from both central chilled water and steam distribution systems. Chilled water and steam are passed through exchangers that serve the primary building systems. Tempered water in these primary systems is then pumped to the individual units. A hot water recirculating system for perimeter radiation is also included.

Maintenance

Before proceeding with any optimization scheme, it is essential to upgrade the existing systems. We determined that the major mechanical components were sound. However both the air and hydronic systems needed to be balanced, the perimeter radiation pump needed to be reworked, and most important, the control air compressor had to be rebuilt. Other maintenance items included cleaning coils, calibrating controllers and insuring that control valves and dampers functioned properly.

Typically, many of the same maintenance problems existed in this building that are seen in others. The control air compressor had been pumping small amounts of oil and some of this oil had entered the system resulting in either improper controller operation or their actual failure. It is paramount that the compressor, the heart of the control system, be in top running order if further efforts to optimize are to be successful.

The balance of the air and hydronic systems is also a critical item. In several zones we discovered that over the years the arbitrary adjusting of distribution dampers had actually created a situation whereby certain areas received little or no air. Obviously this had to be corrected before any modifications could be attempted. System balance was completed by an outside contractor.

Outside air dampers are another frequently neglected item. In one case a piece of wood was found jammed into the dampers in an apparent attempt to increase comfort levels. Other dampers needed either new seals or to be unfrozen. These problems were easily remedied during the normal course of maintenance upgrading.

Environmental Considerations

Before discussing the control changes that were implemented, it is important to address the human factor. Although we've all been exposed to conditioning systems most of our lives, very few understand the complex principles of commercial building operations. When an optimization program is proposed, people seem to immediately develop a negative attitude. Communication is extremely important in alleviating concerns and must be an integral and ongoing part of the project, begun at the earliest possible time.

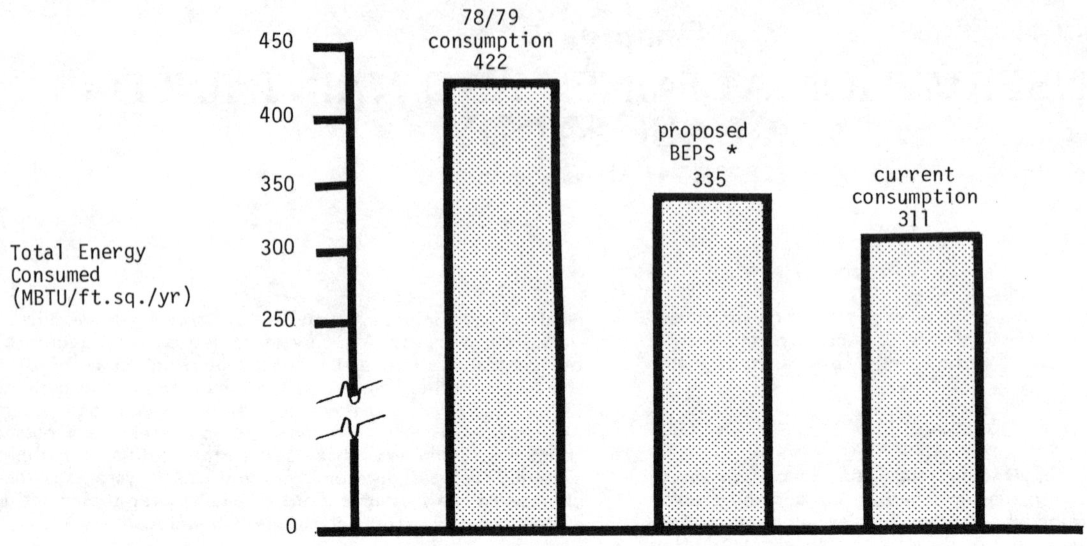

* consumption proposed for hospitals shown for comparison purposes

note: consumption is adjusted for HDD

FIGURE 1 TOTAL ENERGY CONSUMPTION

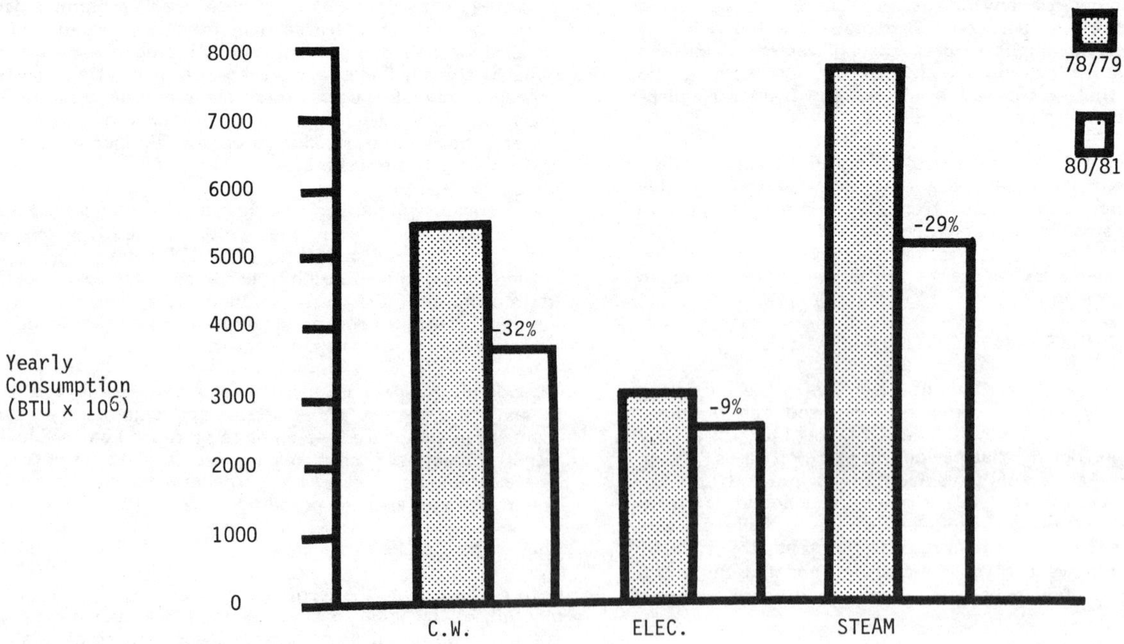

FIGURE 2 ENERGY REDUCTION BY UTILITY

The primary concerns seem to be the amount of fresh air, indoor air quality, resulting space temperatures and the overall building conditions during normally unoccupied hours. Additionally, research buildings have stringent process concerns. Research projects at times are run twenty-four hours per day for several days or even weeks. All of these considerations must be addressed.

Indoor air quality is a concept that defies easy definition. The indoor air quality question arises primarily as one result of techniques which reduce the infiltration of air. After meeting with concerned occupants many times, a decision was made to test the air for traces of ozone, urea formaldehyde and carbon monoxide. An outside certified testing firm was contracted for this purpose. The amounts of contaminants found were well within acceptable federal air quality standards: ozone-0.1 ppm max. (10% of standard), formaldehyde-.014 ppm, in a newly carpeted room only (.5% of standard), and carbon monoxide-not detected.

In addition to the control of potentially hazardous air contaminants, undesirable odors and humidity are certainly components of air quality. Although odors are present on rare occasions, a minimum of between two and three fresh air changes per hour are maintained. This condition provides a safe, healthful environment for people working in the building and also satisfies research requirements.

Fume hood operations is yet another area that will often present unique problems. Frequently, we found that the original hood system had been modified, in an effort to increase the exhaust, without consideration of the overall building operation. Subsequently, the air balance for the entire building was upset and the effectiveness of some hoods was jeopardized.

Cornell's Life Safety Department was requested to perform a hood survey. Currently all hoods are either functioning acceptably or tagged. Also, the distribution system is as close to the original design as possible. However, laboratory practices tend to become somewhat lax over a period of time. The importance of proper lab procedures can best be addressed by research directors within the building.

Controls

The general objective of the control strategy is to provide adequate ventilation, cooling, and humidity control at the lowest total annual cost.

Originally the building was fitted with Honeywell controls. Cost and service availability prompted us to modify the system using Johnson Controls. Major changes included an economizer cycle, cooling changeover capability, return air cooling control, and dehumidification. Calibration was necessary on most of the thermostats as well as the Honeywell controllers that were left in place.

The new control system takes maximum advantage of the local climate allowing outside unconditioned air to be used whenever possible. The cooling changeover system will permit outside air dampers to modulate whenever temperatures are between 25°F and 70°F. The dampers are modulated according to return air conditions.

The economizer cycle will set the dampers to a minimum position whenever outside air conditions are less desirable than the recirculated air. Currently this cycle is activated when the outside air is above 70°F. However, a minimum of 12% fresh air is introduced at all times to maintain air quality standards.

Dehumidification was included to enhance research efforts and to accommodate the NASA photographic storage and computer facilities. The addition of this feature has increased the effectiveness of that area of research. The dehumidification controls will override the chilled water control valve on each unit and use reheat coils to maintain appropriate temperatures. Usually summer humidity levels are kept around 55%. Although this type of control consumes more energy, the overall yearly reduction in utilities is within the goals that were established.

PHASE II - COMPUTER MONITORING, ANALYSIS, AND CONTROL

The second phase of this project involved connecting CRSR's systems to Cornell's central energy management computer. The computer, an IBM System/7, is currently connected to 60 buildings via a hybrid interface consisting of a tie-in to a Honeywell Selectrographic 6, vintage 1964, and a more nearly state-of-the-art audio tele-communications system manufactured by Texas Controls (formerly F+M systems). The Honeywell system permits only stop/start control, while the Texas Controls system expands our capability to permit the monitoring of any analog or digital sensor. Campus-wide we now have nearly 1000 control and sensor points. The system, in its entirety, performs 3 major functions: monitoring, analysis, and control.

Monitoring And Analysis

Our monitoring program is a modified version of an IBM package called "Basic Monitoring". Every five minutes, the program scans each of 700 sensors, including the eighteen that are in CRSR. The computer stores these values on disk. It also processes about 500 "alarm conditions", making use of various combinations of inputs or limit conditions and, in the event of an abnormality, prints a message on a hardcopy terminal which is monitored around the clock. For those alarms considered to be "critical" in nature, an audio-visual annunciator is also activated. The text of the alarm message refers the operator to a graphic display where all of the relevant system parameters can be seen in context. See Fig. 3 for CRSR's AC-3.

A severe limitation inherent to this type of monitoring program is that any resulting alarms are "digital" in nature, i.e. they are either on or off. A knowledge of the current value of a temperature, for example, often sheds little light on how the temperature came to take on its present value. To overcome this difficulty, a data acquisition and graphic display package was written which permits "tracking" a measured quantity over time. The operator enters the hardware addresses of the points to be observed, whether they are analog or digital, the time interval between readings and the length of the data collection period. This program has been an invaluable tool for analyzing both system difficulties as well as permitting the empirical determination of normal system operation.

With the data gathered last winter in CRSR during periods of night setbacks, we now have a means of determining the optimal starting and stopping times of the building systems based on actual plots of interior space temperatures as functions of time and outside air conditions rather than an theoretical calculations. See Fig. 4. Furthermore, on several occasions we have been able to diagnose problems using the computer before sending mechanics into the field.

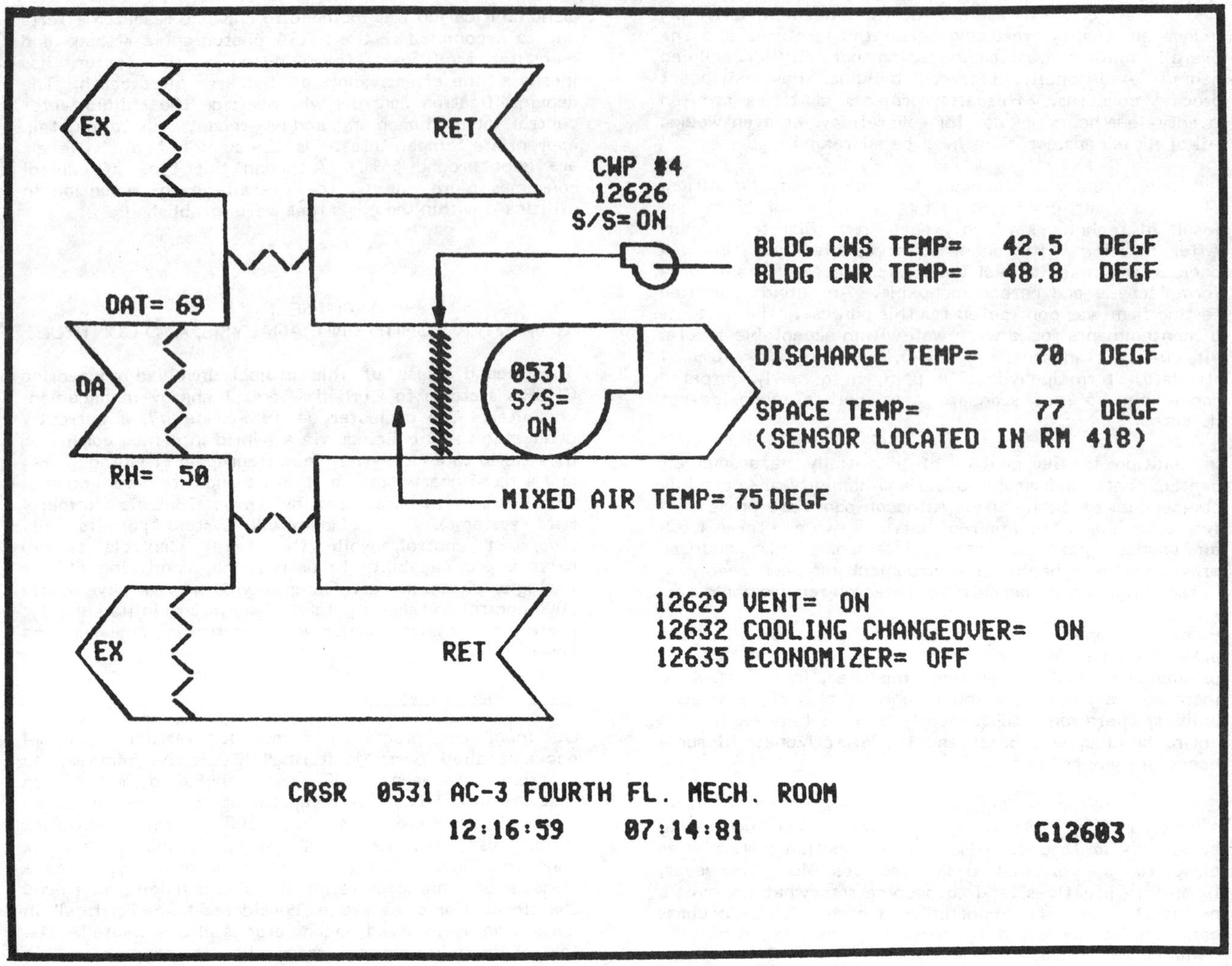

FIGURE 3 BUILDING SYSTEM PROFILE

Graphic displays facilitate the analysis of entire building systems at a glance. Combined with the data acquisition and plotting capability, they provide a powerful tool for optimizing system operation and diagnosing system ills.

Computer Control

The fundamental control strategy in CRSR is the shutdown of the major air-handling system during periods of low occupancy coupled with night temperature setback. Under this scheme, the space temperature is allowed to gradually decrease until, in the vicinity of 60°F, the temperature of the perimeter radiation supply water is increased to the degree necessary to maintain this temperature under local control.

The normal operating schedule was relatively straightforward; however, an innovative plan was needed in order to satisfy off-hours users of the building. It was decided to implement a "purge cycle" during these periods. During the purge cycle, equipment is automatically turned on for one hour according to the schedule in Table 1.

This compromise does little to degrade the energy conversation effort, but it materially improves the environmental conditions during evenings and weekends and allows research to continue essentially uninterrupted.

Additionally, the computer is used to set the overall "mode" of system operation based on outside air temperature. Below 65° F, the chilled water is locked-out and cooling is attained using outside air. Above 70° F, the outside air dampers are set in such a way as to bring in only enough outside air to satisfy the ventilation requirements of the building and the cooler, inside air is recirculated reducing the load on the cooling coils. The use of these computer-aided techniques of monitoring and control has permitted us to observe and regulate this building, as well as to respond to the varying needs of the building's users, in a manner that was not possible prior to the advent of this technology.

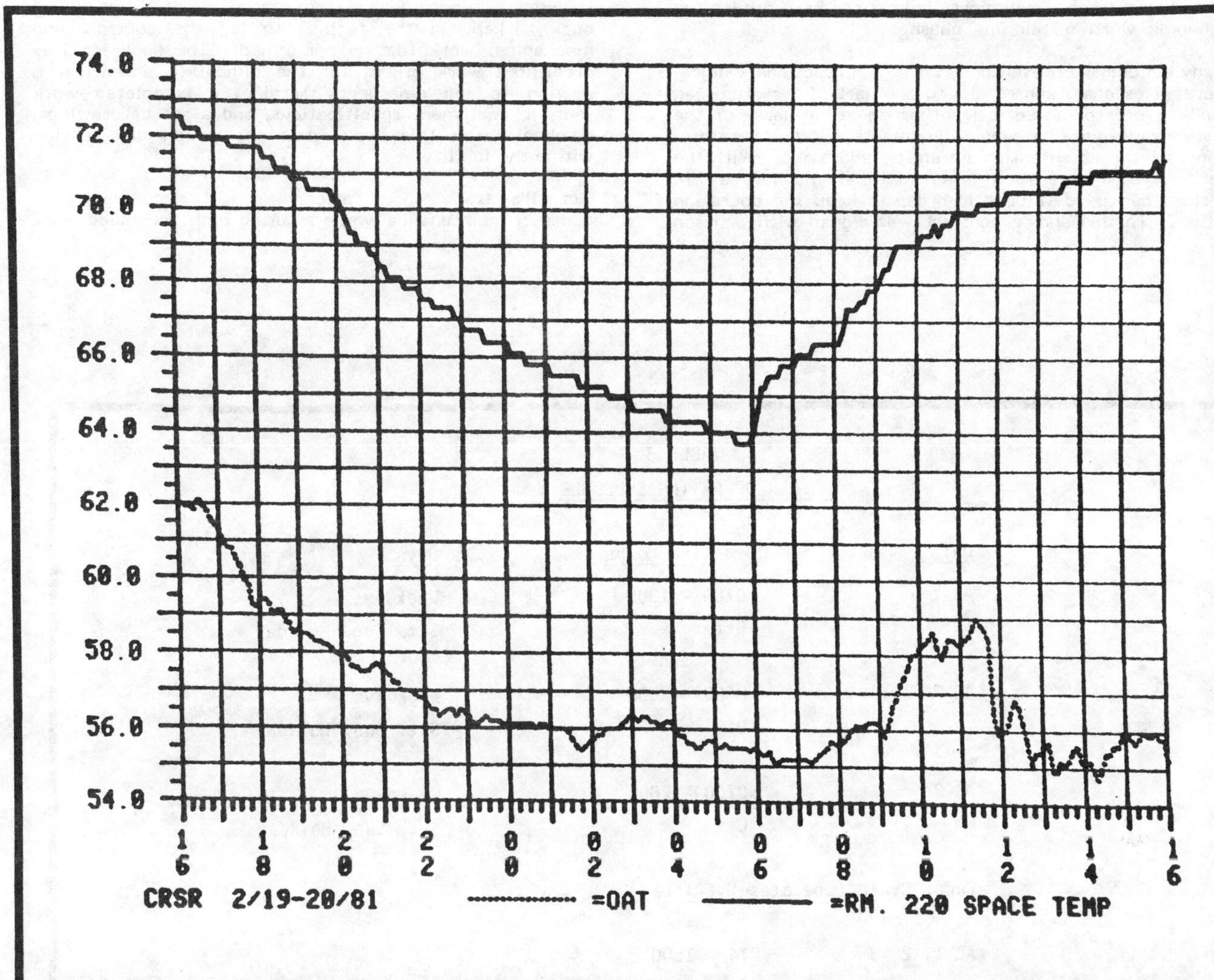

FIGURE 4 SPACE AND OUTSIDE TEMPERATURE PROFILE

Using the computer's data acquisition and display capabilities, a number of temperature profiles for the several building zones were accumulated for various outside air conditions. In the above plot, the air handling unit was shutdown at 2000 and restarted at 0600. In the interim the temperature was allowed to float. To have reached 68°F by 0800, rather than 0900, would have required starting the system about one hour earlier.

PHASE III - PREVENTIVE MAINTENANCE

Implementing a responsive preventive maintenance (PM) program insures that once attained, savings will continue. For purposes of evaluation, we have isolated maintenance expenses at CRSR by issuing a separate two year maintenance contract. Preliminary indications have proven the economic worth of this final phase.

Many buildings will eventually become a reactionary nightmare. Responses and repairs to one part of a system can tend to create problems in other zones or parts of the system or require seasonal adjustments. It is, therefore, imperative that an entire system be addressed. With the entire system operating in an optimal manner, a PM System can be developed in order to keep the operation intact. To this end, a contract was signed with Johnson Controls that includes both scheduled work and emergency service.

The scheduled work is accomplished bimonthly. Our present arrangement ensures that all items are checked and/or calibrated twice a year. Emergency service is needed for contingencies but has not been a severe problem. To document system status and condition, a PM book is kept in the facility to log the controls and mechanical work that is performed. The book includes preprinted work orders for the bimonthly activities, a section to log emergency service, a completed work section, equipment specifications, and a set of drawings. Mechanics use this book when performing any service within the facility.

Cornell's trade shops mechanics will accompany the contract maintenance workers during their scheduled work

TABLE 1

CRSR OPERATING SCHEDULE

Unit	Operating Hours	Days
AC-1	0700 - 1800	Weekdays
	Off	Sat. Sun. Holiday
AC-2	0600 - 2000	Weekdays
	Off	Sat. Sun. Holiday
AC-3	0700 - 2000	Weekdays
	Off	Sat. Sun. Holiday
Purge Cycle (Operates EVERY Day)		
AC 1, 2, 3	14 - 1500	
	20 - 2100	
AC-3 Only Additionally	03 - 0400	

Cooling Changeovers, Chilled Water Pumps on $T > 65$

Economizer on $T > 70^oF$

Vents on $T > 25^oF$

visits. This will provide training for our personnel and eventually allow us to assume inhouse reponsibility for the preventive maintenance of the building.

SUMMARY

Significant results are evident in three areas: energy consumption, environmental conditions and maintenance.

Results

We have attained and maintained a 26% reduction in utilities at CRSR. Energy consumption has been accumulated on a fiscal year basis and analyzed both before and after this project. Actual consumption data are shown in Figs. 1 and 2. During the construction period records were kept but, of course they are not representitive because the building was in a state of fluctuation. This significant reduction has been in spite of increased research activities and the addition of environmentally demanding computer equipment.

Environmental conditions have, for the most part, been noticeably improved both in terms of temperature and humidity. Humidity improvements can be seen primarily in the summer months. Temperature improvements are most noticeable in the winter because of better distribution providing a more even range throughout the building.

A comprehensive Preventive Maintenance program is currently in place. This program will also provide a means of training shop people to deal efficiently with the modified system. During start-up the system required a fair amount of attention from engineering personnel, mechanics, and computer operators. Over the first full operating year, attention has been more in line with a normal facility. Next year we expect the hours will be reduced well below our previous requirements.

Costs

The financial aspect of this conservation project is very attractive. Cornell's investment of about $32,000 was returned in just over one year. Tables 2 and 3 present details of these costs. We have included not only new equipment but also the cost of updating and maintaining existing systems. Hopefully, even further benefits will surface involving reduced maintenance hours and increased equipment life.

Conclusion

Research facilities and similar operations have historically been treated as exempt from energy conservation. However, a careful analysis of the facility can result in significant savings without jeopardizing research. In this particular case, a simple payback analysis has shown a complete payback in approximately one year despite the constraints that were imposed concerning operating hours, humidity control, and increasingly demanding research requirements.

TABLE 2
COST AVOIDANCE

Utility	Current Use	% Saved	Amount Saved	Unit Cost	$ Saved
Chilled Water (ton hr.)	311,219	32	99,590	$.12/Ton Hr.	$ 11,951
Electric (KWH)	781,337	09	70,320	.0515/KWH	3,622
Steam (M#)	4,445	29	1,289	11.60/M#	14,952
				Total Yearly Savings	$ 30,525

TABLE 3
COST TO IMPLEMENT

Initial Investment	
Controls Optimization	$ 22,900
System Balance	3,850
Routine Maintenance	2,000
De-humidity Control	3,000
	$ 31,750

Yearly Savings	
Utility	$30,525
PM	(5,742)
	$24,783

Simple Payback $\frac{31,750}{24,783}$ = 1 yr. 3 mo.

Chapter 10
EFFICIENT LIGHTING FOR LARGE AUDITORIUMS

M. Mytels

It has been stated and documented by many that lighting controls are among the simplest and most effective devices for energy conservation. According to the May 4, 1981 issue of Energy User News, "Purchases of lighting controls that serve as an alternative or a supplement to energy management system have increased substantially in recent years ...". Among the commercially available lighting controls four major types are discernible:

1) Ultrasound or infrared personnel sensors;
2) Photosensitive dimmers and switches;
3) Voltage reducers; and
4) Programmable time of day systems, manual and automatic.

None of the above controllers seem to be ideal for use in large auditoriums, especially in an academic environment. The personnel sensors might be unreliable in the highly complex geometry often encountered in large auditoriums. It might be possible to use many sensors and several types of sensors in one auditorium, but such an installation would be expensive and might end up providing unstable and hard to calibrate control. Photosensitive dimmers would necessarily be manually controlled and would, therefore, suffer from the same disadvantages as manual switching. It is part of our culture not to switch the lights off when one is not certain what will occur next in the auditorium, which essentially means that only custodians or people "in charge" are the ones who often use the switches on the wall, while the general public does not. And even if dimmers would be used, it is important to realize that 10 to 15 percent of power would still be used in a non-incandescent lamp dimmed to its minimum (about 96% dimming). Voltage reducers provide no significant mean of energy conservation in a large auditorium. The programmable controller might be proper if it is very carefully managed, but this would require the employment of a skillful operator/custodian.

After reviewing the foregoing limitations inherent in the commercially available controllers, we decided to develop a controller which would contain the following features:

A. Provide a preset (programmable) time of auditorium illumination.
B. Be fully interruptable and resettable by the user (i.e. lecturer).
C. Near the end of the time of illumination, provide a clear warning that the lights will soon be turned off.
D. All the manual features of a wall switch need to be available.
E. The controller needs to be inexpensive both in hardware and installation.

The controller hardware and strategy is presented in Figures 1 through 3 for the case of an auditorium possessing independent front and rear lights. In brief, the controller has two relays for the ON/OFF switching of the two sets of lights. These are "latching" relays providing the feature of full interruptability from the wall button-type switches. Two additional relays provide the reset signal to the programmable timer controller. The programs commence always from a reset signal which is, of course, simultaneous with the ON command from the wall switch. An OFF signal from the wall switch deactivates the lights, but it does nothing to the program which continues to be executed until the next reset occurs.

It is the function of the timer controller to provide the warning that lights will soon be turned off. One channel in the timer is used for this purpose for the front lights, its program being shown in Figure 2. The second channel of the timer is used to provide a warning when the rear lights are ON, its program being shown in Figure 3.

If it is desired, the system may be made fully automatic through the closure of the B-B Switch. In this mode the wall switches are by-passed (along with the resettability function) and the timer controller executes the ON/OFF programs of the operator. We have used a timer controller which can execute time-of-day functions and so the B-B switch allows execution of this type of "commercial" controls.

Although the system described here is a lighting control system, it probably could be applied in many other situations whenever it is desired to have semi-automation along with human interaction.

The system has been operational in two large lecture halls at Stanford since January 1981. Not a single complaint has been received thus far. Professors and students have accepted the system without any difficulty. Without explaining the system's intricacies, the users of the auditoriums quickly learned that when lights are flickering, the wall switch needs to be reset.

Based on our observations and comparison of intervals the lights are OFF in the two auditoriums equipped with the timer controller, 25 percent saving in energy is achieved on a day on which the auditoriums are heavily used in comparison with similar uses of auditoriums not possessing the controller. On minimal use days, the saving was observed to be as high as 70 percent.

It is worth pointing out that we do not as yet know what is the optimal time interval for which the timer controller should be programmed. For energy conservation reasons, the time interval should be as short as practical. The "practical" should be defined by lecturer's willingness to reset the wall switch in the midst of his lecture. We have set the intervals to one hour and to 90 minutes. In neither case have we received any negative comments. However, in response to requests from the academia, we have not yet tested interval settings below the one hour duration.

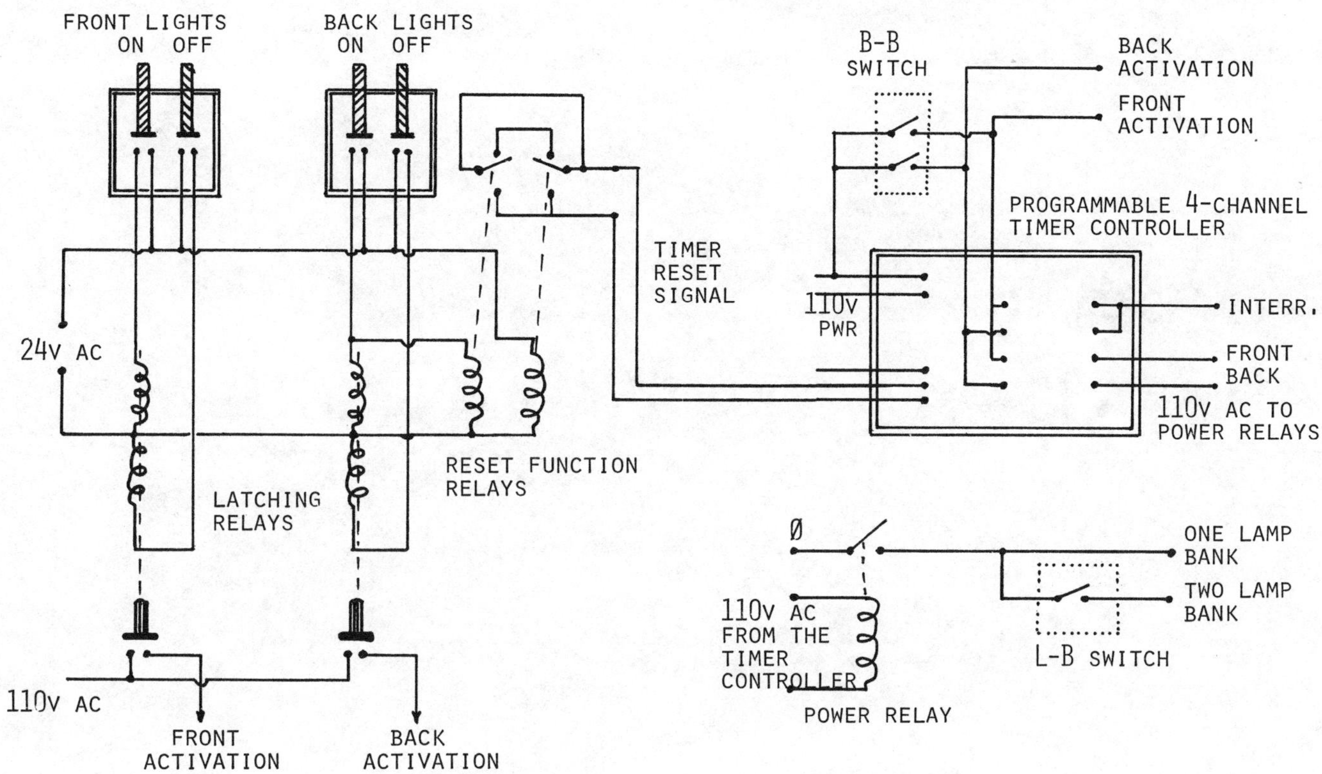

Figure 1: Auditorium Lighting Control Hardware For Front and Back Lights.

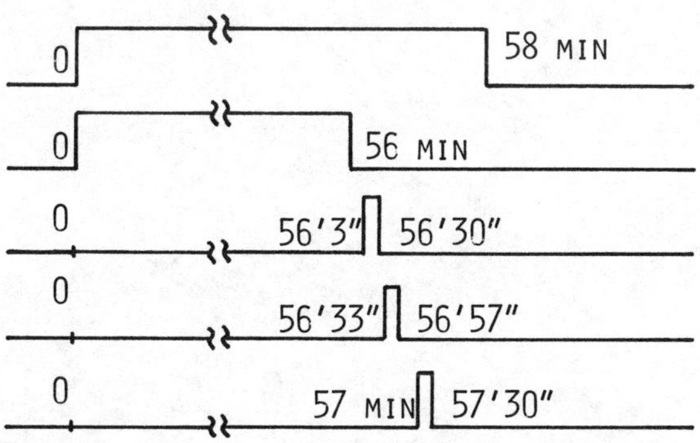

Figure 2: Program For Front Lights.

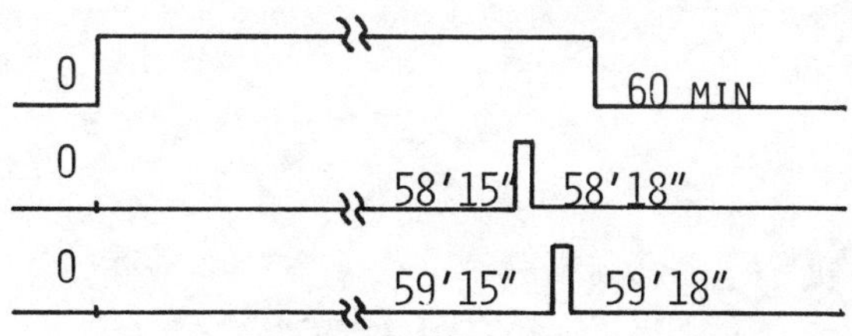

Figure 3: Program For Back Lights.

Chapter 11

THE APPLICATION OF ENERGY MANAGEMENT SYSTEMS AT PRINCETON UNIVERSITY

M. E. McKay, H. W. Smith

Princeton University was founded in 1746 and has grown through the years to the point where it now includes two campus locations; a main campus and the Forrestal campus located approximately five miles away. The University has six million square feet of building floor space in 186 buildings ranging in character from single family homes converted to dormitories up to sophisticated genetic research laboratories. A large number of the laboratory and academic buildings were constructed in the early 1900's and have since been modified considerably including the addition of air conditioning and laboratory hoods.

Energy costs for the University as a whole have risen from less than $3 million in FY73 to in excess of $9 million expected costs for FY82. This three-fold increase is due primarily to tremendous increases in energy prices but also partly to a vigorous campus expansion program that has seen the addition of four new buildings over the past three years with several more scheduled to be constructed in the future.

As energy costs became a larger and larger percentage of the University's operating budget, the administration turned considerable attention to means to reduce these costs. The program that emerged consisted of three levels of actions. The first level included such items as establishing a winter heating policy of 65°F and a summer cooling policy of 78°F, the installation of time clocks on mechanical systems, lamp removal, and generally improved preventive maintenance. Level two included changes to our utility distribution systems, academic and administrative buildings. Examples are the installation of a secondary pumping system in campus buildings connected to our chilled water plant, adding attic insulation, the installation of variable volume on some air handling systems, and some additional utility metering. Level three consisted of more capital intensive projects in the utility area, including the installation of a heat exchanger in the chilled water plant to provide "free cooling" in the winter-time and the installation of an energy management system, which is the subject of this paper.

The first step in the long road to the installation of the energy management system was to perform a feasibility study. A major New York consulting firm (Flack & Kurtz) was retained for this purpose and evaluated all campus buildings to determine the potential savings associated with an energy management system. They then estimated the cost of the system and indicated, using present value and life-cycle costing, that the investment was justified. At that time, which was 1976, the cost of fuel oil was $14.00 per barrel and was expected to escalate at approximately 10% per year which would indicate an expected cost today of approximately $25.00 per barrel. Our current cost for fuel is actually in the range of $35.00 per barrel. Following this step, the consultants prepared a specification which included three phases of construction. The first phase involved the 20 most energy intensive buildings, and the second phase included the next 20 most energy intensive buildings. All of the buildings in these two phases with the exception of one were academic, administrative or laboratory buildings. The third phase included some of the less energy intensive administrative buildings and the dormitories on campus. Bids were solicited from the major vendors of energy management systems at that time, including Honeywell, Johnson, Powers, Robertshaw and one lesser known company, Hamilton Test Systems. All of the proposals were evaluated on a life-cycle basis which included such items as first cost, annual maintenance and operating costs, and the ability of the proposed system to meet the specifications.

When evaluating the bids for system performance, we used a point system. Each of the proposals was assigned a certain point value under six different categories. These categories included the vendor's experience with similar systems, his willingness to accept the turn-key role in the project, his capability to train our personnel, and the terms of the warranty and service contract. The next category was basically a system overview which covered the reliability of the entire system and redundancy. The third category was central station performance in which we looked at such things as computer equipment hardware, information displays, the ease with which commands could be given, difficulty in making software changes -- basically, the human engineering aspect of it from the system operator's point of view. The fourth category concerned energy conservation software in which we evaluated the bids based on the software's capability to perform start/stop, enthalpy control, cycle, load-shed, hot and cold deck reset, and other optimization functions. The fifth category had to do with field hardware where our concern centered on noise immunity, ease with which trouble shooting could be performed, and fail safe operation. The last category covered special requirements such as the ability of the system to handle fire and security information which we eventually deleted.

After this evaluation, Hamilton Test Systems was selected as the contractor for the job and a contract was signed in August, 1977. Phase I construction began at that time and approximately six months later we authorized the beginning of Phase II construction. As the time approached to make the decision on Phase III, we reevaluated the potential savings and found that the ability of an energy management system to save energy in our dormitories was extremely limited and subsequently Phase III was dropped. Only one dormitory which included dining halls, common areas, and which is air-conditioned, was included in the second phase. The entire system installation was completed in April, 1980 and we have been operating successfully since that time.

The energy management system called CSCS, or Central Supervisory Control System, consists of a central computer located in our Physical Plant offices that is linked primarily by direct burial tri-axial cable to 38 buildings, and via modems and leased telephone lines to two other buildings. Within each building is located a RMDM or remote microprocessor data multiplexer. These panels communicate with the central computer and perform a limited control function within the building. All wiring from the RMDM to the individual points, i.e.; motor starters, temperature sensors, valve controllers, is with twisted pair or in some cases twisted shielded pair for sensors and analog signals.

In a typical mechanical system that is controlled by the CSCS, the supply fan is directly controlled by a normally closed relay to the automatic leg of the hand/off/auto switch. The return fan is subsequently interlocked with the supply fan. Temperatures in the space and duct work are sensed using RTD type sensors and the signals transmitted back to the RMDM. Closed loop calculations are performed in the RMDM which then transmits signals to all of the analog control points, including the damper motors and heating and cooling coil valves through a pneumatic control panel. A pneumatic control panel contains three-way solenoid valves and transducers to convert the 4 - 20 milliamp analog signals from the RMDM to 0 - 18 PSI pneumatic signals going to the valves. It should be mentioned that all of our analog controls on air handling and water systems are pneumatic controls.

As I mentioned before, all communications between the buildings and the central computer are transmitted on direct burial tri-axial cable or on leased telephone lines. The RMDM also serves as the communication interface in this link. Information accumulated by the RMDM's such as temperature readings, status of mechanical equipment, steam, chilled water, and electrical meter readings are collected in the RMDM and signals are then multiplexed and transmitted back to the central computer. In some cases, such as with steam, chilled water, and electricity consumption, the information is stored on discs later to be used in reports. Other information such as temperatures and fan status, is available to the operator via the CRT, and is used by the computer in open-loop calculations.

I have already mentioned closed-loop and open-loop calculations and it would probably be helpful to explain what I mean by these terms. Closed-loop calculations are typically those performed within the RMDM located in the buildings and are used to reset valves, damper motors and inlet vanes to maintain a predetermined reference duct or water temperature or static pressure. Open-loop calculations, on the other hand, are performed within the computer itself in our Physical Plant building and establish this predetermined reference. In a typical open-loop calculation, the computer would compare the space or room temperature within a portion of the building to the heating or cooling reference for that building and then calculate the appropriate reference temperature to be maintained on the fan discharge. This information is then transmitted back to the RMDM and resides there for a short period of time (15 minutes) and is used in the closed-loop calculation until the next time the open-loop calculation is performed. Certain control functions are reserved exclusively for the computer, as opposed to the RMDM, including starting and stopping of mechanical equipment, time of day scheduling, and optimal start time calculations. The RMDM has a very limited control capability and cannot stand alone for more than 15 - 30 minutes operating effectively without communication to the central computer.

At any point in time, the operator can interrupt the control loops and take manual control, using the CRT's that are connected to the central computer. He has the ability to start or stop mechanical equipment, open or close valves and dampers, and to interrupt the control logic and establish a reference temperature for any of the closed loops. These options become extremely important in trouble-shooting mechanical systems to determine whether or not the problem is in the computer, any of the control operating devices, or the mechanical system itself. This can save tremendous amounts of time when dispatching mechanics to repair problems. It also becomes important when making software changes to optimize performance of some of the mechanical systems because the operator can simulate what would happen if he changed some of these parameters in the data base and make sure that the system will operate properly afterwards.

Earlier it was mentioned that each of the pneumatic control panels contain three-way solenoid valves in addition to the transducers. These three-way solenoid valves play a very important role in the architecture of our system. When energized, or enabled, which is the normal condition, these solenoids transmit the pneumatic signal from the CSCS transducer to the pneumatic device, such as the valve or the damper motor. When de-energized, or disabled, the signal that passes to the valve or damper motor comes from the local pneumatic control panel. For the most part, our system is a retrofit system and was added on top of existing controls that have been in use in the building for years. A single signal from the RMDM either enables (energizes) all of the solenoids on a system or disables them. Again this becomes important in certain situations. In the event one of the CSCS sensors should fail to operate properly, the operator can de-energize these solenoids and turn control back to the building controls. Of course, this means that we lose our energy management functions, but the building can continue to function comfortably without immediately dispatching a mechanic to solve the problem. Should the communication link between the RMDM and the central computer be broken, we again automatically de-energize these solenoids. In the unlikely event that we had no computer to communicate with the RMDM's, all of the solenoids on campus would become de-energized and the control would revert to original building controls. This provides a great deal of flexibility in that we have a backup control system for all of our computer controls. While on the subject of redundancy and backup, it should also be mentioned that all of the communication links are redundant. We have two cables to each RMDM from the central computer. If one should fail the other one automatically takes over and continues the communications. Within the central computer room, we have redundant computers. If for some reason there should be a problem with the main computer, the backup computer will automatically take over until the problem with the main computer can be resolved.

It is relatively easy to compare the normal pre-energy-management operation for a typical system with the operation after the CSCS is applied to the system. In a typical dual duct system, of which we have a large number, a hot deck and a cold deck temperature are maintained and supply air to each zone is reset by the local zone thermostat and a mixing box. In many of these systems, the cold and hot deck are operated at 55°F and 90-110°F respectively. This is done to accommodate the range of temperature requirements that might be experienced by the different zones. At all times, the system is providing cooling (either with chilled water or outside air) and heating using steam. With the CSCS, typical zone temperatures are sensed in the hottest and coldest zones in the building. These temperatures are then used to determine whether or not any hot deck or cold deck is required and only the minimum amount required is provided. On a typical 30° day the hot deck on a system without energy management would be 100°F, the cold deck would be 55°F. Using the energy management system, the hot deck would be maintained at approximately 70° - 80°F and the cold deck would be 68°F using return air only. Considerably less energy is being consumed in the latter case.

Another typical system is one in which we have an air handler with cooling and terminal reheat. This system is one that is located in one of our office buildings and when it was installed it was designed to operate with a 55°F fan discharge year round and 200°F water to the reheat coils, provided by a steam to hot water converter. Even on a 65° - 70°F day it is likely that this system would be both cooling at the central fan and reheating at the terminal reheats. After application of the energy management system, cooling would only be provided when it was found that one of the zone thermostats connected to the computer was calling for cooling and at that point it is extremely unlikely that any of the zones would be calling for heating, so no hot water would be provided to the reheats. Generally speaking, in a system like this, only the cooling or the heating is in operation at any one time and for a good portion of the year neither one is provided and we operate strictly on return or outside air.

There are, as noted above, a number of energy management functions performed by the computer itself. One of these would be optimal start. The optimal start calculation in the computer is performed beginning at a time several hours before the building is to become occupied. Space temperatures in the building are sampled and compared to the desired temperatures in the upcoming occupied period. Using this information as well as the outside air temperature and certain factors pertaining to the mechanical system and building, the computer determines the optimal time to start the system in order to provide the desired temperature by the time the occupants arrive for work. The mechanical system is then started up earlier if required, but only if required; whereas without the CSCS we would typically have time clocks set for two or three hours before the people come into the building. Occupancy schedules also become important in our facility. Prior to the CSCS, the most sophisticated kind of control we had for occupancy consisted of time clocks. Even with the use of seven day time clocks, most systems were started at 6 or 7 in the morning to ensure that they were heated or cooled to the proper temperatures by 8 or 9 and left on until 5 or midnight depending upon the type of building. The only way to modify these schedules was to send a mechanic out to the building to change the time clock. Using the CRT in the central control room, the operator can, in a few seconds time, change the occupancy schedule for any system for any day and we can thereby schedule the use of our buildings on a daily and hourly basis. A good example of this would be a lecture room that is used only for special events. The operator feeds the schedule into the computer on a weekly basis and can run the system for the two hours required for a lecture or for an entire day if necessary but does have the capability of only running the system when it is actually required.

Load shedding, of course, is an option that is available with the central computer system. The computer receives a signal from our main substation indicating the instantaneous demand level we are drawing from the utility. Since we are on a 15 minute demand interval, the computer continuously calculates the projected demand for the end of each 15 minute demand period. If this calculation indicates that we will exceed a predetermined demand level, the computer then sends out load shed signals. This load shed signal will, in some cases, turn off mechanical equipment that is operating and in other cases, will extend the off time if a piece of mechanical equipment is already in cycle. Cycle, of course, is a brief period every hour when we normally turn off the mechanical equipment, generally, 15 minutes to a half-hour out of every hour. Depending upon the type of facility, load shedding and reduction of demand charges can be significant. In our case, we feel that it has not been as beneficial as hoped because the mechanical equipment is a fairly small percentage of our load and because the cycle program will normally have 25% of our mechanical equipment off at any one time anyway.

Another function referred to previously was the use of the computer to accumulate consumption data in the buildings. With this option, we can provide hour by hour read-out of the steam, chilled water and electricity consumption for each of the buildings on the system. This helps us to determine where, for instance, a large part of our electric demand is going. These reports have been used extensively as an energy management tool.

One of the significant benefits of this system is the ability that it gives to one person to check up to 40 buildings by simply sitting in a chair and pressing a few buttons. We have found a large number of previously unidentified problem areas and sources of energy waste by simply looking at these buildings during unoccupied periods or even occupied periods. A good example of this is steam heating coils. A decision was made early in the design of the system not to include steam pre-heat coils on computer control because they are used to prevent freeze-ups and maintain a constant 45°F leaving air temperature. Therefore, little optimization is possible. Normally, any rise in temperature above that point is done with the heating coil that follows the steam pre-heat coil. In checking the buildings, it was found that most of these steam pre-heat coils have normally open valves that would open whenever the fan shuts down. After identifying these cases, we were able to take corrective action by automatically shutting these coils down when the outside air temperature exceeded 40°F.

No energy management system is without problems, and we have our share of these. The problems stem primarily from the fact that we are trying to operate mechanical systems designed in the days of energy inefficiency in an energy efficient manner. One of the problems that has cropped up is the fact that in many of our mechanical systems we supplied too much air to the occupied spaces. Prior to the energy management system this was not a problem because the supply air would be cooled or heated sufficiently that it did not cause any discomfort. However, after application of the energy management system, we experienced propblems during the heating season in that the temperature of the supply air was considerably reduced, creating a situation wherein the occupants of the building felt drafts. Solutions to this problem ranged all the way from application of variable volume systems and reducing the volume of air during the heating period, to extended cycling of the fans so that when the fan system did run it was supplying a higher temperature air for a shorter period of time and the same total amount of heat was supplied to the space. In a number of our buildings, we have fan coil systems. These were designed so that when the room thermostat called for heat, a valve would open and hot water would flow to the fan coil unit while the fan motor was blowing. Our energy management system stops the pumps when heat is not required in the building. The result is that the fan coil units continue to blow air but at reduced temperatures of approximately 65° - 68°F (recirculated air) and again create a chill or draft problem. The solution to this problem involved trying all of the fan coil unit blower circuits into the same contactor we used to stop the pumps so that the fan coil blower stopped with the pump.

Steam systems do not lend themselves well to control optimization but we did find that the measured savings indicated considerable success. While there is no optimization that can be accomplished on a single valve steam radiation system, it is much easier to maintain a lower level of temperature in the building when centrally controlled than when the building is under control of a weatherstat or other such system.

One of the ways we averted a number of problems with our system was to insist on a point to point check. This consisted of verifying the operation of, or reading from, each point in the field and comparing it to what the operator read on the CRT. This process took several months to accomplish, but the results were well worth the effort in that we found a large number of sensors that were out of calibration, a large number of fan status relays that were not providing true status, and a certain number of analog control signals that were not being properly transmitted through to the valve or damper. It has also served as a valuable tool for our maintenance people in that they were able to see where every single point was and come to a better understanding of how the system operated.

Our single biggest problem with this system was not inherent in the system itself but in our application of it. During the construction of the energy management system, we also had several new buildings under construction and specified that these should be tied into our central system at the same time. As mentioned before, our system consists of 40 buildings with 100% backup for every component and one of these backup modes consisted of the local controls backing up the CSCS controls. As we added two more new buildings to the system, we felt that it was only practical to require local pneumatic controls in these buildings as a backup to the computer to be consistent with the rest of the system. We were wasting a considerable amount of money to purchase controls that would rarely, if ever, be used. We felt, however, that in the event of a computer failure, all of the buildings on our system should be able to operate independently.

This led us to make some inquiries of Hamilton Test Systems and many of the other vendors from whom we had received proposals as to what could be accomplished in the way of energy management on a smaller scale, independent of our present system. We found that most of the vendors were providing distributed processing at this point and after a careful search selected Hamilton Test Systems again for our next generation of energy management. This next generation of equipment, the unit controller, is a much different concept from our original system, primarily in the system architecture. The unit controllers are truly distributed processing in that most if not all, of the actual control functions, calculation, and scheduling, are performed in the building panel which is analagous to the RMDM. In fact, the unit controller can and does operate completely independently of a central computer but can eventually be tied into a central computer.

We have found a number of significant advantages in the newest generation of energy management systems. Installation costs are reduced for a number of reasons. The unit controller is in itself considerably less expensive than the RMDM. Connection to the central computer, if one is installed, is also considerably less expensive because there are lower cabling costs; for the most part communication is via telephone lines. There is increased system modularity, which is a major consideration when looking at system reliability. In our present energy management system, the CSCS, the loss of two major communication lines to any one set of buildings or the loss of both computers results in the loss of any affective control function performed in those buildings. However, with the unit controller system, we can lose the central computer or telephone communications to it and it will still stand alone and operate with 99% effectiveness. Because of our reduced exposure to failure with this system's architecture, we have elected not to install any backup controls in new construction buildings and we are in a position where we no longer feel that it is necessary to have redundant communications between the buildings and the central computer; nor are we required to have redundant central computers.

Present plans are to use the unit controller as a basic building block in expanding our future energy management capabilities. They will be installed in two types of buildings, one will be retrofit in existing buildings where the existing control system is either extremely wasteful from an energy point of view or in need of replacement from a maintenance point of view. The other application will be new construction projects where the unit controller will be applied as the only control system in the building. At this point in time, we do not have a significant need for a large number of these but expect this to develop over the next few years because of our renovation and construction program. Rather than wait until we have 25-30 buildings to put on unit controllers and purchase a central computer along with the field units at one time, we can begin realizing the energy savings in some of our existing buildings by installing unit controllers individually. At some point in the future, when we feel that a central computer is warranted, we can purchase the computer and obtain the additional benefits, but in the interim we will be achieving energy savings. The unit controller also allows us to perform energy management functions in smaller facilities that would not warrant a large RMDM.

We look at the unit controller as a basic building block, as I mentioned earlier. It has stand-alone capabilities and all of the unit controllers are interchangeable because they have the same basic programming and ROM (read only memory). The data base for each particular application is loaded, in our case, by the owner's module. This is important when evaluating different energy management systems. Many systems claim to be stand alone and can operate without a central computer, which is to some extent true. However, many of these require down-line loading from a central computer that you either purchase or hook into by telephone lines at the vendor's factory. The unit controller, on the other hand, has the ability to be loaded individually and we are not relying on anyone else or a central computer to keep a building on line. At least at this point our vendor has indicated that three types of unit controllers will be available, varying in the number of points each is capable of handling.

Some of the other less significant but extremely convenient features of the unit controllers are that the optimization program runs more frequently than in the central computer, providing for closer adherence to scheduled temperatures. In many large computer installations, temperature sensors in the field must be calibrated by the installation of resistors on the field hardware and this becomes an extremely time consuming and inaccurate way to calibrate them. In our unit controller, we can recalibrate temperature sensors simply by punching into the unit controller the actual temperature measured in the field and the bias is adjusted automatically.

The unit controller is a new development for Hamilton Test Systems and like most universities we are fairly conservative in our approach to acquiring newly developed equipment so we purchased and installed one in the Physical Plant building to watch the operation and determine whether or not it truly met our needs. This unit controller has been in operation for approximately six months and we found that it does an excellent job. At this point, we have construction documents out for a new dining hall to be constructed on campus and plan to use the unit controller as the only control system for this building. As much as we had confidence in Hamilton, based on previous dealings with them and our experience with the unit controller, it was additionally reassuring to know that Walt Disney World, Inc. had selected the unit controller as the sole control system for EPCOT (Experimental Prototype Community of Tomorrow) now under construction in Florida near Disney World.

RESULTS

While we have just begun application of the unit controller on campus and have no measurable results, we have spent a considerable amount of time analyzing the results of our several years of operation with the central supervisory control system. Like most institutions that have a large multi-building campus, we did not have the luxury of metering all of our utilities prior to the installation of the energy management system. This makes it extremely difficult to determine whether or not the anticipated savings were achieved. Considerable time and effort was spent on our part recently to determine whether or not the anticipated savings were achieved, and we did this by two routes: the first was an in-house analysis followed by an independent consultant's analysis of our utility plant reports. We found some extremely interesting information.

The first areas that we looked at was electrical because, fortunately, each individual building has its own electric meter. We were surprised to find out that it was almost impossible to determine whether or not the savings were being achieved. For each of the 40 buildings on the system, we analyzed the electrical consumption pattern for the seven years prior to the application of the energy management system and determined the mean and standard deviation of the total annual kilowatt-hours of consumption. In almost every case, we found that our calculated savings were smaller than the standard deviation and therefore determined that we could not reach any decisive conclusions. Table I shows some of the data we obtained from the mean and standard deviation calculated based on fiscal years 72 through 78 (the before data) and compared to the actual FY79 and FY80 consumption. The last column is the original savings figure calculated by the consultants for each of the buildings shown. On a building by buildings basis, it is hard to determine whether or not any savings were actually achieved given the statistical significance of the data available, but on the whole it appears that our consumption has been reduced far more than expected. Of course, there are other energy conservation projects such as relamping, voltage reduction, and rewiring of buildings that have occurred during this period that have also contributed to the savings.

TABLE I

ANNUAL ELECTRICAL CONSUMPTION BY BUILDING
BEFORE AND AFTER THE CSCS

	FY72-78				
Building	$\bar{X}$	σ	FY79	FY80	Estimated Savings
3	603.2	127.0	411.2	352.1	199.5
46	739.5	116.6	532.3	499.8	175.7
62	1,427.0	143.1	1,461.2	1,459.6	122.0
84	1,131.4	423.0	1,306.3	906.6	127.8
98	509.4	175.7	378.0	198.0	323.6
Totals	4,410.5	-	4,089.0	3,416.1	948.6

(All data above is in thousands of KWH)

The next area examined was our cooling load. In this case, we looked primarily at our summer cooling. While we do run our chilled water plant year round, our summer load is the only one affected by an energy management system. The rest of our load is a continuous year-round load for cooling of experimental equipment and computers. Once again, detailed data is not available. However, using the central plant data, we compared our total campus summer-time cooling requirement before and after the CSCS. The before period consisted of the five years from 1974-78, and the after period consisted of the single year 1980. The data was plotted on a graph (Figure 1) where total monthly ton hours of cooling were compared to cooling degree days and we found a significant decrease in FY80 as shown below:

A similar analysis was performed on our steam consumption. If anything, the metering problem was amplified when it came to the steam savings. Whereas most of the buildings that were connected to our central chilled water system were also on the energy management system, only 50% of our connected steam heating load was operated by the energy management system. In the case of chilled water, a 10-20% reduction in chilled water consumption in each building resulted in a 10-20% reduction in our total chilled water usage. In the case of steam, however, the 10-20% reduction in the use of steam for heating purposes in 50% of our connected load would only result in a 5-10% decrease in steam usage produced by the boiler house. By performing an analysis similar to that for chilled water, we found, however,

FIGURE 1
COMPARISON OF COOLING REQUIREMENTS
BEFORE AND AFTER CSCS

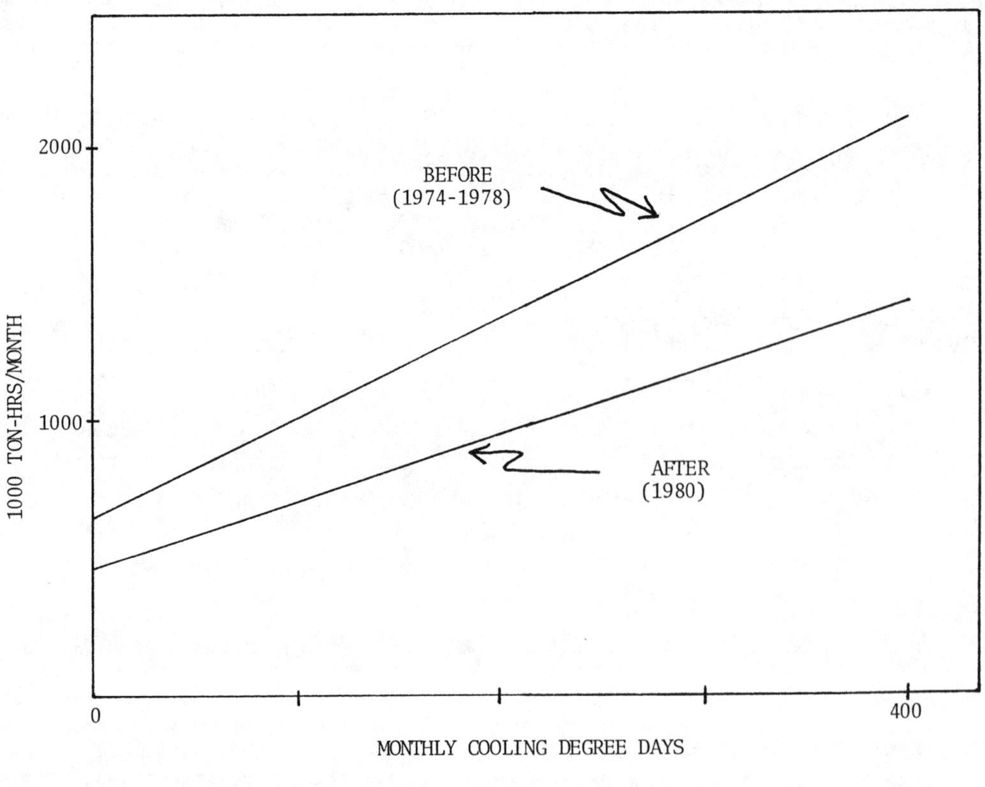

that the expected savings in steam for heating purposes were being achieved. For each of the main steam lines that are separately metered, a comparison was made of before (FY77-79) and after (FY81) data. The data used was heating degree days per month plotted against 1000's of lbs. of steam per month.

Figures 2 and 3 show the least-squares fit to the before and after data for two different steam lines. In Figure 2, the data is for a low-pressure line. Only a small percentage (25%) of the buildings on this line are controlled by the CSCS. Figure 3 is for a medium-pressure steam line on which 50% of the buildings are CSCS - controlled.

FIGURE 2
COMPARISON OF LOW PRESSURE STEAM REQUIREMENTS
BEFORE AND AFTER THE CSCS

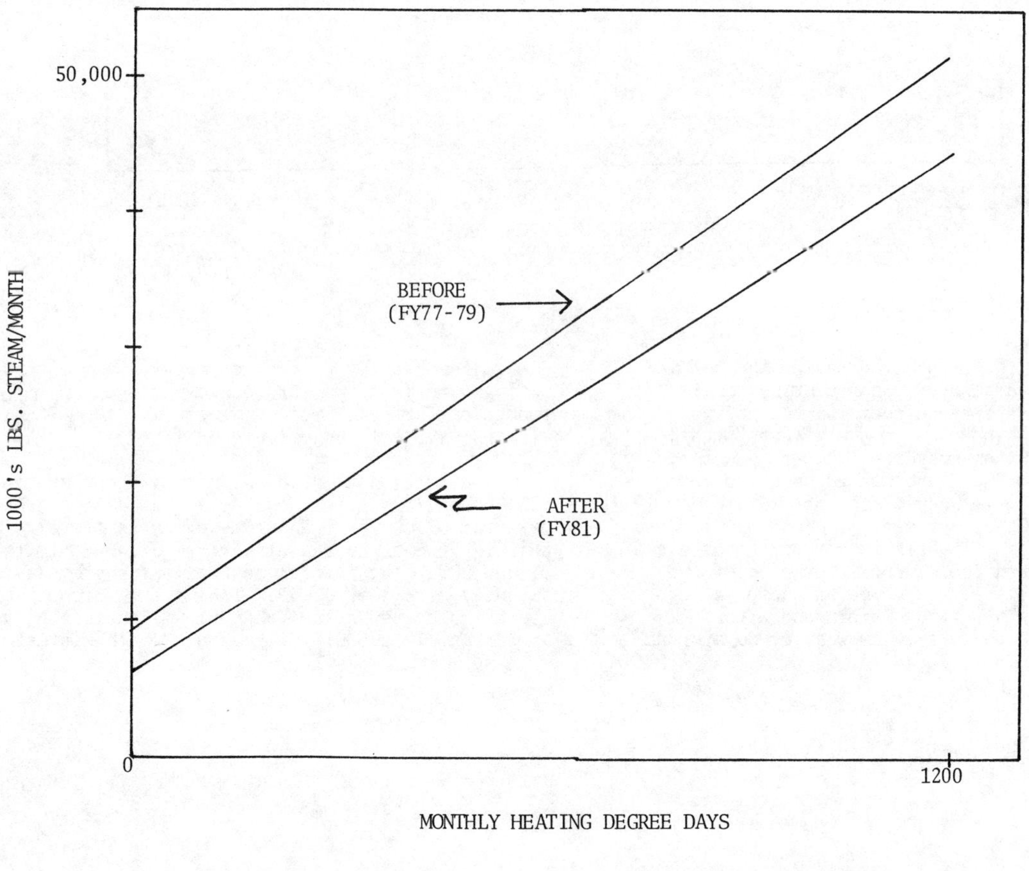

FIGURE 3
COMPARISON OF MEDIUM PRESSURE STEAM REQUIREMENTS
BEFORE AND AFTER CSCS

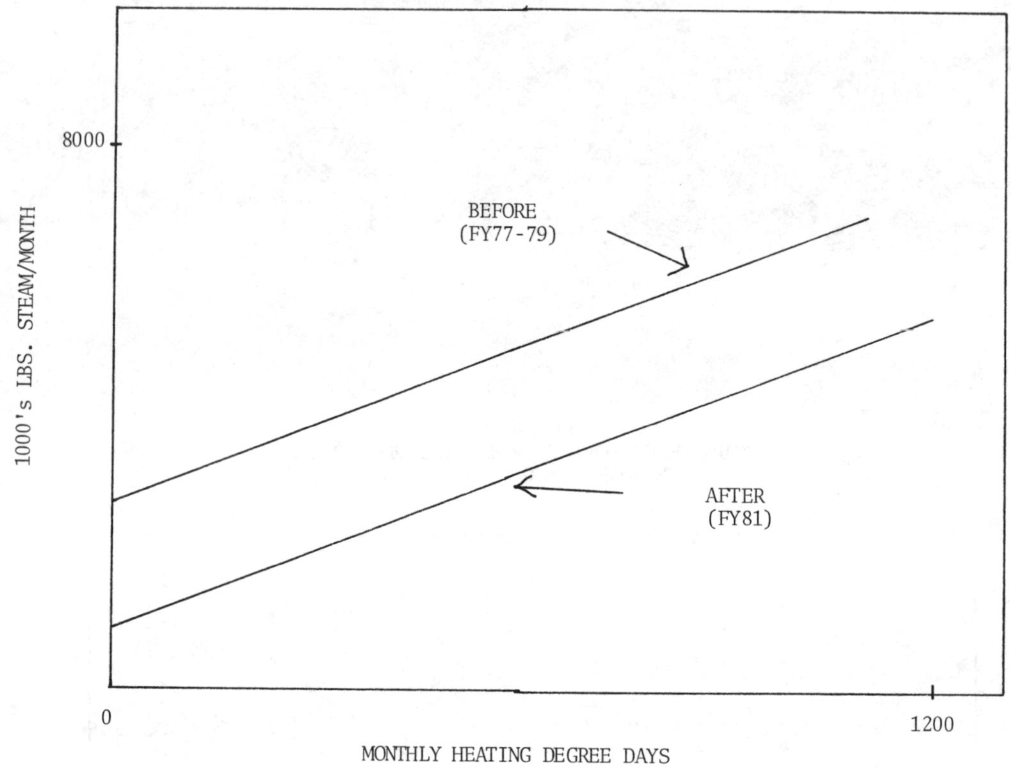

A significant factor when comparing the before and after data for an energy management system is the ability to isolate the effects of the energy management system as opposed to the effects of all the energy conservation measures taken as a whole. As mentioned earlier, we have had an on-going energy conservation program in many different areas beginning in 1973. Other attempts were made to isolate these effects using a BTU balance for our total plant in which we estimated the savings from each of the projects we accomplished and compared those savings to our annual BTU consumption before and after the project. While this is a somewhat crude approach and relies heavily on the accuracy of our estimated savings, we found we were achieving the anticipated reduction for each project, including the CSCS, within 10-15% of the calculated figure.

The total estimated savings from our system were 825,000 gallons of oil and 4,500,000 kilowatt hours per year of energy. In terms of current energy prices, this represents an annual savings of approximately $1,000,000. The initial cost of our system was slightly over $3,000,000. Taking into account operating costs we feel that we will have achieved approximately a four-year payback on the investment.

Chapter 12
ENERGY CONSERVATION WITH 100% OUTSIDE AIR

J. D. Ball

Many facilities of varying types have areas that require 100% outside air. These units are the largest energy users in the facility. During periods of low temperatures the air is preheated to a temperature of 50°F or so, and then channeled down a hot deck and cold deck, or to a reheat coil for final conditioning. During the summer months this air is pre-cooled and then channeled to hot decks and cold decks for final tempering. The system presented here utilizes the chilled water cooling system as a run-around heat exchanger and eliminates the need for conventional preheating during portions of the heating season.

The system utilizes the cooling coils of the return air type units as the heat source and the pre-cool of the outside air units as the heat rejection side (Figure 1). Properly designed and controlled, this system will achieve a balance point furnishing free cooling to the cold decks and free heating in the pretreat units.

Disregarding piping heat losses and other small heat losses, the governing equation for the system is:

$$Q \text{ in} - Q \text{ out} = \text{stored}$$

Q in - heat absorbed by the cold deck coils
Q out - heat rejected in the pretreatment units
Q stored - heat stored in the chilled water which will cause a temperature change

Expanding these terms:

Q in BTU/HR = CFM · 60 min/hr · Cp BTU/lb°F · D lb/ft^3 · [ΔT · Coil Eff]
Q out BTU/HR = CFM · 60 min/hr · Cp BTU/lb°F · D lb/ft^3 · [ΔT · Coil Eff]

CFM = Air volume across the coil
Cp = Specific heat at constant pressure BTU/lb°F
D = Density lb/ft^3
ΔT = Temperature difference, absolute value of incoming air temperature - chilled water supply temperature.
Coil Eff = Correction factor for the heat transfer capability of the coil. Chilled Water Supply Temp/Discharge air temp.

These expressions can then be inserted into the governing equation to calculate the heat that is stored in the chilled water. This heat will change the chilled water supply temperature, and is defined by:

$$CHWS_1 = CHWS_2 + [Q \text{ stored}/\#H_2O \cdot Cp]$$

$CHWS_1$ = New chilled water supply temperature
$CHWS_2$ = Old chilled water supply temperature
$\#H_2O$ = lbs of water in the system
Cp = Specific heat of water

The chilled water supply temperature is then inserted back into the governing equations on the next iteration.

These expressions can then be used as the body of a computer simulation for the system. Computer simulations have varying degrees of complication, depending on the desired results. Practice has shown that calculations done on five minute time increments and outside air temperature changes on an hourly basis give satisfactory results for most applications. Shorter time increments, changing cold deck loads, and actual outside air values could be used if more accurate results are required. The ΔT in the Q in expression is for a wild coil. If a cold deck controller is present,

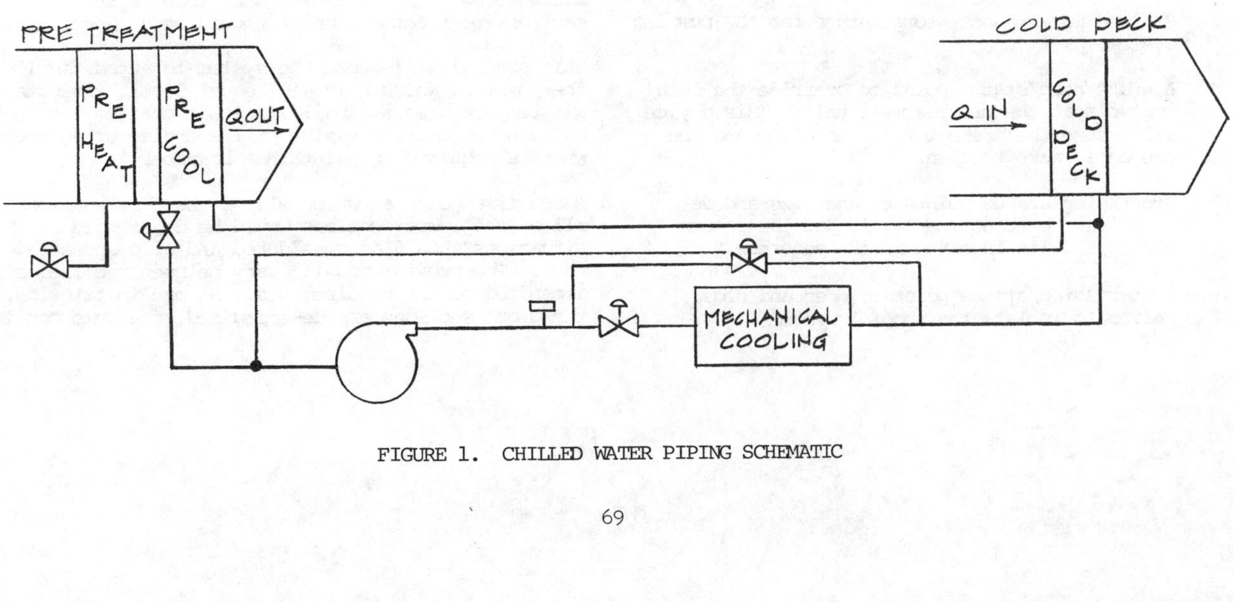

FIGURE 1. CHILLED WATER PIPING SCHEMATIC

this term becomes the mixed air temperature - cold deck temperature, and the coil eff - 1.0.

The simulation should answer the questions:

1] Is this sytem feasible for this facility?
2] At what outside air temperatures can the system operate and produce satisfactory results?

In order to make these evaluations the simulation should output, chilled water supply temp, Q in, Q out, and discharge air temperature of the pretreatment section.

The next step is to examine the economic benefits of such a system at the facility. The BIN method makes calculations of the savings very simple. The number of hours at each BIN during each month are tabulated. Using this in conjunction with the data from the simulation, it will allow the calculation of the energy savings. In buildings that normally preheat the air on the outside air units, and use the outside air economizer to furnish the cold deck on the return air type units, the energy savings would be:

$$Q \text{ saved} = [Q \text{ out} + [T_{RA} - T_{MA} \cdot 1.1 \cdot CFM_{HD}] - Q \text{ pump} \cdot \text{hours at that BIN}$$

Q out - from simulation

T_{RA} - return air temperature

T_{MA} - Set point of the economizer controller

CFM_{HD} - air volume across the heating coil in cubic feet per hour

Q pump The energy required to operate the chilled water pump

This general equation would have to be applied to the system in question to account for all of the energy savings.

Field Devices

For best results the heat exchange cycle should be controlled and monitored by a building automation system. Any computer based system on the market should be able to handle the interlock and monitoring requirements of this control scheme.

The monitoring and control points required are:

1. Pretreatment discharge temperature of each pretreatment unit involved.

2. Cold deck temperature of the return air units.

3. Chilled water supply temperature for the pumping system.

4. A start/stop/status point to override the local controls at the pretreatment unit. This device will cause the preheat valve to close and the pre-cool valve to open.

5. Start/stop/status points to energize and de-energize the economizer cycles of the return air type units involved in the sequence.

6. A start/stop/status point to open and close valves to isolate the pumpint system.

7. A start/stop/status point for the chilled water pump to be used with the sequence.

8. An outside air temperature sensor.

Most of these points are currently available in most building automation systems.

Operation

As the outside air temperature drops to the level at which the heat exchange process becomes economical, the computer will position all of the valves required to isolate the chilled water pumping system. The chilled water pump is then started. The status feedback for the chilled water pump should be a flow switch to ensure that water is actually flowing.

Once water flow has been proven, the cold deck high limit, and the chilled water supply temperature should be compared. If the chilled water supply temperature is higher than the cold deck high limit, then the pretreatment unit controls should be started first. If the chilled water supply temperature is less than the cold deck high limit, then the economizer cycles should be de-energized first. When the chilled water supply temperature reaches a point below the cold deck high limit, but above the pretreatment discharge low limit, the remaining units can be energized.

As the heat exchange scheme operates the building automation system should continuously monitor the chilled water pump status, the cold deck temperatures, the pretreatment discharge temperatures, and the chilled water supply temperature. If the chilled water stops flowing, all units should immediately return to local control. If any cold deck should rise above its high limit, its associated economizer should be energized. This will reduce the heat input to the heat exchange circuit. If any pretreatment discharge should fall below its respective low limit, it should be de-energized. These two conditions represent a system fault. If they should occur the local controls and the limits should be checked, and when the fault is corrected the units manually reconnected.

The main controlling element is the chilled water supply temperature. As this temperature rises and falls, it is a direct indication of the degree of imbalance between the Q in and Q out sides of the system. The chilled water supply temperature should have a series of high and low limits. As each of the high limits is reached, one stage of the Q in section should be de-energized (economizer turned on). As the water temperature returns to normal the Q in sections should be re-energized. Conversely, if the water temperature falls below each low limit, one stage of the Q out section (pretreatment unit) should be de-energized.

This control will cause the system to automatically seek the point at which Q in will equal Q out. As outside air temperatures and building loads change, the building automation system will energize or de-energize what is required to maintain a balance.

There is a point at which all of the Q in sections, or all of the Q out sections would be de-energized. These situations should be considered and allowances made for them. The provisions will vary between facilities depending on the requirements. At one institution, if the Q out sections are de-energized, the pump continues to operate supplying chilled water to the cold decks to handle the small loads. As the water temperature rises a section of the Q out will be energized to cool the water. If the Q in sections are de-energized, an outside air temperature too high for the control scheme to be effective is indicated. At this time the pretreatment units are returned to normal control, and the mechanical cooling is energized.

Thus far the Q out section of the heat exchange cycle has been defined as a pre-cool coil in a 100% outside air unit. In many areas a pre-cool coil is uncommon. In this case, some of the return air units having a large cooling load could be left in economy and thus become the Q out section. A close evaluation of the loads and the system performance should be made to ensure that desired space conditions and energy conservation can be achieved.

The heat exchange system presented here is a simple idea and easy to implement. There are some compromises which may be required before the system can operate. One of these is a lower temperature out of the pretreatment unit. The lower this temperature can be and still maintain the space conditions, the more beneficial the heat exchange cycle will be.

The method for evaluating and implementing a heat exchange cycle using the chilled water system have been outlined in this paper. The next step is for you to use this method to evaluate the feasibility of implementing such a system at your facility.

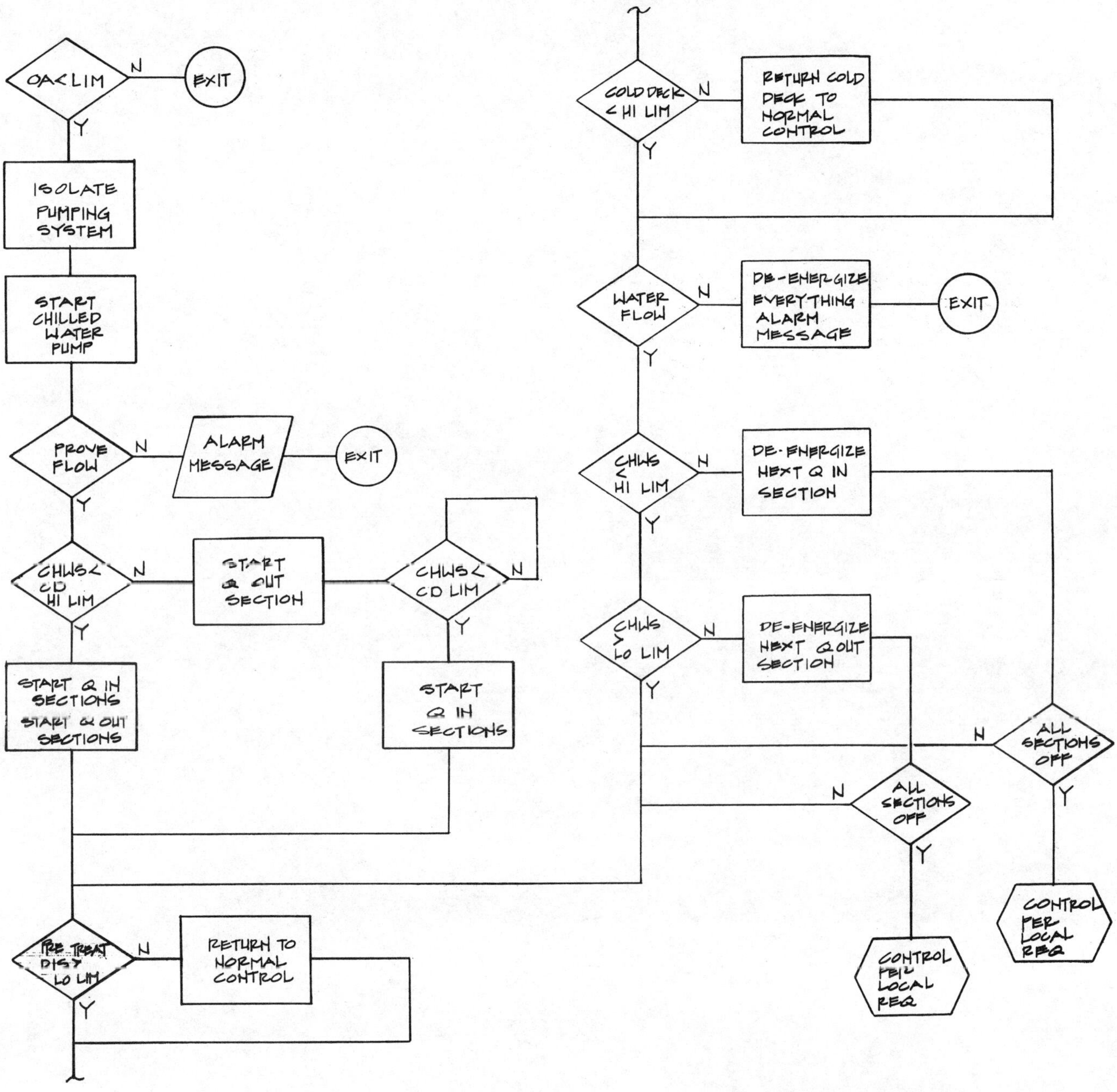

FLOW DIAGRAM OF CONTROL SEQUENCE

Chapter 13

DESIGN AND INSTALLATION OF ENERGY MANAGEMENT AND CONTROL SYSTEMS FOR COMMERCIAL BUILDINGS

D. M. Fowler, J. S. Gossett

INTRODUCTION

All the modern technology and sophisticated hardware have not changed basic concepts. We still save energy by turning energy users off or by "turning down" the energy user to the point it just meets the load or requirement. If we keep these basic concepts in mind, some of the complexity and mystique associated with energy management and control systems disappears. And do not be confused by nomenclature; these systems are variously referred to as building automation systems (BAS), energy management systems (EMS), building management systems (BMS). They all operate in basic principle by turning energy users "off" or "down".

EMCS - WHAT, HOW, WHY?

What

Energy management and control systems are electrical control systems, usually 24 volt low voltage, for monitoring and controlling energy users. They include programmable timers, microprocessors, micro-computers, computers, smart microprocessor remotes, power-line carrier transmitters and receivers, to name the principal varieties.

How

As we said earlier, these systems save energy by turning energy users off, or by scheduling of loads, or by turning them down, based on feedback from the load being served that the design output of 100% is not needed. For example, chillers are designed to provide 42-44° chilled water leaving the chiller, and today most chillers are still set at 42-44°. If the design conditions were for an outside temperature of 100°, it goes without saying that the full capacity is not needed when the outside temperature is 80°. But most of our chillers are running at full capacity whenever cooling is required. An increase in chilled water temperature of one degree saves at least 1%; so an increase of chilled water temperatures in January of 14 degrees will save at least 14%.

Why

Such systems are being installed because it makes economic sense. Would you spend $100,000 now if your investment were returned in savings in three years, and if thereafter you had an annual increase in your cash flow of $30,000 a year? Of course you would - if you could only believe the salesman or the consultant.

But you say you can do the same thing with a good building operating staff and time clocks. Even if you have both - and most today do not have the former and not everyone the latter - the statement is not true. When electrical demand costs reach $5, $10, $20 per KW or KVA, you must have an energy management system of some kind. Time clocks can not monitor your demand meter and shed loads to avoid peak demand charges. And such demand costs are not in the future - today demand costs in Dallas are $7.56, in El Paso $11, and have been rising at a rate much more rapidly than energy or KWH and fuel costs. In addition to demand savings of 10-15%, a good control system will also save 10-15% more in energy costs over a period of time than you can save with time clocks.

PROPOSAL ECONOMICS

Savings

Every owner wants economic justification for a major expenditure such as that involved with these control systems. For an existing building the actual energy bills can be used as a basis for either a detailed energy savings estimate, or as is more often the case, for a percentage saving of the total.

For a new building, we have to be a little more inventive. Here is one way that potential savings can be derived. This is a 460,000 square foot office building with estimated electrical energy use of 75,000 BTU/Sq. Ft./Year. Notice that the demand savings are calculated separately from KWH savings, so that it may be pointed out what portion absolutely cannot be saved by building operators and time clocks.

Consumption

1. $$460{,}000 \text{ GSF} \times \frac{75{,}000 \text{ BTU/Sq. Ft./Year}}{3{,}413 \text{ BTU/KWH}}$$
   $$= 10{,}108{,}409 \text{ KWH/Year}$$
   $$= 842{,}367 \text{ KWH/Month}$$

2. Load Factor for average office building is approximately 0.5

   $$\text{Load Factor} = \frac{\text{Actual KWH Consumption}}{\text{Maximum Possible KWH Consumption}}$$

$$.5 = \frac{842{,}367}{\text{Demand} \times 24 \times 30}$$

Demand = 2340 KW

Estimated Savings

1. Experience indicates such systems can save, conservatively, 10-15% of demand or KW costs and 10-15% of energy or KWH costs.

2. Public Service Company of Oklahoma Rate Code 440 (proposed):

   On-peak (June - October)

200 KW	@ $5.60	$1120	
300 KW	@ $5.00	$1500	
(2340-500)	@ $4.00	$7360	
		$9980	X 5 = $49,900

   Off-peak (November - May)

200 KW	@ $4.60	$ 920	
300 KW	@ $4.00	$1200	
(2340-500)	@ $3.70	$6808	
		$8928	X 7 = $62,496
			$49,900
			$112,396

   Annual demand savings with EMCS = 10-15% ($112,396) = $11,240 - $16,859

   Energy Charge (proposed):

10,000 KWH	@ $.02642	$ 264.20	
90,000 KWH	@ $.02442	$ 2,197.80	
(842,367-100,000)	@ $.02142	$15,901.50	
		$18,363.50	X 12

   = $220,362

   Annual energy savings with EMCS = 10-15% ($220,362) = $22,036 - $33,054

   Total Savings $33,276 - $49,913

Costs

Immediately below are the four general types of EMCS, the number of points each can handle, and the cost of the central equipment for comparative purposes. These costs do not include additional central equipment which can be added, as options, nor do they include the costs of wiring, relays, sensors, and systems installation and check-out.

System Type	No. of Points	Central Equipment Cost
Microprocessor	40	$12,000
Micro-Computer	256 Inputs 256 Outputs	$26,000 (IO cards for 80 points)
Computer	4000	$70,000 (IO cards for 100 points)
Smart Remote or Microprocessor With Memory	256 Inputs 256 Outputs	$ 8,500 (IO cards for 40 points)

The total installed cost of a microprocessor will run from $30,000 to $40,000 and is obviously limited in cost since the number of points is limited to 40. The computer based systems will cost from $350 to $1000 per point including the central equipment cost, depending on the characteristics of the point mix.

Payback - Investment

Assuming a very conservative 3 year payback, an initial investment of 3 ($33,276-$49,913) = $99,828 - $149,739 could be made based on savings of 10-15%. At $500 per point, this amount of money could buy a system with 200 to 300 points. A three building complex of high rise buildings was recently designed with fewer than five hundred points, so one high rise building can be controlled easily with 200 points.

CONSULTANT ROLE

Bid Evaluation

Once the owner has decided to install a system, he has several choices. He can invite vendors to provide a proposal to install a system, either with or without providing them a cost limitation. Most owners do not provide such a cost guideline since they are not familiar enough with costs. With no specific strategy nor design, each vendor will design the system differently and the owner will have a very difficult time comparing the bids and selecting the best one. The owner may wish to call on a consultant to evaluate the bids and give him a recommendation.

Competitive Bidding

Or - the owner can call an experienced consultant and ask for help. Help can be provided in two ways. First, the energy system consultant can design the system, write specifications, provide an input/output point list, and necessary schematics to enable the owner to seek competitive bids. One such project was done in this matter, and all four bids came in under the estimate and within $10,000 of each other at a level of $180,000.

Negotiated Bids

Third, if the owner decides to negotiate with a vendor - say the one that is installing or has installed the pneumatic local controls, the consultant can assist the owner in obtaining the best system for the least money and with the greatest saving. A computer based system for a high rise office building can be purchased for as little as $100,000 to $150,000, or as much as $300,000 to $400,000, with very little difference in energy savings.

A consultant can design a system with only the interests of the owner in mind; he has more time to spend on the design since he is being paid for his time. As a result, he can spend more time educating the owner and his staff, and he can develop alternative pricing packages for the owner's negotiation with a vendor. All of these assume a negotiated purchase. If the local market provides vendors who are truly price competitive, then the consultant can be used to good advantage to prepare a bid package for competitive bidding.

DESIGN CONSIDERATIONS

System Capabilities

The following tabulation lists the energy saving routines or programs which are possible with each type system. This listing is an exhaustive one and includes all those programs which are currently feasible.

	Micro-processor	Micro Computer	Computer	Smart Remote
1. Start/Stop scheduling	X	X	X	X
2. Optimized start/stop	X	X	X	X
3. Duty cycling	X	X	X	X
4. Demand monitoring and limiting (load shedding)	X	X	X	X
5. Night set-back (space temperature)	X	X	X	X
6. Lighting control	X	X	X	X
7. Event initiated program		X	X	X
8. Calculated point		X	X	X
9. Chilled water reset		X	X	
10. Condenser water reset		X	X	
11. Multiple chiller optimization		X	X	
12. Economizer (dry bulb) cycle		X	X	
13. Economizer (enthalpy) cycle		X	X	
14. Ventilation and recirculation (warm-up and cool-down)		X	X	
15. Hot and cold deck temperature reset		X	X	
16. Boiler plant optimization - steam		X	X	
17. Boiler plant optimization - hot water		X	X	
18. Outside air temperature reset - hot water		X	X	

This list is a tabulation of the possible; not every vendor has available standard software packages with all these programs.

In addition, many feel that programs 12 and 13 are more appropriately handled by local controls, rather than by an EMCS or automation system.

Key Assumptions

The two key assumptions that might be recommended to clients are:

- first, that the local controls for the building, whether existing or new, be capable of operating the building in the absence of the EMCS or communications with the EMCS. All-electric systems are available and may in time be adopted in large numbers. But the reliability of pneumatic systems, together with the fact that most building operating personnel have some familiarity with them, indicate that a majority of local control systems will continue to be pneumatic for some time to come.
- second, that energy management, fire and security systems be separate stand alone systems. This design guideline greatly enlarges the competition and with proper design and bidding procedures will result in a considerably lower overall first cost. Because the systems are separate, maintenance on each system will be easier and should be less expensive, again because of increased competition.

Other Design Considerations

With the key assumptions and the knowledge of the capabilities of the various systems in terms of software, the designer is ready to begin his design. Listed here are a few of the considerations which the designer encounters and must resolve.

Electrical System

Number of electrical meters
Number of buildings on one meter
Demand cost
Distance between buildings on same meter
Availability of pulse device for demand meter
Need to sub-meter

Building

Size
Single or multiple
Central control for several buildings
Central monitoring for several buildings with control in each

HVAC

Type of cooling system
- Packaged or roof-top - duty cycle
- Reciprocal - Limit demand using unloaders
  - Reset chilled water temperature
- Centrifugal - Limit demand in steps
  - Reset chilled water temperature

Type of heating system
- Electric Resistance - Disable in summer
  - Set back nights and week ends

Type of air handler
- Use two-speed motors

Sub-metering of chilled or hot water or both

Lighting Strategy

Modular wiring system
Local override switching
Computer monitoring and use calculation

Other Controls

Condenser water reset
Chiller optimization for multiple chillers
Control of outside air supply
Control of exhaust air
Control of electric domestic hot water heaters
Monitoring and back-up control of heat recovery and solar systems

Design

Specifications: Why have a specification for an EMCS? For the same reason we have specifications for any other construction work. Primarily for the protection of the client or owner, to insure that he gets a system which may have been previously described in some manner. But the specification also assists the EMCS contractor; with a good performance specification, input/output point list, and a copy of the mechanical and electrical drawings, the contractor can estimate his cost very accurately.

Many are aware of the military services specification referred to as the tri-service specification. You should not use it for other jobs; it results in costs for a system as high as $1200 per point, when a large well designed and specified system may be purchased for $500 to $750 per point. Portions of that specification may be used if the designer is aware of what the portion he is using is really requiring.

Many of us have seen specifications which were not really specifications; they were merely word descriptions of what the system can do. A good specification must detail the quality and accuracy of the various components of the system in performance terms.

These components include hardware, software, wiring, sensors, and control devices. The state of the art should be used in such specifications, so that the owner does not buy an obsolete system. Here are a few things to watch for and include if a computer based system is being specified. The size of the computer should be specified, together with the type and size of the memory. The computer capability should be expandable as the system grows and as buildings or points are added. The high level language used in the software should be specified. Software is of two kinds. One for the computer operating program, and one for the various energy management programs.

State of the art systems on the market today provide CRTs with all programs provided with English language prompting, to lead the operator step by step through each operation. The basic cost is usually a black and white CRT, with color as an optional added cost. All truly competitive systems also offer a color graphics package for an extra cost. This is a microprocessor based CRT which permits the owner to generate his own graphics, for air handlers, chillers, and the like. In addition, the data being collected in the system is dynamically updated on the CRT color screen as each temperature or other variable changes.

It is no longer necessary to pay the vendor to update software each time a new point or building is added. All competitive systems provide this capability for the owner at no cost, and with no reprogramming of software required. And, as a matter of fact, all truly competitive systems have available, some at added cost, a user development package which permits the user to do his own programming.

Software packages are available today for the list of energy management programs discussed earlier. But not every such package provides for a "calculated point" program and an "event initiated" program. With these two programs the EMCS contractor and the owner can accomplish any monitoring and control desired which can be reduced to a series of logical steps. They are both very powerful tools and should be specified.

Accuracy of the system is important. The owner should know how accurate a system is being purchased. The accuracy of sensors and control devices should be specified. In addition, the accuracy of the data transmission system is important. All competitive systems provide for transmission of data in digital form; hence, analog sensor inputs and analog command outputs must be converted from analog to digital signals. The conversion accuracy of the analog to digital conversion cards used in the system should be specified.

Input/Output Point List: Such a list must be provided for any well designed job. This list provides an exact listing for each floor of each building of both inputs and outputs for the computer system. The latest tri-service specification requires such a listing for the first time, and the Army tri-service technical manual provides a form for this listing. Every major equipment vendor has such a form also, for the first major item of work to be accomplished in a vendor-designed system is the completion of such a form.

Schematics: For the simplest of controls, schematics are not required; but for a multiple chiller optimization package, a schematic for both condenser water and chilled water is desirable. Such schematics show the various types of sensors, controls, and reset devices, as well as their location. Such schematics, together with existing mechanical and electrical drawings, effectively eliminate the need in the design phase for drawings. The specifications, however, must require the EMCS contractor to submit as-built drawings and schematics.

Specification Location: Local controls and control sequences for the HVAC system are normally located in the mechanical section, or Division 15 of a specification. When an EMCS is specified in addition, it may be placed in a separate Division 17. Some owners may indicate a desire for both local controls and the EMCS system to be included in a Division 17, so that one vendor can be held responsible for the requirements of both. Further, fire and security systems may be added to the EMCS functions. The net result is that the engineer or architect may place the specification for the EMCS in any Division, as long as the bidding instructions highlight very clearly the responsibility for installation, and for coordination and supervision of that installation.

Future Installation: The economics of energy use and cost is such today that every building can justify the installation of some type of system. That does not mean that every existing building has such a system, however. It does not even mean that every new building is being provided such a system.

For those new buildings which are not being provided such a control system, there are some specific actions which can be taken at little cost, or at lower cost than may be the case later in a retrofit situation, to permit later installation. A few of these are listed below for consideration by the designer, the architect, and the owner.

1. Include in the base bid a 2" conduit with pull wire from the basement to the roof, with a junction box at each floor.

2. Consider installing low voltage relays or contactors for lighting control. They are more easily purchased as a part of a long term mortgage than from cash flow or a short term loan.

3. Consider installing a central control panel in the mechanical room with remote start/stop and pilot lite for status for all air handler motors, cooling tower, exhaust fans, outside air fans, and the like. Separate this central panel as a bid item so that a proper deduction may be made should you decide to install an EMCS instead. The deduction would also include the 120 volt wiring from the equipment to the central control panel.

4. Install thermometer wells in chilled water and condenser water lines for future EMCS use.

5. Provide auxiliary contacts on all motor starters, control relays, or contactors for future EMCS use as status feed-back.

APPLICATION ENGINEERING & SYSTEM IMPLEMENTATION

General Considerations

Energy management and control systems (EMCS) often do not fulfill their expectations; the reason is not defective hardware, nor defective software, nor faulty installation. It is, instead, lack of proper planning - particularly in the implementation phase of the project. The absolutely vital key to successful implementation of an EMCS is thorough, insightful, and practical application engineering. As many as possible of the potential problems that can affect the project in any way should be identified and solved prior to any hardware being delivered or wire pulled. After the system has been correctly installed, results can then be achieved by properly checking out, starting up, and fine tuning the system.

Assuming for purposes of this discussion that the system hardware and control points have already been specified, the major areas of concern during application engineering relate to two areas; namely, proper interfacing with controlled and/or monitored points, and system installation. Let us briefly look at both.

Specifics of Interface

Systems perform their assigned functions by turning equipment on and off, adjusting control setpoints or position and monitoring status and/or other pertinent variables such as temperatures, pressures, flows, and the like. The application engineer must identify in detail and commit to drawings specifically how the controlled or monitored point is to be connected to the system.

Most of the interfaces are relatively uncomplicated, such as connecting to a motor starter in order to start and stop a pump or fan. The important thing is to get the interface - however simple - correctly on paper with all other operating and safety controls in the circuit shown. Since a discussion of every type of interface is impossible here, this discussion is confined to three of the more involved interfaces, namely lighting controls, automatic capacity limiting of centrifugal chillers for demand control, and remote temperature (or control point) adjustment.

Lighting Control

In deciding how to automatically control lighting in a facility, the application engineer must evaluate several alternatives. Consider the question of how to control lighting on a floor of a high rise building with 16-20 amp lighting circuits fed from a single breaker panel. The most cost effective interface with the lighting circuits will differ, depending on how the circuits are grouped for control. To illustrate this, here are three examples of controlling the 16 lighting circuits mentioned.

Case 1 - Control all 16 circuits as a single group.
Case 2 - Control the circuits in two groups.
Case 3 - Control all 16 circuits individually.

Case 1: The interface to control all 16 circuits as a single group could either be a single 400 amp 3 pole contactor installed ahead of the breaker panel or two 8 pole, 20 amp contactors connected to the load side of the breaker panel. Despite the additional electrical installation labor required by the multipole contactor approach, the fact that the 400 amp contactor costs approximately 10 times more than two 8 pole contactors makes the multipole contactor approach more cost effective.

Case 2: The interface to control the 16 circuits as two groups could either be two 8 pole contactors or a panel with 16 low voltage lighting relays (G.E. RR7 or equal) with necessary auxiliary devices. The total installed cost of the two 8 pole contactors should be less than that for the panel with the 16 individual lighting relays.

Case 3: Controlling all 16 circuits individually can be accomplished with 16 G.E. RR7s. Note that the energy management system will need to provide momentary output pulses to energize and de-energize the RR7s, particularly if individual wall mounted circuit override switches are to be added. Otherwise a maintained contact output to the RR7s is permissible with the inclusion of a half wave rectifier into the control circuit to the RR7s.

Chillers

Shedding of chiller capacity on centrifugal

chillers for purposes of demand control is becoming more prevalent because of escalating demand charges. Since Trane and Carrier are the two manufacturers most widely seen in the field, some comments and precautions to take for proper interfacing with these machines are appropriate.

Trane Centravac: This is a very simple machine to which to add automatic capacity limiting, since it has a 4 position detented switch in the chiller control panel whose function can be automatically duplicated. For one stage of capacity reduction to 80% a single pole double throw relay will be inserted in series with the "Common to 100%" circuit of the chiller control panel. The Common terminal of the relay will connect to the "Common" terminal on the terminal block that feeds to the capacity limiting switch on the chiller control panel. The normally closed relay contact will connect to the "Common" terminal on the switch block of the capacity limiting switch. The normally open relay contact connects to the "80%" terminal on the terminal block connected to the capacity limiting switch. In its de-energized (fail safe) state the relay allows the capacity limiting switch to function normally. It is strongly suggested that the switch be left at its "100%" position and used only if the energy management and control system is down for any reason. In its energized state the relay will limit maximum chiller capacity to 80%. To add capability to shed to 60%, a second relay would be added in series with the "normally open" contact of the first, wired common normally closed (C-NC) to the 80% terminal, common normally open (C-NO) to the 60% terminal. A third relay to provide shedding to 40% could be added and similarly wired. All of this is quite simple and straight forward; however, improper selection of relays can burn out the current transformer (CT) that is feeding this circuit. An "open" in series with the CT circuit for as short a time as 50 milliseconds can cause the CT to self destruct. Therefore, "make before break" relays should be employed to insure against the possibility of a slow transferring relay contact burning out the CT.

Carrier: One method for automatic demand limiting with Carrier centrifugal machines is to step in a set of low value, high power precision resistors in series with the voltage producing resistor in the CT or capacity limiting circuit. If 3 stages of limiting are desired, three resistors are added in series with the voltage producing resistor. In the non-shed (100% capacity) mode, each of the resistors is shunted out of the circuit by a relay wired C-NC in series with the resistor. In the 80% mode, the first relay opens placing the first resistor in the circuit. In the 60% mode the second relay opens and puts the second resistor in the circuit. The 40% mode is identical. Because the voltage producing resistor is such a small value (typically less than 1 Ohm), even with the new resistors shunted out any additional resistance in the circuit added by wire and relay contacts will be very significant. Therefore, if this method is used, be sure to have a Carrier service man recalibrate the full load controller to 100% with the added line and contact resistance in the circuit.

Remote Resets

Many energy management and control systems have the capability of providing outputs for remotely resetting temperatures or other controlled variables. Interfacing with the local loop control system is not as simple as getting in series with a fan starter or contactor because the interface often requires modifying the local loop control system before the reset signal can be accepted.

For example, let's look at a case where remote space temperature adjustment is desired in a space served by a single zone air handling unit controlled by a pneumatic room thermostat. To remotely adjust the room temperature setpoint the pneumatic thermostat must first be changed out to a wall mounted temperature sensor and it must be piped to feed a (new) pneumatic receiver - controller with a pneumatic reset port. The reset output from the EMCS will typically be a DC voltage output to an electronic/pneumatic transducer that will provide a 3-15 psig reset signal to the reset port on the receiver - controller. It should be obvious from this description that what first appears to be the simplest of remote resets can be quite involved and costly. The important thing to keep in mind in planning a remote reset interface is that the sensed variable (e.g., temperature) must be controlled by a controller with a reset port that is available to the EMCS. In most cases this will not be available without costly modifications to the local loop control system.

Installation Considerations

Many installation related problems that are likely to be encountered can be eliminated during the application engineering planning also.

Codes: You must be aware of and follow applicable local and national electrical installation codes. In a retrofit application be sure to become familiar with the current National Electrical Code requirements pertaining to Class II (low voltage) wiring.

Electrical Noise: Another potential problem area to be aware of is electrical interference on the power and control lines that are installed. Because we are typically dealing with some type of microprocessor based system we must reduce the level of electrical noise and switching transients as much as possible to ensure system reliability. Here are some suggestions pertaining to incoming power and control output circuits.

1. Provide power to the central processing unit (CPU) and any remote multiplexers through separate grounded circuits. Do not use any circuit that has wall receptacles, water fountains, or any device other than the EMCS on it.

2. As a minimum, be sure to use some type of RFI filter on the power circuit to the CPU. Consider using a Constant Voltage Transformer if there are a number of inductive loads in the building.

3. It is suggested that 24V AC or DC interposing relays be used to control any line

voltage loads - even though the EMCS contacts can handle line voltage or high current draw. This is to isolate the EMCS from the switching transients induced by the starter or contactor coil being controlled. It is also suggested that a filter be added across the coil of the interposing relay; e.g., blocking diodes for a DC relay or a resistor in series with capacitor for an AC relay.

These measures, though seemingly trivial, are inexpensive insurance and will add greatly to system reliability.

Hard Wiring Vs. PLC

One major decision to be made concerns whether to hard wire all control outputs or to control via PLC (Power Line Carrier) transmitter and receivers. In evaluating the two alternatives you need to consider the following:

1. Most PLC equipment is currently designed for one way digital output control only. If status feedback is required it must be hard wired back to the CPU or multiplexing panel, eliminating some of the savings contemplated through the use of the PLC equipment.

2. Analog inputs and/or outputs cannot be transmitted via PLC; they must also be hard wired back to the CPU or multiplexing panel.

3. PLC transmission can be particularly cost effective where the majority of points in the EMCS are digital output control points, spread across a large horizontal area that would otherwise require extensive hard wiring.

System Start-Up-Checkout

The most important consideration during system checkout is that end to end operation of 100% of the points in the system must be verified. Any type of random sampling is absolutely unacceptable. Verifying end to end operation unavoidably requires a minimum of 2 men with walkie-talkies to go thru the entire system before the installation can be considered complete.

Training

All final documentation as built wiring diagrams and sequences of operation should be supplied before training begins. Emphasis during training should be on becoming comfortable with normal system operation. The operators should be able to make all normal changes (schedules, demand targets, holiday dates) comfortably. Once the operator understands normal operations he can begin fine tuning the system.

Fine Tuning

No system can be set up to provide its maximum energy savings when first installed. The process of maximizing savings is often referred to as "system fine tuning" and is required for a system to perform optimally. Fine tuning simply requires setting objectives (e.g., demand targets and daily KWH targets), implementing measures to reach those objectives, tracking the results, and setting new objectives to further refine the process. Just as in all previous phases of the project, application knowledge and creative thinking will be essential to maximize potential savings. Fine tuning will not be as simple as it first may appear. Take for example, an 80,000 square foot office/manufacturing facility served by 20 DX rooftop units. For demand control you are shedding individual rooftop unit compressors. At first glance the task of maximizing demand savings appears to be no more than lowering the demand setpoint until your demand goal is reached. However, even in this very simple building you will quickly discover that the task is not that simple. Office areas will react differently to shedding of compressors than will manufacturing areas. Interior areas react differently than exterior perimeter areas. Because of these factors, reducing overall demand will require controlling each unique area differently. Compressors serving each area will require unique priority levels for shedding, maximum shed time duration, and minimum run times after being restored from shedding. Only when the requirements for each area are accommodated individually through fine tuning can you successfully lower the total demand target without adversely affecting occupant comfort.

Summary

The principal thought emphasized throughout this discussion is that the business of energy management and control systems is one for application engineering; it is not merely a matter of hardware. Successful implementation of an energy management and control system requires knowledgeable application planning all the way from the system specification to layout, installation, start-up, and fine tuning. Lack of good application engineering at any point in the project will produce mediocre results and a dissatisfied customer or client.

SECTION 3
MEASUREMENT METHODS

Chapter 14

MEASUREMENT TECHNIQUES FOR ENERGY CALCULATIONS AND DOCUMENTATIONS

A. B. Kramer, Jon Darcy

Introduction

This paper will present arguments in favor of engineering, designing, and selecting instrument systems, which are to be used in energy calculations. Prior to 1973, steam, cooling, and electric usage, metered by the utility company, was generally taken at face value. If the central metering station acted as the source point for distribution throughout a complex, no means were available to isolate large energy users within the complex. Accordingly, no allocation for energy use could be supplied.

Recently, there have been attempts to submeter energy so that allocation and billing could be more equitably distributed, and to check the energy supply company. Electical energy offers no particular problems since measurements of watts, KVA, etc. is relatively simple and well defined, however, is subject to some human errors. Measurements of steam, chilled water, and BTU's are considerably more complicated.

Orifice/Flow nozzle/Head type Installations

Head type flow meters rely on restrictions in the pipe line to generate a differential pressure resulting from a change in velocity through the restriction. The flow is a function of this differential pressure the area of the restriction and the velocity. These flows must be corrected for deviation from theoretical by various coefficients and correction factors.
In most cases these coefficients are Reynolds number related and are a function of many other variables not the least of which is a flow pattern. This flow pattern must be uniform turbulent. Coefficients are inaccurate and non-repeatable if certain parameters are not met. These parameters are specified by the American Societey of Mechanical Engineers (ASME), and are as follows:

1. Minimum straight pipe up and down stream of the restriction.
2. Maintainance of uniform turbulent flow pattern.
3. In the case of orifice/flow nozzles, taps for the differential pressure specifically noted as either pipe, flange, or vena concentrate taps.
4. Locations such that no upstream disturbance in the form of valves, tees, elbows are within the minimum upstream pipe diameter limits.
5. That the turndown (maximum to minimum flow range be recognized as a limit. Typically 3 to 4 to 1).
6. Flowing conditions should remain constant or be compensated for.

Examples

A. Steam Flow Metering

Typically, engineers would specify instrumentation to measure steam flow distribution in commercial or residential buildings. The steam conditions are generally saturated and the metering methods are of the head type (orifice) or shunt flow methods. It is extremely difficult, if not impossible, to define the point of saturation in a flowing process of steam, therefore very difficult to measure with any degree of accuracy.

An orifice is a volumetric flow measuring orifice. The specific volume of a saturated steam condition is taken directly from steam tables and is a function of pressure and saturation temperature. The specific volume of a mixture (below saturated conditions in the steam dome of a Mollier chart) is given as:

$$v(ft^3/lb) = xv_g(ft^3/lb) + (1-x)v_f$$

where:
v = specific volume of mixture
v_f = specific volume of liquid
$v_{fg} = v_g - v_f$, the difference between specific volume of saturated vapor (v_g) and saturated liquid (v_f).
x = quality of mixture in %

Consider the following example:

An orifice shunt flow meter being volumetric in nature is a function of specific volume at a given pressure. The following conditions are given:

150 psia saturated steam
$v_g = 3.0139$; $v_f = .01819$;
$v_{fg} = 2.9958$
95% quality

Since mass flow rate is proportional to velocity and velocity is proportional to specific volume, then flow rate is proportional to specific volume.

Let Q_x = flow rate at conditions of less than 100% quality (95%)
Q_s = flow rate at saturated conditions

$$\frac{Q_s}{Q_x} = \sqrt{\frac{f(v_s)}{f(v_x)}} = \sqrt{\frac{f(3.0139)}{f(v_x)}} = \sqrt{\frac{f(3.0139)}{f(2.8639)}}$$

$$= 1.0258$$

$$v_x = xv_g + (1-x)v_f$$
$$= .95(3.0139) + (1-.95).01809$$
$$= 2.863 + .0009045$$
$$= 2.8639\ ft^3/lb$$

The possible flow rate error due to non observance of steam quality is approximately 2.58%, ie: $Q_s = 1.0258Q_x$.

The total instrument system for measuring steam flow consists of the orifice plate or primary measurement element, a conversion element to convert the flow energy to a physical signal and a readout device to observe the signal. Typical accuracies of the orifice is ± 4%, readout ± 1%, conversion ± 2%.

Total system inaccuracy (E_{rms}) based on root mean square values and quality of error of 2.58% is given by:

$$E_{rms} = \left(\sqrt{(.04)^2 + (.01)^2 + (.02)^2 + (.0258)^2}\right) \times 100$$
$$= 5.25\%\ \text{error.}$$

B. BTU Metering

BTU Metering for cooling or heating distribution and accounting costs is becoming increasingly more common. BTU's cannot be measured directly: they must be calculated by the simple heat equation:

$$Q = mc_p(\Delta T) = mc_p(T_R - T_S)$$

where:
- Q = BTU/hr
- m = mass flow rate lb/hr
- c = specific heat BTU/lb °F
- ΔT = Temperature difference °F
- T_R, T_S = supply and return temperatures °F

It can be shown that the error in measurement of a calculated variable as a function of the measured variable is given by:

$$\Delta Q_{rms} = \left[\sum\left(\frac{\partial Q}{\partial u_i}\Delta u_i\right)^2\right]^{1/2}$$

where: ΔQ is the rms error
$\frac{\partial Q}{\partial u_i}$ = is the partial derivative (weighting function) of the calculated variable with respect to the individual measuring instrument
Δu_i is the error in absolute value of the measuring instrument

Consider the following example for BTU cooling accounting:

- mass flow(m): 25,000 lb/hr
- specific heat (c_p): 1.00
- temperature supply (T_S): 45°F
- temperature return (T_R): 55°F
- temperature range: 25° - 75°F

Accuracies:

mass flow Δm = ± 4% of full scale
specific heat c_p = ± .1%
temperature devices, T_S, T_R = .5% of span

Absolute errors of instruments:

$$\Delta Q = \sqrt{\left(\frac{\partial Q}{\partial m}\Delta m\right)^2 + \left(\frac{\partial Q}{\partial c}\Delta c\right)^2 + \left(\frac{\partial Q}{\partial T_R}\Delta T_R\right)^2 + \left(\frac{\partial Q}{\partial T_S}\Delta T_S\right)^2}$$

$$\Delta m_{absolute} = \Delta m_{\%} \times \text{maximum flow}$$
$$= .04 \times 25,000 = 1000$$
$$\Delta c_{absolute} = \Delta c_{\%} \times \text{maximum } c_p$$
$$= .001 \times 1 = .001$$
$$\Delta T_{S,R} = \Delta T_{S\%} \times \text{maximum span}$$
$$= .005 \times (75-25) = .25°F$$
$$\Delta T = 55-45 = 10°F$$

Weighting functions:

$$\frac{\partial Q}{\partial m} = \frac{\partial}{\partial m}(mc_pT_R - mc_pT_S) = c_pT_R - c_pT_S = c_p T$$
$$1 \times 10 = 10$$

$$\frac{\partial Q}{\partial c_p} = \frac{\partial}{\partial c_p}(mc_p\Delta T) = mc_p\Delta T = 25,000 \times 1 \times 10 = 250,000$$

$$\frac{\partial Q}{\partial T_R} = \frac{\partial}{\partial T_R}(mc_pT_R - mc_pT_S) = mc_p = 25,000$$
$$\frac{\partial Q}{\partial T_S} = \frac{\partial}{\partial T_S}(mc_pT_R - mc_pT_S) = -mc_p = 25,000$$
$$\frac{\partial Q}{\partial m}\Delta m = 10 \times 1000 = 10,000$$
$$\frac{\partial Q}{\partial c_p}\Delta c_p = 250,000 \times .001 = 2500$$
$$\frac{\partial Q}{\partial T_R}\Delta T_R = 25,000 \times .25 = 6250$$
$$\frac{\partial Q}{\partial T_S}\Delta T_S = 25,000 \times .25 = 6250$$

Actual BTU recorded:

$$Q = mc(T_R - T_S)$$
$$= 25,000 \times 1 \times 10 = 250,000\ \text{BTU/hr}$$

$$\text{Error } \Delta Q_{rms} = (10000)^2 + (2500)^2 + (6250)^2 + (6250)^2 = 13,578$$

$$\%\ \text{error} = \frac{13,578}{250,000} = 5.4\%$$

Some BTU measuring systems use the same installation for accounting both heating and cooling energy supplied. In this case the temperature devices are typically ranged from 0 to 200°F. If in fact these temperature devices were used in the example cited for BTU cooling, the errors would be calculated using the same procedure.

$$\Delta T_{S,R} = \Delta T_{S\%} \times \text{maximum span} = .005 \times 200 = 1^{o}F$$

$$\frac{\partial Q}{\partial T_S} \Delta T_S = 25{,}000 \times 1 = 25{,}000$$

$$\frac{\partial Q}{\partial T_R} \Delta T_R = 25{,}000 \times 1 = 25{,}000$$

Actual BTU recorded is still 250,000 BTU/hr

$$\text{Error } \Delta Q_{rms} = \sqrt{(10000)^2 + (2500)^2 + (25000)^2 + (25000)^2} = 36{,}827 \text{ BTU/hr}$$

$$\% \text{ Error} = \left[\frac{36827}{250000}\right] \times 100 = 14.7\%$$

C. Monitoring of Liquid and Gaseous Fuels

Liquid fuels are generally heavy to light distillates ranging from about 40API for #1 fuels to approximately 19API for a good quality #6 fuels. The viscosity of these fuels range form less than 2SSU for #1 to greater than 140SSU for #6. It is the viscosity characteristic which determines the methods available to measure the gross volumetric quantity of fuel consumed. The measurement of BTU (heating value) offers a more significant problem in that BTU variation of 10-15% for the same volume and same class of fuel is known to exist.

It is painfully obvious that variation in BTU of 10-15% along with the inherent inaccuracies of the volumetric measurement of the fuel may lead to errors in actual BTU's consumed in a fuel burning plant of between 12 and 17% errors.

An adequate correction for BTU variation due to non hydrocarbon constituents is given by:

$$\text{BTU/lb} = \frac{\text{BTU} \times 100 - (A + B + C) + 40.4(C)}{100}$$

A = % by weight of H_2O
B = % by weight of ash
C = % by weight of sulfer

BTU = heating value at constant volume on a gross basis

Consider a #6 fuel with the following analysis:

Gross BTU/lb = 18,300
% by weight sulfer = 2.3
% by weight ash = .12
% by weight H_2O = 3.5

$$\text{BTU/lb(corrected)} = \frac{18{,}300 \times 100 - (3.5 + .12 + 2.3) + 40.5(2.3)}{100}$$

$$= 17{,}309 \text{ BTU/lb}$$

$$\% \text{ error} = \frac{18{,}300 - 17{,}309}{17{,}309} = 5.7\%$$

The measuring instruments typically range from the "dip stick" to a sophisticated variable orifice (area meter). Instrument accuracies vary from manufacturers stated ±2% to realistic ±4-5% of full scale reading. Keep in mind, however, that these accuracies are a % of full scale and that the inaccuracies of the reading at less than full scale increase.

Consider a meter as follows:

Fuel oil flow range: 0 - 100 GPH
Fuel oil meter accuracy: ±3%

At 100 GPH, the actual reading would be 100 ± 3GPH or ±3% At 50 GPH, The actual reading would be 50 ± 3GPH or ±6%.

D. Gas Flow Metering Systems

Gas flow metering systems are typically wet gas type rotating drum, connected to a series of gears which rotate dials or needles. The typical accuracies of these meters is approximately ±2-4% of full scale, and are volumetric measurements.

However, few of these meters are corrected for Standard Pressure and Temperature (SCFH). The Standard Pressure and Temperatures are 14.7 psia and 60°F. Assuming the molecular weight of the gas is a constant, a correction factor back to standard conditions from ambient conditions is in order to determine actual ft^3/hr consumed. Since ambient temperatures continually vary, either a continuous compensation is required or the errors due to ambient temperature changes must be accepted.

Volume flow rate, V, referred to standard temepratury and pressure conditions is given by:

$$V = \text{SCFM} \times \frac{14.7}{520} \times \frac{T.Z.}{P.}$$

where: SCFM = standard cubic feet per minute
T. = temperature in degrees Rankine
P. = pressure in psia
Z. = compressibility

Consider the following example:

200,000 ft^3/min flow rate

60°F temperature
14.7 psia pressure
Z = 1 compressibily

$$V = 200{,}000 \times \frac{14.7}{520} \times \frac{(460 + 60)}{14.7}$$

V = 200,000 SCFM or no error.

Note that in a volume meter, the actual volume flow rate varies directly with temperature and inversely with the pressure. If the primary reason for gas metering is to obtain fuel containing a certain quantity of BTU/ft^3, a considerable error can exist if metering is not corrected to standard temperature and pressure conditions.

Conclusion

This paper presented some examples and considerations for the measurement of non electrical variables. An error analysis was given, showing that apparent quantities of these variables are not what they appear to be. Considerable errors exist which are predictable.

BTU measurements, not directly measureable, can offer considerable errors if ranges of instruments used in the calculations are not carefully selected.

Typically in large process and power industries and consultant organizations which serve these industries, there exists a division or department which specializes in "Instrumentation and Control Engineering". These consulting organizations which serve the HVAC industry may have a person whose function is instrumentation in addition to a multitude of other duties. Since energy measurement and indeed control of energy conservation processes is becoming increasingly more complex, it becomes incumbent on the HVAC industry to promote a greater understanding of the concepts of instrument and control engineering, and to acquire personnel who are educated in these areas.

SECTION 4
ADVANCES IN COMBUSTION TECHNOLOGY

Chapter 15

RETROFITTING EXISTING BOILERS WITH SOLID WASTE BURNERS

L. G. Desmon

This paper is based upon the results of DOE contract No. DE AC03-78-ET-13148 entitled "Direct Combustion of Industrial Dry, Composite Waste in Suspension". Costs associated with the project were shared in the ratio of 78% DOE, 22% Peabody Gordon-Piatt.

OBJECTIVES

The objectives of the project were to (1) identify classifications of industrial sites producing sufficient (process) industrial wastes which could satisfy on-going energy requirements, and (2) develop the technology necessary for acceptance by industry of systems in which fossil fuel burners on existing boilers in these and at similar industrial sites can be retrofit with solid fuel suspension type burners.

The overall program objective is the adoption by industry of demonstrated retrofit procedures thereby saving oil and gas.

INTRODUCTION

Boilers are used extensively throughout the United States to generate steam for process use and for space heating; more than 95% of these boilers use fossil fuels as their primary energy source. Considerable quantities of dry solid combustible wastes are generated at many plant sites, however, and with the proper retrofit technology, much of it could be used as a replacement for fossil fuels.

Solid wastes have been used as fuel for very many years in saw mills, plywood plants, furniture factories and other forest product plants. These wastes are generally homogeneous and consistent in combustion characteristics, a condition which is not usually true of wastes generated at industrial plants.

The random variations in combustion characteristics associated with these dry industrial wastes must be accommodated for these materials to be used as fuels in retrofit suspension burning systems.

RATIONALE FOR SELECTING SUSPENSION BURNING

1. The potential for saving oil and gas by retrofitting existing boilers with suspension type burners that use dry composite solid (waste) fuels is great.
2. The technology to do so did not exist.
3. Some of the reasons for choosing suspension burners are:

- Retrofit is relatively easy to accomplish. In most cases, the retrofit can be accomplished in the shortest period of time and at the least cost.
- In many cases, the burner will occupy no more space on the face of the boiler than that occupied by the original fossil fuel burner.
- Suspension burners will follow variations in load more closely than comparable solid fuel systems and almost as closely as oil or gas burners.
- Should an interruption occur anywhere in the solid fuel system, the fuel can automatically and immediately be switched to the original oil or gas without interruption to the process needing heat.
- Should the fossil fuel supply be curtailed, the waste fueled system will be unaffected.

TELEPHONE SURVEY

Fifteen (15) plant sites each in twelve (12) Standard Industrial Code (S.I.C.) Groups that appeared to have good potential in meeting the objectives of this project were selected. Table I is a list of the S.I.C. Groups selected, along with the number of establishments in the major group.

TABLE 1

SIC MAJOR GROUP		NO. OF ESTABLISHMENTS IN THE MAJOR GROUP
20	FOOD & KINDRED PRODUCTS	29,454
21	TOBACCO MANUFACTURERS	272
22	TEXTILE MILL PRODUCTS	7,203
23	APPAREL & OTHER FINISHED PRODUCTS MADE FROM FABRICS AND SIMILAR MATERIALS	24,438
24	LUMBER & WOOD PRODUCTS EXCEPT FURNITURE	33,948
25	FURNITURE & FIXTURES	9,232
26	PAPER & ALLIED PRODUCTS	6,038
27	PRINTING, PUBLISHING & ALLIED INDUSTRIES	42,102
28	CHEMICALS & ALLIED PRODUCTS	11,425
31	LEATHER & LEATHER PRODUCTS	3,201
37	TRANSPORTATION EQUIPMENT	8,802
39	MISC.MANUFACTURING INDUSTRIES	15,188

The major objectives were the determination of:

a. The quantity of waste generated and whether or not it has sufficient energy potential to be of significance.
b. Whether or not the company is interested in using the waste as fuel.
c. The company energy needs for firing boilers, etc.
d. Whether or not the company is representative of the industry group in terms of energy needs and the classification of energy uses (comfort heat, process steam, heat requirements, or cooling).

Of the 180 company names selected, we were able to contact 172.

Table 2 is a summary of these data.

TABLE 2

TELEPHONE SURVEY
SUMMARY OF RESULTS

COMPANIES	NUMBER
NAMES LISTED	180
CONTACTED	172
INFORMATION OBTAINED	144
DECLINED TO PARTICIPATE	25
NO LONGER IN BUSINESS	3
POTENTIAL REMAINING	116
HAVE, OR WOULD CONSIDER CONVERSION	67
SEEMING TO HAVE SUFFICIENT WASTE	35

S.I.C. GROUPS SHOWING GREATEST POTENTIAL FOR RETROFIT	RATIO
27 PRINTING, PUBLISHING & ALLIED PRODUCTS	8/15
24 LUMBER & WOOD PRODUCTS EXCEPT FURNITURE	5/11
25 FURNITURE & FIXTURES	6/14
21 TOBACCO MANUFACTURERS	3/11
26 PAPER & ALLIED PRODUCTS	4/15

Analysis of the meaningful data resulted in establishing the 5 S.I.C. Groups seeming to have the greatest potential for retrofit. Group 27, Printing, Publishing & Allied Products had the best "potential ratio". The second and third best groups, 24, Lumber & Wood Products and 25, Furniture & Fixtures were not selected for further activity because:

a. a significant level of energy is currently obtained from wastes in these S.I.C. Groups, and
b. the wastes used as fuel in each plant is homogeneous and consistent, not "composite".

As a result Group 21, Tobacco Manufacturers, and 26, Paper and Allied Products were selected as the second and third groups for further study.

The following five plants were selected as sources of fuel for the full scale burn tests:

Group 21 - Tobacco, Etc. - Brown & Williamson, Philip Morris

Group 26 - Paper, Etc. - Kansas City Star

Group 27 - Printing, Etc.- Kable Printing, Norway Gravure

LABORATORY TESTS

Typical samples of dry composite waste were obtained from all five of the fuel source sites selected and sent to the analytical laboratory, Commercial Testing and Engineering of Chicago.

The Kable mixture runs the gamut from newsprint to slick paper and includes tramp materials such as glue and binding materials. The moisture content is low (4.51%) and the volatile fraction is high (78.74%). Both of these characteristics promote good combustion. Fuels that burn satisfactorily in suspension usually have moisture contents of 15% or less and volatile fractions of 60% or more. The high ash content (11.10%) generates a need for costly removal (in furnace) and emission (at stack) control systems.

The analysis performed on a sample of waste from the Brown & Williamson Tobacco Co. shows a higher moisture content (11.12%) than that for Kable and a lower volatile fraction (57.86%). As can be deduced from previous comments, this fuel is not as "easy" to burn as Kable's. Further, the ash content (19.75%) is nearly twice that of Kable's.

A comparison of the two heating values shows Kable's to be 7261 Btu/Lb (as burned) vs 5089 Btu/Lb for B & W. To generate the same amount of steam with each waste fuel would require a 42% increase in feed rate of the B & W material over that for Kable. The resulting ash removal requirement would jump to 2½ times as a consequence.

Combustion Characteristics of Fuels:

The combustion characteristics of fuels that may seem to be similar can vary widely. When burning a material for the purpose of transferring its energy to another medium for extended periods, it is imperative that more than its heating value, moisture and ash contents be known.

Ultimate analyses were conducted so that the elemental composition of the fuels could be established. From these analyses, the stoichiometric (chemically correct) air/fuel mixture were calculated and the relationship between heat release rate and excess air established. Stoichiometric air flow rates have ranged from 48.5 to 91.6 cubic feet per pound of fuel burned during the course of this project.

Another important difference between fuel components shows up in ash fusion temperature. White wood, for example, may have an ash content of about 1% with an incipient slagging (initial deformation) temperature of 2200°F. Slick paper may have an ash content of 35% and a slagging temperature of 2700°F. The ash from some agricultural products may be expected to slag at 1500°F or slightly higher. Each of these materials has different ash handling and combustion air needs, and these needs must be carefully determined.

Excess air is a very important independent variable in the combustion of fuels, and this is especially true of solid fuels.

To operate a heat exchange process at its best efficiency, the inlet gas temperature should be held at the highest possible level. Theoretically, the stoichiometric air/fuel ratio (zero excess air) will produce this condition. Operating at zero excess air will however, cause some combustibles to pass through the reaction zone so rapidly that they don't burn to completion.

For vaporous or gaseous fuels, a small amount of excess air (10% to 20%) is necessary to obtain maximum combustion efficiency (maximum CO_2 and minimum CO).

When burning solid fuels, the excess air rate must be held at an even higher level because combustion of fixed carbon occurs on the surface of the particles and the presence of an excess of oxygen molecules helps the carbon find a "mate" within a reasonable period of time.

Excess air is also used to control the temperature of the combustion gas in which the ash is being conveyed. If the temperature is too high, the ash will melt, become slag, then solidify on and insulate the first surface it "sees" that is cooler than fusion temperature. All too frequently, this surface is the inlet heat transfer area of the boiler. Combustion control is a key requirement to the satisfactory use of composite wastes as fuels.

TEST FACILITY

Based upon the selection criteria described and the results of the laboratory tests, ten ton samples (approx) were collected from five of the participating sources.

The Test Facility in which the fuels were burned has three major areas: fuel preparation, energy exchange, and instrumentation room.

A. Fuel Preparation Area:

The fuel preparation area occupies the south half (40') of the Solid Fuel Laboratory (SF Lab). The flow path taken by the incoming dry waste is depicted schematically on Figure 1.

FIGURE 1
FLOW PATH TAKEN BY INCOMING DRY WASTE

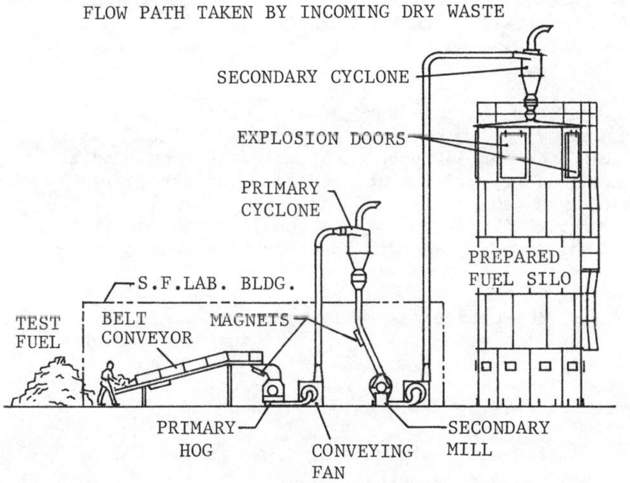

Previous work on similar low bulk density fuels resulted in modifications to the pneumatic conveying systems in the fuel prep area from negative to positive type. This was found to be necessary because the fuel could not pass through the rotary air locks at the product exit of the cyclone separators. The air lock rotors were removed from their housings and stored. These modifications served well for processing the B&W waste. Processed B & W waste produced a low bulk density fuel (about 4 Lbs/Ft^3) after grinding.

B. Energy Exchange Area:

The energy exchange area occupies the north half (35') of the SF Lab. Movement of the fuel from the silo through the remainder of the system is shown schematically on Figure 2.

FIGURE 2
FLOW FROM FUEL SILO THROUGH
THE REMAINDER OF THE SYSTEM

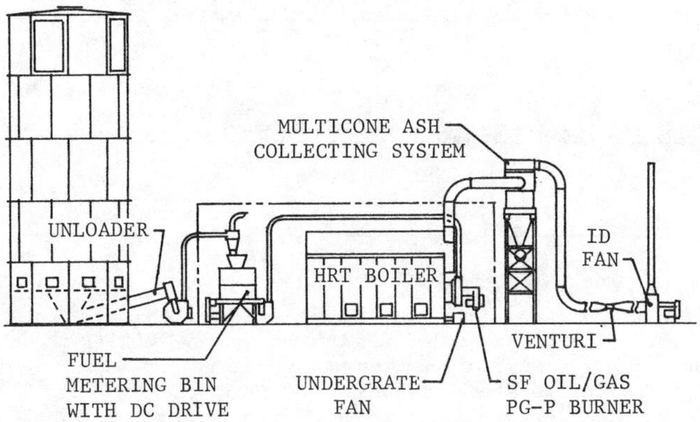

From the silo, the prepared fuel travels to the metering bin on demand as determined by the level control switches in the metering bin.

It was found that although the auger in the bottom of the silo and the short unloader auger would operate satisfactorily with the low density fuel, the material would not drop into, and continue moving along the auger that fed fuel from the unloader to the metering bin. Figure 2.

Augers in the bottom of the metering bin, driven by a variable speed D.C. motor, meter the prepared fuel out of the bin whereupon it falls into the drop box, is picked up by air and sucked into the conveying fan.

Prepared fuel flows through a 4" pipe extending from the conveying fan at the drop box exit to the solid fuel (SF) portion of the burner assembly on the face of the boiler. A combustion air blower located at the side of the SF burner mixes air with the fuel and blows it into the furnace chamber of the HRT boiler.

In addition to the SF portion, there is a standard gas/oil portion to the burner assembly. The gas burner is used to pre-heat the furnace refractory and initiate combustion of the solid fuel. Having established ignition of the solid fuel, the gas burner is turned off, permitting autogenous combustion of the solid fuel. Should there be an interruption in the flow of solid fuel, automatic controls will shift to fossil fuel if desired.

Figure 3 is a section through the HRT boiler. The furnace chamber has a target wall 9'3" from the burner exit and a pin hole grate extending from the target wall to the front of the refractory chamber. A separate undergrate air fan has been installed at the front of the boiler so that undergrate air could be controlled separately from the remainder of the combustion air.

FIGURE 3
SECTION THROUGH HRT BOILER

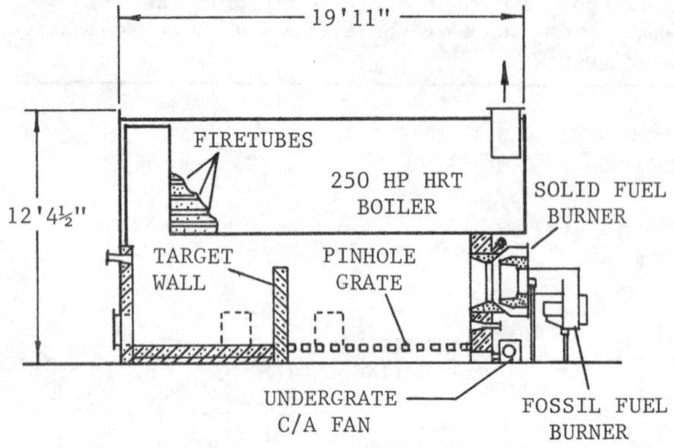

Combustion gases pass from the refractory chamber through the horizontal firetubes of the HRT boiler and into the multiple cyclone. The carryover separated from the stack gas passes through a rotary air-lock into a barrel under the multiple cyclone. The stack gas travels through a venturi flow meter, into the induced draft (ID) fan and then to atmosphere.

C. Instrumentation:

The instrument room contains all controls for operating the burner-boiler system, in addition to manometers, thermocouple readouts, and stack gas composition meters. Exterior to the instrument room are the boiler feed water and (natural) gas flow meters. A fuel analysis room is located south of the instrument room.

SHAKE-DOWN

Some of the problems that surfaced during the project and were resolved during the shake-down operation were:

- The interior of the boiler combustion chamber was examined and photographed prior to initiating combustion. Some cracks in the refractory were observed but declared normal by an agent of the refractory vendor.
- Loads of scrap wood were obtained locally and put through both the fuel preparation and the energy conversion sections of the Facility. Six shake-down tests were conducted using wood as fuel. As a result, additional grids were ordered for the two hammermills and a number of new approaches to the measurement of fuel flow were tested.
- Preliminary efforts at comminution of baled mixed paper for the first DOE burn test revealed problems in conveying the low bulk density materials.
- The rotary air lock hard surface clearances had to be increased to avoid binding by thin platelets of waste materials.
- Cyclone air flow rates had to be reset to suit the bulk density of the solids.
- The metering bin drive ratios had to be revised to suit the different bulk density ranges.
- Orifice and venturi flow meters were calibrated to permit heat and mass balance calculations to be made. Calculation methods were set up to establish the variation of CO_2 and O_2 with excess air for various fuel analyses.

TEST PROGRAM

All five test burn programs were conducted in the full scale Test Facility. Each program consisted of two preliminary and one or two final tests. During the preliminary tests, parameters such as fuel particle size and the ratio of undergrate to overfire were optimized.

Although particulate emission measurements were discussed prior to the award of this contract, the requirement was not included because the cost would have increased very significantly, the results would have been specific to the particular equipment and operating conditions employed, and the technology required to control emissions already exists and little value would be gained for the additional costs incurred.

KABLE PRINTING CO. (SIC No. 2752)

23,000 pounds of waste in 22 bales was received from Kable Printing Co. The mixture consisted of newsprint, slick paper, and included tramp materials that were non-combustible.

Kable produces magazines, catalogs, books, and other publications. They use several types of paper, all of which come in kraft paper covered rolls. In addition to the kraft paper, the waste includes up to 8% of the roll feed stock. Some of the waste also comes from the bindery and so contains glue, stitchery, staples, etc.

Two preliminary and two final tests were made and much was learned during this series about the differences in the suspension burning of composite waste as versus wood waste.

The Test Facility design was typical of the best practice for suspension burning of wood and other materials with essentially "three dimensional" particle configurations. Paper particles do not seem to fit this description.

At the start of this test series, difficulty was experienced in processing the baled paper into a usable fuel. The initial problem occurred in conveying it pneumatically. The processed fuel had a very low bulk density which made it difficult to separate from the conveying air stream. Fines generated during the milling process added to the conveying problems experienced.

Because conventional negative pressure conveying systems did not work satisfactorily with fuel having a bulk density as low as 2 Lbs/Ft^3, all pneumatic conveying systems were converted from negative to positive pressure as previously stated.

The first preliminary test burn was run on 6-6-80 and the three major observations made were: combustion was autogeneous, was better in suspension than on the grate, and appeared to be "cleaner" with finer particle sizes.

Additional fuel was processed after modifications were made to the secondary mill to produce a smaller fuel particle size with fewer fines. The fuel produced had a different bulk density and so the metering system had to be recalibrated.

The second preliminary test burn was run on 6-17-80; the burn looked good and it was decided to proceed with the first of two full scale test burns.

The first of two final test burns was run on 6-27-80. One of the objectives of this test was to determine the minimum excess air ratio for good combustion. This test had to be abandoned after a few hours of running for the following reasons:

- We were in the middle of a heat wave (114°F) and some electronic components in the control system malfunctioned. This problem was subsequently corrected.
- The fuel silo unloading mechanism could not keep up with the firing rate (this was corrected by modifying drives and adding a supplementary pneumatic conveying system, bypassing the original screw augers).
- The combustion chamber had started to fill up with a considerable quantity of ash which was expected to, but did not carry over to the multiple cyclone system, due to the reduced undergrate air flow.

The second final test burn was run on 7-9/11-80. Because of the wide variations in fuel moisture content, (2.8%-7.5% moisture content wet basis), the ash content (5.0%-38% dry basis), it was concluded that firing rate be controlled in the following manner:

- Combustion air flow rate was held constant for a calculated heat release rate of 8 x 10^6 Btu/Hr. and excess air of 95%. Calibrated orifice plates were used to determine flow rate which was adjusted for air density changes due to temperature, relative humidity, and barometric pressure.
- Combustion gas carbon dioxide and oxygen readings were continuously monitored (CO_2 = 10%, O_2= 10%) along with gas and refractory temperatures at strategic locations.
- Fuel flow rate, controlled by the variable speed D.C. drive on the metering bin, was manually adjusted throughout the test to hold CO_2 and O_2 levels and combustion temperatures on setpoint.

The procedure described has proven to be quite satisfactory, was followed in all the succeeding tests, and is the basis for a change in automatic control design.

BROWN & WILLIAMSON TOBACCO CORP. (SIC No. 2111)

14,800 pounds of loose (unbaled) waste was received from the Brown & Williamson Tobacco Corporation.

The waste consisted of "empty" cigarettes, filter tip stock, roll ends of foil/paper, roll ends of polypropolene, cigarette carton stock, corrugated cartons, and a very small amount of tobacco waste.

About 3 weeks after its receipt, evidence of spontaneous combustion in the "pile" was noted. Some of the waste was removed, ground and stored in the silo and one end of the dumpster was allowed to remain open. This seemed to end the problem.

Two preliminary and two final test burns were made using Brown & Williamson waste. The results varied considerably due to the fuel mix.

The following conclusions are based upon these tests:

- The dry industrial composite waste from B & W varies continuously in its make-up. Random samples can vary dramatically from an "average" fuel profile and this variability can give dramatic changes in combustion requirements.
- Grinding of this material, when clear of contaminants, presented no problems. Secrens having 3/8" diameter holes were used in the second stage mill. It is probable that a screen with larger holes could have been used with satisfactory results.
- Combustion was autogeneous, and as observed through sight ports and video equipment, looked very good with most of it occurring in suspension.
- The gas leaving the exhaust stack appeared hazy as was expected. The only stack gas cleanup equipment in the system was a multiple cyclone.
- Excess air rates of 75% - 80% appeared to result in the most satisfactory combustion results. It was found that when excess air rates were lowered to approximately 45% - 50%, the combustible fraction in the ash would increase as would visible emissions. Higher excess air rates increased the combustible fraction and also reduced steaming rates.
- The laminated foil on paper has a very high heating value (9924 Btu/Lb), probably because aluminum can be a fuel. The ultimate analysis did not however, show the presence of aluminum. The standard laboratory procedure used in performing the analysis relegates all elements not specifically identified therein, to the category entitled ash.

NORWAY GRAVURE (TEMPO COMMUNICATIONS, INC.) (SIC No. 2752)

25,000 pounds of waste was received from Norway Gravure. The company is one of the largest producers of beer labels consisting of a thin film of aluminum deposited on a heavy paper backing. Some of the waste was in the form of stacks of 30 inch square sheets of uncut labels and the remainder was in the form of cartons of trimmings that remain after the labels are "punched out".

When the labels were made by an earlier process, the aluminum was 60 "molecules" thick and the scrap was bought by an aluminum recycler. The process now used deposits a layer of aluminum only 2 "molecules" thick and the recycler will no longer accept it.

Although the waste did not suffer from wide and random variations in composition as was the case with the waste from Kable and Brown & Williamson, the high ash content (20%) required careful combustion procedures.

Interlocking characteristics of the prepared fuel promoted a tendency to bridge in the silo. Accordingly, the horizontal auger that sweeps the material toward the center hopper had to be extended by about 3 inches so that it was less than 1 inch from the silo wall. This "fix" eliminated any further bridging tendency.

Conveying the fuel from the silo to the metering bin and from the bin to the solid fuel burner did not pose any additional problems.

Optimum excess air rate appeared to be 62%, combustion was autogenous, and as was the case with all high ash fuels, the stack appeared hazy.

Although the indicated combustible content of the ash collected by the multiple cyclone was between 1 and 2%, the actual value was probably higher and masked to some extent by the aluminum oxidizing in the analytical furnace.

PHILIP MORRIS USA (SIC No. 2111)

37,140 pounds of waste in 23 bales was received from Philip Morris USA. The waste was quite similar to that received from Brown & Williamson, in that it consisted of a composite mixture of empty cigarettes, filter tip material, polypropolene, foil paper, etc. but almost no tobacco.

Satisfactory comminution, conveying, storage, and retrieval of the fuel was demonstrated. As was the case with previously burned fuels having low bulk density, it was necessary to perform all conveying functions in positive pressure systems. Further, new sonar type bin level sensors had to be installed to control the fuel level in the metering bin because of the low fuel bulk density.

Combustion of the fuel as observed through sight ports in the combustion chamber of the HRT boiler looked excellent with most of it occurring in suspension.

KANSAS CITY STAR (SIC No. 2621)

24,820 pounds of waste in 40 bales was received from the Kansas City Star. The Star is a newspaper publishing company and their waste consists of crumpled newsprint, office scrap, veloxes (photographic paper), plastic coffee cups, etc.

There is also a considerable quantity of imperfectly printed but folded newspapers and returned newspapers. This material is sold to a recycler and is currently worth less as fuel than as stock for recycled newsprint. It is probable that the economics will change in the near future.

Although the waste was a composite material, the variation in combustion characteristics was small and so control was not as big a problem as with the previous waste materials.

Satisfactory comminution, conveying, storage and retrieval of the fuel was demonstrated. Combustion as observed through sight ports in the combustion chamber looked excellent and appeared to take place almost entirely in suspension.

SUMMARY OF SIGNIFICANT FINDINGS

As a result of the tasks performed during the course of this project, the following findings are considered to be most significant:

- Industrial dry composite wastes can be used effectively to replace fossil fuels.
- Although the subject industrial wastes are available within many SIC (Standard Industrial Code) Groups, this investigation has shown the following groups typically generate sufficient quantities to meet a significant part of their energy needs:

  Group 21 - Tobacco Manufacturers
  Group 26 - Paper & Allied Products
  Group 27 - Printing, Publishing & Allied Industries.
- Substantial technical knowledge has been gained relative to the handling, preparation, and combustion of dry composite solid wastes as fuels.
- A technical data base has been established to optimize the retrofit of oil or gas burners on existing boilers with suspension burners that use industrial dry composite waste as fuel.
- It is economically attractive to use these wastes as a replacement for fossil fuels in SIC Groups 21, 26, and 27.

The decision as to whether a waste is recyclable or best used as a fuel, is often a matter of economics. If, for example, an industrial plant is currently using oil at \$0.90/gallon (\$6.42/10^6 Btu) as fuel for a boiler that is generating steam at an overall thermal efficiency of 80%, the net fuel cost is \$6.42/0.80 or \$8.02/10^6 Btu.

Assuming the waste has a heating value of 8000 Btu/Lb (16 x 10^6 Btu/Ton and the overall thermal efficiency of the waste heat boiler is 70%, the heat converted into steam is 11.2 x 10^6 Btu/Ton and its equivalent value as fuel is \$8.02 x 11.2 or \$89.82/Ton. Although this illustration does not consider other important costs such as amortization of retrofit equipment and savings in waste disposal, it serves as an example of how the decision to recycle or burn is approached. The payback for retrofit is good as shown on the table below for which a capital investment of \$350,000 has been assumed.

Capital Investment (in thousands)	\$350
Inflation for Costs & Savings	10%
Depreciation - 4 Years Double Declining, then Straight	
Investment Credits, 2 at	10%

YEAR	1	2	3
1. Cost Savings	191	210	231
2. Less Depreciation	100	71	51
3. Operating Costs	48	53	58
4. Savings, Gross (1-2-3)	43	86	122
5. Savings after Income Tax ½ (4)	22	43	61
6. Investment Credit - Machinery, Energy	70		
7. Cash Flow (2 + 5 + 6)	192	114	112
8. Payback	2.4 Years		

ACKNOWLEDGMENT - The author wishes to express his gratitude to Gerry Hottinger, who directed the construction of the Test Facility and conducted the test burns with eclat.

APPENDIX - RETROFIT GUIDELINES

The intent of these guidelines is to provide an indepth check list for those considering the retrofit of existing oil or gas burners with solid fuel suspension burners that use industrial dry composite waste for fuel.

Although the hot combustion gases from solid wastes can be used in almost any process where the combustion gases from oil or natural gas is presently being used, these guidelines relate solely to boiler applications.

1. PRELIMINARY EVALUATION

1.1 Space:

Check adequacy of the following conditions:

- Space on the face of the boiler for attachment of the SF Burner.
- The floor space in front of the boiler for accessibility.
- Area for collection and storage of combustible waste.
- Area for preparation of waste as fuel.
- Location of prepared fuel silo.

1.2 Waste - Fuel:

Check:

- Adequacy of quantity to warrant retrofit.
- Degree of contamination and its effect on cost.
- Safety in storage and preparation.
- Combustibility.

1.3 Combustion System:

There are a number of waste combustion systems presently available for retrofit and the proper selection therefrom is of great importance. Because this document deals with suspension type burners, the advantages of selecting it for retrofit are listed:

- It will follow variations in load almost as closely as the oil or gas burner it is intended to replace.

. In many cases, it will occupy no more space on the face of the boiler than that occupied by the original fossil fuel burner.
. It is easiest to accomplish.
. Should an interruption occur in the solid fuel system, the fuel source can automatically and immediately return to the original oil or gas and maintain the steam load.

2. FUEL

For material that has been classed as waste to become suspension-burner fuel, it must have certain characteristics. These are discussed below.

2.1 Specifications:

. Non combustibles - all metallic objects and materials that fuse below a temperature of 1800°F (it forms slag) must be removed.
. Moisture content - the moisture content (contained water) should not be greater than 15% wet basis. As the name implies, fuel particles burn in a short time while suspended in air. During this period, the contained water is evaporated, the volatile fraction is "boiled off" and finally, the fixed carbon burns.

2.2 Receiving & Storage of Raw Fuel:

. How and where is the waste being collected now (truck, cart, conveyors) and must changes be made to accommodate its moving into the fuel preparation area.
. If a stoppage occurs in the solid fuel system, can the waste be diverted (perhaps to a landfill as might previously have been the case).
. If more waste is to be "imported" to improve economics, will the receiving system accommodate it.
. What is its condition upon arrival. Is it dusty, explosive, contaminated by metal or glass, are edible materials or rodents in it, is it in bulk or baled.
. What is its density, is it subject to spontaneous combustion, does it require covering.
. Is there sufficient storage volume. Is it collected over a 5-day 8-hour shift and used over a 7-day 3-shift basis. Is a 3-day storage volume adequate or will seasonal effecte influence this.

2.3 Handling for Preparation:

. The first step in preparing the waste as fuel is transporting the raw fuel from storage to the preparation area.
. Depending upon its condition in storage, the raw fuel might be transported by a tractor type loader that needs an operator. This might be a solution if the waste is stored loose or in containers.
. Another approach might be pneumatic conveying. This requires a feeding mechanism in the storage area and a cyclone separator at the delivery end. The pneumatic system can be operated at a pressure below or above that of the atmosphere. The usual system is negative because air will leak into the system vs a positive system in which the possibility exists for dusty air to leak out. The negative system requires an airlock to allow the solids collected by the cyclone to drop out. The positive system does not need an airlock. Both systems exhaust dusty air to the atmosphere which will probably require further attention.
. Other transportation methods include belt or chain conveyors, bucket elevators, hoists, etc.

2.4 Fuel Preparation:

The extent of the fuel preparation requirement depends greatly upon the condition of the raw waste. Large pieces of metals and other non-combustibles must be removed before comminution is attempted. Also, if the fuel is in bales, the straps must be removed. This is one area that proponents of other combustion systems point to as an additional source of electrical energy usage. Although this criticism is true, careful design of the comminution system will minimize the energy use and the ability to follow the load makes suspension burning more efficient in its use of fuel.

The objective of the fuel preparation system is to provide particles that will pass through a 1/8 inch diameter screen, have a moisture content of 15% or below, have a minimum non-combustible fraction, and be as homogeneous and consistent in other ways as possible.

If there is a possibility of dirt or sand adhering to the waste, it should, as an initial step, be passed over a screen designed to clean it.

What happens next depends upon whether or not the sizing can be accomplished in one stage. This and succeeding decisions in the area of fuel sizing require the recommendations and guarantees of hammermill manufacturers.

If the raw waste is in pieces larger in maximum dimension than about 3 inches, it is probable that two stages of comminution will be required. The first stage may be a hammer hog and have a grid through which the particles fall. The grid may have openings that are 3 x 1½ inches. The hog should have a magnet to catch ferrous particles before they can enter the rotating hammers. Although one reason for the magnet is indeed minimizing damage to the mill, another reason is to prevent the occurrence of sparks which can cause dust explosions. The inclusion of properly sized explosion doors, rupture discs, or vents will prevent damage to personnel and equipment.

One should be aware that there are materials of widely varying physical characteristics, and size reduction devices of many designs.

To increase the material flow through a hammermill, it is usual to produce a low pressure area below the grid or screen by pulling air across its exterior. The flow of air from the inlet side of the screen to the discharge side air conveys particles that are small enough, through the screen and away from the mill in a duct. The proper air flow rates are important for reasons additional to that described above. The ratio of air mass to solids should be held within certain ranges in order that the solids remain in suspension while being conveyed. This usually determines the volume flow rate of air required. Further, the air velocity must not be so low that the material will not go around bends. A velocity that is too high will impose an unnecessarily high pressure drop and associated fan power.

What has been said about one stage of comminution also applies to a second stage if one is required. If two comminution devices are required, the cyclone separator for the primary hog is usually placed atop a small surge bin which will allow minor variations in flow rate between the first and second stage sizing operation to occur. The shape of the surge bin is important because some particulate materials have a tendency to bridge over a discharge disallowing further flow. The bottom of the surge bin will usually have a variable speed rotating valve to control the feed rate to the second stage mill.

2.5 Storage of Prepared Fuel:

The second stage cyclone (or equivalent) sits atop the prepared fuel silo in which it is usual to have a volume capable of holding a 3 day supply.

Where packing of prepared fuel is anticipated, the silo is replaced by a large rectangular bin with negatively sloped sides. Outside storage of dry, sized fuel is not usually acceptable. Explosion doors are usually installed near the top of the storage device.

In either case, the bottom of the storage device contains an unloader. When the fuel is free flowing and in a silo, a cone shaped vibrating unloader with feeder valve is sometimes used. If the fuel is not free flowing, a "live" bottom is usually employed. This may consist of one or more mechanical augers or conveyors.

The unloader(s) operate to satisfy level switches on the side of a small live bottom negative slope surge bin, whose primary function is to meter fuel to the solid fuel burner.

2.6 Fuel Feeding:

Mechanical augers (usually) cover the entire bottom of the metering bin. These augers are chain or gear driven by a variable speed drive. The speed of the drive is controlled by a modulating signal that is proportional to the need for heat. In this case the signal is intended to maintain constant steam pressure. As steam pressure decreases, the signal amplitude increases causing an increase in the speed of the live bottom augers which, in turn, screw more fuel out of the metering bin. The fuel drops into a mixing box where it is picked up by an air stream and conveyed to the solid fuel burner in a pipe.

If the burner is not more than about 75 feet away, a centrifugal type material handling fan can be used to ingest the fuel and transport the mixture. If the bin-burner separation is more than 75 feet, it is customary to use a positive displacement pump as an air source and the fuel is fed into the conveying pipe through an air lock downstream of the pump. In cases wherein the fuel must be introduced into a chamber at a pressure significantly above atmospheric, the positive displacement pump must be used.

It is important to realize that the live bottom augers are a volumetric feed device and although they may expel an equal volume of fuel for each revolution, the weight flow will vary with bulk density of the fuel. Further, as has been determined in this project, the heating value per pound of dry composite wastes is subject to wide and random variations.

2.7 Combustion:

The character of combustion associated with a suspension type burner is similar to that for oil and gas in that it occurs in space. The fuel particles are small and so the oxidation reaction goes to completion rapidly. Accordingly, as long as the combustion volume can contain the reaction, a suspension burner can follow a load rapidly.

Despite the above similarity, the solid fuel combustion process is much different from burning either gas or oil. When burning a solid biomass fuel, the initial action occurs as heat (from previous combustion), causes the contained water near the particle surface to reach 212°F, and evaporate thereby allowing the surface temperature to rise. As it rises, water vapor is driven from the inner volume of the particle and the increased surface temperature causes the volatile hydrocarbons to distill off at temperatures of about 325°F and burn when it "sees" oxygen.

When all of the volatile matter is gone, there remains about 20% by weight of the original combustible matter (fixed carbon), plus all of the non-combustibles in the form of a porous particle. The heating value Btu/Lb is greater for the carbon than it was for the volatile hydrocarbon and it is importnat that as little as possible of the carbon be allowed to pass through the system without giving up its heat.

Combustion occurs on the surface of the fixed carbon and as a result, it has to mate with an oxygen molecule before it can become exothermic and turn to carbon dioxide. After the carbon is oxidized only the non-combustibles (ash) remains and if it gets hot enough, it will fuse and become liquid slag.

With this in mind and returning to the point at which the fuel along with a minimum amount of air enters the burner, it is apparent that provisions must be made to supply additional combustion air rapidly. Note that neither the solid fuel burner nor the fossil fuel burner it replaces is really a burner. They are both injectors and the burning actually occurs in a combustion volume immediately downstream thereof.

3. BOILER MODIFICATIONS

If the boiler being retrofit has a refractory lined combustion chamber such as an HRT boiler, the solid fuel combustion process will usually go to completion quite satisfactorily. The refractory becomes incandescent and radiates heat to the fuel particles thereby helping the reaction.

If the boiler does not have refractory surfaces it absorbs some of the heat of combustion before it can help the carbon burn. As a result, much of the carbon can pass through the system only partially oxidized. In addition to the loss of energy, this condition can generate sparks (burning carbon particles). The best way to avoid this condition is to install a refractory furnace between the solid fuel burner (injector) and the "black" boiler to be retrofit.

The additional combustion air required is normally added where needed along the length of the combustion chamber. If some of the fuel particles are larger than will burn in suspension, it is customary to provide a pinhole grate in the combustion chamber floor with a supply of undergrate air for the fallen particles.

Chapter 16
ADVANCES IN COMBUSTION CONTROL

U. Bonne, Sr., A. Patani

ABSTRACT

This is a review of our analytical and experimental studies on flame sensing, efficiency improvement and control, and flue emission. In addition, we report on new insights gained by an analysis of flue gas enthalpies dependence on firing rate and flue gas volume, and how this was used to predict and verify efficiencies of boilers fired with different fuels. Other results:

- Geiger-Müller - type UV tubes can detect the UV emission of flames of hydrocarbons, CO, H_2, methanol, and sulfur with air, in addition to H_2-Cl_2 flames.
- Low or medium BTU fuel gases may lead to boiler efficiencies that are not at all low. For example, CO is more "efficient" than CH_4 or H_2 on both lower or higher heating value bases.
- Using cold outdoor air for combustion leads to lower boiler efficiencies.
- The influence on NO_x generation of over 10 different parameters on one combustor is analyzed.

In addition to reporting on spectral flame measurements, the effects of reducing firing rate or steam pressure setting, and emulsifying fuel-oil with water on efficiency are discussed.

NOMENCLATURE

a	Coefficient to Express Influence of Inerts in %/F
A	Oxygen Content of Air in Volume Fraction
B	Atomic Hydrogen to Carbon Ratio of Fuel
B_4	Fuel Inerts in Equivalent Nitrogen to Carbon Ratio
E	Excess Air, in Fraction of Stoichiometric
F	Firing Rate in W or BTU/h
N_9	Total Number of Flue Gas Moles per Mole of Fuel
T_F	Flue Temperature in C or F
x	Mole Fraction
Δ	Difference
ΔH_O	Lower Heating Value in MJ/kgO_2 Consumed
ΔT_F	Flue Temperature Rise in C or F
η_{SS}	Steady State Fuel Utilization Efficiency in % of Input

1. INTRODUCTION

The equipment used to convert chemical energy (or heat of combustion of fuel) into thermal energy may be considered as consisting of the following parts:

- Fuel supply system, e.g., pipes, tanks, mills or fuel generators
- Burner
- Fire box or combustion chamber and heat exchanger
- Venting system, including appropriate dampers and/or air preheaters, and
- Controls for fuel and air flow including pre- and post- purge, ignition, flame proving, combustion emissions and efficiency

The purpose of this paper is to discuss the latter in light of requirements or features of unconventional or new fuels. "New fuels" is interpreted here in the broader sense of including the impact of new regulations or new ways to control combustion of conventional fuels. The discussion below focuses on three aspects only of controlling the combustion process, i.e., efficiency controls, flame sensors and control of combustion generated pollutants.

Although ignitors, sequencers/programmers, and valves are part of any combustion control package, these aspects were not included in the discussion at this time.

2. DISCUSSION

2.1 Efficiency Controls

The efficiency of a combustion system has conventionally been measured in terms of the steady state composite of combustion and thermal efficiencies. The former is a measure for the completeness of combustion, i.e., the combustibles remaining in the flue, ash or equipment walls, while the latter indicates the useful fraction of the heat energy that did not get lost via the stack or jacket.

The above definitions are still useful today and have served us well in the past; they are represented by charts such as those of Figures A1-A4 of ref. (1) in terms of the conventional parameters flue O_2 or CO_2 concentration, flue temperature, and various types of fuel. The new feature of Figures A2 and A2.1[1] (metric) is that they were plotted on the basis of a constant heat release per consumption of unit of oxygen, rather than per unit fuel. This means that these charts can be used for all hydrocarbons. The only inconvenience is that if gross efficiencies are desired, the latent losses, L_W, of Fig. A4[1] have to be added to the sensible losses by an additional operation.

One exception may be the fact that fuel efficiency definitions are generally not consistent with operating cost since they do not allow for the influence of the used electric (input) energy to drive feed water pumps, circulating pumps or fans, etc. To alleviate this shortcoming we defined the concept of Cost Efficiency as the ratio of Value of Delivered Heat/Total Cost of Input Energy.[2]

New fuels, the increasing cost of energy and use of multiple fuels demand that traditional efficiency concepts be fine-tuned and the feasibility of reducing all pertinent efficiency losses be evaluated. According to this need, we discuss below possible improvements in efficiency by controlling:

- Firing rate and excess air
- Flue combustibles
- Fuel interchangeability
- Fuel inerts
- Air inerts
- Fuel additives
- Boiler outlet water or steam temperature
- Miscellaneous items such as pre- and post-purging, steam blow down and soot blowing, intake air temperature, pilot flames, cycling and off-period draft.

2.1.1 <u>Firing Rate and Excess Air.</u> Reducing the firing rate of a combustion system, without changing the air flow rate, e.g., with an atmospheric burner, leads to increases in excess air that are only in part compensated by decreases in flue temperature, and therefore lower the steady state fuel utilization efficiencies, η_{ss}, and is not recommended. Reducing fuel <u>and</u> air flow rate or excess air alone, on the other hand increases η_{ss} by increasing residence time, in spite of reduced flow velocities in the convective transfer section. Just by how much the efficiency is increased can be simply determined by using Figures 1 and 2, since they include the effect of excess air and firing rate on flue temperature. Predicting flue temperature as a function of changes in excess air and firing rate may be regarded as the next step in the development of combustion efficiency charts like Figs. 1A-4A of ref. (1) and a brief description of this development may be in order.

Earlier analyses of how excess air and firing rate change flue gas temperature and furnace or boiler efficiency had resulted in the relation[3]

$$\Delta T_F \sim F^{.81} (1+E)^{(.55 - 4.14\cdot 10^{-4}\cdot \Delta T_F)} \quad \text{---(1)}$$

where

ΔT_F = flue temperature rise in F

F = firing rate, and

E = excess air

Data used to arrive at this relation included both lean and rich combustion conditions, see Figure 3,

- Theoretical computations of temperature vs. time or heat exchanger length, based on radiative heat exchange of CO_2, H_2O, and soot, and natural and forced convection,[4]
- Gas Engineer's Handbook data,[5] and
- Experimental data from
  - Radiant tube high temperature gas furnaces[6,7]
  - An oil-fired boiler[8]
  - A gas-fired boiler and gas-fired furnaces

Fig. 1 was drawn using the direct relationship between efficiency, flue oxygen and flue temperature rise for the solid lines and eq. (1) for the dotted lines, after converting E into flue oxygen.

Figure 2 is a redrawn version of the dotted curves of Figure 1 and helps to predict the amount of simultaneous fuel and air derate (expressed in arbitrary units before and after derate) required to achieve a desired efficiency.

Adjustment of excess air can, when firing a fuel of unknown composition, be a problem if only a CO_2 measuring instrument is available. O_2-measuring devices do not have this problem since stoichiometric combustion or zero excess air conditions correspond to 0% also. Besides advances in ZrO_2 sensor technology, O_2 measurement offers these advantages:

- Requires little or no resetting of the control setpoint when switching over from natural gas to oil or propane, in good approximation, if the excess air is to remain approximately constant.
- Provides more accurate excess air readings at the low end of the scale than CO_2 measurements, if the relative measurement error is the same.
- Easier interpretation than CO_2 measurements.

Depending on the temperature of ZrO_2 sensors, the sensor output may represent actual O_2 at low, or net O_2 at high temperatures.

2.1.2 Flue Combustibles. Flue combustibles have been expressed in a number of ways, such as percent of fuel that is wasted, equivalent flue volume percent of CO or H_2, etc. A plot of the corresponding losses, L_C, (see e.g. Fig. 3A of ref. 1) shows that the Bacharach smoke concentration can be substantial before it hurts financially, and that cost penalties due to combustibles are higher than the numerical value of the equivalent hydrogen flue volume percent.

Flue combustibles are made up by components like H_2, CO, CH_4, aldehydes, solid carbon (soot), etc. with relative concentrations that depend on the type of fuel, combustion intensity, flame cooling rates, etc. It is therefore not a trivial problem to find the best practical way to determine flue combustibles that offer an acceptable compromise between accuracy and sensitivity. While methods based on measuring only hydrocarbons or CO may be sensitive, they suffer from accuracy since they measure only one combustible component, which may not be the most relevant one. For example, small oil-fired systems generally smoke heavily before generating significant amounts of CO. Calorimetric methods, on the other hand, are more accurate or comprehensive, but not as sensitive.

Monitoring and controlling only flue combustibles, without measuring CO_2 or O_2 we do not recommend, since the combustibles may not only be generated near stoichiometric or fuel rich conditions, but also under quite fuel lean conditions as a result of mixing or quenching problems; in addition, every burner/furnace or boiler combination is likely to have combustible vs. excess air profiles of its own.

2.1.3 Fuel Interchangeability

The purpose of this section is to describe how the achievable efficiency of furnaces or boilers fired with new or different fuels can be predicted, to show that our predictions compare very favorably with experimental data, and to demonstrate that low BTU fuels may have boiler efficiencies that are not at all low.

We hypothesized that the above equation (1) may provide a tool to predict the relative efficiency on a boiler when fired with a variety of fuels either at constant flue temperature, constant firing rate or constant output, e.g., in the form of lbs. of steam/h.[5] Equation (1) was used in such a way that changes in flue gas volumes from different fuels are simulated by substituting the total combustion air (1+E) by the total number flue gas moles N_9, per unit of net heating value released, where $N_9 = CO_2 + \frac{B}{2}H_2O + (1+E)(1/A-1)(1+B/4)N_2 + E(1+B/4)O_2$ resulting from burning an equal "amount" of CH_B fuel, where "amount" means released net energy, B = atomic H/C ratio and A = mole fraction of O_2 in air.

The boiler/furnace input is iterated until the output equals the desired level, e.g., the output obtained by firing the boiler with pure methane under 430F flue temperature rise and 0% excess air conditions, as presented in Table 1. Efficiency predictions for combustion of different fuels controlled to a flue oxygen level of 4.65% is shown in Table 2, Columns 8 and 10. Column 8 is for constant flue temperature, and Columns 9 and 10 are for constant output. The verification of these results is based on comparing them with experimental results in Reference 9. Table 3 shows this comparison, which was made by adopting the flue O_2 and temp. conditions for natural gas and comparing the efficiencies for methanol and fuel oil #5 obtained experimentally with those predicted by using the measured excess air (but not the flue temperature!). As shown, the efficiency agreement (81.8, 81.9%) for methanol and (89.6, 89.1%) for fuel oil #5 is excellent.

Table 3 also shows predictions of efficiencies for other fuels. Carbon monoxide, as shown in Tables 1 and 2, turns out to be the most "efficient" fuel, achieving 96.1% efficiency at the conditions listed in Table 3.

Since CO figures prominently in many low BTU gases, one may expect that low BTU gases do not necessarily lead to low efficiency boiler performance. This is indeed so as shown in Tables 2 and 3 and also results from the fact that the presence of N_2, CO_2, or H_2O in the fuel-air mixture does not reduce efficiency very much; (see next section).

The data of Table 1 show that:

- The energy released by combustion per kg of oxygen, ΔH_O is fairly constant compared to other ways of expressing heating value.
- The single value that best correlates with efficiency is heating value/unit volume of flue products (not listed). A better correlation is obtained by a combination of ΔH_O, B and X_{CO_2}.
- For equal fuel costs based on HHV BTUs, the operating cost of a system burning methane under the conditions of Table 1 would be higher by 4.6% if it used H_2. Fuel savings would be about 1.6% in favor of H_2 if fuel costs were equal on the LHV basis. On that basis, CO-firing would save an additional 1.3%. These numbers obviously shift with changes in operating conditions; see Table 2.
- The high value of ΔH_o for blast furnace gas is the result of the 1% of free O_2 contained in the fuel. The high values for CO and formaldehyde are due to the chemically bound oxygen of the fuel, which reduces the need for atmospheric oxygen (and the flue losses associated with heating the nitrogen part of air).

2.1.4 Fuel Inerts. We computed that for fuels containing 1 mole of N_2, CO_2 or H_2O per carbon atom of combustible fuel, the efficiency is lowered by amounts, $\Delta\eta_f$, expressed by the expression $\Delta\eta_f = a\Delta T$ where a is the temperature coefficient determined by linear regression for natural gas. Values for a, typical values of efficiency reduction for an average flue temperature rise of 300F and the standard error are listed below:

	a Coefficient %/F		$\Delta\eta_f$ Δ Efficiency % at ΔT = 300 F		$\Delta\Delta\eta_f$ Std. Error + %
N_2	$0.188.10^{-2}$	$(.44.10^{-2})$	.56	(1.3)	.0018
H_2O**	$0.220.10^{-2}$		.66		.0052
CO_2	$0.264.10^{-2}$	(54.10^{-2})	.79	(1.6)	.021

The listed Efficiency Drops are due to the presence of one Mole of N_2, H_2O or CO_2 per Mole of Fuel (CH_B), for Flue Gas Conditions 20 $\leq$ % Excess Air, $\leq$ 150 and 100 $\leq$ ΔT in F $\leq$ 500. Numbers in parenthesis refer to efficiency reductions when the combustion system is controlled to operate at constant output power rather than at constant flue temperature. Water was assumed to be introduced as a vapor.

This information should be of help to determine price rates or to decide on the level of cost-effective purification of gaseous fuels.

The possibly surprising low values of a (e.g., switching from natural gas to a 50/50 mix of CH_4 and CO_2 will only decrease efficiency by .79% at equal flue gas conditions) are caused by the fact that the fuel volume is still relatively small compared to the volume of N_2 introducted with the combustion air.

2.1.5 Air Inerts. Combustion with air generates more flue losses than combustion with oxygen. This is because the inert components of air (nitrogen, carbon dioxide, water vapor, argon, etc.) cause part of the so-called dry flue losses, similar to the losses introduced by the presence of fuel inerts.

One way to avoid these losses is to remove those inerts before combustion, i.e., "enrich" the air or increase its oxygen concentration.[4,10] Figure 4 shows that if the cost of generating O_2-enriched air was low or zero, substantial cost savings can be achieved, which are highest for systems with high flue temperatures. Note that the fuel savings are always achieved, assuming that no fuel is needed to generate O_2-enriched air; and that it is only the cost savings that depend on the cost of oxygen. Fuels containing unreacted O_2 (or compounds with oxygen bonds weaker than those to CO or CO_2) are attractive for the same reason: less O_2 from the air is required for combustion, resulting in fewer N_2 - fuel losses.

2.1.6 Fuel Additives. Evaluation of the benefits of such additives requires an analysis of their effect on:

- Changes in equipment maintenance requirements
- Fuel savings
- Cost savings, and
- Changes in particulate emissions

Organic and metal-organic additives are advertised primarily for fuel oil firing, and a number of publications analyze their merits or benefits.[11,12,13] A recent Battelle study for EPA found fuel savings ranging from 0 to +2%.[14] Besides the above compounds, claims have been made that addition of water in gaseous or liquid form reduces smoke formation and heat transfer surface fouling rates. The catalytic influence of CO_2 and H_2O in reducing soot formation in flames has been known for many years and was used to develop the blue oil flame systems by recirculating a fraction of the combustion products back to the burner. The key question is whether the additional energy losses and costs due to:

- Evaporating the emulsion water
- Venting the additional hot water vapor
- Amortizing or renting the additional equipment, and
- Operating (energy and maintenance) that equipment, are in a favorable balance with the benefits of:

- Less particulate mass loading
- Less heat transfer surface fouling, i.e., less maintenance and higher efficiency and/or
- Fuel savings due to reductions in excess air, made possible by the reduced smoke.[15]
- Energy savings are realized if excess air reductions can be controlled to values above those shown by the "breakeven curves, indicating that the additional flue losses are compensated by excess air reduction.[15]

The reduction in excess air, needed to compensate for the amount of energy lost due to the addition of water was plotted in Figure 5. It shows that:

- For example, a 20% by wt. addition of water requires an excess air reduction by about 20% to compensate for that loss, for a boiler flue temperature of 400F(204°C).
- For higher water contents and for lower boiler/furnace temperatures, the required excess air reduction increases.
- Lighter fuel oils need smaller excess air reductions to break even.

Figure 5 was drawn under the assumptions that:

- Water additions have no influence on flue temperature.[16]
- Flue temperature decreases, as indicated by the dotted lines on Figure 1, when excess air is reduced.
- There are no long term benefits in terms of less heat transfer surface fouling, because excess air is readjusted,

resulting in similar levels of smoke.

Figure 5 is valid also for systems using combustion air as the vehicle to add water vapor, since the energy to evaporate water is not recovered, and the flue gas losses are again increased by the sensible heat of the added water.

In summary, we concluded that:

- The latent and sensible energy of 10 to 20% water in a fuel oil-water emulsion is difficult to recover in boilers operating at or near 30 to 50% excess air and low flue temperatures, as shown in Figure 5. The required amounts of excess air reduction would lead to operation at undesirably low amounts of excess air.

2.1.7 Boiler Outlet Water or Steam Temperature. Analysis of experimental data from two different boilers showed that the steady state flue temperature increases or decreases with increasing or decreasing water outlet temperature, by about equal amounts, i.e., $\Delta T_F \simeq T_W$.

To help evaluate actual efficiency or fuel savings, in cases where the flue temperature is proportional to changes in saturated steam temperature, Figure 6 helps to make the transition from steam pressure change, ΔP, to flue temperature change, ΔT_F, and to the steady state efficiency improvement $\Delta\eta_{ss}$, for several values of excess air. Note that $\Delta\eta_{ss}$ does not depend on the type of hydrocarbon fuel used (natural gas, propane, butane or fuel oils).

2.1.8 Miscellaneous Ways to Improve Efficiency. Forced combustion chamber purging before and after burner operation may be needed for safety or odor suppression reasons. However, any air flow through the combustion chamber that picks up heat from the combustion chamber or heat exchanger represents a stack loss.[17]

Steam blow-down and soot blowing losses, L_B, in units that can be compared to efficiency losses are obtained by simply forming the ratio of the lost average steam enthalpy flow and the average input firing rate, F:

$$L_B = QM\Delta H_S/F$$

where Q = vapor quality of the released water
M = the average mass steam flow and
ΔH_S= the steam enthalpy

Ideally, one would control the frequency and quantity of the steam/water flow rates to maximize overall, long term equipment efficiency. In addition one should recover as much energy as possible from the released steam.

The higher the intake air temperature can be made, the higher is the heat transfer process efficiency. This was found to be true not only for wood stoves,[18] but also for residential furnaces.[19] Figure 7 shows this graphically for a number of intake air temperatures which resulted in essentially the same flue temperature. The curves were drawn from values of flue gas temperature vs. time as they travel from the ignition point through a plug flow reactor.[20] Fig. 16 shows the results of Fig. 7 expressed in terms of efficiency improvement vs. increases in intake air temperature.

The corollary is that one should avoid low air intake temperatures such as direct vents (sealed combustion) without air preheating, from an efficiency point of view. However, NO_x emissions decrease as the air intake temperature drops, as shown also in Fig. 7.

The problem of having to choose between cycling or modulating mode of operation is not uncommon. The advantages and disadvantages of modulating vs. cycling are listed below:

Modulation	Cycling
• Saves fuel, i.e., fuel efficiency is higher if excess air is kept constant	• Shorter equipment operating time
• Saves cost only if increased electric cost does not mask fuel savings	• Uses less electric energy
• Achievable small load (or room temperature) swings lead to improved process control (or comfort.	• Does not require fuel and air modulation controls
	• More efficient than fuel-only modulation[21]

The key to the understanding of the modulating vs. cycling alternative resides in the higher steady state efficiency achievable with modulation (assuming that excess air can be held constant) on one hand, versus the longer operating hours on the other. Operating cost may be higher if the fuel savings are overshadowed by increased electric costs.[21]

2.2 Flame Sensing

Flames provide direct evidence of the occurrence of combustion. The problem of determining that combustion in proceeding safely and steadily is then reduced to reliably sensing the presence of flames by detecting one or more of their properties:

- Chemical composition of the product gases
- High temperature
- Optical spectral emission
- Optical flicker emission
- Ionization
- ΔP across the combustion front, etc.

Speed of response and reliability requirements have led to the preferred use of optical spectral and flicker emission, and ionization flame properties. We discuss below the optical and electric properties of conventional and new fuel flames, relevant to these sensing approaches.

Optical Sensors. The UV emission of a natural gas flame recorded with a .75m Jarrell-Ash Scanning Monochromator and an 1P21 photomultiplier tube is shown on Figure 8&9. Dominating features are the well-known OH and Co-bands.[22]

The latter stretch all the way to "vacuum UV", where the O_2- Schumann Range bands limit the light transmission in air, below 1900A.

The OH-bands are prominent also in the fuel oil flame (vaporized, premixed with air) spectrum shown in Figure 10, although the short wavelength UV is not as intense as with the natural gas flame. The H_2-air flame spectrum of Figure 11 does not exhibit the CO-bands, as expected. Somewhat surprisingly, these CO-bands were not emitted by a CO-air flame, see Figure 12, although the hydrogen present from impurities was enough to generate some OH and excite the 3064A OH band ($A^2\Sigma^+$-$X^2\Pi$ transition) on top of the continuum attributed to $CO + O \rightarrow CO_2 + h\nu$. Honeywell's ultraviolet light detectors of the gas-filled Geiger-Mueller counter type are sensitive to wavelengths below 2800A. They are therefore "solar-blind" and blind to hot refractories below 2600F or 1400C. They are sensitive to the flame emissions shown in Figures 9-12. One question one can now ask is: how will these detectors respond to flames of fuels other than natural gas, fuel oil, H_2 or CO? We can make the following statements:

- All flames of hydrocarbons with compositions between methane and fuel oil can be detected.
- Unburned, vaporized fuel oil absorbs strongly in the UV below 2300A and for best performance, needs to be removed from the space between flame and UV detector, e.g. by blowing air through the detector sight tube. Results of UV absorption measurements with a small laboratory Meker burner holding a premixed, vaporized fuel oil #2-air flame are shown in Figure 13.
- New fuels containing or generating mixtures of hydrocarbons H_2, CO, CH_4, CO_2 and N_2 can be detected. This includes heavy oil, wood, lignite and coal provided absorbing vapors or coal particles are not in the way.
- We have recently checked and found that methanol and flames of elemental sulfur generate enough UV for Geiger-Mueller type UV tubes.
- While H_2-Cl_2 flames can be detected, the much cooler H_2-Br_2 flames do not appear to generate enough UV.

In general, we found that the emission intensity of hydrocarbon flames is fairly constant in terms of UV generated per carbon atom burned, regardless of the molecular origin of that carbon fuel atom.[22] This is shown in Figure 8 for several hydrocarbons. Note that the sensitivity to CO is significant, but distinctly lower than that of hydrocarbons.

The optical or electrical flicker is another property of a flame used to sense presence. Flicker is indicative of the flow pattern and only depends remotely on the flame chemistry. The detector therefore need not be blind to radiation of hot refractory or daylight if it is steady, but it still has to be sensitive to radiation in the range of the flame emission.

Flame Ionization Sensors. When the energy received by a molecule is very high, e.g., as a result of releasing chemical reaction energy, an electronically excited molecular state may result. Optical emission is generated when the excited electron makes a quantum jump back down to the ground or a lesser excited state, before it gets deactivated by a collision with another molecule. An ion is generated if the exciting energy received by the molecule and transmitted to one electron is higher than the highest excited state, and is therefore sufficient to remove the electron from atomic or molecular orbit. Flame ionization or chemi-ionization thus results in positive, ionic charge carriers and negative, mostly electronic, charge carriers.

Chemi-ionization is similar to the generation of UV chemiluminescence described above in that:

- It is found with all hydrocarbon flames.
- It has been used for flame detection in large burners, as well as for hydrocarbon detection on microburners burning H_2 in gas-chromatography systems, and
- It is independent of the molecular origin of the burned carbon atom, as long as it is from a hydrocarbon.

Flame ionization is initiated by reactions between CO and CH or CH and OH,[24] which may explain why pure CO or H_2 flames show little, if any, ionization. Fuels like heavy fuel oils and coal containing alkali metals, which require a relatively small ionization energy and ionize readily, generate highly conductive flames. However, that type of ionization depends on the amount of alkali-metal present, rather than the fuel/air ratio. The emission and ionization saturation current of a premixed natural gas flame depend on the fuel/air ratio. That dependence is influenced by the degree of premixedness and disappears at zero premix (diffusion flame).

2.3 Emission of NO_x

The maximum permissible flue concentration of these pollutants has been promulgated by EPA for combustion systems in excess of 250.10^6 BTU/h (250 GJ/h) of fuel input; the state of California has proposed NO_x limits for residential furnaces. The feasibility of staying below those and future legislated limits with new fuels naturally depends on the fuel chemistry, and on the type of combustion equipment and controls used.

Means to reduce the emission of carbon particulates via addition of water or flue gas recirculation was mentioned in Section 2.

The emission of SO_2 and SO_3 depends on the sulphur content of the fuel and the flue gas scrubbing equipment. Figure 14 shows how the flue gas dew point depends on the concentration of SO_3 and water vapor. For example, 1 ppm SO_3 and 10% water vapor can increase the dew point by about 60°C. Their emission can be reduced by using fluidized bed combustion with limestone or dolomite. Such combustion equipment also reduces NO_x because of the lower peak combustion temperatures. The discussion here will be limited to a brief review of the result of an earlier computer modeling study(22) on the simulation of the kinetics of NO generation in simple plug-flow combustors. That study did not include the effect of fuel-nitrogen; it was based on the simple Zeldovich mechanism(25) for cases of positive excess air combustion as used for Figure 15 and on the expanded one to represent the kinetics in fuel rich situations.

Figure 15 presents the influence of modifying a 30cm ID combustor operated under the conditions shown at the upper right-hand corner, on the NO emission in ppm. Some of the trends indicated by Figure 15 may be summarized as follows:

- Fuels with a smaller hydrogen content than methane are likely to generate more NO than methane, even if their heating value per unit weight is smaller.
- Reducing the combustor diameter, D, increasing excess air, E, flame luminosity or soot mass fraction, F_2, and increasing recirculation, R. decreases NO emission.
- Increasing swirl did not have a significant influence, because NO generation is completed before reaching the convection-sensitive part of the combustor.
- Increasing pressure, the intake air temperature, T_o, (by preheating) or the wall temperature, increases NO emission.

3. SUMMARY AND CONCLUSIONS

The thrust of this paper is directed towards an analysis of the feasibility, nature, performance and benefits of combinations of new and conventional controls for new and conventional fuels, from the point of view of efficiency control, flame sensing and pollution control. The conclusions may be summarized as follows:

- Fuel composition is generally a given and not a controllable parameter. However, when selecting a fuel, proper consideration needs to be given, not only to its heating value, but also to its effect on the efficiency of a boiler or furnace. For example, buying and burning the same amount of energy for equal amounts of money in the form of a 23624 BTU/lb fuel vs. a 4960 BTU/lb fuel, see Table 1, would seem advantageous because of its higher HHV. However, the former achieves an efficiency of 81.4% at flue conditions of 500F(260°C) and 0% O_2 while the latter achieves 92.76% under the same boiler output conditions. The corollary is that although fuels with higher heating values generally generate smaller relative flue losses, numerous exceptions exist as shown by comparing fuel oil #2 and #6, CO with CH_4, and ethane with methane, etc.
- Oxygen enriched combustion is an attractive way to reduce N_2-flue losses, and save combustion fuel although costs of generating oxygen and the special high temperature burner costs have to be taken into account as well.
- The short term benefits and penalties of using fuel oil-water or air-water additions were analyzed from an energy balance point of view, and a chart was drawn that shows the amount of excess air reduction needed to compensate for the energy to vaporize the water and for the increased flue losses. In order to achieve energy savings, excess air reductions should be controlled to values greater than those indicated by the "break-even" curves shown on Figure 5.
- Boiler water or steam outlet temperature (steady state) have a direct influence on flue temperature. Any reduction of the former will bring about some improvement in efficiency.
- Increasing the intake air temperature was found experimentally to decrease flue temperature rise, i.e., to increase efficiency. We therefore do not recommend taking cold outdoor air for combustion for this and other reasons related to the use of "free" room temperature air and avoiding casing losses from indoors to outdoors.
- Flame sensing can be accomplished with UV detectors for any fuel containing hydrocarbons, H_2 or CO. Flames of elemental sulphur, H_2-Cl_2, and methanol were detected as well. H_2-Br_2 flames could not be detected. The use of flame ionization is limited to hydrocarbon flames and excludes H_2 and CO flames.
- The use of power rather than atmospheric burners increases the System Efficiency of cycling combustion equipment without further modifications. In addition, it also facilitates:
    1. The use of air pre-heaters to increase efficiency still further, and
    2. The implementation of recirculation for controlling NO_x and carbon particulates.

4. REFERENCES

1. U. Bonne and J. P. Langmead, "A Short Method to Determine Seasonal Efficiency of Fossil-Fired Heating Systems

for Labeling Purposes", Trans. ASHRAE 84I, 422 (1978).

2. Bonne, U., "Definition of Efficiency for Hybrid Heating Systems," ASHRAE J. 20(3), 31 (1978).

3. Bonne, U., and Patani, A., "Combustion Controls for New Fuels", IGT Symposium on "New Fuels and Advances in Combustion Technology", New Orleans, La., March 26-30, 1979.

4. Bonne, U., Brosvic, J., and Johnson, A. E., "Improved Monitoring and Control of Metallurgical Processes," IGT Symposium on "Efficient Use of Fuels. II," Chicago, Illinois, December 9-13, 1974, Proceedings, p. 51.

5. Segeler, C. G. (ed.), "Gas Engineers' Handbook", Industrial Press Inc., First Edition, Fifth Print, New York, New York (1977).

6. Bonne, U. and Torborg, R. H., "Combustion Efficiency," The 1977 AGA Industrial Gas Seminar, "Profit from Energy Management," St. Louis, Missouri, November 2-4, 1977.

7. Torborg, R. H. and Bonne, U., "Instrumentation for Adjusting Burners for Improved Efficiency," 1978 Annual Meeting of APCA, Paper No. 78-49.6, Houston, Texas, June 25-30, 1978.

8. Bonne, U. and Johnson, A. E., "Thermal Efficiency in Non-Modulating Combustion Systems," Conference on "Improving Efficiency in HVAC Equipment and Components in Residential and Small Commercial Buildings," sponsored by NBS and ASHRAE Purdue University, Lafayette, Indiana, October 7-9, 1974, Proceedings, pp. 1 and A14.

9. D. R. Miller and R. W. Duhl, "Alcohols, the Now Fuels", Vulcan Cincinnati Co. IGT Symposium on "New Fuels and Advances in Combustion Technology", New Orleans, Louisiana, March 26-30, 1979.

10. Thekdi, A. C. and Hemsath, K. H., "Improvement of Fuel Efficiency in Industrial Heating Equipment," ASME 1973 Winter Annual Meeting, Detroit, Michigan, November 11-15, 1973.

11. Andrade, R. C. and Gelosa, L. R., "Need for Fuel Oil Additives Grows," Power Eng., February, 1977, p. 60.

12. Lewis, D. C. and Goldberg, H. J., "Prescription for a Successful Fuel Treatment Program," Combustion 50(6) 28(1978).

13. Schwarz, R. C., "Superior Fuel Oil Additives," Paper presented at the Industrial Oil Conference, III. State National Association Power Eng., Chicago, Illinois, May 17, 1972.

14. Giammar, R. D., Weller, A. E. and Krause, H. H., "Evaluation of Fuel Oil Additives for Increasing Efficiency and Reducing Emissions of Boilers," ASHRAE 1978 Semi-annual Meeting, Symposium on "Efficiency of Combustion Systems IV," Atlanta, February, 1978 and ASHRAE Trans. 84I, 367(1978).

15. Bonne, U., "Efficiency of External Combustion Systems. Limitations of Firing Fuel Oil-Water Emulsions," Seminar on "Efficiency of Combustion Systems II," ASHRAE 1976 Semiannual Meeting, Dallas, Texas, February 1-5, 1976, Paper No. 970S, ASHRAE Trans. 83I, 28(1977), and invited discussion of Reference 10.

16. Hall, R. E., Private Communications, 1975 and "The Effect of Water/Residual Oil Emulsions on Air Pollutant Emissions and Efficiency of Commercial Boilers," Paper No. 75-WA/APC-1, ASME WA Meeting, Houston, Texas, November 30 - December 5, 1975.

17. Bonne, U., Schuldt, S. B. and Johnson, A.E., "Process Control for Increased Fuel Efficiency" IGT Symposium on "Efficient Use of Fuels," Chicago, Illinois, April 16-19, 1974, Proceedings, p. 103.

18. Shelton, J. W., Black T. Chaffee, M. and Schwarz, M., "Wood Stove Testing Methods and Some Preliminary Experimental Results," ASHRAE Trans. 84I, 388 (1978).

19. Torborg, R. H., Honeywell Residential Engineering, Private Communications, 1978.

20. Bonne, U., "Source Emission Control - Key to Clean Air," Honeywell Computer Journal 7 (3) 199 (1973).

21. U. Bonne, R. D. Jacobson, A. Patani and R. L. Aagard, "Seasonal Efficiency of Residential Combustion and Heat Pump Systems", Internat'l (IASTED) Symp. on Simulation, Modeling and Decision in Energy Systems, Montreal, Canada, June 1-2, 1978, Proceedings Act. Press, Anaheim, Cal., p. 277.

22. Bonne, U. and Berg, R. E., "Hydrocarbon Flames as Reference UV Light Sources," J. Opt. Soc. Ann. 60, 1285 (1970).

23. Bonne, U., Discussion on "Vacuum UV Light Emission," 11th Symposium (International) on Combustion, Berkeley, California, August (1966), Proceedings, p. 1190.

24. Zeldovich, U. B., "The Oxidation of Nitrogen in Combustion Explosions," Acta Physicochimica USSR 21, 577 (1946).

Table 1. Influence of Fuel Composition* on Heating Values, Ultimate Flue CO_2 Concentrations and Steady State Efficiency, η_{ss}. η_{ss} Was Computed for 0% Excess Air and a Flue Temperature Rise of ΔT_F = 430F(239C), for Methane. ΔT_F for Other Fuels was Adjusted to Account for Variations in Flue Gas Flow to Maintain Constant Output Energy Rate.

	B	B_4	ΔH_C Higher Heating Value				ΔH_0 Lower Htg. Value	X_{CO_2} Ultimate Flue CO_2		η_{ss} Steady State Efficiency			ΔT_F Flue Temp. Rise
			Free of Ash and Moisture		As Found		Per Unit O_2			Const. ΔT	Constant Output		
Type of Fuel	H/C Ratio -	"N_2"/C Ratio -	BTU/lb	MJ/kg	BTU/lb	MJ/kg	MJ/kg	% Dry	% Wet	% LHV	% HHV	% LHV	^oF
CH4	4.000	.000	23942	55.65	23942	55.65	12.536	11.70	9.48	90.12	81.20	90.12	430
C2H6	3.000	.000	22371	52.00	22371	52.00	12.749	13.15	10.98	90.40	82.87	90.54	424
C3H8	2.667	.000	21708	50.46	21708	50.46	12.775	13.72	11.60	90.46	83.45	90.64	422
C4H10	2.500	.000	21353	49.63	21353	49.63	12.791	14.02	11.93	90.49	83.75	90.69	421
CO	.000	.000	4348	10.11	4348	10.11	17.687	34.64	34.64	91.81	92.76	92.76	382
H2	∞	.000	61533	143.03	61533	143.03	15.125	.00	.00	91.14	77.47	91.57	409
CO+3H2	6.000	.000	14439	33.56	14439	33.56	15.766	11.70	8.66	91.33	81.26	91.91	402
CK OVN	5.139	.102	24096	56.01	20375	47.36	13.334	10.73	8.41	90.42	80.78	90.58	423
BL FRN	.211	1.789	4960	11.53	1355	3.15	18.257	24.75	24.12	86.25	81.94	83.56	510
NAT OH	3.887	.021	23810	55.34	22825	53.06	12.551	11.86	9.64	90.12	81.30	90.12	430
NAT MN	3.782	.089	23624	54.91	20389	47.39	12.572	11.88	9.70	90.09	81.39	90.07	431
K-T PR	1.000	.124	6547	15.22	5458	12.69	16.767	27.43	24.12	91.28	86.85	91.97	397
U-GAS	1.367	1.831	9098	21.15	2300	5.35	15.141	18.45	16.33	87.43	79.24	85.69	487
LURGI	1.932	.508	13038	30.31	5139	11.95	14.745	24.18	19.60	89.72	80.81	89.68	432
OIL #2	1.800	.000	19500	45.33	19500	45.33	12.626	15.45	13.56	90.49	84.95	90.70	421
OIL #6	1.500	.000	19000	44.16	19000	44.16	12.799	16.16	14.41	90.68	85.94	90.99	417
METHOL	4.000	.000	9763	22.69	9763	22.69	13.294	15.01	11.55	89.65	78.60	89.45	438
ETHNOL	3.000	.000	13981	32.50	13981	32.50	14.195	15.01	12.25	90.91	83.25	91.32	411
FRMALD	2.000	.000	8082	18.79	8082	18.79	16.235	20.95	17.32	91.41	84.89	92.09	397
FRMACD	2.000	.000	2381	5.54	2381	5.54	13.161	34.64	25.73	85.09	67.36	81.45	529
ACTACD	2.000	.000	6271	14.58	6271	14.58	12.288	20.95	17.32	88.65	79.05	87.91	457
WOOD	3.583	.017	8670	20.15	4736	11.01	13.191	20.16	14.81	86.74	71.11	85.28	476
LIGNTE	1.726	.046	12048	28.01	7548	17.55	12.326	19.39	16.61	88.77	80.86	88.30	447
BIT MV	.751	.045	15491	36.01	13610	31.64	12.376	18.58	17.37	90.39	87.42	90.58	422
ANTRCT	.365	.043	14974	34.81	12896	29.98	12.651	19.94	19.24	90.70	89.37	91.05	414
GRAPHT	.000	.000	14104	32.79	14104	32.79	12.294	20.95	20.95	90.75	91.11	91.11	414

*from References 1, 2, 3, 20, 29 and 30

FL26A, D1, B, D2

Table 2. Influence of Fuel Composition on Flue Gas Composition and Temperature, Firing Rate and Steady State Efficiency, η_{ss}. η_{ss} was Computed for 4.65% O_2 and a Boiler which Resulted in a Flue Gas Temperature for Natural Gas of 289F(143C). ΔT_F for Other Fuels was Adjusted to Account for Variations in Firing Rate and Flue Gas Flow to Maintain Constant Energy Rate

	B	X_{CO_2}		E_A	T_F	F	η_{ss}		
		Flue Gas CO_2		Excess Air	Flue Gas Temp.	Relat. Firing Rate	Steady State Efficiency		
							Const.ΔT	Constant Output	
Type of Fuel	H/C Ratio -	% Dry	% Wet	%	F	BTU/BTU	% LHV	% HHV	% LHV
CH4	4.000	9.10	7.70	25.55	305	1.01	93.95	84.24	93.49
C2H6	3.000	10.23	8.87	25.97	302	1.01	94.09	85.79	93.73
C3H8	2.667	10.67	9.34	26.14	302	1.01	94.12	86.34	93.78
C4H10	2.500	10.91	9.60	26.24	301	1.01	94.13	86.62	93.80
CO	.000	26.95	26.95	34.51	281	1.00	94.90	95.09	95.09
H2	∞	.00	.00	22.56	289	1.00	94.73	80.14	94.74
CO+3H2	6.000	9.10	7.15	25.55	287	1.00	94.78	83.85	94.84
CK OVN	5.139	8.34	6.87	25.47	301	1.01	94.17	83.71	93.86
BL FRN	.211	19.25	18.87	60.18	356	1.07	91.38	86.94	88.67
NAT OH	3.887	9.23	7.82	25.64	305	1.01	93.95	84.34	93.49
NAT MN	3.782	9.24	7.87	25.90	306	1.01	93.92	84.45	93.45
K-T PR	1.000	21.34	19.28	31.29	287	1.00	94.65	89.43	94.71
U-GAS	1.367	14.35	13.07	42.40	341	1.05	92.21	83.54	90.34
U-GAS	1.936	10.72	9.71	40.11	332	1.04	92.67	84.34	91.20
U-GAS	1.937	15.30	13.33	28.08	290	1.00	94.57	87.42	94.53
LURGI	1.932	18.81	15.92	30.08	306	1.01	93.78	84.07	93.29
OIL #2	1.800	12.02	10.85	26.68	302	1.01	94.11	87.81	93.76
OIL #6	1.500	12.57	11.49	26.90	300	1.01	94.21	88.72	93.93
METHOL	4.000	11.68	9.47	26.54	308	1.02	93.74	81.87	93.18
ETHNOL	3.000	11.68	9.94	26.54	294	1.00	94.44	85.97	94.31
FRMALD	2.000	16.30	14.01	28.53	286	1.00	94.77	87.43	94.85
FRMACD	2.000	26.95	21.23	34.51	356	1.07	91.18	73.13	88.42
ACTACD	2.000	16.30	14.01	28.53	321	1.03	93.08	82.78	92.05
WOOD	3.583	15.63	12.21	28.35	326	1.04	92.08	75.64	90.71
LIGNTE	1.726	14.98	13.26	28.25	316	1.03	93.09	84.46	92.23
BIT MV	.751	14.36	13.62	27.95	304	1.01	93.98	90.30	93.57
ANTRCT	.365	15.39	14.97	28.42	300	1.01	94.14	92.11	93.84
GRAPHT	.000	16.30	16.30	28.53	301	1.01	94.14	93.82	93.82

FL26A, D1, B, D2

Table 3. Comparison between Experimental and Computed Boiler Efficiencies for Various Fuels Firing into a Boiler with Economizer and Air Pre-Heater, that Resulted in a Gross Efficiency of 85.5% for Natural Gas. Fuel-Independent Steam Output of 430 klb/h.

	Natural Gas		Methanol		Fuel Oil #5		CO	U-Gas*	U-Gas**	Wood	Lignite
	Expt.(1)	Calc.(2)	Expt.(1)	Calc.(2)	Expt.(1)	Calc.(2)	Calc.(2)	Calc.(2)	Calc.(2)	Calc.(2)	Calc.(2)
Atomic H/C		3.8		4		1.3	0	1.37	1.94	3.58	1.73
Ultimate Flue CO_2, % Dry	11.8	11.8	15.3	15.1	16.7	16.7	34.6	18.5		20.16	19.39
Excess Air, %	8.5	8.5	26.3	26.5	35.6	35.5	11.4	14.1	13.3	28.3	27.9
Flue O_2, % Wet		1.81		4.65		5.70	1.81	1.81	1.81	4.65	4.65
Temp. at Air Heater Outl., F	289	(289)	315	308	258	303	264	319	312	326	316
Relative Firing Rate, BTU/h		1.00		1.02		1.01	.99	1.03	1.02	1.04	1.03
Efficiency, % of HHV	85.6	85.4	81.8	81.9	89.6	89.1	96.1	85.3	85.9	75.6	84.5
% of LHV		94.7		93.2		93.7	96.1	92.3	92.9	90.7	92.2

	CH_4	H_2	CO	CO_2	N_2
*Composition in %	4.69	13.84	19.29	9.98	52.19
**	5.21	15.37	21.43	0	57.98

References

1. D. R. Miller and R. W. Duhl, "Alcohols, the New Fuels", IGT Symposium on "New Fuels and Advances in Combustion Technology", New Orleans, La., March 26-30, 1979, Proceedings.

2. This study.

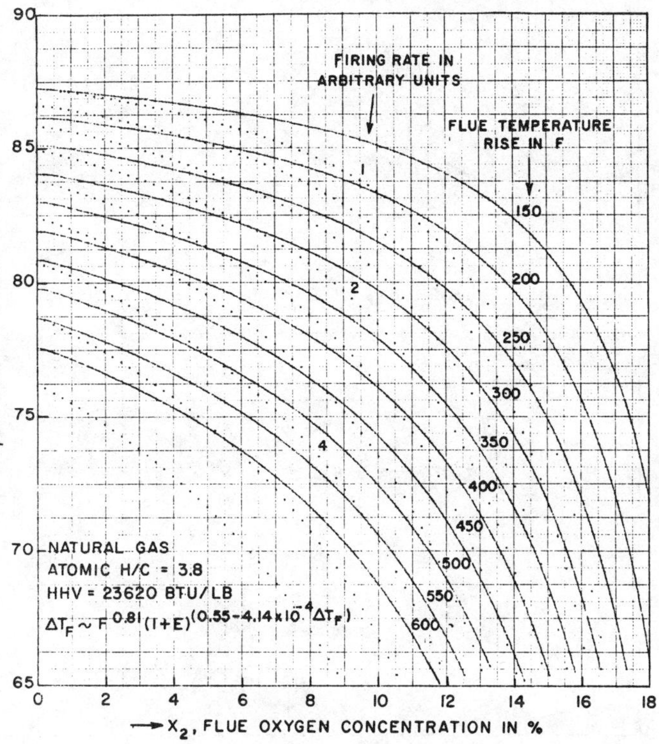

FIG. 1. GROSS EFFICIENCY vs FLUE OXYGEN CONCENTRATION
——— FOR CONSTANT FLUE TEMPERATURE AND
······ FOR CONSTANT FIRING RATE

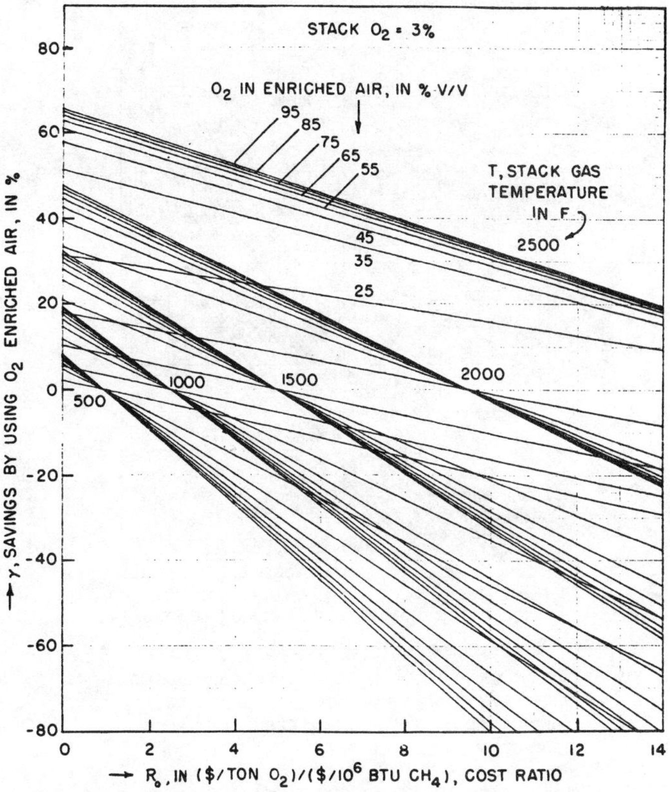

Fig. 4. Savings ($, with respect to combustion with air) realized by using oxygen-enriched combustion vs. O_2/CH_4 cost ratio, stack temperature and oxygen concentration. For $R_o = 0$, γ represents the fuel savings.

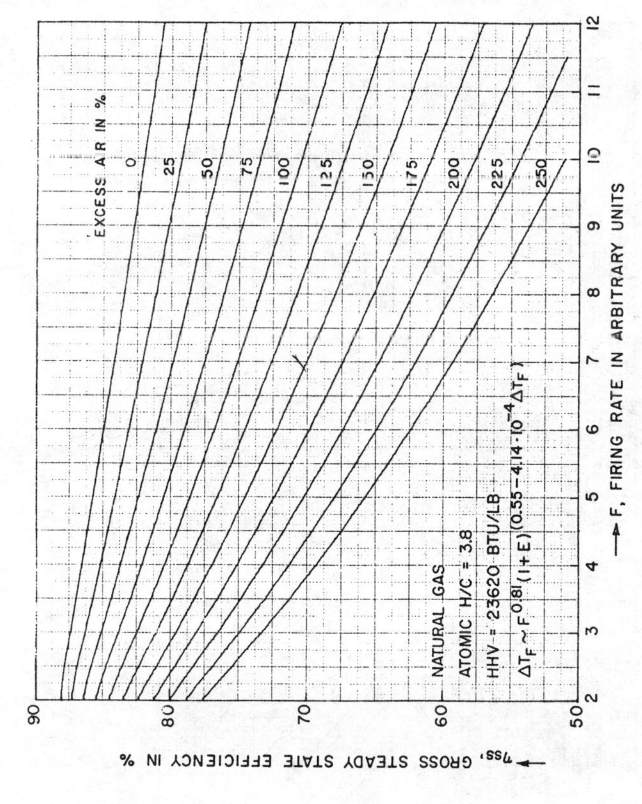

FIG. 2. GROSS STEADY STATE FURNACE EFFICIENCY FOR DIFFERENT FIRING RATES AND AMOUNTS OF EXCESS AIR.

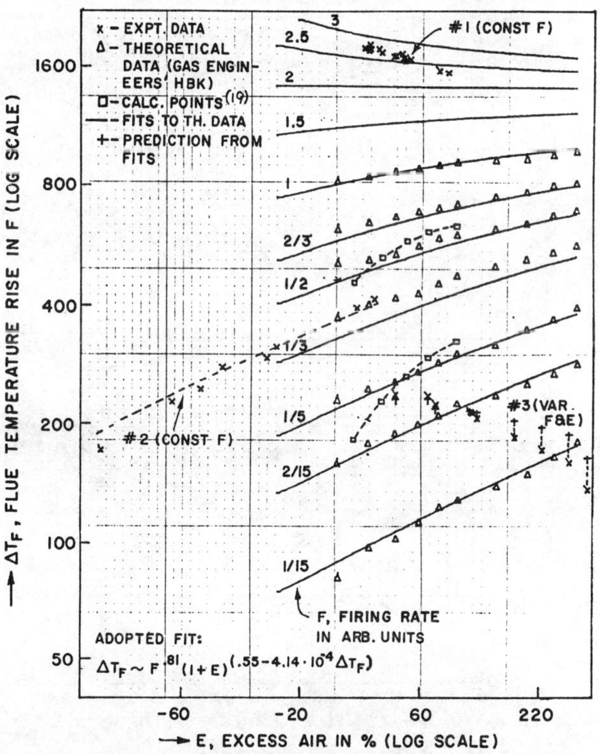

FIG. 3. DETERMINATION OF THE FLUE TEMPERATURE RISE vs EXCESS AIR RELATIONSHIP. COMPARISON BETWEEN DATA AND ADOPTED FUNCTION.

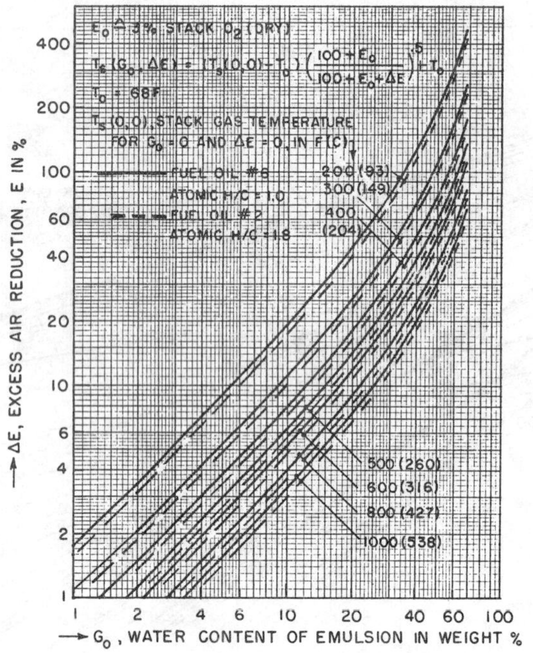

FIG. 5. EXCESS AIR REDUCTIONS REQUIRED FOR COMBUSTION OF OIL-WATER EMULSION FUELS IN ORDER TO MAINTAIN EQUAL COMBUSTION EFFICIENCY WITH RESPECT TO FUEL OILS ALONE. TEMPERATURE REDUCTION CREDIT INCLUDED.

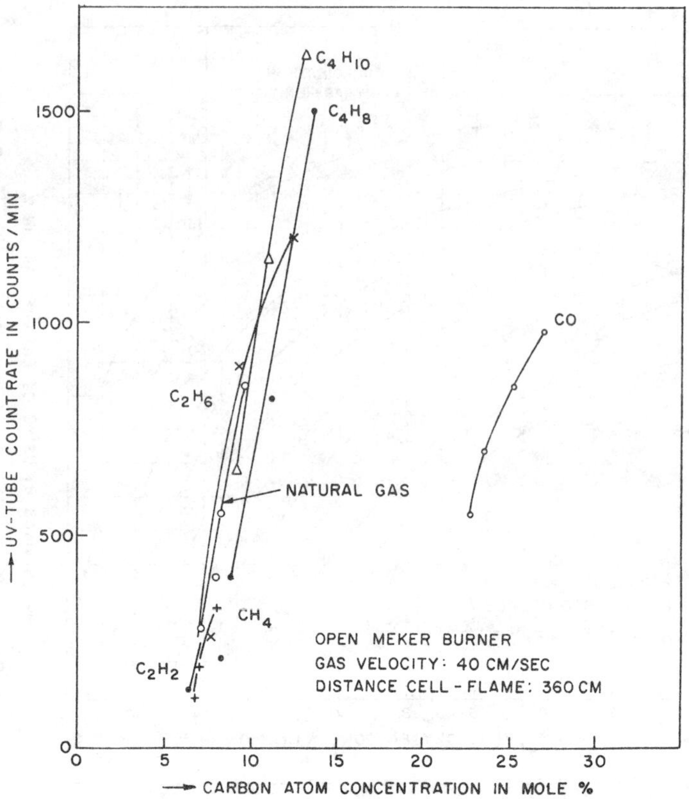

FIG. 8. UV-EMISSION OF AIR-RICH CO AND HYDROCARBON-AIR FLAMES vs CARBON ATOM CONCENTRATION

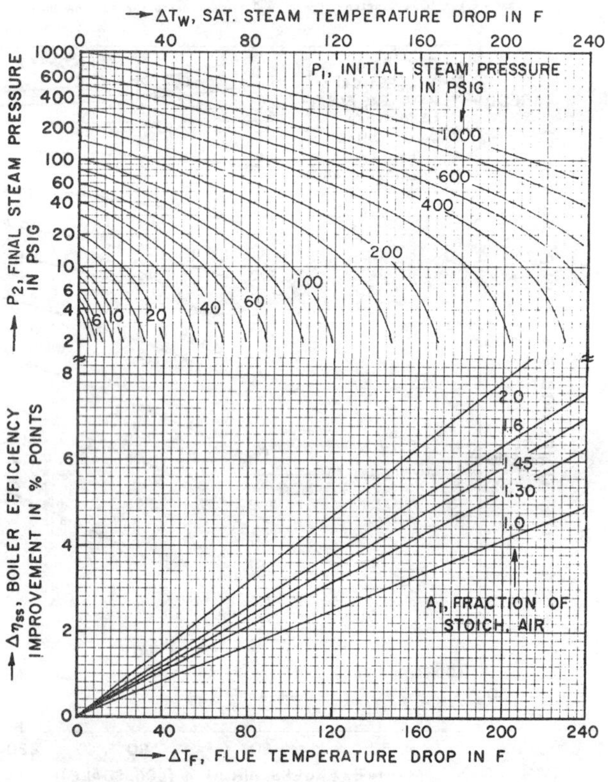

FIG. 6. BOILER EFFICIENCY IMPROVEMENT BY STEAM OR FLUE TEMPERATURE REDUCTION.

FUEL SAVINGS = $\Delta\eta_{ss}/\eta_{ss2}$

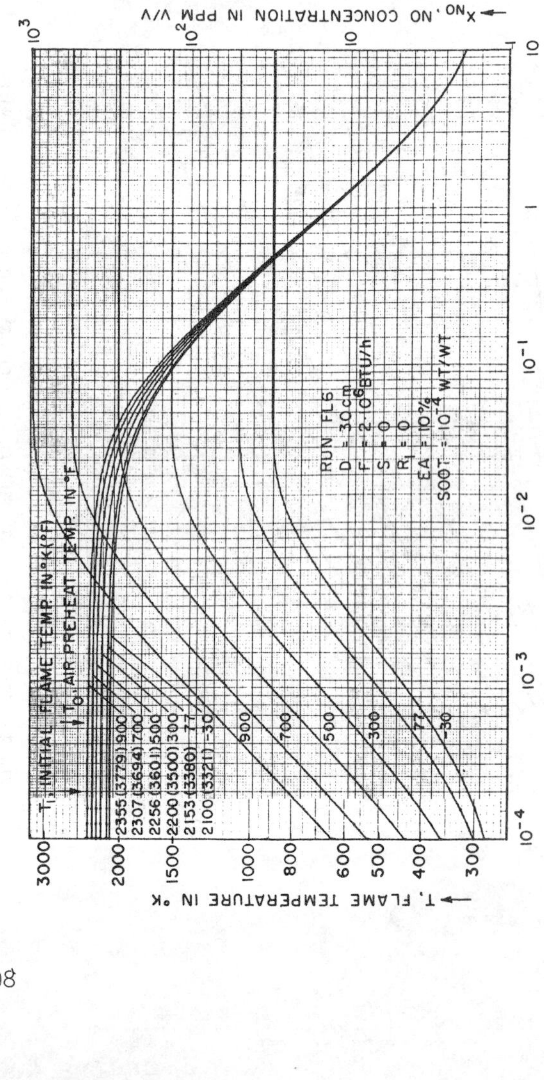

FIG. 7. FORMATION OF NITRIC OXIDE (NO) IN METHANE-AIR FLAMES, AS A FUNCTION OF COMBUSTION AIR PREHEAT.

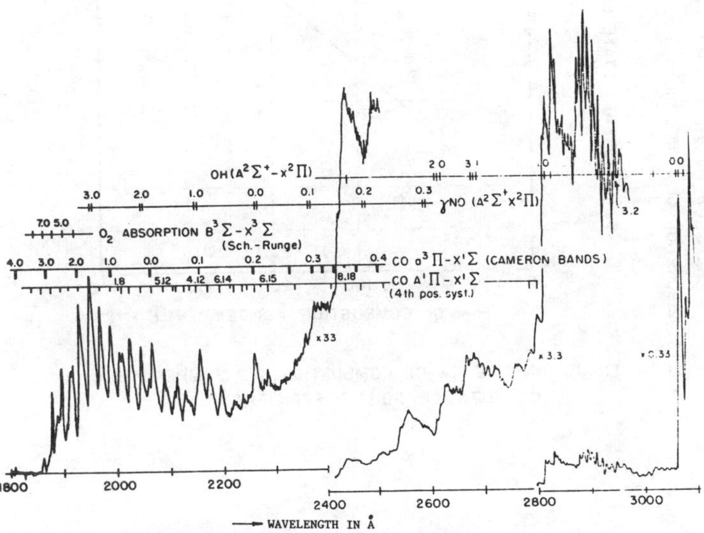

FIG. 9. EMISSION OF A STOICHIOMETRIC, PREMIXED NATURAL GAS-AIR FLAME ($p = 1$ atm, $v_o = 80$ cm/sec, closed Meker burner)

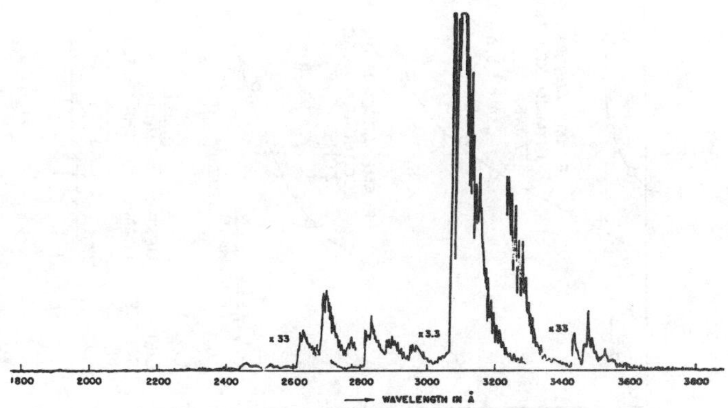

Figure 11. UV Emission Spectrum of a Stoichiometric, Premixed H_2-Air Flame, $p = 1$ atm, $v = 70$ cm/sec, Open Meker Burner of 2.5 cm D.

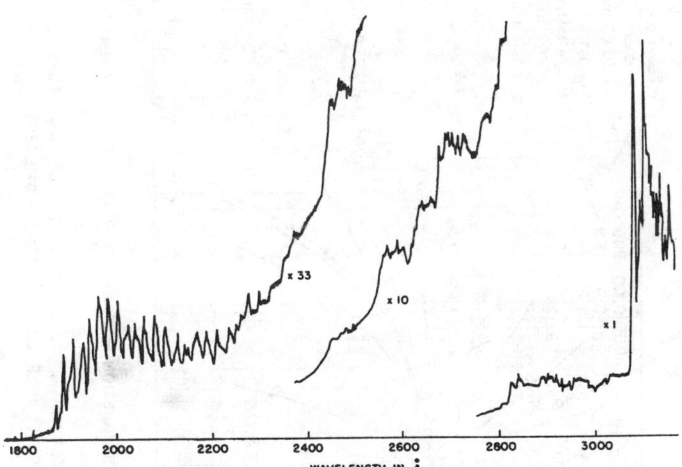

Figure 10. UV Emission Spectrum of a Stoichiometric, Premixed Fuel Oil #2-Air Flame, $p = 1$ atm, $v_o = 80$ cm/sec, Open Meker Burner of 2.5 cm D.

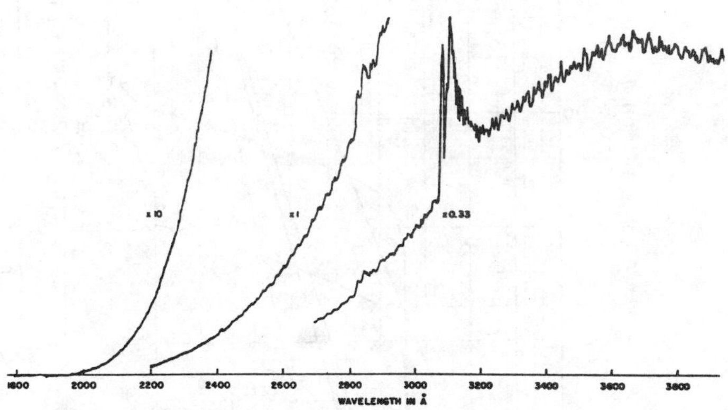

Figure 12. UV Emission Spectrum of a Stoichiometric, Premixed CO-Air Flame, $p = 1$ atm, $v - 80$ cm/sec, Open Meker Burner of 2.5 cm D.

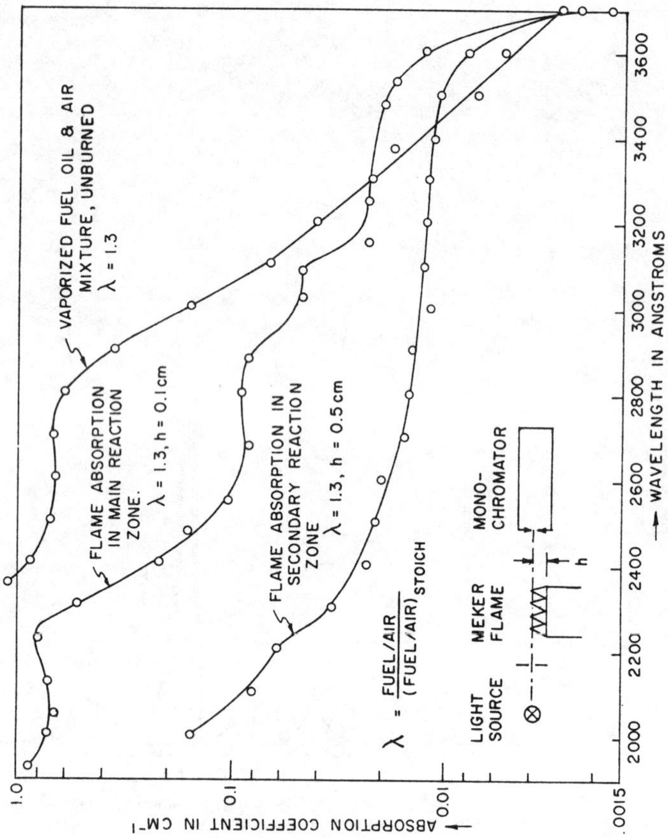

FIG. 13. ABSORPTION OF A FUEL OIL #2 - AIR FLAME

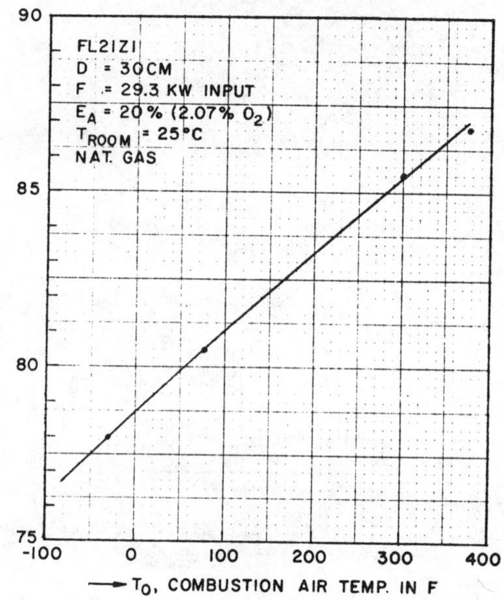

FIG. 16. INFLUENCE OF COMBUSTION AIR TEMPERATURE ON FURNACE/BOILER EFFICIENCY.

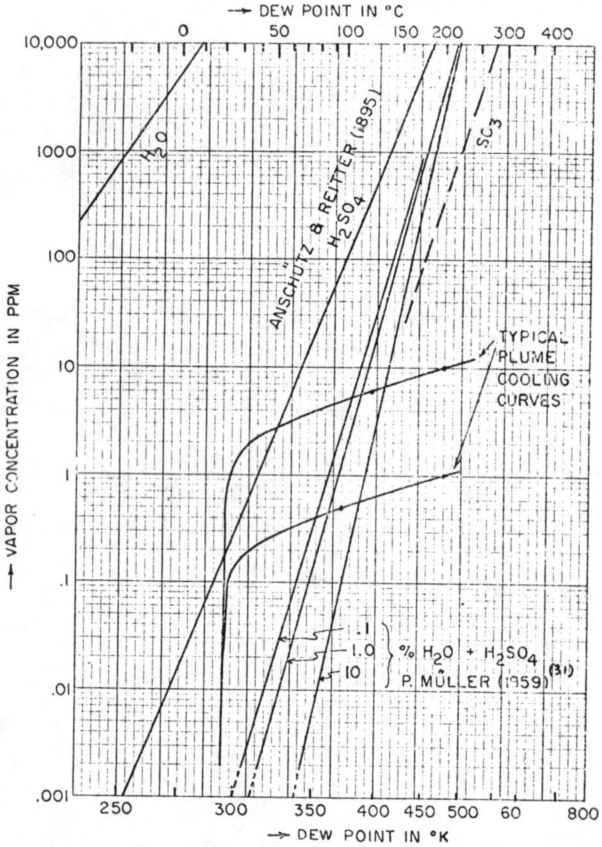

FIG. 14. VAPOR PRESSURE AND DEW POINT TEMPERATURES FOR SOME COMBUSTION PRODUCT COMPONENTS

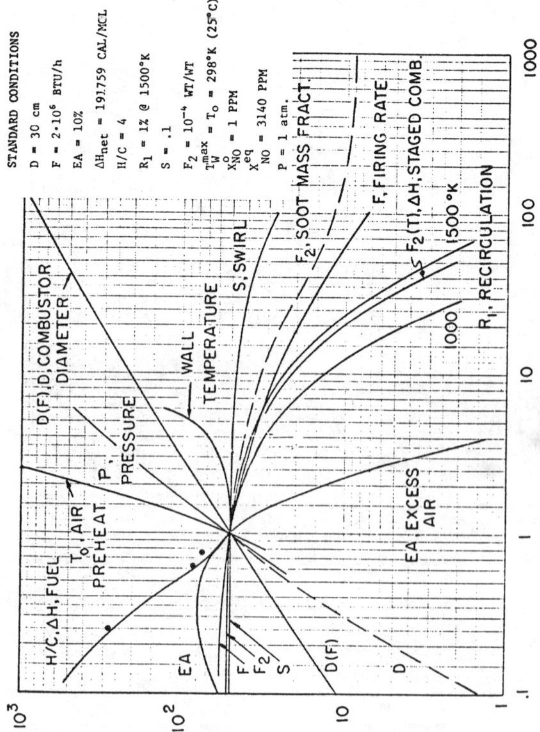

FIG. 15. EXHAUST CONCENTRATION OF NITRIC OXIDE (NO) AS A FUNCTION OF VARIOUS CONTROLLABLE COMBUSTION PARAMETERS

SECTION 5
LIFE CYCLE COSTING FOR ENERGY PROJECTS

Chapter 17
A GRAPHICAL APPROACH TO THE WORTH OF PREMIUM-PRICED ENERGY EFFICIENT MOTORS

D. C. Brewer

ABSTRACT

An extensive set of graphs has been prepared to aid the plant engineer or purchasing agent in the selection of premium priced energy efficient motors. Topics covered include power costs, demand charges, rate structures, machine duty cycles, oversized motors, power factor controllers, and variable speed a-c drives.

Most energy efficient motor lines are in the 1 to 200 hp range of polyphase induction motors. Machines larger than 200 hp may command a higher price when furnished to specifications for energy efficiency. Here, small improvements in machine losses can result in major benefits. These machines also are covered.

Use of the graphs can give a quick evaluation of whether to make further calculations of the benefits. A computer program and a program for the TI-59 calculator with printer are furnished, either of which may be used for further studies.

Introduction

"Energy Efficient Motors" is a buzz phrase designating that class of polyphase induction motors in the 1 to 200 hp range with improved efficiency compared to currently produced standard motors. These energy efficient motors are used in industrial applications which often have long duty cycles. They offer important savings in power costs. These savings are the result of design and material changes that increase the cost of the motors. The question is whether the power savings can justify the increased cost of the energy efficient motor. If true, then, "when buying electric motors, spending more may cost you less".

Technical papers abound giving arithmetic methods for the evaluation of the benefits of energy efficient motors. By use of tables, graphs, calculator or computer programs, or by a combination of any or all of these, the tedious arithmetic may be greatly simplified. Examples of the use of the graphs, the output of the calculator and computer programs are featured. A detailed set of graphs, copies of the computer and calculator programs are available from the author at Armco Inc, Electrical Steel Division, Dept. LE61, P. O. Box 600, Middletown, Ohio 45043.

A First Approach

Before worrying about more complicated financial analysis, a table of values of motor cost and efficiency can speed the motor purchase decision process. After ranking the motors in order of ascending price, comparison of the efficiency values is the basis for further consideration. In examining the values in the table, the rule of the table is that if a motor of higher first cost has a lesser or equal efficiency, reject it. The example table below shows values for a typical group of 25 hp motors. The source of some of these values are from a recent publicity release by a major motor manufacturer.

TABLE I

SELECTION OF 25 hp Motors for Further Analysis

Motor	Price	Efficiency	Search: Motor 1	Motor 2	Motor 4	Motor 5	Study
1	$ 990	85.0	Out —	—	—	—	
2	$ 990	89.5	Try	Retain			X
3	$1000	87.0		Out —	—	—	
4	$1217	91.7		Try	Out —	—	
5	$1217	93.0		?	Try	Retain	X
6	$1316	93.0		?	?	Out —	
7	$1316	93.6		?	?	Retain	X
8	$1350	91.0		?	?	Out —	

Out of eight possible motors, only three are left to be studied further.

Power Costs

The entire process for calculation of savings from improved efficiency of motors depends on the relationship between first costs and power costs.

What are power costs? With every power company in the nation having a different rate structure, the determination of power costs can become overwhelming. A simple solution for an existing installation is to divide the total bill over the last number of months (or since the last rate increase) by the total KWH usage for the period to establish an average price per KWH.

One could, on the other hand, establish an average price for the power used and an average demand charge. Assuming that the motor contributes to the demand charge by virtue of its duty cycle, one could calculate the savings a bit more accurately. If the demand charge is based on peak KW, the power factor of the motor does not enter into the calculation and should be assigned a value of 100%. When the demand charge is based on KVA, the value of the power factors for the motors being compared should be used if the values of power factor can be obtained. Stated simply, the most practical value to use for power cost is the average cost per KWH.

"Duty and the Least"

Since the motor consumes no power when turned off, the duty cycle must be taken into account when making calculations of annual power usage or savings. It will be shown later that operation at reduced load (even at the least-no-load) presents a different condition.

Certain power control devices claim to take advantage of the fact that the motor consumes no power when turned off. Allegedly, the motor is turned off during a portion of the voltage (or current) cycle and operates at maximum efficiency during the balance of the cycle. The evaluation of the benefits of these control devices and their use with energy motors is still being investigated.

Many motors operate at reduced load, either because they have been oversized to provide necessary starting torque, for more reliability, or because their load is variable. To understand all of the benefits of energy efficient motors, one must understand motor losses and their relation to power savings. The following equations are given in consistent units.

By Definition:

$$\text{Efficiency (\%)} = \frac{\text{Power Output}}{\text{Power Input}} \times 100 \tag{1}$$

also

$$\text{Power Output} = \text{Power Input} - \text{Loss} \tag{2}$$

and

$$\text{Efficiency} = \left(1 - \frac{\text{Loss}}{\text{Power Input}}\right) \times 100 \tag{3}$$

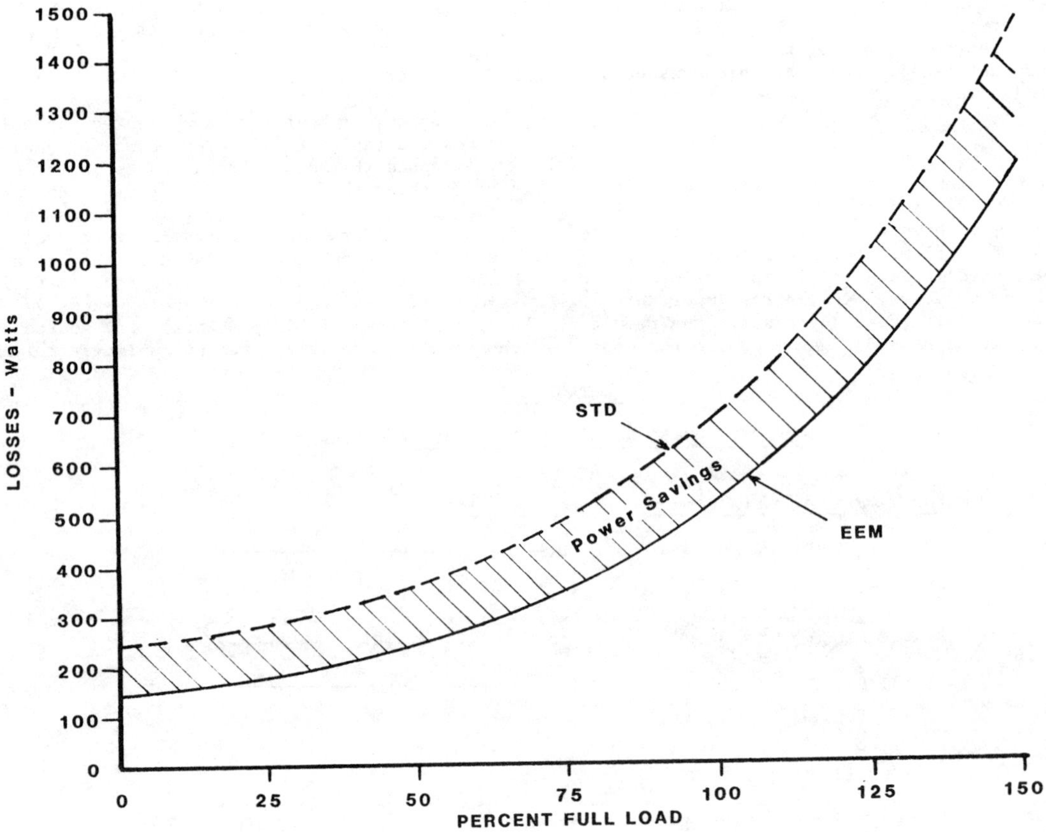

FIGURE 1. A COMPARISON OF THE LOSSES OF A STANDARD MOTOR AND AN EEM AT VARIOUS LOADS (BOTH 5 HP)

When comparing two motors where B is more efficient than A

Power Savings = Power Input A - Power Input B (4)

Since the comparison is being made at equal power output

Power Output A = Power Output B (5)

or

Power Input A - Loss A = Power Input B - Loss B (6)

transposing

Power Input A - Power Input B = Loss A - Loss B (7)

thus, the power savings is also

Power Savings = Loss A - Loss B (8)

Figure 1 shows the loss curve for a standard motor compared to one for an energy efficient motor. Inspection of the curve shows that the power savings is nearly the same at all loads on the motors. Thus, an analysis of power cost savings based on full load efficiencies will reflect the true savings, although total operating costs may be incorrect for the actual load condition.

In order to achieve these savings, the motor designer will have used all the tricks of his trade. These could include use of the maximum core iron (maximum length of core stock), use of CRNO silicon steel for laminations, use of thinner laminations, and use of more or larger diameter copper wire in the windings.

Sample Problem

From Table I above, there are three motors for evaluation. First is the standard motor (#2); second, a motor with improved efficiency (#5); and third, a 3rd generation motor showing even greater improvement in efficiency (#7). The cost and efficiency figures for these 25 hp motors were taken from a recent publicity release by a major motor manufacturer for a new generation of energy efficient motors.

Motor Parameters:

	#2 Std.	#5 Improved	#7 3rd
Cost:	$990.00	$1217.00	$1316.00
Efficiency %:	89.5%	93.0%	93.6%
Power Factor:		not given	

Analysis Conditions

Horsepower:	25
Power Cost:	$.05/KWH
Demand:	not applicable
Motor Life**:	12 years
Interest Rate:	15%
Operating Hours:	8760 hrs/year

**The motor manufacturer may not agree with this figure

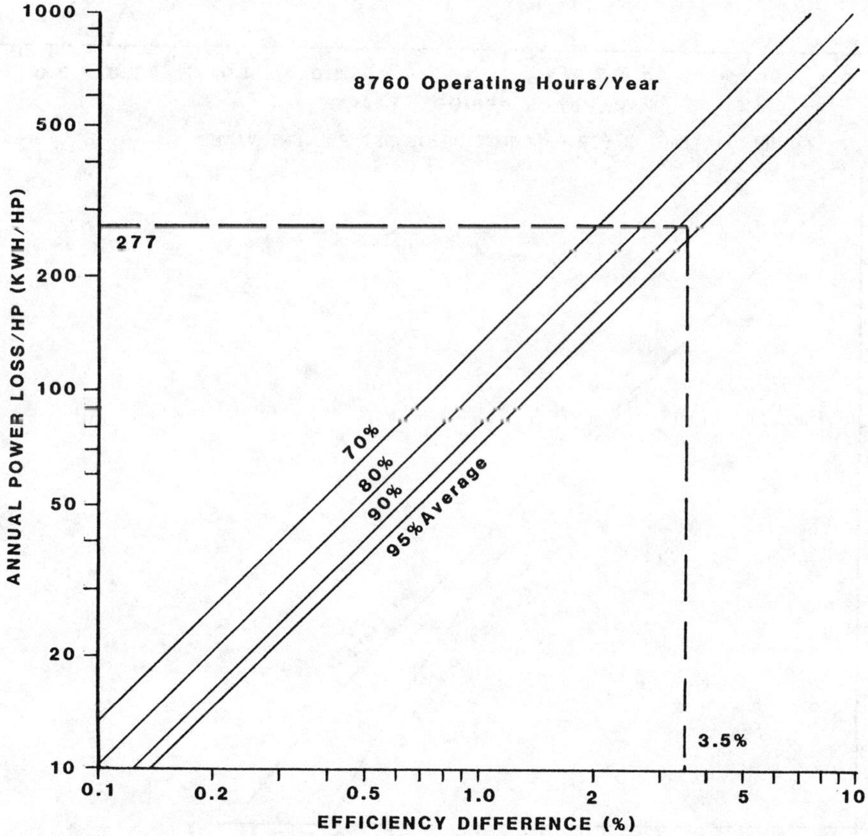

FIGURE 2. THE RELATION BETWEEN EFFICIENCY DIFFERENCES AND ANNUAL POWER SAVINGS. (8760 OPERATING HOURS)

The data shown above will be analyzed by three methods: the graphs, the calculator program, and the computer program.

Use of Graphs

The graphs, Figures 2, 3 and 4, allow the evaluation of power savings, operating cost savings, payback period, net present value of savings, and % ROI (return on investment) through simple calculations.

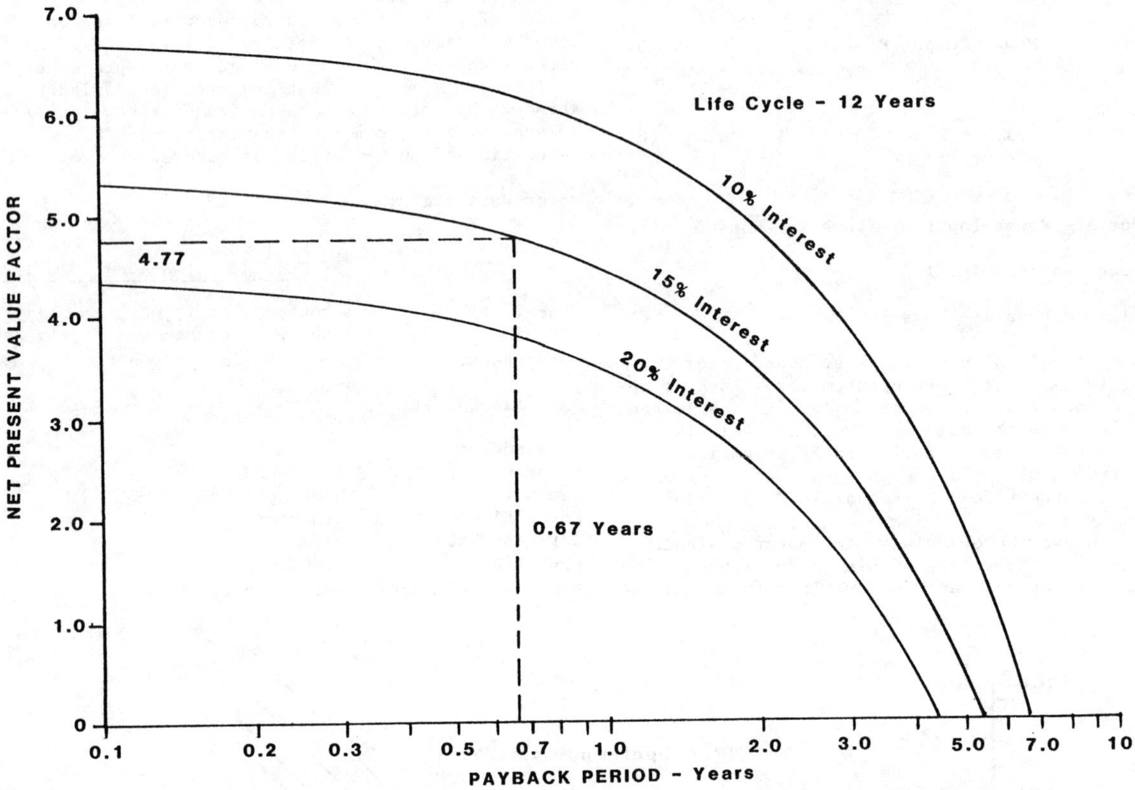

FIGURE 3. FACTOR FOR CALCULATION OF NET PRESENT VALUE

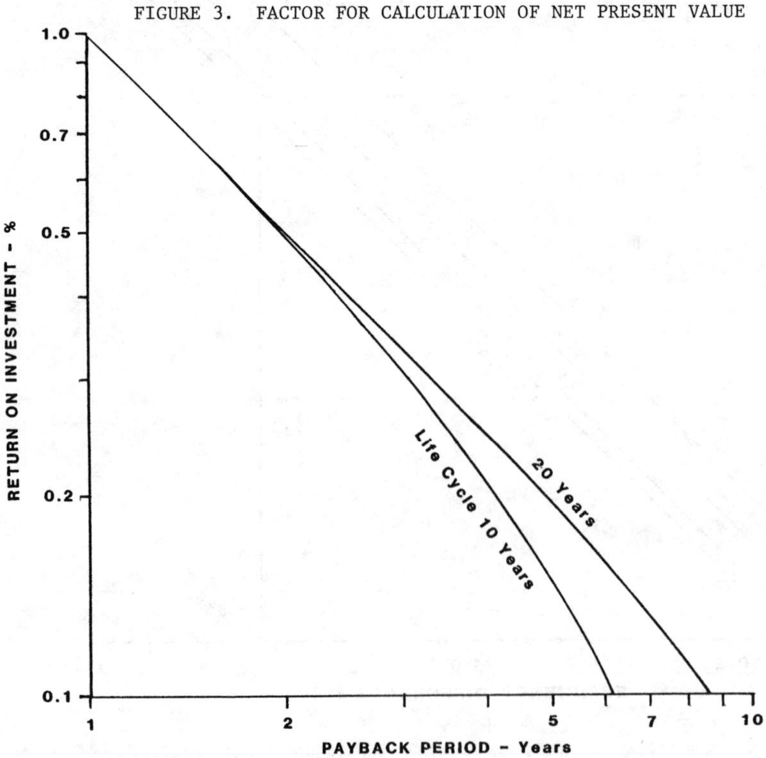

FIGURE 4. PERCENT ROI AS A FUNCTION OF PAYBACK PERIOD

Before continuing, let it be emphasized that the power savings graph is for 8760 hours per year of continuous operation. First, determine the cost difference, the efficiency difference, and the average efficiency for the motors to be compared. Thus for the standard motor (2) compared to the improved motor (5):

Cost Difference - $1217 - $990 = $227

Efficiency Difference = 93.0 - 89.5 = 3.5%

Average Efficiency = 91.25%

The dotted lines on Figure 2 show that for these conditions, the annual power savings is 271 KWH/hp. Multiplying this value by 25 hp and the power cost of $.05/KWH yields an annual power cost savings of $339.00.

Dividing the price difference by the power cost savings equals the payback period in years. In this case 227 divided by 339 equals .670 years. This figure is now used with Figures 3 and 4 to find the net present value of savings.

Using Figure 3, which is for a life cycle of 12 years, and the curve for 15% interest. The value of the factor for net present value of savings is 4.77 (as shown by the dotted line). Multiplying this value by the power cost savings of $339 gives a NPV of savings of $1617.

When the payback period is less than one year as in this case, the % ROI is the reciprical of the payback period multiplied by 100. Thus we have 150% ROI for the extra investment in the more efficient motor. Figure 4 which was not necessary in this case gives values of % ROI for payback periods over one year.

The calculator and computer program confirms the values which have been obtained from the graphs.

Calculator Program

The calculator program is designed for use with a Texas Instruments TI-59 calculator with a printer. Operation of the program is interactive, i.e. when it needs data, it prints the identification of the value needed and displays the current value in memory. This value can be accepted or modified by the user. Computations are printed as the program proceeds. A printout of the sample problem is shown in Table II.

TABLE II

INTERACTIVE DATA ENTRY AND OUTPUT FROM TEXAS INSTRUMENT TI-59 CALCULATOR

```
POWR  RATE

* = DOLLAR SIGN
ENRGY      */KWH
        0.05
DEMND */KW       */KVA
          0.

FINANCIAL  DATA

YEARS LIFE CYCL
            12.
% INT RATE
            15.

MOTOR DATA

MOTOR  A
PRICE
           990.
% EFF
           89.5
% PF
           100.

ANUAL OPER COST
ENRGY COST
       9127.04
DEMND COST
             0.
TOTALANUAL OPER COST
       9127.04

MOTOR  B
PRICE
            1217.
% EFF
              93.
% PF
             100.

ANUAL OPER COST
ENRGY COST
       8783.55
DEMND COST
             0.
TOTALANUAL OPER COST
       8783.55

ANALYSIS

PRICE DIFF
          227.
SVNGS OPER COST
        343.49
PAYBACK   YEARS
         0.669
NPV   OF  SVNGS
       1634.92
% ROI
        151.32
```

Computer Program

A computer program was developed for internal use by Armco (see Acknowledgment) and is written in Fortran IV. It compares up to 12 motors at one time and can do repeated analyses with changed parameters. Table III is the computer output.

TABLE III

MOTOR DECISION OUTPUT FROM ARMCO FORTRAN PROGRAM

```
                              EXAMPLE **  25.0 HP MOTORS

  CHART KEY:(A)=FIRST COST DIFFERENCE (B)=COST POWER COST DIFFERENCE INCLUDING DEMAND CHARGE IF APPLICABLE
     (C)=PAYBACK IN YEARS (D)=NET PRESENT VALUE (PV OF POWER COST DIFFERENCE PLUS FIRST COST DIFFERENCE)
                          (E)=% RETURN ON INVESTMENT (ROI)
                      ALL NEGATIVES ARE COSTS AND ALL POSITIVES ARE SAVINGS.

ANALYSIS CONDITIONS: INTEREST RATE-15.00%  PERIOD(YRS)-  12. COST/KWH-$.05000 HOURS/YR-8760. DEMAND CHARGE/MO $  .0000

                               IMPROVED  3RD
STANDARD        OPERATING COST PER YEAR $ 9127.04,PER DAY $25.01.ANNUAL POWER COST $ 9127.04,ANNUAL DEMAND CHARGE $   .00
1ST COST $   990.00 A   -227.00   -326.00
LOSS/YEAR$   958.34 B    343.49    399.80
EFFIENCY %    89.50 C       .66       .82
POWER FACT.% 99.90  D   1634.92   1841.14
                    E    151.32    122.64

IMPROVED        OPERATING COST PER YEAR $ 8783.55,PER DAY $24.06.ANNUAL POWER COST $ 8783.55,ANNUAL DEMAND CHARGE $   .00
1ST COST $  1217.00 A       .00    -99.00
LOSS/YEAR$   614.85 B       .00     56.31
EFFIENCY %    93.00 C       .00      1.76
POWER FACT.% 99.90  D       .00    206.21
                    E       .00     56.62

3RD             OPERATING COST PER YEAR $ 8727.24,PER DAY $23.91.ANNUAL POWER COST $ 8727.24,ANNUAL DEMAND CHARGE $   .00
1ST COST $  1316.00 A       .00       .00
LOSS/YEAR$   558.54 B       .00       .00
EFFIENCY %    93.60 C       .00       .00
POWER FACT.% 99.90  D       .00       .00
                    E       .00       .00
```

Data and Operating Costs for Standard Motor (STANDARD line)

Comparison of Standard Motor and 3rd Motor (3RD column, STANDARD block)

Comparison of Standard Motor and Improved Motor (IMPROVED column, STANDARD block)

Comparison of Improved Motor and 3rd Motor (3RD column, IMPROVED block)

The first row compares the standard motor with the other two; it yields for motors 2 and 5:

A. First Cost Difference = $-227.00

The negative sign indicates that the improved motor costs more than the standard motor.

B. Annual Operating Cost Difference = $343.49 Or Lost Power Cost Difference

Since the value is positive, it is a savings.

C. Payback in Years - .66

D. Net Present Value of Savings = $1634.92

E. Return on Investment = 151%

Discussion

The above examples show that comparable answers can be developed by all three methods--graphical, calculator or computer. The computer program offers more rapid solutions with greater ease of varying parameters as well as some additional output data.

An important attribute of energy efficient motors is that they offer the same savings in power costs when operated at any load as they do when operated at full load. Thus the calculations of payback in years, % ROI, net present value, and power cost savings are valid for all loads using the full load efficiency values.

The graphs which were developed for net present value of savings vs. payback in years and % ROI vs. payback in years are applicable for any purchase having a payback period with no salvage value at the end of the life cycle.

References

(1) Energy Efficiency and Electric Motors, A. D. Little Inc., U. S. Dept. of Commerce, Publication PB-259, 129, Aug. 1976.

(2) Value Analyses Aids Purchase Decisions, H. Wayne Beaty, Electrical World, Feb. 15, 1978.

(3) Laminations in Energy Efficient Motors, G. H. Johnston, Coil Winding 1979.

(4) Saving Money with Costlier Motors, J. G. Terry, Machine Design, Nov. 23, 1978.

(5) IEEE Standard Test Procedure for Polyphase Induction Motors and Generators, ANSI/IEEE, Std. 112.

(6) Power Factor Controller - An Energy Saver, Frank Nola, MotorCon 81.

(7) The A-C Answer to the Adjustable Speed Drive Question, James D. Johnson, MotorCon 81.

(8) Induction Motor Behavior on Switched A-C Drives, G. R. Phillips, MotorCon 81.

(9) Evaluation of Energy Efficient Motors, David C. Montgomery, IEEE Textile Industry Electrical Conference, May 1979.

(10) Alternating Current Machines, M. G. Say, John Wiley & Sons, TK2711.S3, 1976.

(11) Energy Efficient Electric Motors, DOE/CS-0163, May 1980.

(12) Understanding Efficiency in Squirrel Cage Induction Motors, Austen H. Bennett, IEEE Transactions on Industry Applications, Vol. 1A-16, No. 4 July/Aug. '80.

(13) Principles of Engineering Economy, Eugene L. Grant and W. Grant Ireson, The Roland Press, New York, New York, 1960.

(14) Justification for Energy Efficient Motors, Richard J. Buschart, 3rd Annual Industrial Energy Conservation Technology Conference, April 26-29, 1981.

(15) The High Efficiency Induction Machine of the 1980's, Paul Diamant, 81 SM 508-1, IEEE PES Summer Meeting, 1981.

(16) Induction Motor Efficiency Test Methods, P. G. Cummings, W. D. Bowers, and W. J. Martiny, IEEE Paper PID 78-12B, Sept., 1981.

(17) A-C Motor Control, John Reason, Power, Feb. 1981.

(18) The Effects of Manufacturing Techniques on Motor Efficiency, D. R. Aldworth, Insulation and Circuits, July 1981.

Acknowledgment

The computer program was developed for Armco's internal use by Mr. Jerry L. Slama of the Middletown Works Engineering Department. It was later modified and features added by Mr. Jay G. Terry now retired and the author.

Chapter 18
THE ECONOMICS OF ENERGY INVESTMENTS

C. A. Wieczorek

Economic analysis is important to energy managers when making energy investments. Life-Cycle Cost analysis is becoming an essential evaluation tool as inflation and escalating energy prices impact long-run program decisions. The recurring annual and nonannual costs of many projects are so significant that they cannot be disregarded during the acquisition process.

First, engineering estimates and measurements identify potential energy conservation measures (ECM's) or retrofit projects, and associated energy use and costs for existing and retrofit systems. Life cycle costing (LCC) then provides the methodology for evaluating these ECM's and retrofit projects. This approach takes into account relevant costs and benefits, and assesses the net effect over time of purchasing, installing, maintaining, operating, repairing, and replacing buildings, their systems and equipment. Life-Cycle Costing uses a common set of criteria to evaluate, rank, and select energy conservation retrofit projects for funding and implementation.

The major steps for performing a LCC analysis of a potential retrofit project or energy investment (Figure 1) are as follows:

- Identify alternative approaches to achieve your investment objective within your established constraints.
- Establish a common time basis for expressing costs and a time horizon.
- Identify and estimate the monetary and nonmonetary benefit and cost parameters to be considered.
- Convert costs and savings which occur at different times to a common time.
- Compare the investment alternatives on the basis of their relative economic efficiencies. Select the alternative that will be the most cost-effective.
- Analyze the results for sensitivity to initial assumptions and constraints.

Life-cycle cost analysis is performed after a retrofit project has been identified and the engineering and cost calculations have been made. The LCC analysis can provide a meaningful evaluation method only if these calculations are accurate and complete.

IDENTIFY ALTERNATIVE APPROACHES TO ACHIEVE YOUR INVESTMENT OBJECTIVE WITHIN YOUR ESTABLISHED CONSTRAINTS

Many energy conservation goals can be accomplished by selecting one of several alternative retrofit systems. For example, incandescent lamps can be replaced with fluorescent, sodium, or metal halide lamps, just to name a few alternatives. To determine which of the alternatives is appropriate, you must know any selection constraints which may impact the decision process. Such constraints will include, in this case, the type of space being lit, the ceiling height, the required lighting level, the importance of color rendition, and the kind of work being performed. These constraints will limit the types of lighting you can consider in your retrofit analysis.

Examples of other considerations are budget, siting, code, fuel, and lease constraints. While these are not equipment requirements, they are equally important. Limited funding may impact the number and types of retrofit projects that are implemented. It may be more cost-effective to complete several smaller projects, than one capital-intensive project. Special siting requirements may prohibit a solar project from being evaluated. Considerations regarding the availability of a particular fuel, or a long-term fuel contract, as well as building codes, may impact your decision. Additionally, if the building for which the retrofit is being considered is leased, the investment should not be evaluated over a period of time greater than the length of the lease.

ESTABLISH A COMMON TIME BASIS FOR EXPRESSING COSTS AND A TIME HORIZON

To Establish a Common Time Basis for a valid comparison of benefits and costs that accrue at different times during the life of an energy retrofit project, they must be considered on an equivalent time basis. This is necessary because the significant costs of an energy conservation project are primarily "first costs" associated with the design, purchase, and installation of the retrofit. The benefits, however, accrue over the life of the project, and are normally in the form of energy savings.

When performing economic evaluations of energy conservation projects, inflation should be eliminated from estimates of benefits and costs because the costs are increasing at the same rate relative to each other. However, the prices of fuels, which have increased so rapidly in recent years, are projected with a factor for inflation included.

There are several methods for eliminating the effects of inflation from future prices. One method is to use a discount rate which reflects both real changes in value and the expected inflation rate. The method used in LCC analysis to eliminate inflation is to express all dollar amounts in "constant dollars" in some base year. The constant dollar approach states that if you strip away inflation, the "real" value of goods and services relative to each other will remain constant. For example a pencil costs 10¢ and a pen costs 20¢ today. Five years from now the pencil may cost 30¢ and the pen 60¢, Even though the prices have tripled due to inflation, it will still cost twice as much to buy a pen as it does to buy a pencil, therefore the "real" values relative to each other have not changed.

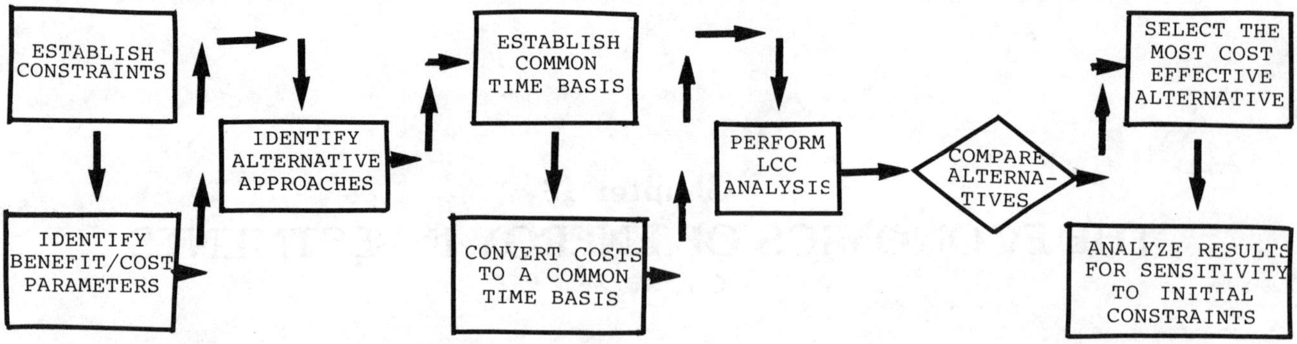

FIGURE 1: LCC ANALYSIS STEPS

Assuming that all costs will be expressed in today's (base year) dollars, a motor which costs $1,000 to replace today, will still cost $1,000 in "real" terms to replace in ten years. This approach facilitates estimating the operating, maintenance, and replacement costs which occur over the life of the investment because we are able to use the dollar values they have today.

The second factor which affects the value of a dollar over time is the ability to invest that dollar to yield a return, exclusive of inflation called interest. If $1 is placed in a risk-free savings account which yields 7% interest per annum, the $1 would be $1.07 in one year. Another approach is to ask how much money must be put in the savings account at 7% interest per annum to yield $1 in one year. The answer, which may be found in the Present Worth tables at the back of most engineering economics and statistics texts, is $0.94. This is the "present value" (PV) of that $1. The rate of interest used to convert future dollar amounts to present equivalant amounts is called the "discount rate." This rate is frequently set by corporate management. Present worth tables may be used to find the present value of a future sum in a specivic year using a 7% discount rate, and to find the present value of a future sum occurring in equal amounts over some number of years.

EXAMPLE: COMMON TIME BASIS FOR EXPRESSING COSTS

Using a 7% discount rate: The present value of a $25,000 component which must be replaced in year 5 is $17,750

$	x	SPW	=	PV
$25,000	x	0.71	=	$17,750

Using a 7% discount rate: The present value of an annually recurring cost of $120,000 for 25 years is $398,000

$	x	UPW	=	PV
$120,000	x	11.65		$1,398,000

The Time Horizon is simply the period of time over which the energy investment or retrofit project will be evaluated. This is normally the number of years for which the costs and benefits accruing to the investment or retrofit are calculated. The time horizon is also called the "study period" and the "life cycle" in life-cycle cost analysis.

The time horizon may be the investment life of the project, the time perspective of the investor, the remaining life of the building, or the investor's planned lease period. In lieu of any preference for one of the previously mentioned selection criteria, the normal maximum time horizon is 20-25 years. The investment life is the period of time over which the investment has value in conserving energy, and is normally the average equipment life. The time perspective of the investor is dependent upon corporate objectives. There may be an established minimum rate of return or maximum length of time within which an investment or retrofit must become cost-effective. The length of time can affect the value of the net benefits accruing from a retrofit project and thereby impact investment decisions.

EXAMPLE: SELECTING TIME HORIZON

- The office complex to be retrofitted has an indefinite life. It is estimated that the new automatic seasonal control system will last at least 25 years. Therefore a time horizon of 25 years is selected.

IDENTIFY AND ESTIMATE THE MONETARY AND NONMONETARY BENEFIT AND COST PARAMETERS TO BE CONSIDERED

To determine whether a retrofit project is cost-effective, the costs and benefits accruing due to the project must be measured. Benefits and costs can have both monetary and nonmonetary values. Monetary benefits from an energy conservation project which reduces fuel oil requirements include the dollars saved from reduced energy use and saved due to reduced O&M requirements. Nonmonetary benefits include reduced pollution, improved balance of payments, and less reliance on foreign oil. Nonmonetary benefits are difficult to place a dollar value on and are therefore not included in the benefit/cost parameters; however, they should be included in the overall decisionmaking process.

The monetary parameters that are considered in an LCC analysis are: investment costs, energy costs, nonfuel operating and maintenance costs, replacement costs, and salvage value.

Investment costs (I) are the "first costs" associated with the system. For an existing system in place, the investment cost is the current salvage value of the system plus any investment required to make it functional. For the proposed retrofit system the investment cost includes the costs of designing, purchasing, and installing the system (i.e. all materials, labor, design plans). For an LCC analysis these "first costs" are treated as if they are all incurred at one time at the beginning of the time horizon.

Energy costs (E) are based upon the quantity of energy delivered to the building. To estimate the cost of energy, multiply the quantity of fuel used by the current (base-year) price for the fuel type. This process is repeated for each fuel type. In a case where fuel is used to generate electricity, the current (base-year) energy price is based upon fuel prices plus electricity generation and distribution costs.

As discussed previously, energy prices are increasing at a rate faster than general inflation; therefore, over the time horizon energy prices must be escalated to reflect this "real" increase in the cost of energy. Additionally, the cost of each fuel type is increasing at a different rate. These "differential energy price escalation rates" are published by groups such as the Energy Information Administration of the Department of Energy, and are available to the public.

Non-fuel operating and maintenance costs (M) are those labor and non-fuel material costs required for routine upkeep. There are two types of O&M costs: annual and nonannual. Annual O&M costs occur on a regular basis each year. Examples of different annual O&M items include changing the oil in your car, boiler tune-up, and changing filters. Nonannual O&M costs occur in specific years and examples include replacing tires on your car, group relamping, and caulking and weatherstripping.

Replacement costs (R) are those costs incurred to replace major system components which become unserviceable during the time horizon, Examples include replacement of the transmission in your car, an electric motor, or a chiller.

Salvage value (S) is the value received, less disposal costs, resulting from the removal or replacement and subsequent sale of any building, system, or component.

EXAMPLE: ESTIMATING MONETARY BENEFIT & COST PARAMETERS

To evaluate the cost-effectiveness of replacing a manual seasonal control system with an automatic system it is necessary to estimate the investment, energy, O&M, and replacement costs and the salvage value for the existing manual system if it were to be kept functional over the time horizon, and for the retrofit system if it were to replace the existing system.

The data and assumptions are as follows:

Existing System

Immediate value (Base-Year $): $260,000
Annually Recurring Nonfuel O&M Costs (Base-Year $): $120,000
Nonannual Nonfuel O&M Costs (Base-Year $):

Year	Amount
5	$25,000
15	$25,000
20	$25,000

Replacement Cost (Base-year $):	Year	Amount
	10	$150,000

Energy Data:
Quantity of Electricity (Base year): 9,982,200 kWh
Quantity of Natural Gas (Base year): 422,600 x 10^6 Btu
Price of Electricity (Base year): 0.065/kWh
Price of Natural Gas (Base year): $3.43/$10^6$ Btu

Energy Cost:

Electricity: (9,982,200 kWh) x ($0.065/kWh) = $648,943
Natural Gas: (433,600 x 10^6 Btu) x ($3.43/$10^6$ Btu) = $1,487,248

Retrofit System

Initial Investment Cost: $1,500,000
Investment Life of the Project: 25 years or longer
Annually Recurring Nonfuel O&M Costs (Base-year $): $145,000
Nonannual, Nonfuel O&M Costs (Base-year $):

Year	Amount
10	$40,000
20	40,000

Salvage Value in 25 Years: $0

Energy Data:
Quantity of Electricity (Base year): 9,190,000 kWH
Quantity of Natural Gas (Base year): 386,800 x 10^6 Btu
Price of Electricity (Base year): $ 0.065/kWh
Natural Gas (Base year): $3.43/$10^6$ Btu

Energy Cost:

Electricity: (9,190,000 kWh) x ($0.65/kWh) = $597,350
Natural Gas: (386,800 x 10^6 Btu) x ($3,53/$10^6$ Btu)= $1,326,724

CONVERT COSTS AND SAVINGS WHICH OCCUR AT DIFFERENT TIMES TO A COMMON TIME

Costs and savings are converted to a common time period (present, annual, or future) through discounting. Discounting is accomplished by applying interest (or discount) formulas to cash flows. There is a specific formula to discount each type of cash flow. The discount formulas used in LCC analysis are: the Single Present Worth (SPW) formula to convert a future amount to its present value; the Uniform Present Worth (UPW) formula to find the present value of annually recurring and equal cash flows; and the Modified Uniform Present Worth (UPW*) formula which permits escalating the "real" price of annually recurring fuel costs over time.

EXAMPLE: CONVERTING COSTS TO A COMMON TIME

Existing System$_{(E)}$

- Investment = \$260,000 = (I_E)
- Salvage Value = \$0 = (S_E)
- PV of Nonannual O&M = \$ x SPW:
  \$25,000 x 0.71 = \$17,750
  \$25,000 x 0.36 = \$9,000
  \$25,000 x 0.26 = \$6,500
  \$33,250

  PV of Annual O&M = \$ x UPW:
  \$120,000 x 11.65 = \$1,398,000

  Total O&M = \$1,431,250 = (M_E)
- \$150,000 x 0.51 = \$76,500 = (R_E)
- PV Energy Cost = \$ x UPW:
  \$648,843 x 11.81 = \$7,662,836
  \$1,487,248 x 14.66 = \$21,803,055

Total Energy Cost = \$29,465,891 = (E_E)

Retrofit System

- Investment = \$1,500,000 x 9.9* = \$1,350,000
- Salvage Value - \$0 = (S_R)
- PV of Nonannual O&M = \$ x SPW:
  \$40,000 x 0.51 = \$20,400
  \$40,000 x 0.26 = \$10,400
  \$30,800
- PV of Annual O&M = \$ x UPW:
  \$145,000 x 11.65 = \$1,689,250
- Total O&M = \$1,720,050 = (M_R)
- Replacements = \$0 = (R_R)
- PV of Energy Costs = \$ x UPW*
  \$597,350 x 11.81 = \$7,054,704
  \$1,326,724 x 14.66 = \$19,449,773

Total Energy Cost \$26,504,477 = (E_R)

*Investment cost adjustment factor

COMPARE THE INVESTMENT ALTERNATIVES ON THE BASIS OF THEIR RELATIVE ECONOMIC EFFICIENCIES. SELECT THE ALTERNATIVE THAT WILL BE MOST COST-EFFECTIVE

Life cycle costing permits comparing an existing system with one or more retrofit alternatives. The different ways of combining cost data to evaluate existing or retrofit systems' economic efficiencies are called "modes of analysis." The modes of analysis are Life-Cycle Costs (LCC), Net Benefits or Savings (B-C), Savings-to-Investment Ratio (SIR), and two which will not be discussed in this article, Internal Rate of Return (IRR) and Discounted Payback (DPB).

Life-cycle cost (LCC) sums the energy costs (E) of the building or system together with purchase and installation costs (I) less any salvage value (S), O&M (M), and replacement costs (R). All amounts are converted to present values using discount factors. The following is a general formula for finding the total life-cycle cost of an existing system or energy retrofit where all costs are in present values and adjusted for taxes and other incentives. The result is the total dollar cost of the existing or retrofit system over the time horizon.

$$LCC = I - S + M + R + E \quad \text{(Eq. 1)}$$

For an energy conservation investment to be cost-effective the LCC with the retrofit (LCC_R) must be less than the LCC without the retrofit (LCC_E).

EXAMPLE: CALCULATE LIFE CYCLE COSTS (LCC)

To calculate the LCC's of the existing manual seasonal control system and the proposed automatic seasonal control system, the variables estimated previously are substituted into Equation 1.

Existing System: $I_E - S_E + M_E + R_E + E_E = LCC_E$

\$260,000 - 0 + \$1,431,250 + \$76,500 + \$29,465,891 = LCC_E

\$31,233,641 = LCC_E

Retrofit System: $I_R - S_R + M_R + R_R + E_R = LCC_R$

\$1,350,000 - \$0 + \$1,720,050 + \$0 + \$26,504,477 = LCC_R

\$29,574,524 = LCC_R

While the LCC with the retrofit is less than without the retrofit, it appears that this particular project will prove to be minimally cost-effective because the total LCC is only 1/15 less over 25 years.

Net Benefits or Savings (B-C) finds the difference between the life-cycle dollar energy savings and the differential life-cycle costs of an energy investment. The differential life-cycle costs are found by calculating the difference between the present value costs of the existing system and those of the proposed system. The result is the net dollar saving attributable to the conservation investment. The general formula for calculating Net Benefits is as follows:

$$(B\text{-}C) = \Delta E - (\Delta I - \Delta S + \Delta M + \Delta R) \quad \text{(Eq.2)}$$

CALCULATE NET BENEFITS (B-C)

The net benefits may be calculated using Equation 2 and the net values between the existing and retrofit systems.

- $\Delta E = E_E - E_R$
$\Delta E = \$29,465,891 - \$26,504,477 = 2,961,414$ (ΔE)
- $\Delta I = I_R - I_E$
$\Delta I = \$1,350,000 - \$260,000 = \$1,090,000$ (ΔI)
- $\Delta S = S_R - S_E$
$\Delta S = \$0 - \$0 = \$0$ (ΔS)
- $\Delta M = M_R - M_E$
$\Delta M = \$1,720,050 - \$1,431,250 = -\$288,800$ (ΔM)
- $\Delta R = R_R - R_E$
$\Delta R = \$0 - \$76,500 = -\$76,500$ (ΔR)

$$(B\text{-}C) = \Delta E - (\Delta I - \Delta S + \Delta M + \Delta R)$$

$$(B\text{-}C) = \$2,961,414 - (\$1,090,000 - \$0 - \$288,800 - \$76,500)$$

$$(B\text{-}C) = \$1,659,114$$

For an energy conservation investment to be cost-effective the net benefits must be positive.

Savings-to-Investment (SIR) is also based on the present value of cash flows; however, the savings and costs are expressed as a ratio rather than as a dollar amount. To achieve a positive net savings, the ratio must be greater than one. The higher the ratio, the more dollars saved per dollar invested. The general formula for calculating the Savings-to-Investment Ratio is as follows:

$$SIR = (\ E - \ M) \div (\ I - \ S + \ R) \quad \text{(Eq. 3)}$$

EXAMPLE: CALCULATE SAVINGS-TO-INVESTMENT RATIO (SIR)

The Savings-to-Investment Ratio for the automatic seasonal control system may be found using Equation 3 as follows:

$$SIR = \frac{(\Delta E - \Delta M)}{(\Delta I - \Delta S + \Delta R)}$$

$$SIR = \frac{(\$2,961,414 - (-\$288,800))}{(\$1,090,000 - 0 + (-76,500))}$$

$$SIR = 2.64$$

While this project is cost-effective, the SIR is low, indicating that other projects with higher SIR's may be more attractive. To select the project which is most attractive other projects may also be considered. The following summary has been prepared to illustrate how this subject might compare with other possible projects.

EXAMPLE: RANK PROJECTS

	Project	SIR	Ranking
(1)	Automatic Seasonal Control	2.64	3
(2)	Power Factor Correction	1.41	4
(3)	HVAC Controls	4.83	1
(4)	EMCS	4.74	2

Having calculated the SIR for each project it is possible to rank the projects by dollars saved per dollar invested. This process is helpful when selecting projects for funding.

ANALYZE THE RESULTS FOR SENSITIVITY TO INITIAL ASSUMPTIONS AND CONSTRAINTS

Because some life-cycle costs and most life-cycle benefits accrue in the future, there may be some uncertainty as to the correct values to use in predicting future benefits and costs. Sensitivity analysis can test how responsive the Life-Cycle Costs, Net Benefits, or Savings-to-Investment Ratio are to changes in the values of key factors. This type of analysis will help predict how the viability of a project will vary as fuel prices increase, and the discount rate, time horizon or other factors change.

To illustrate, Figure 2 presents the sensitivity of a $1,000 base-year energy savings to changes in the time horizon (0 to 25 years), discount rate (D = 0, 5, 10, 15%), and the energy escalation rate (E = 0, 5, 10, 15%). Even though the Btu savings and base year cost savings are the same in all cases, the present value of these cost savings varies significantly depending upon the escalation rate and discount rate selected.

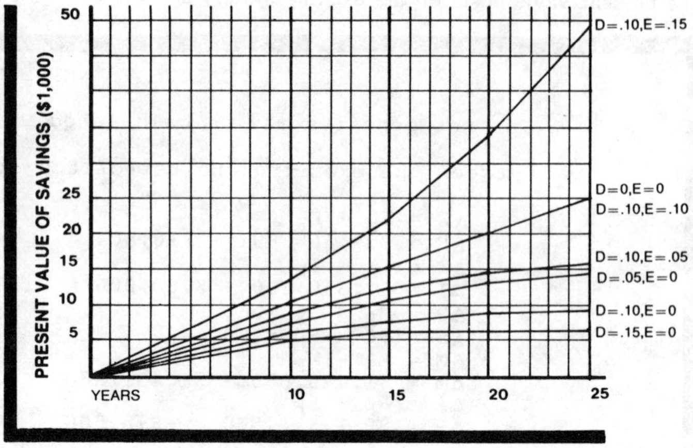

FIGURE 2. SENSITIVITY OF FUEL SAVINGS TO TIME HORIZONS, DISCOUNT RATES AND ENERGY ESCALATION RATES

D = Discount Rate E = Energy Escalation

SOURCE: Department of Energy

CONCLUSION

Too often the decision to invest in energy conservation is based on inadequate information and inadequate economic analysis. Life-cycle costing provides an important tool for making decisions. This methodology will help maximize energy and cost savings over the life of an investment because it provides the necessary logical framework for analysis of potential retrofit projects or building designs. As energy prices, and material and labor costs escalate it is imperative to evaluate the impact of all relevant costs associated with an energy investment.

SECTION 6
INDUSTRIAL PROCESS OPTIMIZATION

Chapter 19
EXPERIENCES AND TRENDS IN TEXTILE INDUSTRY ENERGY PROBLEMS

S. S. Stecco, E. Carnevale, M. Bona, M. Paganelli

INTRODUCTION

The energy costs in textile industries, during the last eight years, have grown rapidly, e.g. petroleum products by 11 times, electricity by 6 times, and also water more than 10 times ([1]) so that, today, they are not far from the 50% of labour costs.

These important problems suggested to EEC to examine possible heat recoveries for textile industries, reaching, in the various phases, the following figures:

a) heat production: employed energy percentage 18%, with 5% maximum theoretical recovery,
b) heat distribution: employed energy percentage 16%, with 10% maximum theoretical recovery,
c) heat utilization: employed energy percentage 66%, with 23% maximum theoretical recovery.

It is to point out that the total possible recovery (38%) is impressive, specially considering that some 15-20% (i.e. half of it) can be performed at reasonable prices with investments reduced to a few per cent of proceeds of sales.

The paper summarizes experience of Institute of Energetics in the field, sponsored by Tecnotessile, showing in some detail the most promising solutions in the various types of operating plants.

STEAM GENERATION PLANTS

Power plants to produce superheated steam and water for dye works can be indicatively listed in four different kinds, namely:

a) steam generation systems for direct heat exchange dye vats
b) steam generation systems for non-direct (surface) heat exchange dye vats
c) steam generation systems with double network plant (steam plus water)
d) steam power station with counterpressure turbine for electricity cogeneration.

Plants of type a) have the great disadvantage of complete reintegration of water in boiler; the convenience of direct heat exchange in terms of efficiency is now totally overcome by increasing costs of water. Moreover, no realistic recovery is here possible.

Plants of type b) have surface exchangers in dye vats, where steam condenses and recirculate in the boiler; a high reduction in water consumption is now possible. Two new kinds of problems are likely to arise: heat losses in exchangers and flash losses at condensed steam draining.

The first one obliges to have large surfaces in exchangers, or to have high enthalpy variations across them (with consequent irreversibility losses), the second one is related to relatively high drainage temperature with losses from 10% to 15%.

In double network plant systems (HTW)[2] steam goes to a reservoir (collector), fig.1, from where all utilizers are supplied, and remaining steam is sent to a 'mixer' to heat water coming from utilizers exchangers.

The HTW system is more complex but heat exchange coefficients are here increased of some 100%, and of course flash losses are absent, so that it appears appreciably more convenient.

In the last type (CPT) with counterpressure turbine steam reaches 750°F and 600 lb/in^2, flowing after boiler through a turbine moving an alternator to produce electrical energy.

([1]) Values referred to Italy, from 1973 to 1981

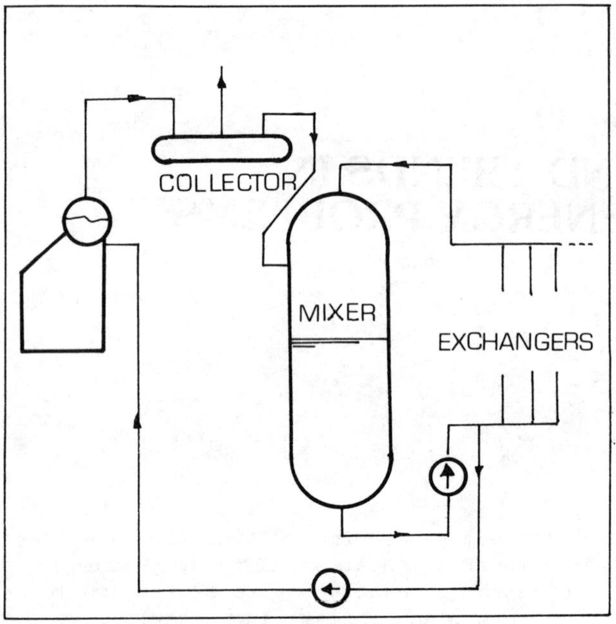

FIG.1

HTW SYSTEM SCHEMATIZATION

Superheated steam from turbine outlet at a pressure of 100÷180 lb/in^2 enters utilizers network. The final cost of steam is here the lowest but, of course, plant costs can be quite high.

According to the preceding different systems, different solutions can be envisaged: for types a) and b), after having transformed direct into surface exchangers, flash losses can be greatly reduced, e.g. by means of ejector recovery(fig.2) ([2]).

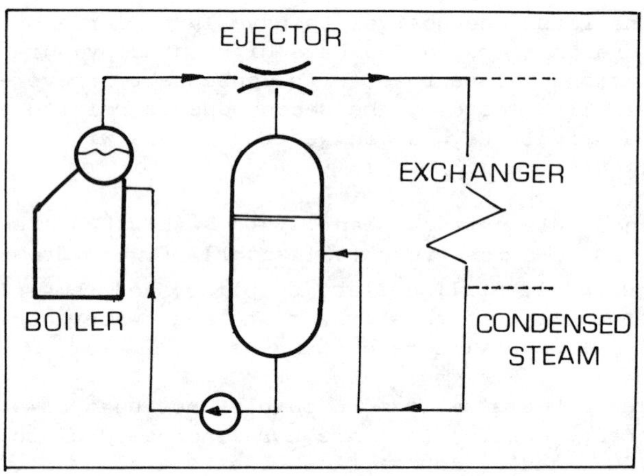

FIG.2

EJECTOR RECOVERY SOLUTION

([2]) Other solutions are possible, but they are less convenient [1,2,3] .

Condensed water from exchangers is sent to a reservoir at 212°F and atmospheric pressure: whence a steam fed ejector compresses flash steam to a suitable enthalpic level, avoiding thus flashing losses.

It is important to notice the relatively minor modifications to apply to existing plants.

In HTW systems the most important inconveniences found in practice are the following: a) in particular working conditions steam generators operate at reduced efficiency, b) steam generator must be choosen for maximum flow, c) when several steam generators operate in parallel it is likely to turn off transitorily one or more of them, with consequent losses.

If a steam accumulator of suitable capacity is inserted in the plant (fig.3), the steam generators can be choosen for the average needs instead of maximum ones.

Details of accumulator operation are available in [1], and a rough indication, for steam production of 10 t/h, indicates a volume of 3500 ÷ 5000 cuft.

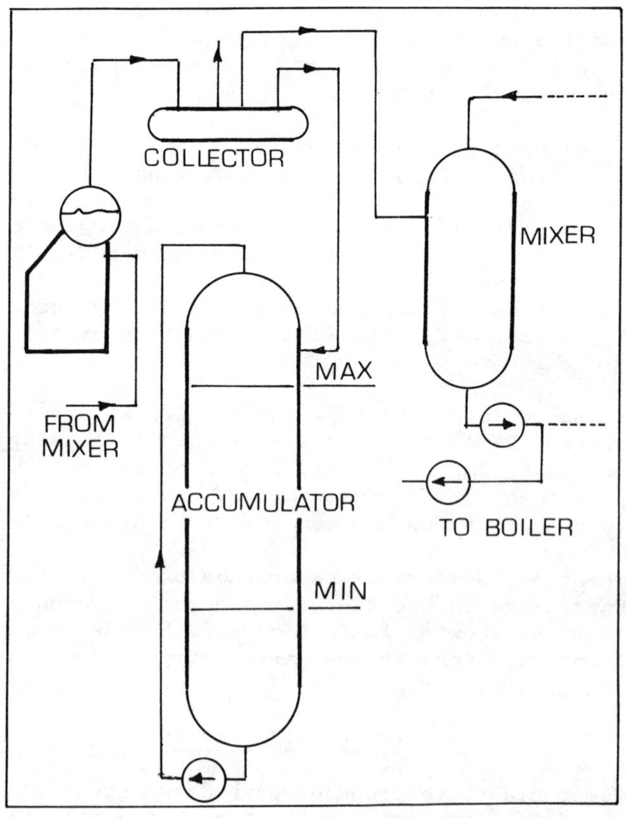

FIG.3

ACCUMULATOR RECOVERY SOLUTION

The same solution applies also to cogeneration systems.

Particular care, in all cases, should be devoted to rational and controlled operation in steam production plants in order to avoid long low-load running conditions and cold air introduction in network; both circumstances, according to authors' experience, can in practice have a decreasing effect in CO_2 contents, increasing sensible heat losses and fuel consumption, lowering the efficiency.

It can be interesting to compare, for the different solutions here proposed, the unit cost of produced steam, according to 1980 values and admitting the following hypotheses (3):

fuel cost	L. 60/lb
electricity cost	L. 50/kWh
stoker cost	L.1.5×10^7/yr
amortization time	5 years
interest allowed	20%
maintenance cost	5% per year
plant operation	2500 hrs per yr

Four types of plants are taken into consideration, namely:

- P1) Indirect heat exchange with saturated steam (paid up capital 150×10^6 liras)
- P2) Indirect heat exchange with saturated steam and pressurized condensed steam draining (paid up capital 185×10^6 liras)
- P3) HTW (paid up capital 200×10^6 liras)
- P4) CPT with steam at 750°F and 600 lb/in^2, and electricity production 60 kWh per ton of steam (paid up capital 300×10^6 liras).

Table 1 shows how and why P3 and P4 are more convenient even for short amortization times.

ITEMS	P1	P2	P3	P4
FUEL	8825	8825	8825	8520
EL.ENERGY	125	130	140	200
LABOUR	1200	1200	1200	1200
AMORTIZATION	1280	1480	1600	2400
INT.ALLOWED	1280	1480	1600	2400
MAINTENANCE	320	370	400	600
TH.RECOVERY	--	-882	-1323	-882
EL.RECOVERY	--	--	--	-3000
FINAL COST	13030	12603	12442	12438

TABLE 1

UNIT COSTS OF STEAM IN LIRAS PER TON

HEAT EXCHANGERS TO REGENERATE WASTE WATER FROM DYE BATHS

Dying processes consist in immersion of textile material into water solution of colour to be absorbed: for synthetic fibers the process is performed at high pressure to reduce high absorption times. This procedure has so many advantages that, nowadays, all new plants are operated at high pressure.

For optimizing water and energy in dying we must know the values and plant positions of hot water requirements and consumptions, together with possible modifications.

Since dying factories taken into consideration have important differences with respect to steam generators, dimensions etc., a parametric analysis seems to be more appropriate.

The following data are then assumed [3]:

steam production	10 t/h
reintegration water	700 cuft per day
waste water for each cycle	
- vats and flows	1050 cuft
waste water temperature	
- vats and flows	208 °F

A double heat exchanger is suggested: the first one, reintegration water exchanger RWE, the second one, collecting waste water coming from RWE at 194°F, warms up clean water feeding (CWE).

Reintegration water temperature passes from 68°F to 140°F. Further details, here omitted, referred to real conditions, are reported in [3].

If preheating of reintegration water and waste water utilization have to be performed, after easy calculations figs. 4 and 5 can be settled the first one referring to RWE, the second one to CWE.

In all cases a very short amortization time can be experienced

ENERGY SAVING IN DRYING

Textile materials drying is generally performed through convection (4), operating with hot air to eliminate water.

In such systems total efficiency, defined as the ratio between theoretical thermal energy to complete the process and the employed ther-

(3) All data refer to italian situation, and then costs are expressed in liras
(4) Data refer to italian situation

mal energy, can vary between 0.40 and 0.60 .

The most recent drying machines (rameuses) have an independent burner for various fuels (normally clean fuels such as GPL or CH_4) in order to get the required tempe r ature, increasing heat exchange efficiency.

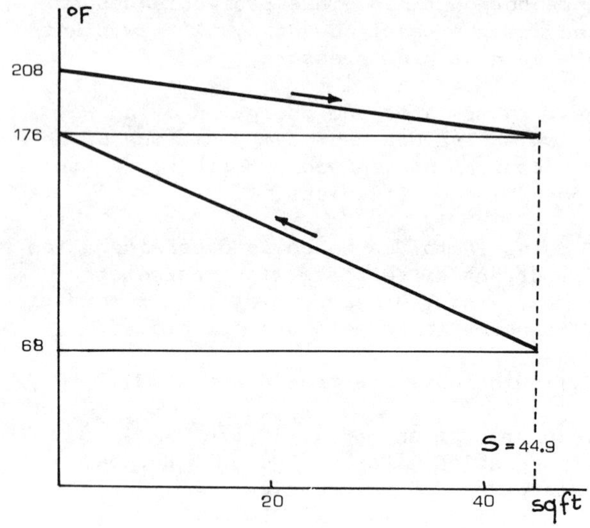

FIG.4

RWE TEMPERATURE-SURFACE PLOT

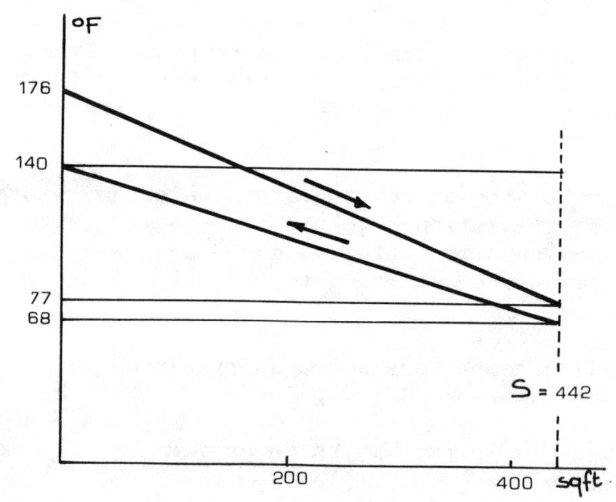

FIG.5

CWE TEMPERATURE-SURFACE PLOT

Recovery possibilities are then the following

a) utilization of air enthalpy drop after operation
b) elimination of all overdrying (i.e. heat supply after total water elimination)
c) regulation of process following air humidity.

For solutions b) and c), which appear stimulating but technologically more difficult to achieve, we are proceeding with experiments to get all involved important information.

For solution a) two kinds of exchangers can be employed, namely air-water and air-air systems.

The first ones can give water at 165°F, whilst the second ones increase air temperature up to 176°F, so that it can be directly employed in the rameuse or for central heating.

WASTE WATER UTILIZATION

In dye processes of textile industry the water normally drained in sink can be several thousands of cuft per day, at temperatures higher than atmosphere and the problems of water availability and pollution arise.

Regeneration possibilities of sink waters to recirculate in the plant can of course reduce consumptions : the most interesting among them, for this particular case (outlet temperatures of 140 ÷ 170 °F), are the following:

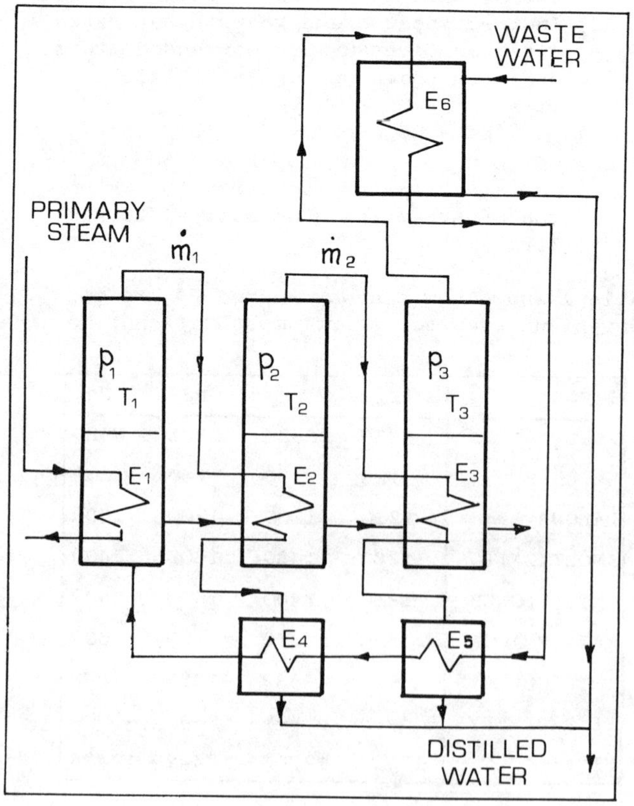

FIG.6

MULTISTAGE EVAPORATION

a) regeneration through multistage evaporation
b) multistage flash distillation
c) vapour compression distillation

All these ways refer to aqueous solutions treatment, normally employed in desalting systems [5] .

Regeneration through multistage evaporation

Fig.6 shows the process.
Waste water, heated to temperature T_1, enters the first stage where, by means of primary steam fed exchanger, the quantity $\dot{m}_1$ evaporates

Waste water enters then the second stage, where steam produced in first stage condenses, driving evaporation of $\dot{m}_2$ (note that $T_2 < T_1$ and $p_2 < p_1$), and then in E_4 preheats waste water to regenerate, etc.

A thermodynamic analysis shows that, in N stages, we can have theoretically N lb of distilled water per lb of primary steam condensed at first stage.

A compromise between clean water production and plant costs suggests a number of 10 ÷ 15 stages in practical applications.

Multistage flash distillation

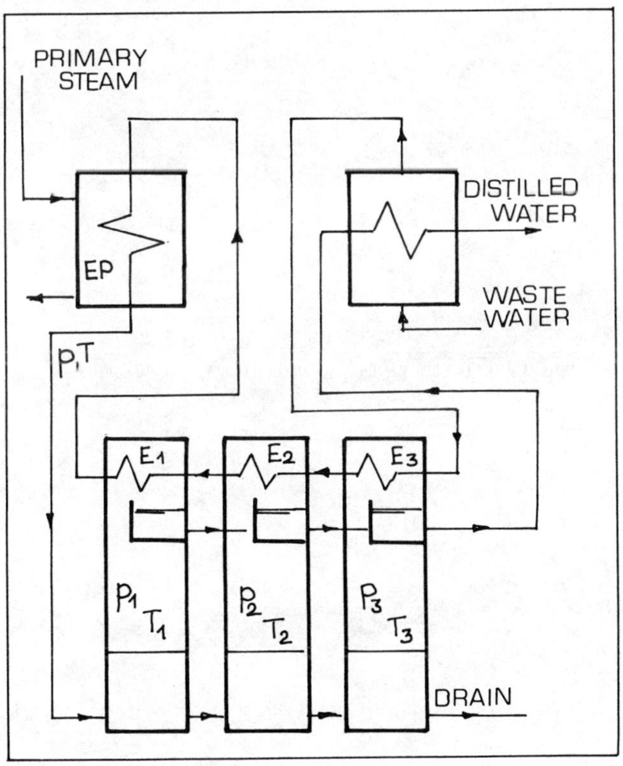

FIG.7

MULTISTAGE FLASH DISTILLATION

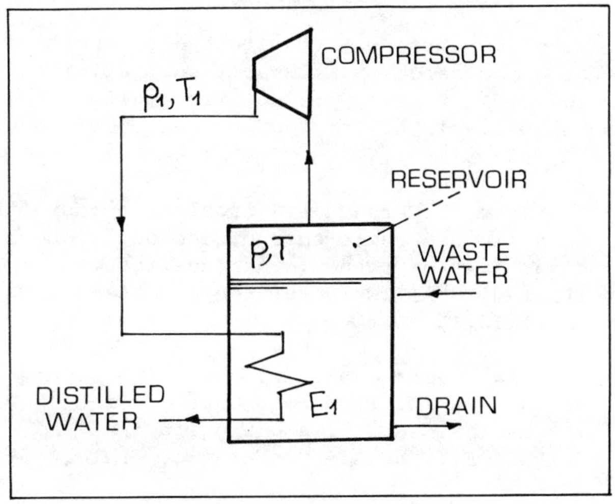

FIG.8

VAPOUR-COMPRESSION DISTILLATION

Fig.7 shows the process.
The waste water, after some preheating by means of produced steam, comes to exchanger ES, where primary steam heats it to a temperature slightly lower than equilibrium temperature at pressure p.

The water enters first stage ($p_1 < p$), undergoes a flashing process of evaporation, and the steam in the upper part is condensed and collected.

The waste water enters then the second stage etc.

In this case not only stage number affects clean water production, but also temperature and pressure drops play an important role in it.

Practical envisaged applications should easily reach 10 lb of distilled water per lb of condensed steam.

Vapour compression distillation

Waste water at temperature T_w enters an evaporating reservoir at pressure and temperature in equilibrium p,T, where $T > T_w$.

Saturated steam is sucked from compressor, and reaches a temperature $T_1 > T$ and a pressure $p_1 > p$ (fig.8).
The superheated steam is then sent to exchanger E_1, where it condenses, heating waste water to be regenerated.
This solution can be interesting for lower productions.

REFERENCES

[1] M.Bona, S.Stecco - Risultati e prospettive di una ricerca sui risparmi energetici nella industria tessile pratese (in italian) - ENERGIA e INDUSTRIA Conference, Prato, nov. 28, 1980.

[2] S.Stecco, E.Carnevale - Problemi eprospettive nelle richieste energetiche delle operazioni di tintura nelle aziende tessili (in italian) - AICTC Conference , Montecatini, april 18,19, 1980

[3] A.Bosio, E.Carnevale, F.Martelli - Alcune soluzioni per il risparmio energetico nelle lavorazioni tessili a umido (in italian) - ENERGIA E INDUSTRIA Conference, Prato, nov. 28, 1980.

[4] S.Stecco - Su alcuni problemi energetici nei cicli di tintura delle aziende tessili (in italian) - INDUSTRIA TESSILE PRATESE DI FRONTE AL PROBLEMA ENERGETICO Conference, Prato dec.10, 1979.

[5] S.Stecco *et al.*- *Energy for Industry* - Ed. P.W.O'Callaghan, Pergamon Press, 1979.

Chapter 20
ENERGY MANAGEMENT IN STEAM PROCESSING KILNS

L. Polisner

BACKGROUND

The steam curing of industrial products constitutes a major use of energy in the country's industrial sector. Considering that industry consumes about 30% of the country's total energy at an annual cost of $87 billion dollars, attention should be directed towards savings in this area.

CONSERVATION:

As applied to Steam Processing, Particularly in the Concrete Industry.

Steam is utilized in various processing industries. As an example, in the field of concrete, concrete masonry units are manufactured with a "no slump" concrete and require heat and moisture for the hydration of the concrete. This means the introduction of live steam into holding kilns where the product is stored for its steaming process. The process is similar for concrete pipe and precast concrete, although the curing temperatures may vary.

I am going to speak this afternoon about how the concrete industry has applied automatic control systems for substantial conservation of energy. Although I will be talking about steam curing in the concrete industry, the analysis and methods employed here will be applicable for any process where direct live steam is utilized without return lines.

The average manufacturing plant has a steam generating plant. This consists of a boiler which is hooked into steam supply lines to individual kilns.

GENERAL ENERGY CONSERVATION METHODS FOR ALL STEAM PROCESSING APPLICATIONS.

The first steps taken are those that should be taken in similar plants throughout the country. Combustion efficiency of a boiler should be studied and the unit should operate at a minimum of 80%. Once the steam generating unit has been brought up to maximum efficiency, steam lines and kilns should have the cost effective investments of complete insulation.[1] Therefore, we first must bring up the boiler to top efficency and add insulation wherever needed.

Obviously, steam leaks wherever found should be immediately corrected. For example, a steam blow that is only two feet can mean a loss of 50 pounds of steam per hour. A six foot steam blow means a loss of 70 pounds per hour. At the end of the day, this means 4,080 pounds lost, and at the end of the year, 1,490,000 pounds lost. Consider an average cost of $4.00 per 1,000 pounds of steam, this means a loss of $6,000.00 in one year.

The reason for perhaps repeating what everyone knows and illustrating same, is that before one attempts to go into sophisticated systems for actually mechanizing steam in the processing, one has to be assured that the basic steps of 80 to 85% boiler efficiency are reached, leaks are eliminated, exposed pipes covered, and kilns, kettles, or holding vessels are completely insulated. Then, and only then, we are ready to reduce energy use in the steam systems that we are monitoring. In this way, and only in this way, we are able to evaluate the actual dollar savings and paybacks that emanate from the system itself. Many industries are being constantly made aware of new systems at substantial investments that have really saved energy through the basic fundamentals of boiler efficiency and insulation. The system itself may not offer a payback which is cost effective.

STUDY OF THE STEAM PROCESS ITSELF IN CURING CONCRETE MASONRY UNITS.

In the concrete industry, and more specifically the manufacture of concrete masonry units, we have studied the entire curing process.[2,3,4] The steam processing system requires a preset period which is nothing more than a holding period of so many hours before the steam is allowed to enter the kilns. Once the preset period is complete, steam is turned on and the atmospheric temperature in the kiln is allowed to rise (not more than a degree a minute) to the set point, dew point, or equilibrium temperature. The moment this temperature is reached, the steam is shut off.

Before the cost of fuel went up to its present level, this entire process was activated by manual labor. Individuals in the plant were given the task of manually opening and closing the steam valves. Steaming periods, rather than being controlled by exact temperatures, were approximated in terms of time. Boilers were on timers for shutdown after the steam processing, while in some plants, foremen simply came back after hours to shut down the boiler.

A careful evaluation of this system showed

serious defects in the steam processing operation. Since the kilns were on time and not temperature, the temperature variation in the kilns caused a variation in the product. There was much over and under steaming since the manual timing process did not take into account differences in ambient temperatures. In addition, time studies within plants showed that it takes a man twelve minutes per kiln to open and shut the steam valves. This timing is conservative as a man is assigned the task of visually supervising the process in watching the time to open and close valves.

Not only is time taken to open and shut valves but following the steaming, it is absolutely essential that the valve be closed tightly because leaking steam after a process is finished, can only affect the material in a deleterious manner. So we see that this type of manual energy serves no real purpose and this manual energy is an extra cost that can easily be eliminated.

AUTOMATING THE STEAM PROCESS SYSTEM TO ELIMINATE LABOR AND REDUCE STEAM.

In order to eliminate the manual energy that is required to open and shut valves and kilns, and in order to make for a precisely engineered curing process based on exact time and temperature, we have introduced a curing control panel and automatic valves. The automatic valves are self-explanatory. They are simple, steam solenoid valves that remain closed until they receive a signal from the control panel for them to open. These valves open on signal from an indicating controller and remain open until the process is completely closed. As stated, we are controlling steam to accelerate the hydration process of the concrete so that the concrete masonry units will have a compressive strength of eight hundred pounds, so that it will be handled and cured. Formerly, the steaming was handled by manual valves and simple timers up to the point that fuel costs reached the present level.

We were asked to design a curing control panel which would eliminate the 'human error' and the human labor associated with opening and closing valves and timing out processes. In addition, and more importantly, a control panel was needed which would eliminate, of course, the aforementioned human error, but would precisely time out all sequences and would monitor the temperature in the kiln, and signal the valve that the process was complete. The panels would have to be graphic in that all logos, indicating controllers, timers, and lights would be on the outside of the panel so that at this position, any plant operator could see exactly what was happening as to what sequence of the processing was going on in what kiln. At the same time, all the setting could be accomplished in a very simple manner. As a result of all this, we have come up with a curing control panel which seems to contain everything that was needed in order to automate the process and eliminate any human factor.

As previously stated, our panel was devised after a study of the curing standards as proposed by the American Concrete Institute. In fact, we made our panel in such a way that these standards could easily be accomplished by putting the information on our panels. In addition, enough flexibility was allowed so that if the plant operator had what he considered a different aggregate or different cement or a different product, and he wanted to make other changes on the panel, anything could be done quite simply right on the face of the panel. He could see what was going to be done when the start button was depressed.

THE ACTUAL PANEL

You will note the panel we came up with. There are five kilns in this particular installation. Each kiln has its own set-up of controls. We will explain exactly what we have for this particular process in light of what you see on the panel.

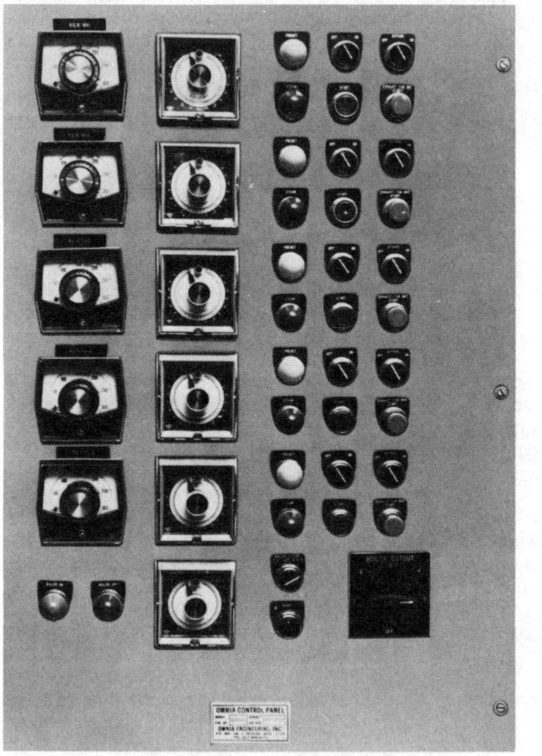

FIGURE 1. ILLUSTRATION OF STANDARD PANEL.

On the extreme left, we have the start button. We have an Eagle Cycle-Flex Timer, which is a plug-in type timer we use on the face of the cabinet. This timer has a number of hours of setting time that we want the material to set before the steam is put on. Each kiln has an indicating controller to show steam temperature in the kiln and set point temperature. Lights to show every period of the steam process operation: amber: preset; red: steam on; green: exhaust. Manual override for valve on panel as well as start button. On bottom of all panels are indicating lights and switch for immediately shutting off boiler once curing period has ended.

The interior of the panel shows the simple

hook-up to TC panel on top and valve panel on bottom as well as exhaust timers.

FIGURE 2. INTERIOR SECTION OF CURING CONTROL PANEL.

In this process, it is absolutely necessary to prevent thermal shock occurring when steam is turned on immediately when the fresh product arrives in the kiln. On the other side the other control that is mounted on the panel is a solid state indicating control, which of course, is plugged in, and when the button is depressed to start this kiln, then the indicating controller immediately registers the present temperature in the kiln. There is a dial and a pointer which is then moved to the objective temperature that is desired for the kiln. This could be 160-degrees Fahrenheit or whatever temperature is desired. The lights on the panel indicate when the preset is on, and when the exhaust is on.

So rather simply stated, for each kiln, dependent upon the type of curing that is going to be utilized, the plant operator sets the preset timer for so many hours, then he sets the indicating controller for the set point temperature when it will shut off. In the interior of the panel, we have timers for the exhaust fans, and these can be set as to their experience for the best exhausting methods and time periods.

The sequence that then follows is as simple as this: the plant operator goes to the panel, depresses the start button once the kiln has been filled and the door shut; the preset timer will time out the preset period; the indicating controller will then come on and open the valves. The valves will remain open until such time as the objective temperature in the kiln is reached. Once it has been reached, the temperature is shut off and a soaking period is started. The product is then allowed to soak overnight. The next morning, before the kiln is to be used, it is necessary, of course, to have a small opening in the front and rear of the kilns in order to allow exhaust fans to pull the steam from the kilns. A man will go to the panel and depress the exhaust button. A green light will come on to indicate to plant personnel the kiln is exhausting and the interior timer will take over and keep the fan operating until such time as the exhausting is completed. Once the kiln has been exhausted of steam, then it is opened and the cured products are taken out, and it is ready to receive the green products being made on the machine.

For sensing the temperatures required in the kilns, we are able to use thermocouple wire which is simply run from the thermocouple terminal in the rear of the panel to the interior of the kiln. The wiring that is used in this operation is Type "J" thermocouple wire.[5] It is unsheathed and the dissimilar iron and constantan wires are braided and fused, and we find that this is an excellent probe for the kiln. We are working in a steam atmosphere, and certain sensors do not react favorably with steam. However, the iron constantan seems to have a long life and should this metal become corroded, it is a simple matter to cut off an inch of the wire, strip it back, braid it, and you are able to have the use of another thermocouple. All instruments have thermocouple break protection[5] in that should the thermocouple wire become inoperative, the the control would go to the other end of the band and shut off the steam so there would not be any wastage of energy.

There are many different procedures that can be put onto a curing control panel such as this. We have added alarm systems with dual lights so that what will happen is this: instead of shutting off the steam once the temperatures are reached, we hold it for one, two, three, or four hours, depending on what the plant operator desires. The steam holding period is generally visible with an amber light. Once this is completed, then a green light and an alarm come on to tell the block plant operator now it's time to exhaust the kiln. On operations which run 24 hours a day, we have automatic door openers and automatic fans so that the exhaust system will come on immediately upon the completion of the steam holding period.

This panel is not only utilized to manufacture concrete masony units, but has been used in concrete pipe production, precast, and other such processes where time and temperature are of utmost importance. The results of the survey taken with over 50 panels throughout the country has shown fuel savings ranged from 20% to 60%. We have had plants with a 200 horsepower boiler which has saved as much as 14,000 gallons of #2 fuel oil in a six months' period by the use of precision controlled temperature and accurate timing of all procedures. The results in other plants have shown similar savings. There is absolutely no way that plant personnel can watch the process without having overruns unless they are using precision control set-ups.

A method that we are exploring at the moment is the use of programming controls where the entire program could be put onto a programmer and then have it processed in this way.

We have been rather slow to adopt a programmer for this operation because of the fact that we are working with small and medium size businesses where they do not have the expertise for working on integrated circuitry which would require a more intensive knowledge of advanced electronics. With the present control system as you see it, any one in the plant can test for accuracy. There is no way of losing the program once it has been set up, and every kiln is individually visible on the panel without having to go into the interior of the panel and call for certain information.

Now, likewise the capital investment. As long as we are working with a minimum number of kilns, that is, under 20 kilns, the cost of the entire set-up is much less than using a programming type unit. However, when we go into 20 kilns or more, then the price of all the controls and hardware may become prohibiting.

The payback is directly attributed to savings in fuel alone. That is, reviewing the fuel consumption with an automated control system in the plant as opposed to the same production using manual valves and timers has averaged 18 months. This system has allowed plant operators to function with lower temperatures which were effective due to the fact that all kilns were uniformly steamed.

SUMMARY

Finally, I would say that the establishment of control panels to monitor, time, and control temperature in the industrial steam processing will yield to plants the following persuasive advantages: 1. A positive direct reduction in fuel consumed. 2. A uniform product that will be cured according to temperature and time with no possibility of human error. 3. Less manual labor and cost for a man to open and shut valves during a productive day or be forced to return after hours to shut down the boiler. 4. An opportunity to visually monitor the effect of temperature on the product and to assess the results based on the quality of the product achieved. This could lead to the use of lower temperatures which would, of course, again mean less fuel consumed.

The challange of the eighties has arrived and it can be met by automated control of processes to bring down production costs and raise quality levels.

REFERENCES

1. Piping Handbook, Fifth Edition, McGraw-Hill Book Company, 1973.

1a. Insulation Performance Data and Assessment Procedures for Steam Kiln Energy Conservation Investments, by James V. Zaccor, United States Department of Energy, September, 1980.

2. American Concrete Institute Report: Recommended Practice for Atmospheric Pressure Steam Curing of Concrete (ACI 517-70), adapted as a standard of the ACI in New York, April 16, 1970.

3. Steam and Boiler Size Requirements for Curing Systems, by Robert Dikkers, National Concrete Masonry Association, 1958.

4. Manufacture of Concrete Masonry Units, by William Grant, Concrete Publishing Corp., 1959.

5. Trouble Shooting Hints: Thermocouple Input, Solid State Controls, Love Controls Corporation, August, 1977.

Chapter 21
DEVELOPMENT OF THE UTILITY SYSTEM FOR THE FOOD INDUSTRY IN 1981

R. R. Cutler

ABSTRACT

The design of a new processing plant has enabled our client to effect conservations of energy and water.

The utilization of coal-fired fluidized bed boilers, turbine driven equipment and heat recovery deems that this new facility will have the lowest utility costs of any of their present operations.

The paper will outline the studies performed, a description of the energy saving features and the resultant savings.

INTRODUCTION

Our firm provides engineering services for a number of industries and I have chosen a current project, for an international food company, to suggest the various opportunities available to industry for conservation of their utilities and the associated savings. The project we are involved in is a new plant and both the clients staff and ourselves have enjoyed the opportunity to implement this program.

It is my intent to outline the overall program and not to dwell on the design engineering details of the various areas involved.

UTILITY REQUIREMENTS

The services required in this plant include:

1. Steam for heating and process.
2. Water for domestic services, process, cooling and various make-up requirements.
3. Electricity for lighting and power.
4. Chilled water for comfort and process cooling.
5. Compressed Air for manufacturing and controls.
6. Natural gas for boiler light-off and manufacturing. As the use of gas is quite limited, no significant savings were identified.

In the following sections we will look at the above services and the alternatives we have considered in arriving at our design. On Figure 1 - UTILITY SERVICES we have shown the various services involved.

STEAM GENERATION

The energy needs for steam production, in an intergrated food plant, is the highest of any other energy requirement.

I feel it is important to know that in the United States, over 80% of industrial steam is produced in boilers, 100,000#/hr. or smaller, utilizing gas or oil for a fuel. This is typical of our clients present situation in their other plants.

A major decision was the selection of the fuel for the new plant. The alternatives available included oil - interruptible gas and coal. The use of gas was discarded as the availability was less than 100%, plus the likelihood that prices would baloon, and possible Federal action prohibiting such use for industrial steam purposes.

The oil picture has not been very good since 1973, and not until recently has there been any settling down of price increases and availability. We still expect prices to increase at a nominal 8%/year. If there should be any disturbance in the Mid-East, this could change overnight.

The availability and pricing of coal is favorable for industry, and should be part of any evaluation of a steam production facility.

On Figure 2 - FUEL COSTS, we have shown the relative costs of the different fuels. For discussion purposes we will use a cost of $6/million BTU for oil and $3/million BTU for coal. We have included $1/million BTU in the price of coal for ash disposal. Any evaluation you may do will have to be site specific.

Coal has been burned for many years in industrial plants, however the availability of low cost oil/gas after World War II saw conversion from coal units to oil/gas operation or direct placement with package type oil/gas boilers.

Two (2) major developments have required the re-thinking of coal utilization, those being the OPEC oil policy and the environmental concerns.

The need to burn coal cleanly, involves consideration of not only the boiler proper but the needed pollution control devices. The burning of coal involves not only particulate removal, but in our case the ability to utilize coals having up to 3% sulfur content. To remove the sulfur dioxide either a tail gas system or a fluidized bed are available.

While field experience is limited, it was decided by our client to proceed with fluidized bed boilers. This decision was arrived at for the following reasons:

1. Ability to burn various ranks of coals.

2. Low temperature - Low NO_x.

3. Relatively simple SO_2 removal.

4. Relatively simple waste disposal as the ash and spent limestone are in the dry state.

On Figure 3 - FLUIDIZED BED BOILER COMBUSTION we have shown the basic parameters of this type of boiler. The introduction of sulphur bearing coal into a fluid-bed of limestone at 1300-1500°F results in combustion of coal and the absorption of the generated sulfur dioxide gases by the limestone. The amount of sulfur in the coal will determine the limestone feed rate. Spent lacquer from can manufacturing will be burned in the boiler eliminating a disposal problem.

On Figure 4 - STEAM PLANT EVALUATION we have shown the present worth of operating an oil fired and a coal fired plant for 25 years. While an oil-fired steam plant may only cost one-third that of a coal plant, look at it over the long run. We cannot afford to continue to appraise our energy decisions on a two (2) year pay-back or a first cost basis. Accordingly the final decision was made to burn coal in fluid bed boilers.

After selection of the boiler fuel, a study was made of various plant schemes to achieve an economical and energy effective system. The major energy users were identified and three (3) progrms were evaluated to meet these needs.

The lowest cost was for a steam plant producing 150 psig steam using package type fluid bed boilers, followed by the same type unit operating at 300 psig, with the highest cost being field erected boilers generating steam at 600 psig which would provide 3000 Kw of co-generated power.

Based on the relatively low steam usage, 150,000#/hr., it was found most economical to generate steam at 270 psig and then reduce it to plant pressure (100 psig) through mechanical drive turbines.

In the boiler plant, over 2000 HP of turbine drives will be available. Start-up and plant steam balance will be obtained by use of electric motors on certain equipment as required.

The boilers selected were three (3) 50,000#/hr. skid mounted Johnston Boilers. The boiler flue gases will pass thru a dust collector, economizer, and baghouse before exiting thru a single stack. Material handling in the plant consists of the following:

1. Coal Handling

   The boiler plant will be capable of receiving coal by rail or truck. A forty-five(45) day supply of coal will be stored outside on site. The coal handling system is traditional with belts, elevator and over bunker conveyor.

2. Limestone System

   Limestone will be delivered by tanker truck and will be unloaded into a storage silo. Limestone will be transported pneumatically to individual boiler lime silos.

3. Ash Handling System

   Bed ash and spent limestone will be drained from the fluid-bed and cooled by water cooled ash conveyor screws. Bed ash and fly ash will be collected in surge hoppers. The hoppers will be periodically emptied by a pneumatic unloading system. Ash will be collected in a silo, from which trucks may be loaded.

On Figure 5 - BOILER PLANT we can look at the major components of this plant.

HEATING, VENTILATING AND AIR CONDITIONING

The central air handling units are equipped with chilled water coils and steam coils for heating. The air handling system in the office areas are of the variable volume type to more efficiently meet the required loads.

To reduce the duty on the heating coils, additional heat recovery coils will be provided in the exhaust and supply ducts to reclaim building heat, this is referred to as a "Run Around System".

Air conditioning has been kept to a minimum, however the total load is still over 3000 tons. An evaluation was made as to the most economical means of satisfying this seasonal load. Flexibility of the chilled water system is necessary to meet partial and full load requirements, therefore a combination of equipment was necessary. While two (2) motor driven compressors would be lower in first cost we found that using one (1) 1000 HP motor driven compressor; one (1) turbine driven compressor and one (1) absorption machine, receiving its steam from the exhaust of the turbine driven compressor, justified the higher investment cost. On Figure 6 - CHILLER SELECTION - 3000 TONS we have shown the summary of these two (2) schemes.

To further reduce the demand on these chillers, the system has been designed with variable flow. If the chilled water return temperature is below the operating setpoint a chilled water pump will drop off the line. By matching the equipment duty to the load requirements, the 100 horsepower circulating pumps and the chiller equipment on line will be kept to the minimum.

A new approach was developed in connection with the cooling for can welding lines. Up to now, chilled water has been circulated through the can coolers and returned to the chillers directly. We have added an outside air exchanger to provide cooling of chilled water when outside air is 60° or less, again reducing chiller duty.

ELECTRIC POWER

In addition to utilization of mechanical drives were suitable, we have required all motors to be "high efficiency". The standards were published in our contract specifications. At full load the high efficiency motors have 4 to 8 points higher efficiency over the standard line of motors, depending on the size.

COMPRESSED AIR SYSTEM

Historic data established that 4000 SCFM of 100 psig plant air would be required. It was agreed that two (2) machines would be installed. One (1) motor driven centrifugal machine and one (1) turbine driven centrifugal were selected in lieu of two (2)motor driven machines. The turbine will operate at 270 to 100 PSIG, exhausting to the plant steam header. We look at the energy as being essentially free, as the steam is required for plant load. The turbine drive will reduce electrical use and demand by 500 KWH. On Figure 7 - COMPRESSED AIR SYSTEM - 4000 SCFM we show the annual energy savings with this arrangement.

HEAT RECOVERY

In the preparation of canned food products some water is contained in the product, however the majority is utilized for cooling services, plant clean up and sterlization.

As we became more familiar with the process our engineers found numerous ways to conserve water and reclaim heat.

In the cooking building we developed a system of booster pumps and reclaim heat exchangers utilizing the incoming treated process water. The process water is heated from 75°F to 150°F in this system. Previously separate cooling water service went to drain and steam heated hot water generators furnished the hot process water. The outcome is that there is a significant reduction in the cost of fresh water, cost of waste treatment and steam for water heating. On Figure 8 - WATER AND HEAT RECOVERY we have tabulated the associated costs and savings. The simple payback is less than six (6) months.

SUMMARY OF THE UTILITY SYSTEM

Now that we have walked thru the various areas of the plant utility system we have put this all together on Figure 9 - SUMMARY OF UTILITY SYSTEMS.

In addition to the dollar savings, we should not overlook other contributions of this program which are the:

1. Utilization of Coal for the industrial scale.
2. Commercial utilization of fluidized-bed boilers.
3. Features of this design are applicable to existing plants and are now being evaluated.

As the cost of energy escalates we, the consultant and you the plant engineer-manager or owner must strive to better his utility systems. When fuel or water were inexpensive, their impact on product cost was of lessor importance, their costs today are now quite significant.

We appreciate the opportunity to pass on our thoughts in regard to industrial utilities.

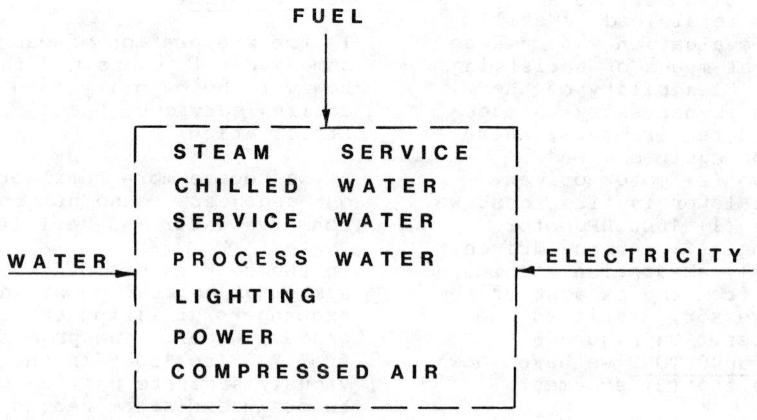

FIGURE 1- UTILITY SERVICES

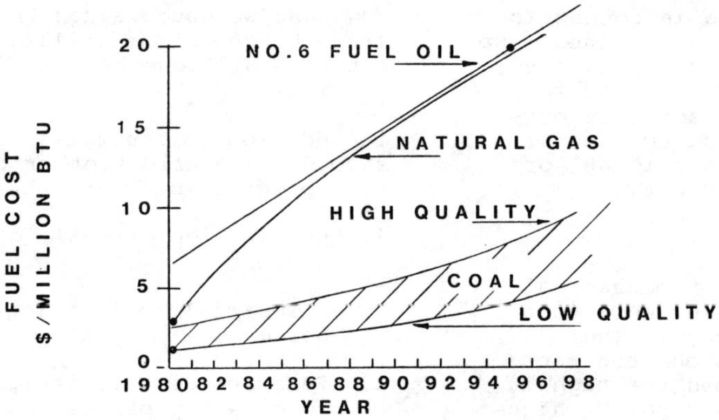

FIGURE 2- FUEL COSTS

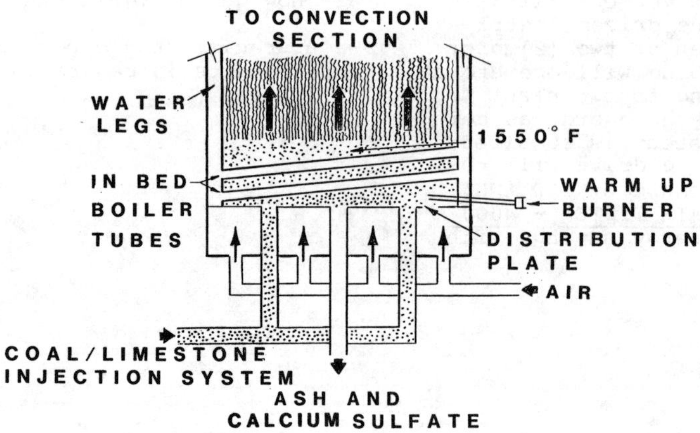

FIGURE 3 - FLUIDIZED BED BOILER COMBUSTION

	OIL FIRING	COAL FIRING
PLANT COST	$4,500,000	$15,000,000
ANNUAL FUEL COST	$6,300.000	$3,150,000
PRESENT WORTH-25 YEAR LIFE AT 15%-ESCALATE OIL AT 8%/YEAR,COAL AT 5%/YEAR	$81,500,000	$44,600,000
SIMPLE PAYBACK	—	3.4 YEARS

FIGURE 4- STEAM PLANT EVALUATION

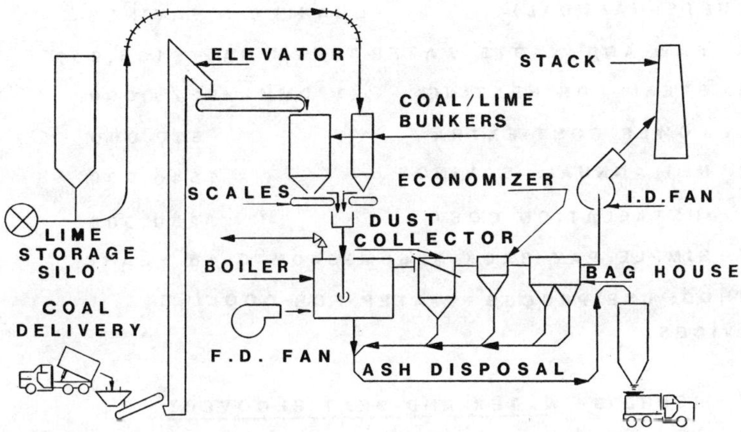

FIGURE 5 - BOILER PLANT

ARRANGEMENT	TWO (2) CENTRIFUGALS	COMBINATION
DRIVERS	3000HP ELECTRIC	ELECTRIC/ STEAM
EQUIPMENT& INSTALLATION	$400,000	$514,000
SIX(6) MONTH ENERGYCOST	$262.000	$239.000
SAVING IN ENERGY		$23,000/YEAR
PRESENT WORTH AT 15%-25 YEAR LIFE -ESCALATE ELECTRICITY AT 7%	$29,665,400	$15,804,190
SIMPLE PAYBACK	5 YEARS	

FIGURE 6- CHILLER SELECTON - 3,000 TONS

	SCHEME # 1	SCHEME # 2
DRIVERS:	2-500 HP MOTORS	MOTOR/TURBINE
EQUIPMENT & INSTALLATION	$150,000	$200,000
ANNUAL ENERGY COST	$175,000	$ 87,500
PRESENT WORTH AT 15%-20 YEAR LIFE-ESCALATE ELECTRICITY AT 7%	$1,936,750	$1,093,375

SIMPLE PAYBACK - $50,000÷87,500=0.58 YEARS

FIGURE 7- COMPRESSED AIR SYSTEM-4000 SCFM

OBJECTIVE: REDUCE WATER AND STEAM USAGE

RESULTS: (ANNUAL)	QUANITY/$ SAVINGS
A. RAW AND WASTE WATER	300MM GAL/$185,000
B. STEAM FOR HEATING	170MM#/$375,000
C. POWER COST EXTRA	-$20,000
D. NET ANNUAL SAVINGS	$540,000
E. INSTALLATION COST	$250,000

F. SIMPLE PAY-BACK $250M÷540M=0.46 YEARS

METHOD: USE PROCESS WATER FOR COOLING SERVICES.

FIGURE 8- WATER AND HEAT RECOVERY

AREA	ANNUAL SAVINGS	ROI
COAL FIRED STEAM PLANT	$3,150,000	3.4 YEARS
STEAM/ELECTRIC CHILLERS	$ 23,000	5 YEARS
STEAM/ELECTRIC COMPRESSORS	$ 87,500	7 MONTHS
PROCESS - STEAM WATER REDUCTION WASTE TREATMENT	$ 540,000	6 MONTHS

SUMMARY: UTILITY COST WILL BE $3.8MILLION/YEAR LESS THAN OTHER COMPRABLE PLANTS

FIGURE 9- SUMMARY OF UTILITY SAVINGS

Chapter 22

ENERGY AUDITING IN FOOD PROCESSING PLANTS ON PRINCE EDWARD ISLAND

J. L. MacLeod, C. Macdonald

INTRODUCTION

Prince Edward Island is Canada's smallest province. Located in the Gulf of St. Lawrence, it has an area of 2184 square miles, a coastline 1100 miles long, and a population of 120,000. Nicknamed "the million acre farm", P.E.I. is primarily a fishing and farming province, with only a small industrial base. Our largest and possibly most energy-intensive industry is food processing which runs from lobster canning and freezing, to vegetable processing, to the production of canned milk and ice cream.

Four years ago the governments of Canada and Prince Edward Island started an energy audit program in P.E.I. called "Enersave for Industry and Commerce". Its purpose is to help businesses of all kinds save energy. Since dollar savings go hand in hand with energy savings, the program has been well received in this province which has the highest energy cost. Over 500 facilities have been audited to date, with total potential savings at $3 million per year identified.

Food processing plants tend to be large energy users, making the search for energy savings there especially worthwhile. In this paper we would like to share with you some of the things we have learned and approaches we have taken in performing energy audits of food processing plants on P.E.I.

THE BASIC APPROACH

Our energy auditors enter a food processing operation solidly grounded in the principles of thermodynamics, but without a lot of detailed information about food processing. In some ways this lack of knowledge is a blessing. We are free to ask "why"?, to ignore sacred cows, and to question the unquestionable. What we do bring to the site is a year's history of the plant's fuel and electrical consumption, our mobile office --- (a GMC Transmode bus equipped with a Data General Nova 3 computer), a rough idea of building size, and some information about plant operating conditions.

Upon arriving we sit down with the production manager, foreman, or other knowledgeable person. Before even looking at a piece of equipment we get this person to describe the process in as much detail as possible. As this description progresses we make the first rough energy flow diagrams. We typically make a separate diagram for each specific production line or product, tying them together into a plant-wide diagram later on.

Each production line diagram starts with the incoming raw product itself. How many pounds per day and per year? What is its temperature upon arriving? Is it frozen? Must it be stored on site for a period of time, or is it processed immediately? What are the storage conditions?

Most food production processes consist of a series of discrete steps or stages. And each of these steps normally involves some transfer of energy. At certain stages; for example, the product will be heated --- often by steam or hot oil. At other stages it will be cooled --- sometimes by city or well water, sometimes by mechanical refrigeration. So the quantity of product, the temperatures, pressures, flow of coolant, quantity of steam used, etc are all noted in the appropriate places on the energy flow diagram. At each step of the process we ask ourselves: "Is there a significant energy interchange in this step?" "What happens to the waste products?" "To the condensate flow?" "Where does the heat entering this step eventually go?"

A typical energy flow diagram appears in Figure 1. The product in this instance is canned evaporated milk. Heating in the various steps is done with steam from an in-plant boiler. Most cooling is accomplished with 50°F water from the city mains. Note that each process stage is energy-isolated from the other stages. Each stage is treated as an independent entity with condensate and city water being sewered, and spent steam being vented to the air. The energy potential of these effluents is wasted. Cool water is going down the drain rather than being used in other stages of production or areas of the plant; and usable heat in the hot effluents is being wasted.

When each process stage is considered by itself it often seems that all this energy waste is unavoidable. But when the operation is viewed as a whole, possibilities for coordinating and interconnecting energy flows become apparent. Figure 2 shows this same evaporation and canning operation as it might be configured to reduce waste:

1. The steam-operated vacuum system has been replaced by a 5 H.P. mechanical vacuum pump. The 500 lbs per hour of steam required by the old system cost $18,000 per year to generate. In contrast running the mechanical pump costs only $2,000 per year. (Savings paid back the capital cost of this change in well under one year.)

2. A portion of the cooling water from the condensor is rerouted to preheat the raw milk by approximately 40°F. This action saves 865 MBH (253 KW) of steam use or $27,000/yr. We are still not finished with this water flow however and we now will use it to precool the hot evaporated milk as it leaves the evaporator. While no energy is saved directly, we have reduced the requirement for city water by 20,000 lbs/hour and saved approximately $7,500/yr. water cost.

3. The cooling water from the can cooler is now used to preheat the evaporated milk as it leaves the holding tank before being sent to drain. This action saves 620 MBH (182 KW) of steam for a dollar saving of about $12,000/year.

4. Finally the waste streams from the preheater and sterilizer are collected and used to preheat boiler makeup water saving approximately 350 MBH which in this case would mean a dollar saving of about $6500 a year.

SPECIFIC CONSIDERATIONS

The primary value of a plant-wide energy flow chart is to make visible the potentials for energy exchanges and the utilization of waste. A flow-chart shows up the non-local possibilities --- where waste energy at one stage of the process can be used in another, or even elsewhere in the building. In analyzing the flow chart, and in walking through the plant, we pay attention to a number of things:

Water Flows

We look at the water flows through the plant. We ask ourselves several questions:

1. What is the water used for?
2. Where is it put to drain,
3. What are the possibilities for cascading its use?

Even water without much heat content may be useful for washup. And simply reducing the quantity of water used saves energy and dollars.

Air Flows

We also look at air flows:

1. How much air is exhausted?
2. What is the temperature of the exhausted air?
3. Does the exhausted air contain appreciable moisture? (Air which is used to dry and cool french fries, for example, contains considerable latent heat which is recoverable.)
4. What is the present source of the makeup air? Should that be change? Should the temperature of makeup air be raised through heat exchange with exhaust air?

Heating and Refrigeration Plants

In creating and analyzing the flow chart we look at how energy is being used. But we also question whether or not that energy is produced efficiently in the first place. We look, for example, at basic steam generation in the central boiler plant. Is combustion efficiency up where it should be? Are stack economizers advisable? We also look at the refrigeration plant. Are insulation levels adequate on lines, tanks, and storage rooms?

Examples

In addition to the example cited in figures 1 and 2 there have been numerous other instances where our audit techniques have proven their worth.

In one case involving a producer of french fried potatoes, we found steam heated blanchers which, in order to control the level of solids in the hot water, had a continuous overflow of approximately 9000 lb/hr which was made up by cold ($50^{o}F$) water from the plant's wells. Plant personnel were considering collecting this waste stream along with others to be used to preheat boiler makeup water. We suggested, however, that heat from the waste stream be exchanged with the cold water makeup at the blanchers. This action resulted in a saving of approximately $40,000/yr in steam cost.

In a small fish processing plant where water flumes were used to move the fish from one end of the process line to the other, we observed that the water was being used on a one pass basis. Well water was pumped into the flumes and put to drain at the end. The water carried all the fish waste to the end of the flume where it was screened out and later processed into fishmeal. The water after leaving the screen was reasonably clear so the suggestion was made to divert at least 50% of the flow back to the inlet of the flume. This action obviously would save at least 50% of their cost of pumping water for the flumes.

While many of the ideas we come up with are obvious, the usefulness of the flow diagrams to the persons conducting the audit can not be denied. When a person has the opportunity to see the processes on paper first, they have a much easier time when they tour the plant. Instead of mysterious machines you see segments of the process whose purpose you understand. This allows the people involved to use the tour to assess the practicality of already formed ideas rather than struggling to understand the operation of the equipment they are seeing and find areas where energy might be saved.

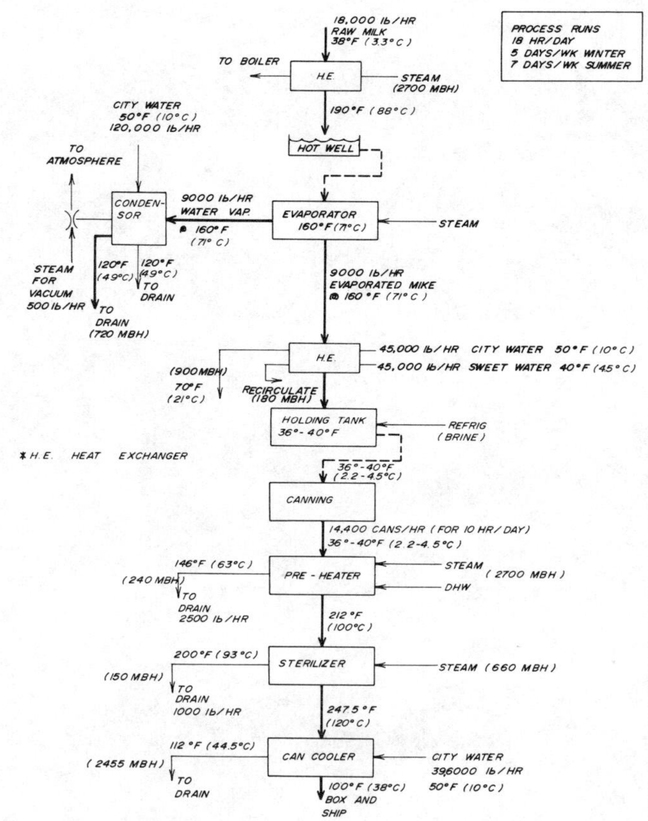

FIGURE 1 - ENERGY FLOW DIAGRAM FOR A MILK EVAPORATION AND CANNING PROCESS.

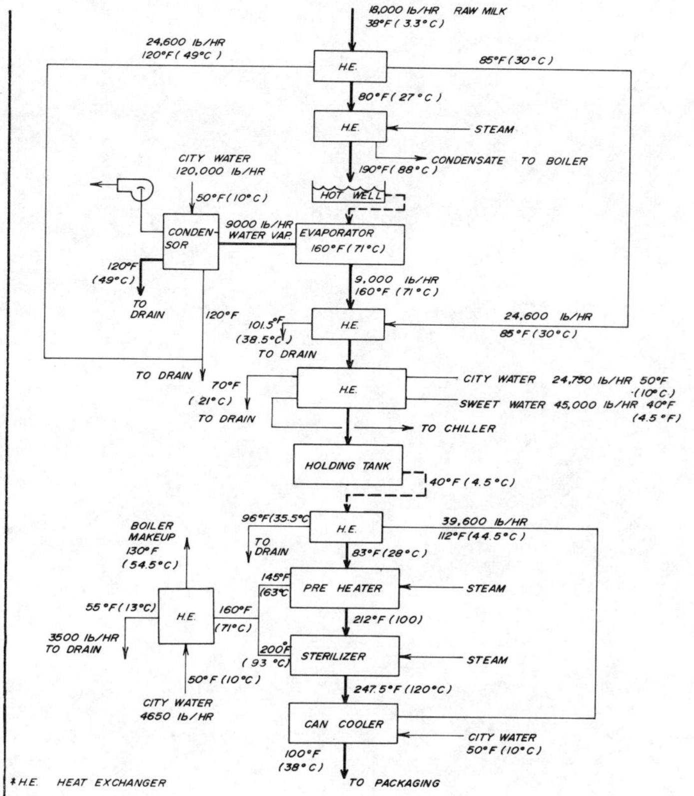

FIGURE 2 - MODIFIED FLOW DIAGRAM

Chapter 23

A SOLID FUELED GAS TURBINE COGENERATION SYSTEM FOR THE MINERAL PROCESSING INDUSTRIES

J. Adams, M. L. Brown

Introduction

The mineral processing industries are a $20 billion per year business in this country. This group as a major industrial class is the fifth largest user of purchased fuels, consuming 1.12 quads of energy (1). Found in this group are non-intensive energy users such as stone products industries and energy intensive users such as the clay and glass industries.

Because of concerns with production cost and product quality, energy intensive plants have tended to utilize natural gas because of its economy, efficient combustion, and low emissions. The rapid price increases of recent years coupled with the speculation of even larger increases should gas be deregulated has caused most segments of this industry to explore alternative fuels. This paper examines an energy system which is capable of utilizing less expensive solid fuels while maintaining compatibility with process requirements.

The energy system to be discussed is particularly well suited to the clay processing industry, thus system specifications are set to meet this criteria.

Clay Processing

The clay processors form a significant subset of the mineral industry. In 1980, clay industries accounted for 50 million tons of production in the United States alone. Revenues from this activity exceeded $920 million (2). Clay products included kaolin, fuller's earth, bentonite, fire clay, and common clays with applications ranging from building materials and refractories to paper coatings and absorbents.

The processing of clays varies depending on the type and end use, but typically will consume over one MMBtu thermal energy and 150 kWh electrical energy per ton of product during processing. Thermal energy is used to dry and/or calcine clay while electricity is used as the driver for mechanical processing, pumps, and other plant equipment. The clay industry is similar to the glass, pottery, and lime industries in that it is unable to utilize low grade, ash bearing fuels for fear of product contamination because drying and calcining operations generally use direct contact between the product and combustion gases.

Alternate Fuel Resources

Because the processing of clay has high purity requirements, past fuels have been limited predominantly to natural gas with high grade distillate fuel oil used as back-up. The energy price escalations of the past decade has had a profound effect on intensive consumers of conventional fuel resources such as the clay industry. A study performed by a consultant to the fuller's earth industry illustrates this point by analyzing the cost of energy as a percentage of the total product cost (3). Before the rapid price increases of the 1973 Arab oil embargo era, energy averaged about 15% of the cost of processing fuller's earth. By 1980, this percentage had increased to 35% and continuing at the current rate will reach 50% in less than six years. This analysis assumes natural gas price increases at a rate similar to the past several years and does not include abrupt disturbances such as another fuel embargo or sudden deregulation of gas prices by the government.

Consumers of natural gas have been aided in holding down their fuel expenditure by the governmental regulation of gas price. Once gas is deregulated, which will occur in 1985 by law or possibly before if felt necessary by the executive branch of government, gas prices will rise to a level comparable with oil. The comparable prices of gas and oil was the case historically before the days of OPEC oil price fixing. The increase of gas prices to that of oil will translate into a doubling of cost. Clearly it could be most difficult for clay industries to maintain their current market segment with energy forming a continually growing percentage of their product cost. One solution to this gloomy outlook for clay processors and other massive energy users is to utilize alternative sources of energy. Several alternate fuel possibilities exist that are plentiful and have displayed impressive price stability over the past several years. Table 1 shows the costs for both conventional and alternate fuels. This table illustrates that solid fuels such as coal and wood may offer the most cost effective option.

Two problems must be overcome before less expensive fuels can be used as a substitute fuel in the clay industry. First, the problem of mineral product contamination must be solved, and, second, the solution must be achievable at an attractive rate of return on investment. The proposed energy system presents a solution that satisfies both criteria. The proposed design is based on utilizing wood fuel instead of coal since it offers a slightly lower capital investment system. This is because wood fuels are very low in ash content and virtually free of sulfur, thus emissions equipment is limited to particulate control.

TABLE 1

AVERAGE 1981 FUEL COSTS

Fuel	Price	Unit Price
Natural Gas	\$0.40/therm	\$4.00/MMBtu
#2 Fuel Oil	\$0.95/gal	\$6.79/MMBtu
#6 Fuel Oil	\$.50/gal	\$3.33/MMBtu
Coal	\$40/ton	\$1.66/MMBtu
Wood Chips	\$12/ton	\$1.40/MMBtu

*MMBtu = 10^6 Btu

One might question that the problem of wood fuel availability has been overlooked. To the contrary, numerous wood fuel studies in the Southeast have indicated a significant overlap in location of clay and forest industries. Studies of forest industry waste and overgrowth indicate adequate waste wood fuel supply for many locations in this country.

Wood Cogeneration System

The processing of clay involves the investment of great quantities of energy, both electrical and thermal. Given these requirements, it is not surprising that a cogeneration system capable of supplying electricity and process heat would prove practical. However, it must be verified that a plant's thermal and electrical demand is compatible with the proposed cogeneration system.

To operate economically, the cogeneration system must be capable of utilizing solid fuel. To guarantee that the mineral product quality is not sacrificed, the heat provided by cogeneration must be clean and low in particulates. This requirement places severe restrictions on solid fuel combustion systems which, by nature, have high particulate emissions due to the ash content of solid fuel.

The prime mover that best satisfies the requirements of the proposed cogeneration system is a gas turbine. Gas turbines are simple to install, have wide fuel flexibility, and produce high temperature exhaust usable in thermal processes. The problem with gas turbines as with clay drying operations is their intolerance of ash bearing fuels. Though solid fuels have been burned in gas turbines for short periods of time, they have proven unfeasible because of the severe gas path erosion caused by suspended ash. Thus, for a solid fuel cogeneration arrangement to be practical, a method to eliminate ash particles in the combustion gas must be found.

One approach to the elimination of particulate has been the utilization of an indirect gas turbine cycle. The conventional simple cycle gas turbine consists of a compressor, a combustion system, and a turbine as shown in Figure 1. Ambient air is compressed and continuously flows into the combustor where fuel is burned to add heat and to increase the temperature of the compressed air. The hot gas then expands to atmospheric pressure through the turbine which drives the compressor and also provides power to drive the load. Since the products of combustion pass through the turbine, only clean fuels such as natural gas or petroleum distillates are normally used in order to avoid particulate erosion of the turbine blades.

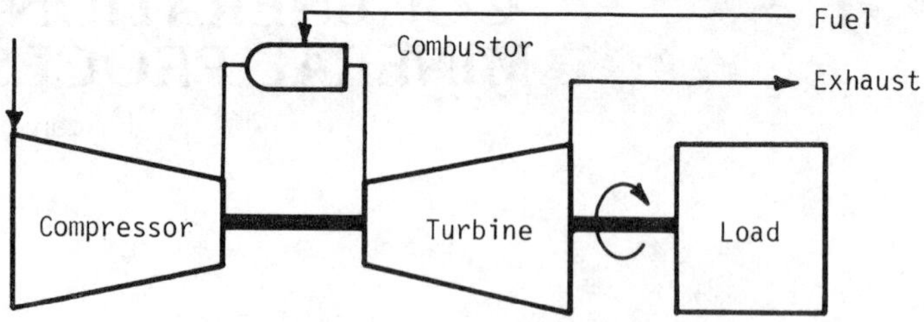

FIGURE 1. SIMPLE CYCLE GAS TURBINE

Figure 2 shows the indirectly heated cycle. The compressor and turbine of the conventional cycle are retained, but the necessary heat addition is achieved by passing the compressed air through a heat exchanger rather than the direct burning of fuel in the air stream. Only clean heated air passes through the turbine and the potentially damaging products of combustion are restricted to the hot side of the heat exchanger.

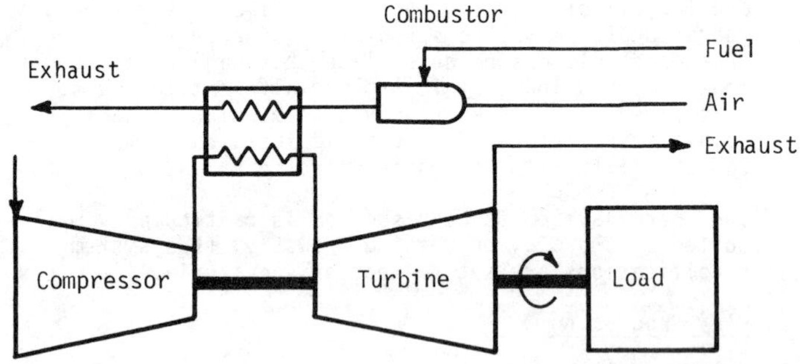

FIGURE 2. INDIRECTLY HEATED GAS TURBINE CYCLE

The indirect heated gas turbine offers several advantages in this instance. This arrangement can accommodate a variety of fuels including wood, coal, or other ash bearing fuels. The cost of adding a heat exchanger to the cycle is less than modifying a conventional gas turbine to accept high ash fuels in a standard combustor. Also, the component life of the indirect fired cycle operating on high ash fuel will not be reduced unlike direct fired turbines. Lastly, this turbine configuration is a good source of thermal energy for process requirements.

Notable disadvantages include the size and life expectancy of the turbine heat exchanger. Because of existing material constraints, the heat exchanger temperature is limited. This restricts the turbine inlet temperature and concurrently the turbine cycle efficiency. Though ceramic heat exchangers hold considerable promise for the future, commercially available units are metallic.

For the proposed indirect fired gas turbine to be viable, several design questions must be addressed including the selection of a heat exchanger and combustor, determination of appropriate operating temperatures, and arrangement of the system to permit maximal use of waste heat. These concerns are addressed below.

A central component of the proposed cogeneration system is the combustor. When utilizing biomass fuel, a two-stage wet cell combustor is one of the foremost options (Figure 3). This device utilizes the high volatile content of biomass to generate temperatures in the range of 2000°F even with high moisture content fuel. In the lower chamber of the combustor, fuel is fed in with an underfeed stoker to be burned on a pinhole grate with less than a stoichiometric amount of air. During the first stage of combustion, all of the volatiles are driven off. The volatiles and low Btu gases then enter a second chamber where they are burned with 10% to 20% excess air. By limiting the amount of excess air, higher temperatures are achieved. Burning biomass on grates in a pile bed gives significantly lower particulate than would be possible in a conventional spreader stoker furnace. These low levels of particulate will help to extend cleaning intervals in the heat exchanger. For proper operation with wet fuel, combustion air preheated to about 500°F is necessary.

A stainless steel alloy heat exchanger will be employed to transfer heat to the turbine. Metallic exchangers have shown much greater resistance to temperature flux than ceramics and are generally preferred for this reason even though they have lower temperature limits. When used with an indirect fired gas turbine, the heat exchanger inlet temperature is limited to approximately 1750°F which corresponds to turbine inlet temperature of 1450°F.

The next step is to arrange the system components and select the operating temperature to best match the thermal and electrical outputs with those required in the process. The heat exchanger inlet temperature and thus the turbine inlet temperature is limited by the materials. Given these system constraints, an arrangement to best utilize the available energy is shown in Figure 4. The model gas turbine system produces approximately 2100 kW of electricity. The clean, high temperature exhaust is used as pre-heated air in the wet cell combustor. At 700°F, the turbine exhaust is too cool to be used directly for clay drying, therefore, using it as pre-heated air to the burner helps reduce the fuel requirements. Operating the wet cell on green wood, it is possible to approach outlet temperatures of up to 2000°F, but this is too high to use in the turbine heat exchanger, thus a portion of the gas is blended with the 795°F heat exchanger exhaust to produce 1750°F inlet gas temperature. An additional amount of 2000°F air is produced to blend with the remainder of the turbine heat exchanger exhaust to produce 1200°F. This gas is fed through a second heat exchanger producing 1000°F air for drying. The net effect of this arrangement is to produce electricity and 1000°F clean process air while rejecting heat at 300°F.

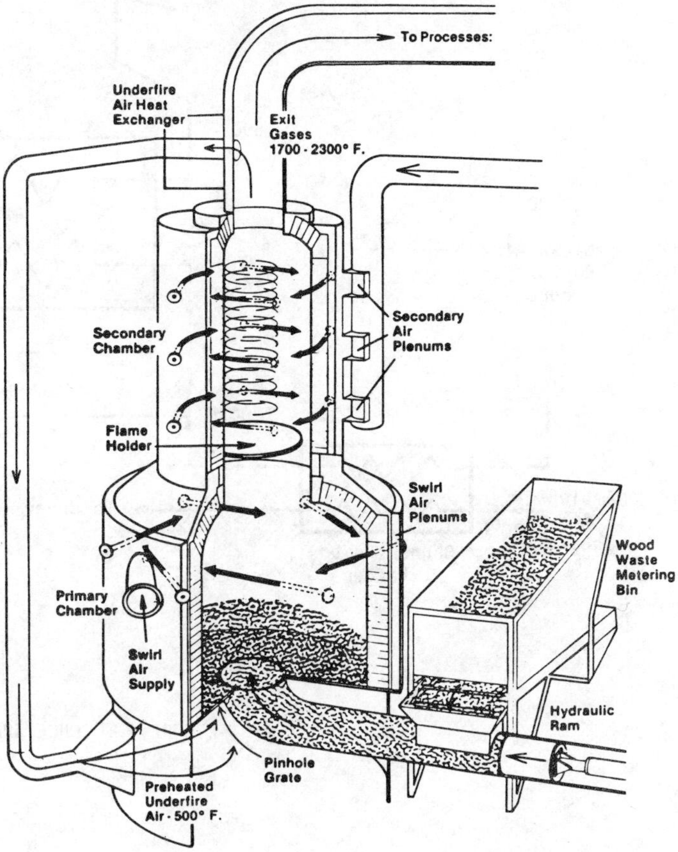

FIGURE 3. LAMB-CARGATE WET CELL

The major advantages of this configuration include high thermal efficiency and accommodation of low quality fuel. The system is flexible so that processes with different operational constraints can be accommodated.

System Economics

The economic feasibility of converting to a wood burning energy system is addressed by an analysis of those factors which are common to any financial investment. Specifically, the analysis will focus on the following topics:

- Capital Investment
- Operating Costs
- Fuel Costs
- Annual Income/Savings
- Payback and Return on Investment

Case studies have shown wood fueled systems are capital intensive. Conventional wood combustion equipment such as steam boilers can cost up to ten times that of a natural gas, fuel oil, or a propane fired equivalent.

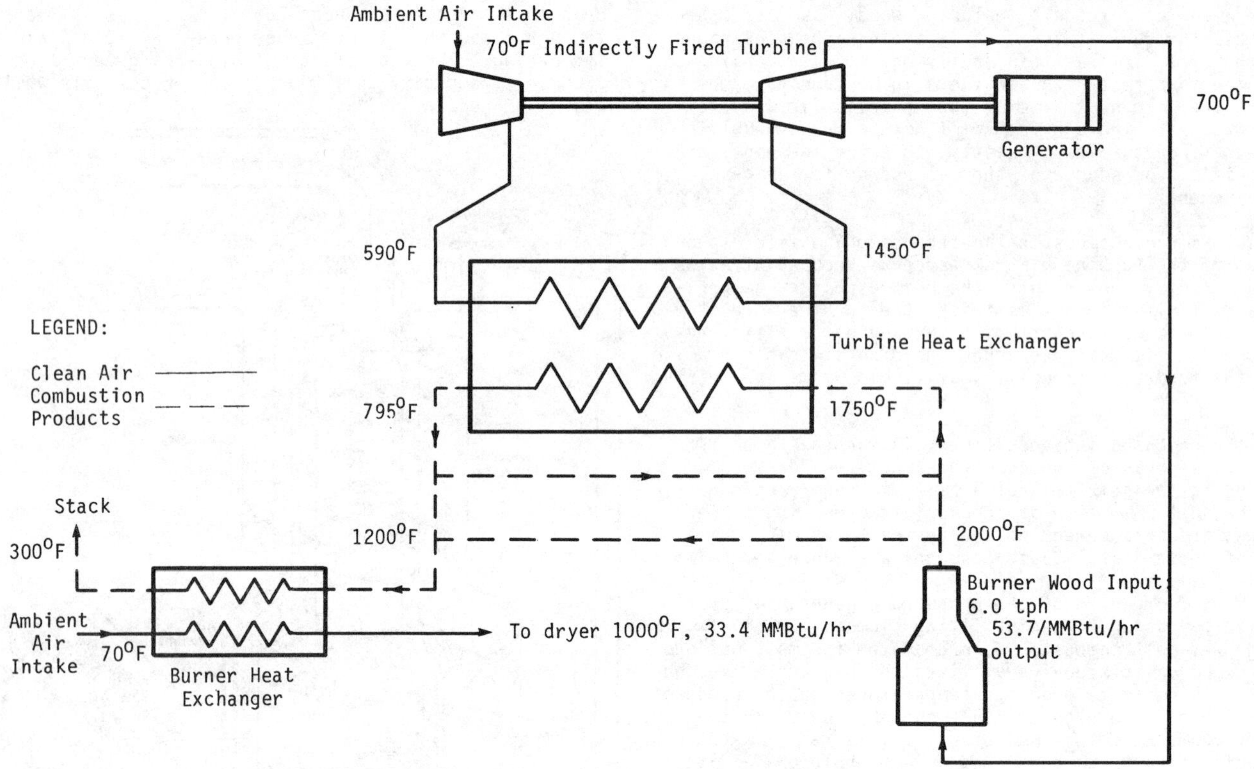

FIGURE 4. CONCEPTUAL COGENERATION SYSTEM

Costs of the system are more expensive due not noly to furnace design, but also to the additional cost incurred for wood fuel storage and handling equipment. Budget capital costs for the proposed cogeneration energy complex is obtained by surveying several vendors of the various system components. No one vendor can supply the entire system so estimations are based on the conjecture of several manufacturers' equipment to produce the complete system. Budget capital cost estimates for the proposed system are shown in Table 2.

TABLE 2

BUDGET CAPITAL COSTS

Turbine	- 2100 kW x $380/kW	$ 800,000
Combustor	- 55 MMBtu/hr	600,000
Heat Exchangers	- high temperature	600,000
	medium temperature	300,000
Fuel Handling		300,000
Miscellaneous, Piping & Controls		200,000
		$2,800,000

Operating costs are broken down into two groupings: (1) fuel costs and (2) maintenance, utilities, and labor. These cost estimates are derived from vendors, quotes, surveys, and industrial experience. Labor and maintenance costs for a comparative natural gas system are assumed to be approximately zero. Maintenance for the wood cogeneration system was calculated by assuming the yearly cost to be 3% of the total capital investment. Operating costs for the conventional and wood energy system are presented in Table 3.

TABLE 3

HOURLY OPERATING COSTS

	Conventional System	Wood System
Fuel	33 MMBtu/hr x $4/MMBtu = $132	6 ton/hr x $12/ton = $ 72
Electricity	2100 kW x $.05/kW = $105	
Labor	0	2.5 men x $10/hr = $ 25
Maintenance	0	$ 10
	$237	$107

Hourly Savings = $237 - $207 = $130
Yearly Savings = $130/hr x 8600 hr/yr = $1,118,000

Wood fuel costs vary greatly in most regions of the country. Generally, transportation costs are the principle determinant of wood fuel pricing to specific customers. A detailed study of actual cost and availability as well as investigations of long term contractual arrangements would need to be performed prior to final design. The economic analysis for this report is based on wood fuel costs of $12.00 per ton. Energy costs for the existing arrangement area based on natural gas at $4/MMBtu and electricity at $.05/kW. Annual savings is derived by comparison to the cost of operating the existing system and is equivalent to income.

Results of a simple payback analysis are presented in Table 4. The net capital cost of the system is derived by subtracting the appropriate tax credits from the total capital. For this case, a 20% tax credit composed of a 10% investment credit and a 10% alternate energy credit, was used.

TABLE 4

ECONOMIC ANALYSIS

Total Capital Cost	$2,800,000
20% Tax Credit	560,000
Net Capital Cost	$2,240,000
Annual Savings on Operation	$1,118,000
Interest on Capital (15%, 20 years)	- 381,500
	$ 736,500

Simple Payback = $2,240,000/$736,500 approx. 3 years

The net yearly savings for the proposed cogeneration system is found by subtracting the interest on the capital cost from the annual operational savings. Interest on the capital expenditure was calculated based on a 15% interest rate and a 20-year amortization.

The simple payback period for the conceptual cogeneration system can be calculated by dividing the yearly savings into the net capital cost. For this case, a simple payback of approximately three years was calculated.

Summary

With energy costs consuming larger and larger portions of the finished product costs, profits and the economic vitality of the industry suffer. Innovative energy systems such as the one discussed above offer great promise to the minerals industry by supplying a long term solution to energy costs and availability.

References

1. Annual Survey of Manufacturers, U.S. Department of Commerce, Bureau of the Census, 1976.
2. Minerals Yearbook, U.S. Government Printing Office, Washington, 1980.
3. Private Communication with Brian Richardson, Kinetics Consulting Group, July 1981.

SECTION 7
PRODUCTS AND APPLICATIONS

Chapter 24

A CASE HISTORY ON RETROFITTING FACTORIES WITH A BOILER SYSTEM FUELED ON COAL, WOOD, AND WASTE

R. D. Pincelli

Dover Mills of Shelby, North Carolina, a major producer of finished goods in the textile industry, with dye house and manufacturing plants approximately one-half mile apart, made the decision in July of 1980 to proceed with a boiler plant locaed between the manufacturing facilities and positioned to serve both plants. This system would be required to burn local wood waste, as well as coal, with backup fuels of gas and oil. Mr. Dave Roberts, Energy Manager at Dover, led the corporation into the strategy of eliminating vulnerability to oil cutoff and in addition, maintain their competitiveness by the ability to generate steam at approximately 50% of the former cost.

Dover Mills has three plants, J & C Dye House, Dover Textiles, and Ora Mills, located within approximately 2,000 feet of each other. The project calls for J & C Dye and Dover Textiles to be tied into a central solid fuel boiler facility near the J & C Dye House, which will require approximately 25,000-30,000 lbs./hr. of steam with the remainder of the 42,000 lbs./hr. capacity being available for Dover and ultimately for Ora Mills when that tie-in is accomplished at a later date under a Phase Two construction program.

The current boilers supplying all three plants are firetube, Scotch Marine, gas/oil only boilers. The three plants have seen massive increases in the cost of their oil and gas over the last five years, as have all textile mills. Current cost of No. 6 oil is approaching \$.90 per gallon and natural gas is approaching \$4.00 per MCF. Coal, equivalent of \$1.95 natural gas and \$.29 per gallon No. 6 oil on a delivered basis. Wood waste will be available at approximately the equivalent of \$1.00 gas or \$.15 per gallon No. 6 oil. After allocation is made for additional maintenance, electrical load for intermittent conveyors, feed systems, etc., as well as some additional labor, the payback period for Dover is projected to be 2.1 years. Substantial tax credits from North Carolina and the federal government of 30% plus were obtained. These tax credits are 10% investment and 10% energy tax credits from the federal government and 15% investment from the North Carolina Department of Revenue. The ROI on this project exceeds 20%. Obviously, each passing year the project becomes more economically attractive to Dover as the costs of gas and oil continue to outpace the cost of coal and wood by 10-15% per year.

Dover Textiles made the decision to install a multi-fuel boiler system capable of burning coal, wood waste (both dry and green), selected textile and plant waste, as well as gas, oil, and propane/air mixtures after observing units already in operation for standard Knitting Mills of Knoxville, Tennessee, Division of Stanwood Corporation in Charlotte; Spring Mills units at Calhoun, Georgia; Mount Vernon Mills of Greenville, South Carolina at the Tallassee Mills Division in Tallassee, Alabama; as well as Steel Heddle Corporation at Greenville, South Carolina currently burning coal, wood, and oil.

The savings from this project at the current cost of gas and oil is anticipated to be \$300,000-\$400,000 per year. However, this number was calculated prior to the surge in oil prices after oil deregulation by President Reagan. The savings figure now would be in excess of \$500,000 per year. Dover generated a positive (profit improvement) throughout the life of this boiler project, because of the vast difference in costs per BTU between gas/oil and coal/wood. This is obvious when one observes the oil/price graph in Table I and the fact that coal has increased approximately 7%, while gas and oil have increased over 31% in the last 12 months as exhibited in Table 2.

The Dover solid fuel boiler system will consist of two T-600P CNB multi-fuel boilers. The design pressure will be 150 PSI saturated and the operating pressure will be approximately 130 PSI. These boilers will generate, on a combined basis, 34,500 lbs./hr. of steam when fired on coal, wood waste, or other solid fuels. When "over-fired" with gas, oil, or propane, the two units are capable of generating a combined steam output of 41,400 lbs./hr. The units are firetube with massive water walls. Efficiency is guaranteed at 80% minimum without economizers or blow-down heat recovery.

Turn-key delivery at current industry activity is approximately 12 months with deliveries extending 14-16 months in the next year due to substantial business increases. Complete EPA compliance was guaranteed by the design and build contractor, as well as operational satisfaction of the material handling system for all types of fuels specified. Site preparation, soil analysis, concrete buildings, material handling equipment, ash removal equipment, water side auxiliaries, as well as boilers, piping, wiring, instrumentation and controls were all supplied through the sole-source contractor eliminating substantial involvement, time and expense on Dover personnel, as well as eliminating problems between equipment suppliers and installing contractors at start-up time.

Relative to financing of this project, Dover elected not to use any unusual financing mechanisms. However, several programs were offered. This program would have essentially worked as follows:

> The investor group would have paid for the entire installation at Dover.

The investor group would retain all investment tax credit benefits.

Dover and the investor group would share equally in the fuel savings as documented by historic gas/oil use and actual coal, wood, and waste use.

The investor group would take 50% of these savings for a period of between 7 and 10 years. At the end of this time the entire project would be turned over to Dover at no cost, meaning essentially that Dover could have had, without capital investment, 50% of the fuel savings for the first 7-10 years and then obtained a multi-fuel boiler system at no cost.

Obviously, these financial programs are structured for the benefits of both parties and in-depth tax and other ramifications must be studied. We're also participating in programs offered by Manufacturers-Hanover Bank and others in which lease-purchase options are employed to maintained positive cash flow throughout the life of an investment such as this. If further details on either of these financial options are of interest, please contact the author at Energy Systems, Chattanooga office.

Environmental considerations on the Dover project were:

1. Particulate Emissions.

2. SO_2 Emissions.

3. Opacity.

Applications were submitted to the North Carolina environmental authorities and permits have been issued to construct without clean-up equipment for the above three categories. This is possible primarily due to the combination of the underfed ram stoker and the conservative furnace design of the CNB Tri-Fuel boiler, which is inherently capable of very low particulate emission rates. To date, over 80 of these multi-fuel boilers have been installed without clean-up equipment of any type, including multicyclones. Obviously, each site must be evaluated on its own individual characteristics, including emission off-sets from prior sulphur content of the oil, ambient air quality conditions, BTU input size, etc. Again, a sole-source contractor taking responsibility for total construction, as well as EPA compliance, has considerable advantage to the client.

Material handling at the Dover site is kept simple as evidenced by the slide here, showing a large bunker filled by front-end loader with coal or wood. The bunker, which has a live bottom unloader to a primary "u-trough" feed screw, which on demand automatically maintains the desired fuel level in each stoker hopper. The Dover system is designed for expansion to a third multi-fuel unit, which will boost capacity to over 50,000 lbs. on solid fuel and over 60,000 lbs. when over-firing with gas, oil, or propane is desired. This over-fire burner capability is <u>not</u> required for EPA compliance. It's purpose <u>is</u> two-fold:

1. To allow rapid and automatic fluctuations in steam demand to be taken with liquid and gas fuels which have a quicker response characteristic than do solid fuels.

3. Should a problem develop in the solid fuel feed system, the liquid/gas fuels can automatically supply needed steam for plant operations during repair of solid fuel systems.

The Dover system will utilize solid fuels, i.e. coal, wood, and selected plant waste to maintain base load conditions anticipated to be about 30,000 lbs. per hr. The liquid and gas fuels, i.e. natural gas and fuel oil, will be used to make up any sudden swings in load on an automatic and quick response basis. The burners are not continually operating, but only come on with a delay in pressure. This feature makes the gas/oil fuel consumption minimal. The fuels which are currently being contracted for by Dover include 50% moisture content green wood in the form of sawdust, chips, and general mill residue, as well as dry wood from local furniture sources; coal, via railroad delivery to existing railroad facilities on the Dover property. Textile waste is also being considered as a fuel. Dover Textiles anticipates being able to eliminate waste haul-away costs, land fill charges, and potential EPA water contamination problems by burning this material in the multi-fuel boiler system. More development work is required, as to the processing and introduction of this fuel source to the boiler system.

Ash removal will be semi-automatic with the operator manually dumping the ash grates, raking ash to a drag chain ash removal system, which will then automatically deposit this material in a 10-yard dumpster. Interest has been shown by a local concrete block manufacturer in obtaining this ash as a filler material for his product, as well as the County and State Highway Department for utilization of the ash as a road anti-ice media during the winter.

Labor on the Dover project will consist of approximately 4 hours per shift from one boiler operator. This will involve dumping of ash one to two times per shift, monitoring of all systems, water softener regeneration, logging of critical temperatures, pressures, etc., routine maintenance, stoker bed maintenance depending on the coking tendencies of the coal. Skill level of the operator would be that of a typical \$5.00-\$6.00 per hour textile worker with normal aptitude and job interest.

Space requirements for the entire Dover project including boiler room, fuel storage bunker, ash removal facilities, etc. will be approximately 100 x 200 ft. outside dimensions. Fuel storage will include approximately 2-3 weeks of storage on site of combined wood, coal, and waste.

Additional reasons that Dover Textiles chose Energy Systems of Chattanooga, Tennessee to do the "turn-key" design and build construction on this central solid fuel boiler facility included numerous North Carolina jobs completed on the same basis and in the same approximate size range. These jobs included RCA's Crown Wood Products Division wood/coal/oil burning 21,000 lb. unit at Mocksville, North Carolina, just outside Winston-Salem; Monarch Furniture Company, High Point, North Carolina, also a 21,000 lb./hr. unit burning wood and coal; and Steel Heddle in Greenville, South Carolina a 12,000 lb./hr. unit burning wood, coal, gas, and oil.

An interesting observation in the simultaneous mixing and burning of wood waste with coal is that at several current locations including Galaxy Carpet Corporations, South Pittsburg, a Tennessee spinning mill, this combined burning has proven that low grade, inexpensive types of coal can be fired in the boiler. The technical reasons include the fact that the wood has "high volatility" and burns out prior to carbon burn-out in the coal, thus leaving a very porous bed in which oxygen from the underfire air fan can readily finalize combustion of the carbon remaining in the coal. Again, many of these technical advantages are available from a "turn-key" contractor experienced with this size and type of installation and fuel utilization that would not be available from other groups.

Our company, in conjunction with Dover, is also exploring the possibilities of mechanical cogeneration. This would involve utilization of steam turbines to replace large continuous running electric motors, such as those found on air compressors, chillers, etc. In this case, 125 PSI steam would be introduced to the turbine, 15 PSI discharge steam would be available from the turbine. This 15 PSI steam could then be used for plant heat, process water heating, absorption air conditioning, etc. with most of the heating benefit still available. The payback on this type mechanical cogeneration is approximately 12-18 months.

A special circular economizer, which can be lowered directly down the existing stack will be considered as a Phase 2 add-on project when the economics dictate. This is illustrated by the slide here showing a similar type economizer being installed in South Carolina on a 42,000 lb. Cleaver-Brooks No. 6 oil fired boiler. Design and construction consideration is given to both very aedquate soot blowing and pressure drops on the water and flue-gas side to prevent any build-ups or corrosion from sulphur or vanadium. In addition, cold end corrosion will also be specifically designed out of the system.

The cost of the Dover system, including over 2,000 ft. of (6 inch) and 500 ft. of (8 inch) steam tieins between J & C Dye House and Dover Mills, was less than $1,000,000 after tax credit impacts are considered. Dover will see approximately a 50% reduction in boiler operating costs and this reduction would include the increased labor, preventive maintenance and electrical requirement imposed by the multi-fuel boiler system. The gross fuel savings will exceed 60%, with 50% being a net savings figure.

These systems make excellent technical and economic sense for intermediate size plants requiring steam and currently burning gas and oil on a 4,000 hr. per year basis or more. The attractiveness of the investment increases each time OPEC, natural gas deregulation, curtailments, and other supply and price problems impact on the gas/oil fuel system. Not only are the economics outstanding, but the competitive edge offered by eliminating the vulnerability to gas/oil supply disruptions means that Dover will not get caught with their "plant" down.

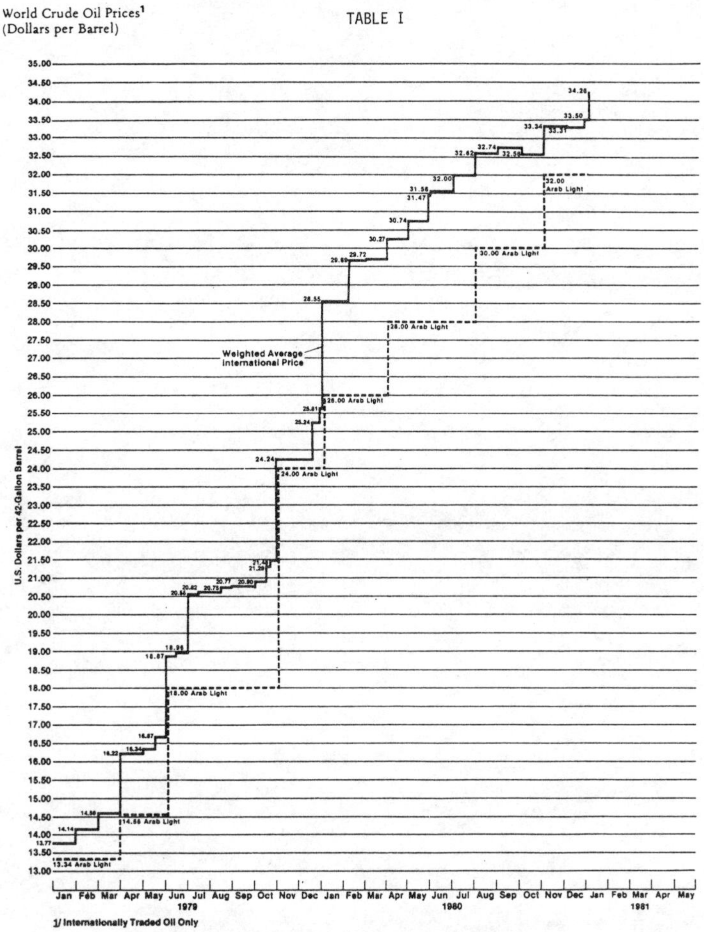

Energy Information Administration Weekly Petroleum Status Report, United States of America, Department of Energy

TABLE I. ENERGY INFORMATION ADMINISTRATION WEEKLY PETROLEUM STATUS REPORT

Energy User News

A Fairchild Business Newspaper

7 East 12 St., New York, NY 10003

ENERGY USER NEWS, MONDAY, FEBRUARY 23, 198

EUN PRICE OUTLOOK SURVEY

Each month, EUN asks some energy users to offer their best percentage estimates of price increases likely to be encountered in the coming 12 months. This month's survey covers expectations for the year ending January 1982.

	Next 12 Mos.	Past 12 Mos.
Natural gas...........	+29%	+31%
Electricity...........	+18%	+18%
Distillate oil.........	+30%	+26%
Residual oil..........	+30%	+27%
Steam coal...........	+10%	+7%

TABLE II. EUN PRICE OUTLOOK SURVEY

Chapter 25

A TEN YEAR OVERVIEW OF CONDENSER ON-LOAD BRUSH CLEANING SYSTEMS

M. S. Gengler

Fouling of heat transfer surfaces has been termed by the Heat Transfer Research Institute as the major unresolved problem in heat transfer. By the application of a unique automatic condenser tube brush cleaning system, the problem of fouling in the heat exchangers of refrigeration machines can now be considered resolved. With over 1,000 installations of Condenser ON-LOAD Brush Cleaning Systems, a broad base of information has been developed to assist the Energy Engineer in reducing the operating cost of centrifugal refrigeration equipment.

In order to best understand the contribution made by the Automatic ON-LOAD Brush Cleaning System, it is important to have a basic knowledge of the problem it answers.

FOULING OF HEAT TRANSFER SURFACES

Because of the complexity of fouling of heat transfer surfaces, it is often a little-understood and neglected subject. Still the importance of fouling and heat transfer cannot be ignored, the designer and operator of plant equipment must have a knowledge of the major characteristics of fouling and its impact on operations.

In general, there are six types of fouling experienced in heat transfer equipment. First is crystallization, the most common type of fouling which occurs in many process streams, particularly cooling tower water. Often interconnected with crystallization is sedimentation which usually is caused by deposits of particulate matter, such as clay, sand or rust. Polymerization is a chemical reaction. Coking occurs on high temperature surfaces and is a result of hydrocarbon deposits. Biological fouling is often complexed with crystallization and sedimentation but may occur upon its own. Finally, corrosion and corrosion by-products can form a layer of fouling quite similar to the fouling experienced in sedimentation.

Of the six previously-listed categories of fouling, we find that four categories - crystallization, sedimentation, polymerization and biological fouling - are often the most important types of fouling because of their ability to form complex compounds with each other. In order to understand these complex compounds, we must first be able to understand the formation of each of these types of foulants. Crystallization fouling takes place when a material, such as calcium carbonate or calcium sulfate, reaches its limit of solubility in the water and comes out in a pure crystalline compound on the heat transfer surface. Sedimentation fouling is caused by the oversaturation of suspended matter within the water causing particulate matter, such as clay and sand, to settle out on the heat transfer surfaces. This is quite often aggravated by the formation of crystallization fouling or crystal growth on the heat transfer surface. Polymerization fouling is a chemical reaction whereby a material carried within the water, sometimes a constituent of the water or a material added as part of the water treatment program, reacts due to the temperature increase of the heat transfer surface and its contact with a reactant. This reactant may be oxygen or it may be something on the order of automobile or plant emissions absorbed into the water stream through a cooling tower. This type of fouling provides a more sticky-type surface which can assist in the bonding of other types of fouling. Biological fouling, or the growth of organic material on the heat transfer surface, is a rather unpredictable type of fouling and most often appears on a random basis. Biological fouling also exhibits the tendency to provide a more sticky surface with which to bond other foulants. This sticky surface tends to aggravate all the other types of foulants that may be present within the heat transfer medium.

From Figure 1 (see next page), we can see that the build up of the fouling resistance is a function of time. Initially, fouling is zero. When we reach the value expressed by the point of Curve A, the rate of deposition is equal to the rate of removal. However, not all fouling levels off. Curve B shows that we do not reach an asymptotic point at anytime within the frame of this figure and that the fouling progresses as a lineal-type of function. Curve A represents the type of fouling resistance that would be expected with sedimentation fouling and polymerization fouling. Curve B, on the other hand, represents the type of fouling resistance that would be expected with pure crystallization fouling and biological fouling. Depending on the severity of the complexing of the four types of fouling previously mentioned, Curve A would tend to shift toward Curve B and form an almost perfect straight line of continued build up of thermal resistance. When we reach Curve B, we find that heat transfer is no longer acceptable and the surfaces must be cleaned.

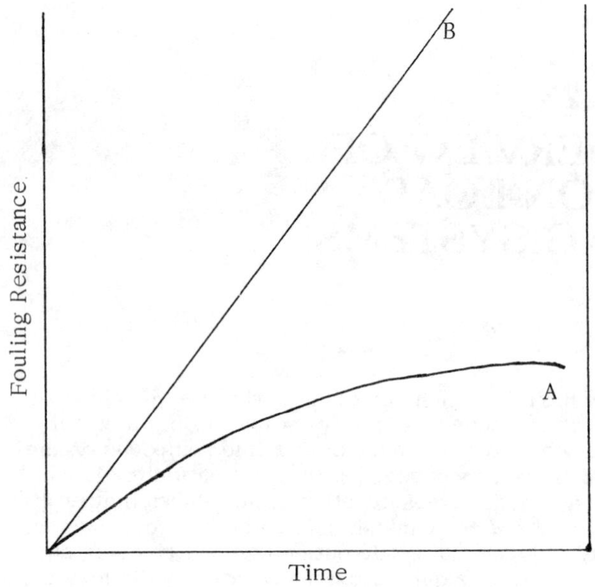

FIGURE 1. BUILD UP OF FOULING RESISTANCE.[1]

EFFECT OF FOULING ON REFRIGERATION MACHINE CONDENSERS

Armed with the basic knowledge of what fouling is and how it takes place, we can now look at its effect upon heat transfer. Foulants on the heat transfer surface act as an insulator; as foulants increase in thickness, the resistance to heat transfer is also increased. The higher the thermal resistance, the lower the heat flow is and, consequently, a lower amount of work is done.

The effect of fouling on heat transfer in a large centrifugal refrigeration machine is shown in Figure 2. From Figure 2, we can see that as fouling or thermal resistance increases from a clean tube to a 0.001, the overall heat transfer coefficient drops from 850 BTUs per hour, per square foot, per degree Fahrenheit temperature difference across the tube to 460 BTUs per hour, per square foot, per degree Fahrenheit temperature differential across the tube. It is interesting to note that the increase in thermal resistance from clean tube to 0.001 has given us a 45% decrease in the ability to transfer heat, caused by the build up of a thickness of scale of approximately 0.012 inches, about the thickness of this paper. In order to offset the build up of this minute amount of scale, we would have to increase the heat transfer surface of this refrigeration machine by 85%. It would be quite impractical to either build in a large amount of excess surface or add surface as scaling takes place; therefore, we must pay the penalty of decreased heat transfer across the tube.

Fouling Thermal Resistance (hr) (ft.^{2}) (°F temp diff)/Btu*	Over-all Heat Transfer Coefficient** Btu/(hr) (ft.^{2}) (°F.temp diff)*		Thickness of Scale*** Approx. (in.)
	Cooler	Condenser	
Clean tubes	400	850	.000
0.0005	333	595	.006
0.001	286	460	.012
0.002	222	315	.024
0.003	182	240	.036

***Assume a mean value for the thermal conductivity of the scale of 1.0 Btu/(hr) (ft.^{2}) (°F/ft.)

**The over-all heat transfer coefficients selected for this illustration are typical for a flooded cooler and a water-cooled refrigerant condenser. However, because it is possible to have different over-all heat transfer coefficients depending on the systems, the effect on the over-all heat transfer by the scale will vary.

*Sq. ft. of inside surface of tube in heat exchanger.

FIGURE 2. HEAT TRANSFER COEFFICIENT VS. FOULING THICKNESS.[1]

Because we do not have the ability to increase the surface area in the refrigeration machine, the temperature differential across the tube wall has to be increased in order to transfer the same amount of heat. The increased temperature differential (or driving force) is accomplished by running with a higher gas temperature on the outside of the tube. The increased gas temperature is the result of a higher lift across the compressor; and, therefore, more work must be accomplished by the compressor.

Figure 3 illustrates the effect of increased condenser fouling on compressor horsepower. It can be noted that this is a linear function, progressing at the rate of 1.1% horsepower increase for every increase of .0001 in fouling factor.

Figure 4 illustrates the effect of fouling on condensing temperature. This again is a linear function that shows why the increase in power in Figure 3 takes place.

FIGURE 3. EFFECT OF FOULING ON COMPRESSOR HORSEPOWER WITH CLEAN COOLER TUBES.[1]

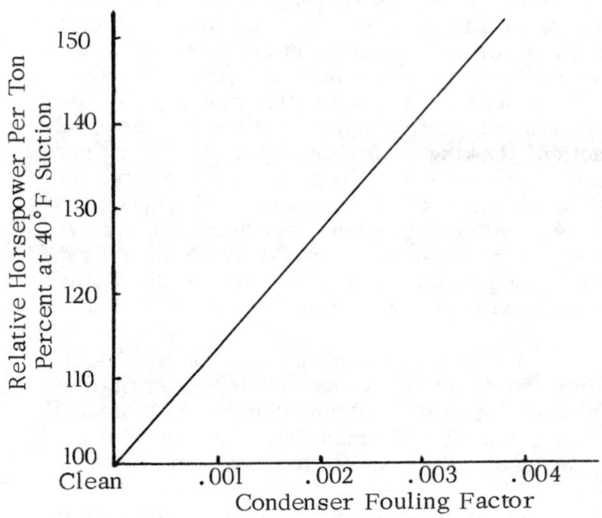

FIGURE 4. EFFECT OF FOULING ON CONDENSING TEMPERATURE [1]

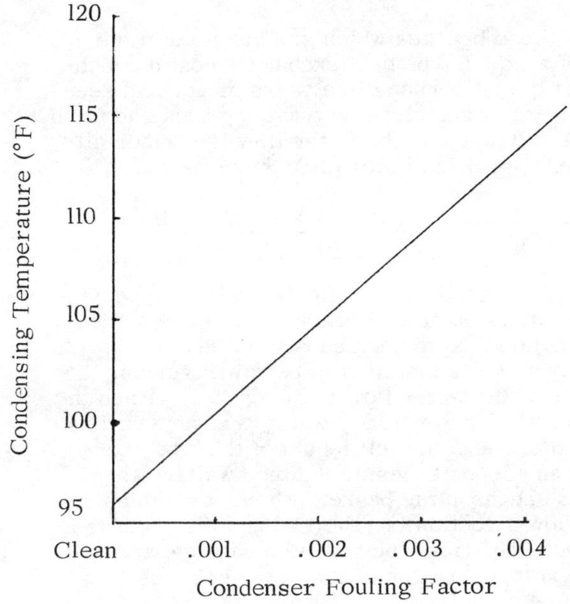

FIGURE 5. EFFECT OF CONDENSER FOULING CENTRIFUGAL REFRIG. MACHINE[1]

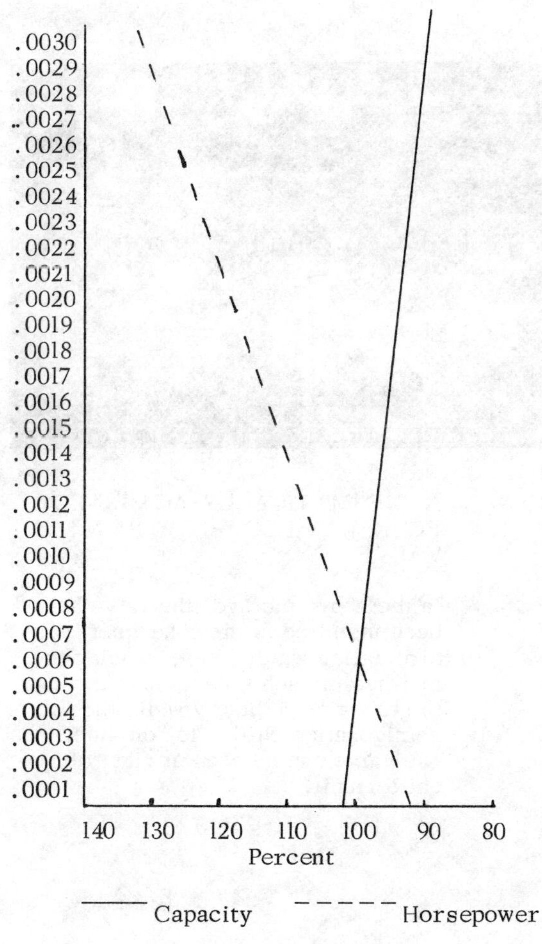

Figure 5 portrays the combined effect of horsepower increase and capacity loss due to condenser fouling within a centrifugal refrigeration machine. This figure is a compilation of Figure 3, Figure 4 and machine design ratings for a typical centrifugal refrigeration machine. It is evident that we must pay a dual penalty for condenser fouling. The first penalty is the increased horsepower, and the second penalty is capacity loss. Capacity loss takes place because we must run the compressor beyond its normal lift efficiencies and the fact that we are returning a warmer liquid refrigerant to the evaporator.

Of great importance to the Energy Professional is the need to know how long it takes normal fouling to reach a point of significant impact. The following is a figure showing actual in-service heat transfer data. This data was developed from a typical cooling tower installation. It can be seen that reduction in heat transfer due to fouling is a rather rapid process: in this case, a 24-day period resulted in approximately a 30% loss in thermal conductivity.

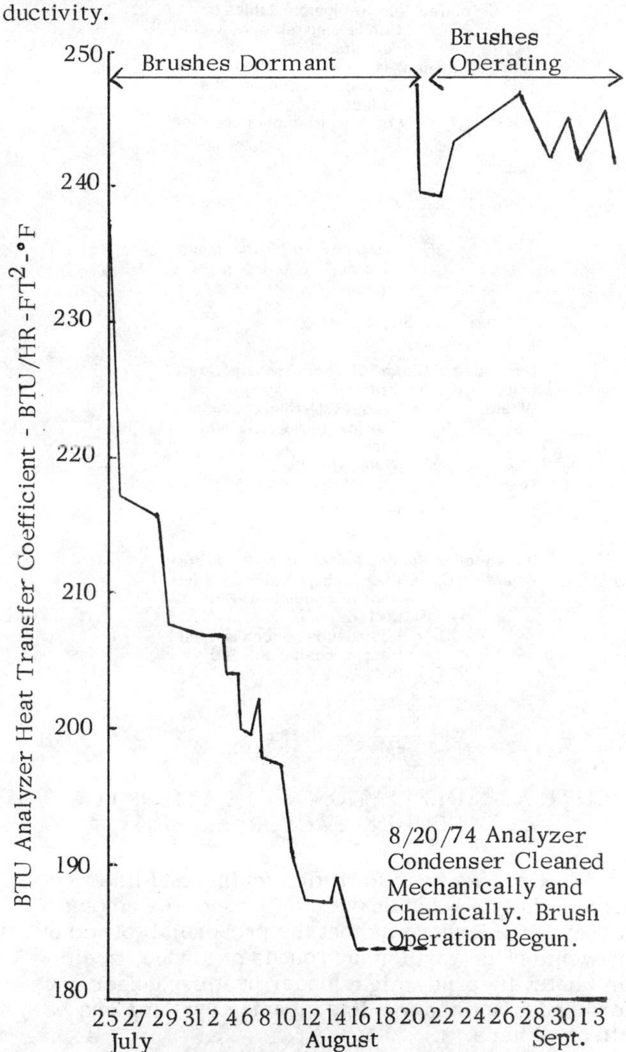

FIGURE 6. REFRIGERATION MACHINE CONDENSER TUBE FOULING STUDY BY BTU ANALYZER-Heat Transfer Coeff. /Time [4]

Having identified the major types of fouling and illustrated their effect on refrigeration condensers, we can now look at a practical method for determining in-service fouling factors.

DETERMINATION OF IN-SERVICE FOULING FACTORS

The determination of actual in-service fouling factors is of itself not a difficult task. As we can see by the following equations, fouling factor is a derivation of the overall heat transfer coefficient for steady state conduction.

Derivation of Heat Transfer Coefficient - U, Btu/hr-ft²-°F

Where: Data acquired is:

T_1 = Entering condenser water temperature, °F

T_2 = Leaving condenser water temperature, °F

q = Condenser water flow rate, gpm

P_c = Condensing refrigerant pressure, psig

Pressure/temperature conversion acquired from refrigerant tables is:

T_c = Condensing refrigerant temperature, °F

Constant is:

A = Outside condenser tube surface area, ft²

Then: ΔT_m = Log mean temperature difference, °F

$$= \frac{T_2 - T_1}{\log_e \dfrac{T_c - T_1}{T_c - T_2}}$$

Q = Heat rejected by condenser, Btu/hr

$$= q(T_2 - T_1) \times 500$$

$$U = \frac{Q}{A\Delta T_m} \text{ Btu/hr-ft}^2\text{-°F}$$

Derivation of Clean Tube Heat Transfer Coefficient - U_c, Btu/hr-ft²-°F

Where: h_i = Inside (water) film coefficient

h_o = Outside (refrigerant film coefficient

v = Water velocity

Then:

$$\frac{1}{U_c} = \frac{1}{h_i} + \frac{1}{h_o}$$

Derivation of Fouling Factor - ff, hr-ft²-°F/Btu

Where: U_c = Clean tube condenser heat transfer coefficient, Btu/hr-ft

U_f = Fouled tube condenser heat transfer coefficient, Btu/hr-ft

Then:

$$ff = \frac{\dfrac{1}{U_f} - \dfrac{1}{U_c}}{SR} \text{ hr-ft}^2\text{-°F/Btu}$$

FIGURE 7. DERIVATIONS & EQUATIONS FOR FOULING FACTOR.[2]

Once we have identified fouling and its effects, what do we do about it? "Consensus among forum participants was that the proverbial 'ounce of prevention' is worth many pounds of sludge, scale, or whatever 'crud' might hinder heat exchanger performance, and that on-line cleaning systems can be effective here.

The best approach? Most members with experience endorsed the use of brushes.

These brushes which are installed in the tubes of a liquid-type heat exchanger cost a couple of dollars, last perhaps five years or so, and keep the heat exchanger clean by traveling back and forth through the tubes as the water flow is periodically reversed, specifically for cleaning purposes."[3]

AUTOMATIC ON-LOAD BRUSH CLEANING SYSTEM

The system is simplicity itself. A special nylon brush is inserted in each tube, and a catch basket is attached to each end of the tube. A special four-way valve is installed to permit reversing the direction of the water flow in the chiller. When the flow direction is reversed, water pressure pushes the brush through the full length of the tube, removing any deposits inside the tube walls. The brush is caught in the basket, where it remains until the flow direction is reversed and the process is repeated. The frequency of the cleaning cycle will depend on the condition of the water being used.

The system has proven to be practical, dependable, efficient and usually pays for itself in savings in 6 to 24 months.

FIGURE 8. A PNEUMATICALLY-ACTUATED FOUR-WAY FLOW-REVERSING VALVE.[5]

In the above picture, the valve has been installed in the condenser circulating water lines, causing the brushes in each tube to pass through the tubes periodically while the refrigeration chiller is "on-line." The valve can be seen at right of chiller unit.

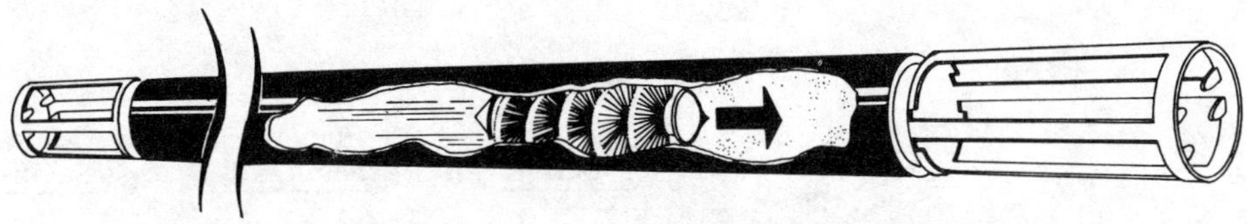

FIGURE 9. ONE BRUSH WITH TWO BASKETS FOR EACH CONDENSER TUBE. [4]

FIGURE 10. CASE HISTORIES - FOULING FACTORS AND ENERGY REDUCTION - COOLING TOWER WATER. [5]

Plant Location	Unit Chiller Capacity-Tons	Fouling Factor Automatic Tube Cleaning Without	With	Energy Reduction By Test °F
Guelph, Ontario	2,000	N/A	N/A	7.4
San Antonio, Texas	2,600	.00097	.00029	8.76
Tucson, Arizona	255	.00136	.00018	13.0
Chicago, Illinois	1,200	N/A	.00039	N/A
Oklahoma	645	N/A	.0001	N/A
New York, N.Y.	2,500	.00245	.00020	25.0
Washington, D.C.	3,000	.00175	.00047	16.0-19.0
North Carolina	2,000	.0018	.0002	26.0

FIGURE 11. CASE HISTORIES - FOULING FACTORS AND ENERGY REDUCTION - RIVER WATER. [5]

Plant Location	Unit Chiller Capacity-Tons	Fouling Factor Automatic Tube Cleaning Without	With	Energy Reduction By Test - %
Alabama	2,000	N/A	N/A	33
Virginia	2,500	.00188	.00026	28
South Carolina	1,175	.00131	.0005	15
New York	1,190	N/A	.00005	15
Virginia	300	.00308	.00052	28.2

Figures 10 and 11 depict a number of typical experiences in energy reduction through the use of ON-LOAD Brush Cleaning equipment. Although this information is quite dramatic, it does not demonstrate the true financial impact of the W.S.A. ON-LOAD Brush Cleaning System. Let us look at one particular case, the Tucson International Airport. In 1974, W.S.A.'s Automatic Brush Cleaning System was installed on a centrifugal and an absorption chiller at the Tucson International Airport. As of May 1980, these systems have yielded $61,400 in savings versus $11,400 in investment in six years. The figure on the right graphically depicts the true benefits of the ON-LOAD Brush Cleaning System in this case.

FIGURE 12. ENERGY SAVINGS COMPUTATION. 285 Ton Centrifugal Chiller -

Year	Hrs/Yr	$/kwh	Energy Savings(1)	Maint. Savings	Total Savings
1974*	1744	.025	$1,700	$1,000	$2,700
1975	2080	.031	2,510	1,000	3,510
1976	2400	.040	3,740	1,000	4,740
1977	2288	.042	3,750	1,000	4,750
1978	1200	.044	2,060	1,000	3,060
1979	1200	.048	2,250	1,000	3,250
1980*	720	.051	1,477	800	2,277
			$17,487	$6,800	$24,287

*Partial year

(1) Energy Savings = 17 1/2 % based on actual tests before and after installation of the W.S.A. ON-LOAD System.

FIGURE 12. (CONT.)

350 Ton Absorption Chiller -

Year	Hrs/Yr	Avg Load %	Gas Cost $/100ft^3	Energy Savings (1)	Maint. Savings	Total Savings
1974*	3100	76	.07	$1,450	$1,000	$2,450
1975	3100	76	.11	2,280	1,000	3,280
1976	3324	76	.14	3,120	1,000	4,120
1977	3180	73	.18	3,681	1,000	4,681
1978	3756	86	.21	5,970	1,000	6,970
1979	4380	100	.25	9,640	1,000	10,640
1980*	2736	62	.28	4,180	800	4,980
				$30,321	$6,800	$37,121

*Partial year

(1) Energy Savings = 12% based on actual tests before and after installation of the W.S.A. ON-LOAD System.

Relying on a wealth of experience, it is now possible to confidently maintain the highest possible performance on refrigeration chiller condensers and other heat exchangers. We no longer need to suffer equipment inefficiencies caused by fouling. Through the use of the ON-LOAD Brush Cleaning System, we can quarantee the highest possible cost effectiveness in the operation of heat transfer equipment.

REFERENCES

1. Carrier Design Manual. Carrier Air Conditioning Co., Syracuse, New York.

2. "Energy Saving System Bought with Actual Savings." Jan./Feb., 1978. Plant Energy Management, Vol. 2, No. 1.

3. "Forum Agrees: Liquid Heat Exchanger Fouling May Be Difficult to Determine But Is Also Easy to Prevent." July 30, 1979. Air Conditioning, Heating & Refrigeration News.

4. Hicks, Neal G. and Frederick Rubel, Jr. Feb./March, 1975. "New Condenser Cleaning System Makes Large Dent in Energy Usage." Application Engineering.

5. Leitner, Gordon F. Feb., 1980. "Controlling Chiller Tube Fouling." Ashrae Journal, pp. 40-43.

Chapter 26

THE GLO-KLEN SYSTEM AND ITS APPLICATION FOR ENERGY EFFICIENCY AND POLLUTION CONTROL

N. D. Cox, C. L. Sallman

Today we are going to discuss chimneys, stacks, and draft, and how they can be improved using new technology; or, more specifically, how they can be improved using a new application of old technology.

We're going to present a number of case histories which show how a new fuel-saving system based on this technology is cutting energy costs by an average of 14 percent for industrial plants, apartment and commercial buildings, schools and hospitals throughout the United States and Western Europe.

This is the patented Glo-Klen Thermal Chimney Fuel Saving System, manufactured and distributed by JEFCO LABORATORIES, INC. of Chicago, Illinois. It's a low-cost installation that, in most cases, is paying for itself in less than one year.

Natural gas, oil and coal burning heating plants are using the new Glo-Klen System which is based on two basic, but often overlooked principles of science:

1. Gas temperature variation and its effect on buoyancy.

2. Catalytic combustion.

Briefly, it's a two-part system:

1. A patented Thermal Chimney, consisting of a flat plate, with a properly sized and centered orifice, positioned on the top of the stack. This fosters more uniform draft conditions in the stack. (See Figure 1.)

2. The Glo-Klen Combustioneer, which performs its function efficiently when uniform combustion conditions are maintained by the Thermal Chimney. Like any catalyst in the presence of a chemical reaction, Glo-Klen Combustioneer shifts the chemical reaction to produce more of the desired product yield - in this case, heat. (See Figure 2.)

But before we launch into a technical explanation of how the Glo-Klen System works, we need to define the terms. At least, let me give you my definitions of terms such as chimney, stack, and draft, so that we can start from the same point of reference. All of these terms are vitally important in an understanding of the Glo-Klen System.

A DEFINITION OF TERMS

A stack is a vertical conduit, or duct, for the conveyance of the exhaust gases of combustion. It serves a two-fold purpose:

a) it delivers the products of combustion and fly ash to a high altitude, sufficient to dilute the polluting effects of the air at ground level;
b) it produces the draft necessary to cause air flow into the furnace. (Current popular theory indicates this is not the case in forced draft firing; however, we will discuss this premise later in our presentation.)

Figure 1. The Patented Glo-Klen Thermal Chimney Cap.

Draft

As for the definition of draft, the gases within the chimney are at a higher temperature than that of the surrounding air. The weight of a column of hot gases in the chimney is less than that of a column of air at outside temperature. This situation sets up the condition known as "draft," or chimney action. Draft is the differential pressure that causes the gases to move. The hot gases in the chimney are light, and they rise. The heavy outside air enters the combustion chamber or damper, to replace existing gases. This draft can be approximated by the following formula:

Formula for Draft

$$D = 0.52H \times P\left(\frac{1}{T_a} - \frac{1}{T_s}\right)$$

where D = draft at base of chimney, in. of water

H = height, ft., of chimney above point where draft measurement is taken.

P = atmosphere pressure, psia (can be taken as 14.7 up to elevation of 1,500 ft. above sea level)

T_a = outside temperature, absolute °F (°F+460°)

T_s = gas temperature in chimney, absolute °F (°F+460°)

This formula represents the static, or theoretical draft. When gases flow, it will be reduced, owing to friction. The amount of reduction will depend upon the size of the chimney, upon the quantity of gases, and, to some extent, upon the material used in the construction of the chimney. The actual available draft may be more realistically considered at 80 percent of the calculated static draft.

Whereas the height of a chimney determines the amount of draft it will produce, the cross-sectional area limits the volume of gases it can successfully discharge. The quantity of gases that a chimney must expell depends upon the amount of fuel burned, and the amount of air used for combustion. Air leaks in the boiler setting, breeching, etc., dilute the chimney gases, increase the volume, lower the temperature, and thus lower the effective draft.

The above information was taken from Steam Plant Operations, written by Woodruff and Lammers in 1967,[1] and is currently regarded as a reliable source of design data.

Expansion of the Formula Interpretation

You will note that the formula makes no allowance for the difference in gas composition inside and outside the chimney. And while this may be a reasonable assumption when discussing products of combustion from burning coal, the hydrogen content in oil and gas (especially gas), and the excess air requirements, suggest a new look into the logic of this assumption.

When hydrogen burns, the product of combustion is water. The more water vapor you put into the flue gas, the lower the dew point. Should the dew point be reached in the stack, the loss of draft can be dramatic. Thus, the importance of the gas composition and its "density" cannot be underestimated. Wet stacks and their associated draft problem plague many boiler operators throughout the country. If you have a wet stack, you had better pay close attention to what we are saying here today.

HOW CAN WE CORRECT THE WET STACK PROBLEM?

The wet stack can be eliminated if we can keep the stack gases above the dew point temperature. What are the factors that work against us in our effort to maintain elevated temperatures in the stack? One is the quantity of gases that a chimney must expel. This depends upon the amount of fuel burned and the amount of air used for combustion.

Air leaks in the boiler setting, breeching, etc., can dilute the chimney gases, increase the volume, lower the temperature, and thus lower the effective draft. Another problem is over-sized chimneys which slow velocity and extend cooling time of the combustion gases in the stack. This excessive cooling creates an increase in specific weight, a reduced draft, and a build-up of flue pressures. This back pressure creates a condition of poor combustion, and wastes fuel.

To digress a moment, let's consider one important reason why your chimney may be over-sized. This is the result when a heating plant is converted from coal to oil or gas - or from oil to gas, without modifying the stack. Gas and oil require less excess air than coal. The volume of air that your chimney must handle, therefore, is reduced when a heating plant is converted.

Figure 2. A Fully Automatic Glo-Klen Combustion Feeder.

Over-sized chimneys can create an over-cooling situation that becomes severe enough for condensate to form in your stack. If you have a wet stack, you are wasting fuel.

One way to solve an overcooling problem (which is most obvious when you have a wet chimney) is to increase your excess air. But that is a costly, fuel-wasting answer. True, an increase in excess air will reduce the chance of saturating the flue gases, and thus lower the dew point and reduce condensation. But, while excess air reduces flame temperatures, it elevates boiler exit temperatures, contrary to what might be supposed, because of poorer heat transfer with higher gas velocities.

A far better answer for keeping stack temperatures up, reducing pulsation and improving draft, is a "Thermal Chimney."

Exhaust gases are cooling and densifying from the time they leave the combustion zone until the time they become completely diluted in the atmosphere. As they densify they occupy less space per unit weight. Upon entering the stack they cool rapidly and start falling back onto the gases that follow. This eddying produces a turbulence in the stack and a non-uniformity of pressure in the exhaust duct system. This impulse or cyclic pressure change not only occurs in the stack but effects the entire heating system, all the way back to and through the burners. Even the fans and air entries into the system are affected. As exhaust pressure increases, the volume of air coming into the combustion chamber decreases. Fuel input, on the other hand, remains nearly constant. Thus in nearly all combustion systems we have a pulsing or cyclic variation of the fuel and air mixture at the point of ignition. This contributes to incomplete combustion! Energy-inefficient combustion! Pollution-causing combustion!

To overcome this pulsation, we have developed the patented Thermal Chimney Cap, a fabricated device that is placed on the top of your stack. The Thermal Chimney Cap produces a trapped blanket of insulating gases against the stack wall. This insulating blanket keeps the gases from cooling. It creates a duct with a frictionless surface, through which the hot exhaust gases can pass, and transforms turbulent flow into smooth laminar flow. The end result, back in the combustion chamber, is the elimination of the cyclic variation of the fuel-air mixture. A uniform combustion condition is maintained. Stack temperatures stay well above dew point, and draft is stabilized.

The draft design criteria of the "Thermal Chimney" do not contradict what has been developed in the past, but emphasize the gas density importance, and provide a means of salvaging costly stacks, which were designed and erected either with inadequate information or for purposes no longer appropriate for today's applications. (You would be surprised to discover how many recently constructed stacks are over-sized.)

Thus, your chimney, with a cap installed, becomes an insulated expansion chamber that provides a number of dollar-saving functions:

1. Condensate no longer forms in the stack to cause corrosion and hasten the necessity for replacement.

2. Particulate matter, seeking the cold surfaces of the stack's outer wall, gets trapped in the dead air space of the insulating wall, and falls to the base of the stack.

3. Smooth exiting of the combustion gases reduces back pressures. Combustion gases no longer recycle into the flame. Boiler vibration and pulsating diminishes even at high firing rates. All contribute to fuel savings.

What is more important, the improvements can be demonstrated in forced-draft boilers as well. Pushing air into the boiler at the burner forces the combustion chamber to pressurize in order to overcome the impediments downstream. The "Thermal Chimney" reduces that impediment by pumping the exhaust gases out with buoyant force.

BENJAMIN FRANKLIN AND "THE THERMAL CHIMNEY" CONCEPT

I am certain that many of you here today have received my presentation with skepticism. That does not come as a shock, for technological change surrounding stack design has not come easy.

Even 200 years ago, the "Thermal Chimney" concept had its proponents. Benjamin Franklin, in his "Observations on Smoky Chimneys,"[2] wrote his remedy for the chimney which was too large, so that it permitted smoke to back down into the building, by stating:

"As different circumstances frequently mix themselves in these matters, it is difficult to give precise dimensions for the openings of all chimneys. Our fathers made them (chimneys) generally much too large; we have lessened them; but they are often still of greater dimension that they should be, the human eye not being easily reconciled to sudden and great changes. If you suspect that your chimney smokes from the too great dimension of its opening, contract it by placing removable boards so as to lower and narrow it gradually, till you find the smoke no longer enters the room."

Mr. Franklin did not elaborate on his reasoning, and did not refer to gas density, draft, or dampers, but the beneficial aspects of his theories could be demonstrated. And so can those of the "Thermal Chimney." We would like the opportunity to demonstrate them at your power plant.

WHEN IS A "THERMAL CHIMNEY" NEEDED?

Several Criteria point to the need for a "Thermal Chimney."

1. If you have converted from coal to oil or gas; or from oil to gas without modifying your stack.
2. If you have a wet chimney.
3. If the smoke rises lazily from the chimney top, or pulsates irregularly.
4. Even if you have none of these visible problems, you may need a "Thermal Chimney." A JEFCO LABORATORIES Combustion Consultant is available to survey your boiler facilities.

Once a chimney cap with a properly-sized orifice is installed, it will pull the flue gases smoothly and uniformly from the stack and combustion chamber, and pull the air smoothly through the fuel at the burner. The result will be smoother, more efficient burning. No longer will the smoke rise lazily from the chimney top, or pulsate irregularly. You will have cooler breeching and warmer, more buoyant, exiting flue gases.

The fuel savings can be remarkable.

CASE HISTORIES

Day Kimball Hospital, Putnam, Connecticut

Consider, for example, the installation of the Glo-Klen/Thermal Chimney Fuel Saving System at the Day Kimball Hospital in Putnam, Connecticut, in August, 1980. (See Figure 3.) Day Kimball realized a 17.8 percent fuel savings in the 1980 - 1981 winter heating season. Chief Engineer Charlie Morkis attributes this savings directly to the installation of the Glo-Klen System, which paid for itself in less than six months at Day Kimball.

FUEL CONSUMPTION REDUCED BY 4.36 GALLONS PER DEGREE DAY WITH THE GLO-KLEN SYSTEM

	BEFORE			AFTER	
1979/80	Degree Days	Gals.	1980/81	Degree Days	Gals.
Oct.'79	**505**	**13,142**	**Oct.'80**	**600**	**16,877**
Nov.'79	**668**	**20,971**	**Nov.'80**	**917**	**20,337**
Dec.'79	**1,060**	**26,192**	**Dec.'80**	**1,348**	**29,373**
Jan.'80	**1,290**	**27,789**	**Jan.'81**	**1,593**	**21,088**
Feb.'80	**1,301**	**24,794**	**Feb.'81**	**978**	**19,991**
Mar.'80	**1,008**	**28,943**	**Mar.'81**	**1,019**	**21,120**
	5,832	**141,831**		**6,455**	**128,786**
Above Figures Average 24.31 Gals. Per Degree Day			**Above Figures Average** 19.95 Gals. Per Degree Day		

Figure 3. **Consumption of Fuel Oil After the Installation of the Glo-Klen/Thermal Chimney System at Day Kimball Hospital in Putnam, Ct.**

Newberry Plaza Apartments, Chicago, Illinois

Similar figures resulted form the installation of the Glo-Klen System at the 52 - story Newberry Plaza apartment complex in Chicago's presigious Gold Coast area. (See Figure 4.) The system was installed there at the end of January, 1980. In its first month of operation, it brought a reduction of 14.5 percent in the building's fuel use, which represented a one-month dollar saving of $5,607.45.

Rust-Oleum Corporation, Evantson, Illinois

A recent article in Chemical Processing reported on the savings realized at the 14-acre Rust-Oleum Corporation plant in Evanston, when the stack cap/ combustion catalyst system was installed on each of the plant's three boiler stacks. According to the article, "Rust-Oleum has experienced an energy savings of 11.43 percent on a therm/degree day basis during the first year. This savings will vary with weather, production rate, and fuel cost. It is expected that the second year savings will be at least 18 percent."[3]

These are but three of over 100 installations of the Glo-Klen/Thermal Chimney Fuel Saving System. Others include the famed Chicago Museum of Science and Industry, the Chicago Stadium, and the United of America Building in Chicago, Illinois. With today's fuel prices, it is difficult to overlook the tremendous fuel savings potential of the "Thermal Chimney."

However, so far I have only spoken of the "Thermal Chimney" and its effect on flue-gas buoyancy, improved combustion, and fuel savings. Now, I will explain its interaction with Glo-Klen Combustioneer, and why the cap-and-catalyst, or the Glo-Klen/Thermal energy-saving system works.

WHAT IS GLO-KLEN?

As you may know, Glo-Klen Combustioneer has been around a long time and has produced fire-side miracles in the boiler room. When boilers were fired with coal, natural draft was used to pull air into the combustion chamber. Coal-firing presented many dirty clean-up problems to the engineer. At this time, Glo-Klen Combustioneer was truly the engineer's friend.

Glo-Klen's ash-modifying performance made boiler clean-up easier, and less frequent. The pungent odors that were always present when the boiler was opened at shut-down did not exist in boilers where Glo-Klen was used. Sooting and oil-coking did not exist as long as sufficient air and Glo-Klen was used. Fuel usage was efficient, and many larger coal consumers were grateful that Glo-Klen was available.

However, when fuels were changed to cleaner burning oil and gas, the boilers closed up and pressurized, and the clean-up problems eased. Glo-Klen Combustioneer more-or-less became "that great stuff we used to use." Fuels have changed, but Glo-Klen hasn't. It is still the same miracle worker it was twenty years ago.

If you have tried Glo-Klen in your present boiler and have been disappointed with the results, you probably need a "Thermal Chimney." You are undoubtedly recycling combustion gases into your flame, and inhibiting combustion and the catalyst effect. Glo-Klen cannot, by itself, correct this condition of poor combustion, but a Thermal Chimney can. As a catalyst, Glo-Klen shifts the reaction to provide more heat. But, even a catalyst must have the other necessary environmental factors present to do its job. That means the proper air (oxygen) and fuel mixture must be maintained. A catalyst is not a substitute for fuel or air.

Then, you might ask - what is Glo-Klen? Glo-Klen Combustioneer is a mixture of alkaline earths, oxides of alkali metals, kaolin and specific oxidizing materials that provide multifunctional benefits. It provides little or no heat in itself, having approximately 1000 BTU's per pound in caloric value.

When used as directed, Glo-Klen Combustioneer will provide:

1. More heat from your fuel than you now obtain.
2. Elimination of carbon or soot deposits within your combustion chamber, tubes, flues and stack.
3. An improvement in the heat transfer characteristics of your boiler in two ways:
   a. By removing or altering the slag deposits in your boiler;
   b. By changing the heat transfer effects of the combustion gases.
4. Elimination of noxious fumes created by products of incomplete combustion.

BENEFITS OF GLO-KLEN

Let us examine each benefit in detail.

More Heat From Your Fuel Than You Now Obtain.

Calorimetry tests conducted on various fuels have indicated that Glo-Klen consistently provides a 3% increase in heat release. This increase can be translated directly into fuel savings. A three percent fuel savings does not seem to be a significant amount, but if you purchase oil at $1.00 per gallon, a 3% savings amounts to 3¢ a gallon. How many gallons do you burn each year? If you burn a million gallons annually, you'll save $30,000.00 a year.

FROM FEBRUARY 28, 1980 TO MARCH 27, 1980, THE GLO-KLEN SYSTEM SAVED $5,607.45 AT THE NEWBERRY PLAZA APARTMENTS

Interval	Degree Days	Therms
Feb. 28-Mar. 27 '80	**969**	**97,568.5**

97,568.5 Therms
÷ 969 Degree Days
100.69 Therms/Degree Day

From Billing Data:

Standard	**117.71**	**Therms/Degree Days**
Actual	**100.69**	**Therms/Degree Days**
Difference	**17.02**	

17.02 ÷ 117.71 = 14.5 percent savings

At 34¢/Therm, this represents a dollar savings of $5,607.45 from February 28, 1980 to March 27, 1980.

Figure 4. **Consumption of Fuel Oil After the Installation of the Glo-Klen/Thermal Chimney System at Newberry Plaza Apartment Complex.**

Elmination of Carbon or Soot Deposited Within Your Combustion Chamber, Tubes, Flues and Stack.

Glo-Klen Combustioneer contains a particular ingredient that is capable of oxidizing carbon at a temperature considerably lower than normal ignition temperature. Therefore, soot and carbon deposits are literally burned off with a low temperature oxidizing action. What's more, the material is regenerative, in that it can, once reduced, reoxidize at a low temperature to react again with other carbon materials. Hence, a little goes a long way.

Those of you who have used Glo-Klen in coal-fired, natural-draft boilers, can probably remember seeing waves of soft blue flames dancing across the walls of the boiler as it cooled. This is characteristic of Glo-Klen's soot removing action. Glo-Klen turns soot and carbon deposits into usable heat.

An Improvement in the Heat Transfer Characteristics of Your Boiler.

By removing or altering the slag deposits: The alkaline metals have a characteristic that causes an alloying action to occur with slag formations both new and existing. That material which remains on the tubes and walls turns gray in color, softens to a cigar-like ash, and either falls off, or is easily removed. In either case, the result is an improved heat transfer efficiency, i.e., less fuel is required to maintain demand loads. This savings can be more substantial than the heat-release improvements, depending, of course, on the initial boiler conditions.

Clean-up maintenance is also minimized: No chipping or scraping should be required. The catalystic action also effects the coefficient of thermal expansion of the slag formation as well. When the deposits are thick, the alloying condition usually affects only the surface. However, when the slag cools, the alloyed surface usually shrinks more than the slag mass underneath, causing the slag mass to crack and pull away, like the clay bottom of a dry river bed. This phenomenon also eases the clean-up operations.

An Improvement in the Heat-Transfer Characteristics of Your Boiler.

By changing the heat transfer effects of the combustion gases: Imagine, if you can, looking down through the tube of a fire-tube boiler. Hot glowing gases fill the center of the tube as the gases travel toward you down the tube. A dark, or clear, halo surrounds the hot gases protecting the tube wall from the high temperatures. That halo may be 1/2 to 3/4 of an inch thick. Now, introduce a catalyst into the combustion chamber and observe the change. Almost immediately, the halo diminishes and the hot gases nearly fill the tube. What has happened? What caused the halo to almost disappear? What was the make-up of the halo to begin with? What did the Glo-Klen do to change the conditions?

This halo is a mixture of insulating gases that are made up of products of incomplete combustion: gases containing aldehydes, nitrous oxides, and other carbon-hydrogen-oxygen compounds that have not given up their heat. These gases have absorbed heat energy, but do not want to relinquish that heat to the tube walls.

Drawn to the tube walls by the thermal pull, these hot, partially oxidized gases impinge on the tube wall, and insulate the heat-transfer surfaces. Some of you have, no doubt, used turbulators to mechanically break down this layer of insulating gas. With Glo-Klen Combustioneer, you not only mechanically scrub off this layer of insulating gas, you also reduce the incidence of occurrence. The presence of Glo-Klen enhances complete combustion and minimizes the formation of partially oxidized hydrocarbons. The bottom line is more steam, with less fuel and cost.

Elimination of Noxious Fumes Created by Products of Incomplete Combustion.

Fuel molecules are made up of carbon and hydrogen atoms. Carbon, when burned completely, forms carbon dioxide. Carbon dioxide is odorless and non-toxic. Hydrogen, when burned completely, forms water. Neither product smells of poisons. If we burn our fuels completely, we can eliminate odor-causing toxic by-products.

Using Glo-Klen Combustioneer in combination with the "Thermal Chimney" provides conditions best suited to give complete combustion. JEFCO LABORATORIES, INC. can offer you fuel saving, labor saving, money saving and environmental saving advantages you can't find elsewhere.

Evidence has been presented that alkaline earth-kaolin mixtures are capable of increasing the ash deformation temperatures, absorbing SO_3, and reducing vanadium corrosion for residual oils. The ability to combat these three problem areas is most promising news to catalyst users. The chemistry is very involved, and more definitive information is needed.

Positive reports from operating personnel are available on the reduction of sulphurous attack on the respiratory system, skin, and clothing. As more precise instruments are added to the engineer's testing procedures, small improvements are being quantified.

One report comes from the well-known Hawthorne Works of Western Electric, in Cicero, Illinois, where the combustion catalyst has produced a 2% fuel savings on gas and gas-coal fired boilers, a significant figure when considering the total fuel bill.

Thank you for your time and attention. If you have any questions, I'll do my best to answer them now.

REFERENCES

1) Woodruff and Lammers, Steam Plant Operations (5th. Ed.), McGraw-Hill, 1967, Chap. 4, Articles 4 - 10.

2) Franklin, Benjamin, Observations on a Smoky Chimney, 1793, Pamphlet privately published in London, p. 15.

3) Olson, James W. and Parr, Gary, L., "Stack Cap/Catalyst System Cuts Gas Use 11.43%," Chemical Processing, May 1981.

Chapter 27
STRATEGICALLY STORED DISTILLATE FUEL AND ITS CHEMICAL MAINTENANCE

P. J. O'Connor

INTRODUCTION

Energy and risk are two different words, but in today's world they have become synonymous. The Middle East is as unsettled as ever, yet the world still relies upon this area for almost 50% of its needs. And even though the United States imports a much smaller percentage from this region, our mutual agreement with other western countries requires us to equally share in the world's burden during oil supply shortfalls. Thus, our exposure to risk of economic hardship is not measured by our import needs. Indeed, some analysts calculate that risk to the western world by the Middle East threat is much greater than the risk associated with the Russian threat to Europe.[1]

Many economic units, such as industries, commercial and institutional facilities, as well as utilities, have countered the risks associated with oil by switching their fuel to natural gas and by conservation. Yet, natural gas supply and price are not absolute and conservation cannot completely eliminate demand for energy.

To counter price fluctuation of energy (mainly oil), one may enter the futures market. But, the futures market's dampening effect is relatively short term and offers no solution to the true problem of energy demand nor to immediate need for energy during emergencies.

Purchasing and storing fuel on site, however, can play an important and, indeed, strategic role as a hedge against price fluctuation and energy supply shortfalls, whether short term or long term. This conclusion is reinforced by the growing number of facilities with fuel in dormant storage.[2]

Flexibility is key at any facility desiring to offset the risks associated with the energy markets. For back-up storage, distillate fuels ranging from #2 oil to kerosene offer the greatest flexibility. During emergencies they may be called upon for immediate response and may be used in a wide range of power generating equipment such as diesel engines, combustion turbines and steam generating boilers. In contrast, residual fuels have usefulness limited to steam generation and require preheating prior to use. However, even though distillate fuels offer greater flexibility, their maintenance is not effortless.

VALUE OF DISTILLATE FUEL

Oil in storage represents economic value, emergency value and/or strategic value. The objective of maintaining or protecting dormant fuel is the preservation of the value of the fuel. As long as a fuel maintains its value, the need for its storage is fulfilled. But, because distillate fuels deteriorate over time, their value will undoubtedly become impaired and, consequently, the purpose for their storage will not be fulfilled.

Plant operators are at the forefront in recognizing that distillate fuels deteriorate over time. They are the individuals confronted with filter pluggage during emergencies and corrosion problems in storage tanks. But concern should not stop at the operator level. Corporate planners should be aware that as a fuel deteriorates, its value decreases. Frequently, firms wish to sell a fuel back to the open market in order to generate cash. If the fuel has deteriorated, the economic value will be impaired and, subsequently, its price will be discounted. Thus, deterioration leads directly to illiquidity of the asset (fuel) and jeopardizes its strategic role.

In addition to the impact from increased importance of stored fuel, considerable concern about deterioration of dormant distillate fuel is growing because of new refining methods. Due to demand and pricing in the oil market, refineries are obtaining more and more distillate fuel from each barrel of crude oil. Sophisticated catalytic cracking methods are now being used to yield more distillate fuel.[3] However, these methods have brought to the market fuels that are less stable than the straight distillate fuels of the past. In response to these factors, government and industry have embarked on numerous studies covering the technology of distillate fuel deterioration.[4] But real problems still exist in having those individuals responsible for a dormant fuel recognize and apply the technology. Fuel maintenance is the burden of individuals at both the plant level and corporate level.

FUEL DEGRADATION

The predominant factor in the deterioration of dormant fuel is oxidation. Oxidation may occur either directly or through catalytic influences. The result of oxidation is increased particulate matter suspended in the fuel, increased gum accumulation, water formation, fuel discoloration and, albeit, minimal, BTU loss. All of these symptoms have one thing in common; they impair the value of fuel.

Oxidation is the process whereby oxygen chemically combines with the hydrocabon molecules that make up fuel. The oxidation process everyone is aware of is combustion. During combustion, oxygen will combine with carbon and hydrogen in the fuel and cause the release of energy as it does. In contrast to combustion, fuel oil will oxidize naturally without being ignited. Without ignition, oxidation is much slower than during combustion. The specific chemical mechanism of distillate fuel is debatable; however, the general mechanism is understood.[5,6]

The chemical reaction that fuel undergoes during oxidation may be seen and is explained in Exhibit I.

Exhibit I

FUEL OXIDATION CHAIN REACTION

1. $RH + O_2 \longrightarrow R + HOO$
2. $R + O_2 \longrightarrow ROO$
3. $RH + ROO \longrightarrow ROOH + R$
2. $R + O_2 \longrightarrow ROO$
3. $RH + ROO \longrightarrow ROOH + R$

Inhibition of Reaction

1. $InH + ROO \longrightarrow ROOH + In$
2. $R + In \longrightarrow RIn$

RH - Fuel Molecule; **R** - Free Radical; **O_2** - Oxygen;
ROO - Peroxy Radical; **HOO** - Peroxide Molecule;
ROOH - Hyperperoxide; **InH** - Oxidation Inhibitor;
In - Inhibitor Radical; **RIn** - Stabilize Fuel Radical

As fuel remains dormant, an oxidation chain reaction occurs. Oxidation Inhibitors can break this chain reaction. Metals may catalyze the reaction, however, special chelants can deactivate their role.

Metals that are suspended or solubilized in the fuel can act as catalysts and significantly influence the speed of the oxidation reaction. Most prevalent as catalysts are transition metals such as iron, copper and nickel, which may enter the fuel during refining, distribution or storage. Other factors affecting dormant fuel oxidation are temperature, pressure (tank head), exposure to oxygen, crude source and refining methods.

Corrosion is a second area for concern. Corrosion of a storage vessel does not directly deteriorate a fuel. Rather, it impairs the emergency value of a fuel by increasing the chance of filter pluggage while simultaneously destructing a tank. For fuels in emergency back-up status, corrosion can be very harmful. Since corrosion feeds metals to the fuel, it also plays an indirect role in increasing catalytic oxidation of the fuel.[8]

Micro-organism growth on the bottom of storage tanks is a third area of concern. Thriving in water at a water/oil interface, bacteria, fungi and other organisms literally eat the fuel and excrete acids and water. Agglomerating into a slime, micro-organisms have been well known to cause failures due to line, filter and burner tip pluggage; an obvious threat to the emergency readiness of any fuel.[9]

In a dormant storage tank, an environment exists in which each area of concern (oxidation, corrosion and micro-organisms), directly influences the other. To explain this relationship, let us look at the following example. When fuel oxidizes, it causes an increase in water on a tank's bottom. This water will improve the medium for micro-organisms which, in turn, will increase excretions of acids. Corrosion, therefore, will increase and feed more metals to the fuel, thereby, increasing catalyzation of the oxidation process. The increased oxidation brings us back to the beginning of the example. Diagram I lays out some relationships among factors, however, little technical data has been generated to quantify the specific relationships.

FUEL ANALYSIS

It should be reiterated that even though chemical deterioration of a dormant fuel is the cause, the impaired value of the fuel is the prime concern. Therefore, the objective of any program designed to maintain a dormant fuel should be to preserve the value of the oil. To accomplish this objective, it is necessary to have a total approach in the fuel's maintenance.

An analysis of the present status and the prediction of the fuel's ability to withstand storage begin with proper sampling of a fuel from its storage tank. To recognize the condition of an entire volume of fuel, it is recommended that two samples be taken; one from the bottom of the tank and another from a few feet below the top of the fuel level. Proper sampling of a tank's bottom is very important. The best sampling devices to use in obtaining a bottom sample is either a Bradford or the Bacon Bomb Tank Thief. When used correctly, either tank thief will yield a good sample showing the depth of water accumulation, dirt and sediment, as well as stratification of the oil.

With two samples (top and bottom), laboratory analyses can be conducted to examine characteristics relevant to storage. Of primary concern are five characteristics: oxidation or aging, corrosion, suspended particulate matter, micro-organisms and water accumulation.

Numerous tests may be conducted to uncover fuel characteristics. However, cost and time consideration must be taken into account when deciding which test to conduct. For example, it would be difficult to justify $500.00 for tests on a sample taken from a fuel volume of only 1,000 gallons.[10] Also, a test that takes 3 months to conduct would be meaningless when results are desired immediately.

Five base-line tests that can be conducted quickly and inexpensively are (1) an accelerated aging test, (2) a NACE corrosion test, (3) a suspended particulate test, (4)

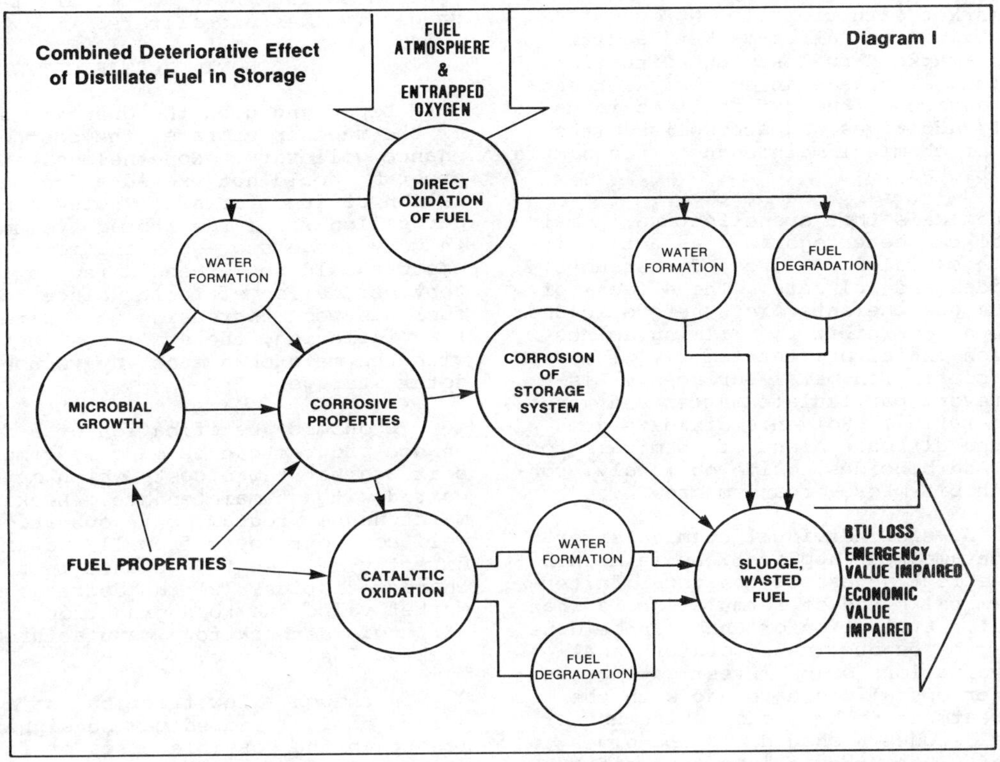

a toluene extract evaluation and (5) a water by distillation test.[11] Each test is a rough analysis which provides information at a minimal cost. More elaborate tests with strict ASTM guidelines may be conducted instead of the five basic tests; however, cost and time may prohibit their use. In addition, tests for ash, tars and asphaltenes, pour point, cloud point, flash point, as well as cetane, BTU value, metals and culture studies for micro-organisms may be conducted to provide total information on a fuel. The cost of conducting every available test, needless to say, would be prohibitive for most storage situations.

The five basic tests may not be ASTM standards, but can be conducted for between $100.00 and $200.00 and will provide basic information within a very short period of time. The evaluation of these tests requires an understanding of fuels and storage practices and is beyond the scope of this article. It is advisable, though, to have an outside subjective evaluation of the fuel based on the tests and storage situation. A laboratory aware of both testing procedures and real-life storage practices should be able to provide a good evaluation. Once an analysis and evaluation of a fuel is reported, it is highly advisable to keep the report on file. This will provide a base from which a maintenance program may be initiated and evaluated in the future.

CHEMICAL MAINTENANCE

With information of a fuel's characteristics in hand, it may be necessary to take steps to maintain the fuel's value. A fuel's maintenance can be divided into two categories, chemical maintenance and physical maintenance.

The chemical maintenance of a dormant fuel is a complex topic that can be very difficult to understand. It encompasses a wide range of organic chemistry from the point of view of the fuel and from the point of view of the chemical additive. However, it is only necessary to understand that chemical maintenance of dormant fuel must accomplish three fundamental objectives in a balanced and inexpensive approach; the inhibition of direct and catalytic oxidation, the inhibition of corrosion and the control of micro-organism growth.[12]

For years, single-purpose chemical additives have been used at the refinery and distribution level. But problems arose when using these chemicals for dormant storage or "after-market" applications. Applying some "pre-market" additives was dangerous because they precipitated when contacted by water or were not capable of dispersing through a dormant fuel. Therefore, new chemical methods were developed to overcome the disadvantages of applying "pre-market" additives to dormant or "after-market situations.[13] New chemical technology brings an "after-market" additive to an "after-market" fuel storage situation. The advantage is to have an economical treatment which overcomes the difficulties in using "pre-market" additives and accomplishes the objectives of chemical maintenance in a simple and compatible way.

The chemicals that are effective in maintaining fuel can be categorized as anti-oxidants, chelants, filming amines, dispersants, micro-biocides and solvents. The effects of antioxidants and chelants are synergistic and are described in Exhibit I. Filming amines are corrosion inhibitors that simply place a protective coating on metal surfaces. Dispersants prevent particulate matter and water from agglomerating. Solvents dissolve gums and carry the ultimate blend of chemicals to the fuel. And biocides, quite obviously, control and inhibit micro-organism growth.

Analyzing the individual chemicals for use in "after-market" applications is meaningless. The effectiveness of the total "after-market" chemical blend or formulation is what is important. Evaluation of this can be done in two ways; by conducting laboratory analysis and evaluations or by investigating the success other operations have had with the chemical treatment. When considering use of any chemical, concern should also be given to O.S.H.A. and E.P.A. regulations that affect each operation.[14]

Limitations on chemical treatment do exist. For example: although fuels may be treated to prevent problems, it is not possible to change the chemical structure of a poor fuel already in dormant storage to the chemical structure of a good fuel. Also, since some chemical agents are used up in the inhibition process, it is necessary to replenish them in the fuel. Thus, it is essential to supplement chemical maintenance of a fuel with physical maintenance.

PHYSICAL MAINTENANCE

One element of physical maintenance that has already been addressed is fuel sampling and analysis. Although chemical treatment slows down the change occurring within a fuel storage environment, it cannot stop it completely. Therefore, it is still necessary to monitor that change; fuel sampling and analysis does this.

If it can be engineered, tanks should have good bottom drains for removal of water formed from oxidation and humidity condensation. Water accumulations should be drained at least annually. In order to prevent undue water condensation inside a tank it is recommended that the tank be kept as full as possible. For large installations, nitrogen blanketing of tanks may be feasible.

Checking the readiness of a fuel system should be done on a quarterly basis. This allows operators to assure themselves of the readiness to utilize the fuel. Areas of concern during these checks are fuel lines, gauges, valves and filters.

ECONOMIC JUSTIFICATION

Depending upon the quantity and quality of the fuel in storage, the cost of maintenance will vary. Nonetheless, a maintenance program should not exceed a few pennies per gallon of fuel for small volumes nor a penny per gallon for large storage volumes. Indeed, in many large storage situations, maintenance costs should not exceed a few tenths of a cent per gallon of fuel. Since the value of fuel has been increasing at a rate faster than inflation, the small cost associated with the maintenance of any volume should not deter storage.

Economic justification of a fuel maintenance program can be made by comparing its cost against those costs which may be incurred without maintenance. Lacking a sound maintenance program, each operation may very well confront costs from (1) power failures directly related to a change in a fuel's characteristics, (2) the decrease in a fuel's market value due to degradation or (3) the strategic setback to a manufacturer losing a back-up fuel source.

Although a few fractions of a cent per gallon of fuel stored may be significantly less than the possible costs of fuel degradation, one needs to be confident that the program is actually preventing the larger costs. To do this, questions should be asked about the program. What chemicals are being used and what do they accomplish? What references are available to verify a chemical maintenance program's success? Are the fuel sample analyses meaningful? Is there a way to check a system to verify that it is functional?

SUMMARY

A total chemical maintenance program plays an important function in distillate fuel storage. But your perspective toward dormant fuel maintenance will depend upon your role. For corporate executives, a sound maintenance program protects a strategic support system that was implemented to remove risks associated with an uncertain energy market. For plant operators and engineers, a sound maintenance program will reduce the headaches caused by the failure of a system to work when it is needed.

REFERENCES

1. Energy and Security, "A report of Harvard's Energy and Security Research Project;" Deere, D., Nye, J., 1981 Ballinger Publishing Co.,Cambridge, Ma.

2. "Commercial and Industrial Storage of Distallate and Gasoline," D.O.E. Supplementary Report, Nov. 1980; 3168-23, 22 pgs., Congressional Information Service, Inc., Bethesda, Md.

3. "Oil Companies are Reshaping Their Refineries," Chemical Week, May 13, 1981, pp. 64-68

4, ASTM Committee D-2, "Distillate Fuel Stability and Cleanliness," June 24, 1980, American Society for Testing and Material, Philadelphia, Pa.

5. Offenhauser, R.D., et al, "Sediment Formation in Catalytically Cracked Distillate Fuel Oils," ACS Dir. Petrol. Chem. Preprints, Vol. 1, No. 3, Aug., 1956.

6. Saver, R.W., et al, "A Mechanism for Organic Sediment Formation in Heating Oils," ACS Div. Petrol. Chem. Preprints, Vol. 3, No. 3, Aug., 1958.

7. Elmquist, E. A., ASTM Spec. Tech. Pub., No. 244 entitled "Stability of Distillate Fuel Oils," Philadelphia, Pa., June, 1958.

8. Stavinoha, L., LePera, M., "Army Experience and Requirements for Stability and Cleanliness of Diesel Fuels," AFLRL No. 128, AD A-088008, U.S. Army Fuels and Lubricant Research, Southwest Research Institute, San Antonio, Tx., July, 1980.

9. Stavinoha, L., LePera, M., "Army Experience and Requirements for Stability and Cleanliness of Diesel Fuels," AFLRL No. 128, AD A-088008, U.S. Army Fuels and Lubricant Research, Southwest Research Institute, San Antonio, Tx., July, 1980.

10. One would reason that very few storage situations exist in which sampling and analysis would be desirable if it was equivalent to 50¢ per each gallon of fuel being stored.

11. a) Accelerated Aging Test - DuPont 300 F aging test.
b) National Association of Corrorion Engineer coupon corrosion test for petroleum products.
c) ASTM D-2275; suspended particulate matter.
d) Nutmeg Chemical Company test. Removal of petroleum base from micro-organism and subsequent identification.

12. Lodwick, J.R., "Chemical Additives in Petroleum Fuels: Some Uses and Action Mechanisms," Journ. Inst. Petrol., Vol. 50, Nov., 1964

13. Pending Patent, "Distillate Fuel Additive and Method for Improving Storage Capability of Distillate Fuel," Paul O'Connor, Nutmeg Chemical Company.

14. Under the Federal Insecticide and Rodenticide Act of 1976, all chemicals intended to be used as biocides must first be reviewed and registered by the E.P.A.

Chapter 28
HUMAN-MACHINE INTERFACE IN BUILDING AUTOMATION SYSTEMS

N. L. Sobczak

The focus of energy conservation and building monitoring and control has changed in recent years. Computerized building automation systems have been developed to conserve energy, maintain comfort, and protect life and property in non-residential buildings. The greater the demand on our energy resources, the greater the need for effective, efficient control of energy consumption. In order to incorporate new energy savings, complex computer control features are employed. As the complexity of these control functions increases, the need for sophisticated computer control systems becomes greater.

Often the more sophisticated the computer, the more sophisticated the operator must be. However, the operators who can most effectively guide the system into an economical and efficient mode of operation often have little or no previous knowledge of the operation of computers. Operators handling the equipment are often maintenance engineers whose experience and aptitudes are mechanical in nature.

Their job has been to keep a building running by manually adjusting damper controllers and switches. Now, they must ask the computer system to adjust the dampers and switches. The human-machine interface they will use must be a conceptual bridge to these operators. It must give them confidence, guidance, and understanding of their new "tools" and how they relate to their job. In order to answer these needs, the manufacturer of building automation systems must create a system with a humanistic approach.

What are the qualities that make up an effective human-machine interface? In a recent article entitled "Human Considerations in On-line Systems Design," Phillip Elam outlined the attributes of a good human-machine interface. One of the most ambiguous items addressed concerns the forces impacting user acceptance, i.e. what are the intangibles that make one system appear better to the user than another. More tangibles issues concerning system response and the hardware interface features are two other items of great concern to users. The type of operator language is one of the issues which relates to the system sensitivity users feel. Finally, the technique used for operators' input and for monitoring and contol type error handling greatly affects the way the users view their interface. All these issues can create the desired symbiosis between users andtheir systems. However, if the items mentioned here are not designed with users in mind, the results will create a negative effect.

In discussing these criteria a building automation system--the JC/85/40--whose design has responded to these operator needs, will be shown.

User Acceptance

The user acceptance of an on-line system can be related to a phenomenon called "subject-object role reversal". This means that operators, because of the way a system communicates with them, can feel that they are the object, and that the control--a normally human attribute--is a function of the computer system. This molding and manipulating by the system can adversely affect the way in which the operators view the system and thus effect the operators' utilization of the system. No matter how sophisticated the energy-saving capabilities of the system, it is of no value if people don't use it. Some of the factors which contribute to the lack of user acceptance are: the lack of "sympathy" on the part of the system, the "peephole" effect, the feeling that the system is "fragile;" and, after the novelty of the system wears off, the belief that it is not responding to the users' needs.

In general, operators may feel that the system lacks sympathy when it does not respond to their needs. Operators have different interface requirements dependent upon the their skill and the frequency of their access to the system. The JC/85/40 has responded to this situation by breaking up the entry method into three modes. The AID mode is used by an inexperienced operator or by an experienced operator who is learning a new function in the system. The DIRect mode is a shorthand entry method for anything that can be entered in the AID mode. The EDiT mode, which is used primarily for data base generation features, is a "fill-in-the-blank" technique which can provide lists to guide operators when they are "stuck" on a particular operator entry.

When operators use the AID mode, they are prompted for each entry field. They then select the next entry from the displayed list. In the DIRect mode, it is assumed that the operators are aware of the entry technique and options available. If operators get "stuck"while entering in the DIRect mode, typing a question mark (?) in the command string will cause the system to switch to the AID mode and display the list for the field containing the question mark.

At each step in the AID mode entry of a command, operators are given an explanation of what the entry is and of what choices are available to them. In each case, the previous entry will be the selection criteria for the next operator-entered parameter. When the command entry is complete and the system has verified the entry's accuracy, it responds with a positive-feedback mode. This serves two purposes: it allows operators one last look at what they are about to do and it shows operators how this command may be entered via the DIRect mode the next time the command is executed. Some of the fields are optional and may be omitted. This allows operators to get a more general view of the contents of the system by specifying only the highest level. For example, if operators wish to view the entire HVAC (heating, ventilating, and air conditioning) system log, only "H,LOG,,,CRT" needs to be entered in the DIRect mode. The display will show all the points in the HVAC system on a page basis. Operators can view the page as long as is required. When further information on subsequent pages is required, a character is placed in the PAGE box and the SEND key is depressed. The next page of points for the requested log is then displayed.

Colors are used to tie the system together. All the colors are significant and have a consistent meaning throughout the system. For example, purple is always used for HVAC and green always signifies an ENERGY application. This feature reinforces the communication and is an important part of the self-training aspect of the system.

If operators feel that only a small portion of the system monitoring a large building is at their disposal at any given time, they may experience the "peephole" effect. The use of access based upon operator need is one way of avoiding this perception. The PASSWORD feature can be used to limit the type of information operators can modify and control. With the use of this feature, the CRT will display to operators only those functions which they can perform.While operators may have only limited access to the information depending upon their type of station and their particular expertise levels (i.e. FIRE, SECURITY, HVAC, etc.), they do not feel left out because only their needs are displayed as valid entries. In general, the "friendly" interface that the JC/85/40 employs gives the operator the control of the system, while the guiding and prompting that the human-machine interface provides will direct operators in the proper direction appropriate to their needs.

Likewise, operators must not feel that the system is "fragile"--that, in the unlikely event of system failure, the data base will be irreparably damaged. This can be avoided by employing a data base backup procedure as part of the normal system. A data base SAVE/RESTORE feature allows operators to keep a backup copy of their data base on the system and in an archive copy external to the system.

When the novelty of the system wears off, the system must be flexible enough to provide information to operators without seeming repetitive or mundane. This challenge has been answered by providing the system with a WEEKLY SCHEDULING feature which allows operators to enter a command once--specifying the days and the time this command is to be executed. For example, if a log showing all the points in the system and their current state is to be available each morning for operators, the system is requested to execute this command through WEEKLY SCHEDULING. This makes operators more efficient by relieving individuals of this task, while automatically providing the information required.

The variety and flexibility of the JC/85/40 design greatly enhances user acceptance of the system because operators have control over the type of, the content of, and the time and place of the information displayed.

System Response

In a discussion of system response, it becomes necessary to view the application of the system and the user interaction with it. If the terminal responds too rapidly, operators will falsely feel pressure to respond with the same speed--thus causing operator tension which affects performance. Likewise, operators tire when the delays are so long that their attention and orientation is lost. The system must respond or communicate with operators in a reasonable length of time, i.e. non-instantaneous, predictable, delay periods seem to give the best operator interface results. The JC/85/40 is designed to respond to the different informational and interactive needs of the specific user in appropriate ways.

When viewing routine information, update delays exceeding one minute can cause an impatient and disoriented user. In the JC/85/40, fields are updated every 30 seconds. Therefore, if an operator has displayed a log of a particular point and is waiting for verification of a command, there is a timely, predictable system response.

Again, the delay situation is a function of the operators level of expertise and familiarity with the system. With the inexperienced user, for example, there are appropriate delays while the AID lists are being displayed. The interactive input of the command string to display the log in the DIRect mode, however, requires much less time. As stated previously, the JC/85/40 has three different entry methods: AID, DIRect, and EDiT. This is important because operator interface to the system is an interactive-specific, user-oriented interface which answers the varying needs of operator. The display of the log, however, is passive relative to operators and, therefore, a uniform predictable response is acceptable.

One situation that can arise in computer applications occurs when the nature of the operation forces a delay in system response, i.e. because of the application or action being requested, certain delays are required. In these cases, it is necessary to inform operators that the appropriate action is being taken and that they should "PLEASE STANDBY". For example, in the SAVE/RESTORE feature, the system takes the backup data base and moves it to the operating system. After this has been accomplished, the system reloads to insure data integrity. During this reload period, operators are informed of the actions being taken by the system. The need to know what the system is doing is answered merely by a message displayed on the screen. This simple message helps to inform and educate users as to the internal procedures which are now being carried out per their request.

There are some instances in which the application requires that the computer delay checking on a point so as not to cause nuisance alarms. For example, start up of a motor may require some time. In the JC/85/40, there is an operator-set time delay in which there is no notification of the status of the motor. When the time interval has passed, the system will verify that the motor is indeed started. If it is not started, a change-of-state message will be given to operators showing them that, after the initial reporting delay, the motor is not responding as commanded. In this case, the design of the system responds to the needs of the users by not informing them of a problem until it really exists. Nothing is more ineffective than a system that is constantly generating alarms which have no real problem associated with them. It is far more effective that operators be informed only when a problem really does exist.

Hardware Used

The only concern operators have with the hardware of the system is the type of media available for them to communicate with the computer. There are two basic types of entry systems--a hardcopy device or a CRT augmented by a hardcopy device. It has been found that operators will respond better to a CRT than to a hardcopy device. The individual will find it easier to make corrections. Also, page formatting as opposed to line formatting is available. The use of a color CRT serves only to enhance this entry media. Now, operators have not only a device they can use for easing entry, but also a more communicative media for the computer to respond to operators. The machine is able to employ multiple colors to produce a variety of sensory images in operators. This increases their attention spans and provides for reinforcement of the message through the consistent use of colors.

In the JC/85/40, there are double-sized characters appearing at the top of the screen. These are used to signify alarm conditions in the system. When an alarm message is displayed these double-sized characters blink at operators. The location of the alarm condition appears between the alarm delimiter lines showing exactly where the alarm has occurred. In addition, the hard copy device which is part of the basic system keeps a log of the points in alarm. When an operator acknowledges the alarm by placing a character in the ACKnowledge box and hitting the SEND key, the hardcopy logs the acknowledgement, the message between the delimiter lines is cleared and--if there are no more messages for this system--the double-sized characters at the top stop blinking. The absence of blinking of these characters signifies that there is still an alarm condition in the system, but that an operator is now aware of it. The absence of the double-sized characters altogether shows that the problems in that system have been corrected and that all the points within the system are operating normally.

Further reinforcement of monitoring and control alarm conditions can be viewed on the system log. If a point is off-normal and unacknowledged, a "U" will preceed it in the log. If a point is off-normal but operators are aware of the condition, a "*" will appear before the point. In some cases, as with analog points, the status of the condition "HI" for high and "LO" for low, will be shown in the log to alert operators as to the specific type of problem.

Additional capabilities for operator control and monitoring include the LOCKING OUT of a point, i.e. change-of-states on the point are not processed by the system. It is possible to lock out messages coming from the point, messages transmitted to the point, or both. This is indicated on the system log by an "I" for input lockout, "O" for output lockout, and "L" indicating both input and output are not being processed. This is particularly useful for time-of-day dependent points and maintenance and servicing type operations.

Again, the JC/85/40 utilizes the consistent use of color. Red is used to signify FIRE systems and ALARM conditions in order to alert operators to a problem. Blue is used for SECURITY operations, green for ENERGY, purple for HVAC, and white for SYSTEM messages and operations. The double-sized characters, the messages between the delimeter lines, the system logs and keywords, and the selection indices all reflect the use of these colors as they relate to operators information requested. The use of color serves to reinforce the on-line training and general user operation of the system. An additional noteworthy feature of this system is the ability for operators toenter commands through the hardcopy device in the DIRect mode. This allows more than one source of inquiry and control. Input via the CRT while a system log is being displayed is also possible. This flexibility allows operators to command the state of a point and watch the system log update to reflect that status change. The system also has the capability to send system logs to the hardcopy device at the option of an operator. This type of flexibility and reinforcement designed with operators in mind contributes greatly to the total individual performance in the use of the system.

Operator Language

Language is our method of communication. To effectively communicate via a computer, the simplest language and techniques available must be used. Users, in general, are oriented to the use of English sentences or phrases as opposed to numbers or numeric entries. Abbreviations are also helpful when accompanied by clearly spelled out expansions. "Menu selection" and fill-in-the-blank type entries aid operators through complex and infrequently done procedures. (This is the basic use for the EDiT mode feature in the JC/85/40 for data base generation and modification procedures.) All the sensing devices are called by alphabetical names. Each sensing device is described as a ten character mnemonic. There are three "levels" or hierarchical breaks in this naming scheme: building (Level 1), system (Level 2) and point (Level 3). Each of these levels has associated with it a twelve-character expansion to describe the location or function of the mnemonic. There are any variety or combination of these points and the owners have the option of naming these in whatever way they feel will best communicate with their operators.

System Sensitivity

Many of the items previously discussed provide operators with system sensitivity. The use of the English language for entering commands and requesting information shows operators that the system provides for their informational needs. The AID mode of entry provides the sensitivity to the very infrequent user. There is the use of the "?" to help operators by prompting them with a list of valid responses. And the very descriptive double-sized characters and eye-catching alarm indicators on the the system logs are all designed to respond to users and their needs as they relate to this system.

Error Handling

Error handling is probably one of the most important aspects of any system. Error messages must be explicit and should tell the user how to make the appropriate correction. This more than anything else must communicate to operators the way to do their job, i.e. show them where and what the problems are. This issue has been addressed from two directions; operator entry errors and monitoring and control alarms requiring operator action. Operator entry errors are flagged by the system and highlighted in a number of ways. In the AID or EDiT mode, the error fields have carets around them. An error message describing the type of error is also displayed at the bottom of the screen. In addition, the cursor "homes" to the field in error to facilitate ease of correction.

In the DIRect mode, the error message also appears at the bottom of the screen, but the caret flashes under the field in error. Again the cursor "homes" on the error field to allow easy correction. If an operator is unclear as to the action to take, inputting a "?" will cause the entry mode to change to AID, and operators will now be able to select the desired entry from the list displayed. For monitoring and control alarms which require operator intervention, the focus of reporting is to alert operators to the type and location of the alarm and to provide a hardcopy record of these conditions and of the subsequent actions taken. The techniques previously described--such as the double-sized flashing characters, exact location of the alarm between the delimeter lines, and alarm indicators accompanying the system logs--are used to alert operators to the type and location of the alarm. The use of the ACKnowledge box demands operator intervention and the hardcopy record of the alarm conditions, time of alarm, and operator name provide the record of the alarm and the action taken to correct it.

Symbiosis

Symbiosis can be described as the interaction between two dissimilar entities in a mutually beneficial relationship. In a building automation system both the system and the user can benefit from an effective human-machine interface. Users can reap the monitoring and energy control savings through the use of the system. And the system--no matter how effective the system's algorithms may be, no one will use it unless the communication mechanism provides for the users' needs. In the JC/85/40, the human-machine interface provides a vehicle bridging the gap between operators and the effective use of their energy management system. The effective and consistent utilization of colors in the display media, the use of prompting to AID users in entering commands, and the self-teaching characteristics of the system make it easy for users to communicate with and accept the advantages of computers.

BIBLIOGRAPHY

1. Dobrohotoff, V. V., Silver, L., Bacher, S., "When User and Vendor Collaborate," Control Engineering, January, 1969.

2. Duncan, J. B., "Anatomy of pc2; Process Control Language," Control Engineering, April, 1974.

3. Elam, P. G., "Human Considerations in On-line Systems Design," Computer World, March, 1980.

4. Gasper, T. G., Dobrohotoff, V. V., Burgess D. R., "New Process Language Uses English Terms," Control Engineering, October, 1968.

5. Kinnucan, Paul, "Systems Designers Strive to Make Software Friendly," Mini-Micro Systems, February, 1981.

6. Martin, James, Design of Man-Computer Dialogues, Prentice-Hall, Englewood Cliffs, N. J., 1973.

7. Sackman, Harold, Man-Computer Problem Solving, Auerbach Pub. Co., New York, New York, 1970.

8. Smith, C. L., "Fill-in-the-Forms Computer Languages For Process Control," Chemical Engineering, March, 1975.

Chapter 29
SOLAR ROOF COOLING BY EVAPORATION

G. V. Patterson

Evaporation is nature's way of cooling. When we exercise, we perspire, and as the perspiration evaporates, it cools our skin. Nothing new there -- simple procedure -- common knowledge. The same principle applies when we try to cool a roof. By spraying a light film of water, in the form of a mist, on the roof of a commercial or industrial building, roof temperatures can be reduced from as high as 165°F to 88°F as the water evaporates.

The first commercial use of evaporative solar roof cooling was in Washington, D.C. in 1934 -- over forty-seven years ago!

Since that initial installation, over 35,000,000 square feet of evaporative roof cooling has been installed on air-conditioned buildings and non air-conditioned buildings.

Articles have appeared in many publications over the past years, such as Architectural Forum, Heating Ventilating and Air Conditioning, Plant Engineering, Industry and Power, and more recently Energy User News, Energy Management, Air Conditioning, Refrigeration News, Textile Industries, and Plant Services, to name only a few of the national magazines that have featured evaporative solar roof cooling.

While these articles in the early years were devoted primarily to non air-conditioned buildings, in the past ten years, with the soaring energy costs, such companies as GENERAL CABLE, IBM, MONSANTO, UNION CARBIDE, CRYOVAC and ABBOTT LABORATORIES have installed evaporative roof cooling. Tests conducted by their corporate engineering staffs have documented savings of 15% to 30%, which resulted in paybacks of eleven to twenty-seven months.

The AMERICAN SOCIETY OF HEATING AND VENTILATING ENGINEERS (ASHVE) conducted extensive tests in their laboratories in Washington, D.C. as far back as 1939. These tests were to determine the effect of ponded roofs versus sprayed roofs to reduce solar heat penetration into the building. The results were subsequently published in the Journal section of Heating, Ventilating and Air Conditioning. (1) They found that peak heat penetration at 3:00 p.m. was reduced by 87%, and total roof heat load was reduced by 94% when using roof sprays.

My presentation deals primarily with air-conditioned buildings; however, it is well to keep in mind that dramatic comfort cooling can be realized in a non air-conditioned building. With evaporative roof cooling, tests have shown a complete elimination of the radiant roof heat load plus actual reductions in inside temperature of from 10°F to 15°F. It is universally recognized that as temperatures rise above 80°F, employees' mistakes increase and productivity decreases.

Let's take a look at the major applications of evaporative solar roof cooling:

1. Many older buildings, particularly in the Sun Belt, have very little, if any, roof insulation. Energy costs in the 50's, 60's and early 70's was so inexpensive that heavily-insulated roofs were not cost effective; however, with electrical costs increasing by as much as 500% in the past ten years, many methods to reduce air conditioning loads are being intensively scrutinized.

On a building with 1" of insulation, peak demand charge can be reduced by as much as one and one-half tons of air conditioning per 1000 square feet of roof surface, and direct energy costs by one ton per 1000 square feet.

2. Many times plants are expanded or new heat producing equipment is added, and the air conditioning is insufficient to meet design conditions.

By eliminating the roof load, we found that in many instances additional mechanical air conditioning is not required. As an example, ALLENDALE YARNS in Ulmer, South Carolina added 20,000 square feet to their existing 76,000 square foot plant. They soon found that when the temperature exceeded 84°F on the outside, the temperature within the plant rose to 86°F at the working level.

After the installation of the evaporative solar cooling system, not only was the temperature reduced to 78°F design, but for the first time the chiller was cycling.

3. In new construction, generally speaking, light insulation is all that is required in the heating season, due to internal heat loads within the plant. The cooling mode is the prime concern of design engineers, particularly in the South.

A major corporation recently conducted an in-depth study on a new 200,000 square foot building in the Austin, Texas area. They were considering a .05 U factor (R-20) equal to 5" of insulation. Their studies concluded that by incorporating an evaporative roof cooling system, they could reduce the insulation to 1" and show a savings of $280,000 in initial insulation cost.

Evaporative roof cooling provides the equivalent of a .03 U factor equal to 7" of insulation in the cooling mode.

Let's now look at the design of an evaporative roof cooling system and how it works.

A gallon of water evaporating absorbs 8,652 Btu of heat. Thus, about 1.5 gallons will provide the equivalent cooling effect of a ton of air conditioning (12,000 Btu).

If we apply the water in a thin film, just enough to wet the roof, then allow it to evaporate, we cool the roof to 86°F to 90°F versus a dry roof of as high as 165°F. This lower roof temperature results in under-roof temperatures of 84°F to 86°F -- thus, virtual elimination of the solar roof load. We accomplish this by careful design of the water pressure and by engineering a system of copper piping on 18-foot centers, supported on redwood blocks, which completely cover the roof surface of the building. The piping is sized to deliver 25 pounds per square inch throughout the laterals from which the spray emerges. With this pressure, we can get a light film of water in the form of a mist, which thoroughly wets the roof without puddling or run off.

Generally speaking, we divide the roof up into zones, with each zone a maximum of 10,000 square feet. Each zone is controlled by a 24 volt solenoid valve activated by a master control, activated by a thermostat set at 90°F, which is imbedded in the roof in Zone 1. The Mark IV master control has timers for the days of the week, hours of the day, and a series of cam timers that control the solenoid valves in each zone. The sequence of operation is as follows:

The controller is energized at 8:00 a.m. and set to turn off at 6:00 p.m. When the temperature on the roof in the morning reaches 90°F, the controller is activated and Zone 1 sprays for a period of ten to thirty seconds, then turns off as Zone 2 comes on.

This sequence follows until all zones have been sprayed, at which time the cycling starts over again, if the thermostat is again calling for cooling.

The major reason for laying the system out in zones is that we can reduce GPM required and the PSI necessary to get adequate coverage. Taking as an example a 100,000 square foot building, we divide this into ten zones of 10,000 square feet each, which would require 45 GPM at 50 lb. PSI at roof level to get adequate coverage for the zone sprayed.

Let's now take a look at the savings we could anticipate using a 100,000 square foot building in Austin, Texas. The roof has 1" of insulation on a steel deck with 4 ply built-up tar and gravel equal to a .23 U factor. Energy cost is .0395¢ per KWH and demand charge is $5.60 per KW on a 100% rachet. Inside design temperature is 78°F and air conditioning run time is 2,520 hours on a two-shift operation. Peak roof temperature will reach 165°F, and average roof temperature between 8:00 a.m. and 6:00 p.m. will be 140°F, with an average 88°F ambient. Water cost is 60¢ per 1,000 gallons.

Using the following formula we can determine the peak demand charge reduction and direct energy savings.

$$((T1-T3) \times U) - ((T2 - T3) \times U) = Y$$

Y = Internal relief provided in Btu's per square foot per hour.

U = Heat transmission coefficient of the roof.

T1 = Roof surface temperature without FANJET.

T2 = Roof surface temperature with FANJET (90°F).

T3 = Design interior temperature of the building (78°F).

Demand Charge Reduction

$((165^\circ - 78^\circ) \times .23) - ((90^\circ - 78^\circ) \times .23) = 17.25.$

100,000 sq. ft. X 17.25 Btu sq. ft. per hr.

= 1,725,000 Btuh ÷ 12,000 Btu (ton of AC).
= 157 tons reduction in roof load.

157 tons X 1.3 KW ton = 204 KW X $5.60 X 12 months = $13,708.00 annual savings.

Direct Energy Savings

$((140^{o} - 78^{o}) \times .23) - ((90^{o} - 78^{o}) \times .23) = 11.5.$

100,000 sq. ft. X 11.5 Btu sq. ft. hr. = 1,150,000 Btuh ÷ 12,000 (ton of AC) = 96 tons.

96 tons X 1.3 KWH ton = 124.8 KWH X 2520 hours AC run time = 314,496 KWA X .0395¢ = $12,422.59 annual direct energy savings.

Since we use approximately 150 gallons of water per day per 1000 square feet, we would use 15,000 gallons per day X 180 days = 2,700,000 gallons annually X .60¢ per 1000 gallons = $1,620.00 annual water cost.

Recap of Savings

Demand charge	$13,708.00
Direct energy	12,422.00
	$26,130.00
Less water cost	1,620.00
Annual Savings	$24,510.00

With an estimated installed cost of 35¢ per square foot, ($35,000) the payback would be seventeen months at today's energy costs.

Water

As long as it's wet, relatively clean and evaporates, we can use just about any water that is available -- waste water from wells, ponds, and of course city water. When using other than city water, we install a pump to deliver it to the roof at 45 gal/min. and 50 pounds per square inch. Normally, city water meets this requirement, so it is only in rare instances that we have to use a pump.

Dr. Frank I. Yellott presented a paper at the 1965 ASHRAE annual convention, titled "Roof Cooling with Intermittent Sprays,"(2) and in his concluding remarks said,

"At a cost of $0.35 per 1000 gallons, evaporating water is by far the most economical refrigerant, since 1 gallon represents about 8650 Btu of cooling capability, and 1000 gallons, used completely, will produce somewhat more than 700 tons of refrigeration. With 50% effectiveness, an expenditure of $1.00 for roof cooling spray water will produce the equivalent of 1100 tons of cooling effect."

Radiant Heat

Radiation of heat from the under side of the roof can not only cause discomfort in a non air-conditioned building but can necessitate a lower temperature setting in an air-conditioned building to achieve a comfortable condition.

Tests run at BIGELOW-SANFORD plant in Landrum, South Carolina showed under-roof temperature reached 98°F with a dry roof versus a maximum 84°F with a damp roof. Since the skin temperature of the human body is 82°F, we have 16°F of radiant heat radiating to the human body versus only 2°F penetrating into the building with a damp roof -- a reduction of 87.5%. The average under-roof temperature from 8:00 a. m. to 11:00 p.m. was 80.8^{o}F under the damp roof as compared to 91.7^{o}F under the dry roof.

This shows a virtual elimination of radiant heat penetration into the building with an evaporative solar roof cooling system.

Let's take a brief look at a few documented Case Histories:

GENERAL CABLE CORPORATION - Clearwater, Florida

A six-zone system was installed on their plant in September, 1979. From August 6 to September 5, the chiller run time was recorded with the roof dry. The chiller ran 85.7% of the time. The evaporative roof cooling system was activated on September 6, and recordings were taken until October 3. Chiller run time was reduced to 50.8% -- a savings of 34.9% in energy cost. The tests were conducted by GENERAL CABLE's corporate engineers, and they calculated an eleven-month payback.

RIEGEL TEXTILE - McCormick, South Carolina

An evaporative roof cooling system was installed in the fall of 1979 at the end of the cooling season. The KW consumption was compared to 1980. SOUTH CAROLINA ELECTRIC AND GAS billings showed an 11.2% reduction in total plant usage in spite of 23% higher degree days -- fourteen-month payback!

SANYO MANUFACTURING COMPANY - Forest City, Arkansas

Roof Protection: In 1964 when this 244,000 square foot plant was built, the engineers incorporated evaporative roof cooling. Through the elimination of 240 tons of air conditioning, the initial savings amounted to $144,000, at $600 per ton, and at present energy costs, they are saving $53,586 annually!

Of equal importance is the fact that the roof is laying as smooth as when it was installed seventeen years ago. There are no blisters, bubbles, or discoloration, and maintenance has been negligible. SANYO engineers attribute the near-perfect condition of the roof to holding the surface temperature of the roof to a maximum of 90°F with evaporative roof cooling versus 165°F with a dry roof.

<u>Fire Protection</u>: Our FANJET Mark IV controller is equipped with a manual switch which overrides the thermostatic control. In the event of fire, when the switch is thrown, the system starts spraying automatically, zone by zone, to keep the roof continuously wet.

<u>Maintenance</u>: Maintenance of the system is negligible as the only mechanical parts are the valves and the controller. In some older systems, spray heads with tiny serrations were used, and these did have a tendency to clog up. We use a system of piercing the pipe at 30-inch intervals at 10:00 o'clock and 2:00 o'clock. To quote W. O. Kind of ITT:

"The type of spray selected was one consisting of pierced pipe, for which the manufacturer claimed no cleaning would be necessary. This has turned out to be the case; in four years of operation, no time has been spent in cleaning." (3)

SUMMARY

FANJET Solar Evaporative Roof Cooling will eliminate an average 94% of the roof air-conditioning load. It turns the under side of the roof into a giant cooling panel, protects and extends roof life, reduces inside temperatures 10°F to 15°F in non air-conditioned buildings, and finally can show a payback of eleven to twenty-seven months. FANJET provides the equivalent of 7" of fiber glass insulation, equal to a .03 U factor. (R33)

References

1. Houghten, F. C., Gutberlet, Carl, and Olson, H. T. Summer Cooling Load as Affected by Heat Gain Through Dry, Sprinkled and Water Covered Roofs, ASHVE Transactions, Vol. 46, 1940, p. 231.

2. Yellott, John I., "Roof Cooling with Intermittent Sprays," presented at the ASHRAE 73rd annual meeting, Toronto, Ont., Canada, June 27-29, 1966.

3. Kind, W. O., Plant Engineer I.T.T. "Roof Spraying Systems Offer Low Cost, Effective Relief from Summer Heat," <u>Heating, Piping and Air Conditioning</u>, January, 1960.

Chapter 30
HIGH TEMPERATURE HEAT PUMP APPLICATIONS

R. C. Niess

ABSTRACT

The high temperature heat pump is increasingly being used to recycle waste heat in many commercial, institutional, and industrial applications. It has been coupled with solar systems to make them more efficient and has been applied as a means to a greater use of geothermal resources.

This paper concentrates on the use of these new technology heat pumps in industry, where the greatest potential for energy-saving heat recovery systems exists. Industry is by far the largest user of energy in the United States. Further, is is estimated that about 25 percent of this energy is actually wasted.

This paper discusses several installations in buildings, plants, and factories that recycle normally wasted heat generated by various processes. The result has been increased efficiency and a substantial savings in energy. Each of these heat pump heat recovery systems has been justified on a purely economic basis. This is the true test of a viable heat recovery system for industry.

THE ENERGY CHALLENGE

The energy crisis is perhaps the greatest challenge our country has ever faced. Seer, skeptics, oil glut, gas "bubble", or rhetoric aside, our nation -- indeed, the world -- will run out of oil and gas.

The extreme cold weather during the winter of 1977 severely reduced natural gas and fuel oil supplies forcing hundreds of manufacturing plants to close. It provided a grim preview of what may happen in the new future when our energy demands exceed our supplies.

Energy use in the past few years, and in the years to come, is and will continue to undergo dramatic changes. Oil and gas are increasingly being regarded as precious commodities -- to be used only in those applications where no other energy source can be readily substituted -- such as chemical feedstocks and aviation.

According to estimates, industry accounts for 37 percent of the nation's energy consumption. About three-quarters of this energy actually performs useful work; the rest is waste heat. Today, the energy costs make it profitable to reclaim much of this waste heat, and with each energy price increase it is becoming more necessary for profitability.

INDUSTRIAL HEAT PUMPS

Because waste heat is found almost everywhere, high temperature commercial- industrial heat pumps can be used for a wide variety of processes in a broad spectrum of applications, or where there is no waste heat in combination with solar or geothermal sources. These heat pumps are now being used for space heating and hot water heating for industrial processes as well as for service hot water.

The major industrial uses for the heat pump appear to be in washing, blanching, sterilizing, and clean-up operations in food processing; lumber and grain drying; metal cleaning and treating processes; recycling heat in distillation operations in the food and petrochemical industries; and for industrial and commercial space and domestic water heating.

The high temperature heat pump employs the same basic carnot cycle as a typical water chiller except that it works at higher temperatures. It offers an enormous savings potential to the large segment of business which use heat at or below 220 degrees F.

Waste heat from the source water is absorbed in the heat pump's evaporator by the unit's working fluid. The working fluid is then increased in temperature and pressure by the compressor and sent to the condenser where the heat is transferred to delivery fluid, usually water.

TEMPLIFIER SCHEMATIC DIAGRAM
95°F Heat Source Temperature
160°F Hot Water Delivered

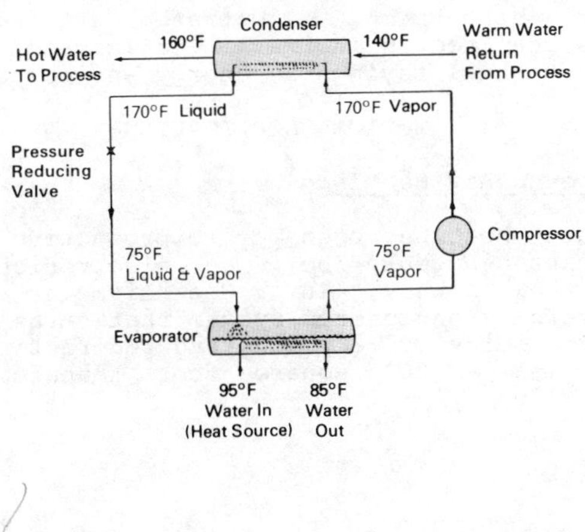

These heat pumps have coefficients of performance which typically vary between three and six, depending on the temperature amplification. This being the temperature difference between the leaving hot water from the condenser and the source water leaving the heat pump evaporator. For many typical applications it averages 4.5.

From an energy conservation standpoint, this is significant. For example, a heat pump with a COP of 4.5 uses only 0.22 units of heat input in the form of electrical energy to deliver one unit of useful heat output. The remaining 0.78 units come from the waste heat source.

PRIMARY FUEL USAGE PER MILLION (10^6) Btu

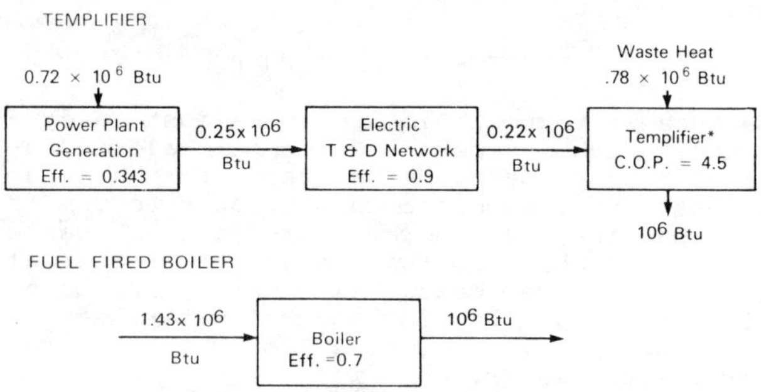

* C.O.P. of 4.5 is typical of 150°F delivery with 95°F waste heat source.

Contrast this with a fossil fuel fired boiler system which consumes twice as much primary energy to deliver the same amount of useful heat.

The heat pump typically recovers the energy in waste heat in the temperature range of 50 degrees F to 160 degrees F. Such low-grade heat is characteristic of industrial cooling water or warm water effluent from plant processes. Other sources include cooling-tower water, water used for quench tanks, as well as for cooling arc welders, extruders, air compressors, furnaces, injection molding and heat treating equipment. Refrigeration and air conditioning condenser water is also an excellent source of waste heat, as are overhead vapors from distillation processes. Solar collectors, water-to-air heat pumps, geothermal resources and warm exhaust stacks are further additions to the seemingly endless list of waste heat sources that can be put to work.

Examples of waste heat pumping at several of the many installations of the high temperature heat pump, illustrate that their use has resulted in increased efficiency and a substantial savings in energy and costs.

TYPICAL APPLICATIONS

Make-up Air Heating

Waste heat has been put to productive use at Robinson-Nugent, Inc., an electronics firm in New Albany, IN. The firm installed a carefully engineered system that saves money and energy by recycling and redirecting heat in its 48,000 square foot manufacturing facility.

All the normally wasted heat -- from injection molding machines, workers, and the lighting system -- is collected by several 20 ton water-to-air heat pumps. This heat pump system, in turn, heats and cools the entire 48,000 square foot manufacturing facility and the adjacent offices; and, in addition, stores excess heat in storage tanks and in an overhead water-loop piping system. Elsewhere, circulating water removes heat from oil coolers for the molding machines. This heat is also added to the water-loop.

HEAT RECOVERY
FROM WATER TO AIR HEAT PUMP SOURCE

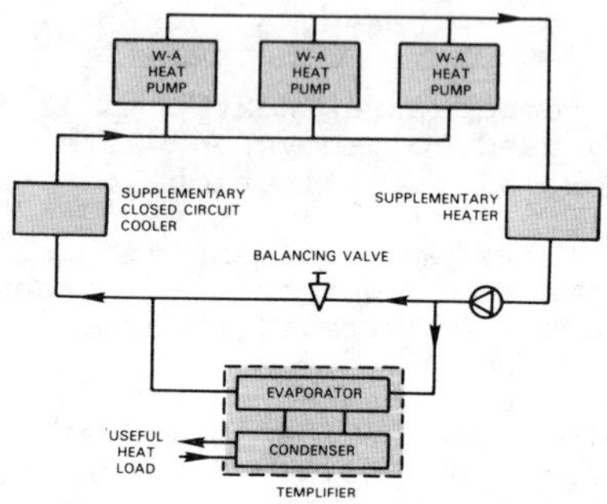

The heat stored in the water and storage loop heats the unoccupied plant on winter nights and during weekends. In the summer, the water-to-air heat pump system provides cool, comfortable air conditioning by removing unneeded factory heat.

The heating system is integrally tied to the pollution control system. So getting air, coming from the cold outside, hot enough to replace air removed by the pollution control system means really squeezing the most out of the plant's waste heat.

The Heat Pump System: A separate high temperature heat pump system was installed to heat the outside make-up air entering the plant.

The pollution control system captures toxic fumes from the plating department and whisks them at the rate of 20,000 cfm, into a scrubber.

The air from outside brought in to replace the air removed by the pollution control system has to be heated. This is done by make-up air heating coils using hot water from the industrial heat pump.

The industrial heat pump removes heat from the plant's water loop and heats glycol up to 150 degrees, which in turn heats the make-up air -- even when the temperature dips to zero degrees, which has occurred several times since the plant first opened in 1977.

The Templifier also saves quite a bit over a more traditional resistance heating system. The Templifier's coefficient of performance -- a measure of efficiency -- is over three.

Hospital Laundry

The Crozer-Chester Medical Center on 67 acres in suburban Philadelphia has 25 buildings and has installed a new energy conservation system that will save countless thousands of dollars in the years ahead. Part of this system uses an industrial heat pump to transfer waste heat from the medical center air conditioning plant to the hot water used in the laundry. This alone is expected to save the hospital over half a million dollars in the next ten years.

In continual daily operation from 7 a.m. to 4 p.m., the laundry processes 5 million pounds of linens a year. There are five special "pass through" washers/extractors that handle 18,000 pounds of laundry a day. Soiled linen goes in one end; clean laundry is taken out the other end. The laundry uses more than 20,000 gallons of hot water a day. This has to be at 160 degrees F so that, when used in conjunction with detergents and bleach, it is hot enough to kill most germs or staph infections.

Before installing the waste heat recovery system, the laundry was the largest user of steam at the medical center. With the new system, the heat pump uses the waste heat in condenser water from the center's centrifugal air conditioning chiller. This waste heat ranges from 90 to 95 degrees F. The heat pump temperature amplifies the incoming water to 150 degrees where it is further heated by steam to the 160 degrees F needed by the laundry.

Daily Cycle Operation: The daily cycle can be looked upon as if the 10,000-gallon storage tank is dumped two times per day. After being drawn down, the tank is filled with city water preheated to 75 degrees and the contents of the tank are gradually heated by circulating 62 GPM through the heat pump's condenser. Thus, for the inital 2-1/2 hours after filling the entire tank, contents will be heated from 75 to 105 degrees and progressively to 150 degrees, requiring approximately 7 hours to recover from 75 to 150 degrees.

Output of the heat pump varies with leaving condenser temperature. The average output per hour and coefficient of performance (COP) on applications of this sort are properly determined by using "mean" leaving hot water temperature (not the final temperature). Thus, the "mean" hot water temperature in this application is 135 degrees. At that mean, Templifier heating capacity is 834,000 btu per hour and the COP is 4.15.

Economic Evaluation: This system is saving a significant amount of money in reduced energy costs. The cost of oil to make steam has become very expensive and is going to become even more expensive in the future. In 1972, for example, the cost of the #6 oil used at Crozer-Chester was 6-1/2 cents a gallon. Eight years later that same oil costs 87 cents a gallon.

A computerized economic evaluation was performed for the engineer to determine the projected cost savings. At the time of the study, electricity cost 4.86 cents per kilowatt-hour, and oil cost $5.72 per million btu. General inflation was assumed to be 10 percent a year, while electricity was predicted to increase by 13 percent a year and oil by 17 percent a year.

With these design conditions, the heat pump system was expected to save $28,000 in reduced energy cost in its first full year of operation. This will increase to over $100,000 during the tenth year of operation. The cumulative net savings over the 10-year depreciation period, after deducting all maintenance cost and costs of ownership, will be $564,000 ($364,000 in terms of 1980 dollars.)

The compounded annual rate of return on investment comes to 119 percent, with a simple payback period of less than one and a half years.

Crozer Chester Hospital Heat Recovery System

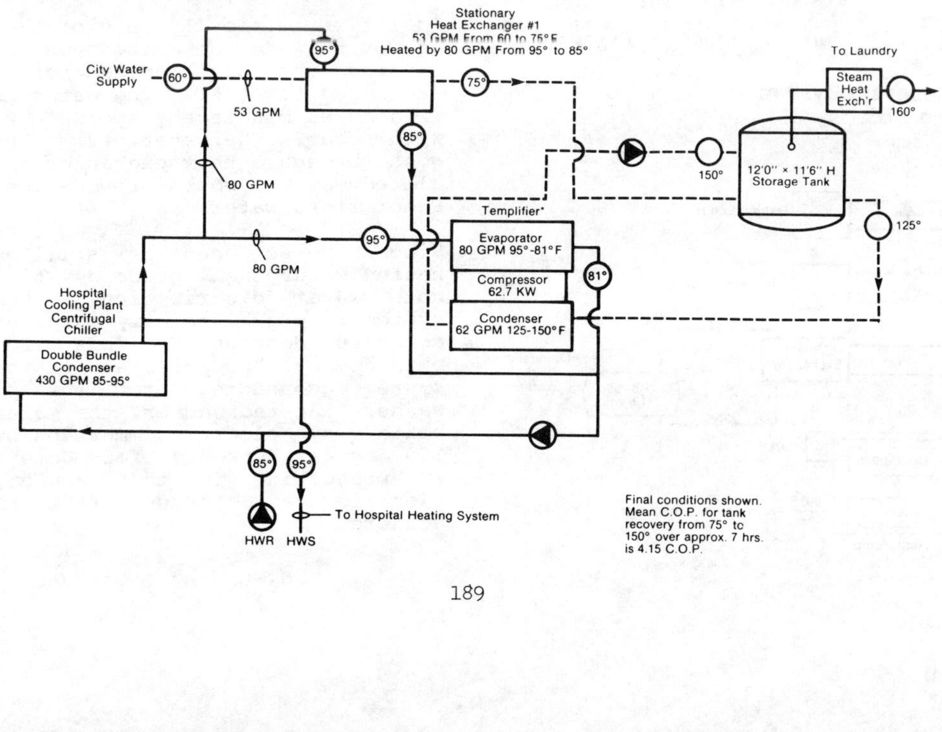

Final conditions shown. Mean C.O.P. for tank recovery from 75° tc 150° over approx. 7 hrs. is 4.15 C.O.P.

Space Heating

Another major commercial application is space heating in office and telephone company buildings. Typically, these structures have a high internal load, plus humidity control requirements where computer or electronic equipment is utilized.

The heat pump taps the waste heat generated by the computers or electronic equipment, gives it a boost in temperature and uses it for space heating, domestic hot water, and even for sidewalk snow removal system in more northern climates.

Telephone Building Installation: A telephone switching center in Pennsylvania is typical. It houses large office areas and each day it used to be filled with scores of operators who were needed to handle calls.

The building's function has now changed. Prior to the installation of the heat recovery system, a steam system was used for heating. Today one entire floor houses sophisticated electronic components and systems which are air conditioned year-round to maintain constant temperature and humidity conditions.

The heat pump heat recovery system eliminates the use of steam by tapping the heat picked up by the 270-ton air conditioning system's chiller.

The year-round cooling load supplies 90 degrees F warm water from the condenser to the heat pump. This source water would have normally been channeled to a cooling tower and the heat dissipated into the air. The heat pump amplifies this heat to produce hot water 150 degrees F and passes it on to the subsystems directly to the coils in the air handlers for space heating and reheating, to radiators for perimeter space heating, to provide domestic hot water, and when needed, to the sidewalk snow melting system.

Energy Savings Achieved: The new heat recovery system in this building has already had a dramatic effect. In November, 1978 -- before the new system was installed -- 230,000 pounds of steam were used. This was cut in November, 1979 to just 55,000 pounds

Waste Heat Recovery System
Bell Telephone Co.

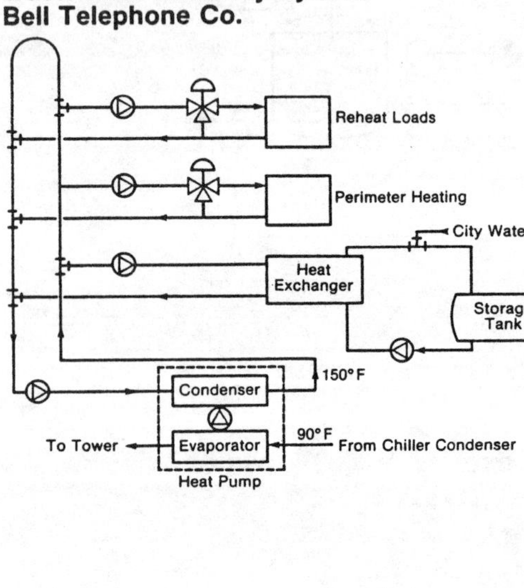

of steam. This energy savings translated into dollars and cents will average over $27,000 per year over the 10 year analysis cycle.

As part of the justification process, a detailed economic analysis of the heat pump system was performed. At the time of the study, steam cost was $5.87 per 1,000 pounds. The estimated savings indicated the installation would pay for itself in about two years and produce almost a 48 percent return on the retrofit investment.

Industrial Process Heating

An industrial heat pump at PPG Industries' glass fabricating plant recovers normally wasted heat from process line air compressors. This recovered heat is used to provide hot water for parts rinsing and other process line applications. The energy recovery system significantly reduces the need for expensive natural gas and currently saves nearly $13,000 a year in reduced energy cost.

Installed in 1979, the industrial heat pump uses waste heat from large air compressors' cooling water. By extracting energy from this waste heat, it heats return water at 120 degrees, boosts its temperature and sends it back to the plant, where it heats demineralized rinsing water, pre-heats wash and domestic water, and even funnels space heat to the plant maintenance area. In all, the system provides much of the plant's current hot water needs.

The Heat Recovery System: A closed loop circulating water system is used to cool the plants large air compressors. The warm water is piped to the heat pump's evaporator where the waste heat is removed, cooling the water to aobut 85 degrees F and returning it to the air compressors for reuse. A second process hot water loop feeds the plant's needs and delivers return water to the heat pump condenser at about 120 degrees F where heat is added and the temperature increased. This hot water returns to the plant to do its work.

The first priority use of the heated water is for the demineralized water needed for rinsing the glass panes before their edges are fused together. The water is demineralized to eliminate any spots that might otherwise occur. The water from the heat pump goes through a heat exchanger in the demineralized water loop to provide the 120 degrees F hot rinse water.

Since excess heat is still available, the heated water then proceeds to the plant's hot water distribution system. In this system it pre-heats city water entering at a cold 50 degrees F. This water is used for both domestic purposes and in the soft water system connected to the process glass washer. At the washer, the water is further heated by electric immersion heaters to the 160 degrees F needed for detergent action. By preheating in this manner, much less electric heat is needed from the immersion heaters.

But there is still more heat that has not been used. In the winter, this excess heat is sent to a fan-coil unit to provide space heat for the maintenance area. In the summer, the heat is simply vented outdoors by automatic duct dampers.

Economic Justification: When operation is increased to the full design load and coefficient of performance (COP) of 4.3, the heat pump will deliver approximately 4.3 btu of energy to the plant (in place of natural gas) while using only 1 btu of purchased electrical energy. The other 3.3 btu is "free" energy recovered from the compressor cooling water.

At the time PPG engineers conducted an economic justification, the cost of electricity was 1.78 cents per kilowatt- hour and the cost of gas was $3.20 per thousand cubic feet.

At the planned part load operation initially, the annual net reduction in gas consumption came to 4150 mcf (or million btu) a year. Subtracting the incremental cost of the electricity to operate the heat pump, and considering the savings in reduced operation of the electric immersion heaters, the net savings were $13,900 a year. At this rate, the heat pump system will pay for itself in less than 3 years.

When the plant expands and the system operates at full design load, two years' energy cost savings will about equal the heat pump system's installed cost.

Summary

A high temperature industrial heat pump can be used to conserve energy and permit oil or gas to be used in applications where no other energy source can be substituted.

In addition to space heating for all types of commercial, industrial, and institutional buildings, the high temperature heat pump can be used in hotels, motels, and apartment buildings to provide service hot water; in office buildings to provide service hot water or hot water for reheat purposes; in hospitals to provide hot water for laundries, kitchen, and wash-up uses; or in industrial plants to provide low cost process hot water as well as boiler and other feed water heat.

It can be used wherever there is waste heat in the form of water, or some similar fluid, and the process hot water that's needed is at 220 degrees F or less.

Speaking of the future, the heat pump makes what most people consider to be the alternative energy sources for the next generation a practicality today. Both solar sources and geothermal sources can be expanded significantly with an assist from a high temperature heat pump.

It is clear that energy conservation and heat recovery systems can make a substantial impact on the energy problems facing our country.

Heat pump installation at PPG Industries Glass Fabricating Plant

Chapter 31
MICROPROCESSOR-BASED COMBUSTION OPTIMIZATION FOR IMPROVED ENERGY CONSERVATION

R. C. Molloy

ABSTRACT

A new hybrid combustion control system has been developed which combines the functions which have traditionally been performed by separate stand-alone measurement and control instruments into one low-cost integrated system. Complete O_2 Trim Control Systems will soon be available starting at less than 6,000 dollars. By utilizing a high performance low-cost microprocessor, both measurement and control functions can now be performed simultaneously. The new systems will feature automatic calibration, self-diagnostics, field programmable memory, and improved operator interface.

By measuring the products of combustion utilizing the latest In Situ Opacity, Oxygen, and CO Monitoring technology, the fuel air mixture ratio of industrial fuel burning equipment can be optimized to insure reduced fuel consumption and improved combustion efficiency. Typical fuel savings of 3 to 5 percent have been experienced on a wide variety of different types of fuel burning sources, including packaged boilers, incinerators, and process heaters.

FORWARD

One of the greatest opportunities for industrial energy conservation today is to improve the combustion efficiency of industrial fuel burning equipment. Up until recently, the boiler room in the typical industrial facility has languished in the "Dark Ages" utilizing few innovations that haven't been available for over fifty years. Most industrial fuel burning equipment, such as boilers, process heaters, kilns, dryers, incinerators, and steam generators, were designed and installed at a time when fuel costs were only a fraction of what they are today.

In the past seven years, the cost of oil has increased from less than two dollars a barrel to over thirty-five dollars a barrel (figure 1). During this same time period, the performance and cost effectiveness of microprocessor based instruments for the measurement and control of combustion efficiency have been constantly improving. A number of recent studies regarding the potential for improving the combustion efficiency of industrial fuel burning equipment have shown that savings of 3 to 5% can be realized. As a result of rising fuel prices and improved technology, the stage is set for a revolutionary improvement in the operating efficiency of industrial fuel burning equipment.

COMBUSTION

Combustion is an exothermic chemical reaction. The most important factor in optimizing combustion is to ensure that the proper mixture of fuel and air (oxygen) are supplied to the combustion process.

Theoretically, ideal "stoichiometric" combustion occurs when the exact amount of oxygen is supplied to ensure complete combustion of the fuel. Unfortunately, stoichiometric combustion is virtually impossible to achieve in the real world as a result of a number of variables which can adversely affect combustion, including variations in: air and fuel pressure, air and fuel temperature, humidity, type of fuel, BTU content, moisture, specific gravity, viscosity, grindability, ash content, burner condition, fan condition, linkage wear, fire box design, etc. In a typical fuel burning application, virtually complete combustion will occur only if there is enough excess oxygen present. Thus, on a practical basis, optimum efficiency occurs when the minimum amount of excess air is supplied to ensure complete combustion.

MECHANISMS OF ENERGY LOSS

When the ideal fuel/air mixture is not present and there is an excess of either fuel or air (oxygen), then significant amounts of energy are lost from the combustion process by the mechanisms outlined below.

Excess Fuel

Combustibles Loss: When unburned combustibles are exhausted from the stack exit, fuel is literally going up the stack with a resulting high loss of combustion efficiency. Also, fuel rich operating conditions can cause safety hazards.

Increased Corrosion: Localized reducing areas in the furnace, caused by poor air or fuel distribution, can lead to accelerated tube corrosion and slagging.

Increased Air Pollution: The air pollution effect of combustion with excess fuel (insufficient oxygen) results in the formation of carbon monoxide (CO), smoke, and solids. Particulate emissions are a major cause of the adverse health effects of air pollution including: respiratory ailments, smog, and gas phase reactions with other air pollutants.

Excess Air (Excess Oxygen)

Sensible Heat Loss: The air (air is 20.95% oxygen by volume) which is supplied in excess of that necessary to complete combustion is heated as it goes through the fire box, and this heat is carried up the stack with the products of combustion in the flue gas.

Heat Transfer Reduction: The excess air supplied, in addition to that necessary to insure complete combustion, increases the volume and thereby the velocity of the flue gas in the fire box, reducing the amount of time the hot products of combustion are in contact with the heat exchanger surfaces. This results in reduced heat transfer efficiency and higher flue gas stack exit temperatures. Thus, excess air contributes not only to a greater volume of flue gas but also to higher stack exit temperatures, resulting in even more severe sensible heat losses.

Fan Power Loss: Significant amounts of electrical energy can be saved by reducing the amount of excess air necessary to be moved.

Increased Corrosion: When sulphur bearing fuels are burned, excess air leads to the formation of sulphur trioxide (SO_3) and, ultimately, sulphuric acid (H_2SO_4). This H_2SO_4 can combine with emitted solids to produce acid smuts that can cause cold end corrosion, air preheater fouling, and severe corrosion to surrounding buildings. Excess air yields higher SO_3 concentrations which results in a higher acid dew point temperature in the flue gas. For every 20 degrees C. that it is necessary to increase the flue gas stack exit temperature to avoid H_2SO_4 condensation in the duct work, approximately 1% in operating efficiency is lost.

Increased Air Pollution: Excess air plays an important factor in the formation of SO_3 and NO_x emissions.

COMBUSTION EFFICIENCY

Combustion efficiency is optimized when the proper mixture of fuel and air is supplied so as to minimize the energy losses associated with excess fuel and excess air (figure 2).

The combustion efficiency curve for typical industrial fuel burning equipment is a bell shaped curve (figure 2). The rate of energy loss is six times greater on the excess fuel side of the curve than it is on the excess air side of the curve. As a result, most combustion control strategies approach optimum from the excess air side of the curve so as to optimize efficiency and also to avoid hazardous excess fuel conditions. It can also be seen that the closer the fuel/air mixture ratio is to optimum, the less potential there is for fuel saving and the greater the risk of the severe energy losses, which occur on the excess fuel side of the curve.

At higher excess air levels, for every one percent reduction in excess oxygen, a one percent improvement in combustion efficiency can be achieved. At lower excess oxygen levels (less than three percent), for every one percent reduction in excess oxygen, approximately a one half percent improvement in combustion efficiency can be achieved. On a practical basis, fuel savings of 3 to 5 percent can be achieved by typical industrial fuel burning equipment.

The selection of the proper combustion efficiency measurement and control instrumentation for a given application is a direct function of the return on investment at the margin for each incremental improvement in system performance. Typically, relatively inexpensive systems can achieve approximately 80% of the fuel saving potential available and thus are ideal for smaller fuel burning equipment. More sophisticated systems which provide a relatively small incremental improvement can still be justified on larger fuel burning equipment where even a saving of ½ to 1 percent can be significant. Also, more sophisticated systems can be used to overcome unique combustion problems where less sophisticated systems are not suitable.

MEASUREMENT AND CONTROL SYSTEMS

By monitoring the products of combustion continuously, the fuel/air mixture ratio can be controlled on a dynamic feedback basis to insure the complete combustion of fuel with a minimum amount of oxygen and thus achieve optimum combustion efficiency.

FLUE GAS MEASUREMENT

An accurate continuous flue gas monitoring system which is properly maintained can provide the source operator with valuable information on both the operation of the emission source as well as the amount and composition of the emissions vented into the atmosphere. A continuous monitoring system can be used to monitor and optimize the performance of air pollution control equipment, including accessing the effects of design improvements and operation variations as well as scheduling maintenance on the basis of need. By maintaining continuous emission records, relations with EPA and other regulating agencies can be improved, and a documented basis can be provided to defend operations based on actual emissions. A continuous monitoring system can provide the dynamic feedback necessary to modify combustion variables so as to minimize harmful emissions while maximizing useful energy output with reduced fuel consumption.

Representative Sample

Any flue gas measurement, no matter the sophistication of the measurement technique, is only as accurate as the degree to which the sample being measured is representative of the actual constituents of the flue gas as a whole. As a result, it is important to avoid sampling locations where there is excessive stratification, such as at sharp turns in the ductwork. Avoid stagnant air pockets. In addition, it is important to avoid high temperature ambient environments, high relative humidity, corrosive gases, excessive vibration, explosion hazards, etc. Typically, sensors are breech mounted in an area which offers reduced installation costs and improved accessibility. In many cases, an ideal location is not available. In these cases, good judgement must be exercised in weighing the various alternatives to pick the best available sampling site and proper measurement technique to be employed. If in doubt, contact a factory application engineer from your flue gas measurement instrument supplier for assistance.

In Situ Measurement

A majority of the continuous flue gas measurement systems used in modern high performance combustion control systems today utilize In Situ measurement techniques such as: In Situ probes for oxygen and wide path visible light and infrared transmissometers (to avoid stratification) for opacity and CO measurements. In Situ measurement insures a more representative sample by measuring flue gas concentrations inside the stack in direct contact with the sample, thereby eliminating the problems associated with extractive gas analyzers such as: high maintenance, unreliable operation, sample transport time lags, possible sample dilution, condensation, sample loss, sample alteration, sample line clogging, sample conditioning breakdowns, etc. By measuring on an In Situ "wet total sample basis", extractive "dry sample" measurement errors caused by combustion variables such as changes in the relative humidity of combustion air, the carbon hydrogen ratio, or the moisture in the fuel, can be eliminated.

THE KEY COMBUSTION EFFICIENCY PARAMETERS

Opacity

Principle of Operation: A beam of visible light in a collimated beam is transmitted across the stack in a single or double pass beam to a photo detector. The output of the detector is proportional to the amount of light incident upon it. The greater the optical density of the emissions, the less light is transmitted across the stack to the detector. The optical density of the flue gas is directly proportional to the mass concentration of the particulates in the flue gas, and thus, an opacity measurement can serve as a relative measure of both air polluting particulate emissions and the completeness of combustion.

Application: Since the earliest days of combustion control, operators have taken the "Casey Jones" approach to optimize combustion efficiency by "operating at a light haze". The smoke opacity curve for a typical liquid or solid fuel burning emission source has a sharp knee as you approach the excess fuel side of the curve, and energy losses start to become severe (figure 3).

Since 1970, with the passage of the Clean Air Act, it has become unacceptable, from an environmental point of view, to operate at high opacity levels. The initial reaction to this new legislation was to provide too much excess air so as to insure low opacity operation.

The oil embargo of 1974 and the resulting explosion in the cost of fuel gave the operators of industrial fuel burning equipment the motivation to purchase both opacity and oxygen measuring instruments so as to reduce excess oxygen to improve combustion efficiency while avoiding opacity levels in excess of EPA guidelines.

Advantages: Opacity monitoring systems are a reliable low cost approach for both avoiding excess fuel conditions on liquid and solid fuel burning equipment and meeting EPA stationary emission source monitoring guidelines. Opacity monitors are useful on small boilers because of their low cost and on large boilers to meet EPA compliance requirements. On some types of coal fired boilers, the maximum allowable EPA opacity limit may be reached prior to reaching optimum efficiency, and thus, opacity may end up as the limiting constraint on the optimum performance of the entire system.

Limitations: On large fuel burning equipment, the relatively high levels of excess fuel necessary to get meaningful opacity readings often occur at unacceptably high levels of energy loss. Also, opacity monitors are ineffective on gas fired equipment where the opacity is virtually zero. Opacity is most useful as a secondary indicator of high excess fuel and as a primary indicator of EPA emission violations. Opacity is typically not sensitive enough to be a primary control parameter with which to fine tune combustion efficiency.

CO_2 - Carbon Dioxide

Application: Another of the early techniques for monitoring combustion is CO_2 measurement, which dates back to the original Orsat test. Today, CO_2 measurements are primarily used in low cost portable instruments and are rarely ever used as a primary feedback parameter in more sophisticated combustion control systems (CO_2 can be used as a secondary parameter).

Limitations: The CO_2 curve (figure 4) has both an excess fuel and an excess air point where the same CO_2 reading can occur. Thus, as a primary measurement, CO_2 is not of much value without a separate reading of another independent combustion parameter to tell you what side of the curve you are on.

CO_2 readings also vary dramatically with C/H (carbon-hydrogen) ratio of the fuel and are thus of little value where there are dramatic fuel quality variations or where multiple fuels are fired simultaneously.

CO_2 requires a greater precision of measurement to be comparable to excess oxygen reading. In addition, whereas an excess oxygen reading approaches zero as optimum efficiency is reached, CO_2 approaches some constant value, which is a function of fuel composition.

CO_2 measurements utilizing extractive analyzers with costly high maintenance sample conditioning systems generally provide a less reliable and less accurate measure of relative combustion efficiency than other more popular techniques.

Oxygen

Principle of Operation: The concentration of oxygen in the flue gas passing through a stack or duct work is measured by detecting variations in the output of a zirconium oxide fuel cell oxygen sensor. As the oxygen concentration varies, an output signal is generated which can be directly related to the oxygen content of the flue gas. An In Situ Oxygen Analyzer is mounted through the wall of the stack so that the oxygen sensor cell can be in direct contact with the flue gas sample to be measured. In Situ measurement eliminates the problems associated with extractive systems which require maintenance prone sample conditioning systems while providing a more representative sample of the gas to be measured.

Fuel cell oxygen sensor consists of a ceramic tube made of yttria-stabilized zirconium oxide. When this cell is heated above 1100 degrees F (600 degrees C), it becomes an oxygen ion conducting solid electrolyte. The Nernst Equation defines the relationship between cell voltage and temperature and the oxygen concentrations on the reference and sample side of the cell.

Combustibles, such as hydrogen or hydrocarbons, if present in the gas sample, will burn on the catalytic platinum surface of the zirconium oxide fuel cell leaving only the more meaningful "net oxygen available for combustion" to be measured.

Application: An In Situ Oxygen Monitoring System can provide the dynamic feedback necessary to enable the operator of industrial fuel burning equipment to supply the proper amount of oxygen to insure optimum combustion efficiency. Oxygen monitors are also an excellent tool for monitoring diluent air leakage when determining the effectiveness of bag houses, electrostatic precipitators, and other types of air pollution control devices.

Oxygen measurement is a direct measure of excess air. Oxygen measurements are not adversely affected by variations in fuel. The oxygen measurement approaches zero in a linear manner when excess air is being reduced (figure 3). Optimum efficiency occurs when the minimum amount of excess oxygen is supplied to insure complete combustion.

Oxygen monitors are used on a wide variety of emission sources.

- Bark & Black Liquor Boilers
- Carbon Black (furnace process)
- Catalytic Crackers
- Cement, Lime Kilns
- Ceramic, Phosphate Kilns
- Chemical Recovery Boilers
- Coal Fired Utility Boilers
- Coke Ovens, Soaking Pits
- Deisel Engines
- Diluent Monitoring
- Driers
- Gas Fired Boilers
- Glass Fiber
- Glass Melting Furnaces
- Heat Treating Furnaces
- Incinerators
- Kraft Recovery Boilers
- Metallurgical Furnaces
- Naval Boilers
- Oil Fired Boilers
- Oxygen Enrichment Systems
- Refinery Process Heaters
- Regenerative Furnaces
- Turbine Engine Test Cells
- Wet Scrubbers
- Windboxes

Advantages: The In Situ measurement of "net oxygen available for combustion" provides the best direct measure of relative combustion efficiency. Approximately 80% of the total fuel saving potential from optimizing combustion efficiency can typically be achieved by oxygen measurement alone.

Solid state In Situ Oxygen Analyzers are relatively low cost and provide reliable, accurate, and maintenance free oxygen measurement. The stable performance characteristics of the zirconium oxide fuel cell oxygen sensor allows for long periods of trouble free operation without need for recalibration. At a 2% oxygen concentration system accuracy is better than $\pm$ 0.1% oxygen. By calibrating the oxygen sensor with a gas calibration standard in the typical operating range of the burning source fuel, more accurate readings can be made.

The zirconium oxide sensor is capable of the selective detection of oxygen with no interference from either water vapor or other typical stack gas constituents. The sensor is insensitive to variations in flow, temperature, or stack pressure. The In Situ probe air foil eliminates dirt accumulation. The In Situ Oxygen Analyzer can accurately measure oxygen concentrations in the combustion efficiency range under severe stack gas environmental operating conditions.

Limitations: Since moisture in the flue gas is a variable and oxygen can coexist with unburned fuel (due to incomplete ignition, poor mixing, atomization problems, burner malfunction, or air infiltration), it is important to use a zirconium oxide fuel cell based oxygen analyzer which measures "net oxygen available for combustion" on a wet basis rather than an extractive paramagnetic oxygen analyzer which measures oxygen by volume on a dry basis without any reference to possible unburned combustibles present in the flue gas sample being analyzed.

Air infiltration (air leakage) has a disproportionate exponential effect on the accuracy of oxygen measurements relative to CO measurements. Since ambient air contains 20.95% oxygen by volume, an air infiltration of 5% (1% oxygen) would cause a 2% oxygen measurement to read 3%, which would be a 50% error. Since there is virtually no CO in ambient air, a similar 5% air infiltration would only have a 5% linear dilution effect on a typical CO reading of 200 PPM causing an insignificant 10 PPM error.

To avoid errors due to air infiltration, select a sampling site upstream of the air preheaters and other known air leaks as close to the source of combustion as possible (while observing probe temperature limitations). Seal all air leaks upstream of sensor. By sampling with an In Situ Probe at an insertion depth of at least 30% of the gas-flow cross sectional diameter, errors due to air infiltration typically found on the surface boundary layer can be avoided.

To avoid errors due to stratification, locate the oxygen sensor in a location where a representative sample can be obtained, and on large breeches, average the output of multiple probes on either a linear or weighted average basis.

The zirconium oxide fuel cell is an ideal oxygen sensor for the majority of combustion efficiency applications. For optimum performance, the following application limitations should be observed: The platinum electrodes on the cell may become contaminated or poisoned in heavy reducing atmospheres in which lead, lead compounds, arsenic, or antimony are present resulting in shortened cell life. The cell cannot be used in process applications to measure oxygen in the presence of high concentrations of combustible background gas. Stack gas temperature limits should not be exceeded, typically 1400 degrees F (760 degrees C) for stainless probes, and 2900 degrees F (1600 degrees C) for ceramic probes. Applications involving repeated temperature cycling, mechanical, or thermal shock should be avoided.

The optimum oxygen set point is a function of both type of fuel and boiler load (figure 5). At low load levels, since smaller quantities of both fuel and air are supplied, mixing is not as good as it is at higher load levels, thus greater amounts of excess oxygen are required to insure complete combustion. This limitation can be minimized by utilizing a programmable set point function generator or by simultaneously measuring CO as outlined below.

CO

Principle of Operation: Modern In Situ CO Analyzers utilize an advanced infrared negative absorption gas filter correlation technique which takes into account the unique signature of the CO absorption band to achieve greater measurement accuracy while avoiding the interference effects of CO_2, water vapor, and other background gases.

A source of infrared radiation is located on one side of the stack, and a high performance infrared detector is located on the other side. A beam of infrared energy is transmitted across the stack. This beam is modulated at a fixed frequency to enable the detector to reject background infrared radiation from the flue gas, duct, or other hot surfaces. The modulated beam of infrared radiation is absorbed over a band of wavelengths as a function of the concentration of CO in the flue gas. Gas filters containing N_2 and CO are inserted in front of the detector on an alternating basis to obtain measurement, reference, and upscale CO calibration signals. The CO concentration of the flue gas is computed as a function of these signals, the pathlength of the sensors, and the average flue gas temperature.

Application: Significant fuel savings can be obtained by using continuous CO readings to determine the fuel/air mixture ratio necessary to ensure optimum combustion efficiency.

CO is a direct product of combustion, and thus, its concentration is a direct measure of combustion conditions. The CO curve rises exponentially as optimum combustion efficiency is reached (figure 3).

An In Situ CO Monitor is essential for large combustion systems which burn gas, oil, coal, or other liquid or solid fuels and where fuel bills reach or exceed $500,000 per year; also, on process applications where air infiltration, variations in fuel type and quality, or variations in load level make it difficult, if not impossible, to utilize a conventional oxygen monitor based O_2 trim control system.

CO monitors are ideal for use on the following applications:

- Power Generation Boilers
- Process Steam Boilers
- Marine Boilers
- Cement and Lime Kilns
- Incinerators
- Bark and Black Liquor Boilers

Advantages: At low excess oxygen levels, CO is the most sensitive and the most reliable measure of combustion efficiency.

As the boiler load level is increased, the amount of excess oxygen necessary to complete combustion is reduced dramatically while the optimum CO set point remains unchanged.

Whereas O_2 measurements are a useful indicator of relative combustion efficiency, CO is the ideal combustion efficiency control parameter because unlike O_2 and CO_2, the optimum CO set point is not affected by variations in boiler load level or type and quality of fuel (figures 6 and 7).

Since ambient air is 20.95% oxygen, small amounts of air infiltration leakage can cause exponential errors in oxygen readings. In that CO is a product of combustion, air infiltration leakage does not have a significant effect on CO readings even when operating at low oxygen levels.

With multiburner systems problems arise in ensuring that the correct balance of load and air distribution to each individual burner is always maintained. It is common for some burners to operate with an air deficiency while the overall oxygen level is correct. This maldistribution of air results in high CO and solids emission undetectable by O_2 control.

In Situ non-extractive open path measurement ensures a more representative sample and helps avoid errors due to stratification. Most systems can operate over pathlengths up to 10 meters (32 feet) in length.

A modulated IR source minimizes the interference effects of background IR radiation and allows for CO measurement in flue gas streams up to 550 degrees C (1020 degrees F) in temperature.

A microprocessor based CO monitor can automatically perform a true check of both system zero and span without interfering with the operation of the emission source. In addition, automatic compensation is provided for zero and span drift, dirty windows, minor misalignment, etc. A self-diagnostics system can alert the operator in the event of system operating problems, including air flow alarm, alignment, and dirty windows. Typical system accuracy under worst case field conditions is better than $\pm$ 20 PPM at 200 PPM CO with a response time adjustable from 3 to 60 seconds.

Limitations: High performance In Situ CO Monitoring Systems complete with protective weather covers and a remote mounting display cost approximately $20,000. In that in most applications 80% of the fuel saving potential can be achieved with an In Situ Oxygen Monitor costing less than $5,000, the fuel burning source must have a sufficiently large fuel bill or unique combustion problems in order to justify this expenditure based on return on investment for an additional fuel saving of ½ to 1%.

Air infiltration, which is very undesirable, cannot be detected with a CO monitor.

Since In Situ CO Analyzers measure CO by IR absorption, possible interference due to the pressure of CO_2 and H_2O, which also absorb IR near the same wavelengths as CO, must be carefully avoided by use of the proper narrow band pass filter and other more sophisticated signal discrimination techniques which rely on the unique "signature" of the CO absorption band.

With In Situ CO measurement, once you are outside of the optimum efficiency region, it works on an on/off basis. The output is very high on the excess fuel side and very low on the excess air side making it difficult to gauge the magnitude of corrective action to take. In Situ oxygen measurement is best for major fuel/air mixture adjustments while In Situ CO is best for fine tuning (figure 8).

Recommendations - Flue Gas Measurement

CO_2: Primarily recommended for use in portable instruments or as a secondary parameter on continuous emission sources; approximate cost $2,000 portable, $5,000 continuous.

Opacity - Single Pass: For use on small fuel burning equipment utilizing liquid or solid fuels as a crude measure of excess fuel and to measure emissions in violation of EPA guidelines; approximate cost $2,000-$4,000. (Sometimes used in conjunction with O_2 on medium sized emission sources in lieu of more costly CO.)

Opacity - Double Pass: For use on large fuel burning equipment to meet EPA stationary source emission monitoring requirements and as a secondary measure of excess fuel conditions; approximate cost $10,000 to $15,000. (Normally used in conjunction with systems outlined below.)

Oxygen: All fuel burning equipment; approximate cost $2,000 to $5,000 for Single Point System, $15,000 to $30,000 for Multi-Point System.

CO: Large fuel burning equipment or where air infiltration, load or fuel variations, or other combustion problems limit the ability of less costly systems to operate properly; approximate cost $20,000 to $25,000.

The Ideal Flue Gas Measurement System: Simultaneously measure opacity, oxygen, and CO. Measure opacity to meet EPA guidelines and as a secondary measure of excess fuel conditions. The simultaneous measurement of excess oxygen provides a direct measure of "combustion efficiency". The lower the amount of excess oxygen necessary to achieve complete combustion, the better the relative combustion efficiency.

Since the final stage of the combustion of fossil fuels is the oxidation of CO to CO_2, the measurement of CO is a direct measure of the "completeness of combustion" without regard to the actual amount of O_2 necessary to achieve it. By simultaneously measuring both oxygen and CO, problems due to air infiltration, load variation, fuel variation, burner malfunction, stratification, etc., can be avoided. By utilizing a combustion control strategy utilizing oxygen measurement as a "rough" trim and CO measurement as a "fine" trim, combustion efficiency can be optimized under widely varying operating conditions.

OXYGEN TRIM SYSTEMS

A typical boiler control system consists of a steam pressure transducer, which senses load demand, and a fuel flow controller, which increases or decreases fuel flow as a function of load demand. Typically, a mechanical linkage or a pneumatic transducer interconnects the fuel valve with the air damper. On a practical basis, it is impossible to properly "characterize" the air damper to account for fluctuations in process variables such as changes in fuel and air quality. Typically, the system is set up to operate with excess oxygen to provide for occasional momentary upset conditions. As a result, a majority of the time the system is operated with more excess oxygen than is necessary.

The most important process variable for optimizing combustion efficiency is the fuel/air mixture ratio. By monitoring the products of combustion on a continuous basis, the operator can manually or automatically adjust the fuel/air mixture ratio to insure optimum efficiency.

Manual O_2 Trim

Currently, the most popular approach to oxygen trim is to make use of the feedback from In Situ flue gas measurement instrumentation on either a continuous or intermittent basis to enable the operator to manually adjust the fuel/air mixture ratio to an optimum setting.

In many cases, manual control is the logical first step in the process of upgrading existing fuel burning equipment to automatic O_2 trim control. It allows the operators to become familiar with and to develop confidence in the new flue gas measurement instruments. Once this confidence building stage has been completed, based on the fuel savings accomplished under manual O_2 trim control, the economics of further upgrading to automatic O_2 trim control can be considered.

Automatic O_2 Trim

An automatic O_2 trim system can provide an automatic fine trim adjustment to the combustion control system to enable it to operate efficiently with less excess air by instantaneously reacting to dynamic fluctuations in process variables on an automatic closed loop basis without operator intervention to insure that the proper fuel/air mixture ratio is maintained at all times.

An O_2 trim system typically utilizes flue gas measurement systems to monitor the products of combustion on a continuous basis. A controller compares the opacity, oxygen, and CO in the flue gas with the desired set points and generates an appropriate output control signal to the air damper draft control to automatically adjust the amount of combustion air to the burner to insure that the desired amount of excess air is maintained.

MICROPROCESSOR BASED FLUE GAS MONITORING SYSTEMS

Most flue gas monitoring instruments which have recently been introduced utilize microprocessors to improve performance. Such features as automatic calibration checks, self diagnostics, high and low alarms, signal averaging, etc., are now provided as standard. The current trend is towards higher performance at lower cost.

HYBRID MICROPROCESSOR BASED MEASUREMENT AND CONTROL SYSTEMS

Combustion control system applications which have traditionally been performed by separate stand-alone measurement and control instruments will soon be performed by low-cost microprocessor based hybrid instruments, which will be capable of simultaneously performing both measurement and control functions.

As the cost of microprocessors continues to decline, their performance and computational capabilities continue to improve. As future combustion control systems make greater use of microprocessor technology, the point where the measurement instrument ends and the control instrument begins will become less and less distinct. Today's system application is rapidly becoming the instrument application of tomorrow. This hybrid approach takes full advantage of the computational capabilities of the microprocessor to produce a higher performance yet lower cost system. Overall system reliability is also improved as a result of the elimination of the redundant components which are required by each stand-alone instrument, such as: power supplies, displays, enclosures, analog outputs, relays, etc. System troubleshooting and repair is also greatly simplified as a result of built-in self diagnositcs and modular solid state plug-in boards. Circuit malfunctions can be readily identified and repaired by simply replacing color coded generic-by-function plug-in electronic printed circuit boards.

By utilizing a microprocessor based O_2 trim system which incorporates a field programmable "characterized" set point function generator, the shortcomings of fixed set point O_2 trim can be avoided. The "characterized" set point function generator utilizes a signal from the boiler master controller proportional to load to compute a corresponding excess O_2 set point. The relationship between boiler load and excess O_2 set point can be easily programmed in the field. The function generator can be used to maintain the desired fuel/air mixture ratio for each level of load demand such that the air will lead the fuel on increasing fuel-load demand and lag on decreasing fuel-load demand.

The new hybrid microprocessor based combustion control systems will feature automatic and manual control, automatic sensor calibration, field programmable memory, self diagnostics, set point display, set point function generators for a choice of two different fuels, computer capability through the use of an optional ASR 232 interface, input and output characterization, adjustable sample transport time lag, and more.

Low cost microprocessor based O_2 trim combustion measurement and control systems complete with oxygen sensors and actuators will soon be available for use on small packaged boilers at a cost from $6,000 to $10,000.

More sophisticated application requirements can already be currently met by utilizing advanced stand-alone microprocessor based In Situ opacity, oxygen, and CO measurement instruments in conjunction with computer based control systems which are available from the traditional combustion control system manufacturers. Excess oxygen readings as low as 0.5% can be achieved. This approach allows one to make use of both state of the art flue gas measurement technology and control system hardware from traditional combustion system suppliers to insure optimum compatability with one's existing combustion control equipment.

SUMMARY

As the cost of fuel continues to increase and as the cost of microprocessor based measurement and control systems continues to decline, the potential return on investment from reducing fuel consumption 3 to 5 percent will soon be such that even the smallest industrial fuel burning equipment will eventually be utilizing the most advanced combustion control techniques.

The selection of the appropriate measurement and/or control system for a specific application is a function of annual fuel consumption and the potential for further performance improvements. Relatively small industrial fuel burning equipment can almost always justify an oxygen or opacity monitor for feedback to optimize manual control. The greater the potential for fuel saving, the greater the justification for a high performance microprocessor based combustion monitoring and control system utilizing the ideal combustion efficiency measurement parameters of In Situ opacity, oxygen, and CO.

Now is the time to start gaining operating experience with the latest combustion efficiency optimizing techniques on larger fuel burning equipment where the fuel saving potential is greatest. Based on this experience, and as the economics for fuel saving become even more favorable, these same techniques can be successfully applied over a period of time to even the smallest fuel burning equipment.

As a result of the rapid escalation in the cost of fossil fuels, the industrial boiler room that for so long has languished in the "Dark Ages" has now become the focal point of the application of the most advanced space age technology. By optimizing the combustion efficiency of industrial fuel burning equipment, a major contribution can be made toward improving industrial energy conservation and reducing our excessive dependence on imported energy.

References

Anson, D.; Clarke, W.H.N.; Cunningham, A.T.S.; and Todd, P.; "Carbon Monoxide as a Combustion Control Parameter", J. Institute of Fuel, April 1971, Page 191.

Dukelow, S.G.; "Combustion Controls Save Money", Instrument & Control Systems, Nov. 1977, Page 35.

Dyer, D.F.; Maples, G.; Maxwell, T.T.; and Jasper, M.T.; "Effects of Oxygen and Carbon Monoxide Control on Boiler Efficiency", In Tech, Jan. 1981, Page 45.

Franczak, D.F.; Paprocki, T.J.; "Locating and Installing Equipment for Continuous Emission Monitoring", In Tech, Nov. 1980, Page 38.

Gilbert, L.F.; "CO Control Heightens Furnace Efficiency", Chemical Eng., July 28, 1980, Page 69.

Gilbert, L.F.; "Precise Combustion Control Saves Fuel and Power", Chemical Engineering, June 21, 1976.

Hendler, H.E.; "The Control of Efficiency in Combustion Type Processes", 70th Annual Mtg. American Institute of Chemical Eng., March 18, 1975.

Jutila, J.M.; "Guide to Selecting Stack Gas Monitors", In Tech, Nov. 1980, Page 43.

Kolmar, J.W.; Neuberger, E.D.; "Programmed O_2 Set Point Optimizes Combustion Control", Instrument and Control Systems, Dec. 1977, Page 33.

KUB, Inc.; "Assessment of the Potential for Energy Conservation Through Improved Industrial Boiler Efficiency", Final Report Volume I, FEA Contract No. C-04-50085-00.

Lord, H.C.; Egan, D.W.; Paules, P.E.; Holstrom, G.B.; "Instantaneous, Continuous, Directly On Stream Boiler Flue Gas Analysis", ISA Power Industry Symposium, NYC, May 17, 1971.

Molloy, R.C.; "Oxygen and Opacity Stack Emission Monitoring Systems to Optimize Combustion Efficiency and Meet EPA Requirements", 1980 Conference On Industrial Energy Conservation Technology, April 1980, From Advances in Instrumentation, Volume 34 15A 1979.

O'Meara, John E.; "Oxygen Trim for Combustion Control", In Tech, March 1979, Page 47.

Opel, A.E.; Gilber, L.F.; "Closed Loop Source Monitoring Saves Energy and Money", ASME Summer Meeting, July 1978, pp 1-8.

Ormerod, W.G.; Read, A.W.; "An improved Method of Combustion Control of Coal-Fired Boilers Using Flue Gas Carbon Monoxide Analysis", Journal of the Institute of Energy, March 1979, Page 23.

Price, Leslie M.; "Maximize Boiler Efficiency and Save Fuel With CO Monitoring", Power Vol. 124, No. 1, Jan-1980, Page 70.

Shore, D.E.; McElroy, M.W.; "Guidelines for Industrial Boiler Performance Improvement - Boiler Adjustment Procedures to Minimize Air Pollution and to Achieve Efficient Use of Fuel", EPA 600/8-77-003A, Jan. 1977.

Shore, D.E.; McElroy, M.W.; "Establish Excess-Air Levels", Power, May 1977, Page 76.

Strange, C.; "Practical Combustion Analysis for Boilers and Furnaces", Energy Utilization Source Book AEE, Fairmont Press, 1980.

Takacs, Laszio; "Improving the Efficienty of Industrial Boilers by Microprocessor Control", Power, Nov. 1977, Page 80.

Vorndran, C.J.; "Using Oxygen Analysis to Improve Boiler Combustion Efficiency", Plant Engineering, May 13, 1976.

Walsh, T.J.; "Controlling Boiler Efficiency with Stack Monitoring Instrumentation", Instrument & Control Systems, January 1981, Page 49.

APPLICATION ENGINEERING HANDBOOKS - Available free of charge from Dynatron, Inc., by calling 1-800-243-3124:

"Everything You Need to Know About Smoke Opacity Monitoring Systems", Dynatron, Inc., 4th Edition 1981.

"Everything You Need to Know About Oxygen Monitoring Systems", Dynatron, Inc., 2nd Edition 1981.

"Everything You Need to Know About CO Monitoring Systems", Dynatron, Inc., 1st Edition 1981.

Adapted from a paper given at the IECTC'81 conference, Houston, Texas, April 26-29, 1981.

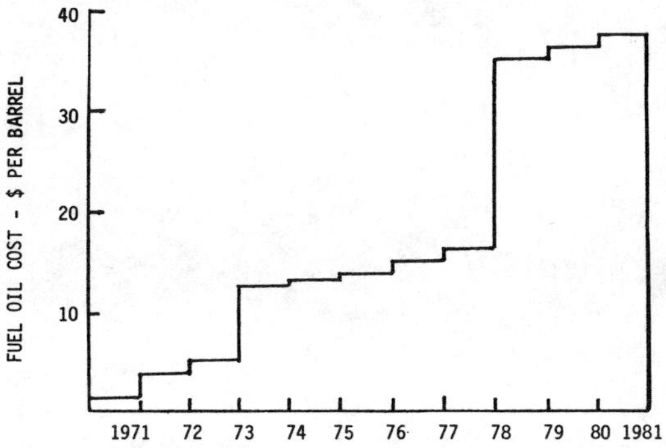

FIGURE 1 RESIDUAL FUEL OIL COST

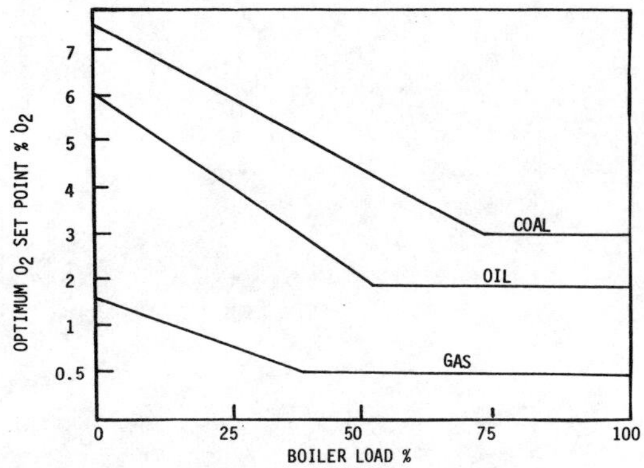

FIGURE 5 EFFECTS OF VARIATIONS OF FUEL AND LOAD ON O_2 SET POINT

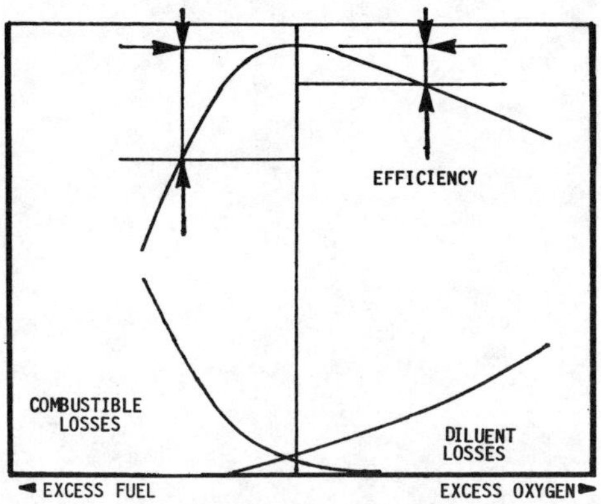

EFFECTS OF EXCESS FUEL AND EXCESS AIR ON COMBUSTION EFFICIENCY

FIGURE 2

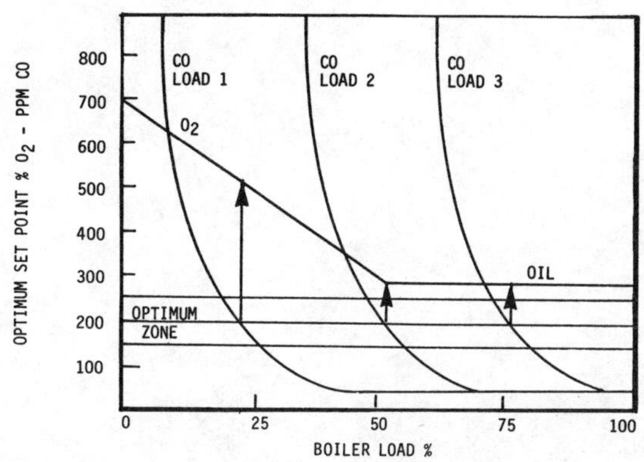

FIGURE 6 EFFECTS OF VARIATIONS OF LOAD ON CO SET POINT AND O_2 SET POINT

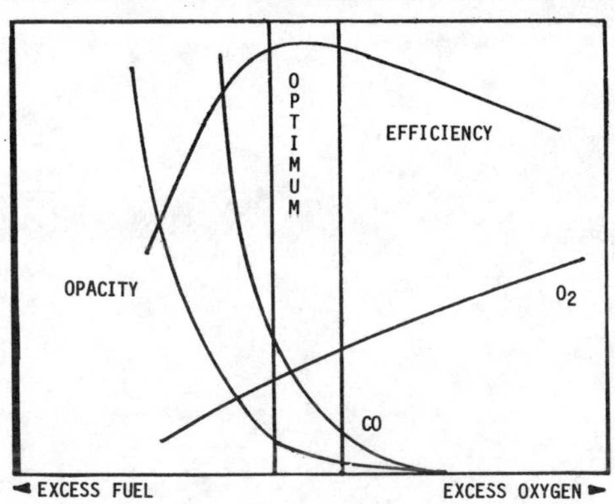

FIGURE 3 FUEL/AIR MIXTURE RATIO

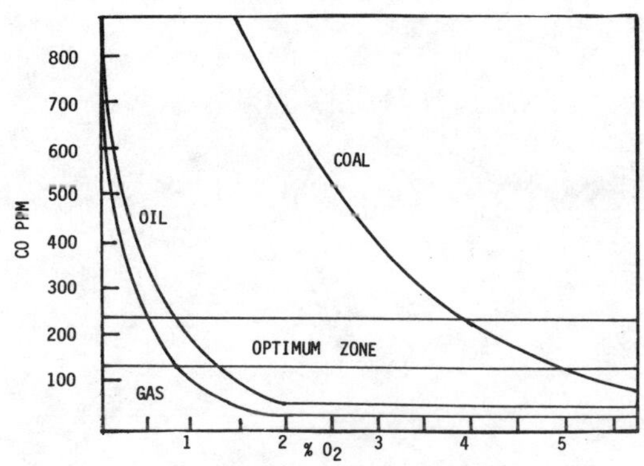

FIGURE 7 EFFECTS OF VARIATIONS OF FUEL ON CO SET POINT

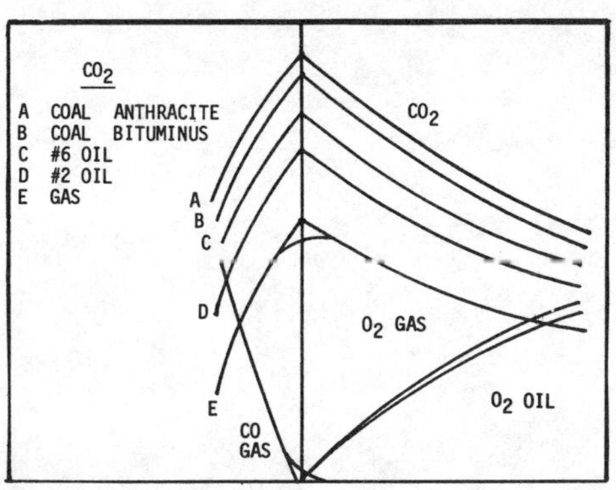

FIGURE 4 PRODUCTS OF COMBUSTION

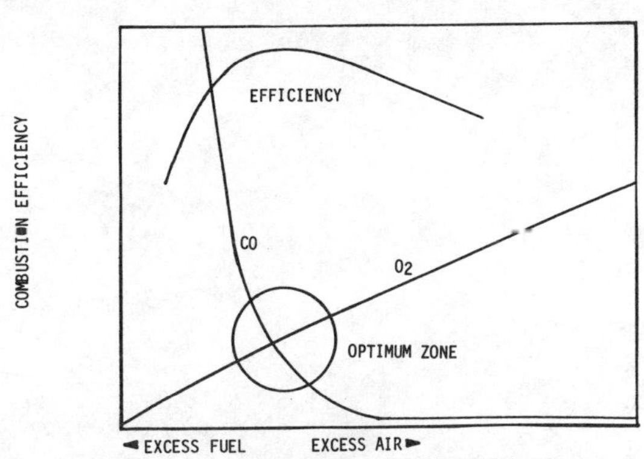

FIGURE 8 OPTIMUM COMBUSTION EFFICIENCY ZONE FOR A GIVEN FUEL AND LOAD LEVEL

Chapter 32

A GENERALIZED APPROACH TO FLARE GAS ENERGY RECOVERY SYSTEM DESIGN

L. C. Hardison, G. J. Nagl, J. J. Pittas

Abstract

As the energy shortage continues to increase daily, the recovery of energy in the form of waste heat and in the reduction of fuel losses becomes more and more necessary and economically attractive. Air Resources, Inc. has been actively involved in waste heat and fuel gas recovery for several years and has developed numerous techniques for expediting the relatively long evaluation and design procedures associated with energy recovery projects of this type. In the examples described herein, recovery of the energy presently lost on a daily basis from flare systems in petroleum refineries and petrochemical plants is explained in regards to flow measurement, establishment of design basis, safety, equipment requirements and economics. It will be shown that even for relatively small systems recovering 500 SCFM of flare gas and having a capital investment of approximately $1,200,000, sufficient energy is recovered to result in a pay out time of less than two years at current prices, and for larger systems recovering 2000 SCFM and costing approximately $1,500,000, pay out times of less than six months result.

Introduction

The primary purpose of a flare system is to provide a means for the safe disposal of combustible vapors generated from scheduled shutdowns of process units, unexpected upsets in process vessels resulting from fires, loss of cooling water, power failures, etc., and controlled vents such as those employed to control a unit's pressure. The flare system generally consists of a header into which is connected various relief valves and vent lines, a knockout drum which removes condensate droplets from the gas stream and a burner which provides safe and smokeless combustion of the vapors. With the exception of process vents, flare systems are designed to operate on an intermittent basis; however, due to corrosion and improper reseating of valve seats, process vapors are generally continuously vented in refineries and petrochemical plants resulting, in some instances, in very large losses of recoverable hydrocarbon products. In addition, flare systems require extensive maintenance to comply with environmental standards regarding opacity and the emission of sulfur and nitrogen oxides. Large flames on flare stacks also result in poor public relations with the public in today's energy conscious world.

Several means of reducing the amount of vapor loss in flare systems, such as the installation of rupture disks and additional relief valves upstream of the main relief valve have been practiced in the past. These methods, however, require considerable diligence and experience in selection of set pressures and valving and additional maintenance for plant personnel. Another method to minimize flare gas loss consists of compressing the vapor and returning it to a fuel gas system or a gas concentration plant. Such a system is illustrated in Figure 1.

Vapor Recovery System

The system shown in Figure 1 consists of three basic parts (the seal drum(s), the compressors(s) and the control scheme) each requiring different design bases. Since a flare system's primary purpose is to provide for safe disposal of vapors during emergency situations, a recovery system must not interfere with this function. The seal drum provides this ability to recover vapor during normal operation and still provide access to the flare stack during emergencies. During normal operations the flare gas is diverted to the vapor recovery compressor by means of a liquid seal. In the event of a major relief, the liquid seal is broken, and the gas is diverted to the flare stack. The system may also be designed such that the seal drum is completely isolated during the event of a major relief. This is beneficial if several flare systems having individual seal drums are tied together into a common vapor recovery compression station. The isolation ability allows continuing vapor recovery from the flare systems which are not experiencing a major relief while the difficulty with the relieving system is being corrected.

The seal liquid may be either a hydrocarbon having a low pour point or water; however, water can present winterizing, corrosion and disposal problems which are eliminated with a properly selected oil. The seal drums must also have a continuous gas purge to the flare stack to prevent air migration down the stack. The quantity of purge gas is dependent on the molecular weight of the available gas, the diameter of the flare header and the type of flare stack. If a molecular or fluidic seal is installed in the system, purge gas rates can be reduced substantially. In conjunction with air migration down the flare stack, the seal drum and upstream flare header must be protected from being depressured below atmospheric pressure and the subsequent leakage of air into the flare leader. This protection is achieved by the installation of low pressure limits which will isolate the seal drum from the recovery compressor.

The type of seal drum is dependent on the individual circumstances under consideration. If space and installation time is available, a relatively small seal drum can be installed immediately upstream of the existing flare knockout drum. Installing a seal drum remote from the knockout drum will result in considerable back pressure in the flare header when the seal is blown. In this case, it is desirable to install a combination seal-knockout drum which is illustrated in Figure 1.

Any type of compressor may be employed; however, reciprocating compressors are generally the most economical for the capacities and pressures associated with a vapor recovery system. Capacity control on the compressor is via suction pressure control consisting of an automatic spillback control and manual or automatic suction valve unloading. As the spillback control valve approaches the fully open or closed positions, the unloaders are adjusted to step capacity, In this manner, power consumption may be optimized throughout the entire capacity range of the compressor. To reduce installation costs, the compressor station can be factory assembled as a "skid-mounted" unit complete with the following items of equipment:

Suction, interstage and discharge K.O. drums
Suction and discharge pulsation bottles
Reciprocating compressor
Suction valve unloaders
Crankcase lubrication system
Induction motor
Intercooler and aftercooler
Spillback system.

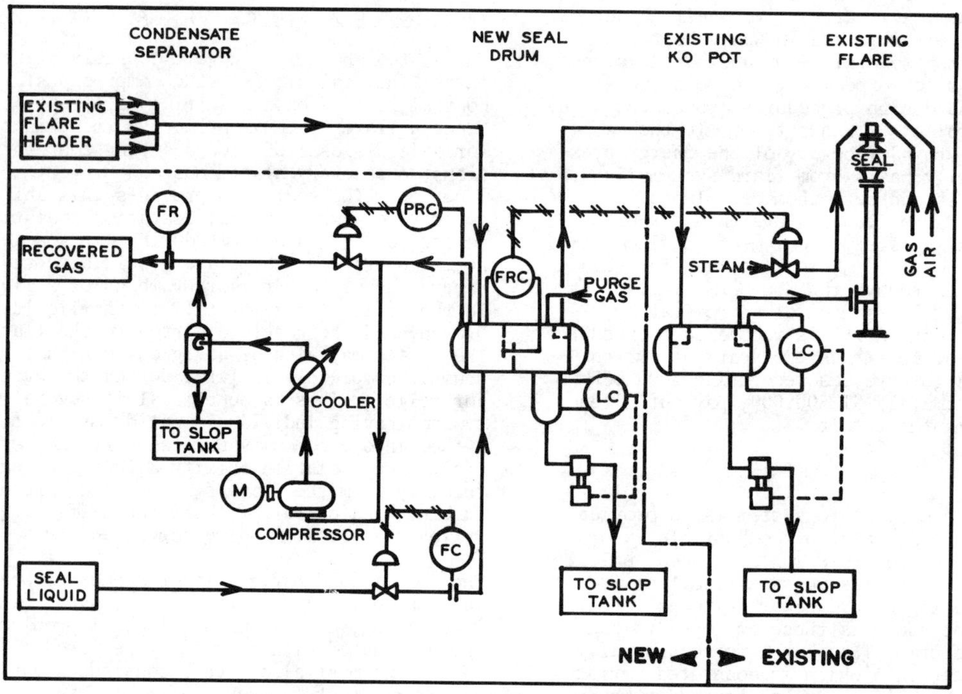

FIGURE I - VAPOR RECOVERY SYSTEM

Design Basis

As previously stated, the design basis for the seal drums and the compressor are quite different. The compressor is sized on the amount of recoverable vapor normally flowing in the flare header while the seal drum must be sized on the anticipated maximum flowrate during major relieving situations. The normal vapor flowrate can be measured by using special techniques; however, determining the maximum flow during relieving conditions is based on considerable design experience and good engineering judgment.

Flare gas rates are not easily measured due to the low velocity heads encountered. For this reason, conventional flowmeters such as orifice plates, pitot tubes, venturis and turbine meters are not useful. Air Resources has developed a technique to measure flare rates which yields accurate and reproducible results. It involves the introduction of a known diluent or tracer gas at a known flowrate into the flare line. Samples of the diluent or tracer and the flare gas before and after the injection point are obtained and analyzed for a key component such as hydrogen sulfide, helium, etc. By performing a component balance, the flowrate is determined. Selection of the diluent or tracer gas is dependent upon the analytical procedures available at the specific location.

Determining the maximum flowrate during major relieving conditions may be a very time-consuming and laborious undertaking which may or may not be required, depending on whether significant additional relieving capacity has been tied into an existing flare system as a plant is expanded. If this practice is not preceded by an investigation of the capacity of the flare system, a potentially dangerous situation may develop. Correction of this type of situation requires obtaining a complete up-to-date inventory of all relief valves and process vents which are connected to the system, preparation of header diagrams showing relative tie-in locations and an engineering analysis of which events such as fire, power failure or loss of cooling water will result in the maximum relieving condition. After this is completed, it can be determined if the existing system is adequate or if it must be revamped to accommodate the additional capacity. the amount of work required to complete this type of investigation is dependent on the accuracy of relief valve inventories and drawings of the flare system.

After the quantity of recoverable vapor has been established, the disposition of the recovered vapor must be determined. This is generally determined by economics and environmental regulations. The gas may be injected into an existing fuel gas system where it will be consumed for its heating value in process heaters and boilers. Prior to injection into a fuel system, removal of hydrogen sulfide may be required depending on local environmental regulations and the quantity of recoverable vapor available. If the recovered vapor contains components which can be employed to produce valuable liquid products such as gasoline, the vapor can be introduced into a gas concentration unit. In most instances, the product values of the liquid components greatly exceeds the value of fuel.

Economics

A typical chemical analysis of a flare gas is contained in Table 1. Estimated capital and operating costs for recovering 500 SCFM and 2000 SCFM of this gas and injecting it into a 65 PSIG fuel gas system and a 125 PSIG gas concentration unit are contained in Tables 2 and 3, respectively. The cost estimates were based on the following conditions:

1. Use of reciprocating compressors.
2. Three flare lines tying into one recovery station.
3. Three new combination seal-knockout drums required.
4. 600 feet from seal drum areas to the compressor area.
5. All utilities available within 300 feet.
6. Explosion-proof electrical classification.
7. Existing pipe racks are available outside of the compressor and seal drum areas.
8. Use of all new equipment.
9. Soil bearing pressure of 2500 PSF.

Table 2 illustrates that the capital cost of a recovery system is relatively insensitive to increases in discharge pressure and capacity; therefore, when deciding to install a vapor recovery system, inclusion of excess capacity in the system to accommodate future plant expansions will result in only a small increase in capital investment. The operating costs developed in Table 3 do not include additional operating labor since this type of system is relatively simple and will have sufficient instrumentation that very little operator attention will be required. Also, purge gas was not considered an operating cost for the vapor recovery system since purging is required for all flare systems with or without recovery. Income taxes, investment credits, etc., were not included in the operating costs since these items vary considerably from company to company and with the specific timing and situation under consideration.

It is very apparent from Table 3 that very fast pay out times of two years or less are realized in all four cases and that recovering the liquid products contained in the flare gas, rather than burning them as fuel, reduces the pay out time by approximately one-half. As the price of fuel continues to increase, the recovery of vapor normally flared will become more and more attractive.

TABLE 1 - TYPICAL FLARE GAS ANALYSIS

Component	Volume %
Hydrogen	35.6
Nitrogen	0.7
Oxygen	0.1
Carbon Dioxide	0.2
Carbon Monoxide	0.1
Hydrogen Sulfide	2.2
Methane	25.4
Ethane	17.2
Propane	9.6
Propene	2.3
Iso Butane	2.6
Normal Butane	1.6
Butenes	0.6
Pentanes	1.1
Pentanes	0.1
Hexanes	0.6
Molecular Weight	20.1
LHV, BTU/SCF	1106
HHV, BTU/SCF	1217

TABLE 2 - ESTIMATED CAPITAL COSTS (1)

SYSTEM CAPACITY, SCFM	HIGH PRESSURE (2) 500	HIGH PRESSURE (2) 2000	LOW PRESSURE (3) 500	LOW PRESSURE (3) 2000
Skid-Mounted Compressor	$ 183,390	$ 389,970	$ 107,820	$ 213,380
Vessels and Pumps	188,880	196,740	188,870	196,740
Instrumentation	145,550	151,610	133,750	139,330
Electrical	71,590	77,500	71,400	77,500
Piping and Insulation	251,100	342,540	256,390	351,760
Painting	30,470	37,060	31,100	38,050
Concrete and site Work	101,550	105,780	95,040	99,000
Mechanical Erection, Shelter Structural, etc.	60,920	63,460	60,920	63,460
Contigency @ 15%	155,020	204,700	141,790	176,880
TOTAL INSTALLED COST	$1,188,470	$1,569,360	$1,087,080	$1,356,100

(1) Each price includes engineering, field supervision, procurement and contractor's fees.

(2) 125 PSIG Gas Concentration Unit

(3) 65 PSIG Fuel Gas System

TABLE 3 - ESTIMATED OPERATING COSTS[6]

SYSTEM CAPACITY, SCFM	HIGH PRESSURE[7] 500	HIGH PRESSURE[7] 2000	LOW PRESSURE[8] 500	LOW PRESSURE[8] 2000
Operating Cost ($/Cal. Day)				
Operating Labor	0	0	0	0
Maintenance [2]	$ 163	$ 216	$ 149	$ 186
Utilities				
Power @ 4.5¢/KWH	70	284	54	214
Water @ 3¢/MGAL	3	13	2	7
Overhead Costs				
Insurance & Property Tax[3]	130	173	119	149
Depreciation[4]	326	433	298	372
TOTAL DIRECT OPERATING COSTS	$ 692	$ 1,119	$ 622	$ 928
Product Values ($/Cal. Day)				
Fuel Gas @ 3.50/MMBTU	$ 1,591	$ 6,363	$ 2,800	$ 11,200
Liquid Products @ $1.00/Gal	3,565	14,255	0	0
TOTAL PRODUCT VALUES	$ 5,156	$ 20,618	$ 2,800	$ 11,200
NET PROFIT[5] ($/Cal. Day)	$ 4,464	$ 19,499	$ 2,178	$ 10,272
INVESTMENT	$1,188,470	$1,569,360	$1,087,080	$1,356,100
SIMPLE PAYOUT TIME (Cal. Years)	0.73	0.22	1.37	0.36

(1) No Additional Manpower Required.
(2) Estimated @ 5% of Installed Cost/Yr.
(3) Estimated @ 4% of Installed Cost/Yr.
(4) Estimated @ 10 Yrs, Straight Line
(5) Before Tax
(6) Estimated @ 8000 Operating Hours/Calendar Year
(7) 125 PSIG Gas Concentration Unit
(8) 65 PSIG Fuel Gas System

Chapter 33
STEAM TRACING OR ELECTRIC TRACING?

J. T. Lonsdale

INTRODUCTION

Both steam and electric energy are used to maintain pipes at desired temperatures. Deciding which system to use involves choosing between a less expensive but less efficient form of energy (steam), and a more efficient but more expensive form of energy (electricity). In the 1960's, the vast majority (over 90 percent) of all pipe tracing systems were steam tracing. Today, a majority of newly installed systems are still steam tracing, but the percentage has dropped significantly (to 70 to 80 percent). Electric tracing has made significant inroads in the petroleum refining industry, which has traditionally used steam tracing.

This paper will consider the trade-offs between the most common forms of steam and electric tracing. These trade-offs have changed significantly due to the dramatic increase in energy costs. Three cases will be analyzed. Energy use will be the main focus but installed costs, maintenance costs, and reliability will also be discussed.

BACKGROUND INFORMATION

Pipes need to be maintained above certain temperatures to prevent a fluid from freezing, to prevent a fluid from becoming too viscous to pump, to prevent a solid from precipitating out of a liquid, to prevent a gas from condensing, or for a number of other reasons. Water pipes are often maintained above their freezing point by burying them below the frost line. In most cases, pipes are maintained above their minimum temperature by a combination of insulation and energy input to the pipe.

Steam energy can be used by running a steam line inside the pipe (gut tracing), surrounding the pipe with a larger pipe and introducing steam in between the pipes (steam jacketing), or by strapping a small pipe or a tube to the pipe and introducing steam into the tube (steam tracing). Steam jacketing and gut tracing are very expensive systems to install. The vast majority of steam systems are steam tracing, and this is the steam system we will consider in this paper.

Electricity may be used to maintain the temperatures of pipes by applying a low voltage current in the pipe itself (pipe impedance systems), by welding a smaller pipe to the pipe and using an insulated conductor to induce a current on the inside skin of the smaller pipe (SECT or Skin Effect Current Tracing), or by installing a resistive electric heating cable on the external surface of the pipe (electric tracing). Resistive electric tracing is by far the most common, and it is the electric system we will consider in this paper.

SYSTEM DESCRIPTIONS

The three major types of resistive electric heat tracing are M.I. (mineral insulated) cable, self-regulating heaters, and zone heaters. Mineral insulated cable designed for pipe heat tracing has been available since the early 1950's. It consists of one or more resistance wire(s) insulated with magnesium oxide (hence the name mineral insulated) and covered with a metal sheath (usually copper or stainless steel). (See Figure 1.) The major advantages of M.I. cable are excellent temperature capabilities (special alloys can be used at over 1,100°F) and good impact resistance. The major disadvantages are that M.I. cable must be factory fabricated to length and nonstandard voltages are often required. It is also not very flexible. These disadvantages make M.I. cable a difficult system to install and a very difficult system to repair or even replace (the exact length required must be obtained from the factory).

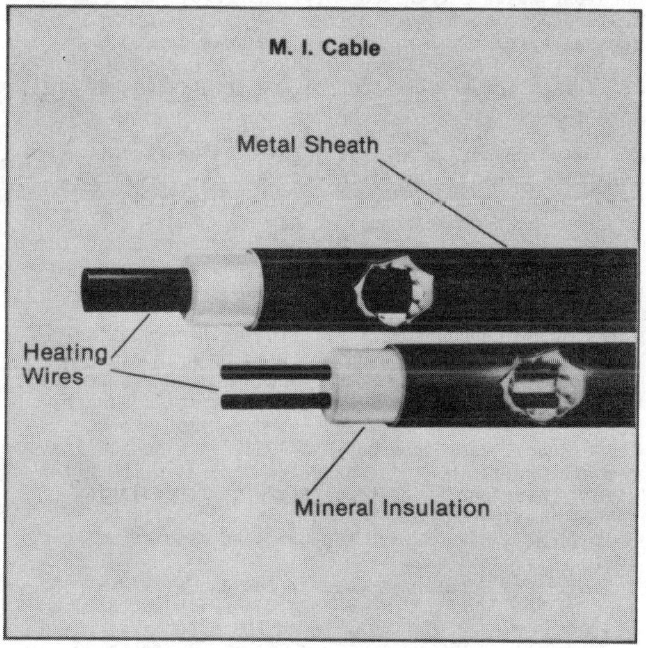

FIGURE 1. TYPICAL M.I. CABLE CONSTRUCTION

Self-regulating heaters were introduced in 1971 by Raychem Corporation. Several other companies have recently entered this field. Self-regulating heaters consist of two current carrying bus wires connected by a conductive polymer. This is covered by a polyolefin or fluoropolymer jacket. A metallic braid and/or a fluoropolymer outer jacket are

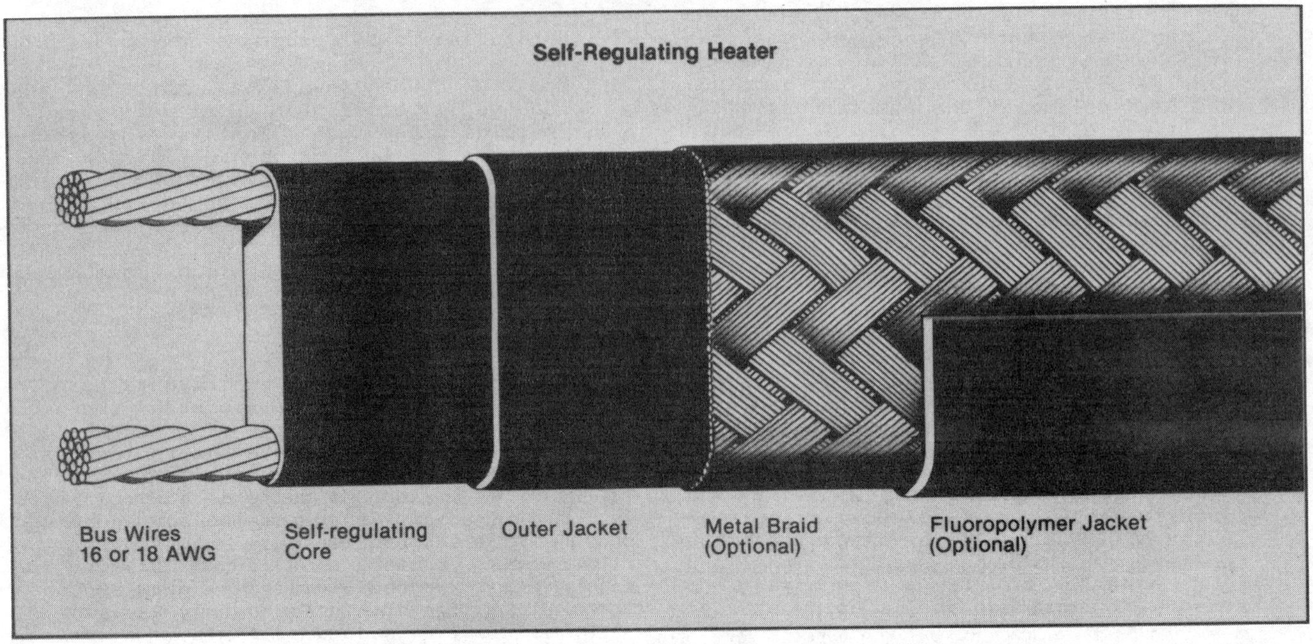

FIGURE 2. SELF-REGULATING HEATING CABLE, TYPICAL CONSTRUCTION

optional. (See Figure 2.) The conductive polymer is the heating element. As the temperature of the polymer increases, its resistance increases, which causes a decrease in the power output. This self-regulating feature means the heater cannot destroy itself with its own heat output as can happen with constant wattage (zone and M.I.) heaters. The self-regulating feature allows the heater to be crossed over itself without excess heat damaging the heater. The self-regulating feature also leads to tighter temperature control and greater energy savings.

The major advantage of self-regulating heaters is that they cannot destroy themselves with their own heat output as can happen with other electric heaters (M.I. cable/zone). Other advantages are that they may be cut to the exact length in the field, they use standard voltages, and have good impact resistance.

The major disadvantage of existing self-regulating heaters is that they cannot be used to maintain temperatures above 250°F and should not be exposed to temperatures above 366°F. (Maximum maintain temperatures vary from 150°F to 250°F and maximum exposure temperatures vary from 185°F to 366°F.) The 366°F temperature is the temperature of saturated 150 psi steam; therefore, high-temperature self-regulating heaters may be used to trace lines which will be periodically purged with 150 psi saturated steam.

Zone heaters were introduced in the early 1970's. A zone heater consists of two bus wires insulated with plastic, wrapped with a 38 to 40 AWG nicrome wire and covered with another plastic insulation. A metal braid and/or an additional outer jacket is optional. The nicrome wire is connected to alternate bus wires at the nodes every 6 to 48 inches. (See Figure 3.) These are the heating zones from which the heater gets its name.

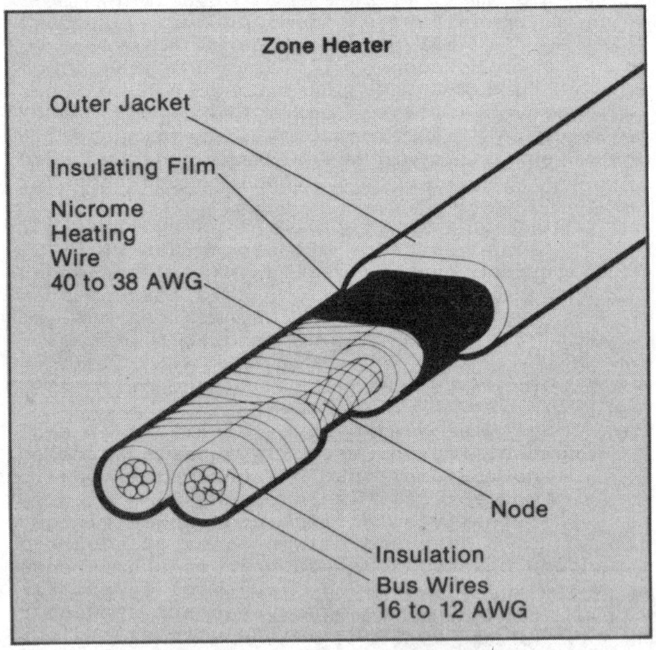

FIGURE 3. ZONE HEATER TYPICAL CONSTRUCTION

The major advantage of zone heaters is that they may be cut to length in the field; however, it is important to be aware of the relationship of the cut to the nearest node.

The major disadvantage of zone heaters is their inability to withstand impact. The very small size of the heating wire (38 to 40 AWG) and the difficulty of attaching the small nicrome heating wire to the copper bus wires are the cause of this characteristic.

Electric systems vary in installed costs, energy use, and reliability. The differences between electric systems in installed cost and energy use are small, particularly when compared to the large differences between electric systems and steam systems. The major difference between electric systems is reliability. This issue will be discussed towards the end of the paper.

Steam tracing systems consist of a supply of steam, a tube (usually 1/2-inch copper) strapped to the pipe, a steam trap to hold back the steam and remove the condensate, and a system to return the condensate to the steam boiler. Both the steam supply line and the condensate return line should be insulated to save energy and prevent burns. The major advantage of steam tracing is that it is a very reliable system. The major disadvantage of steam tracing, as will be seen in the following sections, is that it is very expensive to install and to operate.

ENERGY USE

Table One explains the equations used to calculate the heat lost from a traced and insulated pipe. It is interesting to note the difference between the equation for a steam-traced pipe (Equation 5) and the equation for an electric-traced pipe (Equation 7). Equation 5 has an additional resistance term representing the resistance to heat flow from the steam tracer to the pipe. More importantly, the equation for steam tracing is based on the temperature difference between the steam tracer (essentially the steam temperature) and the ambient air. The electric tracing equation is based on the temperature difference between the pipe and the ambient air. This is due to one of the major operational differences between steam tracing and electric tracing. Electric systems generally control the temperature of the pipe by use of a line sensing thermostat, but it is very difficult to control the temperature of a steam-traced pipe. It is possible to use a thermostat and an automatic valve on the steam supply line to control the supply of steam as a function of pipe temperature. However, these systems are generally very expensive and of questionable reliability. The vast majority of steam tracing systems use no temperature control.

TABLE ONE
HEAT FLOW EQUATIONS

$$Q = \frac{T_2 - T_1}{R_1 + R_2 --- + R_n} \qquad \text{(Equation 1)}$$

Heat flow from Temperature 2 to Temperature 1 with R_n thermal resistors in series between T_2 and T_1. If T_2 is the temperature of the steam tracer and T_1 is the ambient air temperature surrounding a steam-traced and insulated pipe, the resistors are:

$$R_1 = \frac{1}{\pi D_2 h_o} \qquad \text{(Equation 2)}$$

R_1 is the resistance of convective heat flow from the insulation lagging to the ambient air.

D_2 = outside diameter of insulation (ft).

h_o = heat transfer coefficient (BTU/hr ft^2 °F) for convective heat transfer.

$$R_2 = \frac{\mathrm{Ln}\,\dfrac{D_2}{D_1}}{2\pi K} \qquad \text{(Equation 3)}$$

R_2 is the resistance of the pipe insulation to conductive heat flow.

D_1 = outside diameter of pipe (ft).

K = Thermal conductivity of insulation BTU ft/hr ft^2 °F.

NOTE: If K is expressed in units of (BTU in/hr ft^2 °F), it is necessary to multiply by 12 in/ft to keep the equation dimensionally consistent.

$$R_3 = \frac{1}{C_T} \qquad \text{(Equation 4)}$$

R_3 is the resistance to heat flow between the steam tracer and the pipe.

C_T = heat transfer coefficient from steam tracer to pipe (BTU/hr °F ft). (See Kohli[1] for values of C_T and h_o.)

Combining equations 1 through 4, the heat flow from the steam-traced and insulated pipe becomes:

$$Q\ (\text{BTU/ft hr}) = \frac{T_{steam} - T_{ambient}}{\dfrac{1}{\pi D_2 h_o} + \dfrac{\mathrm{Ln}\,(D_2/D_1)}{2\pi K} + \dfrac{1}{C_T}} \qquad \text{(Equation 5)}$$

When the fluid in the line is flowing, heat flows from the tracer to the pipe and is carried away by the flowing fluid. A tracer (or even several tracers) is not usually capable of significantly raising the temperature of a flowing fluid because of the large thermal mass of a flowing fluid. The equation for heat transferred from a tracer to the fluid is:

$$Q = C_T\,(T_{steam} - T_{pipe}) \qquad \text{(Equation 6)}$$

An electric tracer controlled by a pipe sensing thermostat will cycle on and off to maintain the pipe at the desired temperature. The heat flow from the pipe is therefore made up by the heater. The equation for the heat lost from the pipe is:

$$Q = \frac{T_{pipe} - T_{ambient}}{\dfrac{1}{\pi D_2 h_o} + \dfrac{\mathrm{Ln}\,(D_2/D_1)}{2\pi K}} \qquad \text{(Equation 7)}$$

The systems analyzed are shown in Table Two. A broad range of maintain temperatures and pipe sizes was chosen. The total energy used by each system is shown in Table Three. The energy costs shown in Table Three are \$5.80 per million BTU for steam energy and \$14.60 per million BTU for electricity. Electric energy is often twice as expensive as steam energy, and a ratio as high as 2.5 to one (as used in this paper) is not uncommon. As can be seen from Table Three, the high cost of electric energy, in this case, is more than made up for by its high efficiency.

TABLE TWO
SYSTEMS ANALYZED

System (Fluid)	WATER	FUEL OIL	SULPHUR
Maintain Temperature (°F)	50	140	290
Type of Electric Heater Used	Self-regulating	Self-regulating	Zone
Heater Power	8 watt/ft	20 watt/ft	3x10 watt/ft
Pipe Diameter (in)	4	2	8
Pipe Length (ft)	500	400	100
Insulation Thickness (in)	1.0	1.5	3.0
Insulation K Value (BTU-in/hr ft^2°F)	0.25	0.25	0.25
Minimum Ambient Temperature (°F)	-10	-10	-10
Average Ambient Temperature (°F)	50	50	50
Steam Pressure (psi)	150	150	150
Number of Steam Tracers Used (1/2-inch copper tubing)	1	1	4
Distance from Steam Supply (ft)*	30	30	30
Distance from Electric Supply (ft)*	300	300	300

Other considerations: The freeze protection season is five months long. Average heat losses are calculated assuming a 10 mph wind.

* Both the steam and electric supply must be run from the supply point to one end of the line (30 feet or 300 feet, respectively), then along the line to within one circuit length of the end of the line.

TABLE THREE
ENERGY USE
(10^6 BTU/yr)

System	Water	FUEL OIL	SULPHUR
ELECTRIC SYSTEMS			
Tracers (no flow in pipe)	5.5	21.9	24.2
Tracers (flow)	0	0	0
Total (10^6 BTU/yr)	5.5	21.9	24.2
Total $/yr *	$80	$320	$353
Total $/ft yr	$0.16	$0.80	$3.53
STEAM SYSTEMS			
Tracers (no flow in pipe)	52.0	57.0	27.8
Tracers (flow)	108.0	149.0	45.6
Steam Supply Line	40.4	75.6	8.1
Steam Traps Functioning	10.2	24.8	16.5
Steam Traps Failed	16.0	39.0	26.0
Total (10^6 BTU/yr)	226.6	345.4	124.0
Total $/yr *	$1,314	$2,003	$719
Total $/ft yr	$2.63	$5.01	$7.19

* Costs based on electricity at $0.05/Kwh and $5.00/1000# of steam or $14.60/$10^6$ BTU for electricity and $5.80/$10^6$ BTU for steam. (150 psi steam only latent heat of vaporization is utilized.)

In Table Three, the tracer energy uses with no flow in the pipe are calculated from Equations 5 (steam) and 7 (electric) of Table One. This is the energy used to prevent the pipe from cooling down when the line is not flowing. Because of the line-sensing thermostat, the electric systems generate enough heat to maintain the desired temperature and no more. The uncontrolled steam systems put out considerably more heat than necessary in the water and fuel oil cases; however, in the sulphur case, the steam tracer generates only slightly more heat than necessary. The sulphur line is large (8 inches) and being maintained at a high temperature (290°F); therefore, multiple tracers are needed. This means the steam tracing system can be accurately sized by choosing the proper number of tracers. The water and the fuel oil line (and the majority of lines traced) require less than one full tracer. In most cases it is not practical to use lower pressure steam or smaller tracers due to circuit length problems and other considerations. Therefore, most lines are steam traced with one tracer, which is often considerably too much heating capacity. This agrees with Decker's[2] conclusion that as the maintain temperature increases, steam tracing becomes less inefficient.

The energy used when the fluid is flowing in the pipe is zero for the electric system because the maintain temperature is set below the flow temperature. It was assumed that the pipes flowed 50 percent of the time at a temperature of 10°F above their maintain temperature. The steam tracers use more energy when the system is flowing because the nonflowing pipe is heated to a higher temperature by the steam tracer. Since there is a greater temperature difference between the tracer and the flowing pipe, more heat is transferred to the flowing pipe. This energy use can be particularly large in the case of low temperature lines which are flowing a large percentage of the time.

The energy lost from the steam supply line is calculated from Equation 7 (Table One) by assuming the supply line is at the temperature of the steam. It was assumed that a supply of steam was available 30 feet from one end of each line being steam traced. The steam supply line had to be run within one circuit length of the end of the line being traced. The energy used by the sulphur line steam supply is small because only 30 feet of supply line were required. The energy used by the fuel oil steam supply line is considerably greater than that for the water steam supply line because the fuel oil tracing system is on twelve months per year versus five for the water tracing system.

The other energy uses shown in Table Three are the energy used by steam traps. Steam traps use energy when they are properly functioning and use considerably more energy when they fail. Steam traps are designed to fail in the open position in which live steam is allowed to pass through the trap. This is preferable to failing in the closed position and possibly allowing the line to freeze, but a failed steam trap wastes a large amount of energy. The amount of energy lost from failed steam traps depends on the size of the opening, the average life of the trap, and the frequency at which traps are checked and replaced. A 1/8-inch orifice, an average life of three years, and a maintenance program that checked every trap three times a year was assumed. If the traps were checked yearly instead of three times a year, the energy lost from failed traps would be three times that shown in Table Three or the equivalent of $678.50 per year for the fuel oil system which only has three steam traps. This shows the importance of a good steam trap maintenance program.

The total energy use and energy costs shown in Table Three stress the large advantage in energy use that electric tracing has over steam tracing. These results are somewhat sensitive to the percentage of flow assumption. For example, if the flow assumption is changed from 50 percent flow to zero percent flow (this is the most favorable case for steam tracing), the electric energy cost for the water and fuel oil systems will still be considerably less than the steam energy costs. However, if the sulphur line is in the no-flow state 100 percent of the time, the energy cost for the steam tracing system would slightly less than for the electric system ($616 to $708/yr.).

MAINTENANCE COST

The maintenance cost of each system is shown in Table Four. Steam tracing systems require considerably more maintenance because of the high probability of failed steam traps and steam leaks. If the maintenance is not done, the energy lost from leaks and failed traps increases significantly. Some chemical plants check their steam traps every month to minimize steam losses. As shown in Table Four, electric systems require very little maintenance.

INSTALLED COST

The installed cost for each system is shown in Table Five. The installed costs of the steam tracing system vary from almost four times the electric installed costs for the water line to twice the electric installed costs for the sulphur line. It is interesting to note (see Table Five) that the percentage of the total cost which is labor varies from 37 percent to 50 percent for the electric systems and from 80 percent to 85 percent for the steam tracing systems. This confirms the report by Luke and Miserlis[3] that electric tracing systems cost less on an installed basis because they require significantly less labor. The installed cost of each system is very dependent on the length that the steam or electric supply line must be run. In each case, it was assumed that a supply of steam was available 30 feet from one end of the line and a supply of electricity was available 300 feet from the line. This may seem to be unfair to the electric systems, but in a processing plant it is not uncommon for steam lines to be more readily available than electric power. Also, in cases where steam is not readily available, it will usually not be used because of the expense of running and insulating a long steam line.

The installed cost is also a strong function of the number of tracers required. Because of the high maintain temperature and large line size, the sulphur line requires multiple tracers for both electric tracing and steam tracing. Multiple tracers and the need for a long (300 feet) electric supply to trace a short (100 feet) line are the major reasons the installed costs for the sulphur line are considerably larger than the costs for the water and fuel oil line.

TABLE FOUR
MAINTENANCE COSTS

	MAN-HOURS		
	WATER	FUEL OIL	SULPHUR
	(2 circuits)	(3 circuits)	(1 circuit)
ELECTRIC TRACING			
Megger Circuits Once/yr, 0.25 hrs/circuit	0.5	0.75	0.25
Check Thermostat Once/yr	0.5	0.5	0.5
Total Man-Hours	1.0	1.25	0.75
Yearly Maintenance Cost @ $20/hr.	$20	$25	$15

	MAN-HOURS/$ MATERIAL		
	WATER	FUEL OIL	SULPHUR
STEAM SYSTEMS			
Check Steam Traps 3 times/yr, 10 traps/hr	0.9	0.9	0.6
Replace Traps* 33%/yr failure rate	0.3/$20	0.3/$20	0.2/$14
Check and Repair Leaks, Clean Strainers 1.5 hr/yr	1.5	1.5	1.5
Total Material	$20	$20	$14
Labor Hours/Cost @ $20/hr	2.7/$54	2.7/$54	2.3/$46
Total Maintenance Cost $/yr	$74	$74	$60

* Assume in-line replaceable traps, 0.3 man-hours and $20 to replace.

TABLE FIVE
INSTALLED COST

	WATER		FUEL OIL		SULPHUR	
	Labor	Total	Labor	Total	Labor	Total
ELECTRIC SYSTEMS						
Power Supply	2568	3950	2568	4000	1400	2150
Tracers	400	2573	320	3446	300	920
Thermostats	60	463	60	463	60	463
Total	3028	6986	2948	7909	1760	3533
Percent Labor		43%		37%		50%
Installed Cost ($/ft)		$13.97		$19.77		$35.33
STEAM SYSTEMS						
Steam Supply (and Insulation)	10206	11929	7938	9485	936	1105
Tracers	5500	6000	4400	4800	4400	4800
Trap Assemblies	420	853	420	853	280	567
Condensate Return*	5841	8234	4730	6673	330	493
Total	21967	27016	17488	21811	5946	6965
Percent Labor		81%		80%		85%
Installed Cost ($/ft)		$54.03		$54.53		$69.65
Ratio of Installed Costs: Steam Tracing Costs / Electric Tracing Costs		3.9		2.8		2.0

* Pre-insulated tubing was used for the condensate return systems.

Total installed costs are also a function of labor costs (a fully loaded cost of $20 per hour was assumed), labor productivity, material costs, supervision productivity, and many other variables. Most of the data used to cost these systems was taken from some recent work done by a major engineering company. The total cost for any of the above systems could change considerably if different assumptions were made; however, the reader would be correct in assuming that in most cases, electric tracing will be considerably less expensive to install than steam tracing. Several large companies have recently studied this issue to choose the type of tracing to use on large projects. In each case, it was concluded that the installed costs for electric tracing were considerably less than the installed costs of steam tracing.

SYSTEM RELIABILITY

The importance of reliability for tracing systems cannot be over-emphasized. Money saved by installing a less expensive system can easily be lost through one or two poorly timed failures. Before attempting to analyze the reliability of various tracing systems, it is important to define what we mean by reliability. Reliability is best defined as the lack of failure. Failure is defined as the pipe temperature dropping far enough below the desired maintain temperature to cause a problem such as freezing of the pipe, the inability to pump the fluid due to high viscosity, or problems with precipitation or condensation. This may seem to be obvious, but it is not uncommon for operating personnel to say steam tracing is an unreliable system because of the continual maintenance required to repair leaks and replace failed traps. This would not fall within our definition of reliability because a failed steam trap or a steam leak would, in most cases, not lead to a failure.

Steam is actually a very reliable system. Most of the things that can go wrong with a steam tracing system lead to wasted energy but not to a failure. As indicated in Table Six, steam tracing is very rugged, it cannot destroy itself with its own heat output, as can happen with some electric heaters, and it even puts out slightly more heat when the pipe gets cold.

The major types of electric systems differ in installed cost and energy use, but these differences are small compared to their differences with steam tracing. However, electric systems differ considerably in reliability. The destruction of an electric heater by its own heat output, or burnout, is a major concern to potential users of electric heat tracing. Burnout can occur when a constant wattage heater (zone or M.I.) is not in contact with a heat sink such as a pipe. For example, if a piece of the heater is caught inside the insulation, it will continue to generate a constant amount of heat. This can raise the heater to a temperature which destroys the heater. Not only is a burnout likely to lead to a system failure, but the temperature excursion which causes the burnout can cause a fire or an explosion.

Self-regulating heaters lower their heat output independently at every point along the heater in response to an increase in temperature. A self-regulating heater cannot destroy itself with its own heat output. This is a major advantage in reliability. Other reliability differences between the various systems are outlined in Table Six.

SUMMARY

Steam energy is less expensive than electric energy, but electric heat tracing systems cost less to operate because they are much more energy efficient. The reason for the higher efficiency of electric heat tracing is that it is easy to control while steam tracing is difficult to control. As the maintain temperature of the pipe increases, the energy cost advantage of electric tracing decreases, but the electric energy costs for the sulphur example (290°F) were approximately one-half the steam energy costs.

The maintenance costs for a steam tracing system are considerably larger than for an electric tracing system. If the maintenance is not done, the steam tracing system will use even more energy.

The installed costs of a steam tracing system are usually significantly larger than for a comparable electric tracing system even if the steam supply is significantly closer to the line than the electric supply.

TABLE SIX
SYSTEM RELIABILITY

	Steam	Self-Regulating	Zone	M.I. Cable
Impact Resistance	Excellent	Good	Poor	Good
Destruction by Own Heat Output	Will not destroy itself	Will not destroy itself	May destroy itself	May destroy itself
Can be Crossed Over Itself	Yes	Yes	Sometimes*	No
More Heat Output at Low Temperature (to prevent freeze-up)	Slightly	Yes	No	No

* Lower wattage zone heaters may be crossed over without damage to heater

Steam tracing is a very reliable system. The reliability of electric tracing systems varies considerably. The fact that self-regulating heaters cannot be destroyed by their own heat output gives self-regulating heaters a large reliability advantage over other electric heaters.

References

1 Kohli, I.P., "Steam Tracing of Pipelines," Chemical Engineering, March 26, 1979, p. 156.

2 Decker, Roger L., "Electric versus Steam Energy for Pipe-Line Tracing," IEEE Transactions on Industry and General Applications, Vol IGA-6 No. 5 Sept./Oct. 1970, pp. 517-523.

3 Luke, A.G. and Miserlis, C.D., "How Steam and Electric Tracing Compare in Plant Operations," Oil & Gas Journal, November 7, 1977.

Chapter 34

COOLING STORAGE: AWAKENING GIANT

C. D. MacCracken

ABSTRACT

In 1977, Robert Tamblyn, published his ASHRAE prize-winning article, "Thermal Storage: Sleeping Giant". Today this giant is awakening, stirred by the huge market in air conditioning, where cooling storage can cut the cost of the HVAC contract by 30% and the operating cost by 25%-50%.

Also calling to this storage giant are the electric utilities who see the great bargain of saving up to $2,500 per KW in building generating plants, by having their new customers buy storage that will replace the need for additional capacity at no cost, in fact a saving. Even for old customers, the cost of storage will only be $300 to $500 per KW saved.

This paper describes the new all-plastic modular ice storage equipment, which has made these low costs possible, using recirculating glycol brine and standard package chillers. Size and cost of duct systems is cut in half by the use of low CFM, colder air in non-metallic ducts. Actual installations will be shown with performance data and economic savings.

INTRODUCTION

Ice storage for commercial air conditioning and industrial processes is a proven and rapidly growing product line which is expected to become the main vehicle for utility load levelling in the United States in the next ten years. Installed costs are as low as $50 per ton-hour with storage densities of 4 ton-hours per square foot, making storage far cheaper than generating capacity, now estimated as high as $2500 per KW. In fact, in retrofitting an air conditioning system run 10 hours per day reaching a monthly peak condition of 1.5 KWH/ton-hour, the cost of shifting a peak KW is only $333 ($50/ton-hr x 10 hrs divided by 1.5 KWH/ton-hr). In new construction of non-residential buildings, 24-hour operation of small size chillers building ice at night can mean substantial first cost savings in addition to lower demand charges and load levelling.

Old style water/ice systems, such as ice builders and ice shuckers, have been supplemented with lower cost modular solid ice systems in plastic tanks where closely spaced internal synthetic tubes carrying antifreeze liquids provide heat transfer. Standby losses are reduced, control strategy is simplified, and expansion forces during solid freezing have been eliminated by counterflow temperature averaging. Perhaps most important, standard chilled water packaged air conditioning units are used eliminating field refrigerant work and its associated problems.

The lower ice-cooled coil temperatures, producing air down to 40°F, allows for much lower CFM and consequently smaller blowers using less energy. The 5 to 10% lower efficiency (coefficient of performance - COP) of making ice at night at 15°F refrigerant suction and 80°F discharge temperatures compared to more normal air conditioning figures of 35°F suction and 90°F discharge temperatures is offset by the lower blower power. Thus the long time concern of energy loss in ice storage has been shown in actual practice not to occur due also to the very low standby tank loss when ice is frozen solid.

Ice storage has been used for over fifty years in the United States where very short term cooling requirements allow much smaller refrigeration units to be used with consequent first cost and utility demand charge savings. Examples are churches for once a week air conditioning, and dairies for once a day chilling of incoming milk.

The smaller size of ice storage compared to chilled water enabled it to be a factory-made, shippable product reducing installed cost and design uncertainties. The cost advantages to the utility are clear. In new construction costs are lower than standard air conditioning. In retrofitting the payback is two years.

Why then has ice storage not grown rapidly? It seems that the only thing holding it back is the newness of practical equipment and the instinctive worry that ice will expand and break things. It is to these practical concerns that this paper is first focused.

Designed for Low Cost

Direct expansion (unitary DX) systems, where ice is frozen directly on metal refrigerant (such as Freon) evaporator tubes or plates, have been the design of prior years. In some cases the ice is built to depths of up to 3" and retained on the submerged tubes in a water tank. This is referred to as ice building. See Figure 1. In other cases the ice is periodically shucked, or harvested, into a bin and new ice is built on the evaporator. Shucking can be done by scraping, cutting or melting the bond to the metal by several different methods. The water in the tank or bin surrounding the ice is then pumped to the cooling load. See Figure 2.

By using a secondary refrigerant automotive antifreeze brine system, the evaporator is a conventional shell-and-tube chiller and ice can be frozen solid on a closely spaced inexpensive grid of flexible plastic tubes in multiple plastic tanks. Expansion forces are avoided as the ice freezes solid by building ice uniformly throughout by the temperature averaging effect of adjoining counterflow tubes. Since water does not get surrounded by ice and can move sidewards along or upwards between the tubes as ice grows, no damage will occur. See Figure 3.

In addition to the use of plastics and a conventional chiller, costs are low because of small refrigerant volume, no refrigerant field work, ease of shipping and handling, and adaptability to available space by the modular brine concept. PVC cemented header piping as used in many ice rinks is a further cost saver.

Cost and Type of Use

The percent time that cooling from storage is used is critical in its practicality. A church that uses cooling from storage only six hours per week, or 4%, is an extremely practical use of storage and a dairy that chills milk six hours out of twenty-four, or 25% is also very good.

When the use percentage approaches 40 to 50%, a break-even is reached between the cost of a small chiller unit with ice storage and a larger one without storage. See Figure 4.

When a summer peaking utility company designs off-peak rates to induce cooling storage, it should first of all pick a category, like offices, theaters, factories, schools, or stores, where cooling is not used all day; secondly, it should make the ratio of off-peak time to on-peak time as large as possible, preferably 50% or above, and thirdly, it must pick a category that has a large number of operating hours per year over which to amortize the cost.

Thus, residences are a poor choice since they are likely to be used all day and night in the hottest weather and yet used a small number of hours per year compared to commercial applications. To succeed in the residential market time of day rates will need to have high ratios (perhaps 6 to 1) between on and off peak, while commercial uses may well be practical just by the reduction of demand charges.

Ice and Distribution System Costs

Compared to chilled water storage or a 45°F (7°C) phase change storage, ice, while costlier to generate, reduces the cost of ducts, blowers and heat exchangers quite considerably. In some buildings such as schools or offices, conventional ducts can run as high as 60% of the total HVAC cost. When the non-frosting colder coils produced by ice storage are used, ducts and fancoils can be 50% to 75% smaller.

Another advantage of the modular brine form of ice storage is that tanks can be located in various wings or floors of a building, thus shortening ducts still further at the expense of much cheaper PVC piping. Shorter ducts invariably mean quicker response time and lower losses.

Performance of a Commercial Installation

Figure 5 shows a 12-tank installation at an office/factory in Braintree, Massachusetts totalling 400 ton-hours of storage. The three 20 HP Trane chillers produce about 42 tons (70% of normal rating) when making ice at night and freeze the tanks solid in 10 hours. During business hours the chillers run as needed to supplement the ice storage in order to supply the 137 ton design load which requires up to 1000 ton-hours per day. In 10 hours the three chillers can produce 600 ton-hours which plus the 400 ton-hours in storage gives the required peak cooling.

The first cost of the system in this new enlargement of an older section cooled by rooftop units was $178,000 vs. $220,000 bid for a conventional central system. Operating cost per square foot appears to be substantially lower than the rooftops due not so much to demand charge reductions (the local municipal utility charges only $1.20 per KW) as to the following factors:

a) the lower relative humidity caused by the colder air and higher latent heat removal allows thermostats to be set at 80°F (30% R.H.) compared to 75°F (60% R.H.) in the old section

b) the faster response time (17 seconds to deliver 43°F air to room outlets) eliminates the need for precooling before business hours begin

c) air washing due to the fully wetted coil plates and high condensate removal permits less outdoor make-up air to be used for comparable air quality

d) in winter heating is provided by using the chillers in a heat pump mode with heat provided by freezing ice and melting it from waste heat process injection molding equipment.

Based only upon one season's performance with better comfort and lower costs, the management has decided to convert to a central ice system in the old part as well. The rooftop units will be converted into fancoils with PVC piping interconnecting them from central ice tanks and chillers. Alternatively an ice tank can be located at each rooftop fancoil to save space.

Economic Analysis of Typical New Construction and Retrofit Installation

Example - Small Storage: Assume a new 100 ton office building air conditioning ten hours per weekday in a climate where it must be turned on at least once in each of nine months with non-ratcheted monthly demand charges averaging $8.00 per month per KW over the next ten years:

Conventional		
100 tons @ $600/ton installed	$60,000	
Distribution system	60,000	$120,000
With 1/2-sized chiller operating 24 hours per day and ice storage		
50 tons @ $600/ton installed	$30,000	
(50x10) ton-hrs @ $50/ton-hr installed	25,000	
Distribution system	30,000	$ 85,000
First cost savings from ice storage		$ 35,000
Savings on demand charges: 50 tons x 1.5 KW/ton x 9 mos/yr x $8.00 demand x 10 yrs		$ 54,000
Total savings first 10 years		$ 89,000

Payback: Immediate + $35,000

Example - Large Storage: Assume an existing 100 ton office building where chiller is operated 1000 hours per year and at least once a month for nine months with the same $8.00 average demand charge over ten years plus a 10-hours off-peak rate that saves an average of $.05/KWH over the next ten years.

Add Ice Storage for 100% Off-Peak	
10 hrs x 100 tons x 70% x $50/ton-hr installed	$ 35,000

Savings

Demand: 100 tons x 1.5 KW/ton
x 9 mos/yr x $8.00
demand x 10 yrs $108,000

Energy: $.05/KWH x 1000 hrs/yr
x 120 KW x 10 yrs $ 60,000 $168,000

Payback: $35,000 / 168,000/10 2 years

Return on Investment 50%

Conclusion

Ice storage for load levelling (small storage) and load shifting (large storage) has now reached the practical commercial level and will grow very rapidly in the next few years.

Figure 3 - CALMAC'S modular ice storage tanks use anti-freeze liquid to freeze water into solid ice.
Bulk eutectic salts can also be used as thermal storage media.

Figure 1 - Typical unitary DX system in use at a Pennsylvania dairy

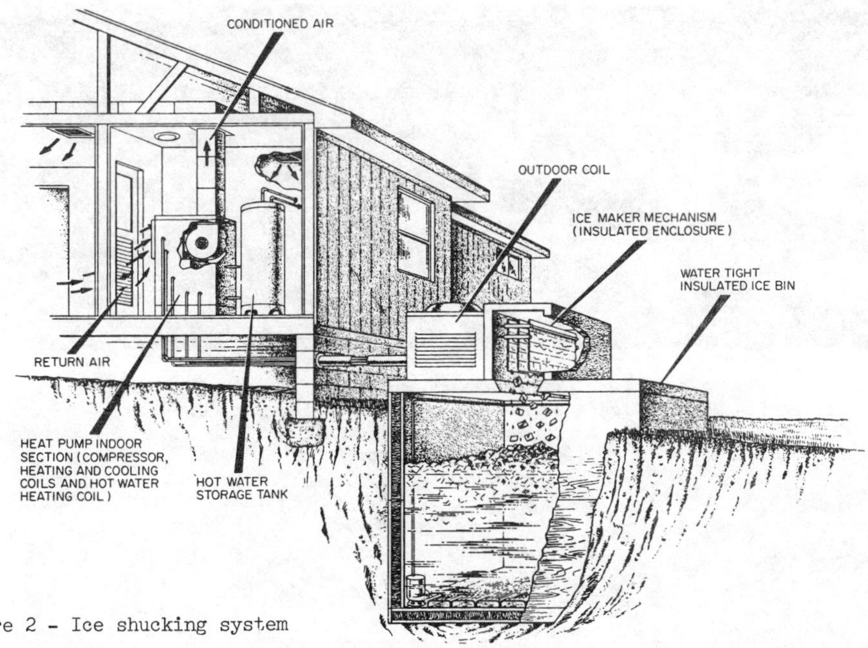

Figure 2 - Ice shucking system

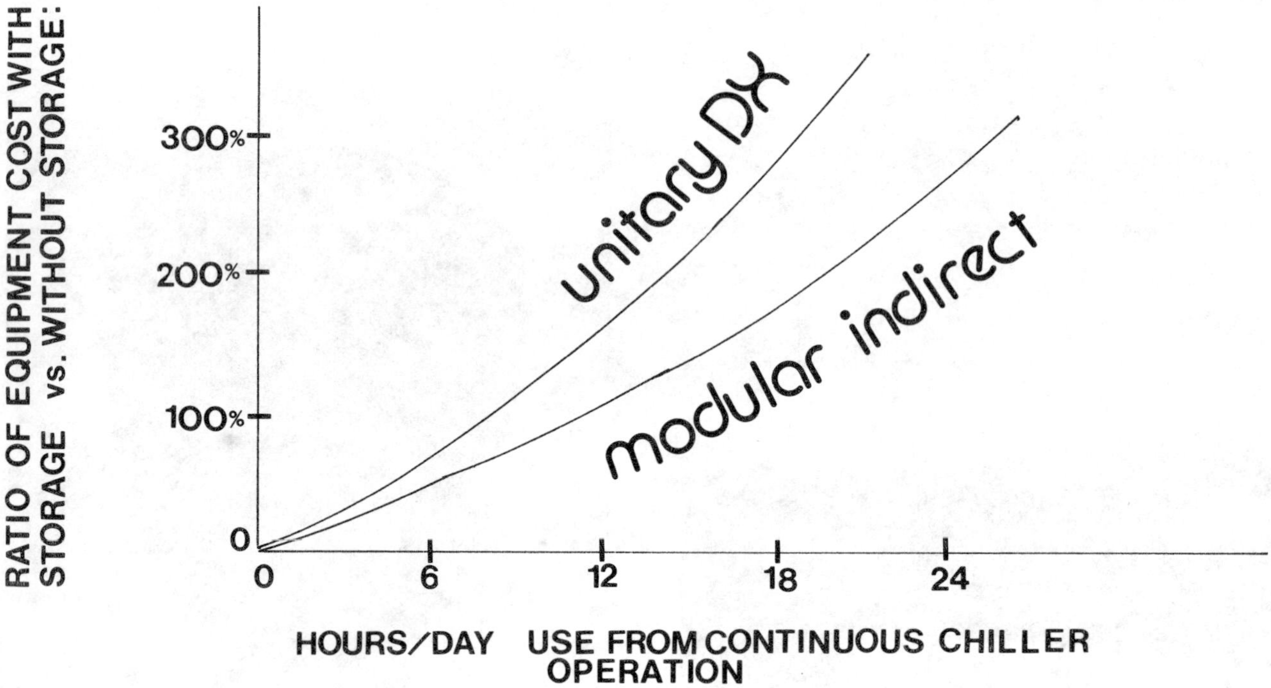

Figure 4

Figure 5 - Modular ice storage installation in Braintree; 12 tanks deliver 420 ton-hours of cooling

Chapter 35

USING OFF-LINE BOILER ENERGY MANAGEMENT COMPUTER PROGRAMS

L. G. Armwood

It has been almost ten years since the oil embargo first began the current spiral of energy prices, but many boilers in the United States are still operated under energy conservation guidelines that were developed during the 1960's. Of even greater concern are the large number of boilers that are operated without any clear indication of what conditions the boiler is operating. It is very unlikely that any boiler whose operations are not constantly monitored is being operated efficiently.

Off-line computer programs can be a useful tool in efficiently managing boiler systems. Off-line computer programs can model boiler operations and prepare detailed reports on the boiler system. Unlike on-line computer systems, which involve costly direct hook-ups from process instruments to the computer, off-line computers can easily use data collected during energy audits to establish boiler operating conditions. The off-line computer can then be used to show trends in boiler operations or model proposed changes in the boiler system.

BOILER ENERGY AUDITS

Boilers may be categorized as boilers that have or have not had a recent energy audit. The boilers that have had an energy audit five or more years ago are probably just as much in need of an energy audit as a boiler that has never been audited. With the rapid changes in the price of energy, the recommendations made in an audit in 1977 will be drastically different from the recommendations made on an audit of the same boiler at the same conditions made in 1981.

If a boiler has not been audited recently, the boiler system may have problems that are unknown to those operating the boiler system. Boilers that have not been audited recently may also have problems that have been recognized but have not been solved. The solution to the problem may be unknown to the boilerhouse personnel or the problem's solution is known, but because the project's economic justification was not properly presented, no action has been taken.

Boilers that have been audited recently need follow-up audits on a regular basis to maintain their current level of efficiency. A follow-up audit is also necessary to prove the value of changes that have already been made in the boiler system.

In spite of all of the benefits from conducting regular boiler system energy audits, they are rarely conducted on a regular basis. A better understanding of why energy audits are not being conducted can be found by examining the nature of the boilerhouse engineer's energy management functions.

BOILERHOUSE ENGINEER

The boilerhouse engineer may or may not have the technical background to conduct an energy audit. Many boilerhouse engineers are actually foremen promoted from the ranks of operators and, although they do a very good job of maintaining boiler operations, they may not have the technical background to evaluate the energy usage inside of a boiler system. Even when the engineer has the technical background to conduct an energy audit, the engineer often has such a limited amount of time because of the other demands of the job, that he is unable to conduct the energy audit and then evaluate the data.

In order to reduce the time required to conduct an energy audit the engineer should be able to use the data that is commonly recorded on log sheets and turn this data into a current picture of the boiler operating conditions. Typical data recorded on log sheets are shown in Table I. These data combined with known operating conditions for the boiler system as shown in Table II, can be used as part of an energy audit to define the boiler system.

TABLE I

TYPICAL BOILER LOG SHEET DATA

Generated Steam Rate

Steam Temperature

Steam Pressure

Boiler Inlet Air Temperature

Flue Gas Temperatures

Flue Gas Composition

Water Analysis (Make-up, Feedwater, Blowdown)

Condensate Return Temperature

Blowdown Temperature After Heat Recovery Units

Fuel Firing Rate

The boilerhouse engineer must be able to take these data and analyze trends to determine problems and find improvements in boiler operations. The engineer also must be able to predict the effect of changes in boiler operating conditions. All of this information must then be combined to form a report on the operating conditions of the boiler. The reports will be both for the engineer's own information and for reports to be presented to plant management.

TABLE II

BOILER OPERATING CONDITIONS

Boiler Capacity

Fuel Analysis

Fuel Cost

Make-up Water Temperature

Blowdown Flash Tank Pressure

Deaerator Pressure

WHY USE COMPUTERS?

The data shown in Tables I and II can be easily recorded in most boilerhouses. The boilerhouse engineer could take this data, sit down with a calculator and steam tables and perform the calculations which are necessary to present a report on the current status of the boiler system. The calculations could include calculations from "ASME Power Test Codes" and the mass and energy balances normally conducted around a boiler system. This usually is not done on a monthly, quarterly, or even a yearly basis because of the other demands on the boilerhouse engineer's time. Solving the day-to-day problems of managing a boiler system leaves the engineer little time to perform energy audit calculations.

Since these data are already being collected in most boilerhouses, performing the calculations is an ideal application for an off-line computer. With little additional time spent by the engineer the computer can be used to provide the complete report on the boiler systems as in the BEMP[1] Computer Program Output shown in Figures 1 and 2. The computer reduces the time required to prepare a boiler audit and allows the audit to become a new tool that can be continually used to maintain boiler performance.

BEMP is a computer program that was designed to provide a detailed description of boiler operating conditions using data that is easily obtainable from most boiler systems. The current data can be compared to baseline boiler conditions to establish long-term trends. This computer program can also perform a sensitivity analysis on many of the parameters in a boiler system to model the effects of changes in the boiler system or its operating conditions.

By providing a complete picture of the boiler systems operation the boiler energy management computer program can present a report that may be presented to plant management with only a short cover letter to highlight pertinent points. Calculations done by hand are more likely to contain errors in data entries and even with the use of strip chart recorders, do not produce reports that can be directly passed to upper management.

The asset of a boiler energy management computer program such as the BEMP is that they can easily model the effect of changes in a boiler system. These changes are made by establishing a new, complete hypothetical case that describes the boiler system after the change. Table III contains a list of some of the most commonly changed parameters. Since a completely new case is created, additional changes can be modelled in the hypothetical case to study the effect of related changes in additional parameters.

THE ENGINEER AND BOILER ENERGY MANAGEMENT COMPUTER PROGRAMS

The engineer in charge of boiler operations will be the person who will reap the greatest benefits from the use of boiler energy management computer programs. Through the use of computers the data supplied by log sheet information may be quickly transformed into usable information about the boilers as shown in Figures 1-A and 1-B, the input data printout. The data shown there is grouped into categories such as steam, water, and fuel data for easy reference.

Using the data recorded on these sheets, the engineer can re-create the conditions of the boiler for further study in the future. This is particularly important when trying to establish meaningful trends in boiler operations.

The Energy Management Status Report, Figures 2-A and 2-B, provides the engineer with information on the overall boiler system efficiency and the combustion efficiency. The Status Report also shows the percent excess air in the flue gas and the flow rate and value for various energy streams in the boiler system. This energy data can be used to explain changes in boiler system efficiency.

Since a complete set of data describing the boiler operating conditions has been recorded, the engineer may compare the current boiler conditions to the conditions of the boiler at an earlier data, the baseline boiler conditions. The engineer can then analyze these cases to determine trends such as an increase in an energy loss. These increased losses could be due to changes in operating conditions such as flue gas temperature, in process condensate flow rate, process condensate temperature or changes in blowdown rates.

Using the computer, the engineer may also project the effect of changes in operating conditions. This will allow the engineer to make recommendations on changes in the operating conditions, such as cycles of concentration or excess air. The engineer may know for instance that running at higher cycles of concentration will reduce energy losses in the blowdown. Using an off-line computer program to model the boiler, the engineer can quickly predict if further increases in cycles would result in significant energy savings to balance increased water treatment costs.

TABLE III

BOILER SENSITIVITY ANALYSIS VARIABLES

Inlet Air Temperature

Stack Gas Temperature

Air Preheater Temperatures

Economizer Temperatures

Flue Gas Composition

Generated Steam Flow Rate

Make-up Water Temperature.

Blowdown Heat Recovery Units

Cycles of Concentration

Process Condensate Flow Rate

In the example shown in Figures 1 and 2, a hypothetical case, Case 4, was created from the baseline conditions, Case 1, by adding blowdown heat recovery units consisting of a blowdown flash tank and a blowdown to make-up heat exchanger. These changes were made while holding constant process steam and process condensate flow rates.

The energy management computer program can be used by the engineer to prepare reports. These reports would not only include those reports for the engineer's own records, but reports to plant management.

PLANT MANAGEMENT AND BOILER ENERGY MANAGEMENT COMPUTER PROGRAM

The major concern of plant management will be how the future cost of steam impacts upon the cost of the company's product. They must be able to take the reports prepared by the plant engineer and project what future steam costs will be based on projected fuel prices. A second concern of plant management is to be able to obtain a concrete decision-making tool for capital investments. Often plant management rejects a project because of inadequate proof of savings.

In order to be useful, the reports for use by upper management must be simple and direct because of the limited amount of time available for decision-making. Plant management commonly has a business background and relies on the boilerhouse engineer's reports as the major decision-making tool.

In these situations the simple, direct approach of being able to point to the net energy savings for a proposed project or a new steam cost such as in Figure 2-A will allow plant management to make sound decisions.

If alternatives are suggested, they can be tested by re-running the computer program. The alternatives can be easily compared using the standard output format.

SUMMARY

Off-line boiler energy management computer programs are being used to help manage more and more boiler systems. The benefits of using off-line boiler energy management computer programs include:

1. A fast, accurate means of doing boiler system mass and energy flow calculations.
2. Complete input data listings defining current boiler operating conditions.
3. Summary outputs for plant management.
4. Justification for energy saving capital expenditures.
5. Detailed output for boilerhouse engineers.
6. The ability to identify trends by comparisons with baseline data.
7. The ability to quickly model changes in the boiler system.
8. The computer printout is ready for inclusion in reports.
9. Off-line computer programs can be used without costly direct instrument hook-ups to the computer.

---

[1] The BEMP, Boiler Energy Management computer Program is available through the Mogul Division of The Dexter Corporation, Chagrin Falls, Ohio - 44022 (216-247-5000).

BIOGRAPHICAL SKETCH

Levin G. Armwood is a project engineer with the Mogul Division of The Dexter Corporation, P.O. Box 200, Chagrin Falls, Ohio (216-247-5000). His responsibilities include the design of environmental and energy conservation facilities.

He holds a B.S. in chemical engineering from Ohio University, Athens, Ohio. He is an active AIChE member having held several positions including local section chairman.

Figure 1A

ABC COMPANY
MAIN STREET
ANYWHERE, USA

SYSTEM:LOW PRESS
BOILER:NO. 1
CASE 1 DATE: 41281
CASE 4 DATE: 0

INPUT DATA

	CASES 1	4
STEAM SYSTEMS		
GENERATED STEAM		
FLOW IN LBS/HR	30000.0	29453.4
TEMPERATURE IN DEG F	350.0	350.0
PRESSURE IN PSIG	120.0	120.0
STEAM USED BY THE BOILER SYSTEM		
FLOW IN LBS/HR	3206.1	2659.5
PROCESS STEAM		
FLOW IN LBS/HR	26793.9	26793.9
RATED MAX. BOILER CAPACITY		
FLOW IN LBS/HR	37500.0	37500.0
FLASH STEAM		
FLOW IN LBS/HR	0.0	229.8
DEAERATOR STEAM VENT		
FLOW IN LBS/HR	6.4	5.3
PRESSURE IN PSIG	5.0	5.0
WATER SYSTEMS		
MAKE-UP WATER		
FLOW IN LBS/HR	14702.7	14414.2
TEMPERATURE IN DEG F	40.0	40.0
CONDENSATE RETURN		
FLOW IN LBS/HR	17297.3	16980.5
TEMPERATURE IN DEG F	197.0	197.0
TOTAL BLOWDOWN		
FLOW IN LBS/HR	2000.0	1941.3
CONTINUOUS BLOWDOWN		
FLOW IN LBS/HR	0.0	1747.2
BLOWDOWN AFTER HEAT RECOV.		
FLOW IN LBS/HR	0.0	1517.4
TEMPERATURE IN DEG F	0.0	50.0
CYCLES OF CONCENTRATION (BASIS FEEDWATER)	16.00	16.17

Figure 1B

INPUT DATA CONTINUED

	CASES 1	4
FLUE GAS DATA		
TEMPERATURE IN DEG F	365.0	365.0
COMPOSITION		
% CARBON DIOXIDE	10.7	10.7
% CARBON MONOXIDE	0.0	0.0
% OXYGEN	9.0	9.0
FUEL DATA:	SOFT COAL	SOFT COAL
FUEL PRICE		
$/ LBS	.0	.0
$/MMBTU	1.3	1.3
FIRING RATE		
LBS /HR	2726.9	2678.3
COMPOSITION		
% CARBON	75.0	75.0
% HYDROGEN	5.0	5.0
% OXYGEN	6.7	6.7
% NITROGEN	1.5	1.5
% SULFUR	2.3	2.3
% ASH	7.0	7.0
% MOISTURE	2.5	2.5
HEATING VALUE		
FUEL IN BTU/ LBS	13165.0	13165.0
ASH IN BTU/LB	0.0	0.0
SPECIFIC GRAVITY BASIS AIR FOR A GAS DENSITY FOR LIQUIDS OR LBS/TON OF SOLIDS	2000.0	2000.0
AIR DATA		
BOILER INLET AIR		
TEMPERATURE IN DEG F	82.0	82.0
WATER IN LBS / 100 LBS OF DRY AIR	.8	.8
ENERGY RECOVERY		
FLUE GAS TEMPERATURE IN DEG F		
LEAVING THE STACK	365.0	365.0
ENTERING THE AIR PREHEATER	0.0	0.0
ENTERING THE ECONOMIZER	510.0	510.0
BLOWDOWN HEAT RECOVERY FLASH TANK		
PRESSURE IN PSIG	0.0	5.0
% OF BLOWDOWN FLASHED	0.0	13.2

Figure 2A

ABC COMPANY
MAIN STREET
ANYWHERE, USA

DATE: JULY 4, 1981
CURRENT PERIOD: HYPOTHETICAL

****
* ENERGY MANAGEMENT STATUS REPORT *
****

SYSTEM:LOW PRESS
BOILER:NO. 1
RATED OUTPUT IN LBS/HR: 37500.
OPERATING PRESSURE IN PSIG: 120.
BASELINE PERIOD:APRIL 12, 1981
COMMENTS: ADDED BLOWDOWN HEAT RECOVERY

SUMMARY

	BASELINE	CURRENT
OVERALL SYSTEM EFFICIENCY		
%	75.7	78.1
FUEL CONSUMPTION		
MMBTU/MMLBS STEAM	1339.8	1316.0
STEAM FLOW, LBS/HR		
GENERATED	30000.0	29448.3
PROCESS	26793.9	26793.9
CURRENT FUEL COST		
$/MMBTU	1.3	1.3
STEAM COST, CURRENT BASIS		
$/MMLBS	1679.3	1649.4
%	100.0	98.2
NET ENERGY SAVINGS		
$/MMLBS STEAM		29.8
$/DAY		19.2

ENERGY LOSS ANALYSIS

	BASELINE	CURRENT
BLOWDOWN		
MMBTU/DAY	15.1	1.9
$/DAY	18.9	2.4
LOST CONDENSATE		
MMBTU/DAY	47.9	47.9
$/DAY	60.0	60.0
COMBUSTION		
MMBTU/DAY	124.0	121.9
$/DAY	155.4	152.8
MISCELLANEOUS		
MMBTU/DAY	14.0	14.0
$/DAY	17.6	17.6

Figure 2B

ENERGY FLOWS

	BASELINE	CURRENT
INPUT, MMBTU/DAY		
FUEL	861.6	846.3
PROCESS CONDENSATE	53.1	53.1
DEAERATOR STEAM	91.1	75.4
OUTPUT, MMBTU/DAY		
GENERATED STEAM	852.5	836.8
BLOWDOWN	15.1	1.9
COMBUSTION		
DRY FLUE GAS	79.0	77.6
MOISTURE - ALL SOURCES	37.8	37.1
RADIATION	7.3	7.3
MISC.	14.0	14.0

ENERGY RECOVERY ANALYSIS

	BASELINE	CURRENT
ECONOMIZER, MMBTU/DAY	43.2	42.4
$/DAY	63.2	62.1
AIR PREHEATER, MMBTU/DAY	0.0	0.0
$/DAY	0.0	0.0
FLASH TANK, MMBTU/DAY	0.0	6.5
$/DAY	0.0	9.6
B.D. HEAT EXCHANGER, MMBTU/DAY	0.0	6.7
$/DAY	0.0	9.8

COMBUSTION DATA

	BASELINE	CURRENT
EXCESS AIR, %	71.8	71.8
COMBUSTION EFFICIENCY, %	85.6	85.6

Chapter 36
ENERGY EFFICIENT FUME AND ODOR CONTROL EQUIPMENT FOR COIL COATING LINE

G. C. Smith

On behalf of the stockholders and management team of the Wolverine Aluminum Corporation, we wish to thank you for inviting us here today to speak before the Fourth World Energy Congress.

As a bit of background about our company, we are this country's largest independent producer of exterior siding building products for residential housing. We are a multistate corporation with headquarters in Lincoln Park, Michigan, employing more than 400 employees. Our annual sales topped $60 million in 1980.

As part of our continuing effort to make major capital improvements at our five major plants, we recently completed the installation of a $2.5 million energy-recovery pollution control system at our Lincoln Park Coil Coating plant.

This new system has eliminated air pollution problems in and around the plant and through a unique energy recovery feature reduces our natural gas consumption by 50%. In 1981 this system is expected to reduce our operating costs by $1 million.

Known as a RE-THERM [R] thermal oxidizer, the unit has incorporated the best available pollution control technology and latest engineering state-of-the-art to conserve energy while meeting federal and state environmental pollution control regulations.

Prior to our installation of the RE-THERM system on our 4 coil coating lines, a natural gas boiler provided the lines with steam and each line contained its own gas operated incinerator which burned solvent laden gases. Unfortunately, these fume incinerators emitted odors and consumed large volumes of gas.

With the escalating costs of natural gas in recent years, we decided we had to go to a retrofit. We investigated several different sources of retrofit equipment and decided on the RE-THERM unit because of its simplicity and suitability to our existing equipment.

REECO designed and engineered a variable energy recovery system which has eliminated the inefficient fume incinerators and directed the fume emissions into one Model G, 56,000 SCFM RE-THERM thermal oxidizer.

Two main ducts were installed with ducts to each zone - one main duct to carry solvent laden air to the RE-THERM unit and the other main duct to return clean, pre-heated air from the RE-THERM unit to the oven zones. There are also ducts from the coater rooms to carry fugitive solvent emissions to the RE-THERM unit.

There are 7 chambers of stoneware bed material with 3 chambers accumulating heat and 3 chambers discharging heat with one idling alternately. The discharged heat is carried through the ducts to the zones with the zone trim burner to level the zone at the exact predetermined temperature. This preheated clean make-up air further reduces our natural gas consumption.

Polluted exhaust fumes from the ovens enter the system and pass over the RE-THERM stoneware bed containing stored energy. The fumes progressively absorb the stored energy. These fumes virtually reach the oxidation temperature, reducing the energy needed for thermal oxidation. Cleaned, super-heated air passes over another stoneware bed where that heat energy is again recaptured and stored.

This clean, warm air leaves the RE-THERM at very near its inlet temperatures and is then reused for process and steam generation - thereby reducing total plant energy usage with no auxiliary fuel required to operate the system.

A variable design feature allows each of these lines to be turned on or off automatically as production demands require, control energy usage, and comply with present EPA Standards. The variable design feature also allows reducing the exhaust flow from each oven to increase the concentration of fumes in each exhaust air stream fed to the RE-THERM [R] unit. This again reduces energy usage, improves productivity, and minimizes emissions to the atmosphere. Our capability to turn on and off each line is controlled at a single control panel.

The installation was quite simple with the foundations, the erection of the RE-THERM unit, electrical and main ducts all done in advance of shutdown. All of this work, other than the electrical, was done outside the plant. At the time of shut-down, the duct work was run from the main ducts to each zone. The insulation of the unit was one of the most time consuming jobs of the entire installation and we went back into production without this work being completed. We were quite pleased to accomplish the installation with very little lost production time.

The REECO ^{R} Variable Energy Recovery System was initially started in December 1980. The RE-THERM ^{R} unit and the 52" line came on stream in early February, 1981. During March the 36" line was tied in. Later the 24" and 18" lines were tied in.

The tie-ins took place over weekends during scheduled production down times.

Upon start-up we had a few minor adjustments to make but were able to go into full production with no interruption of productivity.

The operation of the unit is quite simple and our operating personnel have had no problems assimilating the principle of the unit or the operation of the control panel. The electrical control panel gives the operating personnel a simple quick readout of all functions and this panel does not have to be attended. The paint line foreman merely glances at it from time to time as he passes the control unit. The panel even indicates when the solvent load is high enough that the unit goes on to low fire and the heat is maintained by the solvent load itself. The RE-THERM unit burns at 1,400^{o}F which adequately destroys any hydrocarbons and allows us to be in complete compliance with Air Quality Standards for the portion of air emitting from the stack.

In total we have found the RE-THERM unit to be simple yet reliable. It conserves thermal energy by saving and reusing the heat energy of the oxidation process and through the innovative engineering design we explained earlier, the thermal energy is recaptured and stored to be used again and again.

The eight most positive features we have witnessed with the RE-THERM are:

1. Total system fume and odor control

2. Fuel efficient thermal oxidation through the <u>entire solvent load operating range</u>

3. Minimal amount of re-work to retrofit existing lines

4. Elimination of complicated controllers and instrumentation required to modulate dampers and regulate system air flows to meet changing solvent loads or oven demands

5. System combines fuel efficiency through "regenerative heat recovery" with thermal oxidation of fume and odor in a self-contained unit

6. Inert ceramic recovery media stable to 2,100^{o}F

7. Elimination of tube type pre-heat exchangers

8. Autoignition problems eliminated

The gentleman who will be following me to this dais this morning will be quite able to discuss these points in greater detail, but I will say in conclusion that for the first six months of 1981 we have realized savings of more than $350,000.

These savings will help us to be more competitive in today's marketplace.

And now I pass the technical platform over to Mr. Jim Mueller, President of REECO.

Chapter 37

A CASE HISTORY OF INDUSTRIAL ENERGY CONSERVATION WHILE MEETING EPA REGULATIONS

J. H. Mueller

ABSTRACT

The metal finishing industry, and others, require a quality product paint finish at a practical cost. EPA emission requirements and energy costs have stimulated many alternate solutions. These alternates include water-based compounds, solvent recovery systems, energy efficient incineration and efficient energy recovery systems.

This paper discusses alternate techniques currently available. Included are cost comparisons which show techniques of how an existing paint system can meet the VOC emission requirements while reducing energy costs. A summary will show the economics of meeting VOC control requirements, reducing energy consumption and maintaining product quality.

EXPERIENCE SHOWS THAT many assumptions are erroneous.

One such assumption is that required air pollution control equipment will be expensive to purchase, install, and operate . . . or that process changes may be required which will have a detrimental effect on the quality of the products being produced by the plant.

Industry has had this concern.

The problem, of course, is the solvent-laden fumes released in paint spraying, curing and coating operations. This problem is shared by others who coat aluminum, steel and a variety of other materials including paper and film.

One clean air alternative is a switch to water-based instead of solvent compounds. In many operations, the water-based compounds were satisfactory. An example is an undercoat used by Inland Steel in their coil coating operations. Good! Problem solved.

In other cases, however, the test results were less than satisfactory. The same Inland Steel coil coating operation for instance, continued to use a solvent-based top coat to assure a durable finish.

Another potential solution is carbon adsorption process. Laboratory and full scale models look logical. Air can be cleaned. Solvent can be reclaimed.

Here again, the story, in practice, is somewhat different. Many installations have been quite successful. Companies such as Rexham and Armstrong World Industries, (formerly Armstrong Cork Company) have installed and are benefiting from the reclaimed solvents. Technology, however, isn't yet up to the rigors of all commercial applications. Problems include system complexity and an array of hidden costs. Expenses include the steam required for stripping the bed and carbon bed replacement.

Also, acids sometimes form requiring use of costly metals.

Last, but not least, the recovered solvent often needs further refining.

Fortunately, there is a practical solution which is well within the scope of present-day technology. Solvent emission can be controlled by using a highly energy-efficient form of regenerative incineration that recovers and reuses, within the equipment, 85%, 90% or 95% of the required energy input. Regenerative incineration can offer an excellent return on investment as has been documented by Gregory Smith, President and Chairman of the Board of Wolverine Aluminum in Lincoln Park, Michigan.

Regenerative incineration is simple and straightforward. Energy recovered from the solvent is used for its own incineration and fume self-destruction with very little additional fuel usage. This energy efficiency is true for spray booth cleanup as well as curing ovens. Operation isn't affected by a mixture of water-based and solvent-based exhaust streams. Equipment is simple and reliable. Plant people can concern themselves with production rather than spending valuable time keeping a complex system working.

Incinerators have traditionally had a bad name for fuel consumption. A modern thermal oxidation system with efficient energy recovery and recycling is different! Regenerative incineration can be tremendously efficient both in terms of cleaning the air and in terms of energy self-sustaining operation. Fuel for auxiliary energy consumption can be (and often is) essentially "NIL".

The objective of this presentation is to show the plant manager, engineer and environmentalist the availability of practical, straightforward incineration and heat recovery techniques. These techniques are being used today to reduce plant energy consumption while meeting stringent environmental code requirements. In many cases, complete environmental compliance is possible with the cost of the installation being paid from plant energy savings alone.

The regenerative system is an energy recovery unit that operates on the mass/surface area principle of heat recovery. Figure 1 will give an idea of how it works.

RE-THERM® OPERATION

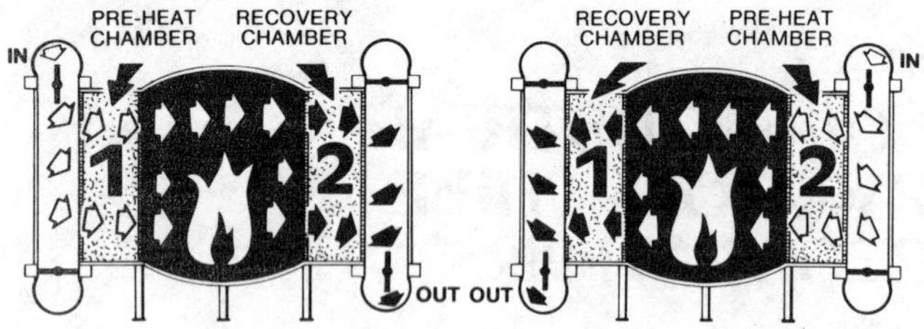

FIGURE 1 - REGENERATIVE INCINERATOR

As hydrocarbon-laden exhaust air travels toward the central incineration chamber, it passes through a smaller chamber containing previously warmed energy recovery elements. The fumes, thus preheated, can be incinerated in a fraction of a second, with little or no auxiliary fuel needed in addition to the fuel value of the solvent fumes themselves.

The superheated clean air is routed through another chamber also containing the energy recovery elements. Most heat energy is stored in these energy recovery elements for use in preheating the next cycle. The minimal amount of heat carryover can be used in the plant or process. Subsequently the flow is reversed and fumes enter through the other chamber. thus, a back and forth cycle is establiished to capture the energy of incineration and reuse it for burning the next cycle. This feature reduces the auxiliary fuel requirement to only 5-10% of the fuel required by a common afterburner, depending on the design and allowing for wall losses. Stated another way, the regenerative incinerator can self-sustain on L.E.L. concentrations as low as 3 to 5%. Typical bake ovens are presently operated in the range of 2 to 25% L.E.L., thus a minimal amount of auxiliary fuel is required with a regenerative incinerator.

Regenerative incinerators are installed at companies such as A. E. Staley, Armstrong World Industries, Champion Spark Plug, Mead, Rexham, Western Electric, Wolverine Aluminum and others. They are controlling fumes containing a variety of otherwise troublesome items such as aluminum and steel coating fumes, plastisols, condensates, clay particulate, coffee roaster chaff, etc. Currently, there are some fifty (50) regenerative systems in or about to be in operation. Their capacity ranges from 2,000 SCFM to 60,000 SCFM each. The solvent/organic concentrations range from a trace to a variable loading of from 8% to 25% L.E.L.

This energy-efficient system has a number of benefits. It offers a total system fume and odor control, and is fuel-efficient through the entire solvent load operating range. It can be installed with a minimal amount of re-work to retrofit existing lines. It eliminates complicated controllers and instrumentation as well as high maintenance tube-type preheat exchangers. Another benefit is that autoignition problems are eliminated.

Another less apparent advantage is the internal temperature profile which, to date, has prevented any condensate or particulate build-up. Nevertheless, a self-cleaning feature (which operates much like a self-cleaning oven) is incorporated.

There are many reports and evaluations regarding equipment costs, full engineering and financial studies. Such evaluations are freely available from my company and from other sources.

A simple comparison of a current auto industry oven is also useful. The following compares an oven as presently operated, versus the same oven with a regenerative incineration system and oven feedback. The oven temperature, solvent concentrations and exhaust volume are as established by a major auto manufacturer.

To simplify this comparison, the following assumptions are used:

1. Both systems will use existing solvent-based materials.
2. There will be no basic change to the process.
3. Energy is the main cost factor and the key of the evaluation.
4. The energy cost for heating the product is the same for both systems - this energy cost may be omitted for this comparison.
5. The energy cost for oven wall losses, solvent vaporization, etc. is essentially the same for both - this also may be omitted for both systems.
6. When the solvents from the oven exhaust are destroyed, 70% of the resulting clean warm air can be used for oven make-up (or a boiler or plant heat or dry off, etc.).
7. The example oven exhausts 53,200 SCFM @ 257°F.

Using the above, a simple cost evaluation for heating the make-up air for the oven and exhausting it to atmosphere gives these results:

OVEN AIR HEATING COST	$373,740/YR
AIR HEATING COST WITH	
REGENERATIVE INCINERATION	
@ 90% THERMAL ENERGY	
RECOVERY AND FEEDBACK	$ 84,240/YR
SAVINGS IN AIR	
HEATING COST	$289,500

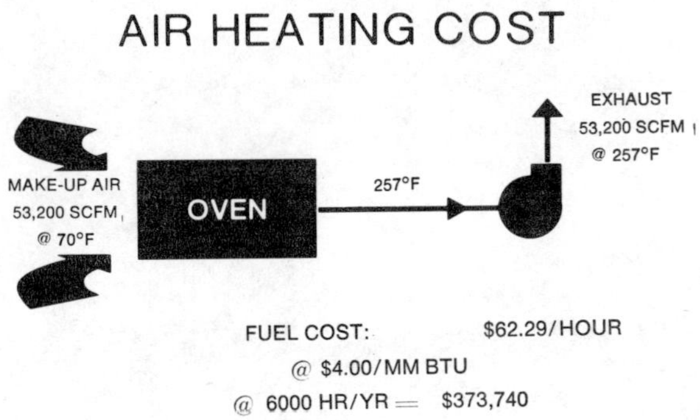

FIGURE 2

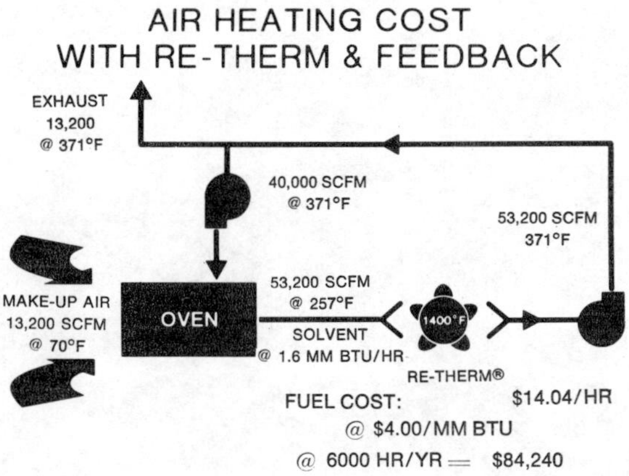

FIGURE 3

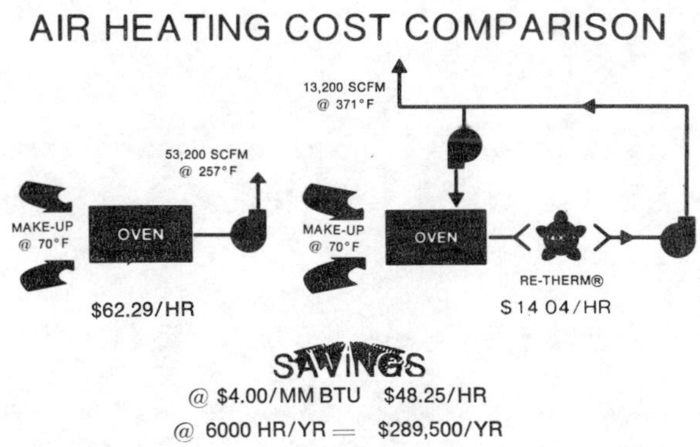

FIGURE 4

The above comparison may seem simplistic. In many respects it is. The results, nevertheless, show that with existing technologies, equipment, paints and most important, existing quality, a substantial savings - energy savings - is possible with regenerative incineration.

As a case history, the Wolverine installation is a combination of seven (7) such ovens covering four (4) separate lines. In addition, the system provides energy to a boiler. The installation, while appearing complex, has the same simple, straightforward approach.

This system has the added advantages of allowing:

1. Variable exhaust volume (ovens on or off)

2. Variable solvent loads

3. Variable energy feedback to the ovens

4. Auxiliary boiler on or off.

The results, at Wolverine, are dramatic. Projected energy savings of approximately $1,000,000 per year with pollution control as a bonus.

In summary, regenerative incineration equipment has been and is being installed which meets the environmentalists needs. This equipment gives industry a financially attractive solution to the ENERGY ENVIRONMENT problem. While each application is different, operators have written that their ROI (Return on Investment) averages form 30% to well over 100%.

I suggest that we . . . as an industry . . . can help ourselves. The government wants us (and OPEC is doing its best to encourage us) to save energy. Is this not the time to muster forces and find a better way to answer all needs? I believe it is!

In conclusion, I suggest consideration of these steps:

1. Establish existing conditions.

2. Exhaust flow management through incineration and recirculation.

3. Installation of quality energy efficient equipment.

4. Obtain government support projects through low interest bonds.

Taking these steps can result in costs and cash flows that are practical. This will give a partial solution to the pressing environmental and energy problems.

And most of all, remember that thermal energy waste is money up the stack and that it is practical to have energy conservation with pollution control as the bonus.

Chapter 38
DIRECT DIGITAL CONTROL... IS IT REALLY BETTER?

R. J. LeBlanc

INTRODUCTION

The automatic temperature control industry is in the midst of what may be the most dramatic technological advance of the last 90 years. Since the late 1800's, environmental control of the HVAC systems in buildings has been the sacred ground of pneumatic controls. Electronic analog controls are also used but have yet to gain the wide acceptance accorded pneumatics. However, with the rapid advance of microprocessor technology, a serious new challenger to conventional pneumatic controls has stepped forward.

The new threat comes from Direct Digital Control (DDC) which has proven its worth in the industrial process control marketplace. In the mid-1970's, DDC challenged pneumatics in the process industry and won. It now commands the lion's share of all process control applications. Yet the commercial HVAC industry offers a much different battle ground for the confrontation of pneumatic controls and DDC. Several issues are raised:

Are the additional capabilities of DDC systems required in HVAC applications?

Is DDC a cost-effective alternative to conventional pneumatic controls in commercial buildings?

To answer these questions the criteria that must be evaluated are: performance, initial cost, reliability, flexibility, maintainability, ease of use and life cycle cost. The main question about DDC is ... Is it really better than pneumatic controls in commercial HVAC applications? The intent of this paper is to evaluate DDC vs. pneumatics and show that on almost every battle front DDC wins.

DEFINITION OF TERMS

Although many of the terms used herein are industry standard, a quick review is given below to minimize any possible conflict of terminology.

Conventional Pneumatic Controls

This is the traditional form of control used in most commercial and institutional facilities for environmental control. The control function is performed by a pneumatic analog controller which receives its inputs from pneumatic sensors (i.e. temperature, humidity) and sends control signals to pneumatic actuators (valves, damper actuators). (See Figure 1) Pneumatic receiver-controllers typically perform a single, fixed control function which cannot be altered unless the controller itself is changed, or other hardware (relays, switches) are added. Conventional electronic analog controls are much the same.

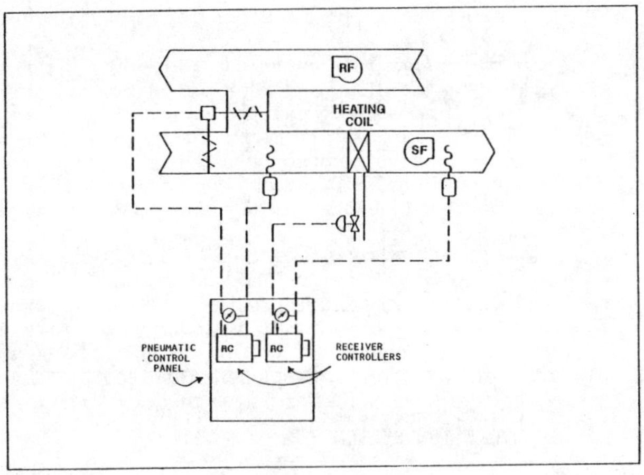

FIGURE 1. CONVENTIONAL PNEUMATIC CONTROLS

Supervisory Automation Systems

These systems are often used when remote control and centralized monitoring capability is required. A central computer communicates with remote electronic panels which can provide central monitoring through electronic sensors which measure environmental conditions. In addition, these systems can provide remote control of fans, pumps, etc.

A supervisory system is often interfaced to the pneumatic control system to allow remote set-point or control point adjustment (CPA) of the pneumatic receiver-controllers. (See Figure 2)

In recent years these systems have been justified primarily as energy management systems.

Direct Digital Control or DDC

This type of control has been in use for years in the process industry. DDC utilizes a programmable digital computer or microcomputer as the primary controller. When controlling the mechanical heating, ventilating and air-conditioning (HVAC) systems in buildings, variables such as temperature, humidity and pressure are measured by electronic sensors. The control function is performed by a programmable microcomputer which sends a control signal directly to the control device (valve or damper actuator).

Direct Digital Control (also sometimes called closed-loop control) is a generic term where the term "Direct" means the microcomputer is directly in the control loop and the term "Digital" means control is accomplished by means of a digital computer. (See Figure 3) Actuators can be electronic or pneumatic.

True Direct Digital Control is not to be confused with electronic controls. Electronic analog controls are much like pneumatic analog controls, in that each controller typically handles one control loop and does so in a fixed manner. A true DDC controller, on the other hand, can control numerous control loops and can be reprogrammed for different control functions without hardware changes.

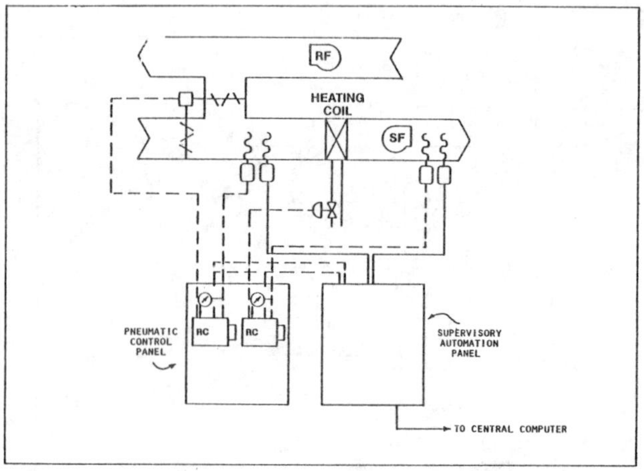

FIGURE 2. CONVENTIONAL CONTROLS AND SUPERVISORY AUTOMATION SYSTEMS

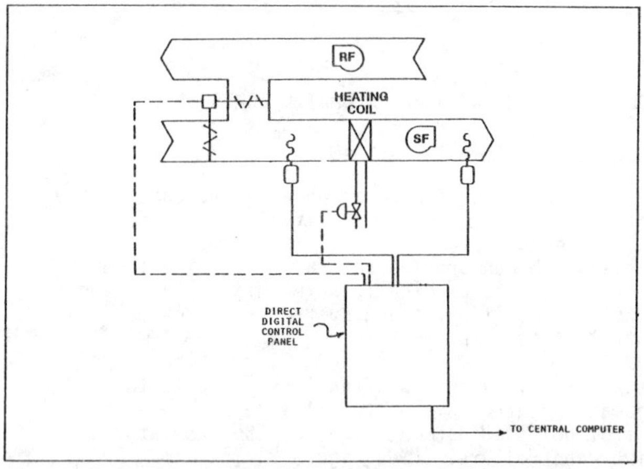

FIGURE 3. DIRECT DIGITAL CONTROL

Distributed Direct Digital Control

The most reliable form of system architecture to accomplish DDC is one in which the DDC controllers (microcomputers) are distributed throughout a building complex, yet remain connected to a central computer. In this configuration, the local DDC microcomputers can operate independently, yet the benefit of central control and monitoring is preserved. (See Figure 4) The measure of a true distributed DDC system is whether the remote DDC controllers continue to perform full HVAC control and energy management, without the central computer. To qualify as a true DDC system, the loss of the central computer or communication to it shall not require the remote DDC controllers to operate in a degraded mode, or rely on separate electronic controls.

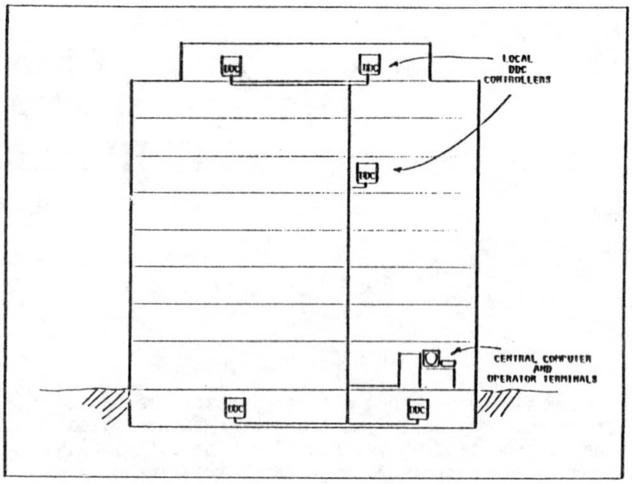

FIGURE 4. DISTRIBUTED DIRECT DIGITAL CONTROL

THE COST/PERFORMANCE COMPARISON

As stated earlier, it is necessary to fully evaluate several criteria to properly compare DDC to pneumatic controls. For the purpose of this evaluation, HVAC control capabilities will be examined first. The comparison will then look at the cost and ease by which energy management functions can be added to a pneumatic system and a DDC system.

Performance

Conventional pneumatic controls have historically provided satisfactory HVAC control. In comparison with the capabilities of DDC, however, conventional pneumatics are greatly overshadowed.

A conventional pneumatic-controller utilizes a proportional control action (i.e. the output of the controller is proportional to the offset (error) between the desired set-point and actual condition. Although pneumatics do provide adequate control in all but the complex control situations, their control action does generate offset and as such, wastes energy. (See Figure 5)

There are various control applications which are not uncommon, yet are difficult, costly or actually impossible with pneumatic controls. Some examples are:

- fan sychronization of VAV systems
- optimizing a chiller plant based on cost of operation
- setting minimum outside air percentage based on temperature ratios

DDC provides much tighter control. With the ability to perform full Proportional, Integral and Derivative (PID) control, the DDC control action can be customized to the specific needs of each loop. This closer control of HVAC systems optimizes the energy consumption for environmental control. (See Figure 6)

With the aid of computer-precision and the ability to perform industrial-type process control, DDC is clearly a superior control system in terms of performance. This becomes an important consideration when the modern energy-efficient mechanical HVAC systems require complex control.

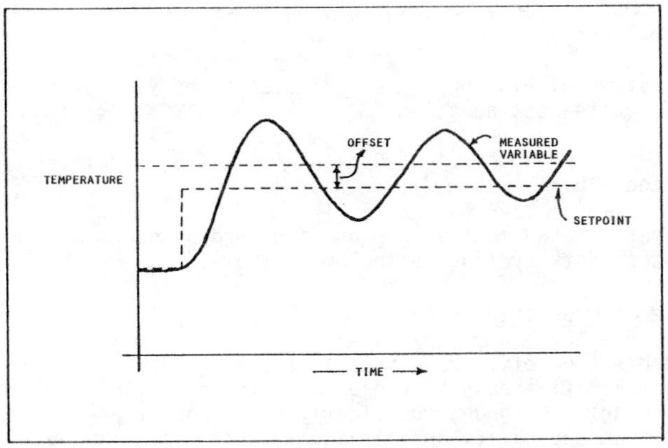

FIGURE 5. CONVENTIONAL PROPORTIONAL CONTROL ACTION

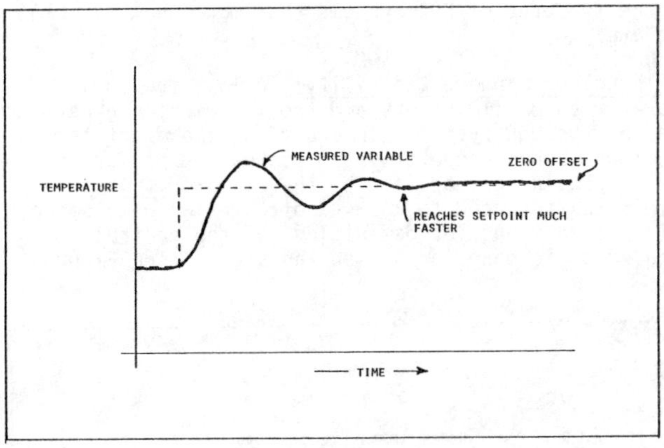

FIGURE 6. DDC CONTROL ACTION

Initial Cost

It has long been conceded that industrial-quality HVAC control is readily available, but often the cost of this higher degree of control has been prohibitive in commercial buildings. With the rapidly dropping costs of solid state electronics, in recent years the cost differential between DDC and pneumatics has shrunk dramatically. While DDC control of simple HVAC functions (i.e. room thermostat control) remains slightly more costly than conventional controls, the more complex HVAC applications can be controlled better with DDC and without a penalty in cost.

As shown in Figure 7, the cost of pneumatic controls tends to rise linearly with respect to the complexity of the control system. Furthermore, when large complex control is required (i.e. VAV system control with fan sychronization), the cost of pneumatics rises even more rapidly. This is largely due to the fact that as more control loops are required more pneumatic controls must be added, installed and calibrated. In addition, normal pneumatic controls are inadequate for many complex control situations, often forcing more expensive controllers (i.e. proportional-plus-reset).

DDC as stated earlier is in fact a multi-loop, programmable controller, and as such is much better suited to the complex control applications. Often additional control functions can be done without any additional hardware, but rather only additional control statements entered at the central and sent to the local DDC controller.

In terms of the initial cost of a control system, the superior capabilities of DDC are often available at no cost penalty over pneumatic controls.

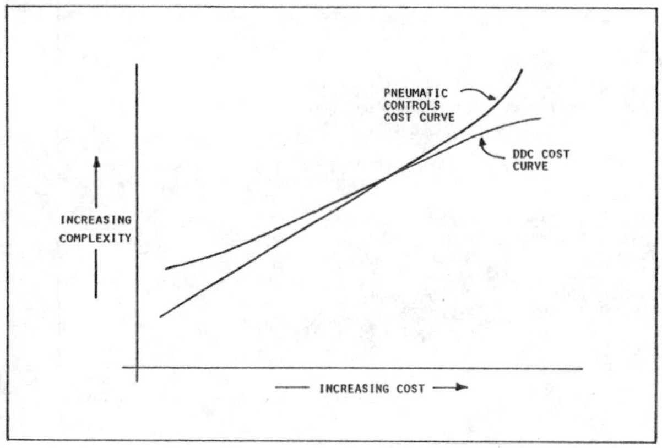

FIGURE 7. COST COMPARISON OF PNEUMATIC CONTROLS VS. DDC

THE NEED FOR RELIABILITY AND MAINTAINABILITY

Reliability

Much of the reason for the success of pneumatic controls over the years has been due to their reliability. It was largely this reliability that helped pneumatic controls to successfully withstand the challenge of electronic receiver-controllers in the later 1960's and 1970's. One of the big tests for DDC systems will be to demonstrate acceptable reliability, when early electronic systems could not.

Most modern pneumatic control systems can demonstrate good reliability, only if they are properly maintained and calibrated regularly. This periodic service is mandatory to prevent contamination of the air lines with water, oil and dirt. In addition, regular recalibration is required since pneumatic controls tend to drift away from their set-point over a period of time.

DDC systems in the process industry have demonstrated reliability for many years, even under harsh industrial environments. Perhaps the reliability of modern electronics themselves is the best proof. Moreover, careful attention to the design of DDC systems can produce truly distributed systems that clearly offer dependable environmental control by allowing each DDC controller to stand-alone.

There exist today enough successful DDC installations to safely conclude that DDC does provide the same degree of reliability as pneumatics.

Maintainability

The ability of building operating personnel to maintain their control system is often as important as the reliability of the system itself.

Pneumatics do allow fairly simple exchange of components to correct a problem, yet the detection of the problem is not always easy, due to lack of as-builts on many jobs.

A DDC system can simplify maintainability by providing automatic as-builts from the computer memory. (See Figure 8)

CONTROL LOOP LOG

CONTROL LOOP: RETURN AIR HUMIDITY
CONTROL STATEMENT: 120, TRUNK 02, RCU 03
STATUS: ENABLED (RCU CONTROL)

	POINT NAME	CURRENT VALUE	ENG UNITS	LOW ALARM	HIGH ALARM
PROCESS VARIABLE	RAHUM	42.0	PCT RH	30	50
CONTROL VARIABLE	HUMCV	6.5	PSI		
SET POINT	HUMSP	42.0	PCT RH		

CONTROL FORM: PROPORTIONAL
CONTROL ACTION: REVERSE
GAIN: P = 5, I = 0, D = 0
BIAS: 10
SAMPLE TIME (SEC): 5

FIGURE 8. RECORD OF ALL CONTROL LOOP PARAMETERS

In addition, DDC controllers tend to have fewer components making the inventory requirements smaller, and component replacement easier. Lastly, built-in diagnostics make it easier to locate faults.

DDC controls makes maintainability easier since they don't drift over a period of time like pneumatics, rather they fail totally and report the type of failure and its location. Repair is then quickly accomplished by a circuit board replacement.

THE NEED TO BE EASY TO OPERATE AND MODIFY

Flexibility

With the ever increasing need to optimize the operation of HVAC systems, building owners are often faced with the need to modify or expand their controls. Control systems must allow operators the ability to make these changes easily with in-house operating personnel.

Pneumatic controls are fairly easy to understand and control changes are relatively simple if as-builts are available and the required control strategy is not too complex. However, each change in the sequence of operation or control action requires new controllers, re-piping and recalibration. This all involves physically going to the HVAC system often with a control mechanic or electrician or both.

DDC offers considerably more flexibility since it is a programmable controller. The control strategy of the DDC controller can be changed by the operator without even going to the remote controller. In many cases, new control can be added without installing additional hardware. Once a DDC system is connected to the fans, pumps, valves, dampers and environmental sensors of an HVAC system, various additional control dequences can be easily added by the operator by simply entering control strategies at the central computer. The superior flexibility of DDC allows full ability to modify or add:

- control loops (i.e. control algorithm, control action, gain, bias)
- equipment interlocks (i.e. interlock chilled water pumps, condenser pumps, cooling tower fans and chillers).
- custom calculations (i.e. calculation of flow using a square root equation).
- custom sequences (i.e. Boolean functions, specialized algorithms).
- energy management functions (i.e. programmed start/stop, duty cycling, enthalpy control).

Ease of Operation

One great benefit to conventional controls has been their relative ease of use by building operating personnel. Although DDC clearly has superior performance capability and flexibility, it will only be accepted by commercial building owners if it is easy to learn and use by their existing personnel.

Some of the advanced DDC systems available today, offer many features that make operating personnel able to fully utilize the vast capabilities of DDC. Some of the features of DDC systems that make them easy to use are:

Full English Lanuage Capability: Allows the building operator to easily understand the information generated by the DDC system. In addition, the exact terminology used to report system status and performance can be chosen and modified by the operator. This eliminates any need to know a software language since all "programming" (or definition) of the control strategies is done in English language. (See Figure 9)

10:08 10/29/81 MON MASTER CONSOLE DJM 65.0 DEG F
PTDISPLAY

USER/ENGLISH NAME	POINT DESCRIPTION	CURRENT STATUS/VALUE	ENG UNITS	ALARM STATUS
ADM BLDG AIR HANDLER ONE 00	SUPPLY FAN	ON		NRML
ADM BLDG AIR HANDLER ONE 01	RETURN FAN	ON		NRML
ADM BLDG AIR HANDLER ONE 02	MIXED AIR	68.0	DEG F	NRML
ADM BLDG AIR HANDLER ONE 03	RETURN AIR	78.3	DEG F	NRML
ADM BLDG AIR HANDLER ONE 04	HOT DECK	97.6	DEG F	"ALML"
ADM BLDG AIR HANDLER ONE 05	COLD DECK	65.0	DEG F	"ALML"
ADM BLDG AIR HANDLER ONE 06	RETU AIR RH	50.4	PCT RH	NRML
ADM BLDG AIR HANDLER ONE 07	OUTSIDE AIR	54.0	PCT RH	NRML
ADM BLDG AIR HANDLER ONE 08	FILTER	CLEAN		NRML
ADM BLDG AIR HANDLER ONE 09	FREEZE STAT	NORMAL		NRML

FIGURE 9. ENGLISH LANGUAGE DISPLAY OF BUILDING STATUS

Color Graphic Displays: These displays take the information generated by a DDC system and present it on a color schematic of an actual HVAC system being controlled. This feature greatly enhances the ability of building operating personnel to understand the status of a complete HVAC system. (See Figure 10)

Modern DDC systems can gather building operational data and display it in a meaningful way. For example, a DDC system can be instructed to monitor a given HVAC system parameter (temperature, energy consumption) and record the data over a period of time. The information can then be presented using a color graphic CRT in a chart format. (See Figure 11)

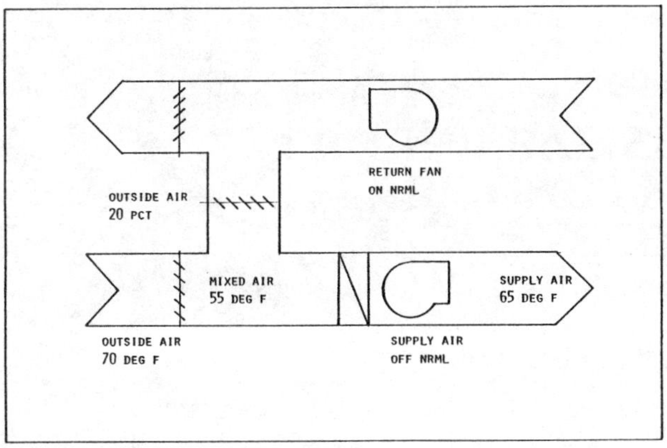

FIGURE 10. COLOR GRAPHIC DISPLAY OF AN HVAC SYSTEM

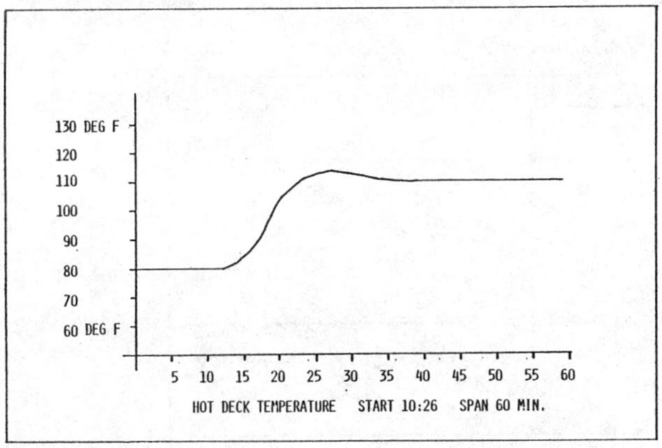

FIGURE 11. COLOR CRT DISPLAY OF HISTORICAL DATA

Automatic Records of All Control Strategies: One of the real benefits of a Distributed DDC system is that when control strategies are defined or modified by an operator, the system automatically stores the new control sequences. This greatly improves the ease of record keeping by ensuring that the as-builts are kept automatically.

THE LIFE CYCLE COST OF A CONTROL SYSTEM

Responsible building owners with an eye toward the future are showing concern about the total cost of a control system over its lifetime. The life cycle cost of controls is affected by several factors including: initial cost, maintenance cost, repair and replacement cost, and the cost to modify.

As stated earlier, in many applications the superior performance of DDC can often be obtained for a cost which is comparable to pneumatics. It was concluded before as well, that the maintainability of a DDC system was easier because of fewer components, built-in diagnostics and the reduced need for regular recalibration. These features also can greatly simplify any repair and replacement activity. Indeed, in many cases when a pneumatic controller is replaced it is substituted with a controller that has a different control action, one that is better suited to the application. With a DDC system the change in control action could be done via the operators terminal without using any additional hardware. Lastly, it has already been discussed how the flexibility of DDC greatly reduces the cost to modify or adapt controls to changing building conditions.

DDC appears to offer a much better altnerative over the lifetime of a control system in terms of its performance versus its cost.

ADDDING ENERGY MANAGEMENT TO A CONTROL SYSTEM

An environmental control system is a necessity in a facility and provides comfortable working conditions for the building occupants/workers. The need for energy management is a financial one, however, building owners and managers struggle to optimize energy consumption without causing undue discomfort for any occupants. When selecting an environmental control system for a building, great care should be taken to ensure that energy management functions can be added easily and economically.

It is in this area that DDC has a distinct advantage over pneumatics. To perform the necessary environmental control in a building a DDC controller must be connected to all HVAC equipment under control (fans, pumps, etc.) and also be directly controlling dampers, and valves. The DDC controller will also be monitoring outside air conditions, discharge and space conditions. In short, the DDC system is probably monitoring and controlling everything that a separate supervisory system would have to be connected to, to perform energy management.

Many energy management functions that can be performed by a DDC system, but are costly to add to a pneumatic control system, are shown in Table 1. However, the same functions can usually be done by simply re-defining control strategies in the DDC panel.

CONCLUSIONS

The preceding evaluation of pneumatic controls and Direct Digital Control is summarized in Table 2. This detailed evaluation clearly shows that DDC is a better control system in almost every category. This is a fact even when one only compares DDC with an environmental control system. If one requires several energy management functions and adds a supervisory automation system to the pneumatic controls, the benefits of DDC are even more obvious.

The time for DDC in commercial HVAC applications is obviously at hand. Building owners who are selecting a control system for their new or retrofit building would be well advised to consider Direct Digital Control. By selecting DDC as their control system, owners can rest assured that they will be able to provide occupant comfort while minimizing energy consumption. DDC is clearly the modern control system for the modern building owner or manager.

ENERGY MANAGEMENT FUNCTION	ADDING ENERGY MANAGEMENT TO	
	PNEUMATIC CONTROLS	DDC
Economizer Cycle (Enthalpy Optimization)	• Add sensors, relays • Add or re-pipe controllers OR • Add separate economizer panel	• Define Control Strategy
Reset of Supply Air, Hot Water, Chilled Water	• Add relays, sensors • Add receiver-controllers	• Define Control Strategy
Programmed Start/ Stop of Fans, Pumps, Lights	• Add timers • Wire timers to equipment	• Define Control Strategy
Duty Cycling	• Add separate duty cycle panel or load programmer • Rewire equipment	• Define Control Strategy
Start Time Optimizer	• Add separate optimizer panel or controls • Rewire equipment	• Define Control Strategy
Demand Limiting OR Load Shedding	• Add separate demand controller • Rewire equipment	• Define Control Strategy
Lighting Control	• Add separate lighting controllers • Rewire lights	• Wire lights to existing DDC panels. • Define Control Strategy

TABLE 1. ADDING ENERGY MANAGEMENT TO A CONTROL SYSTEM

COMPARISON CATEGORY	CONVENTIONAL PNEUMATIC CONTROLS	DIRECT DIGITAL CONTROL	BEST CONTROL SYSTEM
Performance	• Proportional control only. • Single loop controllers. • Complex control is difficult or costly. • Adequate control.	• Full PID control and more. • Multi-loop controller. • Easy to define complex sequences. • Closer control.	DDC
Initial Cost	Cost rises with number of control loops. • Complex control is very expensive.	• Once cost of DDC controller is absorbed, cost rises with number of sensors and actuators. • Capable of most complex control.	COMPARABLE
Reliability	• Proven reliability over many years, however, control system must be well maintained and recalibrated regularly. • Relies on air supply.	• Proven reliability in process industry and many commercial HVAC applications. • Each DDC controller can stand-alone.	DDC
Maintainability	• Relatively easy to maintain. • Require regular recalibration due to drift.	• Automatic as-builts. • Built-in diagnostics. • Fewer components. • No drift. • Service by board replacement.	DDC
Flexibility	• Changes or additions require new or different controllers re-piping and often wiring, and then re-calibration.	• Programmable controller. • New control strategies defined at central. • New control easily added.	DDC
Ease of Use	• All operator interaction at local control panels. • Can read temperatures and change set-point.	• Full English language reports. • Color Graphic Displays • Automatic Records of all control strategies.	DDC
Life Cycle Cost	• Requires regular recalibration. • Modification and expansion require additional controllers.	• Easy to maintain. • Easy to modify. • Easy to expand.	DDC
Cost to Add Energy Management	• Each new function usually requires additional equipment and labor.	• New functions are easily defined by operator.	DDC

TABLE 2. COMPARISON OF PNEUMATICS AND DIRECT DIGITAL CONTROL

SECTION 8
STATUS OF ENERGY ENGINEERING EDUCATION

Chapter 39
ENERGY ENGINEERING AT THE UNIVERSITY OF ARIZONA

R. L. Seale

In recent years there has been an increasing demand for individuals who bring a broad perspective to the problems of energy generation and usage. Whether it is a utility, a state planning or regulatory agency, a division of the federal government, or an industrial firm, the need exists for people who are familiar with many facets of the "energy problem." Decisions made by these organizations are almost never made on the basis of narrow technical grounds, but rather involve a complex set of considerations. Economics, environmental impact, available resources, conservation, options for generation and distribution of electricity --these are just a few of the areas that must be considered. Thus, while the reactor physicist, power line siting expert, or economist have definite roles to play, decisions are most often made using all of the many disciplines that impact upon the energy problem.

While people certainly exist who can integrate these various aspects of the energy situation, the typical path for achieving this perspective is long and difficult. Typically, people with a background in one of the basic engineering disciplines must undergo several years of on-the-job training in government or industry to achieve the more general outlook required. While there certainly is no substitute for experience, it may be possible for the educational process to produce people who can more quickly assimilate the range of information necessary to make decisions or advise those who do. It was in response to this challenge that the Energy Engineering program was developed at the University of Arizona.

There are other aspects of the problem which also influenced this decision. We are all quite aware of the chronic shortage of competent engineering talent that promises to extend into the foreseeable future. This shortage is expressed most vividly in terms of the rather frantic recruiting of qualified graduates in engineering that seems to be the general rule across the country these days. Evidence that this shortage is real and not the result of a fad that is present in American industry may be drawn from the fact that while slightly less than 6% of the bachelors degrees conferred this past year were in engineering in the U.S., the fraction in most of the rest of the industrialized world was above 20%! Indeed, in the Eastern European countries 40% of the bachelors degrees conferred were in engineering. As a head count, there were approximately 10,000 more engineers graduated in Japan than in the U.S. this last year. While a portion of this statistical disagreement is a reflection of the fact that a much larger portion of the college offerings in the United States may be considered to be in non-professional areas, the fact remains that on the basis of total population, engineering training in the U.S. is lagging far behind any of the other industrialized nations. Parenthetically, I might also note that even in the so-called professional areas, the less litigious societies do not indulge in the wholesale education of individuals to enter the legal profession; yet much of today's engineering effort in the U.S. is directed to satisfying requirements generated by lawyers.

By any measure the need for qualified, capable individuals in engineering is real. To meet this need it is necessary to attract individuals from outside the traditional sources of incoming engineering students. We need look no further than the very great appeal that systems and computer oriented engineering curriculum have had for the incoming female engineering students to appreciate that there is some reluctance of some individuals to enter the more traditional engineering fields such as civil, mechanical, and the like. Also, many of us have had the experience of talking to the young science or mathematics major with excellent grades who is not attracted to the traditional engineering disciplines. There must be some appeal for these students in engineering. It should be emphasized that the need for additional engineering talent will not be met by reshuffling student enrollments among the various disciplines. Rather it is mandatory that there be programs which are attractive to an as yet uncommitted constituency of capable people. Such programs must have the quality and rigor necessary to assure that the graduate is fully capable of functioning as an engineer. It was with these points clearly in mind that the Energy Engineering program at the University of Arizona was designed.

The program was designed to provide a strong background in basic sciences. Basic mathematics, physics, and chemistry are obvious prerequisites. All undergraduate engineering students have these introductory courses in their curricula, and this degree program is no exception. Thirty-three hours of mathematics, physics, and chemistry are required in the first two years of study. In addition, the first two years contain two hours of course work designed to give the student a broad-based introduction to the study of energy systems and four hours of computer programming and application.

Beyond the introductory sciences, there are several areas in which a student should be knowledgeable, which are common areas of consideration for any method of energy generation:

1. Thermodynamics and heat transfer
2. Engineering mechanics
3. Engineering materials
4. Electrical circuits
5. Chemical engineering processes
6. Fluid mechanics

Each of these has been incorporated as a course or courses in the curriculum.

Finally, the principles learned in the first two areas of study should be applied to the individual sources of energy which may be available. Specifically, students should be exposed to courses oriented toward nuclear, solar, and fossil fuel power plants--their design, economics, and environmental impact. Thus a series of more specialized courses is required in the senior year. The emphasis is on the integration of information from earlier courses and the application of that information to design problems.

In addition to the broad background in science and engineering described above, a student should have some more specific academic experience oriented toward a particular area of expertise. Otherwise, the possibility exists that the student's knowledge will be so broad that he or she will be unable to make a contribution in any particular field. Thus, the technical electives traditionally left open to student choice are somewhat more constrained in this program, so that the student will be exposed to some technical area in more detail.

The entire program is shown in Table 1 on a semester-by-semester basis. There are prerequisite requirements that impose a heavy discipline on the student's progress as is the case in all other engineering programs.

Table 1 Study Program for the Bachelor of Science Degree in Energy Engineering

	Units
Freshman Year	
First Semester	
Analytic Geometry & Calculus I	3
Fundamentals of Chemistry with Laboratory I	4
Introduction to Nuclear & Energy Engineering	1
English Composition I	3
Humanities or Social Science Elective	3
Second Semester	
Analytic Geometry & Calculus II	3
Fundamentals of Chemistry with Laboratory II	4
Physics-Mechanics with Laboratory	4
FORTRAN Programming	2
Introduction to Energy Engineering	1
Humanities or Social Science Elective	3
Sophomore Year	
First Semester	
Vector Calculus	4
Physics-Electricity & Magnetism with Laboratory	4
Chemical Engineering-Introduction	4
English Composition II	3
Computer Applications	2
Second Semester	
Linear Algebra & Differential Equations	4
Physics-Optics, Acoustics, & Heat with Laboratory	3
Engineering Thermodynamics	3
Engineering Statics	3
Engineering Economics	3
Junior Year	
First Semester	
Chemical Engineering-Mass Transfer	3
Physics-Structure of Matter	3
Fluid Mechanics	3
Electrical Power Systems	3
Technical Elective*	3
Second Semester	
Fundamentals of Materials with Laboratory	4
Analysis of Deterministic Systems	3
Heat Transfer	3
Electronics & Instrumentation	3
Humanities or Social Science Elective	3
Senior Year	
First Semester	
Nuclear Energy & Power	3
Solar Energy Engineering	3
Water Resources in Energy	3
Technical Electives*	6
Humanities or Social Science Elective	3
Second Semester	
Fuels Processing in Engineering	3
Power Systems Analysis	3
Systems Design	3
Technical Elective*	3
Humanities or Social Science Elective	4

*Technical electives to be selected from a specific topic option.

Basic Sciences and Mathematics

The freshman and sophomore level courses are, for the most part, common to all of the undergraduate engineering curricula. Fourteen units of mathematics are required, which should give a student a firm grounding that includes analytic geometry, calculus, linear algebra, and differential equations. Eight units of lecture and laboratory work in chemistry are required. The physics sequence gives the student eleven units which will cover standard topics--mechanics, electricity and magnetism, optics, acoustics, and heat. In addition, four units of engineering-oriented computer programming are required. All of these courses are the offerings also designated as required for the major in that field.

A key feature of the freshman year is a new course which will give an introductory overview of the program in energy engineering. The main purpose of this introduction is to set a context for subsequent courses. This course will be characterized by analysis of world and national energy problems and a broad review of possible solutions. Faculty from all engineering disciplines and guest lecturers will be featured. Some hands-on experience with equipment used in later courses will sharpen the interest and anticipation of the students. Hopefully, by the time two units have been completed, the students will have a clear idea of where the next three years are taking them and of the career opportunities awaiting them afterward.

In addition to mathematics and the physical sciences, economics is another discipline which should be considered "basic" for the program envisioned here. To that end, a new course, "Economics of Energy" will be instituted for the sophomore year. The comparative cost of various energy options will be stressed, along with the principles underlying those estimates. Of particular interest will be an analysis of the way in which technological change can influence costs. A goal of this course will be to produce an engineer who is "literate" in both technology and economics and has an appreciation for the interaction between the two disciplines.

Engineering Sciences

To obtain the broad perspective we hope to create in this curriculum, the student must have knowledge and skills in several different engineering fields:

1. Thermodynamics and heat transfer - six units which cover the basic laws of thermodynamics and conductive, convective, and radiative heat transfer.

2. Engineering mechanics - three units covering traditional analysis of force systems.

3. Materials - seven units which cover both microscopic and macroscopic approaches to the study of matter. Three units will deal with modern physics, including atomic structure and nuclear physics. The other four units will cover engineering applications with hands-on laboratory experience.

4. Electrical circuits - six units will include an introduction to electrical engineering emphasizing problems of power generation coupled with a survey of electronics.

5. Chemical engineering - three units of introductory material with emphasis on mass transfer processes.

6. Fluid mechanics - three units of mechanical engineering oriented toward fundamental properties of fluids including both static and dynamic analysis.

Much of the work described above will be given in the junior year, building on and extrapolating from the basic science program of the first two years. In addition, the junior year contains three units of work oriented toward the analysis and design of deterministic systems which should be applicable to the study of control systems for any kind of power-generating facility.

Advanced Courses and Engineering Design

The senior year is largely devoted to courses in which the student integrates the knowledge gained in the basic science courses and in the study of particular energy generation systems and major activities associated with energy generation. Six units will be devoted to revisions of courses currently offered by the Department in the overall design and operation of nuclear and solar power systems. The nuclear course will stress engineering analysis of current power plants and include environmental effects, siting, radioactive emissions, thermal pollution, and waste management. The solar course will stress solar collector design, solar heating and cooling, and solar systems for generating both electrical and mechanical power. A pair of related courses will be taught in power systems analysis and energy systems design. Such topics as gas power cycles and thermionic, thermoelectric, magnetohydrodynamic, photovoltaic, and fuel cell power systems will be covered, as well as modern techniques in economic evaluation, system modeling, optimization, and decision analysis.

The basic program core will include six units which deal with the problems of natural resource utilization and conservation. One is a fuel-processing course in which the steps for obtaining clean fuel from coal, petroleum, natural gas, oil shale, tar sands, and geothermal sources will be studied. The other course, of particular interest in the western United States, will involve the study of water resources as they are required for energy generation.

Technical Elective Options

Finally, there are twelve units of technical electives included in the program. To provide a focus for the individual student, option sets have been identified. These include:

1. Electrical: Courses in Electrical Engineering which give a student a more thorough understanding of circuit theory and problems of electrical power transmission. These include
   - Electric Circuits Analysis - 2 semesters
   - Electric & Magnetic Fields
   - Energy Conversion
   - Electric Power Systems

2. Thermo-Mechanical: Courses in Mechanical Engineering which delve more deeply into fluid flow and heat transfer. These include
   - Advanced Fluid Mechanics
   - Advanced Thermodynamics
   - Problems in Fluid Mechanics & Thermodynamics
   - Wind Energy Conversion Systems

3. Materials: Courses in Metallurgical and Nuclear Engineering which give more detail in materials problems, particularly for nuclear reactor applications. These include
   - Physical Metallurgy - 2 semesters
   - Nuclear Fuel Cycles
   - Radiation Effects

4. Modeling and Simulation: Courses in Nuclear and Systems Engineering providing more detailed work in the modeling of complex systems not amenable to simple analytic approaches. These include
   - Dynamics of Nuclear Systems
   - Probability and Statistics in Design
   - Economic Analysis of Engineering Systems

5. Environmental Effects: Courses in Nuclear and Civil Engineering giving more detail into problems of air and water pollution, including radioactive wastes. These include
   - Environmental Analysis of Power Systems
   - Environmental Quality Management
   - Air Pollution

The combination of engineering science, an overview of various energy systems, and specialized training should produce a graduate capable of making a direct contribution in the professional world.

Current Status of the Program

The program in Energy Engineering was formally submitted to the university administration in the spring of 1980. It was subsequently approved by the Board of Regents in January of 1981. Of course, prior to formal approval, the word was out that it would likely be available, so during the 1980-81 academic year, ten students indicated their intent to pursue the major in Energy Engineering. Based on preregistration for the fall 1981 term, we expect to have 25-30 freshmen entering the program.

Based on interviews with these students, it would appear that between one-half and two-thirds would not have selected another engineering discipline, but rather would have chosen to major in one of the other programs in the university. The applicants have ATC scores which would indicate a high probability of success, based on the performance experience in the college. Further discussions in which the rigor of the program is stressed appear to be productive in screening out weaker students at the onset.

Additional Comments on the Program

From the very beginning it was recognized that in the current climate on the nationwide university scene, any proposal for a new degree program could expect very serious and probing review. At the same time, the situation in engineering does deviate from the university-wide norm in that enrollment is going up, attrition is higher to the extent that some less capable students are attracted by the popularity of engineering, and the demand for engineering talent is increasing even faster than the popularity of the programs. Thus it was the intent from the beginning that the Energy Engineering program be designed to have a high potential of attracting a new constituency into engineering. Of course, this latter requirement is dependent both on the nature of the study program and on the way in which the program is publicized and administered. These goals are as much a part of the objectives of the program today as they were during the initial development of the program proposal.

An important part of the program development was input from individuals in both government and industry. As might be expected, that input was varied and bimodal. The one thing the proposal did not generate was apathy. On campus the concept had been germinating for several years, going back as far as 1970 to an internal memorandum written by Professor Rocco Fazzolare of our department. It took a long time to develop the program concept to the point where it was essentially complete in detail and, hence, ready for formal presentation to the faculty of the college and the university. In that interim, considerable spade work was done in selling the program. As a result, the general reception within the university was favorable, with some reservations which appear to have been accommodated in the program as it now stands.

Outside the university the arguments against the program are all of those you would expect. They were largely based on general considerations of principle rather than detailed criticism of the program itself. Certainly the requirement for a specialized, in-depth option helped defuse criticisms based on too much generality or a lack of depth.

The support received was, on the other hand, more specific. Individuals from utilities, architect/engineering firms, consulting firms, and equipment suppliers were all a part of that group. Persons involved in planning and in the evaluation of energy resource alternatives were strongly represented. The nature of the comments received included a desire for more economics and the importance of a comprehensive understanding of energy source alternatives rather than a narrow doctrinaire view of one option at the expense of all others.

The program is now under way and promises to be quite dynamic. Indeed, some of the details in some of the courses at the advanced level are still in the formative stage. It will be interesting to see how things progress over the next few years. There are several points, however, that will continue to be emphasized. In the curriculum area, the program will be tough and every effort will be made to assure that the basic package of lower division courses will allow the student the maximum flexibility should other options appear attractive. Secondly, an effort will be made to identify those students who come into the program who might be considered "new" in the sense that they would not have otherwise entered engineering. Finally, we will bring people from industrial organizations to the campus to participate in a special seminar series and, incidentally, to become better informed about the program and the capabilities of its graduates. Ultimately, of course, it is in the marketplace that the soundness of this effort will be judged.

SECTION 9
ENERGY
AUDITS

Chapter 40
ENERGY AUDITS AT 48 U. S. HOSPITALS

E. Hirst

ABSTRACT

Staff at the Oak Ridge Associated Universities (ORAU) conducted energy audits at 48 hospitals in four states (New York, Pennsylvania, Virginia, Tennessee) between 1978 and 1980. Staff at the Oak Ridge National Laboratory (ORNL) and ORAU developed and organized a computerized data base containing information from these audits. This paper describes the ORAU audit process; summarizes the data collected from these audits on hospital characteristics, annual energy use, and the audit recommendations; and analyzes the audit data in terms of cost-effectiveness, type of recommendations, and the relationship between potential energy saving and characteristics of the individual hospital.

INTRODUCTION

Hospitals are the most energy-intensive buildings in the commercial sector, consuming about two-thirds more energy per unit floor area than the average of all commercial buildings.[1] This high energy use occurs because hospitals are occupied 24 hours a day for 365 days a year and have high ventilation requirements. Estimates for 1980 suggest that the 2.3 billion ft^2 (210 million m^2) of hospital floorspace accounts for about 1.2 Quads (1.3 EJ) of primary energy, equivalent to 12% of total commercial sector energy use.[2]

Identifying operation and maintenance (O&M) practices and energy conservation measures is particularly important in hospitals because they are so energy intensive. A variety of programs designed to audit hospitals and implement energy-efficiency practices and measures is underway. Perhaps the most important of these is the federal Institutional Conservation Program, authorized by the 1978 National Energy Conservation Policy Act.[3] In addition, private organizations such as the American Hospital Association[4,5] and Blue Cross/Blue Shield[6] have related programs.

The Oak Ridge Associated Universities (ORAU) conducted energy audits at 48 hospitals in four states (New York, Pennsylvania, Virginia, Tennessee) between November 1978 and April 1980.[7] Data collected by ORAU and analyzed at ORNL include information on hospital location, climate, floor area, number of beds, preaudit energy use and costs by type of fuel, and on each audit recommendation (type of measure, estimated energy saving by fuel type, and estimated payback period).

The audit data and their analysis provide important information on ways to improve energy efficiency in hospitals and the economics of doing so. This information can be used to design programs that encourage energy audits, O&M changes, and capital improvements in hospitals. In addition, the data provide information on a portion of the commerical sector, that sector of the U.S. economy for which we have little energy-related information.

THE ORAU AUDIT PROCESS

Between late 1978 and mid-1980, ORAU audited 40 hospitals in Pennsylvania, four in Virginia, three in New York, and one in Tennessee.

The process used by ORAU includes eight basic steps:[7]

1. Pre-audit checklist: The pre-audit checklist is sent to the hospital administrator at least one week before arrival of the auditors. The checklist informs the administrator of the facilities, information, and cooperation that the auditors require to provide an effective audit.

2. Entrance conference: The entrance conference is scheduled on the first day of the audit. This is the initial meeting between hospital representatives and the auditors. The entrance conference allows ORAU to explain the audit and to gauge how extensively the engineer and administrator want to be involved in the process.

3. Walk-through of the facility: The walk-through of the facility usually takes place immediately after the entrance conference. In the walk-through, the auditors become familiar with the facility and the location of all major energy-using equipment. The auditors question the guide (the engineer or maintenance foreman) on equipment capacities, hours of operation, maintenance, and operating procedures.

4. Data collection and training: The data collection process takes place over a period of several days or weeks. It is desirable to provide training for the hospital maintenance staff while collecting data. ORAU offered training in steam trap maintenance, boiler combustion efficiency, HVAC controls, efficient lighting, and water treatment.

5. Pre-exit conference: The pre-exit conference is held at the conclusion of the data collection. The auditors report initial findings to the administrator and the engineer at this time.

6. Report writing: Depending on the complexity of the hospital and the recommendations, a report

*Research sponsored by the Office of Conservation and Renewable Energy and the Energy Information Administration, U.S. Department of Energy under contract W-7405-eng-26 with Union Carbide Corporation.

requires one to two weeks to prepare. The reports are then produced at ORAU.

7. Report submittal: Reports are usually submitted to the hospital within 60 days of the pre-exit conference.

8. Exit conference: After the hospital examines the report, an exit conference is scheduled. The purpose is to allow the hospital staff to ask questions and obtain clarification or additional interpretation of the audit recommendations, and to motivate the hospital staff to implement the recommendations.

ORAU and ORNL staff developed a data base structure for the information collected from these 48 hospitals (Table 1). The data base includes information on the hospital itself; on baseline fuel consumption and costs, and climatic factors (monthly data for at least one year); and on each audit recommendation made at the hospital. To permit analysis of the audit recommendations, ORAU developed a scheme to classify each recommendation into one of five types; each type is further subdivided into several codes.

Table 1. Structure of the ORAU/ORNL energy audit data base[a]

Hospital ID
Zip code
Date of audit
Number of beds
Floor area
Number of fuels used at hospital
Number of billing periods for fuel use
Number of audit recommendations
Pre-audit (baseline) fuel consumption[b]
Fuel code
Start date for billing period
Monthly fuel consumption
Monthly fuel cost
Monthly heating degree days
Monthly cooling degree days
Audit recommendations
Recommendation code and type
Simple payback period
Number of fuels in recommendation
Fuel code
Fuel saving

[a]The data base has an hierarchical structure, in which indentations refer to repeating groups.

[b]Data are given for at least 12 months, for each of the fuels used at the hospital.

AUDIT RESULTS

This section presents descriptive statistics from the ORAU audits. We first discuss data on baseline energy use at these hospitals and then discuss the audit recommendations.

The hospitals vary considerably in terms of size and energy use (Table 2). The average hospital has 215 beds, but the range is 43 to 595 beds. Similarly, the range in hospital floor area is 36 - 614 thousand ft^2 (3.3 - 57.0 thousand m^2). Average annual energy use across these 48 hospitals is 480 kBtu/ft^2 (5.4 GJ/m^2); the range is 240 - 840 kBtu/ft^2 (2.7 - 9.5 GJ/m^2).*

On average, the fuel used to produce electricity accounts for slightly more than half the total energy use at these hospitals and fossil fuels account for the remaining share (Table 2). The variation among hospitals in electricity use and in fossil fuel use is larger than the variation in total energy use (Figs. 1-3). This occurs because two hospitals are essentially all-electric, while the other 39 use fossil fuels as their primary heating fuel.

Table 2. Characteristics of the 48 hospitals audited by ORAU

	Mean	CV (%)[a]
Number of beds	215	61
Floor area		
ft^2	197,000	61
m^2	18,300	
Annual energy use		
Electricity		
kBtu/ft^2	248	42
GJ/m^2	2.8	
Fossil fuels		
kBtu/ft^2	232	33
GJ/m^2	2.6	
Total energy		
kBtu/ft^2	480	25
GJ/m^2	5.4	
Degree days		
Heating		
^{o}F	6430	36
^{o}C	3570	
Cooling		
^{o}F	700	37
^{o}C	390	

[a]The coefficient of variation (CV) is the ratio of the standard deviation to the mean.

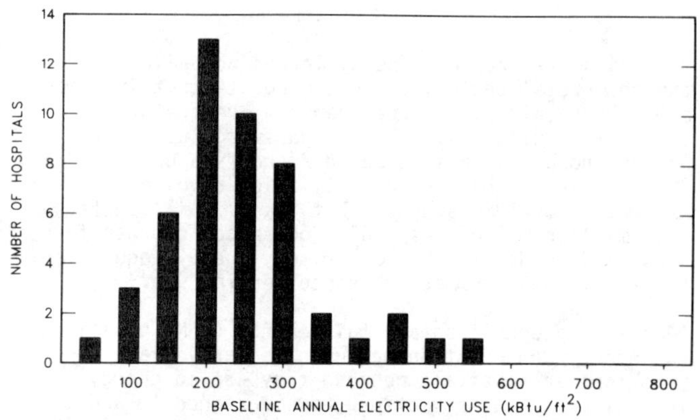

Fig. 1. Variation in annual electricity use among 48 hospitals.

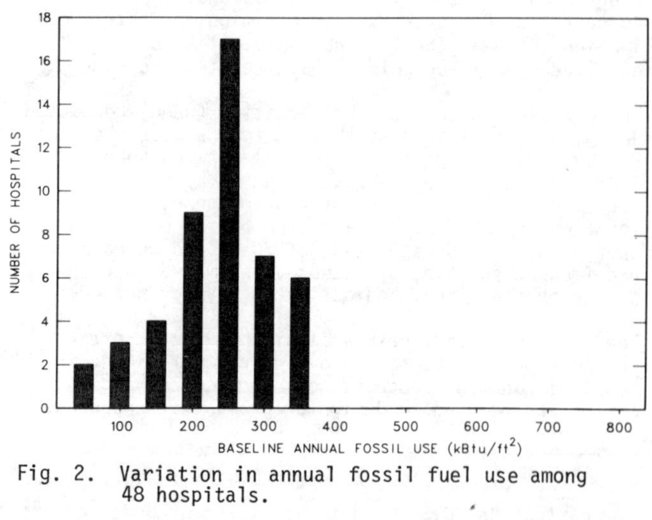

Fig. 2. Variation in annual fossil fuel use among 48 hospitals.

*Electricity is treated in terms of primary energy (11,600 Btu/kWh, 3.4 J/J). Conversion factors for other fuels are in ref. 8.

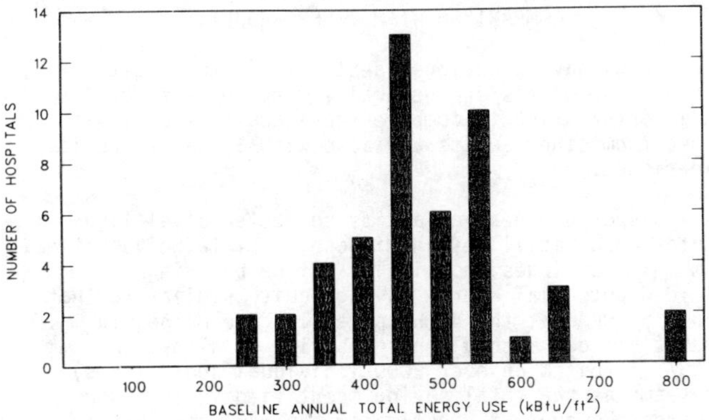

Fig. 3. Variation in annual total energy use among 48 hospitals. The two high energy use hospitals both use residual oil for primary heating and reported high values of electricity use.

Averaged across the 48 hospitals, implementation of all the audit recommendations would cut annual energy use by about 20% (Table 3). Almost two-thirds of the energy saving potential is fossil fuels and one-third is electricity. The percentage reduction in fossil fuel use is greater than in electricity use, probably because all but two of the facilities use fossil fuels as their primary heating fuel. The average total energy saving is almost 100 kBtu/ft^2 (1.1 GJ/m^2), with a range of 27 - 200 kBtu/ft^2 (0.3 - 2.3 GJ/m^2).

Table 3. Summary of the audit results at 48 hospitals

	Mean	CV (%)[a]
Potential energy saving		
Electricity		
kBtu/ft^2	36	85
GJ/m^2	0.41	
% of total elec. use	15	74
Fossil fuels		
kBtu/ft^2	63	59
GJ/m^2	0.72	
% of total fossil use	27	45
Total energy		
kBtu/ft^2	99	46
GJ/m^2	1.13	
% of total energy use	20	40
Capital cost		
$/ft^2	0.25	92
$/m^2	2.65	
Payback period (years)	1.0	60

[a]The coefficient of variation (CV) is the ratio of the standard deviation to the mean.

The estimated cost of implementing these conservation changes averages $0.25/ft^2 ($2.65/m^2). These costs are likely to be repaid within one year, on average, by avoided utility bills. This very rapid payback period is a direct consequence of the ORAU focus on low-cost changes, rather than capital-intensive changes.

As with baseline energy consumption data, the estimates of potential energy saving and implementation cost vary considerably among the hospitals (Fig. 4). The ratio of the maximum energy saving to the minimum is seven and the ratio of the maximum implementation cost to the minimum is 27. In general, those hospitals with higher implementation costs also have higher potential energy savings; however, as Fig. 4 shows, there are many exceptions to this. In particular,

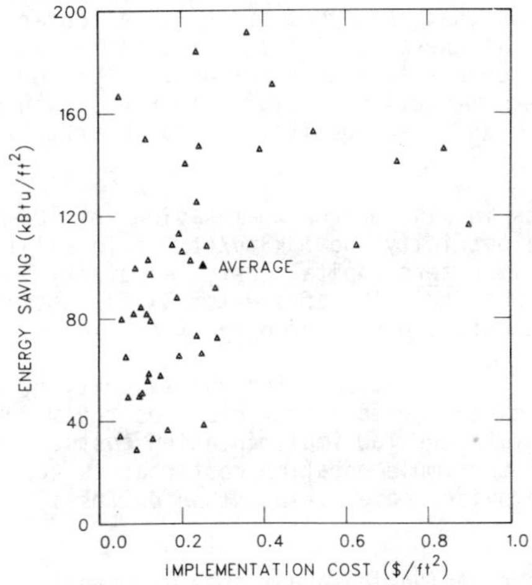

Fig. 4. Total potential annual energy saving as a function of implementation cost at 48 hospitals.

audit results at 16 hospitals yield an estimated total implementation cost of about 20¢/ft^2 (17 - 25¢/ft^2); however, the range in potential energy saving is from 40 to 180 kBtu/ft^2 for these facilities.

The relationship between baseline energy use and potential post-retrofit energy use also shows considerable variation. These results suggest that hospitals are unique facilities, each with its own energy-related characteristics, operating procedures, and potential energy savings.

Despite these large variations among hospitals, certain generalizations concerning audit recommendations can be made. Figure 5 is a plot of the cumulative estimated energy saving per unit floor area as a function of the cumulative implementation cost per unit floor area. Curves are shown for electricity, fossil fuels, and for total energy (the sum of electricity and fossil fuels). The curves are drawn by first ordering all the audit recommendations from all the hospitals (881 recommendations in all) in terms of increasing implementation cost per unit energy saving. The final results are curves with a well-defined shape. Energy savings increase with increasing capital investment, but at an ever-declining rate, a manifestation of the law of diminishing marginal returns.

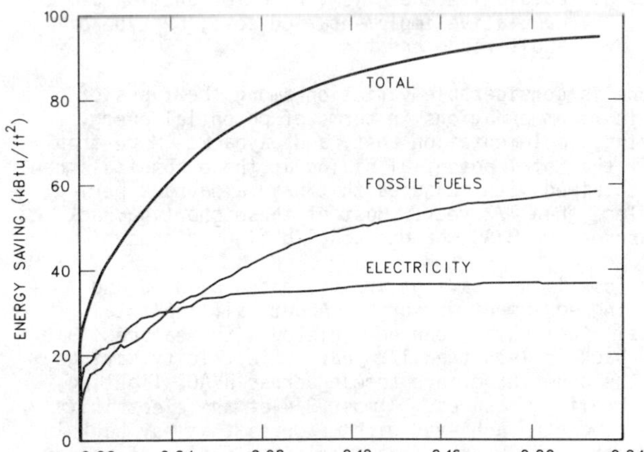

Fig. 5. Cumulative energy saving as a function of cumulative implementation cost for all 881 audit recommendations at 48 hospitals.

The curves show that about 60% of the total potential energy saving at these hospitals is fossil fuels; electricity accounts for the other 40%. Implementation of all 881 recommendations included in Fig. 5 would cut energy use at these hospitals by 96 kBtu/ft^2 with a cost of 23¢/ft^2.

The intercepts in Fig. 5 show that savings of 12 kBtu/ft^2 of electricity and 7 kBtu/ft^2 of fossil fuels are possible with zero capital cost. A potential saving of 40 kBtu/ft^2 (42% of the total) is possible with an implementation cost of only 1¢/ft^2.

Although fossil fuels account for the majority of the total energy saving, electricity provides the majority of the saving at very low implementation costs. Indeed, until the implementation cost reaches 4¢/ft^2, electricity provides more savings than do fossil fuels.

This relative contribution and its shift with increasing capital cost is a consequence of the types of measures recommended and their costs. Figure 6 is similar to Fig. 5 except that Fig. 6 shows energy saving by type of audit recommendation rather than by type of fuel. Changes to the heating, ventilating, and air conditioning (HVAC) system account for more than half the total energy saving. Changes to the central heating plant account for another quarter; lighting, cooling, and other changes account for the rest.

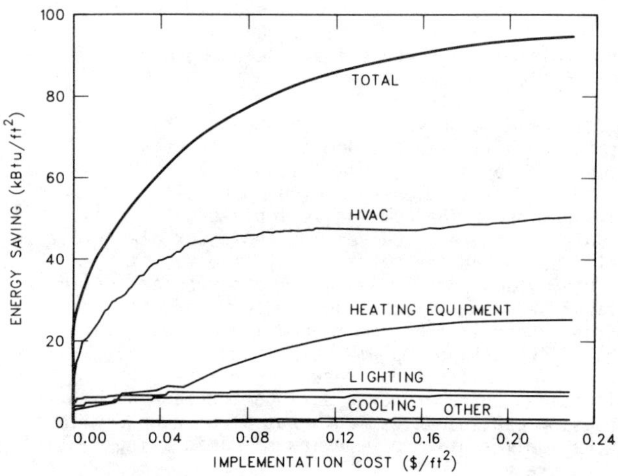

Fig. 6. Cumulative energy saving as a function of cumulative implementation cost, by type of audit recommendation.

There is considerable variation among the types of audit recommendations in terms of potential energy saving, implementation cost, and payback. More than half the total potential saving at these hospitals can be obtained with measures that have a payback period of less than 1/2 year. Most of these short-payback measures are HVAC changes and lighting changes.

The fossil fuel savings are dominated by HVAC and heating equipment measures. About half the total fossil fuel saving can be achieved with measures that pay back in less than 1/2 year. Electricity savings, on the other hand, are spread across HVAC, lighting, and cooling measures. Almost 3/4 of the electricity saving can be achieved with measures that pay back in less than 1/2 year. Thus, the distribution of energy savings by payback period is skewed more heavily towards short periods for electricity than for fossil fuels (Fig. 6).

COMPARISON WITH OTHER RESULTS

The energy saving options identified in the audits of these 48 hospitals are generally very cost-effective. It is interesting to compare these audit results with those from other audits and also with actual retrofit experience.

Analysis of engineering audits conducted at 41 large state-owned facilities in Minnesota (state universities, community colleges, hospitals, office buildings) yielded potential energy savings quite similar to that found by ORAU at the 48 hospitals.[9] The Minnesota audits yielded a total energy saving of almost 100 kBtu/ft^2 (based on over 2000 individual audit items), the same as the total saving identified in the ORAU audits. However, the estimated capital cost for the Minnesota audit recommendations ($1.90/ft^2) was far higher than the ORAU cost estimate. This difference occurred because of the different approaches chosen. The Minnesota audits included considerable attention to envelope measures that are expensive and have long payback periods. The average payback period for the Minnesota envelope measures was 23 years, compared with 3.2 years for all the other measures.[9] The ORAU audits, on the other hand, generally ignored expensive measures. Only 10 of the 881 conservation items recommended in the ORAU audits dealt with the envelope. These ten measures had an average payback of three years, compared with the average payback of one year for all the other measures.

A prototypical hospital located in Kansas City was analyzed as part of ORNL's commercial sector energy use model development.[1,10] Energy use at this hypothetical 180 bed hospital (100 thousand ft^2) was analyzed using the NECAP computer code to analyze hour-by-hour thermal loads. The base case, mid-1970s annual energy use was estimated at 560 kBtu/ft^2. Thermostat setback, reduction in ventilation levels, and some delamping yielded an estimated 32% reduction in annual energy use. The ORNL analysis yielded higher estimates for both baseline energy use and estimated energy reduction than did the audit data discussed here.

A recent review of actual retrofit experience (based primarily on case studies reported in *Energy User News* and *Air Conditioning, Heating and Refrigeration News*) yielded data on 40 commercial buildings that were retrofit during the 1976 - 1980 period.[11] The actual energy saving experienced by these buildings averaged almost 30% of the preretrofit baseline. The average investment was about 30 ¢/ft^2, yielding an average payback period of about two years.

A group at DOE assembled data from various sources on 230 commercial buildings.[12] The focus of their effort was on changes in actual metered energy use after retrofit, rather than changes in estimated energy use. The average actual reduction in energy use after retrofit (including both O&M and capital-intensive changes) was 19% of preretrofit baseline energy use. (The average actual saving for ten hospitals in their data base was 17%.) The average cost of these energy-efficiency improvements was 65 ¢/ft^2, which yielded an average simple payback period of 1.4 years.

DOE obtained data on sixty buildings for which both the audit estimate of energy saving and the actual energy saving were available.[12] Roughly 60% of these buildings saved more energy than was estimated and 40% saved less.

Comparison of the audit results presented here and actual experiences suggests that the ORAU audit estimates are probably realistic.

SUMMARY

This report discussed data collected from energy audits conducted at 48 hospitals in four states (New York, Pennsylvania, Virginia, Tennessee). The audits were conducted by ORAU staff between late 1978 and mid-1980. ORAU and ORNL staff developed and organized a computerized data base containing information from these audits. These data were then analyzed at ORNL during early 1981.

There is considerable variation in energy-related characteristics among the hospitals. Although the mean value of annual energy use is 480 kBtu/ft^2 (5.4 GJ/m^2), the range is from 50% of the mean to 175% of the mean. Similarly, there is considerable variation in the potential energy saving at each hospital. The average saving amounts to 20% of the baseline energy use; however, the range is from 6% to 49% of baseline energy use.

A second feature of the audits is the large potential energy saving at these hospitals. The average total saving is almost 100 kBtu/ft^2 (1.1 GJ/m^2), equivalent to 20% of baseline energy use. Not only is the potential saving large, but the cost to obtain the saving is quite small. The average implementation cost to achieve all the savings identified averages 25 ¢/ft^2 (\$2.65/m^2) across the 48 hospitals. The average payback period for these measures (at the fuel prices that prevailed at the time of the audits) is only one year. Thus the audits reveal the existence of large potential energy savings that are extremely cost-effective.

REFERENCES

1. J. R. Jackson and W. S. Johnson, *Commercial Energy Use: A Disaggregation by Fuel, Building Type, and End Use*, Oak Ridge National Laboratory, ORNL/CON-14, February 1978. Also S. M. Cohn, K. R. Corum, J. Kurish, and C. Emerson, *A Commercial Energy Use Model for the Ten U.S. Federal Regions*, Oak Ridge National Laboratory, ORNL/CON-40, March 1981.

2. D. M. Hamblin, personal communication, Oak Ridge National Laboratory, April 1981.

3. 95th Congress, *National Energy Policy Conservation Act*, PL 95-619, November 9, 1978.

4. P. E. Mehl, personal communication, American Hospital Association, September 1980.

5. M. Gough, "Hospital Energy Management Information System," *Overview: Programs, Projects and Activities*, The Hospital Research and Educational Trust, 1980.

6. R. Maturi, personal communication, Blue Cross/Blue Shield, October 1980.

7. J. M. Stelson, *Energy-Use Data for Audited Hospitals*, Oak Ridge Associated Universities, prepared for the Oak Ridge National Laboratory, February 1981. Also E. Hirst et al., *Analysis of Energy Audits at 48 Hospitals*, Oak Ridge National Laboratory, ORNL/CON-77, 1981.

8. U.S. Department of Energy, "Energy Measures and Energy Audits Grant Programs for Schools and Hospitals and Buildings Owned by Units of Local Government and Public Care Institutions," *Federal Register*, 44(64) April 2, 1979.

9. E. Hirst, R. Tyler, C. Eastes, and J. Dumagan, "Engineering Audits at State-Owned Buildings in Minnesota," *Energy*, 5(2), March 1980.

10. W. S. Johnson and F. E. Pierce, *Energy and Cost Characteristics of Commercial Building Shell Characteristics and Operating Schedules*, Oak Ridge National Laboratory, ORNL/CON-39, April 1980.

11. H. P. Misuriello and R. M. Bily, Jr., "A Study of Actual Metered Energy Savings for Energy Conservation Retrofit Measures Reported for Commercial Buildings," *Proceedings of the Conference on New Energy Conservation Technologies and their Commercialization*, Berlin, West Germany, April 1981.

12. H. Ross and S. Whalen, *Building Energy Use Compilation and Analysis, Part C - Conservation Progress in Commercial Buildings*, Buildings Division, U.S. Department of Energy, draft, May 1981.

Chapter 41

SUMMARY OF RESULTS OF SCHOOL ENERGY CONSERVATION STUDIES

J. Baumgartel, T. P. Cummins

INTRODUCTION

Following are the results of energy analyses of 29 public schools located in New Mexico. The schools contain 67 separate buildings and are located in the southeast, northcentral, and northwest regions of the state. A total of 234 Energy Conservation Measures (ECMs), were recommended for these schools by the staff of Energy Management Services, Inc. A compendium of these energy conservation measures is presented.

Division by Discipline

Of the 20 types of ECMs recommended, nine were architectural, seven mechanical, and four electrical. Of the total 234 individual projects, 101 (43%) were architectural, 82 (35%) were mechanical, and 51 (22%) were electrical. The total estimated project cost is comprised of $563,763 (43%) architectural, $496,375 (37%) mechanical, and $267,812 (20%) electrical.

The average architectural ECM costs $5,582 and saves $483 in the first year. Average mechanical projects cost $6,053 and save $1,148 and average electrical project costs and savings are $5,251 and $544, respectively.

Frequency

Specific Energy Conservation Measures were recommended 234 times. The most frequently recommended ECM is ECM #1 (Night Setback) which occurs 42 times. The next most frequently recommended ECMs are ECM #2 (Insulate Glass) and #3 (Timers and Exhaust Fans), each appearing 23 times. Following, are ECM numbers 4 (Lighting Modifications), 5 (Wall Insulation), and 6 (Roof Insulation) with frequencies of 22, 18, and 14 respectively. The frequencies continue to drop to the final four ECMs with two recommendations each.

The most frequently recommended ECM is a mechanical project, ECM #1. Overall, the architectural, mechanical and electrical projects are distributed uniformly with regard to frequency.

Table 1, "20 Energy Conservation Measures Ranked by Frequency", displays all 20 ECMs listed in order of decreasing frequency. The discipline associated with each ECM is also shown.

BACKGROUND

New Mexico Schools Energy Program

New Mexico became involved with energy conservation in schools through the National Energy Conservation Policy Act of 1978. The act contains provisions for a three phase energy grant program for schools: Preliminary/Energy Audits, Technical Assistance Programs, and Construction of recommended ECMs. The state used Phase 1 funds to complete Preliminary/Energy Audits of all public schools in New Mexico. Subsequently, various school districts applied for Department of Energy matching grant funds to accomplish Phase 2, Technical Assistance Programs. The information presented in this report is the product of Phase 2, Technical Assistance Programs conducted by EMS.

Methodology

Energy Management Services has developed a procedure for conducting Technical Assistance Programs and has used this procedure successfully on over 50 facilities. Since Phase 3, or Construction Project Grants, are awarded competitively on the basis of simple payback, the procedure is designed to provide accurate energy cost savings and project costs for each ECM. Due to the overwhelming number of articles published describing energy audit techniques, the EMS procedure will not be presented within the confines of this paper.

The result of the analysis of each school is a Technical Assistance Report. This report complies with all requirements of the grant program regulations and is complete with text designed to aid the district and school staff in understanding the school's energy operation and in implementing the recommended projects.

ECM Criteria

Criteria for ranking grant applications for implementation of ECMs have been established by the D.O.E. (FR April 17, 1979). These criterion are necessary for the acceptance of the grant application and include:

o ECMs must be recommended in a Technical Assistance Report prepared by a licensed professional engineer. Results from the study must comply with the regulations for Technical Assistance Programs.

o Simple payback for each ECM is less than 15 years and has been calculated with specified values of btu contents of fuels. Interaction between ECMs must be accounted for.

RESULTS

Overall

In total, 234 ECMs (energy conservation retrofit projects) were recommended for 29 schools containing 67 buildings in New Mexico. The total cost of all projects has been estimated to be $1,327,950 with a total first year energy cost savings of $170,707. Assuming all projects are completed, the aggregate simple payback on this investment will be 7.8 years. Payback for the school districts will be less than half this value due to D.O.E. grant participation. Additionally, as the cost of utilities increase, payback periods will decrease correspondingly.

Twenty separate types of projects were recommended. They have been numbered ECM #1 through ECM #20. The ECM numbers represent, in declining order, the frequency with which the ECM was recommended. ECM #1 was recommended 42 times and ECM #20 was recommended two times.

A complete list of all ECMs and a brief description appears in the Appendix.

[NOTE: There are two peripheral factors that influence definition of ECMs in all of the buildings presented. They are; 1) there is very little air conditioning and; 2) schools are used only 180 days each year. Facilities utilizing mechanical refrigeration and/or year round operation would tend to reflect greater energy savings and faster simple payback periods.]

Average Simple Payback

Perhaps the single most valuable comparison of these 20 ECMs is simple payback. This parameter, although it does not account for a great many important factors, does yield a rough comparison of the investment value of each ECM.

Payback periods range from 3.3 years for ECM #16 (Improve Switching) to 14.4 for ECM #17 (Storm Windows). The top five ECMs (Shortest Payback) are comprised of two electrical (ECM #16 and #15) and three mechanical (ECM #13, #18, and #1) projects. Of the five projects with the longest payback, four are architectural and one mechanical. Generally, the projects with short paybacks are mechanical and electrical and those with long payback are architectural.

TABLE 1. 20 ENERGY CONSERVATION MEASURES RANKED BY FREQUENCY.

ECM* Number		Frequency	Discipline**
1	Night Setback	42	M
2	Insulate Glass	23	A
3	Timers on Exhaust Fans	23	E
4	Convert Lighting	22	E
5	Wall Insulation	18	A
6	Roof Insulation	14	A
7	Dampers on Vents	11	M
8	Reduce Infiltration	11	A
9	Door Treatment	11	A
10	Reduce Pump Operation	10	M
11	Radiator Insulation	10	A
12	Skylight Improvement	10	A
13	Reduce U.V. Ventilator	9	M
14	Reduce Outside Air	6	M
15	Convert Lighting	3	E
16	Improve Switching	3	E
17	Storm Windows	2	A
18	Night Pumps	2	M
19	Plug Soffit Vents	2	A
20	Replace DHW Tanks	2	M
	TOTAL	234	

*Definitions of ECMs can be found in the Appendix.

**A = Architectural, M = Mechanical, E = Electrical.

Average ECM Cost

Of the 20 ECMs recommended, the average cost ranges from $591 to $12,306. The four ECMs with the highest costs are ECM #6 (Roof Insulation), $12,306, ECM #9 (Plug-Soffit Vents), $10,612, ECM #4 (Lighting Conversion), $10,307, and ECM #1 (Night Setback), $9,265. ECM numbers 6 and 19 are architectural, #4 is electrical and #1 is mechanical.

Table 2, "20 ECMs Ranked by Average Project Cost", displays the recommended ECMs in decreasing order of average project cost. The discipline for each ECM is also indicated.

The 20 ECMs are listed in order of increasing payback periods in Table 3, "20 ECMs Ranked by Average Simple Payback". The discipline for each project is also indicated.

CONCLUSIONS

o There is one ECM that appears in the top five for both frequency and payback. This is ECM #1 "Night Setback". Night Setback controls should be installed in all schools not having them now.

o There are three additional ECMs appearing in the top ten ECMs for frequency and payback. They are ECM #3, "Timers on Exhaust Fans", ECM #7, "Dampers on Vents", and ECM #10, "Reduce Pump Operation". These three projects should be seriously considered for all schools for which these recommendations apply.

TABLE 2. 20 ECMs RANKED BY AVERAGE PROJECT COST.

ECM* Number		Average Project Cost (Dollars)	Discipline**
6	Roof Insulation	$12,306	A
19	Plug Soffit Vents	10,612	A
4	Convert Lighting	10,307	E
1	Night Setback	9,265	M
5	Wall Insulation	7,329	A
2	Insulate Glass	6,092	A
14	Reduce Outside Air	4,820	M
15	Convert Lighting	4,710	E
12	Skylight Improvement	4,030	A
18	Night Pumps	3,715	M
7	Dampers on Vents	2,790	M
10	Reduce Pump Operation	2,109	M
8	Reduce Infiltration	1,904	A
13	Reduce U.V. Ventilator	1,869	M
17	STorm Windows	1,863	A
11	Radiator Insulation	1,705	A
9	Door Treatment	1,474	A
20	Replace DHW Tanks	1,150	M
3	Timers on Exhaust Fans	1,094	E
16	Improve Switching	591	E

*Definitions of ECMs can be found in the Appendix.

**A = Architectural, M = Mechanical, E = Electrical.

TABLE 3. 20 ENERGY CONSERVATION MEASURES RANKED BY AVERAGE SIMPLE PAYBACK PERIODS.

ECM* Number		Average Simple Payback (Years)	Discipline**
16	Improve Switching	3.3	E
13	Reduce U.V. Ventilator	6.4	M
18	Night Pumps	6.6	M
15	Convert Lighting	6.9	E
1	Night Setback	7.0	M
19	Plug Soffit Vents	7.1	A
3	Timers on Exhaust Fans	7.2	E
20	Replace DHW Tanks	7.8	M
7	Dampers on Vents	8.5	M
10	Reduce Pump Operation	8.5	M
8	Reduce Infiltration	8.9	A
11	Radiator Insulation	9.0	A
4	Convert Lighting	10.7	E
2	Insulate Glass	10.8	A
12	Skylight Improvement	11.6	A
9	Door Treatment	12.0	A
6	Roof Insulation	12.7	A
5	Wall Insulation	13.6	A
14	Reduce Outside Air	14.0	M
17	Storm Windows	14.4	A

o Architectural projects generally have longer paybacks. Mechanical and electrical projects generally have shorter payback, although average project costs are distributed among disciplines.

o There are many projects with paybacks less than 15 years and with D.O.E. grant funds, these projects have much better paybacks. The National Energy Conservation and Policy Act grant program can have a large impact on utility bills paid by school administrations.

DEFINITIONS OF ECMs

ECM #1 Night Setback: This ECM is directed at reducing the amount of heating occurring in a school during nighttime and weekends. It is done through a variety of methods all of which, however, center upon the controls for the heating system. (Mechanical)

ECM #2 Insulate Glass: Much heat is lost through windows in schools, and this ECM provides additional R-factor to reduce losses through the windows. It is accomplished through two ways. First, addition of a second layer of glass; or second, replacing the glass with structural wall and insulation. (Architectural)

ECM #3 Timers on Exhaust Fans: Schools have many exhaust fans in toilet rooms and in gymnasiums. These fans often operate all the time (24 hours a day) or at all times when the school is occupied. The installation of twist timers allows the fans to be operated only when the space is actually being used. The timer shuts the fan off when the space is unoccupied. (Electrical)

ECM #4 Convert Incandescent to Fluorescent: Many schools were built before the introduction of fluorescent lighting into schools and, therefore, have large areas lit only by incandescent lights. This ECM calls for the replacement of incandescent light fixtures with fluorescent light fixtures. (Electrical)

ECM #5 Wall Insulation: Although wall insulation does not payback in many instances, there are certain cases in which it does payback. One such case is the walls in multi-purpose rooms and gymnasiums. These high walls can be improved by spray insulation applied on the interior surface. (Architectural)

ECM #6 Roof Insulation: Roof insulation only pays back when the insulation is quite easy and inexpensive to apply. The cases in this report in which roof insulation was recommended are cases in which the roof insulation could be spray-applied to the exterior of the roof. (Architectural)

ECM #7 Dampers on Vents: Many schools were built before the days of forced ventilation and used gravity ventilators and windows to supply the necessary fresh air. Many of these gravity ventilators have no back draft dampers, and many fans and other ceiling/roof openings have no back draft dampers in them. This ECM called for the installation of dampers in those ceiling/ roof openings. (Mechanical)

ECM #8 Reduce Infiltration: This is a general ECM covering application of caulking and weatherstripping. In addition, a few projects have been added that include closing of openings in the exterior of the building. (Architectural)

ECM #9 Door Treatment: Doors in schools, of course, get very heavy usage and often the previously applied weatherstripping has failed to provide the necessary seal. This ECM calls for reapplication of weatherstripping, and in some cases, rehanging the doors and/ or adding astrigals to improve closure. (Architectural)

ECM #10 Reduce Pump Operation: Many domestic hot water pumps and heating water pumps operate 24 hours a day. This ECM calls for the installation of time clocks to reduce the hours of operation of these pumps. Freeze protection is, of course, taken into account. (Mechanical)

ECM #11 Radiator Insulation: Many schools have radiators mounted under the windows adjacent to exterior walls. This ECM calls for installation of insulation behind the radiators to reduce the high heat loss occurring in that area. (Architectural)

ECM #12 Skylight Improvement: Schools with skylights loose a tremendous amount of heat through the low R-factor in skylights. This ECM calls for the reduction of the losses through the skylight through application of insulation or additional glazing layers. (Architectural)

ECM #13 Reduce Unit Ventilator Operation: Schools with unit ventilators were often built with no capacity to shut off the unit ventilators when the building is unoccupied. Installation of such controls is the purpose of this ECM. (Mechanical)

ECM #14 Reduce Outside Air: Since a great deal of energy is required to heat outside air on a cold winter day, this is an obvious place to reduce energy waste. Installation of control dampers or repair of existing dampers is called for in this ECM. (Mechanical)

ECM #15 Convert to H.I.D. Lighting: In certain applications existing incandescent lighting can be changed to H.I.D. lighting with very high energy savings. That is the purpose of this ECM. (Electrical)

ECM #16 Improve Switching: Many illumination systems have inadequate switching at the present. This ECM calls for improvement of the switching to allow only certain banks of light to be used at certain times, or in some very poor cases to allow one room to be shut off when it is unoccupied without affecting other spaces. (Electrical)

ECM #17 Storm Windows: This ECM calls for the installation of storm windows on certain high-loss, exterior windows. (Architectural)

ECM #18 Night Pumps: Some schools have large, high horsepower pumps operating all night long. This ECM calls for the installation of a lower horsepower pump to maintain the circulation of water at night while accruing large electrical savings due to the reduced horsepower. (Mechanical)

ECM #19 Plug Soffit Vents: Certain schools have large losses through the attic due to unnecessarily large numbers of soffit vents. This ECM calls for the blocking of the excess soffit vents. (Architectural)

ECM #20 Replace DHW Tanks: Some schools have been using the same domestic hot water tanks for many years and these tanks are no longer serviceable. This ECM calls for their replacement. (Mechanical)

SECTION 10
MODEL
ANALYSIS

Chapter 42
PROCESS MODELS FOR INDUSTRY

D. Hill, S. O. Howe, B. L. Pierce

INTRODUCTION

The National Center for Analysis of Energy Systems at Brookhaven National Laboratory is developing and managing a set of mathematical programming models of energy-intensive industries in the United States. Currently, models of the iron and steel industry and the pulp and paper industry are complete, and a model of the aluminum industry is in the final stages of development. These industry-specific models are based on data from economic and engineering analyses of manufacturing processes and are intended to represent energy flows within an industry in considerable detail. Given a projection of the demand for an industry's manufactured goods and a scenario for the basic raw materials and energy prices, the models are designed to select those processes and levels of production that minimize the cost of meeting the projected demands. To the extent that an industry operates to minimize costs and maximize profits, these process models are an analytical aid to understanding the consequences of alternative scenarios. They are useful analytical tools which can be used to project energy demands, assess new technologies and analyze energy policy.

The purpose of this paper is to describe and illustrate how the process models developed at BNL are used to analyze industrial energy use. A model of the U. S. pulp and paper industry is described and discussed as a specific application of process modeling methodology. Case study results from the pulp and paper model illustrate how process models can be used to analyze a variety of issues. Applications discussed include projections of energy demand, conservation technology assessment, energy-related tax policies, and sensitivity analysis. A subsequent discussion of these results supports the conclusion that industry process models are versatile and powerful tools for energy end-use modeling and conservation analysis.

Table 1 gives the current status of industry models at BNL. The final step in the developmental sequence outlined in the table is the completion of a user-oriented version of each process model. These user-oriented versions are based on data obtained from a variety of public sources as documented in References 1 and 2. More highly developed versions of the models with data from other sources are, however, possible. Because relationships and data in the models are represented in terms and units conventionally understood within an industry, the models are amenable to improvement through criticism and review by industrial engineers and other outside experts. A goal of the industry modeling program is to establish relationships with private industry in order to enhance the validity and improve the utility of the models. Dialogues have begun with representatives of the paper and steel industries, as well as other sources of technical expertise.

*Present affiliation, Department of Nuclear Energy, Brookhaven National Laboratory

	Stage of Progress				
Industrial Sector	Identified Existing Model	Data Base Development	Model Development	User-Oriented Version Development	User-Oriented Version Complete
Iron & Steel*					X
Pulp & Paper*					X
Aluminum*				X	
Cement*			X		
Petrochemicals*	X				
Plastics and Polyesters*	X				
Petroleum Refining*	X				
Copper			X		
Textiles		X			
Dairy			X		
Meat			X		
Glass			X		

*Current Version in-house at BNL.

Table 1. Status Report of BNL Industry Modeling Program.

THE PULP AND PAPER PROCESS MODEL

The process model of the U. S. pulp and paper industry is both temporally dynamic and regional. It considers the production of pulp and paper in four regions of the United States over a 30 year time horizon divided into three time periods (a five period, five years per period version is available and is documented in Reference 2). Each ten year period is modeled as a single representative year within the period, with the base year or first representative year being 1975. Periods are linked in that production capacity existing or purchased in earlier periods is available in later periods after appropriate adjustments for physical depreciation. Outputs or inputs other than production capacity, however, may not accumulate in current periods for use in future periods. That is, no inventories of raw materials, semi-finished goods, or final outputs are allowed.

Five general types of activities are included in the model: purchases of raw materials, sales of by-product energy, material/energy conversion, capacity expansion, and transportation. Specific elements or characteristics of these activities considered in the model are listed in Table 2. Figure 1 gives a flow diagram of the major materials conversion processes within a single region. It shows the sequential processing relationships between the three types of fiber, eight types of pulp, and eight types of final paper products.

Directly or indirectly associated with each activity in the model is a cost. As indicated before, the model is designed to select those levels of activities that minimize the costs of meeting a specified demand for final products. In other words, the objective function to be minimized must contain all appropriately

1. Purchases:
   (1) Fiber sources
      (a) roundwood
      (b) sawmill waste
      (c) waste paper
   (2) Btu sources
      (a) residual oil
      (b) distillate oil
      (c) natural gas
      (d) coal
      (e) bark and hogged wood
      (f) electricity
      (g) low pressure steam
      (h) waste paper
2. Sales of by-product energy:
   (a) low pressure steam
   (b) electricity
3. Material/energy conversion:
   (1) Pulping: To convert roundwood, sawmill waste and waste paper to the following pulps:
      (a) groundwood
      (b) defibrated/exploded
      (c) semi-chemical
      (d) unbleached chemical
      (e) semi-bleached chemical
      (f) bleached chemical
      (g) waste paper stock
      (h) thermomechanical
      (The chemical pulps may be made by either the sulfate or sulfite process.)
   (2) Papermaking: To blend and treat pulps to produce the following kinds of paper:
      (a) newsprint
      (b) writing/printing and related
      (c) packaging and industrial converting
      (d) sanitary
      (e) containerboard
      (f) boxboard
      (g) construction paper and paperboard
      (h) dissolving pulp (a natural fiber feedstock)
   (3) Cogeneration: To convert purchased and recovered fuels, i.e., oil, natural gas, coal, spent liquor, bark and hogged wood, to process steam and/or electricity by the following processes:
      (a) boilers (7 types)
      (b) steam turbines (2 types)
      (c) gas turbines (2 types)
4. Capacity expansion: To expand one process or to convert one to another at some cost.
   (1) Pulping capacity
   (2) Papermaking capacity
   (3) Boiler capacity
   (4) Turbine capacity
5. Transportation: To move the material inputs or the products produced from pulping or papermaking from one region to another.
   (1) Fibers
   (2) Pulps
   (3) Papers

Table 2. Model Activities

represented costs of production. The objective function of the pulp and paper model, therefore, is the sum of the present value of all after-tax costs and can be written as

$$\sum_t \frac{(1-\tau)\cdot E(t) + K(t) - \tau\cdot D(t)}{(1+r)^t}$$

where E(t) represents the sum of operating expenses in t, K(t) represents the sum of capital expenses in t, D(t) is the total financial depreciation in t, τ is the corporate tax rate (assumed 48%) and r the discount rate (assumed 10%). All costs are in 1975 dollars. Normally, the model is run subject to the assumption that the industry acts collectively to minimize the cost of producing an exogenous set of demands. However, minimizing production costs is only one of a number of industry objectives. As discussed below, other objective functions can be introduced to explore the near minimum production cost space.

A comprehensive discussion of the industry's structure, sources of data, and results of validation case studies is included in Reference 2.

Scenario Description

Figure 3 gives the scenario description assumed for the cases to be discussed in this analysis. Energy prices are based on the CONAES Demand Panel's Scenario B (3). In this scenario, energy prices increase linearly in time and regional prices equalize in the year 2010. The demands for specific paper products are based on results in Reference 4. The rate of increase in aggregate demand averages about 3% per year.

In the cases presented here a discount factor or after-tax rate of return on investments of 10% is assumed. (For the eight paper companies in Reference 5, after-tax returns on investment ranged from 6.3 to 12.1% in 1977). Projected levels of investments are somewhat dependent upon the discount rate selected. Figure 4 shows that total investment in energy conserv-

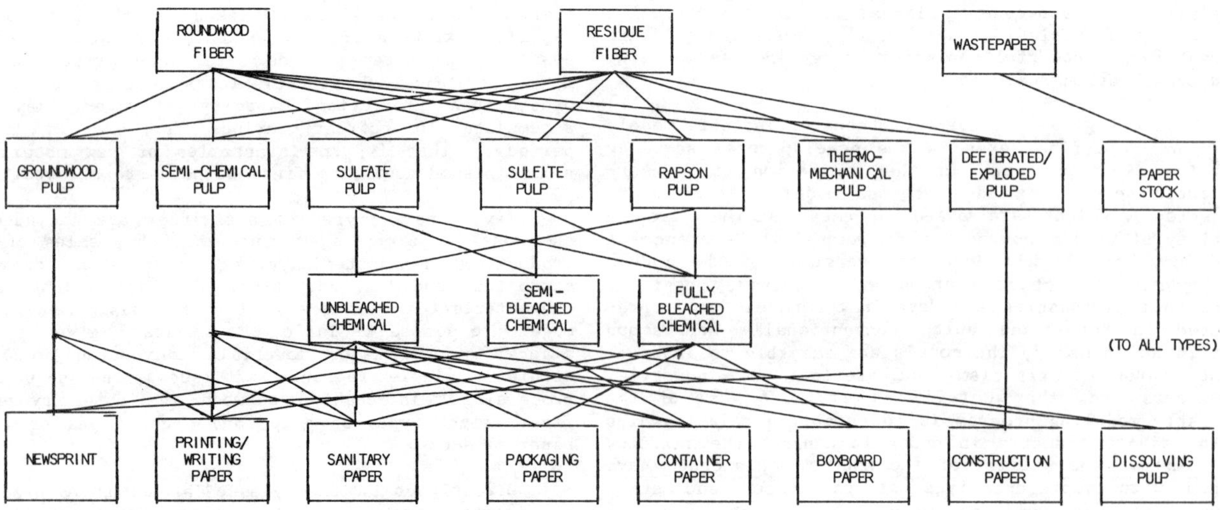

Figure 1. Relationships Between Fibers, Pulps, and Papers.

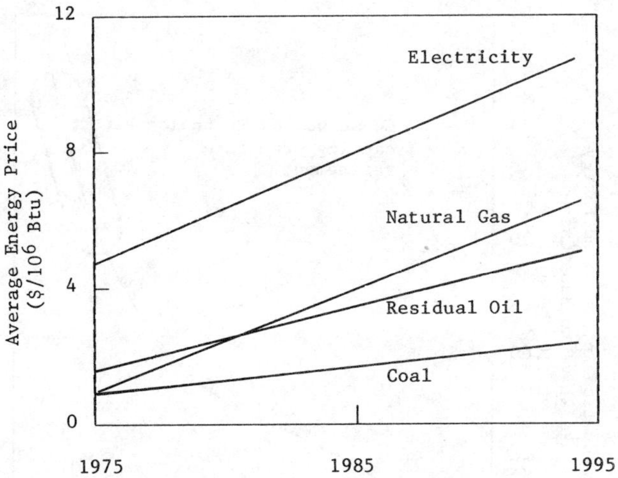

Figure 2. Energy Price Scenario

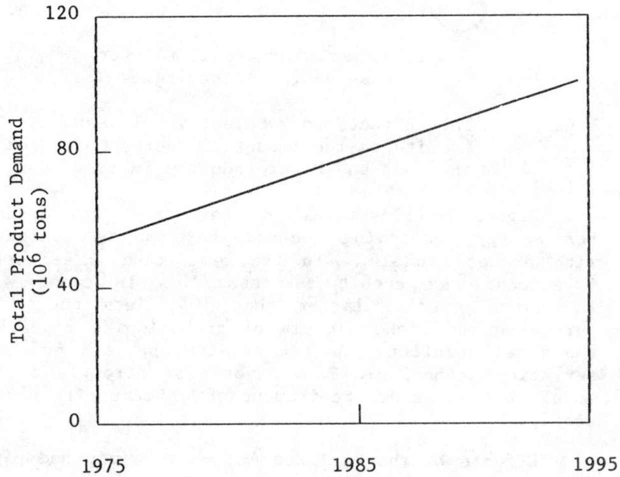

Figure 3. Demand Scenario

ing equipment declines as the required rate of return increases. It varies by a factor of 2 over a reasonable range of rates which could have been selected.

Technology Assessment

Unlike most process models, the pulp and paper model explicitly accounts for risk perceptions within the industry concerning unproven energy conservation technologies. New technologies attract interest if they promise greater returns over time than conventional technologies. Associated with an investment in a new technology is an element of perceived risk; the new (risky) technology may not deliver returns according to expectations. A comparison of initial capital costs for conventional and new technologies, without adjustment for risk, does not reflect the reluctance to invest in a technology whose success is unproven. However, a reluctance to invest in a new technology may be overcome if the rate of return it offers is sufficiently high. The pulp and paper model attempts to quantify the required rate of return related to the risks associated with new new technologies using the concept of a risk-adjusted discount rate. For reasons related to the methodological problems of directly introducing the risk-adjusted discount rates, these rates are used to adjust upward the initial costs of investments in new technologies to mimic their higher required rate of return.

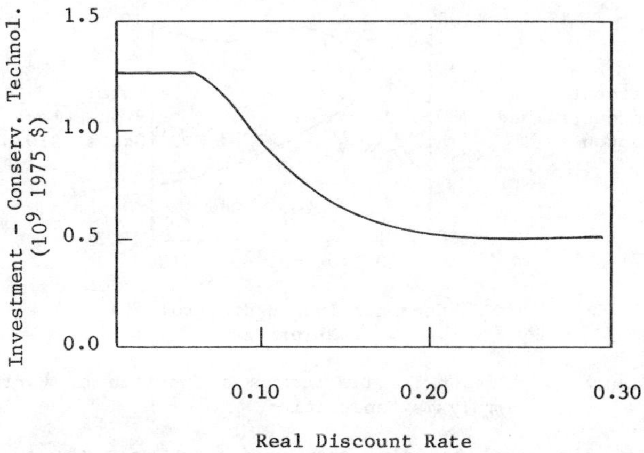

Figure 4. The Relationship Between Investments in Conservation Technologies in 1990 and the Real Discount Rate.

Table 3 gives each of the energy conserving technology options in the paper model with the assumed probabilities of success. To demonstrate the dependency of market penetration on assumed risk, the probability of success of a proposed black liquor recovery system known as freeze crystallization was varied in a set of model runs. Figure 5 gives the 1990 market penetration of this technology as a function of its perceived risk. At very low probabilities of success, the technology is not purchased. For higher probabilities of success the market penetration increases until the market is saturated with this technology.

Technology	Probability
Hydropyrolysis	0.20
Vapor Recompression	1.00
Freeze Crystallization	0.25
Mach Nozzles	0.10
Headbox	0.20
Rapson Closed-Cycle	0.80
Sludge Drying	0.80
Gas Turbines	1.00
High Electrical Output Cogeneration	1.00

Table 3. The Probability of Success for New Energy Conservation Technologies

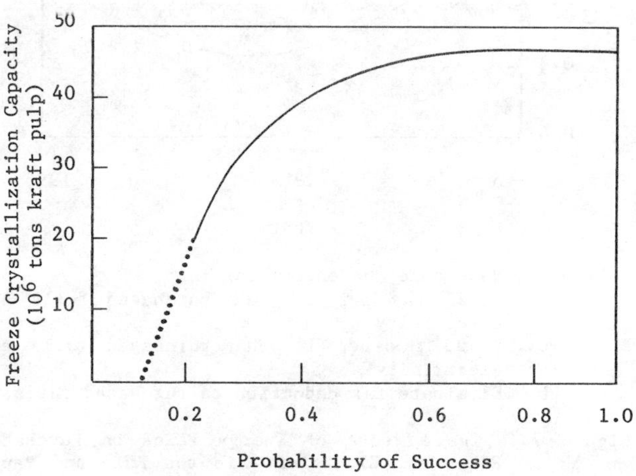

Figure 5. Freeze Crystallization Capacity in 1990 as a Function of Probability of Success.

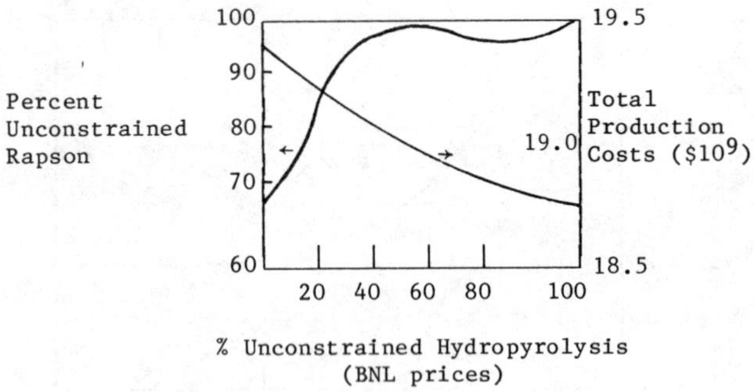

Figure 6. Rapson Penetration as a Function of Hydropyrolysis Penetration

The model has also been used to examine the penetration of new technologies in relation to fuel prices and technological interdependencies (6). This synergestic effect is demonstrated in Figure 6, where the results of constraining the penetration of hydropyrolysis below its optimal value and the concomitant effects on Rapson penetration are given. Allowing hydropyrolysis to be used causes a 50% increase in the use of the closed-cycle Rapson process. More importantly, most of this increase is achieved if hydropyrolysis achieves only 40% of its potential market share. Low penetration of hydropyrolysis has a marked effect on the total penetration of the Rapson pulping process, and under other price assumptions different synergestic relationships appear (6).

Energy Policy Analysis

Figure 7 shows the effect of various Btu taxes on the amounts of energy purchased between 1975 and 1995. In case D, where the purchase of fuels and electricity was not allowed as an operating expense for income tax purposes, the price of purchased energy is effectively doubled relative to base case A with no Btu tax on energy.

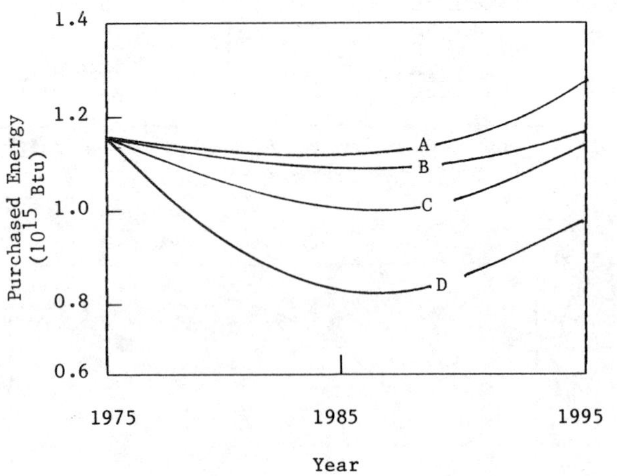

Policy A...Base case; no energy tax
B...$0.25 tax per 10^6 Btu purchased fuels and electricity
C...$1.00 tax per 10^6 Btu purchased fuels and electricity
D...Eliminate tax deduction on purchased fuels

Figure 7. The Effects of Energy Taxes on Purchased Fuels and Electricity in the Pulp and Paper Industry

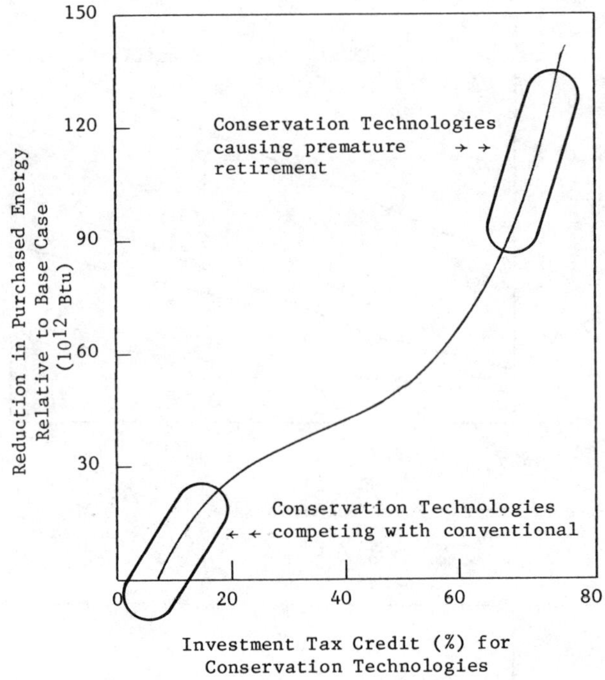

Figure 8. The Effects of Selective Investment Tax Credits on the Amount of Energy Purchased by the Pulp and Paper Industry in 1990

Figure 8 illustrates how investment tax credits for energy conserving technologies can be evaluated with process models. In the base case a selective investment tax credit is assumed. In cases with investment credits larger than 10%, less energy is purchased and higher levels of technologies appear in the model solution. As tax credits approach 60% conservation technologies are favored so strongly in the model that premature retirement of old capacity occurs widely.

Effects of the Fuel Use Act on the pulp and paper industry have been examined using the pulp and paper process model (7). The impact is modelled by not allowing the model to construct any new oil or natural gas boilers. The degree of impact was found to be determined primarily by the ratio of coal to oil prices. If coal prices are low relative to the price of oil (on a Btu basis, including environmental control costs), then the Act had no impact on the industry, since it would convert to coal on its own initiative. If coal prices are nearer oil prices, regional shifts shifts in production and fuel use occurred, causing regionally specific changes in air pollution. The total use of natural gas in the Fuel Use Act case increased by 1995, despite the fact that the use of gas as boiler fuel decreased. This is because the industry responded to the Act, at least in part, by investment in gas fired combustion turbines that cogenerate steam and electricity, equipment that is specifically excluded from the FUA prohibitions.

Alternative Objective Functions

Generally the objective function of a process model is a cost minimization function. The results discussed so far have all been from such models. It is possibble, however, to define other objective functions and explore system behavior when the constraint on cost minimization is relaxed. Figures 9 and 10 show results from cases where the costminimizing objective function was replaced by an alternative function (8).

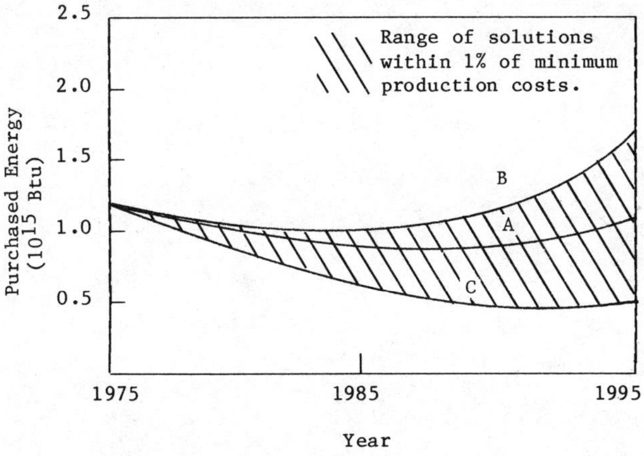

Projection
A...Base case results - minimum cost solution
B...Maximize purchased energy } Allowing the cost
C...Minimize purchased energy } production to increase 1% above the minimum

Figure 9. Purchased Energy Projections for the Pulp and Paper Industry

The projection of purchased energy labelled A in Figure 9 is the standard cost minimization base case. Case B is a case in which purchased energy is maximized subject to the constraint that costs in future periods do not exceed by more than 1% the costs in the cost minimization case. Similarly, in case C, purchased energy is minimized subject to the constraint that costs in future periods do not exceed by 1% the total production costs in the costminimization case. The spread between B and C represents the range of purchased energy projections of all industry operating modes within 1% of absolute minimum cost. About 60% of the difference is related to activities for the burning of feedstocks and self-generated sources of energy in case C, the energy minimization case, that did not enter the solution in case B, the energy maximization case. Nevertheless, the results do show that long run energy demands may vary significantly and yet have little impact on total production costs.

Results are summarized in Figure 10 from cases in which the utilization of particular technologies was either maximized or minimized subject to the constraint that costs do not exceed by more than 1% the total production costs of the cost-minimization case. In all cases except one, it was possible for a particular technology to be eliminated from the solution. In cases maximizing particular technologies, the use of one technology increased by a factor of 9.6 compared to the base case. These results show that the market penetration of any of the energy conserving technologies in the pulp and paper model may vary considerably and yet have little effect on total production costs.

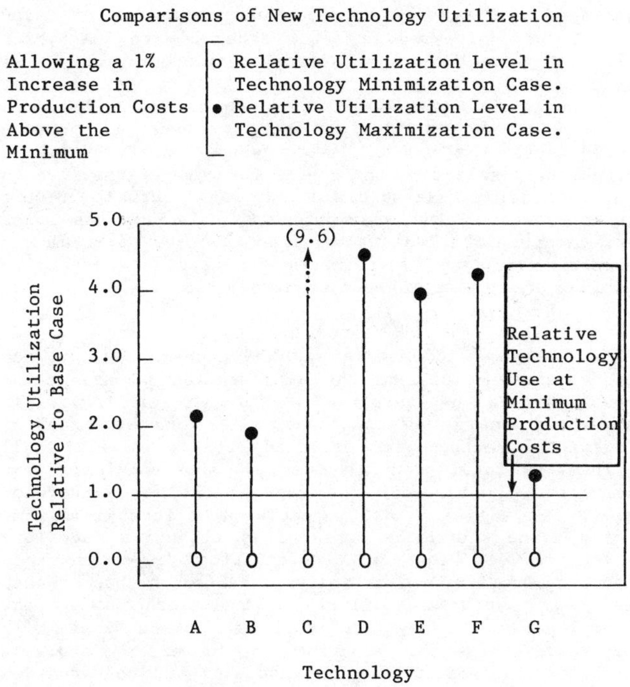

Technology A...Rapson closed cycle
B...Headbox
C...Sludge drying
D...Mach nozzles
E...Vapor recompression
F...Freeze crystallization
G...High efficiency turbines

Figure 10. Range of Market Penetration Results with Production Costs within 1% of the Minimum (for 1990)

DISCUSSION

The case study results from the pulp and paper model illustrate the range and power of the process modeling approach. The cases focussing on energy-demand projections show how future energy use is dependent on both the availability of particular energy conserving technologies and on the price of energy. The models are designed so that specific technological options available to an industry can be readily assessed and compared. Because technologies are explicitly represented, the effects of investment tax credits on specific technologies can also be readily evaluated. However, one limitation of process models is that they cannot foresee technological innovation; technologies considered in the models must be introduced explicitly. This fact could lead to underestimates of factor price sensitivity, especially in cases of large changes in prices over long periods.

Industry process models, by design, offer a systems approach to analysis of energy demand and conservation issues. Results from applications of the pulp and paper model have identified technological interdependencies within the industry that could significantly affect the estimates of the potential of conservation programs. In one instance, the model was used to investigate the conservation potential of three alternatives to a conventional technology. Given the model results for the use of each option, a separate calculation was made to find the engineering energy savings (alternative technology units in use multiplied by the energy savings per unit). The energy savings in the model results were about one to three times the engineering result, depending on the which alternative technology was being considered.

It is also possible that the opposite result could occur; introduction of an energy conserving process could cause greater overall energy use. Analyses of the effect of a selective tax credit for cogeneration led to the conclusion that such a policy could actually result in increased energy consumption (6). In this case, the tax credit effectively subsidized the cost of steam and electricity, and investments in energy conserving options other than cogeneration were reduced.

Finally, the case studies with alternative objective functions illustrate why the accuracy of projections based solely on the minimization of production costs should be questioned. Large variations in projected energy demands and market penetration of individual technologies occurred with relatively small effects on total production costs. This conclusion has important implications for energy modeling methodologies. It suggests that objective functions minimizing costs alone should be replaced by objective functions which reflect the unique realities in which each industry operates. This analysis has shown that process models are extremely flexible, both structurally and functionally. They are well suited to deal with alternative objective functions and to meet the evolving challenges of energy end use modeling and conservation analysis.

REFERENCES

1. F. T. Sparrow, D. Pilati, T. Dougherty, E. McBreen, L. L. Juang, "The iron and steel industry process model," BNL 51073, Brookhaven National Laboratory, Upton, New York 11973, Jan. 1980.

2. D. A. Pilati, J. Change, F. T. Sparrow, S. O. Howe, C. Balzer, B. McBreen, "A process model of the U.S. pulp and paper industry," BNL 51142, Brookhaven National Laboratory, Upton, New York 11973, April 1980.

3. Demand and Conservation Panel of the Committee on Nuclear and Alternative Energy Systems, "U.S. energy demand: some low energy futures," Science, April 14, 1978.

4. H. D. Nguyen. "A study of demand for paper and board products," Oak Ridge National Laboratory, draft report.

5. Ernst and Ernst, "Costs of capital and rates of return for industrial firms and class A & B electric utility firms," Report prepared for U.S. Department of Energy, Economic Regulatory Administration, June 1979.

6. D. A. Pilati and R. Rosen. "The use of the pulp and paper industry process and model for R&D decision making," BNL 50839, Brookhaven National Laboratory, Upton, New York 11973, March 1978.

7. S. Rogers, S. Brown, R. Shypailo, and P. M. Meier, "The impact of the national energy plan on the pulp and paper industry: An energy-environmental assessment," BNL 51311, Brookhaven National Laboratory, Upton, New York 11973, in press.

8. S. O. Howe, D. A. Pilati, and F. T. Sparrow, "Process modeling of industrial energy use," BNL 28838R, Brookhaven National Laboratory, Upton, New York 11973, April 1981.

Chapter 43

APPLICATIONS OF AN ANALYTICAL MODEL TO DETERMINE THE EFFECTS OF AIR SUPPLY RATES ON ENERGY CONSUMPTION

D. P. Mehta

ABSTRACT

In this paper, an analytical model has been applied to study the effects of the air supply* rates on the dynamics of an occupied space as related to energy consumption and occupant's comfort. The concept of Air Diffusion Performance Index (ADPI) is reviewed and used to derive air movement control strategies to reduce energy consumption in buildings. The relationship between the air movement control strategies and passive thermal control strategies for energy efficient buildings is discussed. Procedures to calculate the terminal air velocity for a proposed diffuser and to optimize the selection of diffusers are described. Derivation of the air movement control strategies has shown that a design criterion based upon the dynamic conditions can easily be coordinated with comfort criteria, load criteria and ADPI criteria.

INTRODUCTION

Energy required to operate and maintain a building system depends directly on the desired level of indoor air quality. Air quality considerations of indoor climates must include thermal and mass air quality, lighting, and noise levels. Air supply rates can have a significant influence on the mass air quality, thermal air quality, and noise levels in an occupied space.

Appropriate mass air quality is necessary for safety and health of the occupants. Thermal air quality determines the thermal comfort of the occupants and is dependent upon many factors including building envelope characteristics, occupants, internal thermal processes, weather, air supply rates, conditioned air velocity, and location of the supply air terminal device(s). These factors interact with each other on a dynamic basis and result in a net energy transfer between the interior and exterior environments. To reduce energy consumption in buildings, it is very important to understand the dynamics of these interactions. The intent of this paper is to report an analytical model which describes the dynamic interactions among the factors having influence on thermal air quality and to apply the model to determine the effects of air supply rates on the thermal air quality as related to thermal comfort and energy consumption.

BACKGROUND

Mass Air Quality: Historically, interest in the effects of building air supply on safety, health and comfort of the occupants can be traced back to as early as 1500 when Leonardo da Vinci designed and built a water driven fan to ventilate the bedrooms of his patron (1).** This was probably the first mechanical ventilation system.

An account of the use of natural ventilation is given when Charles the First of England, in 1600, ordered that no house could be erected whose rooms were less than 10 ft. in height (2). He also specified that window height must exceed the width. Subsequent understanding of the relation between "bad" air, ventilation, safety, and health was based on the discoveries of O_2 and CO_2, Dalton's law of partial pressures, description of the composition of air, and description of the respiratory function. Emphasis on the effects of organic impurities in indoor climates on the health of the occupants started in 1872 when Billings, a physician, recommended 30-40 cfm of outdoor air per person (3). The American Society of Heating and Ventilating Engineers (ASH&VE), one of the predecessors of American Society of Heating, Ventilating, Refrigerating, and Air-Conditioning Engineers (ASHRAE), organized in 1895, adopted the view that engineers were ready to accept the idea of hygienists and physiologists (4). The hygenic and physiological communities had adopted the 30 cfm minimum as being required for adequate ventilation. This necessitated mechanical means. Therefore, ventilation to control mass air quality was essentially an engineering problem. Consequently, it was the engineers who were held responsible in future years for wasting large sums of money in over-designed ventilating systems.

Engineers initiated work to save money by correcting over-designed ventilation systems as early as 1930 when a major change in ventilation standards resulted from the experimental work reported by Yaglou, et al. (5). Conducted under controlled experimental conditions, the Yaglou studies have served as the primary reference in codes and standards for the last forty-five years.

In response to demands for energy efficient buildings, attempts have been made to prescribe lower and lower ventilation rates (6,7). For example, ASHRAE Standards 90-75 "Energy Conservation in New Buildings" states that the minimum recommended values of ventilation rates prescribed in ASHRAE Standard 62-73 "Standards for Natural and Mechanical Ventilation" shall be used for design purposes for each type of occupancy. The implication of all such past methods to control mass air quality is that ventilation air quantity, by itself has been used as a reliable gauge of indoor mass air quality. In practice, many factors can influence mass air quality and if sufficient ventilation rates are not prescribed to meet forseeable variabilities, the resultant indoor air could be unhealthy. Conversely, if sufficient ventilation rates are prescribed to meet all contingencies, the energy requirements for thermal control will be excessive.

*Air supply includes recirculated and ventilated air, see Figure 1 on next page.

**Numbers in parentheses refer to references.

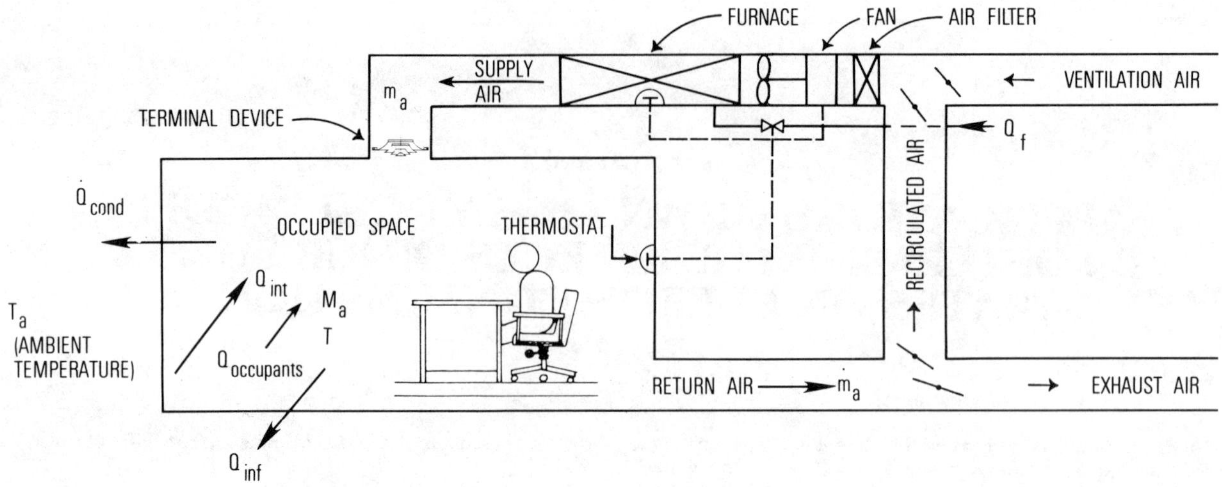

FIGURE 1. SCHEMATIC REPRESENTATION OF A FORCED AIR HEATING SYSTEM.

Therefore, to provide acceptable indoor mass air quality at minimum rates of energy consumption under various occupancy conditions will require environmental control systems that can dynamically respond to changes in occupancy load and to outdoor air quality. However, reliable analytical models which can predict the dynamic conditions in an occupied space and their effects on the occupants, and which can be used to design and analyze the control systems have not yet been reported.

Thermal Air Quality: Thermal air quality of controlled environments can be assessed in terms of human thermal comfort. Thermal comfort effects must be evaluated from two standpoints. First, the thermal environment must be such that the occupant will be thermally neutral for a given condition of dress and activity. This condition can be termed "whole-body" comfort. Second, the thermal environment surrounding the individual should be completely uniform so that conditions of local discomfort do not exist.

The conditions for "whole-body" comfort are determined from four physical factors: dry-bulb temperature; water vapor pressure; air motion (relative air velocity) and mean radiation temperature and two reciprocative factors which are associated with each occupant, namely activity (metabolic heat production) and clothing (thermal insulation). Indoor design temperatures, as specified by building codes, standards, or design criteria, provide conditions which normally satisfy "whole-body" comfort for the occupants of a given space. This, of course, assumes that adjustments have been made for the type of activity and type of clothing to be worn in the space. The works of Rohles and Nevins (8) and McNall et al. (9) provide one source of data from which these criteria can be formulated. Fanger's comfort equation (10) provides another excellent means to predict the comfort response of occupants or to determine the preferred environmental conditions for a given set of variables. Fanger's equation is based on the assumption that there is a unique relationship between skin temperature and metabolic rate and sweat rates and metabolic rate for thermal comfort. A simple model of human physiological response of Gagge et al. (11) can also be used to predict thermal comfort and to provide detailed physiological data such as wetted area and skin temperature.

Of particular interest for this paper is the influence of air motion on thermal comfort. Subjective reactions to velocities from 0.1 to 1.0 m/s have been studied in the laboratory and thermal sensations of "comfortable" or "thermally neutral" have been achieved at velocities up to 1.0 m/s (12). In practical applications, air motion or air velocity will range from 0.1 to 0.2 m/s to a probable maximum of 0.5 or 0.8 m/s, except in those cases using "spot" ventilation. Based on the work of several investigators, Table 1 shows the preferred or comfort temperatures for the range of air velocities given above. Ostergaard et al. (13) have found that the direction of the air motion appears to have no effect on the comfort sensation, at least for velocities up to 0.8 m/s. It must be remembered that these data are for air velocities which envelope the occupant uniformly, and they do not apply to the localized or jet cooling of small areas of the body.

STUDY	VELOCITY, m/s				
	0.2	0.2	0.4	0.8	1.0
Fanger: Comfort Eqn. (10)	25.7	26.7	27.2	27.8	--
Ostergaard et al. (13)	25.1	--	--	27.4	--
Houghten, Yaglou (14)	--	26.7	27.2	28.1	--
Olesen, Fanger, Bassing (17)	24.5	--	--	26.5	--
Rohles, Woods, Nevins (18)	25.9	26.6	27.0	27.9	--
Fanger, Nevins, McNall (19)	--	--	--	--	30.0

TABLE 1. PREFERRED AMBIENT TEMPERATURE (^{o}C) FOR COMFORT Sedentary activity, MRT=DBT, I_{cl}=0.6, RH=50%

Even though the requirements for "whole-body" thermal comfort are satisfied by the air supply system, there may exist local drafts which can cause discomfort within an otherwise comfortable environment. Houghten et al. (14) defined a draft as:

> "Any local sense of cooling of a portion of the body caused either by excessive movement of air of normal temperature, by air having a normal velocity but a lower temperature, by excessive radiation to cold surface, or any combination of these three effects."

It is obvious from the work of Houghten and others that sensitivity to draft varies with different parts of the body. The effective draft temperature can be used to analyze the effect of velocity and temperature variations on the local feeling of comfort or discomfort. The effective draft temperature was based on the work of Rydberg and Norbach (15) and is defined as:

$$\theta = (T - T_c) - 7.66(V_x - 0.15)^* \qquad (1)$$

From Houghten's relationship for the values of local velocity and local draft temperature (temperature of the moving air stream minus the average room temperature) at which 80% of the subjects were comfortable, the limits of θ are -1.67 and +1.11, for V_x between 0 and 0.36 m/s. Figure 2 illustrates the comfort criteria resulting from Eqn. 1 and the limits developed from Houghten's data. Also shown are the conditions at which 90% of the subjects reported comfort when exposed to drafts on the ankle and on the neck, and the conditions at which 80% of the subjects reporting comfort when exposed to drafts on the neck.

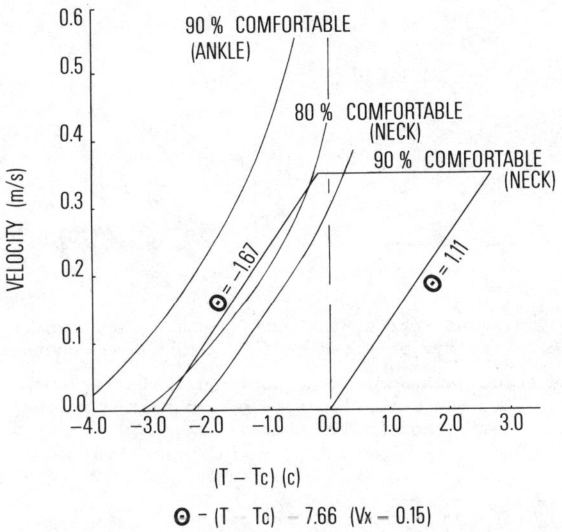

FIGURE 2. SUBJECTIVE RESPONCE TO DRAFTS AND CRITERIA FOR THE AIR DIFFUSION PERFORMANCE INDEX, ADPI (after reference 12)

Air Diffusion Performance Index, ADPI: The ultimate test of an air distribution system is the response of the occupants to conditions in the occupied space. In an effort to provide a single number rating system for the evaluation of air distribution systems; the Air Diffusion Performance Index, ADPI, was proposed in 1964 (2). Using the comfort criteria described above, ADPI was defined as:

> "The number of measuring positions, uniformly distributed in a vertical center-line plane passing through the diffuser or throughout the occupied zone, at which the comfort criteria are satisfied, expressed as a percent of the total number of positions at which measurements of temperature and velocity are made."

ADPI was found to be a valid, sensitive, single number index relating outlet performance, the space, and occupant's comfort.

*See list of nomenclature at the end of this article.

ADPI was shown to be a function of the type of terminal device, the room load, the supply air flow rate and the room geometry for all systems except the air distributing ceiling systems.

Miller and Nash (16) combined the parameters affecting ADPI for terminal devices and presented ADPI as a function of the ratio of the isothermal jet throw distance (X) to a characteristic room dimension (L). Room load was found to be a parameter. The throw of a jet was defined as the distance from the outlet to a point in the isothermal air stream where the maximum velocity occurring in the stream cross-section has been reduced to a selected "terminal" velocity. Throw data cataloged by most diffuser manufacturers, are taken under isothermal conditions.

Throw coefficients vary with the diffuser type. The relationship between throw (X) and outlet velocity (V_o) is:

$$\frac{X}{\sqrt{A_e}} = K\frac{V_o}{V_X} = \frac{KQ_o}{V_X A_e} \qquad (2)$$

Typical values of K for several diffuser types are given in Table 2 which has been adapted from reference 12.

DIFFUSER		K
High Sidewall Grille		
Vane Angle	0.0 deg	4.7 - 8.5
	22.5 deg	3.6 - 6.5
	45.0 deg	2.6 - 4.7
Circular Ceiling Diffuser		0.8 - 1.2
Slot Diffuser		0.9 - 3.1
Light Troffer Diffuser		1.5 - 4.0

TABLE 2. TYPICAL THROW COEFFICIENTS (adapted from reference 12)

The characteristic room length (L) is, in some installations, an easy dimension to determine and, in other installations, more difficult. For example, for a high side wall grille located in the one end of the test room, the characteristic room length would be the length of the room in the direction of the throw. A characteristic dimension for slot diffusers located on the center-line of the ceiling would be half of the width of the room. In some installations the air pattern will be directed towards an adjacent device such that the primary jets will interact. This configuration occurs in large spaces served by multiple outlets. In this case, L would be taken as half the distance between units plus the distance from the ceiling to the top of the occupied zone. A terminal velocity of 0.25 m/s is normally used.

ANALYTICAL MODEL FOR DYNAMIC THERMAL CONDITIONS

The thermal processes to describe the heat transfer rates in an occupied space of a building are shown in Figure 1. The following assumptions are made:

1. The properties of the thermal process involved in heating the occupied space are "lumped" at the location at which the occupants experience the controlled conditions. This assumption permits us to describe the process with ordinary linear differential equations instead of partial differential equations.

2. Changes in the stored energy of space air are assumed to be due only to changes in dry-bulb temperatures; those due to humidity changes are considered negligible.
3. Heat transfer by radiation is not included in this analysis.
4. Infiltration losses are assumed to be a function of air supply flow rate, occupant activity, internal heat sources, room temperature, and construction details in a given occupied space.

An energy balance on the occupied space of Figure 1 yields: Rate of energy in = Rate of energy out + rate of change in stored energy, or

$$\dot{Q}_f + \dot{m}_a c_{pa} T_i + \dot{Q}_{occ} + \dot{Q}_{int} = \dot{m}_a c_{pa} T + \dot{Q}_{cond} + \dot{Q}_{inf} + M_a c_{pa} \frac{dT}{dt} \quad (3)$$

If we define

$$\dot{Q} = \dot{Q}_f + \dot{Q}_{occ} + \dot{Q}_{int} \quad (4)$$

and

$$\dot{Q}_{cond} = UA\,(T-T_a) \quad (5)$$

then we may rewrite Eqn. 3 as:

$$\frac{dT}{dt} = \frac{\dot{Q}}{M_a c_{pa}} + \frac{\dot{m}_a c_{pa}}{M_a c_{pa}}(T_i - T) - \frac{UA(T-T_a)}{M_a c_{pa}} - \frac{\dot{Q}_{inf}}{M_a c_{pa}} \quad (6)$$

For most applications, relatively small fluctuations in net heat gains and space temperature will be experienced within the occupied space. For this reason, only small perturbations about the operating or set point will be considered. The operating point may be found by setting the time derivative contained in Eqn. 6 to zero:

$$\frac{\bar{\dot{Q}}^{*}}{M_a c_{pa}} + \frac{\dot{m}_a c_{pa}}{M_a c_{pa}}(\bar{T}_i - \bar{T}) - \frac{UA(\bar{T}-\bar{T}_a)}{M_a c_{pa}} - \frac{\bar{\dot{Q}}_{inf}}{M_a c_{pa}} = 0 \quad (7)$$

Eqns. 6 and 7 may be combined to yield an equation in terms of perturbations about the operating point:

$$\frac{d(T-\bar{T})}{dt} = \frac{\dot{Q} - \bar{\dot{Q}}}{M_a c_{pa}} + \frac{\dot{m}_a c_{pa}}{M_a c_{pa}}[(T_i - \bar{T}_i) - (T - \bar{T})] - \frac{UA}{M_a c_{pa}}[(T-\bar{T}) - (T_a - \bar{T}_a)] - \frac{(\dot{Q}_{inf} - \bar{\dot{Q}}_{inf})}{M_a c_{pa}} \quad (8)$$

Eqn. 8 may be rewritten as:

$$\frac{d\Delta T}{dt} = \frac{\Delta \dot{Q}}{M_a c_{pa}} + \frac{\dot{m}_a c_{pa}}{M_a c_{pa}}(\Delta T_i - \Delta T) - \frac{UA}{M_a c_{pa}}(\Delta T - \Delta T_a) - \frac{\Delta \dot{Q}_{inf}}{M_a c_{pa}} \quad (9)$$

From Assumption 4, we may write

$$\dot{Q}_{inf} = f(\dot{Q}, T) \quad (10)$$

Linearizing Eqn. 10 about the operating points yields:

$$\Delta \dot{Q}_{inf} = K_1 \Delta \dot{Q} + K_2 \Delta T \quad (11)$$

* Bar (-) notations indicate values at the operating point.

The constants, K_1 and K_2, are defined as follows and are evaluated at the operating point:

$$K_1 = \left.\frac{\partial \dot{Q}_{inf}}{\partial \dot{Q}}\right|_{T = \text{constant}} \quad (12)$$

$$K_2 = \left.\frac{\partial \dot{Q}_{inf}}{\partial T}\right|_{\dot{Q} = \text{constant}} \quad (13)$$

Substituting Eqn. 11 into Eqn. 9, taking the Laplace transform, and rearranging, we obtain for step inputs:

$$\Delta T(s) = \frac{K_{\dot{Q}} \Delta \dot{Q}(s)}{S(1 + \tau_{os} S)} + \frac{K_{T_i} \Delta T_i(s)}{S(1 + \tau_{os} S)} + \frac{K_{T_a} \Delta T_a(s)}{S(1 + \tau_{os} S)} \quad (14)$$

where

$$K_{\dot{Q}} = \frac{1 - K_1}{\dot{m}_a c_{pa} + UA + K_2} \quad (15)$$

$$K_{T_i} = \frac{\dot{m}_a c_{pa}}{\dot{m}_a c_{pa} + UA + K_2} \quad (16)$$

$$K_{T_a} = \frac{UA}{\dot{m}_a c_{pa} + UA + K_2} \quad (17)$$

$$\tau_{os} = \frac{M_a c_{pa}}{\dot{m}_a c_{pa} + UA + K_2} \quad (18)$$

A block diagram representation of Eqn. 14 is shown in Figure 3. It may be noted that τ_{os}, the time constant of the occupied space, is a function of the enclosed mass of air, the mass flow rate of air, the envelope characteristics and infiltration.

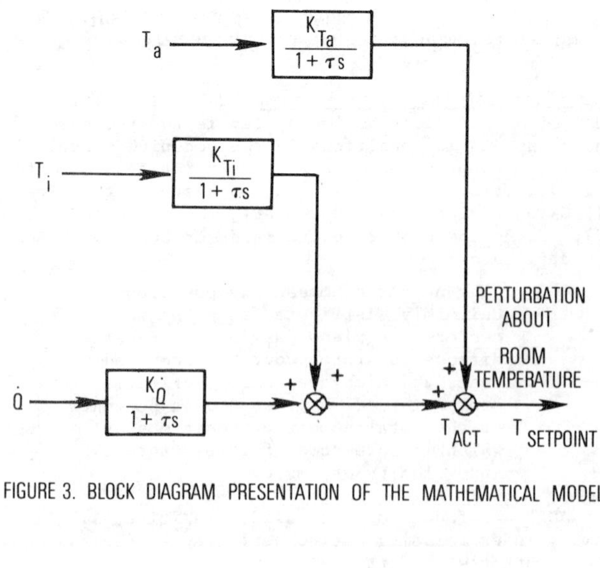

FIGURE 3. BLOCK DIAGRAM PRESENTATION OF THE MATHEMATICAL MODEL

Eqn. 14, represented by the block diagram of Figure 3, is in a form which can be coupled to the models of other components, including the occupant, in an environmental control system to yield an overall transfer function for the building system. This can be subjected to the techniques of control theory for design purposes.

PASSIVE ENERGY EFFICIENT CONTROL STRATEGIES

Passive Control Strategies: The parameter K_2 as described by Eqn. 13 can be used to develop criteria for evaluating energy consumption characteristics of envelopes within a given category of buildings. For example, a number of single family residences with equal floor areas can be compared to study the influence of different types of structures, shapes, infiltration characteristics and life styles of the occupants on energy consumption. Such studies can be used to derive passive control strategies for energy conservation. To illustrate this application, it may be noted that

$$K_2 = \left. \frac{\partial \dot{Q}_{inf}}{\partial T} \right|_{\dot{Q} = \text{Const.}} \qquad (13)$$

and

$$\dot{Q}_{inf} = \dot{m}_{ainf} C_{pa} (T - T_a) \qquad (19)$$

Under the conditions when heating or cooling is not required,

$$K_2 = \dot{m}_{ainf} C_{pa} - \dot{m}_{ainf} C_{pa} \left(\frac{dT_a}{dT}\right) \qquad (20)$$

Therefore, the value of K_2 depends upon the mass flow rate of infiltration air, $\dot{m}_{ainf}$, and the factor (dT_a/dT). This factor depends on the type of construction of the house, heat transfer and infiltration characteristics. Under passive conditions its value may range from +1 to ∞ as shown in Figure 4. A value of (+1) for dT_a/dT means a very loose construction from an infiltration point of view or no resistance to heat transfer across the envelope. Substitution of these two extreme values of (dT_a/dT) in Eqn. 20 reveals that K_2 can vary from 0 (no envelope) to −∞ (perfect envelope for a building with passive control. The numerical value of K_2 is one single parameter which combines the elements of infiltration and heat transfer characteristics of a structure. If the infiltration losses can be defined to include losses due to door openings or window openings, the numerical value of K_2 becomes a parameter which can be used to compare life styles. To use K_2 as a criterion to evaluate energy consumption characteristics of envelopes within a category of buildings, the following procedures are suggested.

Procedures for Existing Buildings:

1. Select buildings within the same category on equal floor area basis.
2. Record hourly indoor temperatures and outdoor air temperatures during spring or fall when heating or cooling is not needed.
3. Measure infiltration rates at the time the temperatures are being measured using tracer gas techniques.
4. Calculate the values of $|K_2|$ using equation 20.

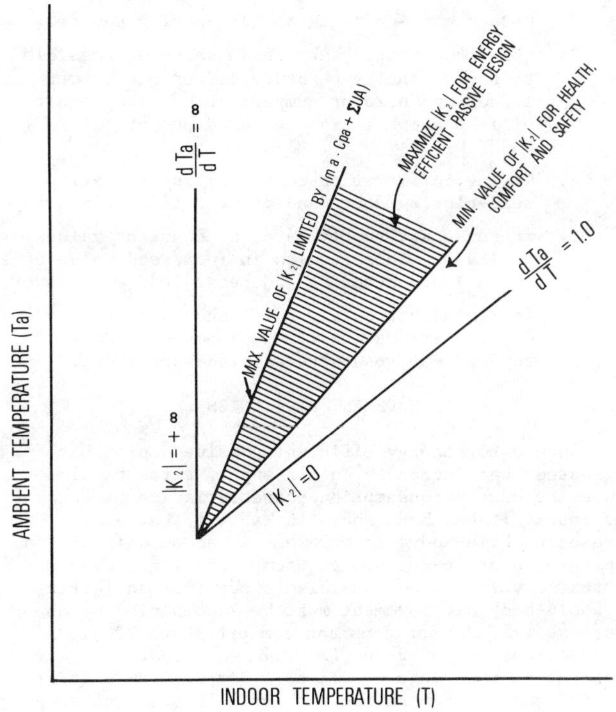

FIGURE 4. RELATIONSHIPS BETWEEN AMBIENT AND INDOOR TEMPERATURES FOR PASSIVE DESIGN

5. Structures within the same category designed to provide equal floor areas and which have an equal number of air changes per hour will be more energy intensive as the value of K_2 decreases. Occupants in structures with equal floor areas, identical envelope characteristics and designed for the same number of air changes per hour will be identified as having more energy intensive life styles as the value of $|K_2|$ decreases. In structures with equal floor areas but with different numbers of air changes per hour, the one will be more energy intensive as the value of (dT_a/dT) decreases. Some corrective steps to control energy intensive passive buildings are suggested in the discussion below.

Procedures For Energy Efficient Passive Design: It has been shown above in Step 5 of the procedures for existing buildings that $|K_2|$ should be as large as possible. In the limit $|K_2|$ has to be less than $(\dot{m}_a C_{pa} + \Sigma UA)$ because a value of $|K_2|$ equal to $(\dot{m}_a C_{pa} + \Sigma UA)$ yields an infinite value for the time constant for the occupied space, see Eqn. 18. Physically an infinite time constant of an occupied space in response to inputs from weather, internal sources or from heat exchanger has no meaning and hence a limit on the maximum value of $|K_2|$ is shown in Figure 4. Thus, codes should be developed to specify maximum and minimum value of $|K_2|$ for health, comfort and safety, and energy considerations. The following procedures are suggested to maximize the value of $|K_2|$ within the limits shown in Figure 4.

1. Increase the thermal mass of the structure. The results will be shown by increased values of (dT_a/dT) and of $|K_2|$.

2. Reduce over-capacity of the heating source.

3. Use space temperature reset whenever possible to reduce the temperature difference between indoor and outdoor temperatures. The result will be shown by increased values of (dT_a/dT) and $|K_2|$.

4. Reduce infiltration losses by incorporating self-closing doors and tight fitting windows.

5. Increase air supply rates at constant values of (ΣUA). This results in increased value of $(\dot{m}_a C_{pa})$ and hence an increase in $|K_2|$. However, increased air supply rates should take into considerations the limits imposed by discomfort due to drafts as discussed above.

AIR MOVEMENT STRATEGIES

Procedure 5 for energy efficient passive design has suggested that increases in air supply rates can improve the energy consumption characteristics of the envelope. It has been shown in Table 1 that an increase in whole-body air movement is accompanied by an increase in preferred ambient temperature for constant comfort levels. Woods has also shown that an increase in whole-body air movement must be accompanied by an increase in SET for constant comfort level (20). Calculations in reference (20) indicate that, for air movement in the range of 0.2 to 1.0 m/sec, a shift of $1C^o$ is required for an increase of 0.38 m/sec to keep the same comfort level. However, when air movement is less than 0.2 m/sec (i.e., still air), the calculated shift is significantly greater: namely $1C^o$ for an increase of 0.04 m/sec. So a very low velocity of air is desired for winter conditions to provide comfort at lower space temperatures to save energy. Alternatively, high air velocities in the summer can permit higher acceptable space temperatures resulting in reduced air conditioning loads and energy savings. Thus, air movement strategies derived from comfort considerations are in agreement with those derived from energy efficient passive design considerations for summer conditions whereas they conflict for winter conditions. Figure 5 has been constructed using equations (16)-(18) and can aid the designer to seek an optimum solution. The following steps are suggested:

1. Select supply air temperature, T_s, for a certain design load and a desired control temperature, T_c, so that $\dot{m}_a$ calculated from the following equation:

$$\dot{Q}_L = \dot{m}_a C_{pa} (T_s - T_c) \quad (21)$$

lies to the right of line XY in Figure 5. This step will prevent the selection of a value of K_{Ta} which is too high and will result in an optimum combination of ΣUA, $\dot{m}_a$ and K_2.

2. Determine the outlet air velocity for the proposed diffuser from the equation:

$$\dot{m}_a = A_e V_o \rho \quad (22)$$

3. Calculate V_x, the local velocity at the location of the occupants, for a specified occupied space temperature, T, using Eqn. 2. The values of K to be used in Eqn. 2 can be read from Table 2.

4. Calculate the effective draft temperature, θ, from Eqn. 1.

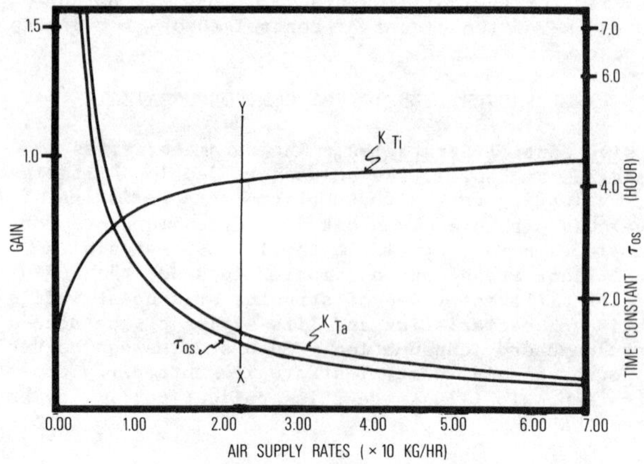

FIGURE 5. CALCULATED EFFECTS OF AIR SUPPLY RATES ON THE DYNAMICS OF OCCUPIED SPACE.

5. Check that the values of θ lie between (-1.67) to (+1.11) when velocities are m/sec and temperature in ^oK.

6. If the calculated values in step 4 do not fall within the limits stated in step 5, choose a new effective area for the diffuser and repeat steps 1-5.

A design procedure based on steps 1-6 should coordinate load, dynamic performance and air diffusion performance index (ADPI) criteria.

CONCLUSIONS

Acceptable indoor thermal and mass air quality can be maintained with minimum energy consumption if environmental control systems respond dynamically to the changes in those variables which have influence on energy consumption in buildings. Analytical models which can predict dynamic conditions in occupied spaces can prove to be useful tools in the design and analysis of such control systems.

A mathematical model which describes the dynamic interactions among the thermal energy characteristics of a building has been presented in this paper. The model has been applied to derive some energy efficient control strategies. Effects of air supply rates on the dynamics of an occupied space as related to energy consumption and occupant's comfort have been discussed.

ACKNOWLEDGMENTS

Some of the concepts reported in this paper are the result of studies funded through the National Science Foundation. Special thanks are due to Dr. Max A. Wessler, Chairman, Mechanical Engineering Department at Bradley University, for his support and for his help in editing this paper. Appreciation is also extended to Mrs. Annelle Fletcher for her interest and efforts in typing this paper.

REFERENCES

1. Ingels, M. Father of Air Conditioning, Country Life Press. Garden City, New York, 1952.

2. Nevins, R. G. Air Diffusion Dynamics. Business News Publishing Company. Birmingham, Michigan, 1976.

3. Billings, J. S. Ventilation and Heating, The Engineering Record, New York. 1893. p. 120.

4. Proceedings of the First Annual Meeting. ASHVE Transactions 1. (1895) p. 40.

5. Yaglou, C. P., E. C. Riley and D. I. Coggins. Ventilation Requirements. ASHVE Trans. 42. (1936), p. 133.

6. Woods, J. E. and R. G. Nevins. Ventilation Requirements and Energy Conservation. Proceedings of the 6th International Congress on Air Conditioning, Heating and Ventilating. Milan, Italy. 1975.

7. Woods, J. E. Ventilation, Health and Energy Consumption: A Status Report. ASHRAE J. (7): 1979. p. 23.

8. Rohles, F. H. and R. G. Nevins, The Nature of Thermal Comfort in Sedentary Man, ASHRAE Transactions, Vol. 77, Part I, 1971, p. 239.

9. McNall, P. E., J. Jaax, F. H. Rohles, R. G. Nevins and W. Springer, Thermal Comfort (Thermally Neutral) Conditions for Three Levels of Activity, ASHRAE Transactions, Vol. 73, Part I, 1967, p. I.3.1.

10. Fanger, P. O. Thermal Comfort, Danish Technical Press, Copenhagen, 1970. Reprinted by McGraw-Hill Book Co., New York, 1973.

11. Gagge, A. P., J. A. J. Stolwijk and Y. Nishi, An Effective Temperature Scale Based on a Simple Model of Human Physiological Response, ASHRAE Transactions, Vol. 77, Part I, 1971, p. 247.

12. Nevins, R. G. Variables Air Volume Systems and Air Diffusion Performance. Proceedings of the 6th International Congress on Air Conditioning, Heating and Ventilating. Milan, Italy, 1975.

13. Ostergaard, J., P. O. Fanger, S. Olesen, Th. Lund-Madsen, Thermal Comfort for Man Exposed to Uniform Air Velocity from Five Directions, ASHRAE Transactions, Vol. 80, Part II, 1974.

14. Houghten, F. C., C. Gutberlet and E. Witkowski, Draft Temperatures and Velocities in Relation to Skin Temperatures and Feelings of Warmth, ASHVE Transactions, Vol. 44, 1938, p. 289.

15. Rydberg, J. and Per Norback, Air Distribution and Draft, ASHVE Transactions, Vol. 55, 1949, p. 225.

16. Miller, P. L. and R. T. Nash, A Further Analysis of Room Air Distribution Performance, ASHRAE Transactions, Vol. 77, Part II, 1971, p. 205.

17. Olesen, S., P. O. Fanger and J. J. Bassing, Physiological Comfort Conditions at Sixteen Combinations of Activity, Clothing, Air Velocity and Ambient Temperature, ASHRAE Transactions, Vol. 78, Part II, 1972, p. 191.

18. Rohles, F. H., J. E. Woods and R. G. Nevins, The Effects of Air Movement and Temperature on the Thermal Sensations of Sedentary Man, ASHRAE Transactions, Vol. 80, Part I, 1974, p. 101.

19. Fanger, P. O., R. G. Nevins and P. E. McNall, Predicted and Measured Heat Losses and Thermal Comfort Conditions for Human Beings, Thermal Problems in Biotechnology, ASME, New York, 1968, p. 61.

20. Woods, J. E. "Thermal and Non-Thermal Human Responses to Air Movement and Relative Humidity." ASHRAE Transactions, Vol. 84, 1978, p. 75.

NOMENCLATURE

Variables

A	area: (m^2) or (ft^2)
K	throw constant, dimensionless thermal conductivity: (J/hr.m.^oC) or (Btu/hr.ft.^oF)
L	length: (m) or (ft)
M	mass: (kg) or (lbm)
Q	energy: (J) or (Btu); volume flow rate (m^3/s)
T	air temperature: (^oC) or (^oF)
U	overall heat transfer coefficient: (J/hr.m^2.^oC) or (Btu/hr.ft^2.^oF)
V	air velocity; (m/s) volume: (m^3) or (ft^3)
X	distance from outlet to occupant; (m) or (ft)
c	specific heat: (J/Kg.^oK) or (Btu/lbm.^oF)
m	mass: (Kg) or (lbm)
t	time: (hr) or (s)

Symbols and Subscripts

K	gain, proportionality constant
L	load
X	local, at the occupant's location
a	air, ambient
c	control
Cond.	Conduction
e	effective, outlet
i	inlet
inf	infiltration
int	internal; integral
m	meter
o	outlet; outside

occ occupant

os occupied space

s sensor, supply

x local, at the occupant's position

Greek Symbols

Δ perturbation of a variable

θ effective draft temperature (^{o}C) or (^{o}F)

ρ density: (Kg/m^3) or (lbm/ft^3)

τ time constant: (hr)

Chapter 44

ENERGY TECHNOLOGY EVALUATION UTILIZING LINEAR PROGRAMMING MODEL

D. Hill

INTRODUCTION

One of the objectives of the International Energy Agency (IEA) is to reduce member-country dependence on imported oil by undertaking long-term cooperative efforts in energy research and development.[1] In order to establish its own priorities for energy technology R&D, the IEA established in 1976 a multinational systems analysis project to assess the relative merits of candidate new technologies. Studies were conducted by national representatives of fifteen* IEA nations located at either Brookhaven National Laboratory, U.S.A. (BNL)**, or Kernforschungsanlage Jülich (KFA), Federal Republic of Germany. In 1980, the IEA published its report, A Group Strategy for Energy Research, Development, and Demonstration, based largely on this analysis.[2]

Initially, "reference energy systems" were developed for each of the several nations in which a balance was shown between available energy resources and final energy consumption, with feasible flows of energy carriers from sources through conversions to destinations. While this was a major step forward in the description of national energy systems, it was insufficient to define the shared energy prospects of the several nations. To describe the potential of each national energy system and to evaluate the effects of energy R&D policies on each of them, it was necessary to formulate a mathematical model.

The model developed, MARKAL (for MARKet ALLocation), is a time-phased linear program (LP) that each nation could use to describe the potential changes in its energy system during the period 1980 to 2020.[3,4] It was an outgrowth of the BESOM, DESOM series of models at BNL and a similar model in use at KFA in which many of the characteristics typical of European energy systems--district heating, night heat storage, coupled production of electricity and heat--were more adequately represented.

At the same time, a major effort went into the characterization of potential new energy technologies at each laboratory, and a final set of "reference technologies" was compiled after preliminary descriptions had been reviewed by technical experts in many of the countries.[5,6] In the model, these new technologies, together with the existing stock of energy technologies, compete with each other in the production, conversion, and consumption of energy to meet projected demands.

Using the MARKAL model and an appropriate set of candidate technologies, each nation then prepared a model of its energy system, as it exists and as it may develop. Common assumptions were made for a set of scenarios in which there were variations in projected oil prices, rates of introduction of new technologies, emphasis on renewable resources, restrictions on the total use of nuclear and fossil energy, and the premium that might be paid to reduce oil imports. Projected final energy demands in each nation were assumed the same in all cases.

The solutions to the MARKAL model in each of these scenarios showed for the 1980-2020 time period the energy resource requirements, the preferred energy technologies and the extent to which they will be needed, the energy system cost, and the level of oil imports. These were compiled for the fifteen nations as the basis for the IEA energy technology R&D strategy.

Although we found that a formal mathematical model was essential to make the evaluation of technologies required by the IEA, questions persist about the value of such large models. Are they necessary? Do they obscure the issues rather than render them transparent? Must they be so large? Can they be believed?

In this paper, the merits of such models will be discussed, using as illustrations some of the results of the U.S. case in the IEA project, which is described fully elsewhere.[7]

THE MARKAL MODEL

The MARKAL model consists of a set of inequations that describe the constraints on the energy system; e.g., each energy service demand must be satisfied, the consumption of each specific fuel cannot exceed its supply, the amount of electricity that can be produced on a summer day cannot exceed the total installed capacity of electric generating plants taking into account probable unavailable capacity, the growth in capacity from one time period to the next cannot exceed a

---

*Austria, Belgium, Canada, Denmark, Germany, Ireland, Italy, Japan, New Zealand, Norway, Spain, Sweden, Switzerland, United Kingdom, and the United States.

**Work at BNL was supported by the U.S. Department of Energy.

specified rate, the amount of oil that can be produced by enhanced recovery techniques cannot exceed present estimates of this potential, etc. The variables in the inequations describe for each time period the capacity of each technology that may be installed, the flow of energy through them, and the production of electricity, heat, and fuel products or energy services. There are essentially seven types of variables and fourteen principal types of inequations. With all of the energy forms and candidate technologies and time periods represented in the U.S. case, however, the linear programming matrix consists of about 3,500 rows and 4,000 columns.

The program first of all finds a set of feasible nonnegative values for the variables; i.e., a solution that does not violate the set of constraints. Second, among the feasible solutions, it finds an optimal solution that minimizes or maximizes an objective function; i.e., some function of the variables that is of particular interest. Three types of objective functions were used:

P = total energy system cost, discounted to its "present value" in 1978

S = total amount of oil imports during the nine five-year time periods

$$Q = P + \left(\frac{\Delta P}{\Delta S}\right) S$$

where

$\left(\frac{\Delta P}{\Delta S}\right)$ = the value of the premium that the nation might be willing to pay to reduce oil imports, which may be interpreted as a surcharge on imported oil.

The program serves three purposes. First, by showing a feasible soluton, it adds up the numbers, a job that is not necessarily satisfied by any amount of rhetoric. Second, by giving an optimal solution, it points in a desirable direction. Third, because it is easy and relatively inexpensive to run the model once it has been formulated, a wide range of possible future situations can be examined by altering the assumed conditions to answer the question, "What if...?"

MODEL MISCONCEPTIONS

It is perhaps equally important to explain what the MARKAL model is not.

It is not a forecast of the future. It describes what ideally should be done in the model world abstracted from reality; it does not predict what will be done in the real world. Thus it is a normative model, not a simulation. As a long-term representation of the national energy system, its most useful results are the trends it exhibits in how the ideal ("optimal") situation changes over time. Since it is a representation of the energy system, it should show some resemblance to reality in the short run, but it is unrealistic to expect the solutions it finds in the short term to be a precise extrapolation of the recent past.

COST AND OIL-IMPORT TRADEOFFS

The way the numbers add up under a variety of different assumptions about the future is illustrated in Figure 1 which shows the total cost of the energy system (discounted at 6% and expressed in 1979 dollars) and the total oil imports during the 45-year period for each of sixteen scenarios.

Scenario PS-1 is a cost minimization scenario with new technologies assumed to enter at a modest rate and with a growth in oil prices assumed to be moderate. The assumption as to future oil price trajectories proves to be quite critical. The cost-minimization result when higher and lower rates of increase are assumed is illustrated by the broken line connecting PS-1/OIL C and PS-1/OIL A. In each of these three scenarios, the cost of the energy system is minimized without regard to the level of oil imports, which proves to vary by almost a factor of two in the extremes. Each of these scenario results represents a different mix of energy technologies and resource use to satisfy the same set of energy demands, with the difference determined by the justifiable cost of alternatives to importing oil. With cheap imported oil, few alternatives are justified; when oil becomes more expensive, more expensive alternatives can replace it.

Scenario SP-1 represents the absolute minimum level of oil imports that can be achieved regardless of cost assuming the same moderate oil price trajectory and modest rate of introduction of new technologies. Oil imports are reduced by about 40% from scenario PS-1, but at a great cost; the marginal cost of the final substitutions in the technology mix to supplant the final drops of imported oil are astronomical. This is a totally unrealistic picture of the future, of interest only as a limiting case.

On the line connecting PS-1 and SP-1 in Figure 1, three other scenarios are shown--QP-1/0.5, 1.0, and 1.5--in which it is assumed that the nation will be willing to pay a premium to reduce oil imports below their level at PS-1. These premiums may be regarded as a surcharge on imported oil which is mathematically equivalent to higher and higher exponentially increasing future oil price trajectories. The effect of this surcharge is to permit more expensive technologies in the mix if they are effective in reducing oil imports. The distribution of these points suggests that a relatively modest surcharge (scenario QP-1/0.5) can result in almost a 30% reduction in oil imports, but that further increases in this surcharge have little additional effect. Up to a point, the substitution of more expensive energy technologies can substantially reduce oil imports, in other words; but beyond that point the gains do not seem to justify the additional cost. This substitution of technologies is determined by the absolute level of the cost of oil imports, so that the amount of the justifiable surcharge becomes less as actual oil imports increase in price.

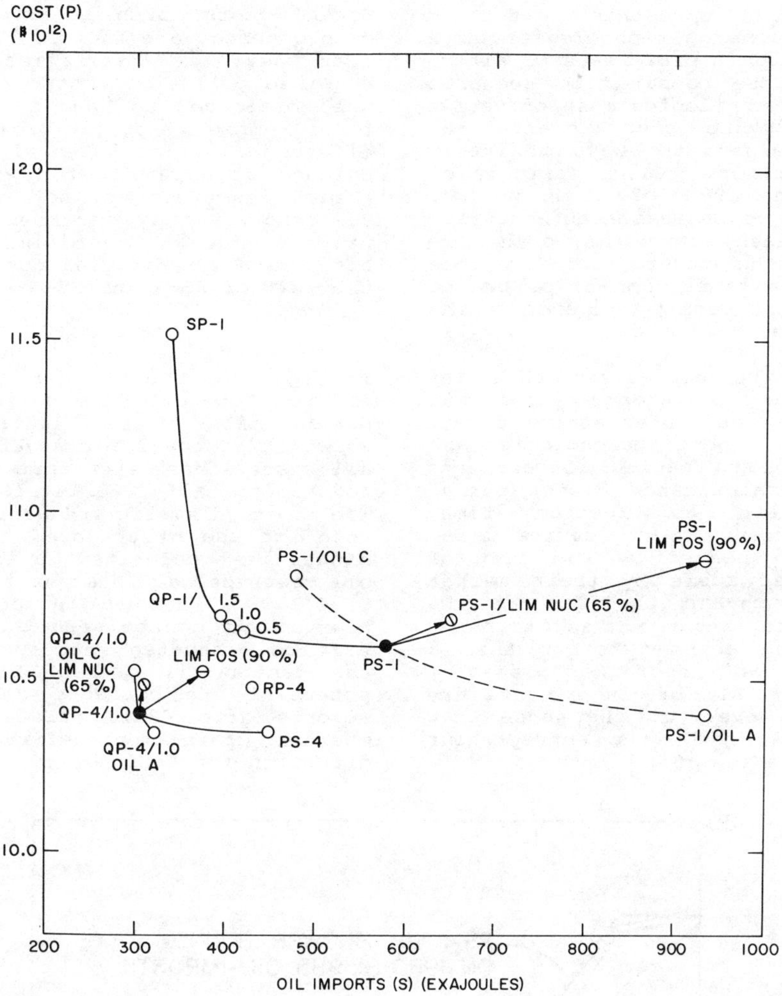

FIGURE 1. TOTAL COST OF ENERGY SYSTEM AND TOTAL OIL IMPORTS

An alternative to changing the mix of energy technologies is the possibility of accelerating the rate of introduction of the most desirable technologies. In scenario PS-4, almost all the technologies are permitted to enter the market at the maximum feasible rate, again assuming the moderate future increase in imported oil prices. In comparison to PS-1, with its modest rate of introduction of all technologies, oil imports can be reduced by about 22% with an actual reduction in the energy system cost of about 2 1/2%. (This reduction would be partially offset by any increases in cost needed to foster the acceleration of technologies, but this would undoubtedly be trivial compared to the absolute savings of 2 1/2%.)

If an oil import surcharge is assumed in addition to the possibility of accelerating technologies as in scenario QP-4/1.0, further reduction in oil imports can be obtained with little increase in energy system cost. Oil imports in this scenario are about 47% less than at PS-1 and the energy system cost is lower. It is worth noting that the marginal cost of substitution of oil-saving technologies is the same at QP-4/1.0 as it is at QP-1/1.0--i.e., not extraordinary--but the level of oil imports is less than at SP-1, the minimum that could possibly be achieved with unaccelerated technologies. Thus, it seems that the most effective technologies should be allowed to prevail rather than encouraging a little bit of everything.

At QP-4/1.0, with accelerated technologies and the surcharge, there is very little slack left in the system. QP-4/1.0 is based on the assumption of the moderate future increase in oil prices. If the high and low oil price trajectories are assumed (QP-4/1.0 OIL C and OIL A), there is little difference in the level of oil imports and even the energy system cost swing is very small compared to that under the PS-1 conditions.

Similarly, if nuclear power is limited to 65% of the level found optimal under the QP-4/1.0 conditions, there is little change by comparison to the same limitation in the PS-1 scenario. (See the LIM NUC scenarios indicated by a diagonal line through the circle.) If fossil energy is limited to 90% of the optimal levels in the two cases (see the LIM FOS scenarios shown by a horizontal line through the circles), the effect is greater and again far more pronounced in the PS-1 situation.

Finally, scenario RP-4 represents a situation in which renewable energy technologies as a class are accelerated in preference to alternatives. This can be compared to scenario PS-4 in which the technologies most effective in reducing energy system cost are accelerated. There are perhaps two ways of interpreting this comparison. Based strictly on their merits in the 1980-2020 time period, there does not seem to be reason to encourage renewables as a class; some are worth more than others. On the other hand, if this period is seen as a transition to an era of renewable energy, the penalty of encouraging renewables as a class is not out of sight.

It should perhaps be emphasized that the costs comprising the total energy system cost in these results are estimated as the direct cost to the nation. Energy technologies are costed by the required capital investments and operating and maintenance costs, taking into account their construction time, availability, efficiency, and service life. Domestic fuels are costed by the cost of extraction; imported fuels by their market price. Taxes, such as the tax at the gasoline pump, are regarded as transfer payments within the economy rather than a cost to the nation, and insofar as possible, they are excluded. All of these costs are "private costs," however, in the sense that they are paid directly by the entrepreneur when the expenditure is made.

Social costs, such as adverse environmental consequences, are included only to the extent that they are "internalized" as the cost of required pollution controls. If oil imports are considered to have a social cost beyond their market price--because of adverse effects on national security or the national balance of payments--the PS-1 and PS-4 scenarios described are not truly minimum cost scenarios. They represent the minimum of private costs. The minimum of the total of both private and social costs is somewhere to the left of PS-1 and PS-4 in Figure 1.

OIL IMPORT FUTURES

In Figure 1, the total amount of oil imports in the 1980-2020 time period is shown. The distribution of oil imports over time is shown in Figure 2 for several scenarios. In all cases, these show that oil imports can be reduced to half 1980 levels or less by 2020. The most dramatic reductions are obtained when the use of enhanced oil recovery techniques is accelerated in exploiting domestic oil reserves, as in scenarios PS-4, RP-4, and QP-4/1.0. When the numbers are added up, however, it can be seen that these resources will be exhausted shortly after the turn of the century if present estimates of their potential are correct. The upturn in oil imports after 2010 raises the question of whether drain-America-first is a desirable direction for U.S. policy.

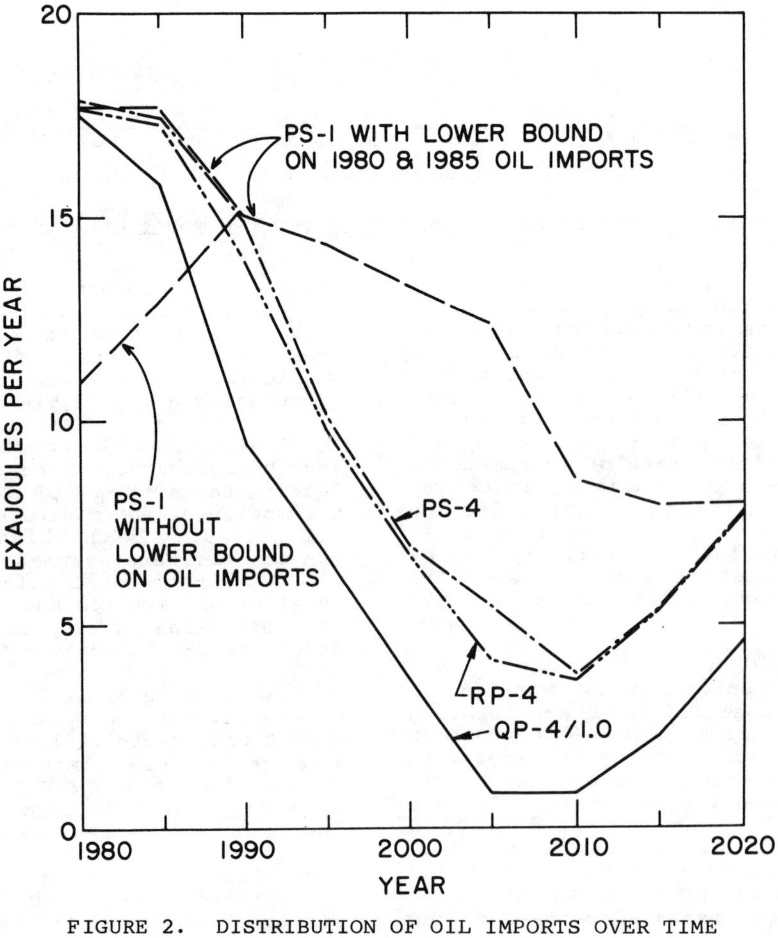

FIGURE 2. DISTRIBUTION OF OIL IMPORTS OVER TIME

Figure 2 also illustrates one of the quandaries facing the energy modeler. Except as noted there, minimum levels of oil imports were specified for 1980 and 1985 to achieve verisimilitude with actual 1979 experience. (Long-term normative models should not be judged by short-term extrapolations, but they are.) In the model, 1980 represents the average of a five-year period centered on 1980, and we therefore still do not know what the "actual" level of oil imports in model year 1980 will be. Since 1979, of course, the direction of oil imports has been in the direction of the unfettered model.

The discrepancy between the PS-1 solution with and without "realistic" lower bounds on oil imports in 1980 and 1985 may be due to two things. One is the necessarily greater efficiency in reducing oil imports of an idealized, simplified, mathematically optimized model energy system than of the reality it represents. A second possibility is that the existence of counterproductive public policies may make the reality even less efficient than it has to be. Either way, even the unattainable 1980 model can serve as a standard by which to assess the reality. Thus, a "least-cost energy strategy," represented in Figure 2 as PS-1 without a lower bound on oil imports, can be said to lead to lower oil imports than those being experienced even without special measures to reduce them. (See, for example, Reference 8.) It seems more consistent, however, to draw inferences from the model by comparing model results from one year to another or one scenario to another, recognizing the inevitable discrepancy between the model and reality.

COAL TECHNOLOGY FUTURES

The possibility of reducing oil imports as shown in Figure 2 depends largely upon the possibility of increasing the use of coal and drastically changing the ways in which it is used, as shown in Figure 3. For the minimum cost scenario shown, PS-1, the annual amount of coal consumed must nearly quadruple by 2020. Whereas most coal is burned to produce electricity in 1980, most will be turned into products or heat in 2020 with only a small percentage used to produce electricity. This change depends in part upon a major buildup in nuclear power (not illustrated in the figure), but even in a limited nuclear buildup (represented by scenario QP-4/1.0 LIM NUC), only about one fifth of the total consumption of coal is used to produce electricity. The main reason for this is that coal is needed to be turned into liquid fuel and to supplant oil as a feedstock for the production of petrochemicals.

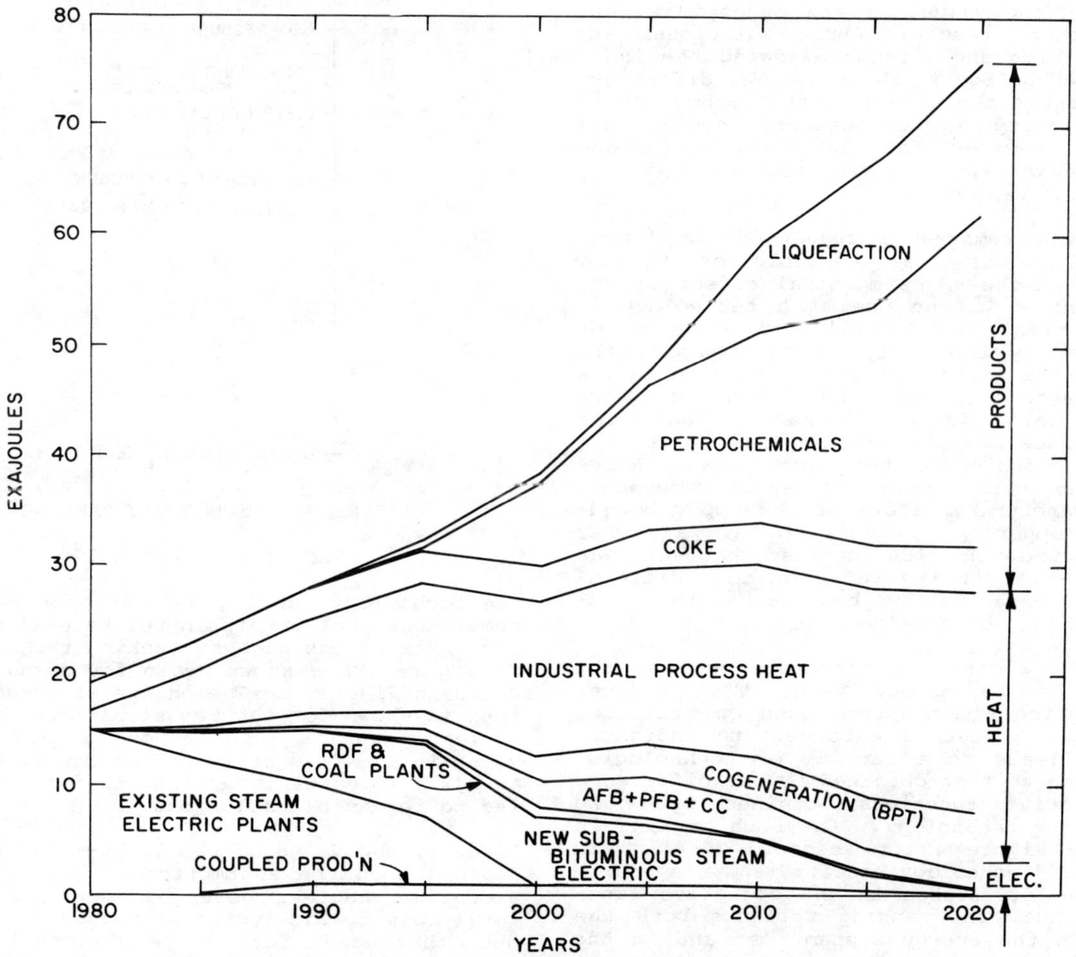

FIGURE 3. USE OF COAL

In scenario PS-1, new sub-bituminous steam electric plants supplant existing coal-fired electric plants as they are phased out during the next twenty years. The more advanced coal-burning technologies--atmospheric and pressurized fluidized bed and combined cycle --do not make greater inroads simply because they become available too late. In a limited nuclear growth scenario, however, these advanced technologies do continue to grow and by 2020 dominate the coal-steam electric plants.

ENERGY TECHNOLOGY COMPARISONS

The merits of individual energy technologies can be assessed from the linear program solution in several ways. First of all, of course, by their appearance in the solution; when several scenarios are run and a technology appears in only some, the cirumstances favoring it are apparent. Most commonly, the value of a technology is measured by the amount of energy passing through it. However, this measure may be limited by upper bounds on installed capacity set by the analyst, and it does not reveal anything about the margin of superiority of the technology; a very slight cost advantage, for example, may favor one technology over a close substitute for it.

An ideal measure is the difference that occurs in the values of the objective functions that results when solutions are obtained with and without allowing the individual technology in the mix. The difference clearly shows the effect of the technology in reducing energy system cost and level of oil imports. However, for many technologies and many scenarios, this is too expensive a procedure.

One of the remarkable results of a linear program is the dual variable or "shadow price" that shows the marginal effect of any constraint. If the use of a technology in any one time period is limited by an upper bound set by the analyst, for example, the shadow price tells what the difference in the objective function would be if the bound were set one unit higher or lower. Thus, the shadow price reveals the marginal value of a unit of capacity of the technology. Moreover, the shadow price tends to compensate for the arbitrary nature of the upper bound: a lower upper bound tends to have a higher shadow price. The IEA interest, however, was not primarily in the value of an additional unit of the technology, but whether the technology should be developed at all.

In the U.S. case, we therefore used as the measure of technology value the product: shadow price times upper bound on installed capacity. A test of this measure indicates that it leads to a ranking of technologies very close to that obtained by the difference in objective functions obtained with and without the technology. One such ranking is shown in Figure 4 for scenario QP-4/1.0 in which the "technology benefit" shown approximates the difference in the objective function. Thus, the ranking reflects both the change in the energy system cost and in the level of oil imports due to each technology.

The scale is logarithmic with the triangular ticks showing factors of ten. Clearly, there is a very great difference in the total value of new energy technologies, and in this scenario, the light-water reactor (LWR), shale oil, coal liquefaction, and enhanced oil recovery all have greater value than the total amount of hydroelectric power available, for example.

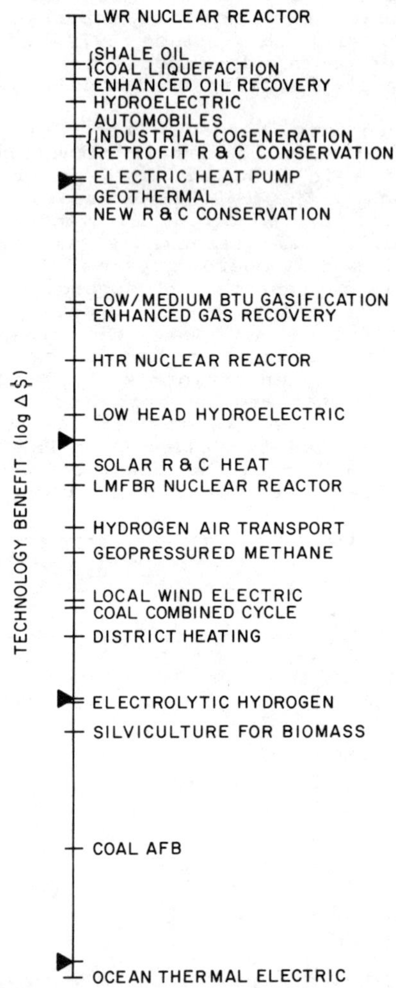

FIGURE 4. TECHNOLOGY RANKING

SENSITIVITY ANALYSIS

The technology ranking depends upon the circumstances that obtain in the several scenarios. Six of the highest ranking technologies in Figure 4 are shown again for scenario QP-4/1.0 in Figure 5. The change in these rankings is shown for the two situations in which nuclear power is then reduced to 65% of its value in QP-4/1.0 or fossil energy is reduced to 90%. (Greater reductions in fossil energy led to infeasible solutions.)

Clearly, the value of the light-water reactor depends upon the assumptions. If the total amount of nuclear power is restricted, its total loss to the system is not as great. On the other hand, if the use of fossil energy is restricted (for example, by concerns about the accumulation of carbon dioxide in the atmosphere), nuclear is all the more important. With the cost and performance characteristics assumed in these runs, nuclear power is the highest ranking technology under almost all circumstances. On the other hand, we have seen in Figure 1 that the effect of sharply curtailing nuclear is not catastrophic. Thus, nuclear may be the best answer, but--as long as there is coal--it is not the only answer.

The importance of industrial cogeneration is also illustrated in Figure 5. One of the highest ranking technologies in any event, its great efficiency in using the energy content of the fuel becomes even more important when either the total amount of nuclear or fossil energy is curtailed.

CONCLUSIONS

In the light of these few illustrations, we can examine some of the questions raised about the use of large LP models.

Are they necessary? The experience of the IEA energy technology systems analysis project was that the model was necessary for a responsible comparative analysis of the energy systems of fifteen nations; it led to results that could be compiled with confidence that likes were being added to likes. However, models are not sufficient. They are simplifications of reality, and their results need to be tempered with consideration of what is left out.

Do they obscure the issues rather than render them transparent? In fact, they should serve to sharpen issues and perhaps reveal possibilities that should be issues. Figure 1 makes clear the extent to which a "minimum-cost strategy" deviates from one in which the reduction of oil imports is regarded as an important objective. Figure 2 indicates that a goal of minimizing U.S. oil imports by the turn of the century may rebound with a growth in oil imports thereafter.

Must such models be so large? They would be more helpful if they were even larger. The technologies represented in this model are typical of the great variety that are candidates for future development, but they are not necessarily optimized technologies. Within limits, the present model can be used as a context within which a variety of candidate technologies can represent one sector,

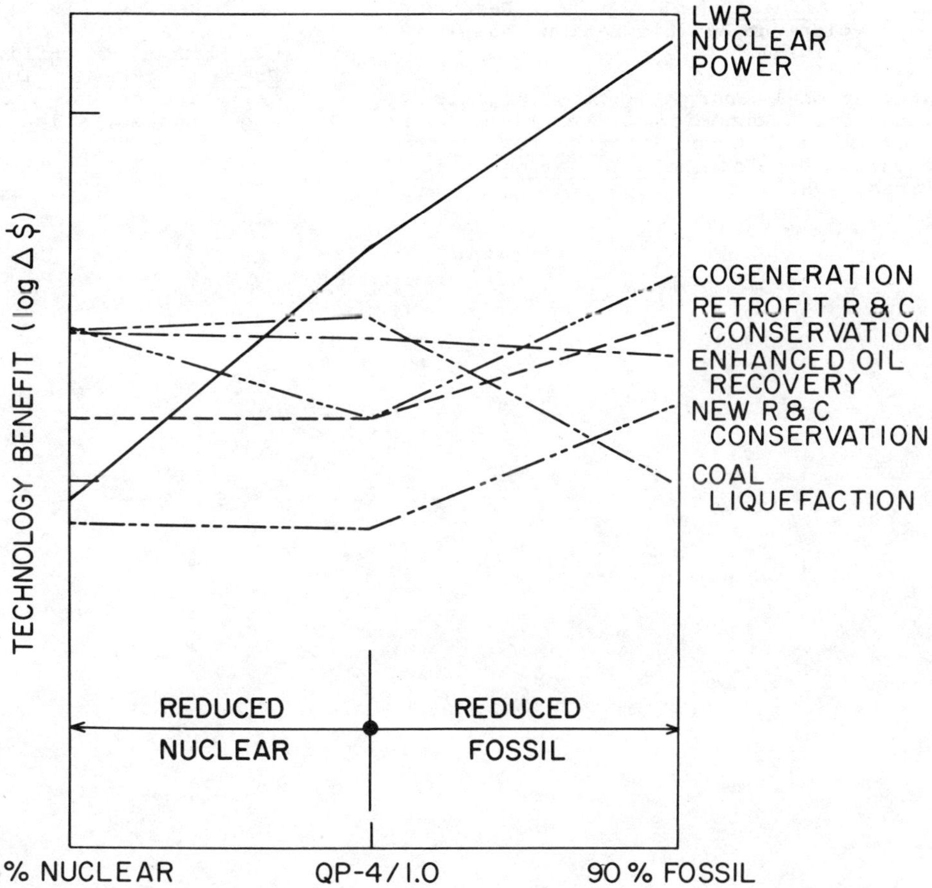

FIGURE 5. TECHNOLOGY RANKING--REDUCED NUCLEAR OR FOSSIL CASES

thus optimizing the choice rather than suboptimizing it within a smaller arbitrary framework. It was not possible in these runs to represent the variety of candidate technologies for individual industries, however. (The use of a linear program to analyze a specific industry is illustrated in another paper at this conference.[9])

Can they be believed? Certainly, the model results should not be taken on faith. The model itself should be scrutinized; the MARKAL model has been tested by fifteen nations. The results themselves should be subject to the test of plausibility. It should be possible to explain surprises. Is it plausible that the major source of petrochemicals must become coal because oil is needed for liquid fuel as suggested by Figure 3? Is the total value of the advanced coal-burning technologies for electric generation --typified by coal combined cycle and atmospheric fluidized bed (AFB) in Figure 4-- rather surprisingly low if nuclear power becomes dominant after the turn of the century? The answer lies in adding up the numbers, which after all is what the model does.

REFERENCES

1. Office of International Affairs, International Energy Agency Background and Current Affairs (U.S. Energy Research and Development Administration, Washington, DC, 1976).

2. International Energy Agency, Organization for Economic Cooperation and Development, A Group Strategy for Energy Research, Development, and Demonstration (Paris, 1980).

3. H. Abilock and L. G. Fishbone, User's Guide for MARKAL (BNL Version) (BNL 27075, Brookhaven National Laboratory, Upton, NY, 1979). The KFA model is described in G. Giesen, H. A. Hymmen, and K. Leimkühler, User's Guide for MARKAL (KFA Version) (Kerforschungsanlage-Jülich, Germany, 1980).

4. L. G. Fishbone and H. Abilock, "MARKAL, A Multiperiod Linear Programming Model for Energy Systems Analysis: Technical Description," to be published in Int. J. of Energy Res.

5. V. L. Sailor, ed., Technology Review Report (BNL 27074, Brookhaven National Laboratory, Upton, NY, 1979).

6. C. Manthey, ed., Energy Technology Data Handbook (Jül-Spey-70, Kernforschungsanlage-Jülich, Germany, 1979).

7. D. Hill, V. L. Sailor, and L. G. Fishbone, Future U.S. Energy Technologies: Cost and Oil-Import Tradeoffs (BNL 28798, Brookhaven National Laboratory, Upton, NY, 1980, to be published in Energy.)

8. R. W. Sant, The Least-Cost Energy Strategy: Minimizing Consumer Costs Through Competition (Mellon Institute, Arlington, VA, 1979).

9. B. Pierce, D. Hill, and S. O. Howe, Process Models for Industry (paper delivered at 4th World Energy Engineering Congress, Atlanta, 1981).

Chapter 45

MODELING SOLAR ENERGY SYSTEMS FOR ECONOMIC OPTIMIZATION

R. W. Leigh

1. ABSTRACT

In the paper I present a simple analytic model of solar energy systems which can describe how the performance of the system depends on collector area and storage capacity with surprising accuracy. The model can then be used to determine easily the optimal (lowest cost) configuration of collector area and storage capacity, and to see how this configuration will change under various possible fuel, collector, or storage costs.

2. INTRODUCTION

Solar energy is becoming increasingly practical in a wide variety of applications, from domestic hot water in parts of the United States to agricultural and industrial process heat in some less developed countries, and a variety of experiments and demonstration projects are now underway both here and abroad. As these technologies become economically attractive and join the engineer's stable of energy supply systems, several questions that were premature or of somewhat less interest during the developmental stages of the technology move to the foreground, focusing on issues such as reliability, availability, and economic design. I will deal with the last of these in the paper, describing a method by which, if the technical characteristics of a solar energy system are known, the engineer can easily find that combination of collector area and storage capacity which will provide the lowest cost ("optimized") system when fuel costs and all other expenses are taken into account on a life-cycle basis. These optimal system configurations will depend on the costs assumed for the components and for fuel over the life of the system, and these are all fairly uncertain quantities today. Therefore, the method is also designed to allow rapid evaluation of the effects of variations in any of the costs involved in the solar energy system, so that the risks and benefits of over or under estimation of these quantities can be estimated easily.

I will be discussing solar energy systems which are designed to meet well-defined loads under specified weather conditions. Although a wide range of components can actually be used in such systems, a configuration consisting of collector area, storage capacity, and a backup system will prove sufficiently detailed for this discussion. The annual fuel consumption, for specified weather conditions and load, will depend on the collector area and storage capacity. We will assume that the fuel consumption has been determined for a set of points in this collector area -- storage capacity (c-s) space either from detailed simulation or from experiment. The discussion applies to a wide range of solar energy systems, including those delivering hot water, space heat, process heat, or electricity. The range of possible storage devices is correspondingly broad.

It should be emphasized that this method does not help in determining the energy which will be obtained from a given solar energy system; this will depend on the load being met, the weather, the temporal correlations between the two and many other factors. Once the load, weather device characteristics, and other inputs have been specified to the degree of accuracy possible, the fuel consumed in the backup system must be determined by estimation, detailed computer simulation or, in the last resort, by experimenting with similar systems meeting similar loads with varying amounts of collector or storage. Simulation is far less expensive and moderately reliable. Available codes include TRNSYS[1] for hot water and active space heating, DOE-2[2] for active and passive space heating, DEROB[3] for building energy demands, and BALDR[4] for industrial process heat, as well as a myriad of other models which can be found in Ref. 5, perhaps the best single source of information on performance estimates and modeling, or by perusing the proceedings of meetings of the International Solar Energy Society. Unfortunately, substantial familiarity with these models is often a prerequisite to obtaining meaningful results. Estimating techniques such as F-CHART[6] are also available and are easier to use but ultimately less accurate. In most cases, it is extremely important that the time dependence of the load and of the weather[7] be modeled on an hour-by-hour basis since it is this temporal interaction of supply and demand which determines the need of the system for storage.

After carrying out these simulations, one will find that both the amount of fuel and the total cost of the system will vary as the collector area and storage capacity are changed, and in general there is a specific pair of areas and capacities for which the system cost is a minimum, given the load, weather, and other exogenous variables. This is an "optimized system." A well-known result, discussed in a succeeding section, is that in an optimized system, the cost of an additional unit of collector or storage is just equal to the value of the fuel it will save when all outlays are levelized over the life of the system. The main purpose of this paper is to present a language in which these optimized system may be easily discussed.

In order to do this I have developed, and here present, an analytic description of the annual operating characteristics of solar energy systems in terms of collector area and storage capacity. This consists of a four parameter equation which describes fuel consumption as a function of collector area and storage capacity, and which can then be solved to yield

*This work was supported by the Technical and Economic Analysis Branch of the Office of Energy Systems Research of the U.S. Department of Energy under contract No. DE-AC02-76CH00016.

closed-form expressions for the sizes and characteristics of optimized systems and various other relations of interest. This description will prove useful in many areas of analysis, since a few (expensive) simulation runs will yield the values of the necessary parameters, and economic and technical analysis for a variety of possible economic futures may then be carried out by evaluating simple expressions.

I have not dealt in detail with the behavior of systems which can sell some of their energy back to a larger entity, such as electricity to a utility. The fuel consumption behavior of such systems will be quite different from isolated systems, and if electricity is counted as fuel, they may actually be net producers. A more general formulation is needed to describe the economic behavior of such systems.

3. DESCRIBING THE TECHNICAL PERFORMANCE OF SOLAR ENERGY SYSTEMS

In this section I define and justify a four parameter equation which describes the annual technical performance of solar energy systems incorporating storage and a fuel-fired backup system in terms of total annual fuel consumption. The backup fuel could be oil or gas, but it could also be electricity from the grid, as long as the possibility of sell-back is excluded. (The equation, with the addition of a fifth parameter, can describe the sell-back situation if the capacity of the grid to absorb power is limited so that for attainable collector areas some collected energy must be stored.)

The parameters are weather and load specific, which is to say they must be obtained anew for systems operating in different climates or under different circumstances. They are not specific to particular amounts of storage capacity or collector area since they describe the dependence of fuel consumption on these variables. We will see in the next chapter that this is enough technical information to permit a complete economic description of all interesting configurations of the system. (A configuration is a specified pair of collector area and storage capacity; in more complex systems than are considered here, a configuration might involve the specification of the areas of several different types of collector, the capacities of several different types of storage and so on.)

Let F(c,s) represent annual fuel consumption as a function of collector area c and storage capacity s. Clearly as c and s increase, F(c,s) will decrease. This behavior is demonstrated by the points plotted in Figure 3-1, which represent the fuel consumption of a solar total energy system operating in Albuquerque.[8] All figures in this report were generated using the parameters from this system, listed in Table 3-1.

TABLE 3-1		
F_o	=	4070.70 GJ/yr
a_o	=	2652.57 GJ/yr
m	=	2.25901 GJ/m^2-yr
m'	=	650.369 GJ/GJ(cap)-yr
λ_F	=	8.75 \$/GJ
λ_c	=	14.46 \$/m^2-yr

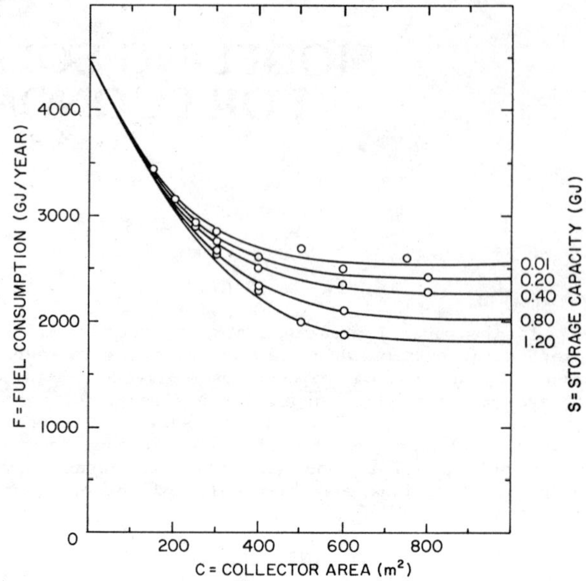

FIGURE 3-1. FUEL CONSUMPTION VS. COLLECTOR AREA FOR VARIOUS STORAGE CAPACITIES

Economic analysis of solar energy systems can be performed by calculating or measuring fuel consumption at each of a set of discrete points (c,s) as shown here, but the process rapidly becomes tedious and expensive, especially if the simulation is detailed. Thus an analytic representation of this dependence of fuel consumption on c and s would be quite useful. I assert that an adequate analytic representation is:

$$F(c,s) = F_o - (F_o - a(s))\tanh(mc/(F_o - a(s))), \quad (3\text{-}1)$$

where

$$a(s) = a_o(1 - \tanh(m's/a_o)), \quad (3\text{-}2)$$

and the four parameters F_o, a_o, m and m' have the following interpretations: F_o is the fuel consumed if c = s = 0; a_o is the fuel consumed if s = 0 and c is very large; m is the fuel savings attributable to the first unit of collector, and m' is the fuel savings attributable to the first unit of storage for very large collector areas. The curves are plotted in Figure 3-1 for storage capacities corresponding to the plotted data points, and the corresponding parameters F_o, a_o, m, and m' are shown in Table 3-1. The root mean square error is on the order of 3% at each of 28 points. Fits to two other different systems have been equally accurate.

The choice of this functional form was based on the nature of solar energy system operations. Consider the zero storage case. The first few units of collector will be used to the greatest extent practical, since the arriving solar energy can always displace fuel. Thus fuel consumption declines linearly with collector area for small areas. This characteristic is independent of storage capacity, as long as collected energy need not be cycled through storage if it can be used directly. The expression for fuel consumption reflects this, since Eq. (3-1) reduces to $F(c,s) = F_o - mc$ for small values of c, independent of the size of s. This expression is not adequate for systems where all collected energy must be cycled through storage of less than 100% efficiency, even if the energy is to be used immediately. For these systems (solar ponds might be an example), the behavior at small c may still depend explicitly on s. We will not consider such systems here.

Returning to consideration of the zero storage case, as more collector is added, it eventually can no longer be used effectively since some solar energy is arriving when it is not needed and is dumped. The fuel consumption curve bends up as a function of collector area and eventually becomes horizontal when the energy supplied by a new increment of collector area cannot be used at all. Systems with storage operate similarly: for large collector areas, the first few units of storage are all used to the fullest and fuel consumption declines linearly as storage capacity is added. Eventually, however, capacity is added which is used only part of the time and the curve representing fuel consumption versus storage capacity turns up. At very large storage capacities the fuel used at infinite collector area, a(s), goes to zero and we have a system which is 100% solar. Under some circumstances, systems with the potential for net sales may be represented by adding a (negative) constant to Eq. (3-2) for a(s).

For both the dependence of a(s) on s and the dependence of fuel consumption on $mc/(F_o - a(s))$, the hyperbolic tangent forms used in Eqs. (3-1) and (3-2) were the simplest analytic forms which would have the appropriate behavior: a linear decrease in fuel consumption at small arguments, and an asymptotic approach to a limiting value for large arguments, with simple monotonic behavior between the limits. If greater accuracy is desired, and sufficient data are available to permit fitting more constants, a more flexible functional form could easily be created by adding to Eqs. (3-1) and (3-2) a series expansion in complete orthogonal functions having zeros for zero and infinite arguments.

If the solar fraction σ is defined as that fraction of maximum fuel consumption displaced by solar energy,

$$\sigma(c,s) = 1 - F(c,s)/F_o \quad , \tag{3-3}$$

then it is easy to show that

$$\sigma(c,s) = (1 - a(s)/F_o) \tanh(mc/(F_o - a(s))) \tag{3-4}$$

It is also straightforward to solve the equations for c, so if we know s, the storage capacity, and the desired solar fraction, σ,

$$c = ((F_o - a(s))/2m) \ln \left| \frac{1 - a(s)/F_o + \sigma}{1 - a(s)/F_o - \sigma} \right| \tag{3-5}$$

Eq. (3-5) may be used to calculate isosolar curves (curves of constant solar fraction) such as are shown in Figure 3-2. Examination of the equations above shows that for a given solar fraction σ, the requirement $c > 0$ means that the argument of the logarithm in Eq. (3-5) must be greater than one. This in turn implies that

$$a(s) < F_o(1-\sigma), \tag{3-6}$$

or

$$s > (a_o/2m') \ln(2a_o/(1 - \sigma)F_o - 1) \quad . \tag{3-7}$$

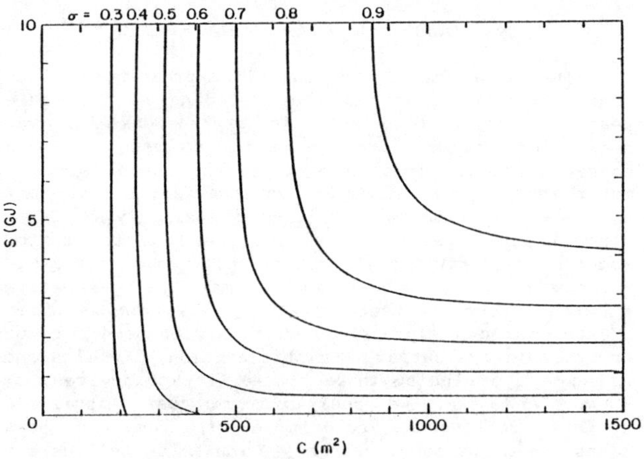

FIGURE 3-2. STORAGE CAPACITY VS. COLLECTOR AREA FOR CONSTANT SOLAR FRACTION σ

Similarly, as s gets very large a(s) approaches zero and Eq. (3-5) indicates that

$$c > (F_o/2m) \ln((1 + \sigma)/(1 - \sigma)) \quad , \tag{3-8}$$

where equality would hold in Eq. (3-8) if $a(s) = 0$. These expressions are the asymptotic values of the "isosolars" plotted in Figure 3-2. Within the limits prescribed by Eqs. (3-6) to (3-8) any solar fraction may be obtained by any of the (c,s) pairs on the appropriate curve. The detailed shapes of the curves, and in particular the numerical values obtained from them depend strongly on the parameters, obtained in turn from the simulations or experiments to which the expression for F(c,s) was fitted. The general form of the curves, however, is remarkably similar for all the systems examined to date.

Note that Eqs. (3-1), (3-3), (3-4), (3-5), (3-6), and (3-8) do not depend on the exact form of a(s). Examination of some systems indicated that the simpler form,

$$a'(s) = a_o(1 - m's), \quad s < 1/m', \qquad = 0, \quad s \geq 1/m', \tag{3-9}$$

would give an adequate representation of the data. In this case, Eq. (3-7) would be replaced by

$$s > (1-(1 - \sigma)F_o/a_o)/m', \quad s < 1/m', \qquad 0 < (1 - \sigma)F_o/a_o, \quad s > 1/m', \tag{3-10}$$

the latter being no restriction. However, this form of a(s) gives perverse economic behavior due to the discontinuity in its slope, and so will not be exploited further in this report.

I have described the heuristic meaning of the four technical parameters F_o, a_o, m and m' above, and in many cases this understanding will be enough to permit estimation of their values. If more precise data are available, such as the points plotted in Figure 3-1, standard least squares techniques can be used to find the values of the four parameters which allow Eqs. (3-1) and (3-2) to fit the data best. The parameter values listed in Table 3-1 and used to generate the figures in this report were determined from data from Reference 1 (the points shown) using a Newtonian search subroutine (ZXMIN) on the CDC 7600 at Brookhaven National Laboratory.

4. OPTIMIZED SOLAR ENERGY SYSTEMS

The existance of the analytic expression for fuel consumption described in the preceding section makes possible closed form analytic expressions for many quantities derived from economic analyses of solar energy systems. This is particularly useful since it means that expensive simulation programs, or even more expensive experiments, may be used sparingly to determine the four parameters above, and that the performance characteristics of systems optimized under a wide variety of economic assumptions may then be derived from the technical representation. In this section I present various expressions which may be used for such an economic evaluation. In this section, I will assume all costs, including those of the storage devices, are known, and that we seek minimum cost (optimized) systems. All costs, including ongoing fuel and operations and maintenance charges (O and M) as well as lump sum capital outlays, are assumed to have been levelized in constant dollars over an assumed economic life of the system. All the equations are equally useful if one assumes that the quantities are all net present values of the outlays, over a similar economic life. I include no prescriptions for levelizing or finding net present values, since this subject has been treated in detail elsewhere.[9-10]

The total annual system cost (or NPV, if you will) is given by:

$$T = c\lambda_c + s\lambda_s + \lambda_b + \lambda_F F(c,s) + b_c + b_s, \qquad (4\text{-}1)$$

where:

λ_c = incremental cost of collectors per unit area, including O and M, replacement costs, and/or salvage value;

b_c = buy in costs for collectors;

λ_s = incremental cost of storage per unit of capacity, including O and M, replacement costs and/or salvage value;

b_s = buy in costs for storage; (4-2)

λ_b = cost of balance of system, including backup, power conditioning and transmission, demand charges, and any costs which do not vary with collector area or storage capacity; and

λ_F= cost of backup fuel per unit of energy purchases.

The buy-in costs for collectors and storage are those costs incurred only when the collector or storage is installed. This would include piping or wiring, scaffolding or other rigging which could be used at no cost once installed, and control systems and power conditioning equipment. These costs are associated with equipment which would not be needed if no collection or storage were present and which does not have any costs depending on the actual area of collectors or capacity of storage. Wires or pipes leading up to the collectors might have to be increased in capacity as more collectors are added, and similar costs can be imagined for storage devices. I assume that these incremental costs are included in λ_c and λ_s, while the initial costs of putting in wires or pipes of capacity sufficient to handle the smallest unit of collector are counted in the buy in costs. As has been pointed out,[5] the incremental cost of the first unit of storage is actually $b_s + \lambda_s$, while the cost of the second unit of storage is simply λ_s. This is admittedly simplification of actual decision-making processes, where a choice is probably based on smearing the buy-in costs over the whole project, but it will turn out to be a useful simplification in the discussions to follow. We will return to treat the effects of buy in costs in more detail at the end of this section.

In reality, cost functions more complex than the linear expressions used here will give a more accurate representation of the dependence of component costs on component size. For example, the costs of insulated tanks for storing hot water will more nearly scale with the surface area than with the volume, leading to an $s^{2/3}$ dependence on storage capacity. The use of more complex functions makes most of the closed form solutions to be derived below impossible, although the general approach remains valid. I will not pursue this avenue here, since the linear formulation may be used to represent the more general case in the following way: In investigations of realistic systems, one generally has or soon acquires a reasonable notion of the component sizes an optimized system will have. Then, if more complex cost/capacity curves are available, the two parameters b_s and λ_s (or b_c and λ_c) may be adjusted to make the expression $b_s + \lambda_s s$ approximate both the value and slope of the more complex cost curve in the regime of interest. b_s will no longer represent the "real" buy in cost, but results based on the linear approximation will closely mirror those which would be obtained from more complex component cost expressions.

An optimal system is one for which the total costs are a minimum, a condition which obtains when:

$$\left.\frac{\delta T}{\delta c}\right|_{c^o,s^o} = \left.\frac{\delta T}{\delta s}\right|_{c^o,s^o} = 0. \qquad (4\text{-}3)$$

Here c^o and s^o are the values of c and s for which the minimum occurs. Substituting (4-1) into Eqs. (4-3), we find:

$$\left.\frac{\delta F}{\delta c}\right|_{c^o,s^o} = -\lambda_c/\lambda_F, \qquad (4\text{-}4)$$

and

$$\left.\frac{\delta F}{\delta s}\right|_{c^o,s^o} = -\lambda_s/\lambda_F. \qquad (4\text{-}5)$$

Note that multiplying both sides by λ_F shows that the costs of increments of collectors or of storage are just equal to the fuel that they save in these optimized systems.

We now seek explicit expressions for c^o, s^o, and $\sigma(c^o, s^o)$ or $F(c^o,s^o)$. Some definitions make the algebra simpler:

$$\alpha = \lambda_c/m\lambda_F, \qquad 0 < \alpha < 1, \qquad (4\text{-}6)$$

$$\beta = \ln(\sqrt{1/\alpha} + \sqrt{1/\alpha - 1}), \qquad \beta > 0, \qquad (4\text{-}7)$$

$$\delta = \sqrt{1-\alpha} - \alpha\beta, \qquad 1 > \delta > 0, \qquad (4\text{-}8)$$

$$\mu = \lambda_s/m'\lambda_F. \qquad 0 < \mu < 1, \qquad (4\text{-}9)$$

The range of values of each parameter is also shown; note that α gives incremental collector costs and μ incremental storage costs, measured in terms of the value of the fuel saved by the first unit of collector or of storage. Examination of Eq. (4-6) shows that if α were greater than one, even the first few units of collector would not be worth their cost in fuel and no optimized system would be possible. Note also that only μ involves λ_s, the cost of storage.

Equation (4-4) above may be solved for c^o:

$$c^o = (F_o - a(s))\beta/m. \qquad (4\text{-}10)$$

This may be substituted into Eq. (3-4) to yield

$$\sigma = (1 - a(s)/F_o)\sqrt{1 - \alpha}. \qquad (4\text{-}11)$$

These, however, still depend on s and so are not yet solutions. Substituting (4-10) into (4-5) does yield:

$$s^o = (a_o/m')\ \ln\left(\sqrt{\delta/\mu} + \sqrt{\delta/\mu - 1}\right), \qquad (4\text{-}12)$$

which is an expression for s^o which depends only on the technical and economic parameters. c^o may be found by substituting this value for s^o back into (4-10):

$$c^o = (F_o - a_o(1 - \sqrt{1 - \mu/\delta}))\beta/m. \qquad (4\text{-}13)$$

Values for the fuel consumption, solar fraction, and total cost of optimized systems may then be obtained by substituting the expressions for c^o and s^o back into Eqs. (3-1), (3-3), and (4-1). The results are:

$$F(c^o,s^o) = F_o - (F_o - a_o(1 - \sqrt{1 - \mu/\delta})) \\ 1 - \alpha, \qquad (4\text{-}14)$$

$$\sigma(c^o,s^o) = (1 - (a_o/F_o)(1 - \sqrt{1 - \mu/\delta}))\sqrt{1 - \alpha}, \qquad (4\text{-}15)$$

and

$$T(c^o,s^o) = c^o\lambda_c + s^o\lambda_s + \lambda_b + \lambda_F\, F(c^o,s^o) + b_c + b_s. \qquad (4\text{-}16)$$

Note that while Eqs. (4-10) through (4-14) explicitly depend only on the four technical parameters and on λ_c, λ_s, and λ_F, Eq. (4-15) must have various quantities from the preceeding equations substituted into it.

Examination of the definitions in Eqs. (4-6) through (4-9) and the results in Eqs. (4-12) through (4-16) reveals the following dependencies: In order to have a viable system with $c^o > 0$, we must have $\alpha < 1$ (the first unit of collector must save fuel equal to or greater in value than its cost). This gives $\beta > 0$ and $\delta > 0$. We must then have $\mu \leq \delta$ for these equations to apply, and if $\mu = \delta$ we have a system with non-zero collector area but zero storage. For $\mu > \delta$ the results obtained from Eqs. (4-12) through (4-16) for $\mu = \delta$ apply.

So far we have tacitly assumed that as collector prices fall or fuel prices rise there is a smooth transition from a non-solar system to one with arbitrarily small amounts of collectors or storage. In practice this is not the case, due to the existance of the buy in costs b_c and b_s, which will inhibit the purchase of solar equipment until the savings are sufficient to cover these initial outlays. In short, we must have

$$T(c^o,s^o) < T(0,0) \qquad (4\text{-}17)$$

which can be reduced to

$$b_c + b_s < \lambda_F\delta(F_o - a_o(1 - \quad 1 - \mu/\delta)) \\ - \lambda_F a_o\mu\ \ln(\sqrt{\delta/\mu} + \sqrt{\delta/\mu - 1}). \qquad (4\text{-}18)$$

If Eq. (4-18) is not satisfied, then the optimal system will have c = s = 0, rather than collector area c^o and storage s^o from Eqs. (4-12) and (4-13). If $\mu > \delta$, Eq. (4-18) will reduce to the prescription of constraints for the transition from a non-solar system to one with collector but no storage is b_s is taken as zero.

The Eqs. (4-12) through (4-18) constitute a complete description of the relevant characteristics of optimized systems (fuel consumption and cost) in terms of F_o, a_o, m and m' and the economic parameters defined in Eq. (4-2). In this section, these have been the independent variables (Table 4-1) from which the characteristics of optimized systems were derived. We may now, for example, examine the dependence of total system cost or fuel consumption on the fuel prices, collector costs, or even discount rate just by varying the appropriate parameters in (4-10) through (4-18). For example, two fuel price scenarios may be developed and the characteristics of optimized system under each of the two possible futures determined from these equations. Then the economic consequences of guessing wrong may be examined by using the component sizes determined from each scenario with the fuel prices from the <u>other</u> scenario to determine whether it is safer (less costly) to overbuild or to underbuild. The practicing engineer will be able to think of many more such applications.

TABLE 4-1

Independent Variables:	F_o, a_o, m, m', λ_c, b_c, λ_s, b_s, λ_b, λ_F
Dependent Variables:	F, σ, c, s, T

REFERENCES

1. "TRNSYS, A Transient Simulation Program," Solar Engineering Laboratory, Univeristy of Wisconsin, Madison, Engineering Experiment Station, Rport 38, 1978.

2. "DOE-2 Users Guide," Lawrence Berkely Laboratory and Los Alamos Scientific Laboratory." LBL 8689, Feb. 1979.

3. F. Arumi-Now, "The DEROB System: Vol. 1, Users Manual and Vol. 2: Explanatory Notes and Theory," School of Architecture, University of Texas, Austin, TX, Jun 1979.

4. J.G. Fingold and F. Ann Herlevitch, "BALDR-2: A Solar Thermal System Simulation," Solar Energy Research Institute, Golden, Co, 1980.

5. J.F. Kreider and F. Kreith, "Solar Energy Handbook," McGraw Hill, New York, 1981 (several chapters are relevant).

6. Chapter 11, Ref. 5.

7. Chapter 2, Ref. 5.

8. Leigh, R.W., <u>Storage in Residential Solar Total Energy Systems</u>, BNL 51230, Brookhaven National Laboratory, Upton NY 11973, Dec. 1980.

9. Grant, E.L., Ireson, W.G., and Leavenworth, R.S., Principles of Engineering Economy, 6th ed., (John Wiley and Sons, New York NY, 1976),

10. Office of Technology Assessment, Congress of the United States, Application of Solar Technology to Todays Energy Needs, Vol. 2, Chapter 1 (U.S. GPO, Washington DC, Sept. 1978)

11. Dr. D. Caskey, Sandia Laboratories, private communication.

SECTION 11
SENSIBLE EVAPORATIVE REFRIGERATION

Chapter 46

APPLYING EVAPORATIVE REFRIGERATION TO COMMERCIAL AND INDUSTRIAL AIR CONDITIONING SYSTEMS

W. Hansen

The Sensible Evaporative Refrigeration (SER) process is the first hybrid solar cooling process that we, as Consulting Engineers, have felt comfortable in presenting to our clients. In analyzing a project for the applicability of SER, it is necessary to pay a little more attention to the climate and the outside air requirements of the facility in question.

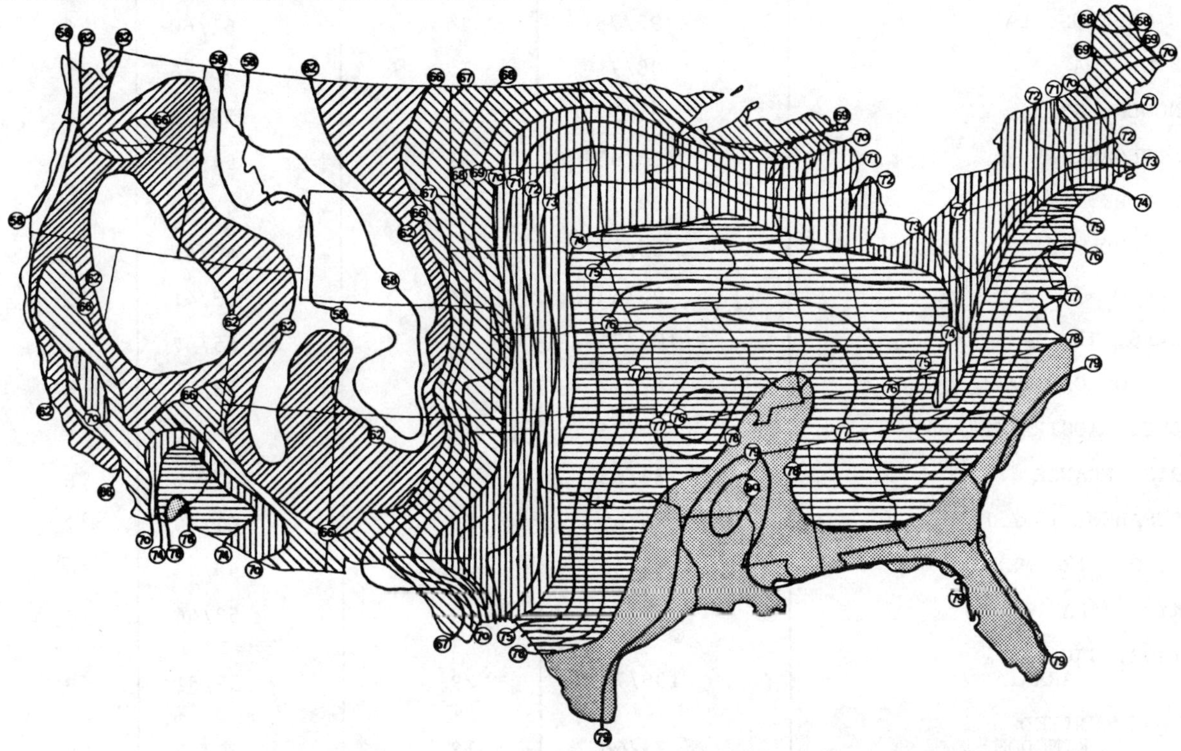

FIGURE 1: WET BULB TEMPERATURE ISO-LINE MAP OF THE UNITED STATES AT THE 5% LEVEL

As one would expect, SER is much more applicable in the Western part of the U.S. than it is in the Eastern and Southern parts (see Figure 1). This does not mean that SER cannot be successfully used in these parts, however. A high wet bulb temperature does not necessarily mean SER technology won't be economically feasible. The weather data shown in Table 1 will help to illustrate this. The important quantity to consider when looking at a site for potential use of SER is the ambient dry bulb/wet bulb temperature differential. Las Vegas, Nevada for example, is a very attractive site because of the low wet bulb and high T_d, while Tokyo, Japan is a poor choice because of the high wet bulb temperature and low T_d. It would be a mistake to eliminate all sites with high wet bulb temperatures, however. Let's compare Riyadh, Saudia Arabia with London, England. While the wet bulb temperature in Riyadh is 11°F higher than London, it actually is a much better candidate for SER due to the high T_d and high cooling demand.

LOCATION	SUMMER DESIGN		AVERAGE WINTER	
	2.5% °Fdb/°Fwb	Td °Fdb-°Fwb	Fdb/°Fwb	Td °Fdb-°Fwb
PHOENIX, AZ	107/71	36	59/53	6
LOS ANGELES, CA	89/70	19	60/57	3
SAN FRANCISCO, CA	77/63	14	53/48	5
DENVER, CO	91/59	32	38/36	2
WASHINGTON, D.C.	91/74	17	46/44	2
ATLANTA, GA	92/74	18	52/49	3
HONOLULU, HI	86/73	13	74/68	6
NEW ORLEANS, LA	92/78	14	61/60	1
BOSTON, MA	88/71	17	40/36	4
LINCOLN, NB	95/74	21	39/35	4
LAS VEGAS, NV	106/65	41	54/46	8
NEW YORK, NY	89/73	16	43/40	3
PORTLAND, OR	85/67	18	46/41	5
EL PASO, TX	98/64	34	53/41	12
DALLAS, TX	100/75	25	55/47	8
TORONTO, CANADA	87/72	15	38/34	4
SYDNEY, AUSTRALIA	84/73	11	58/52	6
PARIS, FRANCE	86/68	18	52/46	6
JERUSALEM, ISRAEL	94/69	25	61/49	12
MEXICO CITY, MEXICO	81/60	21	59/45	14
TOKYO, JAPAN	89/80	9	53/46	7
RIYADH, SAUDIA ARABIA	106/77	29	59/51	8
LONDON, UNITED KINGDOM	79/66	13	40/32	8

TABLE 1: Design weather data for selected cities

The type of facility under consideration plays an important part in determining the level of applicability of SER technology. Specifically, what is the facility used for, and how much outside air/make-up air is required? While it would not be feasible to air condition a typical office building in New Orleans with SER, it would be feasible to use this system to reduce the sensible outside air load. Any facility with high outside air/make-up air requirements, such as laboratories, would be a good candidate for SER.

LOCATION	SUMMER			WINTER		
	T_d	BTUH	EER	T_d	BTUH	EER
PHOENIX, AZ	81	140,000	70.2	55	21,600	10.8
LOS ANGELES, CA	76	70,200	35.1	58	10,800	5.4
SAN FRANCISCO, CA	67	54,000	27.0	50	16,200	8.1
DENVER, CO	69	118,800	59.4	37	5,400	2.7
WASHINGTON, D.C.	79	64,800	32.4	45	5,400	2.7
ATLANTA, GA	79	70,200	35.1	50	10,800	5.4
HONOLULU, HI	77	48,600	24.3	70	21,600	10.8
NEW ORLEANS, LA	82	54,000	27.0	61	--	--
BOSTON, MA	76	64,800	32.4	37	16,200	8.1
LINCOLN, NB	80	81,000	40.5	36	16,200	8.1
LAS VEGAS, NV	77	156,600	78.3	48	32,400	16.2
NEW YORK, NY	78	59,400	29.7	41	10,800	5.4
PORTLAND, OR	72	70,200	35.1	43	16,200	8.1
EL PASO, TX	74	129,600	64.8	45	43,200	21.6
DALLAS, TX	83	91,800	45.9	49	32,400	16.2
TORONTO, CANADA	77	54,000	27.0	35	16,200	8.1
SYDNEY, AUSTRALIA	76	43,200	21.6	54	21,600	10.8
PARIS, FRANCE	73	70,200	35.1	48	32,400	16.2
JERUSALEM, ISRAEL	77	91,800	45.9	53	43,200	21.6
MEXICO CITY, MEXICO	66	81,000	40.5	49	54,000	27.0
TOKYO, JAPAN	83	32,400	16.2	48	27,000	13.5
RIYADH, SAUDIA ARABIA	86	108,000	54.0	53	32,400	16.2
LONDON, UNITED KINGDOM	70	48,600	24.3	34	32,400	16.2

TABLE 2: Pre-Cooler performance for selected cities (nominal 5,000 CFM unit, 2.0 KW input).

Another class of facilities that would be candidates for SER are those which cannot be cooled with ventilation fans or direct evaporation units, and to cool with conventional refrigeration would be prohibitively expensive. This would include the following:

1. Machine shops and manufacturing areas.
2. Textile plants and clothing manufacturing.
3. Indoor athletic facilities, such as handball courts and basketball courts (where good air change is important).
4. Electronics assembly plants (soldering areas, light painting, fiberglass and epoxy, etc.).
5. Automotive repair facilities.
6. Large kitchens (hospital and military bases, etc.).

LOCATION	SUMMER			WINTER		
	T_d	BTUH	EER	T_d	BTUH	EER
PHOENIX, AZ	78	156,600	71.2	54	27,000	12.3
LOS ANGELES, CA	74	81,000	36.8	57	16,200	7.4
SAN FRANCISCO, CA	66	59,400	27.0	49	21,600	9.8
DENVER, CO	65	140,400	63.8	36	10,800	4.9
WASHINGTON, D.C.	77	75,600	34.4	44	10,800	4.9
ATLANTA, GA	78	75,600	34.4	50	10,800	4.9
HONOLULU, HI	76	54,000	24.5	69	27,000	12.3
NEW ORLEANS, LA	81	59,400	27.0	61	--	--
BOSTON, MA	74	75,600	34.4	37	16,200	7.4
LINCOLN, NB	78	91,800	41.5	36	16,200	7.4
LAS VEGAS, NV	73	178,200	81.0	48	32,400	14.8
NEW YORK, NY	76	70,200	31.9	41	10,800	4.9
PORTLAND, OR	71	75,600	34.4	42	21,600	9.8
EL PASO, TX	71	145,800	66.3	43	54,000	24.5
DALLAS, TX	80	108,000	49.1	49	32,400	14.8
TORONTO, CANADA	75	64,800	29.5	35	16,200	7.4
SYDNEY, AUSTRALIA	75	48,600	22.1	53	27,000	12.3
PARIS, FRANCE	72	75,600	34.4	47	27,000	12.3
JERUSALEM, ISRAEL	74	108,000	49.1	51	54,000	24.5
MEXICO CITY, MEXICO	64	91,800	41.7	48	59,400	27.0
TOKYO, JAPAN	82	37,800	17.2	47	32,400	14.8
RIYADH, SAUDIA ARABIA	83	124,200	56.5	53	32,400	14.8
LONDON, UNITED KINGDOM	69	54,000	24.5	34	32,400	14.8

TABLE 3: Positive Energy Reclaim Cycle (PERC) performance for selected cities (nominal 5,000 CFM unit, 2.2 KW input).

SER systems fall into three general categories:

1. Pre-Cooler. In this system, ambient outside air is used on both the wet-side and dryside. The heat exchanger efficiency for one manufactuer (Everest Pacific, Inc.) is quoted at 70%.

2. Positive Energy Reclaim Cycle (PERC). In this mode, ambient outside air is drawn across the dryside, and building return/exhaust air is drawn across the wetside. The manufacturer quotes an 80% efficiency level for this cycle.

3. Tandem. This is simply connecting a pre-cooler in series with a PERC unit. The manufacturer quotes a 98% efficiency level in this mode, or the discharge temperature, ^{o}Fdb nearly equals the ambient wet bulb temperature.

Tables 2, 3 and 4 show the discharge temperature (T_d), cooling effect (BTUH), and Energy Efficiency Ratio (EER) for the 3 categories of SER. This data is based on a nominal 5,000 CFM machine as applied to a task that requires 5,000 CFM of outside air. Naturally, if one was to compare a conventional system and a SER system as applied to a typical project where a combination of return air and outside air was utilized, then the EER for the SER system would drop considerably, but it would still be far higher than the EER for the conventional system.

Giampaolo and Associates, Inc. has completed a number of projects using SER in recent years; laboratories, department stores, gymnasiums, auditoriums, offices, and athletic facilities.

Location	Summer			Winter		
	T_d	BTUH	EER	T_d	BTUH	EER
PHOENIX, AZ	71	194,400	55.5	53	32,400	9.3
LOS ANGELES, CA	70	102,600	29.3	57	16,200	4.6
SAN FRANCISCO, CA	63	75,600	21.6	48	27,000	7.7
DENVER, CO	59	172,800	49.4	36	10,800	3.1
WASHINGTON, D.C.	74	91,800	26.2	44	10,800	3.1
ATLANTA, GA	74	97,200	27.8	49	16,200	4.6
HONOLULU, HI	73	70,200	20.1	68	32,400	9.3
NEW ORLEANS, LA	78	75,600	21.6	60	5,400	1.5
BOSTON, MA	71	91,800	26.2	36	21,600	6.2
LINCOLN, NB	74	113,400	32.4	35	21,600	6.2
LAS VEGAS, NV	65	221.400	63.3	46	43,200	12.3
NEW YORK, NY	73	86,400	24.7	40	16,200	4.6
PORTLAND, OR	67	97,200	27.8	41	27,000	7.7
EL PASO, TX	64	183,600	52.5	41	64,800	18.5
DALLAS, TX	75	135,000	38.6	47	43,200	12.3
TORONTO, CANADA	72	81,000	23.1	34	21,600	6.2
SYDNEY, AUSTRALIA	73	59,400	17.0	52	32,400	9.3
PARIS FRANCE	68	97,200	27.8	46	32,400	9.3
JERUSALEM, ISRAEL	69	135,000	38.6	49	64,800	18.5
MEXICO CITY, MEXICO	60	113,400	32.4	45	75,600	21.6
TOKYO, JAPAN	80	48,600	13.9	46	37,800	10.8
RIYADH, SAUDIA ARABIA	77	156,600	44.7	51	43,200	12.3
LONDON, UNITED KINGDOM	66	70,200	20.1	32	43,200	12.3

TABLE 4: Tandem Performance for selected cities (nominal 5,000 CFM unit, 3.5 KW Input).

One of our first projects to be completed was for James W. Montgomery Engineers Corp. in Pasadena, California. This was a remodel project consisting of expanding an existing first floor laboratory into adjacent areas. When we did our field survey, we discovered that they had experienced some air conditiong problems with the building hydronic heat pump system in that:

1. Make-up to the lab hoods was unconditioned outside air, which was increasing the room AC load beyond the capacity of the heat pumps.
2. The building ventilation air was provided by a roof mounted fan supplying unconditioned outside to the supply plenums where individual heat pumps were located

During peak summer condition, personnel reported that the general building averaged 80° F, while the laboratory would average over 90°F.

Our design incorporated a SER pre-cooler on the building ventilation air system, and PERC units to provide lab hood exhaust and make-up air. The personnel at Montgomery now report no difficulty in maintaining comfort conditions in all areas.

The second project to be completed was the gymnasium/auditorium for California Lutheran College in Thousand Oaks, California. This existing metal building is the only large auditorium in Thousand Oaks and is used for community events as well as school classes and activities. This structure has clerestory windows as its only means of ventilation. The campus wanted to install air conditioning so that this facility could be used in the summer, but the cost of installing conventional refrigeration and bringing in new electrical service for it was prohibitively expensive. Our solution was to install PERC units on the roof of the adjacent locker room to supply the gymnasium. This past summer was the first in the school's history that this facility could be comfortably used.

Our first new construction project is the recently completed 130,000 square foot department store for Bullocks, located in San Mateo, California on the San Francisco Bay. This building had a number of innovative features:

1. A translucent fiberglass fabric roof was used over approximately 70,000 sq. ft. of the building area
2. The store connected lighting load was less than half of conventional designs loads.
3. The air conditioning system was about 500 H.P. less connected load than the originally proposed conventional system.

Here we elected to design the HVAC system around tandem units, feeding a supply air plenum. Besides the energy savings with this system, we were able to design for 70°F space temperatures, rather than the 78° mandated by the California State Energy Commission. This equipment was located at grade level in protected enclosures for ease of maintenance, and to eliminate any visual distractions from the striking roof structure.

The Rockwell International Science Center in Thousand Oaks, California presented us with some interesting challenges. We were originally called in to this 100,000 sq. ft. high technology laboratory to perform an energy audit. We discovered that because of special needs by individual research groups at the facility, exhaust fans had been added to the point that there was a need for 50% make-up air throughout the building. Our solution was to incorporate pre-cooler and PERC units into the basic air conditioning system to reduce this outside air load. This project is in the final stages of construction and should be on line within 30 days.

Another project nearing completion is the second floor completion of Goudchaux's Cortana Department Store in Baton Rouge, Louisiana. Our initial involvement with this project was to do the second floor air conditioning and lighting design for the expansion. As we became more involved with this project, we asked for, and received authorization to do an energy audit. This audit showed that the need for additional chillers and cooling towers could be eliminated, and that the total connected lighting and HVAC load would be over 300 KW less after the expansion, provided that the following steps were taken:

1. Redesign first floor lighting to reduce lighting load.
2. Retrofit enthalpy economizers on existing air handlers and incorporate on new air handlers.
3. Incorporate SER pre-coolers sized for minimum required outside air with the economizers for all air handlers.
4. Reduce first floor air quantity and rebalance air outlets.
5. Reduce existing air handlers motors.
6. Perform complete preventive maintenance on chillers, towers and air handlers.

The pre-coolers work in this application to increase the effective annual hours of beneficial economizer operation. It is anticipated that during the two coldest months, the chillers will not need to operate where before they did.

We recently completed the design for the new Duarte Civic Center, a three building complex for the city of Duarte, California. This complex consists of:

1. City Administration Building; General office space.
2. Passive Recreation Building; This building has a large multi-purpose room that will be used for open council meetings, receptions, dances, and other similar gatherings. Other rooms will be used for hobby craft, study groups, ping pong and card parties.
3. Active Recreation Building; This building houses handball, racketball, shower and locker rooms and equipment check out.

We selected SER Tandem units for the office building and Passive Recreation Building, with space design temperatures of 70°F-75°F. We selected PERC units for the active building to meet design temperature conditions of 74°F - 78°F. We feel that this high ventilation rate of "tempered" air will provide more comfortable conditions in the active areas and locker rooms than less air changes of cooler air. This facility is scheduled to open in the spring of 1982.

As we gain more experience with SER systems, we become more convinced that this technology is truly an idea whose time has come.

Chapter 47

PRECOOLERS FOR GAS TURBINES — ENERGY CONSERVATION WITH A POWER BOOST

M. B. Dubey, L. A. Schlom

INTRODUCTION

Thousands of stationary gas turbines installed throughout the world operate at depressed power levels and with higher fuel consumption when the ambient air temperature rises above the standard design value (59°F or 15°C). If the hot inlet air is precooled, it becomes denser, performance recovers, and efficiency rises. This characteristic in the performance of gas turbines suggests an excellent opportunity for energy conservation.

Precoolers have often been used for power boost in hot climates. However with fuel costs escalating to as much as $40/barrel, attention is focusing on precooling to improve thermal efficiency rather than to boost hot weather power output. With retrofit, precooling systems can provide immediate fuel savings for most powerplant installations. If integrated into a new design, the principle can be applied in several other ways to increase energy conservation, such as cooling the compressor interstage heat exchangers or the generator windings. In many instances, the savings resulting from fuel economy can recover the installation costs in less than two years.

THE ADVANTAGE OF PRECOOLING

An inherent characteristic of thermodynamic machines is their sensitivity to the inlet temperature of the working fluid. For gas turbines operating with the Brayton cycle, the temperature of the ambient air entering the compressor is a key factor in determining performance and efficiency. Figure 1 shows a simplified temperature-entropy diagram of a typical cycle. Air, entering at condition (1) is compressed to a higher pressure (2). Fuel is injected (3) and burned to heat the air; then, the hot compressed gas is expanded through the turbine (4) to (5). The power extracted by the turbine is greater than that absorbed by the compressor; the excess is converted to electric power by the generator. If the inlet air at (1) is cooled (1'), the work of compression is reduced because the two constant pressure lines (P_1) and (P_2) converge at lower temperature (and entropy) conditions. Thus, for a constant load condition, the turbine performs less work, (4') to (5'). Most of the time, however, the power station will be run at design turbine inlet temperature (4) to utilize the increase in available power and to operate closer to the optimum design condition where thermal efficiency is highest. In either case, thermal efficiency is improved compared to uncooled operation, and fuel savings can be realized.

To illustrate this characteristic, the performance curves for a typical gas turbine power plant are shown in Figure 2. The unit is the FT4 Power Pac manufactured by United Technologies Corporation. The data is from their product brochure[1]. Notice that the heat rate decreases with decreasing inlet air temperature. (The heat rate is inversely proportional to thermodynamic efficiency. A 100-percent-efficient system would show 3414 BTU/kilowatt hour.) Also notice that power diminishes with rising temperature. This is because, in most gas turbines, the limits on turbine inlet temperature and rotational speed prevent operation at conditions where additional power must be extracted to drive the compressor under the more adverse condition of higher inlet temperature.

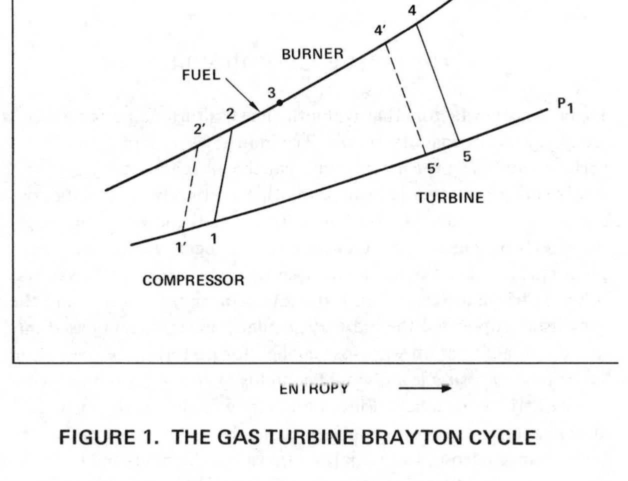

FIGURE 1. THE GAS TURBINE BRAYTON CYCLE

The above characteristic is common with all gas turbines. Figure 3 shows the variation of heat rate with inlet air temperature for two other power units as well as the FT4. Data for the STAL Laval GT 35 and Solar MARS turbines were obtained from Gas Turbine World.[2, 3] The data is plotted for power levels corresponding to base load conditions at each temperature. Notice that the slopes of the curves beyond 70°F are almost identical, averaging 32.3 BTU/KWH °F. Although the three turbines represent a very small sampling, the trend is

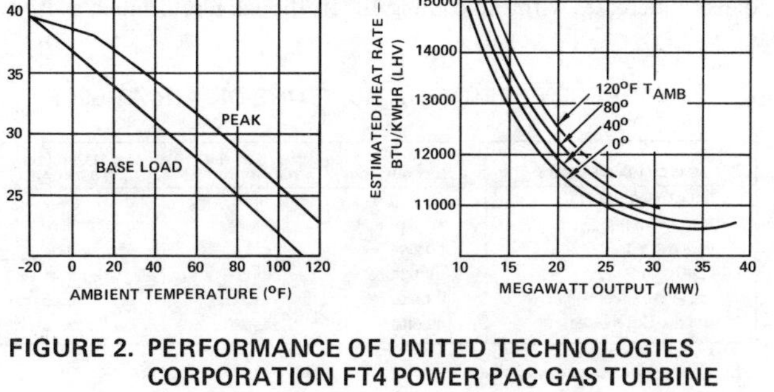

FIGURE 2. PERFORMANCE OF UNITED TECHNOLOGIES CORPORATION FT4 POWER PAC GAS TURBINE

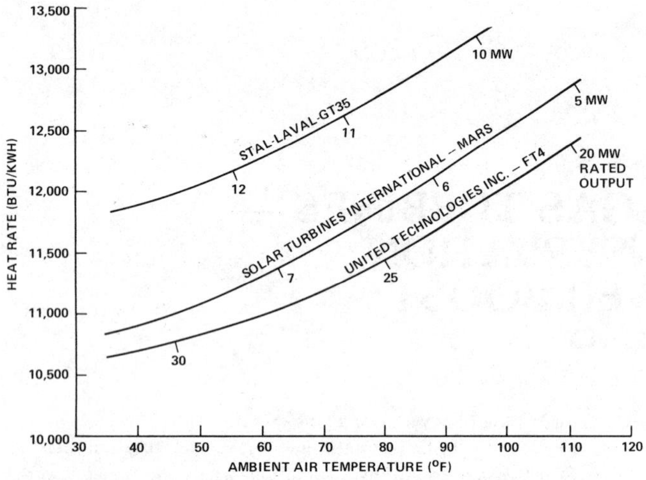

FIGURE 3. EFFECT OF INLET AIR TEMPERATURE ON GAS TURBINE EFFICIENCY AT BASE RATED POWER

obvious, and the average heat rate per degree of temperature can be used for making a rough estimate of feasibility for the precooling concept.

THE INFLUENCE OF HUMIDITY

There are other factors that will influence gas turbine performance, as predicted by the manufacturers. The industry standard is to compute performance using the assumption that the air is absolutely dry, and all gas constants, specific heats, and other thermodynamic data are based on the dry air premise. In 1950, Samuels and Gale[4] investigated the effects of moisture content on turbojet engine performance. Their study and test data showed that the presence of water vapor resulted in trivial influences on thrust, fuel consumption, etc., and their conclusion supported the industry standard; namely, data based on the dry air assumption was close enough for performance prediction. To include moisture levels would complicate computations and contribute little to accuracy. Since evaporative cooling is the most economical process for precooling inlet air, the influence of moisture on performance becomes an important question. Samuels and Gale show that the gas constants for air change slightly when moisture content increases, as shown in Table 1. These factors must be considered when conducting thermodynamic performance computations. In Figure 4, the specific volume is shown to increase with moisture addition for any given dry bulb temperature (hence, the air density decreases). Since the compression cycle is basically equivalent to a constant volume process, moisture addition without reducing air temperature causes a reduction in density, therefore mass flow rate, and the overall power output drops. However, evaporative precooling reduces the air temperature at the same time moisture is added, due to the water's latent heat of evaporation. The cooling process rides up the wet bulb temperature line towards saturation (100% relative humidity) and air density increases with the cooling effect, though not as much as it

TABLE 1. EFFECTS OF MOISTURE ON AIR AT 90°F

GAS CHARACTERISTICS	DRY AIR	0.01 LB WATER /LB DRY AIR	0.02 LB WATER /LB DRY AIR
RELATIVE HUMIDITY	0	33%	66%
R, FT-LB/LB°F	53.3420	53.6460	53.9500
C_p, BTU/LB°F	0.2399	0.2418	0.2438
C_v, BTU/LB°F	0.1713	0.1729	0.1744
γ, RATIO OF SPECIFIC HEAT	1.4000	1.3990	1.3980
MOLECULAR WEIGHT	28.9700	28.8320	28.7550

REF: SAMUELS & GALE[1], KEENAN & KAYE[5]

would if cooling were accomplished strictly by removing sensible heat (no changes in moisture content). If the criteria is to strive for optimum precooling, then the best precoolers would be sensible; i.e., temperature reduction by removing heat with a heat exchanger; the next best would be an evaporative precooler where the cooling is latent; i.e., water is evaporated directly into the air stream. In either case, higher air density results in a power boost. The improved thermodynamic efficiency of the cycle can be explained partially by the reduction in inlet air temperature causing an increase in Carnot cycle efficiency $(T_4 - T_1)/T_4$ (see Figure 1), where the temperature is in the absolute scale (Rankine).

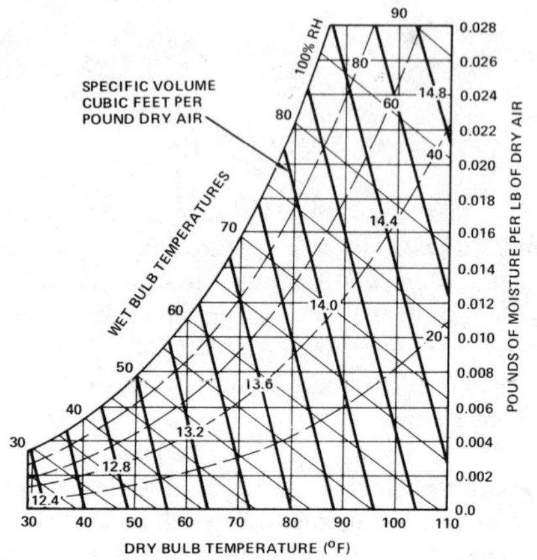

FIGURE 4. EFFECT OF HUMIDITY ON SPECIFIC VOLUME OF AIR

PRECOOLER OPTIONS

The most common choice for precooling is the evaporation process where the air is passed through wet filters before entering the compressor. The air is cooled between 60% and 75% of the difference between dry bulb and wet bulb temperatures depending on ambient conditions and the cooler design. One difficulty is that water droplets and hard water mineral particles sometimes get entrained in the air stream. These erode the compressor blades or deposit on the blade leading edges, thus changing the blade contours and causing compressor performance to deteriorate. Two other methods are proposed here. Both reduce the air temperature to values lower than ambient wet bulb temperature, and they contain the unique feature where the water used for evaporation is obtained by distillation using the hot turbine exhaust gases to operate a waste heat boiler, and in this way, the problem of mineral particles is avoided.

Depressed Wet Bulb Evaporative Cooler

In this first system, the air is first precooled by removing sensible heat, using an indirect evaporative process. The cool dry air with its depressed wet bulb temperature is then passed through a wet spray where it is cooled further by the direct evaporation of water. The schematic diagram is shown in Figure 5. The process is further illustrated on the psychrometric chart in Figure 6. In the example, the entering air (A) is at 90°F dry bulb and 70°F wet bulb. The indirect evaporative cooler consists of a heat exchanger with vertical thin wall plastic tubes. The insides of the tubes are coated by a thin film of water. The tube walls are treated with a hydrophilic substance to assure a highly wetted surface. Air passes up through the tubes, evaporates the water film and cools the tube walls down to the wet bulb temperature. Dry air passing across the outside of the cold tubes has its heat extracted, (A) to (B). It then passes through the wash and is cooled to (D), which is less than the wet bulb temperature of the inlet air (A). The temperatures and specific volumes of the air at the different stages of the process are shown in the table in Figure 6. Note that the evaporative

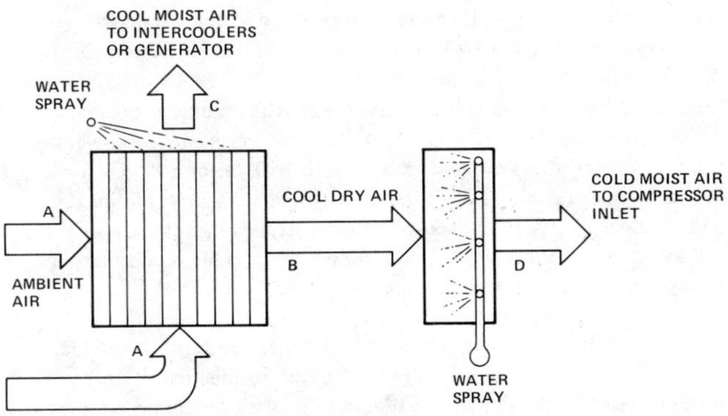

FIGURE 5. DEPRESSED WET BULB EVAPORATIVE PRECOOLER

cooling process on the wet side of the heat exchanger does not follow a true latent heat transfer. This is because the enthalpy given up by the air on the dry side, (A) to (B), is transferred to the air on the wet side, (A) to (C), heating it despite the cooling effect of evaporation. This process decreases the air temperature from 90°F to 68°F whereas the commonly used evaporation process would drop the air temperature down to about 75°F.

		T_{DB}	T_{WB}	V_S
A	AMBIENT AIR	90.0	70.0	14.10
B	DRY AIR EXIT	78.0	66.3	13.78
C	WET AIR EXIT	75.0	73.4	13.85
D	COMPRESSOR INLET	68.0	66.3	13.58

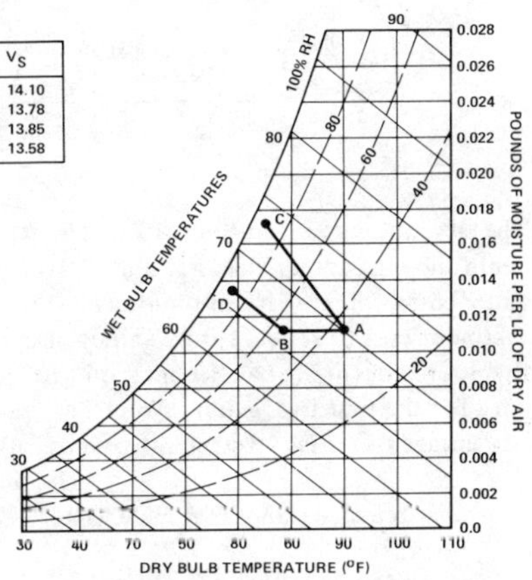

FIGURE 6. PSYCHROMETRIC PROCESS OF DEPRESSED WET BULB EVAPORATIVE PRECOOLER

Tandem Precooling System

The second process can depress the inlet air temperature even further as shown in Figures 7 and 8. Here, an indirect evaporative cooler is placed in front of the prior system, thereby reducing the dry and wet bulb temperatures before entering the depressed wet bulb precooler stage. The air leaving the final evaporation stage (F) is cooled to 65.5°F. This is 4.5° below the wet bulb temperature of the inlet air (A). If mixed with the air leaving the wet side of the second heat exchanger (E), the temperature would be 67.7°F entering the gas turbine compressor (G). A more detailed analysis is necessary to determine if this second concept is more cost effective than the simpler one. It would have to take into account the individual characteristics of the gas turbine being equipped and the climatology at the installation site.

The installation is shown schematically in Figure 9. Notice that the waste heat in the exhaust gases is used to operate a simple water boiler. The generated steam is condensed using the cool moist air from the wet side of the indirect evaporative cooler to cool the condenser. This cold air supply is available only when an indirect evaporative cooling process is used. The distilled water is then used by the precooling system, thus avoiding the problem of mineral particles entrained in the air. Because distilled water is available for the final evaporation stage, the air can be thoroughly saturated with a fine mist to obtain the highest temperature depression, and, at the same time, it is washed clean of any entrained dust. Contrast this feature with the conventional kinds of evaporative precoolers where thorough saturation with a fine mist spray is avoided because it would add mineral particles. As a rule, such precoolers provide only 60% to 75% of the total possible temperature depression.

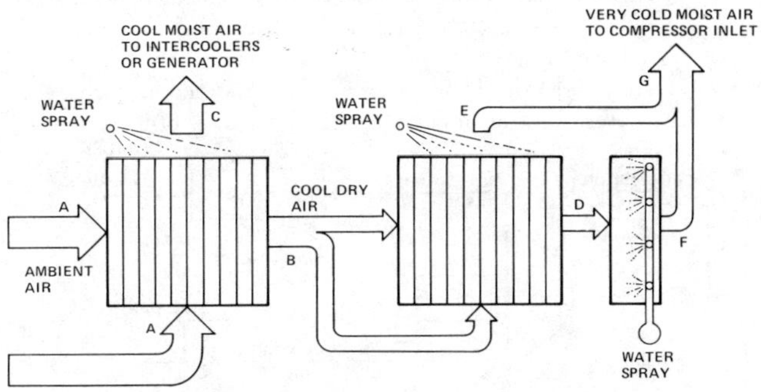

FIGURE 7. TANDEM PRECOOLING SYSTEM

		T_{DB}	T_{WB}	V_S
A	AMBIENT AIR	90.0	70.0	14.10
B	1ST DRY EXIT	78.0	66.3	13.78
C	1ST WET EXIT	75.0	73.4	13.85
D	2ND DRY EXIT	71.0	64.0	13.60
E	2ND WET EXIT	70.0	68.6	36.66
F	AIR WASH EXIT	65.5	64.0	13.50
G	COMPRESSOR INLET	67.7	66.3	13.58

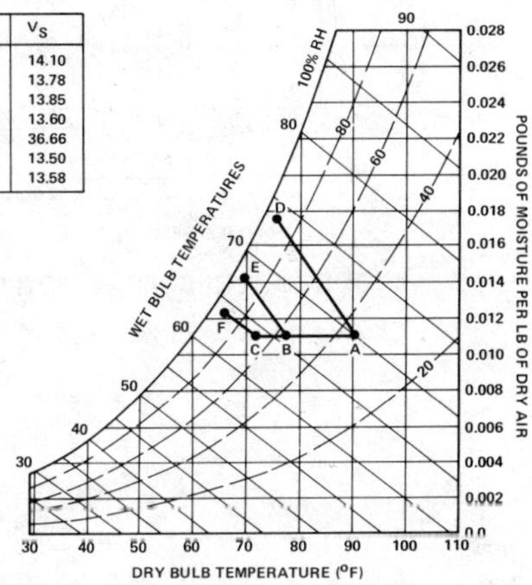

FIGURE 8. PSYCHROMETRIC PROCESS OF TANDEM PRECOOLING SYSTEM

The air exiting from the wet side of the heat exchanger can also be used to cool the generator windings on hot days to improve mechanical power conversion efficiency. There is plenty of such air available. The waste heat boiler system requires only a small amount of the air for cooling the condenser, and it distills far in excess of the water needed for evaporative cooling.

Note that a small amount of power is required to drive the fans that move the wet side air; they use about 2 to 3 percent of the energy saved. The air moving on the dry side is drawn through by the gas turbine. The inlet pressure loss is less than is normally experienced with the air filters and silencers now used in gas turbine installations.

CLIMATOLOGY CONSIDERATIONS

Precoolers depending on the evaporation of water work best, obviously, in low humidity climates. There is no dirth of gas turbines installed in such regions. Indeed, if all the gas turbines located in the drier areas of the United States were to be equipped with precoolers, their contribution to energy conservation would be a significant value. To explore the feasibility of the concept, Las Vegas, Nevada was chosen as a typical site. Data on climate conditions gathered for years throughout the country is presented in the Weather Data Handbook published by McGraw-Hill.[6] In particular, the wet and dry bulb temperatures for Las Vegas are presented in Figure 10. The two curves show the number of hours per year and the $\Delta T^{o}F$ between wet and dry bulb thermometers experienced for each dry bulb temperature. The data is statistically reduced from at least 10 years of recording. They are inclusive for the daytime hours between 9:00 A.M. and 4:00 P.M., when intermediate and peaking power plants employing gas turbines would be furnishing power to match the diurnal demand generated by air conditioners and industrial activity. The mean temperature is 75.5°F and the mean ΔT is 22.3°F. Thus, in the Las Vegas area, depending on the type of precooler used, one could expect to precool the inlet air from 15°F to 25°F, about 2500 hours during the year, on the average. Similar analyses can be made for other locations.

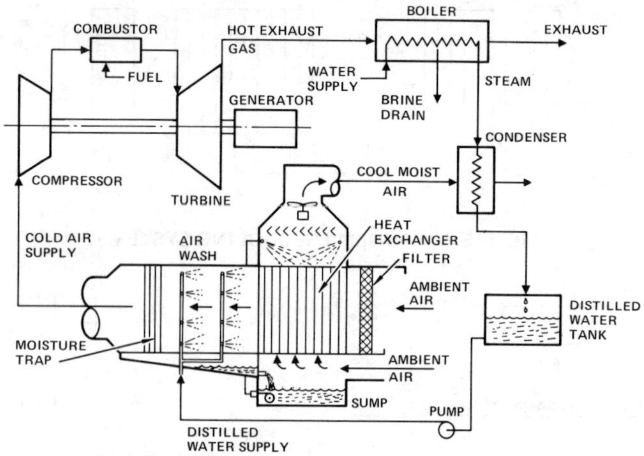

FIGURE 9. PRECOOLER FOR GAS TURBINES

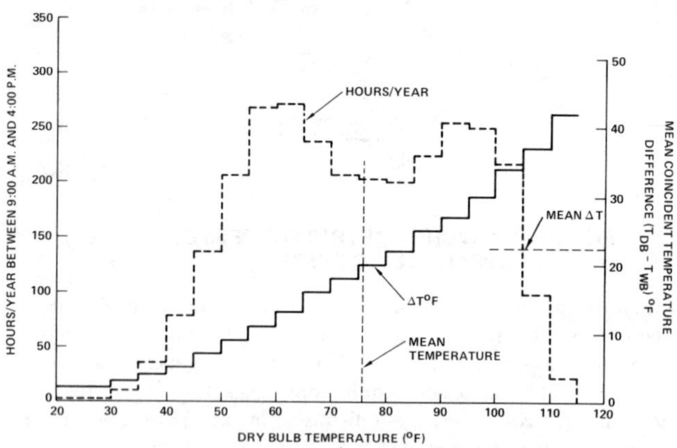

FIGURE 10. DISTRIBUTION OF MEAN COINCIDENT TEMPERATURE DIFFERENCES AT LAS VEGAS, NV, 9:00 A.M. TO 4:00 P.M.

For example, the average ΔT in Phoenix, Arizona is 19.9°F. In Los Angeles it is 8.0°F, but one must note that the data was recorded at the international airport which is adjacent to the ocean. At the city's civic center or in the nearby San Fernando Valley, the average difference between wet and dry bulb temperatures is much higher. To design an installation, the data at the actual site should be available to predict performance.

ECONOMIC FEASIBILITY

The improvement in operating efficiency indicated in Figures 2 and 3 when inlet air temperatures are reduced can be transformed into equivalent cost savings if a price is assumed for fuel oil. Figure 11 was drawn with the premise that the average slope of the curves in Figure 3 is representative of all gas turbines (although the sampling of data is small); that is, the heat rate decreases 32.3 BTU/KWH °F as inlet air temperature is reduced. Assuming the turbine is operated 2500 hours/year, the savings in fuel costs per megawatt of power are shown for fuel prices ranging from \$20 to \$50 per barrel. For Las Vegas, with a ΔT of 22.3°F, the annual fuel savings would be about \$11,800 per megawatt if fuel were priced at \$40/barrel. A 100 MW power station would save \$1,180,000 per year. This is certainly an attractive incentive for precooling.

Installation costs were estimated at \$1000 per thousand cubic feet per minute of airflow, a price closely related to the cost of air conditioners based on the indirect evaporative cooling process, sometimes called sensible evaporative refrigeration (SER). The airflow requirements for 23 gas turbines selected randomly from the specification tables published by Gas Turbine World[7,8] were averaged using regression analysis. The resulting value, 9.5 pounds per second of dry air per megawatt or rated power was used to size the precoolers. The data's correlation coefficient was 0.988, showing remarkable consistency regardless of manufacturer. The average air flow is equivalent to about 7500 cfm per megawatt, or \$7500 per megawatt for a precooler installation. The return on investment is shown in Figure 12. The computation is based on the savings in fuel cost being considered as revenue.

$$\text{ROI} = \frac{\left[\begin{pmatrix}\text{Annual}\\ \text{Fuel Cost}\\ \text{Savings}\end{pmatrix} - \begin{pmatrix}\text{Annual}\\ \text{Depreciation}\end{pmatrix} - \begin{pmatrix}\text{Annual}\\ \text{Operating}\\ \text{Cost}\end{pmatrix}\right] \times (1 - \text{Tax Rate})}{(\text{Investment Cost})}$$

The tax rate is assumed to be 50%. The ROI is averaged over five years allowing for an inflation rate of 10% per year (in the numerator) and a cost of money based on interest rates at 20% (applied on the investment cost to reflect future value). Depreciation is computed using the double declining balance method (20% of the undepreciated balance for the first five years). Annual operating costs are estimated as a percentage of initial investment costs as follows:

Maintenance (labor and materials)	5.0%
Insurance	0.4%
Overhead and miscellaneous costs	2.0%

Figure 12 shows that for a particular inlet air temperature drop characteristic for an installation site, the return on investment will become

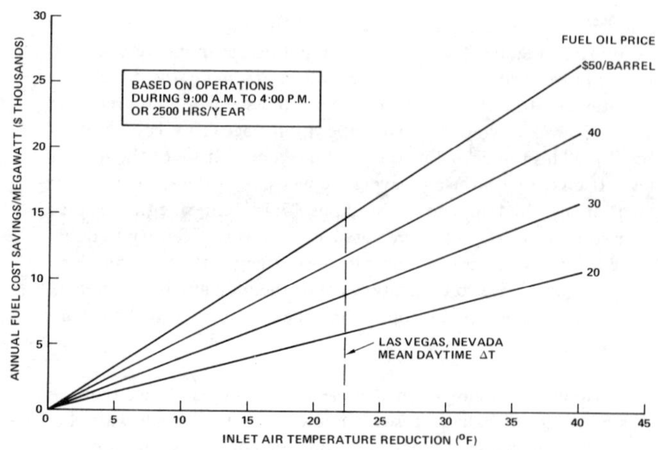

FIGURE 11. POTENTIAL ANNUAL FUEL SAVINGS WITH PRECOOLED INLET AIR

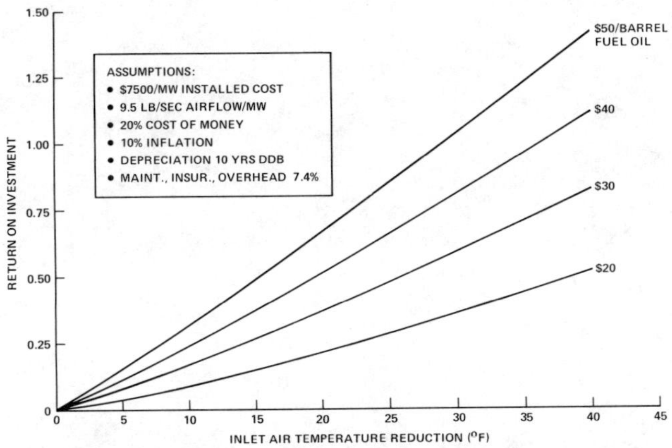

FIGURE 12. RETURN ON INVESTMENT WHEN INSTALLING INLET AIR PRECOOLER

worthwhile when the fuel cost reaches a high enough level to make the investment profitable. An ROI of 0.33 implies a payback within three years. This could be a reasonable threshold for an energy conservation decision. For Las Vegas, the investment would seem well worthwhile, because the investment payback, when the fuel cost is $40/ barrel, would be within 20 months.

CONCLUSIONS

Because of the rapid rise in fuel costs and their direct influence on the cost of electricity, techniques for conserving fuel are receiving serious scrutiny by the power industry. Gas turbine precooling offers an excellent opportunity for such conservation. Cooler air at the compressor inlet results in boosted power and higher thermal efficiency. The fuel cost savings can quickly offset the cost of the precooler installation, thereby making the process economically attractive as well. Several synergistic features of the process as described here make the concept even more desirable:

> The final evaporative cooling stage uses mineral free water to reduce erosion and deposits of minerals in the gas turbine. Also, it thoroughly washes the air free of grime. This reduces the costs of gas turbine maintenance associated with such conditions.
>
> Cool air leaving the wet side of the indirect evaporative heat exchanger can be used to cool compressor interstage gases or the generator windings. These innovations also contribute to higher overall efficiency and improve fuel economy.

The first demonstration of the concept should be made in a hot, dry region where the cost of fuel is high, so that the fuel cost savings would be able to offset the higher cost of the development prototype.

BIBLIOGRAPHY

1. Performance of FT4 Power Pac, Packaged Electric Power, United Technologies Corporation, Power Systems Division, Farmington, Connecticut.

2. STAL-Laval GT35 Performance, Gas Turbine World, Vol. 8, No. 3, May 1978, page 25.

3. Solar Turbines International MARS Performance, Gas Turbine World, Vol. 8, No. 5, September 1978, page 23.

4. Effect of Humidity on Performance of Turbojet Engines, J. C. Samuels and B. M. Gale, NACA Technical Note 2119, June 1950.

5. Gas Tables, Joseph H. Keenan and Joseph Kaye, John Wiley & Sons, New York, 1956.

6. Weather Data Handbook for HVAC and Cooling Equipment Design, Ecodyne Corporation, McGraw-Hill Book Company, New York, 1980.

7. Performance Specs 1981, Gas Turbine World, Vol. 10, No. 6, December 1980.

8. Gas Turbine World Handbook 1980-81, Vol. 5, Gas Turbine World, Pequot Publishing Inc., Fairfield, Connecticut.

SECTION 12
ENERGY MANAGEMENT PRACTICES

Chapter 48
ENERGY CONSERVATION AND MANAGEMENT AT DEERE & COMPANY

D. F. Darby

Introduction

Deere & Company appreciates the opportunity to discuss energy management with you. Energy is a topic that will occupy our thoughts for many years to come, and certainly will be in the forefront in the near future. It is a subject that has become increasingly complex. Fuel availability is a fluctuating phenomenon. Technology to enable energy conservation, and to supply efficient alternative non-depleting sources is complex. Government intervention is becoming increasingly pervasive in energy management. National and certainly industrial responsibilities in wise energy management is a commitment. But it is difficult to determine how one finds a reasonable path to energy security.

Deere & Company faced the issue -- what can we contribute to energy conservation? As one reviews the initiation of an energy management program, one finds that he is faced with a problem that encompasses a large and varied array of disassociated data.

It soon became apparent that an energy breakdown and analysis of major manufacturing processes, and environmental uses of energy was necessary.

Continued study indicated increased refinements were necessary, and that energy consumption would have to be analyzed by fuel type.

With the accumulated data, a reference point is established. This reference point is based upon interpretation of fuel mix requirements, analysis of fuel use for process, and for environmental, and cognizance of fuel costs and availability.

Additional work is then required to organize an energy management program. Direction must be given to channel resources into productive areas of energy management and conservation.

To recap, energy data is required to establish a mile marker, to define the existing situation. Definition must be in sufficient detail to enable an effective first effort. So the message is to start somewhere in an energy conservation program. As a result, the process will gel, and goals will come into focus.

Key Elements of Program

Deere & Company progressed painfully through this process in establishing an energy management and conservation program. I would like to share with you our experience in organizing such a program.

In searching back and examining the various aspects that first originated with, or evolved from our energy program, it is possible to identify several key elements that we believe have contributed the most to the success of our program. These key elements are also identifiable in the majority of successful programs in other industries that we have examined. Therefore, we shall focus on them. These elements are, we believe:

Top Management Support

Energy Audit

Full-Time Energy Coordinators

Direct Responsibility and Control at Plant Level

Realistic and Objective Yardstick

Developing an Energy Ethic

Feedback

I'd like to first comment on these elements so you understand what we mean by them, and why they are important to the success of an energy program.

Top Management Support

Each of you is aware of the importance of having the backing of your superiors in the accomplishment of any particular goal you may seek to achieve. A conservation program and long-range energy program is impossible without top management support. I have discussed "program" - quote unquote - with several firms and have found that the firms with top management support always succeed if they have some form of the other elements listed. The rest fail because manpower for the program does not exist.

We have been fortunate at Deere & Company to have had the active support of many key executives of our corporation. This support has further been reinforced at the top management levels in our factories. Without this support our program could not have achieved the results that it has, because a fundamentally sound and ultimately successful program begins with this single most important element, "Top Management Support".

Energy Audit

A well organized and ultimately successful program begins with a comprehensive energy audit. An energy audit is a planned effort aimed at identifying, categorizing, and quantifying all the various energy forms and energy uses that exist in a factory, or for that matter any business, commercial or public building or your home.

We conducted such an audit at each one of our factories at the outset of our program. We ended up with an analysis of yearly energy consumption by fuel type of each factory, and this then also permitted a composite profile for the whole corporation.

We were able to determine the magnitude of our energy use, the energy mix and conversion losses. This is important since areas possessing the greatest savings potential were identified. Each of Deere's units were individually surveyed to determine the corporate compilation.

Upon breaking this usage down into function we determined that on a yearly basis, environmental needs comprised one-third of our usage and process two-thirds.

To give the energy audit added dimension it is useful to subdivide all energy into three areas -

Environmental energy

Process (value added) energy

Peripheral energy

Environmental energy is that energy required to maintain proper working conditions excluding that required because of process contamination.

Process (value added) energy is that energy required to perform the operation only such as energy to drill a hole, weld a part, or apply the paint.

Peripheral energy is that energy required to negate the impact of process contamination - meet OSHA, EPA and other regulations.

I recognize that this subdivision of energy leaves some "gray areas," but it serves a very useful purpose. The reason for this is that process energy so defined is usually rather inflexible and many times is much less than the peripheral energy requirement. Peripheral energy, on the other hand, is normally highly flexible partially because there are a number of ways to handle the problem and equally important, many times is the result of misconceived requirements. I recognize the fact that this in and of itself, is not really a new concept. But separating it out to show the magnitude, the recognition that it's highly flexible, and that it is not a given quantity will assist in the development of a good conservation program.

Full-Time Energy Coordinator

A full-time energy coordinator on location is essential to obtain maximum success in a conservation program. In a company with multiple plants, an energy coordinator should be established at each location. He must be full time so he knows that his only measure of value to the company is his effort on conservation. In every case where this is his part time job, his other assignments dominate.

He may be associated with production. In that case his job is to get the product out the door. There are always continued problems or so called "fires to put out", so he is constantly distracted from conservation. His time is so consumed with other activities that he generates few if any ideas, and those he does generate are not sold to management because he does not have adequate preparation time.

He must be an energy expert, management consultant, public relations agent, and labor relations man, all in one person.

We believe that anything less than a full-time energy coordinator falls short of full commitment, and maximum potential will not be reached.

The next key element is closely related.

Direct Responsibility and Control at Plant Level

This element is very closely related to the previous element we discussed. Each plant should be responsible for developing its own program. No one is more familiar with a plant's energy requirements, is more capable of refining existing energy audits, of establishing priorities, or of scheduling work than one of the plant's own engineering personnel.

At Deere & Company, the individual energy coordinators report to their unit's management and are responsible for administering the local effort. These local people are directly responsible for the success of their individual programs.

At our corporate office in Moline, an Energy Management Division coordinates the effort on a company-wide basis.

Realistic and Objective "Yardstick"

We feel that it is crucial to develop a realistic and objective "yardstick" to measure progress. A method that is doomed to failure is to calculate or even measure the energy reduction of say a heat recovery device, do the same for other energy reduction programs, add them up, and state this sum as the conservation effort. Firms have done this and then found to their dismay that this was not reflected in their energy consumption. The monthly energy consumption taken from their gas, oil, electric and coal bills divided by ton of product, standard hour, or some other yardstick indicated no conservation. The reason for this has many possibilities. -- Worker aborted modifications, poor maintenance, other neglected areas that have become more energy intensive.

John Deere uses a rather complex computer program to calculate energy efficiency. It takes into account tons of product, degree days and square feet. Nevertheless, the bottom line for us is BTU/ton and since the start of our program we have achieved a significant reduction in the energy required to produce a ton of product.

These kinds of yardstick are objective - tons out the door is a known figure - the utility bills are known amounts - the results are easily audited.

From the analysis of our energy audit data, a computer program was developed as the tool or "yardstick" by which our energy conservation progress is measured.

Essentially our program is a stepwise regression analysis which predicts expected energy usage had there been no energy management and conservation program.

The actual consumption is then compared to the predicted consumption to determine energy savings. This tool has turned out to be the backbone of our reporting system for evaluating each factory's energy savings techniques.

Two more key elements merit attention. These are "energy ethic" and "feedback".

Energy Ethic

The people that comprise a company can enhance a conservation program if they are convinced it is for their good as well as the company's. We strive to develop an energy "ethic".

Once a program has been established, this energy ethic becomes the key element to substantial improvements in energy efficiency. This energy ethic is a very complex element. The engineers (management's commitment) play a very large role and they must have the correct attitude.

At the heart of this attitude must be the belief that it is more cost effective to save a BTU than it is to supply a BTU. I recognize that any statement like this is not always true, but my experience is that it is true in the vast majority of cases.

In addition, the average employee is also contacted through posters, articles, films and meetings. Although energy costs comprise only a small portion of our overall manufacturing costs, we are convinced that a strong "energy ethic" is not only vital but "good business" as well.

One of our factories has had an animated movie produced which features a polar bear hero. He represents the central theme of an energy conservation awareness program. The program continues through the use of posters and newsletters to keep the energy conservation theme and Lester in the minds of employees.

Feedback

The last key element I will mention is termed feedback.

The success of an energy program is dependent upon how effectively the progress and activities are communicated among the various unit and corporate organizations.

At Deere & Company this takes the form of monthly energy usage reports which are sent from each factory to the Corporate Energy Management group. This monthly data is put into a computer data base, and eventually transformed into a quarterly report which is sent to corporate and plant managers, energy coordinators and plant engineering departments.

This report allows for recognition of the unit's energy coordinator, gives him incentive to meet his goals, and gives him an incentive to sell programs to achieve the goal.

The computerized data base (which goes back to 1972) is printed out in the form of our "black book" both by fiscal and calendar year and is updated each year so we have at our fingertips a picture of our conservation program.

We also have an annual meeting of all energy coordinators to disseminate energy conservation information throughout the cor poration. This meeting takes the form of presentations by outside as well as Deere & Company personnel, and by round table discussions between the unit energy coordinators and the corporate energy management staff.

In summary then, these are the seven key elements that we have found to be most essential for a successful program.

Each element, we feel, is independently important and therefore essential to the group of elements.

Energy Conservation Projects

I will now present some of the actual operating and hardware changes that have been implemented as a result of our commitment to save energy. These projects basically include modifications to the systems providing environmental comfort and modification of manufacturing processes to conserve energy.

I categorize the Deere & Company Energy Conservation Program into three phases. Phase I is the housekeeping phase. Phase 2 is the conservation phase, and Phase 3 is the conversion phase. These phases are not a set progression since there is overlap and interrelationships with projects occurring concurrently.

During the housekeeping phase, we implemented a program of lower maintained building temperatures and reduced but sufficient lighting levels. We also initiated a review of building insulation policies.

Deere attempts to maintain a temperature of 65°F in occupied buildings and a lighting level of 50 footcandles in offices and manufacturing areas.

The results are that workers dress accordingly both in the shop and in the office areas.

We have rewired our lighting system to enable zone lighting and have in some cases removed up to 50% of fluorescent lamps and in the shop and warehouse areas have been able to eliminate alternate fixtures.

Concurrently, each unit reviewed the use of insulation to reduce energy requirements. The use of insulation has contributed considerably to the reduction of energy requirements in our buildings.

A program at one unit included a drastic reduction in window area. Extensive window areas in an existing building have been covered with insulated metal siding (1-1/2" thick semi-rigid fiberglass blanket covered with a corrugated metal skin).

In another program, the windows were actually bricked in.

"Casablanca" fans have been installed in many areas to reduce stratification, thereby reducing heating requirements.

"Fan jet" systems have also been installed in buildings which have a make up air deficit. These units bring in untempered outside air into the truss space where it is tempered by mixing with the warm air already in the truss space.

These systems enable us to correct negative pressures within the buildings, thereby eliminating drafts and other objectionable conditions at the work place.

In a factory large outside doors are required to move material in and out of the building. When these doors are open, they result in considerable energy waste.

Some of the things we have been doing to cut down on this type of energy waste are as follows:

- Door temporarily closed
- Plastic strip doors
- Vestibles
- Air curtains

This housekeeping phase requires relatively small capital investment and enables quick and sizeable energy savings. This is the cream off the top of the energy consumption.

Phase 2 includes a series of projects which are more technically complex and more capital intensive.

Basically, these are waste heat recovery systems and modification of building utility systems.

A popular program of relighting has been continuing which involves relighting with more energy efficient lighting sources.

Another continuing program is the modification of direct fired HV units from 100% outside air to units with return air capability.

An example of this was constructed at one of our factories. Existing HV units were dismantled and a return air duct and mixing box were added to the units. The burner plates were revised and the controls remodeled to enable automatic damper operation.

In the process area, use of waste heat recovery has developed into a sizeable program. On forge furnaces with sealed combustion chambers, incoming combustion air is preheated with the waste heat exhaust that was previously exhausted.

Air filtration in areas such as welding where air quality can be a problem is another application of energy reduction. Filtration of the air to remove contaminants enables the recirculation of the filtered air.

Since one of the major energy intensive operations at a factory is heating make-up air, any reduction in the amount of make-up air to be heated is a large contribution to energy conservation.

Another application of the same principle is illustrated by source capture of the air contaminant. Weld fumes at individual weld stations are captured directly and filtered from the air.

A paint system is another energy intensive process and much can be done to reduce energy usage without compromising performance.

At our Davenport Works the paint system was running at 80,000 cfm fresh air which was heated, passed through the booth, and then exhausted.

In the revised system, we have installed the painters in hoods which have their own air supply (heated and air conditioned as required) and we now recirculate 80,000 cfm and bring in only 1200 cfm of heated make-up air! This amount is that required to keep the contaminents and paint fumes within the booth to 25% of the lower explosive limit.

Computerized energy management systems have been installed in most of our factories. These systems allow us to precisely control and monitor environmental and process systems, and in most cases have resulted in payback periods of 1 to 3 years.

This is a review of Phase 2. Some applications of equipment and process revisions to achieve energy conservation.

Phase 3 is a category of energy reduction projects that are most difficult to implement. I separate these since they are projects which press the existing technology or which have unique economic problems.

John Deere is interested in solar energy. We have a solar installation that provides 75% of the domestic hot water for 1000 employees on two shifts. The system provides water flow averaging 75 Gal/Min with maximum demand of 2,700 Gal at 140°F for lavatories and showers. This was an experimental installation to give us "hands on" experience in this low temperature application. It did not have a good return on the investment.

We also have recently completed, an industrial training center that uses solar energy as a heat source.

A passive/hybrid solar system with 8550 Sq. Ft. of exposure to the sun supplies 40 percent of the annual heating requirement of the shop area. (382 x 10^6 BTU/year).

Solar energy impinges on a 35-foot high hollow core concrete wall through a glass facade, heating the wall and the enclosed air space.

Air is drawn out of the shop area, and circulated through, or in front of the wall where it is warmed and subsequently returned to the shop area.

An active solar system with 945 Sq. Ft. of heating mats augments the heating efficiency of a series of heat pumps used for both heating and cooling of classroom areas. This system reduces the electric consumption of the heat pumps by 20,100 KWH/year and is being modified to supply domestic hot water during the summer.

A third system "deep heat" consists of electrical resistance mats buried in the concrete and earthen mass under the floor in the open shop areas. This system will be charged at night when electric demand is lowest, to release its energy during the daytime.

One of Deere's objectives is to determine with this trio of matched systems, how energy efficiency might be improved in future facilities construction.

The economic functions, however, do not favor a solar solution in the majority of cases. We recognize the economic roadblock will decrease as energy costs increase. We see increased use of solar in industry for hot water heating for low temperature applications such as: ovens, parts washers and domestic hot water. One must recognize, however, that these low temperature applications represent only a small part of the energy used in much of industry.

Deere & Company has four waste incineration facilities in operation. Each project has unique features that entered into economic and engineering analyses.

For one application the requirement of a paint oven closely matched the energy available from the plant's solid waste. For this application, the hot flue gas passes through an air-to-air heat exchanger providing heated air for the paint oven.

Another application of waste incineration provided the incineration of 35 tons of solid waste and 2,400 gallons of waste liquid each production day. This is waste material that required expensive landfill and contract hauling for disposal. The energy produced from the waste incineration is low pressure steam which supplements an existing steam system relieving a potential steam energy deficit.

Program Results

This has been an overview of the Deere & Company Energy Conservation Program. Much has been done to improve our energy situation, however, much more needs to be done. Energy conservation and management remains a major concern for Deere & Company. The results of our efforts have been measured and quantified. Let's review the results.

A comparison of actual energy consumption compared to predicted consumption without any energy conservation demonstrates the success of the program. In fiscal year 1980, 13.8 trillion BTU were consumed compared to 24.5 trillion BTU had there been no conservation program. This is a 44% reduction compared to predicted values, and translates directly into dollars thereby maintaining a competitive position in the marketplace. These reductions have provided a buffer against rising energy costs.

Our energy costs per million BTU have risen 358% over the last eight years. The average increase per year is 22% in terms of dollar cost.

Over the same period BTU/ton of production has decreased 44%. The Federal Department of Energy had set a target goal of 15% reduction of our industry to be reached by 1980 when compared to a 1972 base, and we are aiming for a 50% reduction in BTU/ton by 1984.

Another statistic of interest, I think is the fact that for the last six years, 1975 through 1980 we used less total energy companywide than in 1972 despite some 62% increase in tonnage, and 51% increase in manufacturing space over that period of time.

In 1972 we were 61% reliant upon natural gas in our manufacturing operations. In 1980 that number has fallen to 32%. As you can see from this slide we are moving towards coal and electricity as our major energy sources.

This reflects the beginning of a self-imposed trend within our company. We have established a long range energy plan with one of its chief cornerstones being the elimination of natural gas for environmental purposes, and reductions in process usage where feasible.

The continuation of our energy management plan will require large capital expenditures, and perhaps in some cases a reliance on unproven technologies. The experience and data gathered over the past several years has been an essential tool in allowing us to embark on a positive course of action.

Yes, energy conservation is a challenging arena. The wise selection of energy mix and the efficient use of energy is a goal to which Deere & Company is committed. In conclusion, ladies and gentlemen, our message is to encourage the adoption of vibrant energy management programs and to work together to achieve wise use of energy.

Chapter 49
"AST" CONSERVATION STORY

M. A. Mozzo, Jr.

ABSTRACT

American-Standard is a major worldwide manufacturer in four diversified areas: Transportation Products, Building Products, Security and Graphic Products, and Construction and Mining Equipment. Energy supply and costs play a dominant role in our manufacturing process. In 1975, the Corporate Energy Conservation Program was begun in the U.S. only. This was expanded in 1979 to include all American-Standard facilities worldwide in recognition of the fact that energy is a global problem and the corporation is a global manufacturer.

The initial energy problem was one of supply. Curtailments of certain forms of energy, especially natural gas, interrupted production. The Energy Conservation Program initially was aimed at reducing consumption, thereby meeting production requirements with allocated supplies. Today, energy supplies to industry for the most part are adequate, and the energy problem has swung to one of high prices. Our Energy Conservation Program is now geared towards reducing energy consumption and thus energy costs to meet this new problem.

The Energy Conservation Program reviews facilities, equipment, processes, and products in an effort to reduce energy consumption and costs. Many major energy projects have been accomplished worldwide since 1975. 1980 results, the best year yet, showed a savings of over $5½ million (annual) from actions taken in the year, about 8 percent of the almost $70 million spent in 1980 on energy.

INTRODUCTION

Like most industries, the energy problems of the 1970's had significant impacts on our manufacturing operations worldwide. We have had supply interruptions which have curtailed production or at least reduced it. The natural gas shortage of 1977 in the U.S. caused several of our Eastern and Midwestern plants to be shut down. One Brazilian facility was subject to a reduced fuel oil allocation beginning in 1979 which necessitated new equipment purchases converting from oil to electricity. The rapid rise in energy prices has also affected operating income as now energy has become a significantly higher portion of cost-of-goods sold. Government involvement in rate-structuring, such as incremental pricing in the U.S., has had a significant impact worldwide in our energy costs, not consumption.

In 1975, American-Standard responded to the energy problem by establishing a Corporate Energy Conservation Program. This Program is now 5½ years old and has been quite successful. This paper reviews the energy conservation organization utilized, the impact of the energy problem on our operations, the program undertaken to reduce energy consumpiton, and the results achieved.

THE COMPANY

American-Standard is a major worldwide manufacturer in four diversified product areas, each area represented by a global business group. The Building Products Group manufactures plumbing fixtures and fittings, steel doors, pre-fabricated fireplaces, and conference room folding doors. The Security and Graphic Products Business Group manufactures safes, vault doors, safety deposit boxes, and prints bank checks, business forms, and commercial printing publications. Our Transportation and Industrial Products Group manufactures automotive parts such as truck braking systems, railroad braking systems, signalling and switching systems for railroads and mass transit systems, and a line of heat transfer and fluid power components. The fourth business group, Construction and Mining, makes large off-the-road trucks used for coal and mineral mining operations. All business groups have facilities located worldwide. The importance of energy supplies and prices vary within the four groups; however, all groups and facilities are expected to be full participants in the Energy Conservation Program.

ORGANIZATION

In 1975, a Corporate Energy Conservation Program was started in the U.S. only. Initially energy supply was the key problem for American-Standard. A number of U.S. plants were experiencing natural gas curtailments, and several plants on occasion were shut down. Every U.S. group was asked to participate and to develop specific energy cost reduction projects at plant level. In 1979, recognizing that American-Standard is a global corporation, and that the energy problem is a global one, the Energy Conservation Program was expanded to cover all American-Standard groups and facilities.

Figure 1 shows the organization structure used by American-Standard to carry out energy conservation goals. At all plants worldwide, a plant energy coordinator has been appointed. His responsibilities for his facility are: (1) develop and enforce an Energy Awareness Program, (2) prepare and monitor energy budgets, consumption, and costs, (3) plan energy conservation projects and insure they are budgeted, and (4) implement approved capital energy projects.

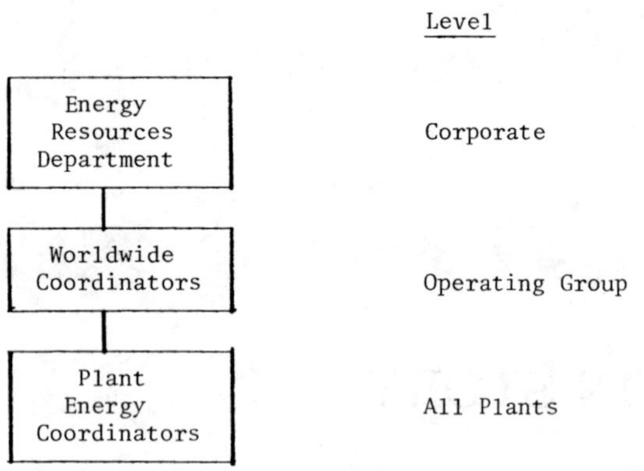

FIGURE 1. ENERGY ORGANIZATION, AMERICAN STANDARD INC.

The plant energy coordinator has a functional reporting responsibility to an energy coordinator at the operating level, a suborganization to the global business group. We currently have 19 operating group energy coordinators located worldwide. The operating group energy coordinator's responsibilities include: (1) monitor plant energy programs to insure that energy conservation is practiced, (2) insure compliance with corporate energy policies and guidelines, (3) coordinate all plant efforts and provide focal point for exchange of ideas within the group, and (4) review Operating and Strategy Plans to insure inclusion of major energy projects which require budgeting of capital funds.

At the corporate level, we have an Energy Resources Department to provide policy, guidance and engineering to establish and accomplish energy programs. The Energy Conservation Program has two individuals in the Department, a Director, Energy Management and an Energy Engineer. Responsibilities for these individuals include: (1) reviewing status of Energy Conservation Programs at operating group and plant levels, (2) collecting and analyzing energy consumption and cost data, (3) forecasting future energy costs, (4) review impacts of regulations, rate structures, and so forth, and plan corporate strategies accordingly, (5) review Operating and Strategy Plans for inclusion of sufficient funds to accomplish major energy programs, (6) provide in-house engineering consultant capabilities, and (7) review energy impacts of planned corporate moves, such as new facilities or major equipment.

THE PROBLEM

As stated previously, American-Standard formally started an Energy Conservation Program in 1975, and expanded it worldwide in 1979. The initial energy problem was one of inadequate supplies to run our facilities without disruption. Currently, energy supplies are adequate, and the shift of the energy problem has been one from inadequate supply to one of high prices adversely affecting manufacturing costs.

Our Energy Conservation Program is now more than 5 years old. Table 1 shows the global energy costs and annual savings for the corporation during the period of 1976-1985. The annual savings shown is for new energy conservation projects completed during the year, and it does not reflect any carry-over from year to year. 1980, the first full year of our global operation, has been the best year to date with an annual savings of $5.7 million from new projects. As indicated in Table 1, our plan is to continue the Energy Conservation Program, with the goal of achieving a minimum of 7 percent per year conservation savings from new projects accomplished each year.

TABLE 1

ENERGY COSTS AND CONSERVATION SAVINGS ($000's)

Year	Global Energy Costs	Annual Savings	Percentage Savings
1976	37,982	2,651	7.0%
1977	45,501	3,156	6.9
1978	48,028	3,266	6.8
1979	54,471	4,109	7.5
1980	67,065	5,709	8.5
1981	82,300	6,842	8.3
1982	104,333	8,034	7.7
1983	119,569	8,968	7.5
1984	131,940	9,500	7.2
1985	145,150	10,160	7.0

Our energy costs have almost doubled since 1976, and will more than double again in 1985. We forecast a relatively constant consumption of natural gas, electricity and fuel oils in developing these costs. Our conservation savings will be offset by increased energy requirement due to increased sales, productivity improvements, EPA and OSHA requirements. Thus the "problem" that we now face in our Energy Conservation Program is primarily one of price. In forecasting our energy costs, we have assumed that global crude oil prices will rise to $60/barrel by 1985, and that natural gas will be priced the same as oil on a BTU basis by 1984.

THE PROGRAM

In order to meet the annual savings goals set out in Table 1, all levels of organization must be involved in our Energy Conservation Program, and plans must be continually developed so that facilities can be more energy efficient. The Energy Conservation Program reviews facilities, equipment, processes, and products to reduce energy consumption and costs. Table 2 lists some of the considerations all operating groups evaluate under each of these four headings.

TABLE 2

PROGRAMS

Facilities

- Design new facilities for low energy use
- Update existing facilities during renovations to improve energy efficiency
- Consider energy costs in the site selection process
- Design efficient distribution systems of all utilities
- Use new efficient lights and control intensity
- Install meters to regulate energy usage
- Provide adequate storage facilities for primary and alternate fuel supplies
- Conduct regular audits to review energy management procedures

Equipment

- Specify energy efficient equipment in new CEARs

- Consider replacement of old, less efficient equipment
- Develop preventative maintenance programs for existing equipment
- Increase efficiency of existing equipment
- Install controls on equipment to insure energy efficient usage
- Provide dual-fuel equipment where economically justified
- Install heat recovery equipment on present equipment
- Investigate new equipment technologies

Processes

- Utilize good process flow layouts to minimize required areas
- Maximize equipment utilization rates
- Reduce BTUs required to produce chinaware
- Use lower temperature finishes
- Change pickling specifications
- Investigate new processing technologies
- Reduce rework and scrap

Product

- Design for simplified manufacturing
- Change material specifications to increase energy efficiency
- Reduce weight and mass

During the past five and one-half years, several capital energy projects have been accomplished to reduce energy costs and consumption. Some of these projects are:

Facilities

- Replace single-pane windows with double-pane or other type insulating panels
- Added insulation to walls and roofs of existing structures
- Build new, energy efficient buildings, replace older existing facilities

Equipment

- Replace old boilers with new, higher efficient units
- Rebuild enameling furnaces with ceramic insulation
- Install recuperators on heat treat furnaces
- Install recuperator preheater on foundry cupola
- Install waste heat recovery systems on sanitary ware kilns
- Install better control equipment and burners on boilers, furnaces, and kilns
- Replace existing energy-inefficient kilns with modern, direct-firing kilns

Processes

- "No-Pickle" enamel steel tub processing
- Single coat enamel steel sanitary ware
- Low mass kiln car for sanitary ware

Products

- Machined product versus foundry casted product

THE RESULTS

The objective of our Energy Conservation Program is to insure adequate energy supplies to our facilities while reducing energy consumption and costs. Our goal is to maintain at least 7 percent annual savings per year from new products. We have achieved through the first five years at least 6½ percent annual savings per year.

One very desirable result from the Program is the cumulative effect of annual energy savings. Most energy conservation projects continue to save energy year after year. Without savings from the Energy Conservation Program during the past 5 years, the 1980 global energy costs would have been $89 million as compared to $67 million actual (see Figure 2). Forecasted costs for 1985 is $145 million with conservation versus $216 million if no Conservation Program is in place, a difference of $71 million.

CONCLUSION

American-Standard has had a very active Energy Conservation Program for the past 5½ years, and plans are in existence not only to continue, but also increase the emphasis of the Program. The Program involves all operating groups worldwide and is primarily geared to making facilities, equipment, processes, and products more energy efficient. The continuing reduction in energy costs and consumption that we have achieved in the past 5 years is the prime reason for management acceptance and support of the strategy reflected for the next five years of the American-Standard Energy Conservation Program.

AST ENERGY COSTS
WITH AND WITHOUT CONSERVATION

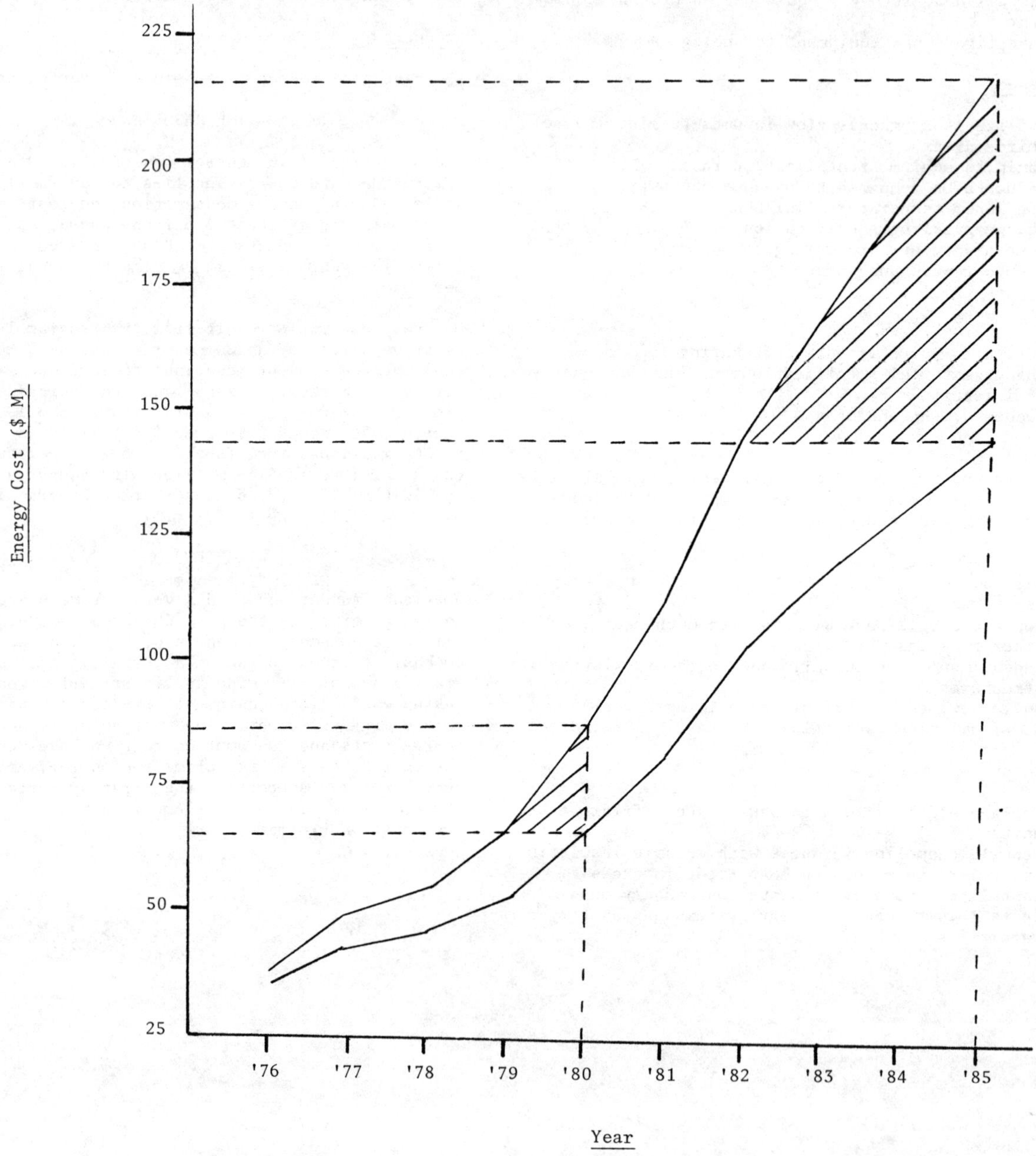

Figure 2.

Chapter 50
LONG RANGE ENERGY ECONOMIC PLANNING

J. S. Gilbert

ABSTRACT

The economic measures we all have been using do address the relative financial attractiveness of individual energy conservation strategies. However, they do not indicate the contribution that an conservation opportunity may have to stay competitive, reduce margins, or provide cash flow to the firm. These issues are of pivotal importance when energy costs represent a significant portion of operating costs and can even be important when energy costs are only 5% of gross revenues.

Because we have concentrated on the multitude of quick payback areas, we have put off the large, significant impact projects, thinking that eventually we'll get around to these also. The author's contention is that in industries where poor margins currently prevail and/or where large capital requirements are normally involved to be in business this may be putting the cart before the horse. Furthermore, if your firm would be nervous when your competitor went "beserk" with conservation projects, you may be in this situation.

The focus of the method is the economic measure of long term profit planning. Simple formulas are presented to put your own firm in perspective. The conclusion is that our current lethargic attitudes about energy conservation and planning come from using economic measures that do not accurately reflect the future earnings capabilities and characteristics of industrial firms.

BACKGROUND

Long range planning has an air of mystique or number mania associated with it. Statisticians and economists scratch their heads and pencils in a search of the most accurate projections of the future. Unfortunately much of this analysis is like the "Seer Sucker Theory," that is, for every seer there is at least one sucker. Or as one CEO put it: "Our planners have predicted 8 out of the past 3 recessions."

But while it is true that predictions are difficult, we have a serious matter at hand. The simple fact is that most firms have not considered energy conservation to be anything more than a cost cutting avenue. Many plants still have leaking steam lines, faulty traps, and on and on. These same plants boast of 20-30% reduction in energy consumed over the past five years.

When you question plant management in these same plants the reply is typically: "We take energy conservation very seriously! Why Joe is our energy conservation manager and he has an on going program to....In fact we've just installed..." When you check into it, Joe is usually either the Chief Engineer, the Plant Manager or the Environmental Compliance expert. Usually he devotes less than 20% of his time to energy management.

The fact of the matter is that most large conservation projects have their roots in either an EPA mandate, public (or internal) relations tokenism or are being used as trial balloon efforts.

Part of the problem has been the fact that there have generally been many more projects with acceptable paybacks than corporate cash made available, or time on hands of the conservation engineer. Quite naturally then, the energy manager has prioritized the energy fixes to bid for the cash. As a rule we have noticed that most projects with an obvious simple payback under one year eventually receive funding. But why not every project with acceptable payback's right now? And what is a legitimate payback that reflects your company's ability to stay in business?

There seems to be a persistent air of energy euphoria that convinces management to believe that you can gradually squeeze the energy budget consistent with the economics.

As an attempt to probe deeper into a possible explanation of these characteristics, this paper will consider the difference between the way large projects are currently authorized and the alternative developed by the author.

Currently: Project savings must payback the invested capital at an acceptable rate.

Alternative: Project must strengthen the ability of the company to maintain or improve profit margins.

Presently, firms will persist in the gradual tightening of the energy budget because they can still recover the energy fat in the selling price of the product. Energy cost increases have historically been passed on anyway.

But now that energy is the fastest rising part of the cost structure of a firm, there is going to be a time in the next few years when this energy cost will affect the ability of a firm to stay in business, and by that time it may be too late. You see there is one aspect that large energy conservation projects have....they take a long time to implement. Typically the minimum time from the decision to implement to the time that the equipment is operational is at least a year. Given that it typically takes 4 to 6 months and up to a year to decide what to do, the total elapsed time is between 1 1/2 and 2 years. So all you out there who don't have the gumption to get serious about large projects could find yourself with 2 years of reduced margins. It isn't that unlikely that some of you could be driven out of business.

As this paper will develop, the handwriting is on the wall. All you have to do is know how to interpret it. It will be tempting for some of you who are advanced econometrics types to take shots at the methods. Please hold your fire until you see the target that these materials address.

Corporate financial strategies differ widely so the methods presented here will be representative of the principles rather than an endorsement of the detail. The first method to be presented is the conventional after tax return on investment method. The second presentation is, for lack of a better term, the marginal impact method.

CONVENTIONAL AFTER TAX RETURN METHODS

The measures used to summarize the attractiveness are after tax payback period and the project internal rate of return (IRR). Another equivalent to the IRR is the project discounted cash flow (DCF).

In general, an energy conservation project may be described as an investment in capital equipment that produces savings beyond the operating costs yielding a return on the use of funds. The following parametric factors were included in this financial model:

Capitalized Cost. The total cost of energy conservation equipment including installation, process modifications and related costs.

Energy Savings. The annual value of the energy saved including credits for reduced cooling water or increased plant production capacity.

Depreciation Charges. The depreciation charges based upon equipment life and standard schedules.

Operating Costs. The annual costs associated with operating and maintaining the equipment including labor, materials, taxes and insurance. The rise in operating costs is also considered. Operating costs have historically risen more slowly than energy costs.

Table 1 illustrates a typical economic analysis for a \$1 million conservation project that is anticipated to return \$350,000 in savings in the first year of operation. These returns have been assumed to rise at 10% per year. The operating and maintenance costs are estimated at 7% of capitalized cost (i.e., \$70,000) and are expected to rise at 8% per year. This company's depreciation method is sum of the years digits with a 10-year asset life. The project is to be evaluated over a 10-year period.

Notice that the 20% investment tax credit is taken in the year the equipment is purchased. The next year, \$350,000 in energy savings was achieved, depreciation charges of \$182,000 were allocated and \$70,000 in operating and maintenance charges were incurred yielding earnings before interest and taxes (EBIT) of \$98,000. Assuming a marginal income tax of 50%, the after tax profits are \$49,000. Cash flow consists of after tax profits plus depreciation charges totaling \$231,000 in this year.

The discounted cash flow represents that percentage satisfying the following equation:

$$\Sigma \quad \text{Investments (year i)} \left(\frac{1}{(1.0 + \text{DCF})^{i}} \right) =$$

$$\Sigma \quad \text{Returns (year i)} \left(\frac{1}{(1.0 + \text{DCF})^{i}} \right)$$

which for this example can be simplified to (for 4 years):

$$800 = \sum_{i=1}^{4} \text{Returns (Year i)} \left(\frac{1}{(1.0 + \text{DCF})^{i}} \right)$$

$$= 231\left(\frac{1}{1.078}\right)^{1} + 236\left(\frac{1}{1.078}\right)^{2} + 244\left(\frac{1}{1.078}\right)^{3}$$

$$+ 253\left(\frac{1}{1.078}\right)^{4} = 214 + 203 + 195 + 188 = 800$$

Consequently 7.8% is that discount rate that equivalences that current expenditure of \$800,000 with the first four years of after tax savings. Notice that the DCF rises very quickly at first and then very slowly. This rapid rise can cause a project that is attractive on a 10-year horizon to be shown quite unattractive in the 6-8 year time frame.

Another common measure of financial attractiveness is the after tax payback period, or that period of time in which the cumulative cash flow is zero. DCF is more indicative of long-term comparative performance. The after tax payback period is related back to a simple payback and relative energy escalation rate in Table 2. Simple payback is defined as the savings (net of operating and maintanance charges) divided by the investment. Notice how inaccurate simple payback calculations are at both extremes.

MARGINAL IMPACT METHOD

The underlying reason for investigating an investment in this manner is that may large firms use this type of analysis to decide upon new processes, plants, or even whether to spend R&D money to develop new products or processes.

Assuming that you would like to understand the way a corporation generates or consumes cash as it grows, please bear with the academics that follow. For those of you who have no love lost for this type of analysis, please skip over to the results and conclusions.

For most industries the price of the product is set by market conditions but normally reflects the actual cost of production. We will assume that you and your competition are manufacturing the product using basically the same approach and that for the moment you are all selling it for the same price. The total cost of manufacture which we will call cost of goods is composed of the following basic elements:

- Materials consumed including components or feedstocks. Energy is normally lumped in here but we will separate it for discussion.
- Labor expended including the administrative, sales, and distribution elements.
- Overhead or burden which includes the other costs of doing business such as the depreciation of plant and equipment.
- Energy consumed which will include all purchased fuels and utilities. This conventionally is lumped in with materials categories.

The sum of all these costs is usually about 80% of the selling price of the product.

TABLE 1

SAMPLES INVESTMENT ANALYSIS

Year	Eqpt. Bal	Savings	Deprec.	Op. & Mnt	Pretax	After Tax	Cash Flow	Cumul. Cash Flow	DCF (IRR)
0	1000	-	-	-	-	200	(800)	(800)	-
1	818	350	182	70	98	49	321	(569)	-
2	655	385	163	76	146	73	236	(333)	-
3	509	424	146	82	196	98	244	(89)	-
4	382	466	127	88	251	126	253	164	7.8
5	273	512	109	95	308	154	263	427	15.8
6	182	564	91	103	370	185	276	703	20.9
7	109	620	73	111	436	218	291	994	24.2
8	55	682	54	120	508	254	308	1297	26.5
9	18	750	37	130	583	292	329	1626	28.2
10	-	825	18	140	667	334	352	1918	29.4

TABLE 2

AFTER TAX PAYBACK PERIOD AS A FUNCTION OF SIMPLE PAYBACK AND ENERGY ESCALATION

Simple Payback (%)	Energy Escalation (% per year) 0	5	10	15	20
10	7.1	6.5	6.0	5.7	5.4
15	5.5	5.2	4.9	4.6	4.4
20	4.6	4.3	4.1	4.0	3.8
25	4.0	3.8	3.6	3.5	3.4
30	3.5	3.3	3.2	3.1	3.0
35	3.1	3.0	2.9	2,8	2,7
40	2.8	2.7	2.7	2.6	2.5
50	2.4	2.3	2.3	2.3	2.2
60	2.1	2.0	2.0	2.0	1.9
70	1.8	1.8	1.8	1.7	1.7
80	1.6	1.6	1.6	1.6	1.6
100	1.4	1.3	1.3	1.3	1.3
140	1.0	1.0	1.0	1.0	1.0

Assumptions: 10 year asset life, 50% tax rate, 20% investment tax credit.

Simple payback defined as first year's net savings ÷ total investment.

For most American industries energy is still a relatively small part of the cost of goods (usually less then 5%). It is almost always the fastest growing element in the cost picture. Consequently, it will eventually become a significant contributor if unrestrained. Current estimates (which vary wildly according to the region of the United States) are that energy costs will rise about 6 to 10% per year faster than general inflation for the next five years. That means that a firm with energy at 5% of the cost of goods will see energy to be between about 9 and 12% in ten years. That is indicative of the euphoria mentioned earlier.

Consider for a moment those energy intensive industries where energy represents 20% of the current costs. This same ten years will see energy move to between 31 and 39%! A firm with a 20% energy edge on the competition could literally hold a 6% edge on margins. Condisering that most firms have margins in the 12 to 16% range, energy conservation will clearly play a crucial role.

So now that you all believe that the energy cost situation is not going to get better, let's look at the management decision strategies for energy conservation the same way a firm might investigate the purchase of new process equipment. When a new piece of equipment is considered, the after tax advantage of the hardware is compared to current planned operations. This sounds just like the analysis from before, except that the focus is on cash flow characteristics, not payback alone.

New equipment offers between a 10 and 40% investment tax credit, and typically has between a ten an fifteen year useful life. Consequently, more than 40% of the cost of the equipment can usually be recovered in the first year through tax credits and accelerated depreciation methods.

Corporate cash flow as shown before is after tax profits plus depreciation and tax credits. After tax profits are "sheltered" by depreciation as well as any interest expense on financed capital items. Working capital and capital required to fund new fixed assets fill in the remaining blanks to define cash flow.

As an illustration of this analysis, Table 3 has been prepared to show a hypothetical industrial firm that makes an investment in 1985 of buildings and equipment and additional equipment in 1986. The rational for the equipment was that the cost of manufacturing could be brought from 70% of revenues down to 60% of revenues. This had been modeled by a series of steps as indicated.

The plant modification required about two years at a cost of $2.5 million (see Uses of Funds - Fixed Assets). The cash flow of the firm and cumulative cash flow indicate that this firm would have to go outside (or into the past retained earnings) for $1.324 million but would recover all of the investment sometime in 1988.

Evaluating a complex decision like this requires a great deal of surmising about the future. It can also bury you under boxes of computer output. Through numerous investigations like this we have found a simple formula that reflects the performance of a firm over the long haul to decisions like the one discussed.

The cash flow generating capabilities of a firm affect the ability to grow. For all firms there is a growth rate that if exceeded will require the firm to get outside cash sources. When this percentage is high it is not difficult to get these funds since investors want to put money into this type of industry. When the percentage is low it indicates industries with poor margins, high capital requirements and poor returns to investors. It can be very difficult to get funding in those situations.

The maximum sales growth rate (%) that can be achieved over the long haul without external funding is:

$$\text{Max Sales Growth Rate (Percent)} = \frac{\left(\begin{matrix}\text{after tax}\\ \text{profits)}\end{matrix} + \begin{matrix}\text{(yearly}\\ \text{depreciation)}\end{matrix}\right)}{\left(\begin{matrix}\text{(total}\\ \text{working capital)}\end{matrix} + \begin{matrix}\text{(undepreciated}\\ \text{fixed assets)}\end{matrix} \\ -\begin{matrix}\text{(after tax}\\ \text{profits)}\end{matrix} - \begin{matrix}\text{(yearly}\\ \text{depreciation)}\end{matrix}\right)}$$

Using the data from Table 3 and knowing that the total working capital was $700 thousand in 1984:

$$\text{Max Sales Growth Rate (Percent)} = \frac{(88) + (503)}{\left(\begin{matrix}(700) + (4395)\\ -\ (88) - (503)\end{matrix}\right)} = 13.1\%$$

According to this company's predictions the situation will improve by 1988 through the addition of the equipment. The calculated maximum sales growth rate in 1987 would be:

$$\text{Max Sales Growth Rate (Percent)} = \frac{(806) + (659)}{\left(\begin{matrix}(2075) + (7796)\\ -\ (806) - (659)\end{matrix}\right)} = 17.4\%$$

That is, the addition of the plant and equipment made a 33% improvement in the cash flow generating capability of the firm.

This formula is more easily used if you take each element as a percent of revenues. Some typical industries where energy is a significant element of cost are shown in Table 4 along with their financial characteristics. Notice that while profitability and capital requirements can vary significantly, most firms can grow at better than 20% per year and "pay their own way." The right hand two columns indicate the relative contribution made by profitability and depreciation.

Now we'll take this same formula and look at the impact of energy conservation decisions. We are basically adding capital equipment, to improve after tax profits. But now we can take on a different over all business strategy. Assume for the moment that energy is 20% of your cost of goods and you can cut that by 40% through an agressive conservation retrofit design. This would amount to an 8% improvement in pretax margins (for the moment neglecting the depreciation expense). Let's assume that your company is normally accustomed to profits of 7% and could now achieve something closer to 11% through this measure. You could choose to lower your price to maintain an 8 or 9% profit, meanwhile forcing your complacent competitor to accept 2 or 3% less profit for their products.

Some of you will no doubt raise the point that firms do not face energy conservation decisions in this way. That is unfortunately very true. With very few exceptions, firms have been using these strategies to frame their choice of process equipment. Energy conservation strategies are going to become (for some

TABLE 3

PRO FORMA FINANCIAL STATEMENT FOR AN INDUSTRIAL FIRM

	Constant Dollars						
	1984	1985	1986	1987	1988	1989	1990
Sales (Tons)	3600	5400	7800	10200	12000	12000	12000
Revenues	3000	4500	6500	8500	10000	10000	10000
Cost of Manf.	2100	2925	3900	5700	6000	6000	6000
General	300	450	650	850	1000	1000	1000
Sales	300	450	650	850	1000	1000	1000
R&D	150	150	150	150	150	150	150
Total Expenses	2850	3975	5350	6950	8150	8150	8150
Profit Before Taxes	150	525	1150	1550	1850	1850	1850
Interest							
Income Tax	72	252	552	744	888	888	888
Invest Tax Credit		75	75				
After-Tax Profit	88	348	553	806	962	962	962
Current Assets	1500	2250	3250	4250	5000	5000	5000
Fixed Assets							
Land	100	100	100	100	100	100	100
Buildings	1250	2250	2250	2250	2250	2250	2250
Equipment	2600	3350	4100	4100	4100	4100	4100
Accum Dep.	1055	1508	2245	2904	3485	3986	4409
Net	2895	4192	4205	3546	2965	2464	2041
Total Assets	4395	6442	7455	7796	7965	7464	7041
Current Liabilities	750	1125	1625	2125	2500	2500	2500
Debt							
Equity	3645	5317	5830	5671	5465	4964	4541
Sources of Funds							
Profit	88	345	553	806	888	888	888
Depreciation	503	453	737	659	581	501	423
Debt							
Capital		1324					
Use of Funds							
Working Capital	250	375	500	500	375	0	0
Fixed Assets		1750	750				
Dividends	341		40	964	1168	1463	1385
Cash Flow	341	(1324)	40	964	1168	1463	1385
Cumulative Cash Flow	-	(1324)	(1284)	(320)	848	2311	3696

industries) the key to being able to stay in business.

Now let's turn the equation around and look at a marginal investment decision. To keep the analysis simple we'll make the following assumptions:

- After tax profits due to an energy conservation measure: 50% of pretax value (after operating and maintenance costs), variable used is "P".
- Depreciation of the retrofit will be made over ten years on a straight line basis, variable used is "D".
- Investment tax credits are 10%. That is, the capital justified by the equation below has been corrected to allow for the equivalent of this credit.
- No changes in working capital are required due to the energy conservation retrofit.
- The retrofit is to strengthen the cash flow of the firm as well as improve the profitability.

The following formulas predicts the maximum allowable capital that can be expended to accomplish the goal of improved profitability and growth.

P = Percent Profitability, percent of revenues

D = Current Years's Depreciation as a Percentage of revenues.

W = Working Capital as a percentage of revenues.

F = Current Fixed assests as a percentage of revenues.

Y = Maximum growth rate in revenues that can be sustained without external funding.

$$Y = \frac{P + D}{W + F - P - D}$$

The incremental investment (Δ F) justified by an incremental change in profitability (Δ P) is approximately:

$$\Delta F = \frac{(F + W)Y - (P + \Delta P + D)\ (1 + Y)}{(.125 - .875Y)}$$

This formula has been used to calculate the results shown in Table 5 which illustrate the percentage of revenues that could be justified for energy conservation given a 1% improvement in profitability. While some of the results seem absurd, they indicate some interesting characteristics about certain firms. The oil company in question was so profitable an operation in relationship to the current capital requirements that only about a percent for percent exchange between capital and profitability seems justified. This situation for the company in question while unusual does reflect the attitude today as to where several large oil companies are spending capital dollars.

More typical of the behavior is the pharmaceutical company. While 16 times the after tax payback is justified by this criteria (a 32 year simple payback!), these firms have fewer opportunities for significant energy conservation because of the batch nature of much of their operations. It does indicate that a major breakthrough in cost can dramatically impact the company's ability to do business.

CONCLUSIONS

While the results can not be used as specific guidance for any one firm, they indicate that the future for much of American industry holds a very serious energy outlook. A corporation's posture now toward process innovation, revamp, and even abandonment could dictate whether they will be in business in the next decade.

Energy, while now usually a relatively small element in the cost structure of a firm could be the Achilles' heel of an industry that experiences poor margins and large capital requirements.

TABLE 4

EXAMPLES OF ANNUAL GROWTH LIMITATIONS IN INDUSTRIAL FIRMS

(All of These Indeces are Percentages of Revenues)

Example Firm	After Tax Profits(%)	Yearly Deprec(%)	Working CAP(%)	Undeprec Fixed Asset(%)	Maximum Growth	Growth Due To Profit	Growth Due To Depec
Oil Company	9.3	9.1	8.9	9.6	>100%	101.1	96.8
Chemical Company	8.4	5.7	27.6	30.1	32.2	17.0	11.0
Pharmaceutical	8.4	2.5	30.0	28.3	23.0	16.8	4.5
Metal Forming	6.6	2.8	19.5	12.2	42.2	26.3	9.7
Railroad	5.6	8.0	5.2	59.1	26.8	9.5	14.2
Manuafacturer	5.3	1.7	21.4	30.1	15.7	11.5	3.4
Paper Company	7.2	5.4	14.0	35.9	33.8	16.9	12.1
Metal Mining	7.4	5.9	16.2	54.1	23.3	11.8	9.2
Steel Mill	2.7	3.4	13.7	26.4	17.9	7.2	9.3
Food Processor	3.8	1.8	14.8	15.8	22.4	14.2	6.2

TABLE 5

CAPITAL JUSTIFIED WITH IMPROVED PROFITABILITY

Example Firm	Current After Tax Profits (%)	Current Maximum Growth (%)	New Capital Justified with 1% Profit Improvement
Oil Company	9.3	>100%	1.2
Chemical Company	8.4	32.2	8.5
Pharmaceutical	8.4	23.0	16.1
Metal Forming	6.6	42.2	5.8
Railroad	5.6	26.8	11.7
Manufacturer	5.3	15.7	94.6
Paper Company	7.2	33.8	7.8
Metal Mining	7.4	23.3	15.9
Steel Mill	2.7	17.9	37.7
Food Processor	3.8	22.4	17.2

Chapter 51

A LOW COST ENERGY MANAGEMENT PROGRAM AT ENGELHARD INDUSTRIES DIVISION

T. S. Brown, R. Michalek, S. Reiter

INTRODUCTION

Engelhard Corporation, established as an independent company on May 20, 1981, consists of the Engelhard Industries and the Minerals and Chemicals Divisions of the former Engelhard Minerals and Chemicals Corporation. Engelhard is a world leader in technology related to precious metals and nonmetallic minerals. It manufactures high-performance chemical and precious metals products, including catalysts for the petroleum and automotive industries.

Engelhard's energy costs have risen dramatically over the past few years. Increases over 19% were experienced both in 1979 and 1980. Energy cost in 1980 was over $7.8 million. Engelhard Industries management realizes that with the continuously escalating cost of energy, energy management has become extremely important to the profitability of the division.

ENERGY MANAGEMENT PROGRAM

Prior to 1978, energy management was carried out at the plant level by plant engineering personnel. In that year, the position of Energy Co-ordinator was established with responsibility of energy matters in the division, and an Energy Management Program was established.

Energy Co-ordinator

The energy co-ordinator has the overall responsibility for the development and implementation of the Energy Management Program. The work divides itself into two areas: technical and communications. Technical help is given to the plants in locating, planning and implementing energy cost reduction projects. Reports are prepared for each plant on a quarterly and semi-annual basis providing energy use and cost data. Plant energy bills are reviewed and approved. Energy awareness among plant managers is promoted by discussions, suggestions and reports. An informed energy-conscious management is the first and most essential requirement of a successful energy management program.

The energy co-ordinator must have a strong technical background in the production process under his responsibility, in general plant engineering, and in energy management. He must also have good communication skills. Energy reports must be carefully organized and clearly written to receive their fair share of attention from busy managers.

Plant Energy Engineers

An energy engineer is assigned at each plant or profit center, with the responsibility of carrying out the Energy Management Program at the plant level. He is generally a process engineer, maintenance supervisor, or plant engineer, assigned part-time to energy matters. They are trained in energy matters by the energy co-ordinator. Energy publications are supplied for study. NBS Handbook 115, "Energy Conservation Guide for Industry and Commerce" has proven particularly useful. Literature on energy-saving equipment is distributed to the energy engineers. All Engelhard reports on plant energy use and cost are sent to them as well as plant management.

Energy engineers carry out the Energy Management Program at plant level by promoting energy awareness and by carrying out energy projects. They distribute pamphlets, hang posters, work with production people on reducing energy use and keep abreast of plant energy trends. Projects for energy cost reduction originate either with them or the energy co-ordinator. Energy engineers carry out the actual engineering, purchase and installation of the equipment. They are periodically brought up to date on Engelhard Industries's Division energy efforts by the energy co-ordinator.

Energy Management Report

The Energy Management Program was initiated by an Energy Management Report which proposed the program described in this paper and presented its financial justification. The data was prepared for each plant and for the division, and was presented to the management at each location. Energy cost reduction goals were proposed for each year from 1979 to 1985 as follows: 4% in 1979, 1980, 1983, 1984, 1985, 8% in 1981 as an incentive for new energy conservation equipment, and 6% in 1982 as a follow-on to 1981. The goals were proposed as targets for the Energy Management Program. Energy costs were predicted to escalate to $21,500,000 in 1985. Using the cost reduction goals above, cumulative savings of $20,000,000 were predicted to 1985. Price escalation of 15% and business expansion of 5% annually were used. The data for the division is shown in the following figures. Figure 1 shows the division energy cost by utility. Electricity is the largest, followed by gas.

ENERGY COST PERCENT BY UTILITY

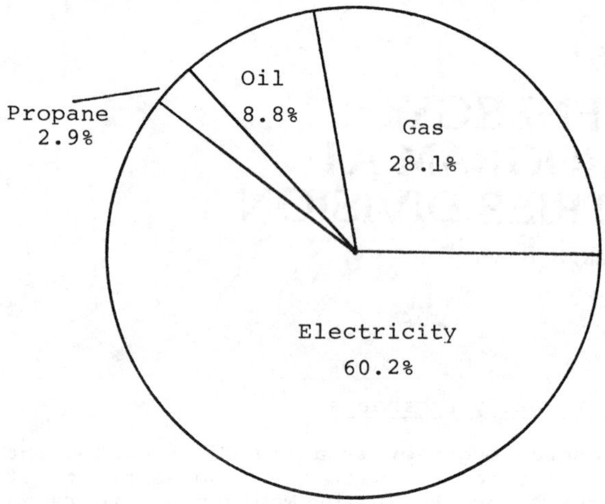

Figure 1

The predicted annual energy costs are shown in Figure 2, and the predicted annual energy savings are shown in Figure 3.

Predicted Annual Energy Costs

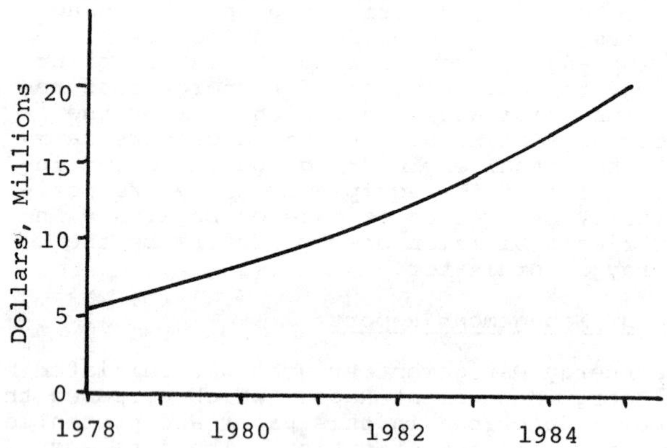

Figure 2

Cumulative Annual Savings

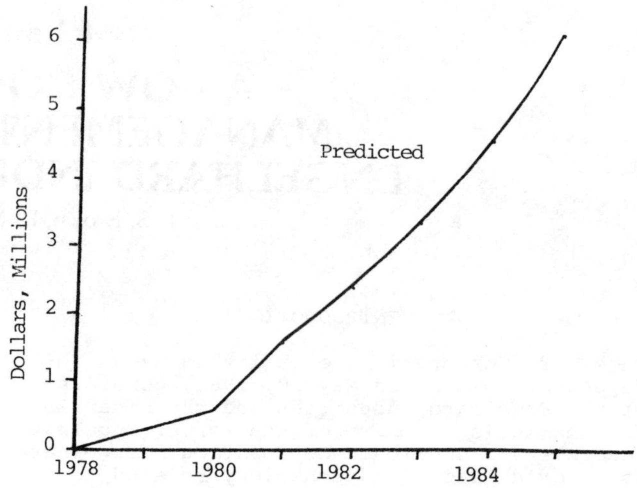

Figure 3

The favorable reception given the Energy Management Report by both plant and division management has provided the basis for the implementation of the Energy Management Program.

PLANT ENERGY MANAGEMENT PROGRAMS

The energy engineers at each plant are carrying out a wide variety of energy programs and projects as their part of the Energy Management Program. These programs and projects are briefly reviewed in this section.

Energy Audit

An energy audit was carried out at each plant. The audit covers each piece of energy-using equipment. It lists:

Identification	Gas or Steam Consumption
Location	Operating Schedule
Motor Size or KW use	

The audit is a valuable tool in estimating process and department use. It pinpoints areas of large energy use, worthy of study for energy cost reduction, and familiarizes the energy engineer and energy co-ordinator with the plant energy-using equipment. It must be updated as plant facilities change.

Energy Awareness

Energy awareness is promoted by the energy engineer through discussions with production supervision, setting of thermostats in accordance with State regulations, putting up energy posters, encouraging better use of lights, heat and machine time, checking on air and steam leaks, and many other normal jobs. Quarterly energy use and cost reports are discussed with plant management.

Billing and Purchasing Review

All utility bills are reviewed by the energy co-ordinator before payment. The amount and cost are entered on a graph, which is sent to plant management each quarter. The graph also retains the previous year's data. Comparison between the two years shows any trends needing investigation or correction. The reviews have turned up billing errors, changes in plant operation and changes in utility rates leading to suggestions for revisions in plant energy services.

Reviews are now being instituted for energy-using equipment purchases. Purchase requisitions are forwarded by the purchasing department to the energy co-ordinator for review and approval before placing the order. The energy co-ordinator can contact the originator of the requisition if he has a suggestion for improving the equipment's energy use. This is valuable where energy efficient motors and heat recovery equipment are involved.

Energy Capital Budget

The Energy Capital Budget was prepared for each plant prior to the submission of the annual capital budget. The energy engineer and energy co-ordinator proposed the budget, listing the projects suggested, cost, saving and required financial ratios. These projects were reviewed by management. Those chosen for implementation were included in the division capital budget. The Energy Capital Budget is updated as new projects are discussed, and serves as a data source for future energy conservation projects.

Energy Cost Reduction Projects

Each plant has a continuing program of energy cost reduction projects to meet its energy goals. The list below is a partial summary of types of projects.

1. Waste heat boilers in incinerators.
2. High pressure sodium lamps in production areas.
3. Programmable controller for lighting and heating controls.
4. Refrigeration heat reclaim on computer room air conditioner.
5. Conversion of boilers from oil to interruptible gas.
6. Ceiling fans in factory area.
7. Part-load control of centrifuge chiller.
8. Substitution of smaller oven for partially used large oven.
9. Timed start-up of electrical equipment.
10. Solar shading of windows.

ENERGY MANAGEMENT AND CONSERVATION REPORT

An annual report is issued to local and division managements, accounting for energy used for production, heating, cooling, pollution control, and other purposes. The report compares the efficiency of energy use in each category with that of a previous base period, and calculates the conservation due to changes in efficiency of energy use. Energy used for uncontrollable factors, such as meeting pollution control requirements, building changes or new uses are accounted for separately. The computation is our adaptation of the Carborundum system (1) of energy accounting. The basic information required is: plant production data, utility bills and government weather data for each plant location. The initial reports were hand calculated to avoid the expense of computer time. However, the time to prepare the reports became excessive. The resulting costs were too high and time was being siphoned from other energy projects. A compromise was found in leasing a terminal and renting time in a time-shared computer network. Terminal lease payments vary with length of lease and buy-out arrangements. A possible range is $75 to $150/month. Time sharing costs peak during months of report preparation. The 1980 annual cost was $2,572. A sample of the report format is shown in Figure 4.

The upper half of the print out compares the annual energy history of the plant or division against the previous year. The comparison is both for total energy and the use categories listed above. The lower half of the print-out analyses the change or variances in these categories as a result <u>only</u> of efficiency of energy use. The conservation is the sum of these variances, and is given in current dollars as well as base period dollars. The current dollars are used to calculate percent conservation. The variance due to fuel and electricity price changes, and the total cost and use changes from the base period are also shown.

The Report compares the amount of energy conservation achieved by each plant and the division against the annual goals set forth in the Energy Management Report. It informs plant and division management of the changes in energy use, cost and price from the previous year. Management uses this data for preparing annual operating budgets. The report summarizes energy cost reduction efforts at each plant and for the division which resulted in the energy cost reduction shown in this report.

The computation of the production energy and the production variance assumes that the energy use for production is proportional to the rate of production. The production quantity used in the computation has been selected for each plant on a trial-and-error basis, as the quantity most closely related to production energy. Quantities used in various plants include: weight of raw material, weight of finished product, dollar sales, and hours of building use. Many production processes involve a base loading and an operating use. Only the latter is proportional to amount of product. Examples of base loading are: continuous ovens, pumps, and heated vessels. Using the energy audit, the base load can be estimated and placed in the "other" category where it will not contribute to the production variance.

The Carborundum system requires the breakdown of total energy use into production, heating, cooling, pollution control, and "other" categories. In order to use the Carborundum system, we have had to use several methods of estimating energy use. Pollution control and "other" energy uses are estimated by establi-

ENERGY MANAGEMENT AND CONSERVATION REPORT

DEC 1980

PLANT:

FUEL TYPE

DESCRIPTION	ELECTRICITY		OIL		L.V.G.		I.S.G.		PROPANE		TOTAL	
	YR-80	YR-79	YR-80	YR-79	YR-80	YR-79	YR-80	YR-79	YR-80	YR-79	YR-80	YR-79
ENERGY 000 UNITS												
UNITS												
ENERGY MBTU												
ENERGY $												
UNIT COST $/MBTU												
UTIL COST $/UNIT												
<u>PRODUCTION</u>												
PROFIT CENTER:												
PRODUCTION UNIT												
QUANTITY X1000												
ENERGY MBTU												
ENERGY $												
MBTU/1000 UNIT												
<u>HEATING</u>												
DEGREE DAYS												
ENERGY MBTU												
ENERGY $												
MBTU/DD												
<u>COOLING</u>												
DEGREE DAYS												
ENERGY MBTU												
ENERGY $												
MBTU/DD												
<u>POLLUTION CONTROL</u>												
ENERGY MBTU												
ENERGY $												
<u>OTHER</u>												
ENERGY MBTU												
ENERGY $												

VARIANCE ANALYSIS
YR-79 TO YR-80 IMPACT

DESCRIPTION	ELECTRICITY	OIL	L.V.G.	I.S.G.	PROPANE	TOTAL
<u>VOLUME/MIX EFFECT</u>						
TOT. PLANT PRODUCTION						
MBTU						
$						
<u>WEATHER EFFECT</u>						
HEATING						
MBTU						
$						
<u>COOLING</u>						
MBTU						
$						
<u>POLLUTION CONTROL EFFECT</u>						
MBTU						
$						
<u>OTHER EFFECT</u>						
MBTU						
$						
<u>ENERGY CONSERVATION EFFECT</u>						
MBTU						
$(1979)						
$(1980)						
%(MBTU)						
%($1980)						
<u>FUEL PRICE EFFECT</u>						
$						
%						
<u>NET IMPACT</u>						
MBTU						
$						

FIGURE 4

shing a base energy use in non-heating and non-cooling periods and subtracting the base energy use from the total energy use during heating and cooling periods. The assumption is made that the base energy use is approximately constant during the heating and cooling period. Production energy use is then calculated as the remainder after the above categories are subtracted from the total energy use. Further refinement of the Energy Management and Conservation Report required more accurate energy use data, particularly at the Newark plant where many different products are manufactured and which is the largest single energy consumer in the division. The metering system at the Newark plant is described in the next section.

ELECTRICAL METERING SYSTEM AT NEWARK PLANT

The Newark plant is the division's largest energy consumer, accounting for the majority of our energy costs in 1980. Electricity represented the largest portion of these costs. Metering is required for accurate measurement of energy use in the categories described in the previous section. Departments within the plant are charged for their monthly electricity consumption on an allocation basis. Metering would provide accurate monthly billing, leading to increased energy accountability and conservation. Location of high demand areas could result in peak load shifting to off-peak hours, with substantial reduction in peak demand cost.

The electrical metering system originated at a time when an on-site computer was available. Transducers were purchased at $200 each for each departmental sub-station or electrical service entrance. The transducers converted the measured watts to an instrument signal. The signal was to be transmitted over wires to the computer for data processing. However, the on-site computer was phased out and its replacement moved to a different plant. An investigation was made of mini-computers and micro-processors to replace the lost computer. Proposals were developed with all programming and installation carried out by a single source to avoid interface problems between hardware and software. The cost of the proposed system, which included steam and gas metering, was excessive.

Alternative lower cost data acquisition systems were evaluated. Dataloggers capable of handling the mathematical requirements were located and found to be easily programmed. The metering project was changed to handle only electricity, use a datalogger and handle all wiring, installation and programming internally. The datalogger, an Acurex Autodata Ten/10, was leased through an instrument leasing firm. The standard rental fee is $1,500 per month for 12 months with a final buy-out of $2,500. The wiring from the transducers to the datalogger was handled by the plant engineering department. The datalogger was programmed internally. The transducers initially purchased were used with the datalogger. Several additional transducers were later purchased for expanded plant facilities.

The datalogger scans the transducers and converts their output to watts every 10 seconds. The data is summed as kilowatt-hours over 24 hours and divided into department totals for daily billing. Point limits can be pre-set to warn of excessive demand or to locate areas of potential electrical service overloading. Utility demand contact pulses are converted to plant demand by the datalogger. Demand values above a pre-set limit initiate a scan of both points and department totals, allowing for corrective action and study of load-shedding procedures. Data can be printed out both by an on-board printer or remotely at a leased terminal.

Through the use of telephone modems, the energy metering system can be expanded to other plants. Similar dataloggers can feed into a central terminal to provide centralized energy billing and data input into the Energy Management and Conservation Report data file. Local print-out would alert the energy engineer to his peak demand conditions.

ENERGY MANAGEMENT PROGRAM RESULTS

The Energy Management Program results are shown in Figure 5. The Cumulative Annual Savings is the sum of the savings of the present and past years. The predicted curve contains the goals of the Energy Management Report. The actual curve is the sum of energy conservation for past and present years, adjusted for price increases since the year of conservation. The goals have been exceeded in every year since the program's inception. The cumulative savings to the end of 1980 have been $1,192,901. Program administrative costs have been less than 10% of the savings.

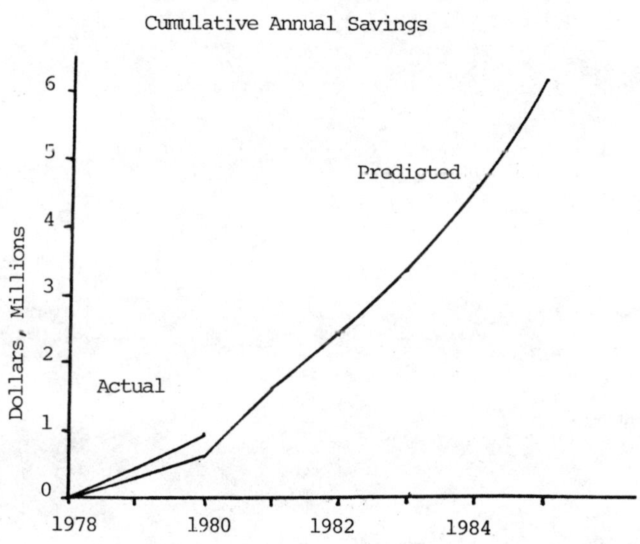

Figure 5

Engelhard Industries Division management has wholeheartedly supported the Energy Management Program. Its low cost and excellent results have made the Program a valuable management function.

References:

1. Carborundum Energy Accounting and Analy-

sis System Workshop
Carborundum Co., Niagara Fall, N.Y. 14303
August 3, 1979

Chapter 52
TOTAL ENERGY MANAGEMENT FOR THE DAIRY AND FOOD INDUSTRY

F. Spollett

INTRODUCTION

H.P. Hood Inc., a Fortune 500 Company, has three major Divisions, Northeast Dairy Division with eleven plants located in four New England States; Citrus Division with one plant in Florida; and the Clearfield Cheese Co., Inc. with four plants located one each in Pennsylvania, New York, Missouri and Utah. These sixteen plants range in size with present utility budgets from three million dollars per plant down to the smallest with a hundred and fifty thousand dollar utility budget. The total company utility budget for this fiscal year is nineteen million dollars.

A Total Energy Management Program (TEM) was established in June 1980. The following gives the process of establishing this program and an evaluation of the results after one year.

TOTAL ENERGY MANAGEMENT PROGRAM (TEM)

- The success of this program must be in direct relationship to the level of top management's commitment.
- Management must recognize that energy conservation is compatible with other objectives such as survival and profitability, and also is necessary for their accomplishment.
- Human nature is such that most employees' time and effort are directed toward those aspects of their job that are reviewed closely and about which management is concerned.
- If management recognizes that energy conservation is a sound business strategy, implementing TEM will be easier and it will be successful.
- Energy conservation must be made a part of the corporate policy and should be treated as one of the primary functions necessary for accomplishing corporate goals.

CORPORATE ENERGY POLICY COMMITTEE

The committee composed of the Director of Engineering, Vice President of Operations, Vice President of Administration, Comptroller and the Corporate Energy Manager meets and acts upon all company policy matters related to the energy conservation program. It first formulated a "Corporate Energy Policy Statement" which was approved by the President and C.E.O. of the Company and distributed to all levels of management and plant supervision. Having complete approval and commitment from the highest level of management of the Company is an essential first step for a program's success.

CORPORATE GOAL FOR ENERGY CONSERVATION

Reduce use of fuel, electrical energy, and water for plants and vehicles a total of 25% by end of Fiscal Year '83, using Fiscal '80 as a base year. A 10% reduction is estimated from optimization of procedures and a 15% reduction from retrofit projects.

Based on the established time schedule for energy audits and capital funds for retrofit projects for FY 81, 82, 83, the following must be the expectation to reach this goal.

Fiscal Year 81	5% Optimization of Procedures and No Cost-Low Cost Projects 3% From Retrofit Projects
Fiscal Year 82	4% Procedures 6% Retrofit Projects
Fiscal Year 83	1% Procedures 6% Retrofit Projects

ENERGY COORDINATOR AND PLANT ENERGY CONSERVATION COMMITTEE

A written position description was approved by the C.E.P.C. and this position was filled at each location in agreement between the Operations Manager and Corporate Energy Manager.

ENERGY COORDINATOR

This position will be established at each operating plant and vehicle location of H.P. Hood Inc.

The Energy Coordinator, a member of supervision/management, will report for all energy matters functionally to the Corporate Energy Manager, H.P. Hood Inc., maintaining all present operating responsibilities to the location manager.

He/she, with the assistance of the Corporate Energy Manager will have the following responsibilities.

1. Establish weekly and monthly reports for all utilities used at the location. The basis of these reports will include both BTU use per production unit, The Energy Utilization Index, and utility cost per production unit.
2. Assist in establishing and implementing an education program to encourage participation of all employees in a conservation program, using employee meetings, newsletter, bulletin board posters, etc.
3. Be a contributor in developing and writing energy policies and procedures.

4. Establish or if established, bring up to date, a listing of all operating equipment giving utility units of electric, steam, refrigeration, gas and water.

5. Conduct an energy plant inspection, listing and correcting all obvious misuse and waste. This program would spend money which would come under maintenance/repair items to correct problem areas of waste, i.e., faulty steam traps, poor or no insulation of utility pipes, leaky valves, improper boiler firing, etc.

6. Establish an Energy Conservation Committee, (approved by Plant Manager). Hold monthly or bi-monthly meetings for purpose of ideas, suggestions, necessary changes etc. for purpose of conserving energy. The Energy Coordinator should chair this Committee.

7. Submit the monthly "Energy Conservation Report" to the Energy Manager, copy to Operations Manager, Location, listing all activities of the energy conservation program for your location.

8. The energy coordinator shall assist the Operations Manager in establishing realistic energy goals which should be compatible with the overall company goals.

9. When approved and scheduled for your location, assist in the complete Energy Audit; following this Audit be a member of decision making team to plan, cost, submit C.A.R. and install energy retrofit projects.

An effective program must have a closed-loop communication system for all input and feed back both positive and negative at each local plant level. Communication with all employees through articles in the company newspaper, posters, suggestion systems, awards, etc. is proving successful at the individual plants. The Energy Coordinator and Energy Conservation Committee play an important role in all these programs.

A two day "Energy Workshop" for all Energy Coordinators was held. The workshop was presented by both "in house" personnel and outside presentations relating to our total program. Planned as an annual affair which will recognize and assist the plant coordinators to meet and maintain the goals for this total program.

ENERGY UTILIZATION INDEX

This index is the total BTU use per unit of production. It is tailored to the particular process and production requirements at each plant, using gallons, pounds/finished product, pounds/solids, pounds/powder as the production unit. Depending upon the size and complexness of the plant, sub metering of the utilities is often required, separating production from non-production areas and systems, for proper reporting. Before a utility can be controlled its use must be measured.

Once the EUI is established for a plant it is used to establish the utility budget for the fiscal year for the total use of energy based on a production volume; the cost then can be determined using the most accurate cost data available. This EUI is reported on a weekly report, then combined into a monthly (period ending) report. The monthly report shows all variances from budget and plan both in utility use and cost. It is a valuable tool in showing energy savings which have been accomplished by means of optimizing operations and no cost - low cost projects.

TABLE I ENERGY UTILIZATION REPORT

Month of March, 1981 | Prepared by: [signature] | Date 4-3-81

Production		Plan		Actual		Cost		MMBTU
Volume	Gals.	4,700,000	Gals.	5,057,420				
Electric	Kwh	1,246,000	Kwh	1,315,310		$95,360	MMBTU	4488
Fuel Oil	Gals.	56,400	Gals	56,430		54,173	MMBTU	8233
Water	Gal(000)	12,200	Gal(000)	12,029		18,590		
BTU/Gal	BTU	2,655			BTU	2,515		
Cost/Ga.		$ .0322				$ .0339		
Variance from Budget					BTU	- 140/gal		
						$+8,598		

****

Non Production		Plan		Actual	Cost		MMBTU
General Office							
Electric	Kwh	145,500	Kwh	144,200	$10,455	MMBTU	492
Fuel	Gal	22,800	Gal	23,410	$22,474	MMBTU	3416
Degree Days - Heating							843
- Cooling							----
Water	Gal(000)	470	Gal(000)	459	$ 709		
Blow Mold Dept.							
Electric	Kwh	161,200	Kwh	156,900	$11,375	MMBTU	535

This Table shows a savings for this period of 140 BTU/Gallon and a cost overage of $8,598. The cost overage relates to an increase cost of the utilities not established when budget was made. The savings of 140 BTU/Gallon comes from energy savings by means of optimizing operations and any no cost-low cost projects accomplished during that period.

Table II shows the method of quantifying these savings.

TABLE II

Production Volume	5,057,420 Gallons
Energy Savings	140 BTU/Gallon
Energy Cost - Electric	$21.97 MMBTU
Fuel	$ 6.08 MMBTU
Electric Power	48% of Use
Fuel Power	52% of Use

5057420 x 140	=	708 MMBTU
Percent Electric	=	340 MMBTU
Percent Fuel	=	368 MMBTU

Electric Savings	-	340 x 21.97	=	$7,470
Fuel Savings		368 x 6.08	=	2,237
		Total Savings		$9,707

PROGRAMS TO CONSERVE ENERGY

There are now three basic programs in effect.

A. Optimizing the Operation

Review and study the present operating systems, methods and hours. Can savings be made by changing procedures, staggering start-up operations, reducing light levels, heating and A/C loads, shutting off all equipment not needed (particularly on off hours), the continued motivation of all employees to conserve energy. This part of our program is and must be automatic and continuing.

B. Energy Survey

By means of an energy survey, a careful study throughout the plant/building by technically trained people listing all energy losses which can be corrected by a good, continuing maintenance program on a "no cost-low cost" basis. Repair of all leaks in each energy system, steam, air, water, condensate etc., survey of all steam traps, repair and/or replacement of defective ones, repair/replacement of insulation on steam lines, condensate lines, refrigeration lines, time controls/switches on HVAC and lighting systems, reduce temperature of hot water (one of our plants reduced temperature from 195°F to 175°F and saved over $18,000/year), reduce steam pressure on boilers, the salvaging and reuse of water. This type of program in all plants is endless and must be monitored on a continuing basis.

C. Capital Retrofit Projects

A complete Energy Audit of a plant/location will show a large number of retrofit projects requiring capital funds, engineering expertise and major installations, which will have the best long range pay back on the investment of our money.

METHODS FOR DOING ENERGY AUDITS

A. Our three year Corporate Energy Goals outlines a complete plan and method for this important phase of our program.

Based on the size of the utility budget together with any unusual circumstances at a particular location, the following time frame for audits was established: three plants in FY '80, six plants in FY '81 and three in FY '82. An allotment of capital funds (total of four million dollars over three year period) is made in yearly phases to the various plants based on results of the audit.

B. Criteria for bidding and awarding contracts for an energy audit.

As our company does not have the necessary engineering staff to do major energy audits it has been necessary for me to contract this important phase using outside consultants. My experience has shown that there are many well qualified energy consultants; how do we pick the best one? Circumstances will determine what is best for your individual operation; there are companies that along with the audit will give you a "turn key" cost for the installation of retrofit projects together with projected savings; others at the completion of the audit will supply engineering on a contract fee basis and/or act on your behalf as project manager; or you can use the results of the audit, do the engineering "in house", contract the work and be responsible for the retrofit project.

In order to compare all quotes for conducting an energy audit I use the criteria outlined in Table III.

TABLE III ENERGY AUDIT - PLANT/OFFICE COMPLEX

An Energy Audit will consist of the following:

1. Engineering analysis of current energy as it relates to the architectural envelope, electrical distribution and mechanical equipment, this analysis will include:

   A. Collect energy consumption and cost records for at least the last 12 months.

      Compile and analyze for each energy source total annual cost and consumption per product unit.

   B. Assemble relevant energy rate schedules and structures. Assemble relevant data on energy prices and supplies.

   C. Collect and review processing, architectural, mechanical, lighting, and electrical plans and specifications for all operating areas as available.

   D. Compile appropriate information on process flow diagrams, equipment specifications, operating and maintenance procedures as available.

   E. Consult with H.P. Hood Inc. on these initial findings.

2. Develop Thermal Performance Models

   A. Hours of plant/processing equipment and systems operation.

   B. Product Energy Requirements

   C. Buildings - Thermal conductivity, heating/ventilation.

   D. Submeter as necessary to determine actual energy use of major areas and/or processes.

   E. Prepare overall thermodynamic model of all plant functions, compare actual performance with this model to identify current energy waste.

3. Recommendations for energy savings in facilities, equipment, systems and processes.

4. Recommendations of points within the facilities which should be controlled and type of equipment best suited for need.

5. Estimates for cost of all recommended retrofit projects.

6. Economical analysis for each individual project recommend using H.P. Hood's system for basic discounted cash payback period and interest rate of return.

7. Prepare a written energy conservation report documenting the methodology findings, engineering calculations and all recommendations of the audit.

8. Participate in review of report with H.P. Hood Inc. in evaluation of total report.

9. H.P. Hood Inc. agrees to fully cooperate by:

   A. Supplying all information reasonably requested.

   B. Providing access to all systems, equipment, mechanical and electrical distributions, all buildings and areas within or relating to the above defined facility.

10. Total cost of energy survey, (including all travel and expenses), method of payment, time frame for completed project.

C. How to Use the Completed Energy Audit

A professional energy audit gives not only a complete written energy report but an analysis of all projects to consider. In this way rather than trying to do "one or two great projects" which every plant believes they have, the audit usually shows there are better projects, in terms of return on investment, than these pet projects.

With the audit completed we are able to make decisions based on all factors which show the best return on investment for the capital funds available and then formulate a continuing plan for the next three years for the additional retrofit projects which again, have the best return.

D. General Results from Energy Audits of Three Plants

1. First plant audited, the largest in the system with an annual utility budget now at 2.9 million dollars, listed 18 electrical retrofit projects, 9 steam/fuel projects and 6 water projects. Cost of installation of total projects was $1,355,000 with an annual savings of $594,900 (1980 figures) for a simple payback of 2.28 years.

   From this study we have completed five projects with an investment cost of $503,000 and annual savings of $284,200 for a simple payback of 1.77 years (discounted cash payback period @ 15% of 2.91 years).

2. The second plant completed with an annual utility budget of 2.5 million dollars, the audit listed 26 steam/fuel projects, and 11 electrical projects. Total cost of these projects was $2,645,000 with an annual savings of $1,045,600 for a simple payback of 2.4 years. Plans call for doing six projects at a cost of $620,000, annual savings of $402,600 for a payback of 1.54 years (discounted cash payback period of 2.66 years).

3. For the third plant with an annual utility budget of 1.3 million dollars, the audit listed 18 steam/fuel projects and 10 electrical projects. Total cost of the project is $1,045,000 with annual savings of $403,800

for a simple payback of 2.6 years. Plans at this location call for doing five projects at a cost of $448,000 with an annual savings of $257,200 for a payback of 1.74 years (discounted cash payback period of 2.71 years).

E. Conclusion From These Energy Audits

Although these three plants are different in product line, facilities and geographical location the following general areas for Energy Conservation was highlighted. Removal of obsolete steam mains and systems, repair/replacement of insulation, boiler replacement of old oversize boilers with package boilers, dual fired for gas/oil, heat recovery from refrigeration systems, also other areas of large volume low grade heat recovery, condensate return systems, retrofit of lighting systems, programmable controls of ammonia refrigeration plant and retrofit of large HVAC system.

The first year of this program has met the established goals set by the Energy Conservation Policy and should accelerate and exceed the goals set for the second year.

SECTION 13
SOLAR TECHNOLOGY UPDATE

Chapter 53
PASSIVE SOLAR CONSTRUCTION HANDBOOK

E. Levy, D. Evans, C. Gardstein

ABSTRACT

Through the sponsorship of the Southern Solar Energy Center and under U.S. Department of Energy contract No. DE-AC02-79CS30166, a Handbook of construction details, using commercially available products, has been developed for passive solar residential applications. The Handbook contains over 150 working-level details, indepth discussions of basic passive solar principles and systems, appropriate Rules of Thumb for design and sizing, a series of relevant case studies, in-depth reviews of specific passive construction materials, and an extensive bibliography.

The Handbook is intended to effectively bridge the gap between the technical state-of-the-art in passive solar design and standard practice in single-family residential construction.

OUTLINE OF HANDBOOK

Section 1. Introduction

This section outlines the overall purpose and scope of the Handbook and describes its organizational framework.

Section 2. Passive Solar Fundamentals

In this section a broad overview of the basic elements and concepts of passive solar design are presented. Five basic passive systems are reviewed:

- o Direct Gain
- o Thermal Storage Wall
- o Attached Sunspace
- o Thermal Storage Roof
- o Convective Loop

Each is described in terms of its characteristic organization of the five basic passive components: Collector, Absorber, Storage, Control and Distribution. Small scale drawings are included to illustrate the operation of each system.

Section 2 is intended to act as a brief introduction to the range of passive solar options available to the reader. The section also provides basic criteria for selecting among the passive systems introduced.

In addition to an introduction to passive systems, this section also reviews the types of energy use analysis tools available to the reader for sizing a passive system and estimating the energy use of a passive home. Four generic types of analysis tools are reviewed: manual methods; programmable calculator programs; micromputer methods; and computer programs. This review is expanded upon in Appendix C.

Passive Solar Fundamentals concludes with a discussion of methods for assessing the overall economic impact of passive solar features in the design of the home. This discussion is general in nature and is intended as an introduction to the various types of tools available for passive solar cost-benefit analyses.

Section 3. Passive Solar Construction

This is the first of two "working" sections of the Handbook and includes over 150 passive solar construction details. The section is subdivided according to passive system type, following the organizational breakdown introduced in Section 2. The initial focus on Direct Gain, Thermal Storage Wall, and Attached Sunspace systems is intended to reflect the state-of-the-art in construction technology and the systems anticipated to be preferred by most builders.

Each system discussion is divided into subsections which include: (1) a general "Design Overview" which discusses the important design and construction issues that apply to the particular system under review; (2) an indexing system for the construction details based on "Key Pages"; (3) the actual "Detail Pages"; and, (4) "Case Studies" illustrating designed and built examples of each system type.

The first subsection, Design Overview, expands upon the discussion of the system introduced in Section 2. The basic operation of the system on a seasonal basis is described, major advantages and disadvantages are reviewed, and general Rules of Thumb for system design and component sizing are provided.

The Key Pages present drawings of specific materials and material combinations which can serve as construction components (e.g., walls, floors, roofs) in passive systems. Because the Detail Pages which follow generally illustrate the intersection of two or more of these components, these Key Pages are designed as an indexing system: they identify those Detail Pages which contain the components illustrated on the Key Page. For example, Key Page 3.1150 (see Figure 1) presents a Brick Cavity Exterior Wall which can be used for Direct Gain storage.

Direct Gain
Storage Component
Brick Cavity Exterior Wall

3.1150
Key

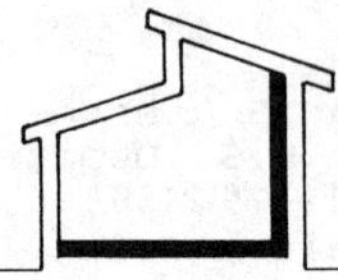

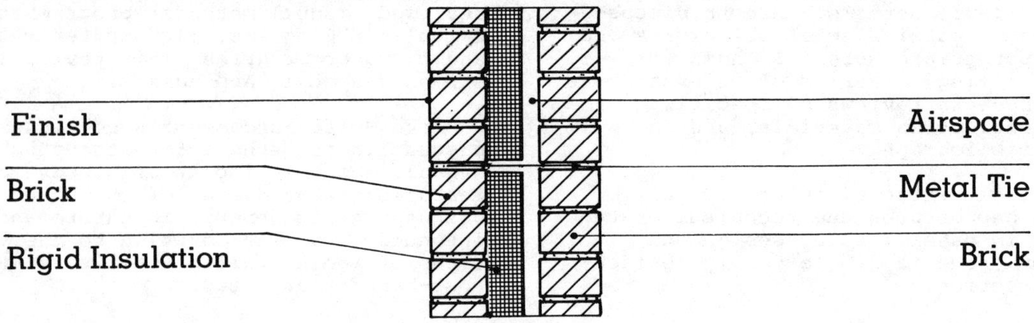

Index to Details

CONNECTION DETAIL AT:

Brick Paver/Slab-On-Grade	3.1484
Brick Paver/Wood First Floor	3.1486
Concrete/Wood First Floor	3.1488
Brick Paver/Wood Upper Floor	3.1490
Concrete/Wood Upper Floor	3.1492
Roof	3.1494

Figure 1 - Key Page

The notes on the back of this page give construction and general design information for such a wall, while the index under the drawing gives page numbers for a variety of construction components which intersect this wall. From the index the location of a detail illustrating the intersection of this wall and a Brick Paver on Wood first floor can be found. Detail Page 3.1486 (see Figure 2) presents this intersection.

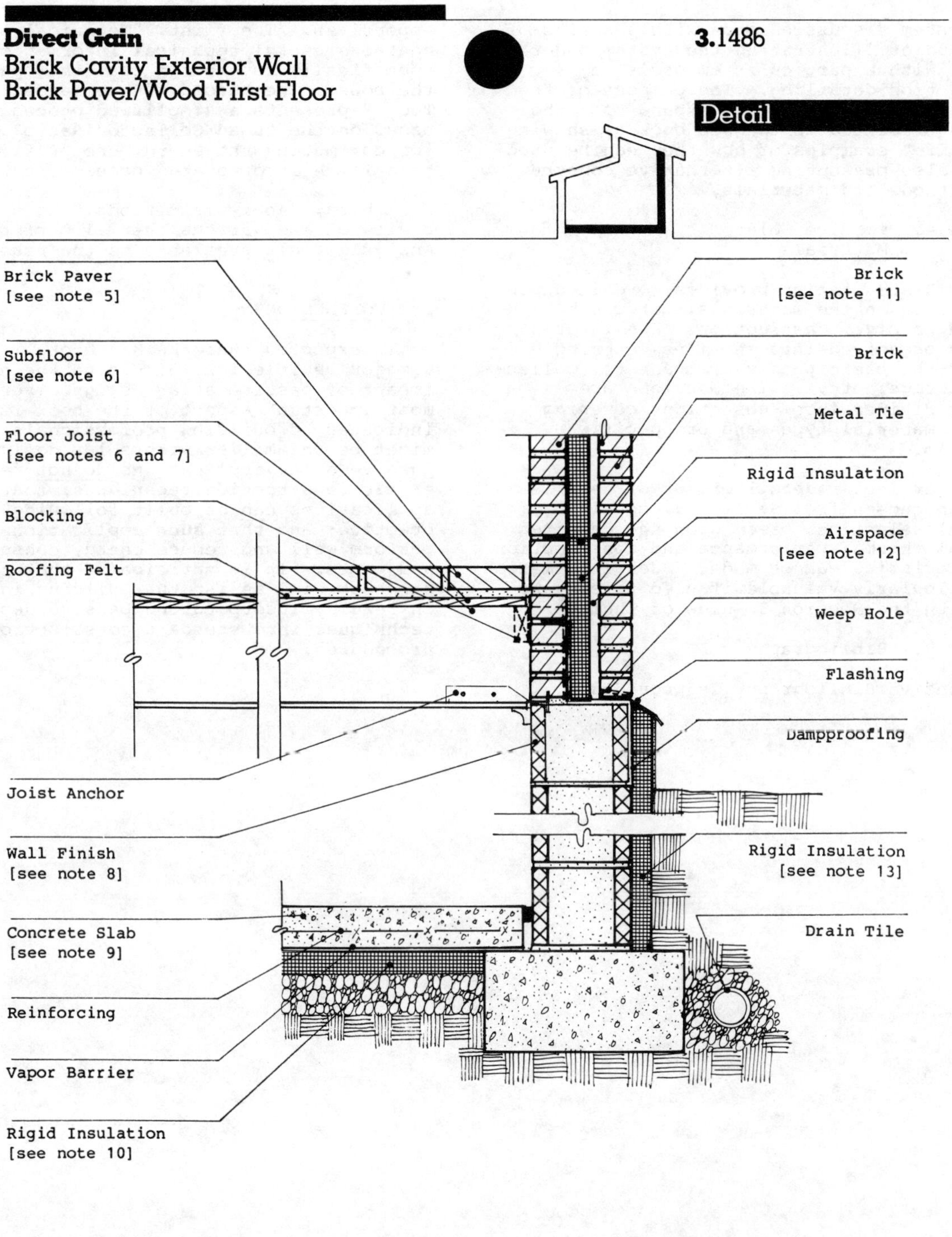

Figure 2 - Detail Page

The construction techniques illustrated in the details conform, wherever possible, to standard construction practice. Notes on the back of each detail explain specific departures from standard practice or elaborate on aspects of the detail which are particularly important from a passive solar point of view. Where applicable, alternative configurations for the details are also presented.

Each system discussion ends with a series of Case Studies illustrating the system under review, with a particular emphasis on construction detailing. The purpose of the Case Studies section is to expand upon the details presented in the Handbook by showing "real life" examples of how they can be used and by also presenting alternative construction methods and materials.

Section 4. Passive Solar Materials

The MATERIALS section provides more in-depth information on the materials that can be used in passive construction. The information is organized into sections covering each of the basic passive components (Collector, Storage, etc.). The sections are further divided into subsections covering generic material types and/or specific materials.

Once a particular detail or set of details has been chosen from Section 3, specific material selections based upon the required physical thermal performance and construction characteristics can be made. Such information is particularly valuable if used in conjunction with the Section 3 Rules of Thumb.

Section 5. Bibliography

An extensive Bibliography is provided to guide the reader to further sources of information on passive solar design and construction.

Section 6. Glossary A detailed glossary of passive solar and energy conservation terms is also included.

Appendices

At present the Handbook includes three appendices. The first, "Technical Data," contains useful technical information not specifically covered in the other parts of the book. The second, "Recommended Analysis Tool," presents a simplified procedure, based on the "Load Collector Ratio" method, for estimating the energy use profile of a home which incorporates passive features.

The third, "Analysis Methods," is a brief review of the various thermal performance analysis tools available to the reader.

2. CONCLUSIONS

It is expected that the Handbook will become a major vehicle for increasing the nationwide impact of passive solar design. Perhaps the most important aspect of the book is that it indicates to building professionals who might be unfamiliar with solar design that passive solar applications do not require exotic construction techniques; that such applications can be built following standard practice; and that such applications do perform well and reduce energy consumption and costs. It is anticipated that this Handbook will assist the building industry in readily incorporating passive design techniques into standard construction procedures.

Chapter 54
THERMAL PERFORMANCE OF SOLAR SYSTEMS IN THE NATIONAL SOLAR DATA NETWORK

S. M. Rossi

ABSTRACT

The paper presents performance results from thirty-three sites in the National Solar Data Network. These sites are being monitored throughout the country and have domestic hot water, space heating and/or space cooling applications. The solar fraction and energy savings are discussed in detail for each application. Various ratios of load to a specific parameter are also used for comparison. Also particular sizing parameters used for some of the sites will be discussed.

The National Solar Data Network is a data bank of system performance factors for solar sites. For the past four years, data has been obtained on a 24-hour basis. The National Solar Data Network is the most extensive data base in the United States.

INTRODUCTION

Vitro Laboratories operates the National Solar Data Network (NSDN). The overall purpose of the NSDN is to assist in the development of solar technologies for buildings by providing data and information on the effectiveness of specific systems, the effectiveness of particular solar technologies and the areas of potential improvement. Vitro Laboratories collects data daily and analyzes it. System performance for each site is analyzed in monthly reports based on guidelines provided by "Thermal Data Requirements and Performance Evaluation Procedures for the National Solar Heating and Cooling Demonstration Program," NBSIR 76-1137.[1] Vitro Laboratories then disseminates information to interested users via the U.S. Department of Energy, Technical Information Center.

This paper will present performance results of hot water, space heating and space cooling systems in the NSDN.[2,3,4] The NSDN has gathered detailed performance data on over one hundred instrumented sites. The purpose of this paper is to discuss the analysis results of the various solar applications.

DOMESTIC HOT WATER SYSTEMS

One of the more practical applications of solar energy is the use of solar energy to meet hot water requirements. Since the demand for hot water exists year round, significant energy savings and reasonable pay back periods may be realized. Several configurations are being monitored in the NSDN:

- Direct use of heated water from collectors (including thermosiphon and drain back),
- Heat exchangers in storage or in the collector to storage loop,
- Anti-freeze collection loop with direct use of preheated water, and,
- Heat exchanger in the collection loop of an air system.

Examples of applications monitored include single and multifamily residences (including apartment buildings, townhouses, and condominiums), schools, hospitals, restaurants, laundries, and other public buildings.

In this section seventeen domestic hot water sites will be discussed. A summary and key design information for these sites are contained in Table 1.

The actual operational performance of some domestic hot water sites was similar to the design objectives. Figure 1 shows the performance of these sites. The design solar fraction vs. actual solar fraction is plotted for the sites for which design information was available.

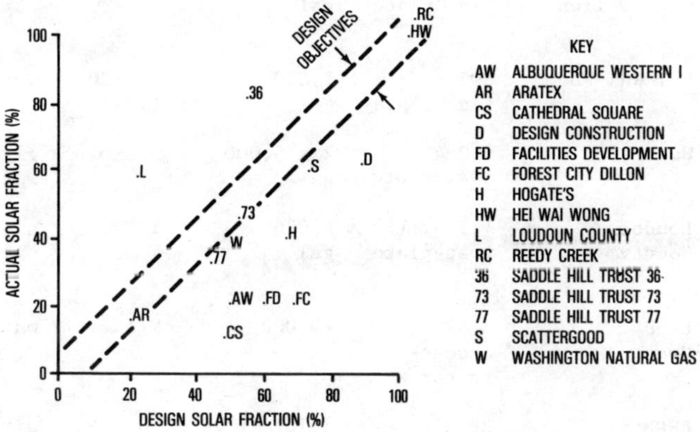

FIGURE 1. ACTUAL MONTHLY AVERAGE VS. DESIGN SOLAR FRACTION

The highest average solar fraction was 73% for the institutional commercial sites (Loudoun County, Reedy Creek and Scattergood School). The industrial commercial sites (ARATEX and Hogate's) had the lowest average solar fraction, 28%. For the residential sites, aver-solar fraction was 52% for single family and 34% for multifamily residences.

Each system performed well and provided significant solar fractions. Many of the sites performed near design expectation or better. Well designed and installed systems provided a substantial percentage of the annual water heating requirements.

For the sites which performed significantly less than the design objectives, there are several reasons:

Table 1. MAJOR SITE CHARACTERISTICS

Site Name	COLLECTOR AREA (FT^2) TYPE	STORAGE/PREHEAT CAPACITY	HOT WATER CAPACITY GALLON	COLLECTOR FLUID	APPLICATION	TYPE AUXILIARY	COLLECTOR TILT/AZIMUTH	LOCATION	RESIDENTIAL/ COMMERCIAL
Albuquerque Western I	2,478 ft^2 concen-trator	1,740 gal	N.A.	water drain-down	DHW	gas	35/0	Albuquerque, NM	Res
ARATEX	6,528 ft^2 flat-plate	12,500 gal	4,000	water	Process Hot Water	gas	30/0	Fresno, CA	Comm
Cathedral Square	1,798 ft^2 flat-plate	2,699 gal	875	glycol/ water	DHW	gas	45/+42	Burlington, VT	Res
Columbia Gas	2,978 ft^2 concen-trator	5,000 gal	50	47% dowtherm SRI	DHW SH SC	gas	variable/ -5	Columbus, OH	Comm
Design Construction	792 ft^2 flat-plate	1,400 gal	65	water drain-down	DHW SH	electric fireplace	45/0	Bigfork, MT	Res
Facilities Development	520 ft^2 flat-plate	1,000 gal	(31) 52	water	DHW	electric	42/0	San Diego, CA	Res
Forest City Dillon	2,592 ft^2 flat-plate	3,200 gal	400	hydro-carbon solution	DHW	oil	40/0	Washington, DC	Res
Hei Wai Wong	807 ft^2 flat-plate	1,230 gal	(6) 30 (1) 85	water	DHW	(6) elec (1) gas	24/0	Honolulu, HI	Res
Hogate's	6,254 ft^2 flat-plate	2 x 5,000 gal	N/A	glycol/ water	Service Hot Water	gas	55/45	Washington, DC	Comm
Loudoun County	1,225 ft^2 flat-plate	2,056 gal	1,000	silicon oil solution	DHW	electric	37/+15	Leesburg, VA	Comm
Reedy Creek	3,820 ft^2 concen-trator	10,000 gal	N/A	water	DHW SH SC	none none fossil	variable/ variable	Lake Buena Vista, FL	Comm
RHRU Clemson	385 ft^2 flat-plate 333 ft^2 greenhouse	1,161 ft^3 rock 2 x 42 gal	30	air	DHW SH SC	electric	60/0	Clemson, SC	Res
Saddle Hill Trust (36)	315 ft^2 flat-plate	750 gal	80	60% glycol	DHW SH	electric	58/0	Medway, MA	Res
Saddle Hill Trust (73)	45 ft^2 flat-plate	80 gal	40	60% glycol	DHW	gas	45/0	Medway, MA	Res
Saddle Hill Trust (77)	78 ft^2 flat-plate	120 gal	40	air	DHW	gas	38/0	Medway, MA	Res
Scattergood School	2,496 ft^2 flat-plate	130,000 lbs rock 2 x 120 gal	52	air	DHW SH-grain drying	electric propane	50/0	West Branch, IA	Comm
Washington Natural Gas	591 ft^2 flat-plate	273 ft^3 rock 80 gal	50	air	DHW SH	gas	57/0	Kirtland, WA	Res

DHW: Denotes domestic hot water SC: Denotes space cooling SH: Denotes space heating

- Recirculation and standby losses were not considered during design.
- Allowing solar energy to enter the hot water tank only when there is a draw.
- Poor matching of predicted load to actual load conditions - oversized solar system especially when designed with a space heating system.
- Long piping runs between storage/preheat tanks to conventional tanks.
- Mismatching of components to functions - undersized heat exchangers.
- Control anomalies - causing unnecessary collection startup and shutoff problems.

Figure 2 illustrates the relative magnitude of the average savings for six residential sites. Some of the sites had relatively large solar operating costs which nullified all or most of the solar savings. Electrical operating energy is an important consideration in predicting net savings and should not be neglected in solar design or evaluation. The monthly average energy savings for single family residential sites ranged from 0.31 to 2.15 million BTU/month. Saddle Hill Trust 36 had the highest monthly savings for this group (2.15 million BTU) while RHRU Clemson had the lowest (0.31 million BTU). The average monthly savings were 1.15 million BTU/month. Due to the interaction of all of these hot water subsystems with heating systems, except for Saddle Hill Trust 73 and 77, specific conclusions cannot be drawn on this variation

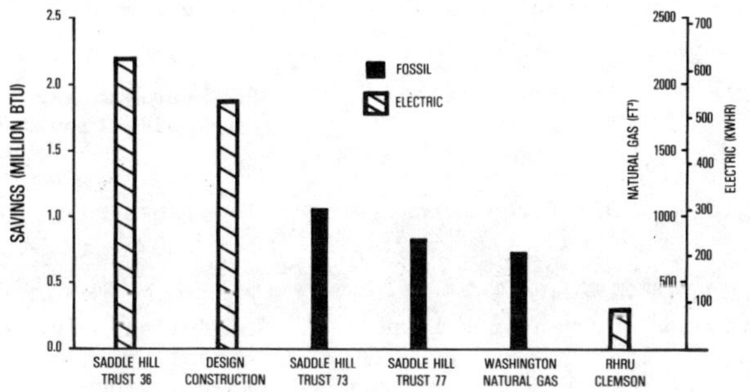

FIGURE 2. MONTHLY AVERAGE ENERGY SAVINGS, SINGLE FAMILY RESIDENTIAL SITES

Fossil savings may be calculated for Saddle Hill Trust 73, Saddle Hill Trust 77, and Washington Natural Gas by converting the BTU figures to cubic feet of natural gas. Electric savings are found for Saddle Hill Trust 36, Design Construction, and RHRU Clemson by conversion of the BTU values to electric power. The right hand vertical axis in Figure 2 approximates the conversions.

Figure 3 illustrates the relative magnitude of the electric and fossil energy savings for the multifamily, residential sites analyzed in depth. Forest City Dillon exhibited the largest absolute value for energy savings, 35.03 million BTU or 1,025 kwh. Facilities Development had the lowest, 9.36 million BTU of fossil fuel. Magnitudes of these quantities reflect a slightly wider distribution of savings than for single family systems. The differing sizes of the loads exhibited by these sites explain the variance of savings. The average savings were 18.26 million BTU/month, over ten times the average value for single family residences. The hot water load for these sites also averaged nearly 20 times the average load for single family sites. No specific interpretation is given for the differences in savings between these two groups.

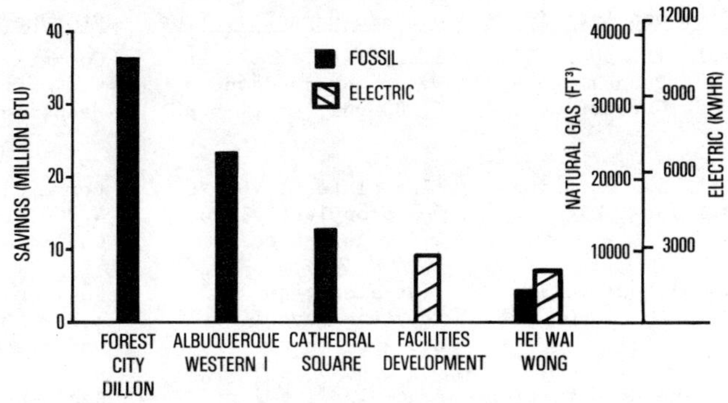

FIGURE 3. MONTHLY AVERAGE ENERGY SAVINGS, MULTIFAMILY RESIDENTIAL SITES

In general, industrial and institutional sites had widely variable savings, as shown in Figure 4. ARATEX and Hogate's are industrial hot water users, and the magnitude of savings averaged over 120 million BTU/month. Reedy Creek had a small load and correspondingly the smallest savings of the institutional sites.

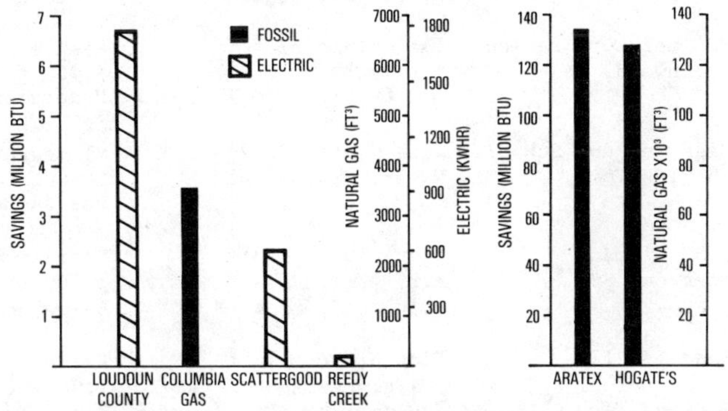

FIGURE 4. MONTHLY AVERAGE ENERGY SAVINGS, COMMERCIAL SITES

SPACE HEATING SYSTEMS

The space heating systems presented in this section were analyzed during the 1979-1980 heating season. Sixteen sites were selected for comparison. The systems are either liquid or air based with various auxiliary systems (natural gas, propane, oil, electric),

A summary table and key design information are presented in Table 2.

Figure 5 shows a plot of actual solar fraction (measured) vs. predicted solar fraction for sixteen NSDN sites. Predicted values were those estimates of performance provided to the U.S. Department of Energy when the system design was presented for review. They are based on a variety of methodologies. A general trend exists in terms of the achievement of design goals as shown by the plot. Nine of the systems nearly achieved or exceeded design goals, while another group of seven performed below design, in terms of solar fraction. Some of these latter sites actually may be cost-effective; however, the design expectations were higher than actual performance.

Table 2. MAJOR SITE CHARACTERISTICS

Site and Location	Collector Type, Size Collector Fluid, Freeze Storage Capacity, Type	Building Type Heated Floor Space Storage Location	Auxiliary 1 Auxiliary 2 Solar Delivery	Comments
Bell Telephone of Pennsylvania West Chester Co, PA	Flat-plate, 2,112 ft^2 Water, drain-down 6,000 gal, steel	Commercial 9,800 ft^2 Mechanical room	Heat pump solar (20) Electric boiler Duct heat exchanger	2-story brick Collectors on roof 62% design solar fraction
Billings Shipping Billings, MT	Flat-plate, 1,968 ft^2 50% propylene glycol 2,500 gal, steel	Commercial, large 4,000 ft^2 Underground	Gas fired hot water Space heat coil - AHU	Freight distribution facility
Bond Construction Gladstone, MO	Flat-plate, 465 ft^2 Ethylene glycol 800 gal, steel	Residence, single 1,932 ft^2 Unfinished basement	Gas fired furnace Fireplace Furnace coil	DHW system, 75% solar contribution
Design Construction Bigfork, MT	Flat-plate, 792 ft^2 Water, drain-down 1,400 gal, concrete	Residence, single 1,800 ft^2 First floor, central	Fireplace, hydronic Electric boiler	DHW preheat
M. F. Smith Jamestown, RI	Flat-plate, 512 ft^2 Water, drain-down 3,150 gal, concrete	Residence, single 1,752 ft^2 Basement	Heat pump - solar Electric element Heat exchanger	Overheating, 1/2 losses to structure DHW system
Matt Cannon Gainesville, FL	Flat-plate, 597 ft^2 Water, none 1,000 gal, steel	Residential, 4 units 2,420 ft^2 Adjacent shed	Electric strip AHU's in each apt.	DHW preheat
Moulder Corporation Greenwood, IN	Flat-plate, 704 ft^2 Air 850 ft^3, 81,000 lb rock	Residence, single 1,769 ft^2 Underground	Electric strips Electric furnace	DHW preheat Large infiltration
Page Jackson School Charles Town, WV	Flat-plate, 11,215 ft^2 Water, drain-down 20,000 gal, 2 tanks	Elementary School 50,600 ft^2 Unattached	Oil fired boiler Hydronic coils	85% design solar fraction
RHRU Clemson Clemson, SC	Flat-plate, 338 ft^2 Air 1,161 ft^3, rock	Residence, single 1,800 ft^2 Below house	Electric strips	DHW heat exchanger in collection loop
Saddle Hill Trust (36) Medway, MA	Flat-plate, 315 ft^2 50% glycerine 750 gal, steel	Residence, single 1,944 ft^2 Basement	Oil fired furnace	DHW preheat
Scattergood School West Branch, IA	Flat-plate, 2,496 ft^2 Air 130,000 lb rock	Institution, school 7,966 ft^2 Adjacent shed	Propane gas furnace	75% design solar fraction Grain drying DHW preheat
Sir Galahad Virginia Beach, VA	Flat-plate, 640 ft^2 15% ethylene glycol 1,500 gal, steel	Residence, single 1,604 ft^2 Attached shed	Heat pump Strip heaters Heat exchangers, AHU	DHW system
South Dakota School of Mines Keystone, SD	Flat-plate, 2,000 ft^2 50% ethylene glycol 3,000 gal, steel	Commercial, large 6,000 ft^2 Equipment room	Oil furnace Collection to load Multizone AHU	45% design solar fraction
T. E. Moseley Lynchburg, VA	Flat-plate, 400 ft^2 Water, drain-down 2,000 gal, steel	Commercial, small 1,780 ft^2 Warehouse	Heat pump - solar Fossil - natural gas Duct heat exchanger	Site built collectors Well insulated structure DHW system
Telex Communications Incorporated Blue Earth, MN	Flat-plate, 11,520 ft^2 Water, drain-down 20,000 gal, steel	Commercial, large Office 97,000 ft^2 Machinery Room	Electric strips Electric unit heaters Multizone AHU	70% design solar fraction
Washington Natural Kirkland, WA	Flat-plate, 591 ft^2 273 ft^3, 27,300 lb rock	Residence, single Basement	Gas fired furnace	DHW preheat

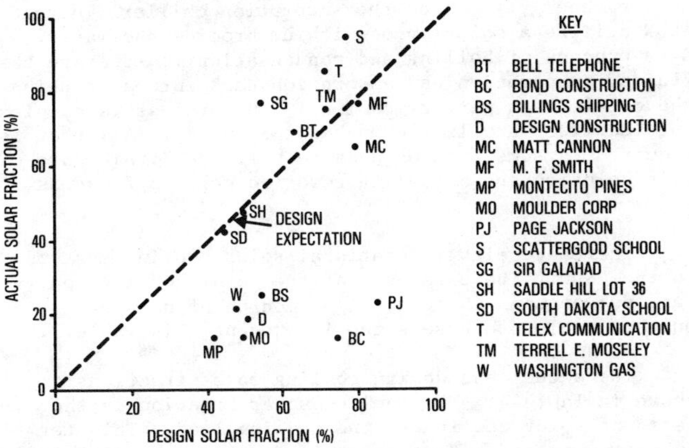

FIGURE 5. ACTUAL SEASONAL VS. DESIGN SOLAR FRACTION FOR SELECTED HEATING SITES.

Many of the designs are at least 50% below expected performance. There are several probable reasons for this result:

- Over optimistic design may be the result of over estimation of available long term weather data or the inappropriate use of solar design tools.
- Poor matching of components to functions - undersized heat exchangers, undersized piping, oversized storage, etc., high transport losses.
- Unconventional design strategy - for example, addition of auxiliary energy to storage caused degraded performance at one site.
- Poor construction practices - improper construction at an air site caused significant leakage to the environment. Uninsulated high temperature piping caused large losses at a number of liquid sites.
- Control anomalies - causing unnecessary energy rejection, mode mixing, collection startup and shutoff problems, etc.
- Large losses from storage vessels.

The energy savings for the sixteen sites is presented in Figures 6 and 7. For the residential sites the energy savings ranged from 0.86 to 3.50 million BTU/month. Due to an increased load, there is a greater range of energy savings for the commercial sites. The range of energy savings for the commercial sites was 5.88 (minus 1.40 million BTU/month for electricity) to 70.13 million BTU/month. As with the domestic hot water systems, operating energy for the solar components must be considered when calculating the net savings.

Since total load varies from site to site, comparison based on total floor area and the number of days heating was required is presented here. Four parameters are plotted for each site: total load (includes all system loads), the ratio total solar to total load, total heating load (equipment heating load) and the ratio total solar to total heating load. The variations between sites are due to such effects as weather conditions, site size, load patterns, energy consumption by other nonheating loads, and site-specific characteristics.

Liquid systems with electric auxiliary are shown in Figure 8. It shows that these sites had the lowest average energy use, 65.2 BTU/ft^2-day. The total load/solar contribution ratio was nearly the same for heating load/solar contribution ratio. The magnitude of the total energy consumed for heating would be expected to vary from building to building. Telex has a smaller surface to floor area ratio (building surface area/building floor area). It has less 'surface area' per square foot of floor space, which requires a smaller contribution from the heating equipment than a site having a larger ratio. For example, each square foot of floor 'sees' a much larger area of wall spare therefore utilizing greater energy quantities.

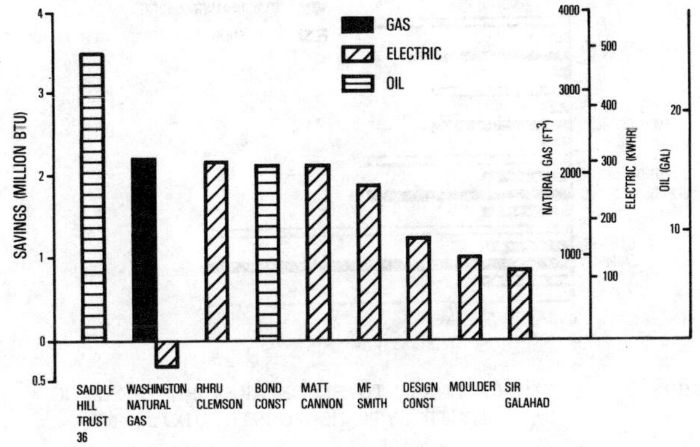

FIGURE 6. MONTHLY AVERAGE ENERGY SAVINGS, SINGLE FAMILY RESIDENTIAL SITES

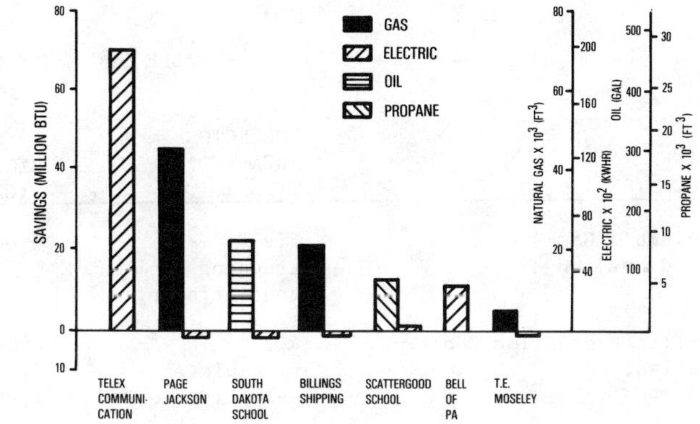

FIGURE 7. MONTHLY AVERAGE ENERGY SAVINGS, COMMERCIAL SITES

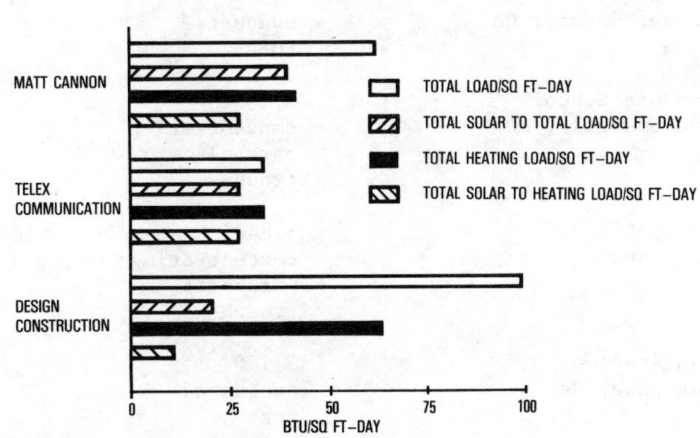

FIGURE 8. RATIO OF LOADS TO FLOOR AREA PER HEATING DAY LIQUID SYSTEMS-ELECTRIC AUXILIARY

Figure 9 shows the ratio of loads to floor area per heating degree day for liquid systems with fossil auxiliary. In this case the difference in system sizing is significant. Heat pump systems have slightly smaller normalized loads and greater solar contributions. The fossil auxiliary sites had less DHW loads than other site groups. Large residential or commercial systems had a larger average load per square foot (average 216 BTU/ft^2-floor).

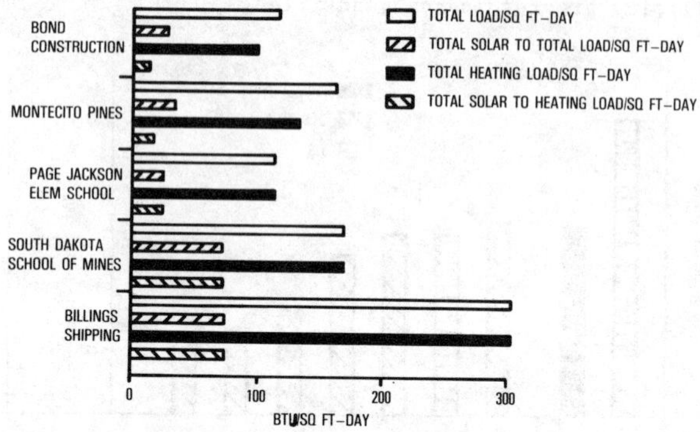

FIGURE 9. RATIO OF LOADS TO FLOOR AREA PER HEATING DAY LIQUID SYSTEMS-FOSSIL AUXILIARY

SPACE COOLING SYSTEMS

The solar cooling systems utilize a system of solar collectors and absorption chillers. The absorption chiller is used to convert solar energy into useful cooling energy. All of the absorption chillers in the NSDN utilize a solution of lithium bromide and water. By a process of boiling and condensation to separate the fluids, evaporation and absorption back into a solution, the chiller can perform cooling. The process is complex but requires very little mechanical energy. All that is required is some form of heat to boil the lithium bromide solution and a cooling tower to carry off excess heat.

There are eight operational solar cooling systems in the NSDN. The data base of the solar cooling applications contains three to eight months of data. A summary table of these sites is presented in Table 3.

The actual and design cooling solar fractions are shown in Table 4. The cooling solar fraction is the ratio of solar cooling to the cooling load. This term is useful in indicating the solar system operation and whether it is properly sized to support the loads. Cooling solar fractions at all NSDN sites, except Dallas Recreation Center, are much below the design cooling solar fraction. This is due to poor performance of the collector or chiller subsystems. The reasons for poor performance are: inefficient control strategies, malfunctioning controllers, undersizing of the collector array and poorly insulated storage.

Dallas Recreation Center has the highest cooling solar fraction but performance is still poor. It was within 60% of design load. It is quite possible that this site could achieve design load if the control problems with cold storage and the cooling coils can be resolved. In fact, during October the cooling solar fraction was 42% (79% of design).

TABLE 3. MAJOR SITE CHARACTERISTICS

SITE LOCATION	COLLECTOR AREA (FT^2) TYPE	STORAGE (GAL)	ABSORPTION CHILLER (TON)	BUILDING TYPE
Columbia Gas Columbus, OH	2,978 tracking concentrating	5,000 hot	100	25,000 ft^2 office
Dallas Recreation Center Dallas, TX	4,258 flat-plate selective coated	6,000 hot 2,000 chilled	15	16,000 ft^2 recreation 8,000 ft^2 office
Fort Polk Fort Polk, LA	11,727 evacuated tube	100,000 hot or cold	150 75 (2)	70,000 ft^2 shopping center
Irvine School Orange County, CA	4,950 evacuated tubes	none	100 (2)	41,109 ft^2 school
Padonia School Cockeysville, MD	3,185 concentrating parabolic trough	9,300 hot 9,700 cold	25	45,000 ft^2 school 10,000 ft^2 office
Reedy Creek Lake Buena Vista, FL	3,840 concentrating parabolic trough	10,000 hot 10,000 cold	25	6,100 ft^2 office
San Anselmo San Jose, CA	3,740 evacuated tube	2,175 hot	25	34,000 ft^2 school
S.D. School of Mines Keystone, SD	2,000 flat-plate concentrating	3,000 hot	3 (3)	6,000 ft^2 visitor center 2,000 ft^2 office

TABLE 4. SELECTED PARAMETER DATA OF NSDN SOLAR COOLING SITES

Site	Collector Type	Array Efficiency (%)	Operational Efficiency (%)	Collector Area Per Ton	Cooling Solar Fraction (%)	Design Cooling Solar Fraction	Solar Cooling Efficiency (%)	Chiller Coefficient of Performance[1] (COP)
Columbia Gas	Concentrator	20.8	29.1	N/A	1.5	7	-0.4	.32
Dallas Recreation Center	Flat-plate	23.8	29.7	283.9	32.4	48	4.8	.49
Fort Polk	Evacuated Tube	15.2	17.3	66.5	2.1	59	-1.4	.19
Irvine	Evacuated Tube	29.0	33.0	N/A	17.8	55	1.1	.35
Padonia School	Concentrator	12.2	19.0	125.3	3.2	75	0.4	.44
Reedy Creek	Concentrator	10.3	15.7	153.6	14.4	N/A	1.6	.53
San Anselmo	Evacuated Tube	19.0	19.2	149.6	22.0	72	2.1	.58
South Dakota School of Mines	Flat-plate	17.2	28.3	222.2	17.5	53	2.5	.46
Average		20.1	23.1					

[1] A coefficient of performance has been created to help in understanding how well the system is using energy. The COP is simply the useful energy output divided by the energy required to operate the chiller processes.

A comprehensive review of these sites has provided several important observations.

- Either evacuated tube or flat-plate collectors may be used for solar cooling. Whichever collectors are used, they must be capable of sustaining operation at 180°F when the ambient temperature is 80°F. Collector efficiencies greater than 25% are desirable.

- The size of the collector to chiller cooling capacity is critical. NSDN data indicates that an effective cooling system requires 200 to 250 square feet of collector per ton of chiller capacity. Reference 5 suggests that with their temperature optimizing controller and a 190°F generator inlet temperature, 170 square feet of collector per ton of capacity would be sufficient. This temperature may be difficult to maintain, therefore, extra collector area is recommended to compensate for the lower operating temperatures presently attainable.

- Excessive hot storage losses should be avoided when designing a solar cooling system. Simply adding more insulation is not enough. Solar heated water should be utilized as soon as possible after collection to reduce storage losses. Two-stage hot water storage systems are being considered by Reference 5 for possible improvement in solar utilization.

- Controls should be set for 160°F chiller startup temperatures.

- Avoid the use of oversized pumps or fans since the operating costs often significantly reduce energy savings.

- Do not use a cold storage tank.

Of the five sites which had hot storage tanks, only Dallas Recreation Center had storage and transfer losses less than 17%. This site has a bypass around the hot storage tank which permits solar heated water to reach the chiller directly. Any excess solar heated water is then directed to storage. There are several advantages to this type of system.

- There is a high utilization of solar energy.

- There is a short residence time of very hot water in the solar storage tank because the chiller is started at 160°F.

- The extra thermal capacity of storage permits greater chiller capacity to meet the peak loads.

- The chiller operates during periods of peak cooling demand so cold storage is not a necessary element in the design.

Energy savings are shown in Table 5. The column marked "Total Energy Saved" is computed from the cooling energy produced by the chiller divided by 2.5, the COP of a hypothetical vapor compression chiller. Operating expenses to run the chiller, condenser, generator and chilled water pumps and the cooling tower fan are included under "Chiller."

Reported cooling savings of electrical energy consist of the useful solar cooling subsystem divided by 2.5 minus the operating energy of the solar cooling subsystem. A value of 2.5 is the assumed conversion efficiency (COP) of a conventional electric chiller. For Columbia Gas and Irvine School, useful cooling and solar cooling operating expenses are proportioned by the cooling solar fraction.

Present applications of solar cooling do not save enough energy to pay for their operating costs in most instances. Two sites, San Anselmo and South Dakota School of Mines have net energy savings which average 0.15 kwh/ft^2 collector and 0.07 kwh/ft^2 collector, respectively. Most sites have high operating costs which significantly reduce energy savings. High operating costs appear to be due to oversized equipment and poor operational control. Poor operational control can be further subdivided into poor control strategies, or control failures and building operation. The worst problems with building operational control come during the spring and fall. These are transition months, times when buildings may need heating during the morning and cooling during the afternoon. Presently, some systems don't have controls which adequately anticipate these changes. A fair amount of overcooling then occurs.

TABLE 5. ENERGY SAVINGS (BTU/FT^2 COLLECTOR-MONTH)

SITE	TOTAL[1] ENERGY SAVED	OPERATING EXPENSE CHILLER	OPERATING EXPENSE COLLECTOR	NET[1] ENERGY SAVED	COOLING SAVINGS ELECTRICAL	COOLING SAVINGS FOSSIL
Dallas Recreation Center	2,016.9	1,653.9	461.5[3]	-98.5	-36.2	-
Fort Polk	513.8	476.8	1,250.8[3]	-1,213.8	349.4	-
Irvine School	1,258.6	1,483.1	1,051.7	-1,276.2	-224.5[4]	13,636
Padonia School	399.1	163.5	454.4	-218.8	-502.0	-
Reedy Creek	661.4	379.7	349.8	-68.1	-342.1	-
S.D. School of Mines	789.7	453.3	107.9[3]	228.5	336.4	-
San Anselmo	1,137.9	355.4	262.0	520.5	782.5[4]	8,240
Columbia Gas	1,258.6	1,483.1[2]	1,051.7	-1,276.2	-69.8[4]	2,227

[1] At evaporator outlet, i.e., no cold storage losses.

[2] Does not include a condenser or generator pump.

[3] Portion of collector operating power required to collect solar energy for cooling.

[4] Not reported but computed by standard NSDN procedures.

SUMMARY

The performance of the DHW sites varied widely when compared to the design solar fraction. Sixty percent of the sites performed better or near the design objectives. For many of the DHW sites, there were seasonal variations in performance. This is because the winter consumption, demand and load, were generally higher than in the summer. During the summer, the solar energy contribution to the DHW subsystem was greater than in the winter since most sites were sized for a space heating load. All DHW sites showed an energy savings. Magnitudes of these savings varied due to site configuration and sizing.

For the space heating sites, fifty percent exceeded the design expectations and eight performed poorer than the design goals. Probable reasons for design goals not being achieved were improper control strategy, system sizing, construction practices and insulation techniques. Significant energy savings were realized for all sites. Also the ratio of operating energy to energy savings was smaller for the space heating sites than the DHW sites. The smaller ratio is attributable to a larger load.

Of the eight solar cooling systems, over eighty percent performed below the design expectations. Reasons for performance being poorer than designed are attributable to improper controls, undersized collectors and poorly insulated storage. Six cooling sites incurred an energy expense rather than energy savings. High operating expenditures due to oversized equipment and poor operational controls are the probable causes for incurring an expense.

REFERENCES

(1) "Thermal Data Requirements and Performance Evaluation Procedures for the National Solar Heating and Cooling Demonstration Program," NBSIR 76-1137, U.S. Department of Commerce, Washington, D.C., 1976.

(2) Comparative Report: Performance of Active Solar Space Cooling Systems, 1980, SOLAR/0023-81/40, Vitro Laboratories, Silver Spring, Maryland, 1981.

(3) Comparative Report: Performance of Active Solar Space Heating Systems, 1980, SOLAR/0025-80/42 Vitro Laboratories, Silver Spring, Maryland, 1980.

(4) Comparative Report: Performance of Active Solar Domestic Hot Water Systems, 1980, SOLAR/0024-80/41. Vitro Laboratories, Silver Spring, Maryland, 1980.

(5) Newton, Alwin B., "Designing for Lower Solar Costs," York Division of Borg-Warner Corp., Form 155.16-TD1 (1980), York, Pennsylvania, 1980.

ACKNOWLEDGMENTS

This work was sponsored by the U.S. Department of Energy, Office of Solar Applications for Buildings and is part of the National Solar Heating and Cooling Demonstration Program. This paper and its supporting analysis is the result of many months of data collection; analysis and problem solving. The author is indebted to the Energy Systems Department of Vitro Laboratories for the data processing, analysis, engineering, and installation and servicing of the monitoring equipment.

Chapter 55

ADVANCED SOLAR THERMAL TECHNOLOGY: POTENTIAL FOR WATER DESALINATION

H. G. Pham, T. M. Liu, L. P. Leibowitz

INTRODUCTION

The demand for new water resources to satisfy human, agricultural and industrial needs has increased substantially in the last few decades. Because of this, desalinated water has become a potential supplement to, and competitor with, fresh water resources. However, desalination technology is still in a developmental stage and considerable effort must be made to improve desalinating methods, reduce equipment cost, reduce energy requirements, and above all, bring down its cost. Currently, the opportunity to couple solar thermal power systems to desalination plants seems particularly attractive. Firstly, water-poor areas have a typical abundance of sunshine. Secondly, there are potential markets for desalinated water both at home and abroad. Presently, it is in short supply in many areas of the world, such as the Middle East. In addition, there is also a potential market for desalination in the U.S.

The U.S. Department of Energy is supporting the development of advanced, low-cost, long life and reliable solar thermal power systems that will supplement and may eventually replace current fossil fuel electricity generating plants. Solar thermal systems such as point focus dish, central receiver, line focus distributed receiver and salt gradient solar pond are being developed to produce electricity power and process heat. Dish concentrators are particularly attractive for remote power generation, industrial and commercial co-generation, and for desalination. Figure 1 illustrates a solar-powered small community electric power plant using a 12-meter parabolic dish concentrator under development as part of the Solar Thermal Power System Project at the Jet Propulsion Laboratory (JPL). Currently, the joint United States-Saudi Arabian Program for cooperation in the field of Solar Energy (SOLERAS), is applying available solar thermal technology for solar desalination experiments.

This work is aimed at identifying the potential of advanced solar thermal technology for desalinating water. By identifying solar thermal desalination schemes that are able to attain high conversion efficiency and possible cost competitiveness with other water sources, this paper is a first step towards the application of solar thermal power to this market.

POTENTIAL DESALINATION MARKETS

Domestic Markets

Domestic markets for desalination may develop due to national and local water shortages, industry needs, and return flow clean-up. In addition, projected growth in energy demands for major industries can also create potential markets for desalination. The total potential desalination market in the U.S. is projected at 29 billion gallons per day (BGD) by the year 2000 (Ref. 1), approximately 10 percent of the 1975 total water consumption in the U.S.

Potential Market in the Foreign Countries

The potential foreign market for desalination depends on the geographic conditions of the country, its current water supply, and the impact of its lifestyles, living standards and industrial growth on the available water supply.

The Middle East is expected to be a significant market for desalination. Countries like Saudi Arabia and other Arab states are typically desert areas with no permanent streams. Therefore, avoiding water shortages in these countries is of critical concern.

Current industrial development and agricultural growth in these countries suffer from the shortage of water. It may be reasonable to assume that for the countries being discussed above, desalinating water is a very important option being considered for survival, agriculture, urban growth, and industrial development during the period of 1985-2000.

A worldwide increase in demand for desalinated water is expected by the year 2000. By this time the total potential desalination market outside the U.S. is expected to be about 5.5 BGD (Ref. 1).

ATTRACTIVE DESALINATION CONCEPTS

Distillation

The advantage of distillation over other desalting processes is that it produces pure water independent of feedwater quality. Several distillation methods are available and are summarized below.

FIGURE 1. ARTIST'S CONCEPT OF A ONE-MEGAWATT SOLAR THERMAL ELECTRIC POWER PLANT.

Multi-stage Flash Distillation (MSF) and Vertical Tube Evaporation (VTE): These methods use low grade thermal energy which can be provided by a solar collector to evaporate water vapor from a saline solution. The vapor produced in any flash chamber or stage in the MSF process is used to preheat the incoming sea water in that chamber. For the VTE process, water is introduced into the evaporation chamber, or effect, which is kept at low pressures. This allows the water to boil at lower temperatures. Water vapor produced from one effect would be condensed in the next effect to produce more vapor. These methods are independent of the feedwater salinity and the product water is of a high degree of purity. Energy consumption, then, is independent of salinity, allowing seawater distillation to cost about the same as brackish water.

With the improvement in performance associated with these processes, lower capital costs and lower fresh water costs can be expected (Ref. 2).

Solar Still: The solar still consists of a body of water that captures solar radiation, causing vaporization and condensation on the inside of a sloped transparent sheet. This arrangement enables the condensed water droplets to run down and be collected in a catch basin. Approximately 1 lb per 1000 Btu of the absorbed solar energy of attainable water is obtained. The efficiency of a single effect solar still is estimated between 35% to 50%.

Solar Pond

The solar pond is also a body of still water that captures solar radiation. Instead of distilling water directly, the solar pond absorbs and stores solar energy for use in thermal processing or for use with a heat engine such as the organic Rankine engine to generate power to drive electric processes. The depth of a solar pond is considerably deeper than that of the solar still. The average depth of a typical solar pond is about one meter. Non-convective solar ponds create a density gradient which prevents the heated, salt-concentrated water from rising to the pond surface and therefore reducing the heat losses. Some means are required to maintain the salt-concentration gradient.

Vapor Compression

The vapor compression process for extracting fresh water from sea water in large quantities has been used only sparingly to date. The concept is based on the principle that if vapor from salt water is compressed slightly above atmospheric pressure, its temperature will rise enough to evaporate more sea water when condensed on the outside of heat exchanger tubes, thereby giving out fresh water without adding more heat. Solar thermal systems can provide the shaft power directly from a heat engine or can generate electric power to operate the compressor. Research and development efforts have been directed towards maintaining low temperature in the main exchanger. This limitation is imposed to minimize scale formation - a technical limitation restricted by current technology.

Membrane Processes

Reverse Osmosis (RO): In this process, impurities are separated from brine by applying a relatively high pressure on the water through a semi-permeable membrane. Solar thermal systems can supply power for the pumping in the process. This method normally requires a nominal pressure between 400-600 psi. Approximately 10% of the dissolved minerals remain with the water. Results of this method are highly dependent upon the quality of the salt water as well as the type of membrane to be used (Ref. 2). Due to the high applied pressure associated with the process, reverse osmosis cells must be designed to support an extremely fragile membrane under very high pressure. Although commercial RO units are currently available on the market, further research and development is needed to produce more efficient and more reliable membranes.

Electrodialysis (ED): In this process, dissolved impurities are removed by passing them through two types of membranes; one will pass anions (anion-permeable) and the other will pass cations (cation-permeable). These membranes are alternately stacked between the anode and cathode. Ions of salt water, passing through the membranes create alternate channels of purified product water and more concentrated salt water. This process requires electric power corresponding to the amount of ions that must be transported through the membranes, and is highly dependent on the salt concentration of the water used. Electric power needed can then be provided by solar thermal systems.

Freezing

The freezing process has advanced steadily since the late fifties. In comparison to the membrane processes, the freezing process has the advantage of having its resistance to corrosion and fouling. Due to the nature of the process, no pretreatment of the water is required, and the equipment used in this process last longer than that of the membrane processes. Vacuum freezing is attained at low temperatures by the use of a vapor compressor to evaporate part of the feedwater. The temperature is reduced and the remaining feedwater is frozen due to the heat of vaporization. Solar driven compressors can be used to power the cycle. The process produces both vapor and ice of nearly pure water (except for a small portion of brine that may be entrained). The process requires shaft power to drive the vapor compressor and refrigeration equipment. Methods to separate the ice from the brine, and to wash and melt the ice are needed to improve efficiency and lower labor and associated costs. Current efforts are directed toward the vacuum freezing and indirect freezing processes. Two vacuum processes are being developed; the Absorption Freezing Vapor Compression Process (AFVC) and the Vacuum Freezing Ejector Absorption Process (AFEA). These vacuum processes have the advantage of direct heat transfer and of using water as refrigerant. In addition, neither of the processes require a mechanical compressor.

Research work during the past few years has concentrated on bulk freezing. This effort is aimed at simplifying the freezing process by eliminating air, refrigirant removal and entrainment problems that are usually associated with direct contact processes. Bulk freezing also helps enhance the scaling effects of desalting plants to larger sizes (Ref. 3).

DESALINATION CONCEPT CHARACTERIZATION

The characteristics of the various concepts are summarized in Table 1 and discussed below.

Membrane Processes

Reverse Osmosis (RO): The high operating pressure associated with the process has created a demand for improved membrane materials, higher efficiency, and longer life. In addition, the power conversion system that produces the high pressure needed to overcome the osmotic pressure must have a high efficiency in order to reduce the energy cost. The efficiency and cost effectiveness of this process is sensitive to the salt concentration characteristic, and therefore does not have the universal applicability for different types of salt water. At present, the energy consumption and the cost of equipment associated with this process are the lowest of those compared. It is expected that reverse osmosis will be used with greater frequency to desalinate water for a variety of purposes.

Electrodialysis (ED): Electric power is the primary energy requirement for electrodialysis units. Performance is sensitive to conversion efficiency, pumping power requirement, and transport losses. Since the electrical power is proportionate to the amount of ions to be transported across the membranes, the process is economically more attractive for low salt concentration water. Like the reverse osmosis process, electrodialysis uses reasonably low energy consumption to produce fresh water. The compactness and simplicity of desalination equipment helps make the water cost economically competitive with other desalination processes.

Pretreatment Requirements for Membrane Process (Reverse Osmosis and Electrodialysis): All membrane processes are commonly subject to fouling and scaling problems. Thus, pretreatment of feed water is required. For the same saline content, surface water sources that contain small, solid particles (minerals such as calcium and magnesium) would require much more sophisticated techniques than for ground water sources. This problem can mean an additional amount of equipment, and associated cost is needed for a pretreatment plant to minimize potential membrane fouling and sealing problems.

Freezing

The freezing process utilizes a mechanically or electrically driven vapor compressor to evaporate part of the feedwater. The efficiency of the power conversion subsystem becomes important since energy requirements contribute significantly to the cost of producing fresh water by this process. Generally speaking, conversion efficiency is improved with increased operating temperature. However, corrosion problems associated with high temperature operation need to be overcome before this becomes feasible.

The energy assumption associated with the process is equivalent with the vapor compression distillation method.

Distillation

Distillation With Vapor Compressor: For discussion purposes, a schematic arrangement of the vapor compression distillation system is shown in Fig. 2 (Ref. 4). As previously stated, current effort is directed towards the low temperature range (less than 250°F), due to problems associated with heat exchangers and scale formation. However, there has been some discussion of operating the system at higher temperature (Ref. 4). The influences of high temperature operation seems to be in the following areas:

(1) Effect on Compression Work. Assuming that steam entering the compressor is always saturated at the temperature T_{S1}, the saturated temperature of the steam leaving the compressor is T_{S2}. The isentropic compression work can be calculated as a function of $\Delta T_S = T_{S1} - T_{S2}$ for a given T_{S1}. Figure 3 shows the compressor work as a function of ΔT_S for various T_{S1}. Since compressor work drops with increasing T_{S1}, it is advantageous to increase the compressor inlet saturation temperature.

(2) Effect on Main Heat Exchanger. The temperature difference between the condensing and evaporating streams, ΔT, is a function of the boiling point elevation (BPE), which is a factor to account for the higher boiling point of mixed solutions, expressed as $\Delta T = T_S - \text{BPE}$.

This implies that the actual temperature difference available for heat transfer will be reduced by the amount of BPE, which is expected to increase with temperature. Thus, saving on compression work will be associated with a size penalty on the main heat exchanger part.

(3) Effect on Equipment Size: Increasing the evaporation temperature will help improve the heat transfer rate and therefore reduce the heat exchanger size. On the other hand, compressor size varies directly with specific volume of the steam at the compressor entrance for a given mass flow rate. As an example, at 200°F, the specific volume of the steam is 33.64 ft^3/lbm where at 600°F, the specific volume of the steam is 0.2668 ft^3/lbm. Thus higher evaporation temperature operation has a significant effect on the compressor size.

Some disadvantages occur along with high temperature operation. By increasing the evaporation temperature, the operating pressure and energy losses will increase. This would lead to higher initial cost for equipment and insulation.

Multiple Effect Distillation Coupled with Solar Pond

In this process, thermal energy from a solar pond is used as the main energy source to drive the multiple effect distillation process. Condensed steam from a solar pond provides the latent heat needed to evaporate an equivalent amount of saline water at a lower temperature than the steam. This is called the first effect. The vapor formed by the first effect is used to evaporate an equivalent amount of saline water at a lower temperature than that of the first effect. A number of consecutive effects are used until the vapor from the last effect is condensed in a condenser. The condensate from the condenser as well as from the effects is collected as product water except for the condensate from the first effect which is returned to the pond. Salt water is fed to the condensed vapor from the last effect and is preheated before entering the first effect.

The system is limited by the delivery temperature and available energy of the solar pond. Product water cost is sensitive to the efficiency of the evaporator and the cost to produce steam used in this process.

APPLICATIONS OF ADVANCED SOLAR THERMAL TECHNOLOGY TO DESALINATION SYSTEMS

The desalination processes described in the previous section are powered either by mechanical, thermal or electrical energy. Solar energy supplies energy under three categories: steam, direct heat, and electricity. Solar energy is collected by concentrating solar ray into a receiver that either heats the water to produce steam or converts the thermal energy to electrical power which can be used in the desalination processes. There are several schemes to couple a solar thermal system to a desalination process. A summary of different arrangements is shown as follows:

Shaft Power Conversion

Steam attained by heating the water through a receiver is fed through a turbine that drives a compressor set in a vapor compression distillation process. A schematic of this arrangement is shown in Fig. 4. The shaft power output can also be used to drive pumps or convert into electricity for a reverse osmosis or an electrodialysis desalination process.

Direct Coupling

A direct coupling between a solar thermal system and a desalination process can be accomplished by taking the steam from the solar receiver directly to desalination process (multiple effect, multiple stage flash evaporation). A schematic of this arrangement is shown in Fig. 5.

Solar Pond

Solar energy can be used directly in a solar pond to provide three important functions in a solar thermal system: collecting, storing, heat supplying. Thermal energy from a solar pond is used as the heat source to produce steam that can be used in a desalination process. A schematic diagram of this arrangement is shown in Fig. 6.

A summary of solar thermal components, their operating temperatures, and the associated efficiencies for each desalination process is shown in Table 2. The operating temperature for each solar thermal desalination scheme is based on the projected operating conditions of the power conversion subsystem as well as of the desalination process. The cost of producing fresh water using solar thermal energy depends on the power conversion efficiency, and concentrator, receiver, and power conversion capital costs.

COST ESTIMATE OF SOLAR DESALINATING PROCESSES

This section summarizes cost estimates of each solar desalinating process for comparison purposes. The economics of each desalinating process were based on projections of the cost of producing power and steam by a solar thermal power plant. Desalination equipment costs were based on estimates by Fluor Engineers and Contractors, Inc., for the Office of Water Research and Technology, U.S. DOI (Ref. 1). The groundrules of the analysis are described below.

Plant Size of 50 MGD

The cost of water is affected by the plant size assumed. Increasing plant size can substantially reduce water cost. As an example, a small distillation plant (about 1 MGD capacity) has an estimated water cost of $4.55 per thousand gallons whereas a plant of 50 MGD capacity or greater is expected to have the water cost of $2.72 per thousand gallons. Approximately 59% of the total water cost is for the cost of producing steam.

The estimates shown above indicate that effort must be directed towards constructing of larger desalinating plants and developing methods to improve the power conversion efficiency. This will allow the production of steam/electricity at lower costs.

Desalinating Equipment is Fully Developed

"Fully developed" desalination equipment is used for the cost estimates. This is defined as the point at which the ratio of incremental R&D development cost to process improvement is too high.

Steam and Power Costs Based on JPL Analysis

In Ref. 1, the power cost of $0.080/kW-hr and steam cost of $2.50/million BTU were used. The solar thermal energy costs used in this study are based on the recent JPL analysis (Ref. 5), and are $0.088/kWhr for the power cost and $5.40/million BTU for the steam cost. A preliminary estimate of $3.20/million BTU is used for the steam cost generated by the solar pond. With energy costs increasing every year, the cost of solar energy will soon be competitive with conventional energy costs.

Results

Based on the above ground rules, the total projected water costs, using solar thermal technology, are estimated and presented in Tables 3-9. Figure 7 shows the cost comparison of solar/desalination methods for seven different desalination processes. The error band associated with each process is primarily due to the uncertainty in the power conversion subsystem efficiencies.

Results of these cost estimations indicate that at salt concentrations greater than 2500 ppm, the reverse osmosis and freezing process are the most cost effective alternatives. Solar pond costs are approximately 40% higher. For salt concentrations less than 2500 ppm, the electrodialysis and reverse osmosis seem the two most cost effective processes. Solar steam generated distillation process for sea water is not competitive with the reverse osmosis and freezing processes.

CONCLUSIONS AND RECOMMENDATIONS

Solar thermal advanced technology currently under development such as point focus parabolic dish Stirling and Brayton systems can be effectively applied to desalination processes. Besides the prospect of being cost-effective, solar thermal/desalination systems show an attractive market potential and offer an alternative way for the conservation of domestic fossil fuel resources. This effort will help reduce our dependency on foreign supplies.

It is recommended that efforts be pursued to integrate advanced solar thermal technology with desalination processes. Particular attention is needed for the power conversion subsystem for the desalination processes with the highest potential growth (reverse osmosis, electrodialysis, solar pond). A detailed system trade-off study should also be made among the solar thermal power systems and desalination processes. This effort will serve as a basis for component development and hardware integration. Feasibility testing to characterize the system should then be performed.

ACKNOWLEDGMENT

The research described in this paper was carried out at the Jet Propulsion Laboratory, California Institute of Technology, and was sponsored by the Department of Energy through an agreement with NASA.

References

1. Fluor Engineers and Constructors, Inc., Final Report, "Desalting Plans and Progress and Evaluation of the State-of-the-Art and Future Research and Development," prepared for Office of Water Research and Technology, Sept. 1978.

2. North Texas State University Institute of Applied Sciences, et al., "A Feasibility Study of the Use of Solar Energy for Desalination of Water," prepared for Texas Water Development Board, May 31, 1977.

3. Johnson, W. E., "Current Development in Freezing Processes," Proceedings of Solar Desalination Workshop, Denver, Colorado, March 23-25, 1981, pp. 113-130.

4. Shouman, A.R., "A Desalting Process with Potential," Journal of Power, May 1965.

5. Leibowitz, L. P., and Liu, T. M., "Point Focus Solar Collectors for Industrial Process Heat Applications," presented at 5th Annual Solar Industrial Process Heat Conference, Houston, Texas, Dec. 16-19, 1980.

Table 1. Desalination Concepts Characteristics

PROCESS & ENERGY CONSUMPTION (Btu/lb H_2O)		CHARACTERISTICS
Reverse Osmosis	(10.00-20.00)	o Low Temperature application, compact components. Sensitive to feedwater concentration. Commercial units available. o Areas of development: high electrical conversion efficiency, methods of increasing membrane lifetime and reducing replacement costs.
Electro-dialysis	(18.00-30.00)	o Sensitive to feedwater concentration. Commercial units available with compact components. o Area of development: low resistance, high efficiency membranes. Components for high temperature operation, reductions of pumping power and transport losses.
Freezing	(21.00-35.00)	o Sensitive to equipment efficiency and costs. o Areas of development: freezing processes, low temperature heat transfer characteristics for salt water. Washing and melting ice processes.
Distillation (Vapor Compressor)	(34.00)	o Sensitive to steam cost. Involves equipment with moderate complexity. o Area of development: - component efficiency improvement - operation at high temperature - corrosion problems at high temperature
Distillation (Multiple Effect & Solar Pond)	(34.00)	o Sensitive to the number of effects. o Temperature limited. o Selection of an efficient evaporator plays an important role in the cost of water.

Table 2. Solar Thermal Desalination Technology

PROCESS	COMPONENTS NEEDED	SOLAR OPERATING TEMPERATURE	EFFICIENCY
Reverse Osmosis	Concentrator		0.90-0.94
	Receiver	1200-2000°F	0.80-0.9
	Power Conversion Unit	1200-2000°F	0.30-0.45
	Generator	---	0.94
Electrodialysis	Concentrator		0.90-0.94
	Receiver	1200-2000°F	0.80-0.90
	Power Conversion Unit	1200-2000°F	0.30-0.45
	Generator	---	0.94
Freezing	Concentrator		0.90-0.94
	Receiver	1200-2000°F	0.80-0.90
	Power Conversion Unit	1200-2000°F	0.30-0.45
	Generator (option)	---	0.94
Distillation (Multiple Effect & Solar Pond)	Solar Pond	180°F-220°F	0.20
	Heat Exchanger	180°F-220°F	0.80
Distillation (Vapor Compressor)	Concentrator		0.90-9.94
	Receiver	700-2000°F	0.8-0.90
	Power Conversion Unit	700-2000°F	0.20-0.40

Tables 3-9. Solar/Desalination Water Cost Estimates

Table 3

Distillation Process	$/Kgal
1. Desalting equipment capital recovery	0.52
2. Site related capital recovery	0.11
3. Indirect capital recovery	0.12
4. Operation and maintenance labor	0.03
5. G&A supplies and maintenance materials	0.02
6. Chemicals	0.02
7. Power (6.5 kWhr/kgal)	0.21
8. Steam	0.81
Total Water Costs, $/kgal	1.84

Table 4

Freezing Process	$/Kgal
1. Desalting equipment capital recovery	0.19
2. Site-related capital recovery	0.03
3. Indirect capital recovery	0.03
4. Operating and maintenance labor	0.02
5. G&A supplies and maintenance materials	0.02
6. Chemicals	0.03
7. Power	0.03
Total Water Costs, $/kgal	1.15

Table 5

Reverse Osmosis Process (Brackish Water, < 2500 ppm)	$/Kgal
1. Desalinating equipment capital recovery	0.04
2. Site related capital recovery	0.02
3. Indirect capital recovery	0.01
4. Operating and maintenance labor	0.01
5. G&A supplies and maintenance materials	0.01
6. Chemicals	0.05
7. Electric power	0.10
8. Membrane replacement	0.04
9. Membrane replacement labor and labor G&A	0.05
Total Water Costs, $/Kgal	0.33

Table 6

Electrodialysis Process (Brackish Water, <2500 ppm)	$/Kgal
1. Desalinating equipment capital recovery	0.11-0.15
2. Site related capital recovery	0.03-0.04
3. Indirect capital recovery	0.02-0.03
4. Operating and maintenance labor	0.02
5. G&A supplies and maintenance materials	0.01
6. Power	0.14-0.30
7. Membrane replacement	0.03-0.04
8. Chemicals and filters	0.02
9. Membrane replacement labor and labor G&A	0.05
Total Water Costs, $/Kgal	0.38-0.61

Table 7

Reverse Osmosis Process (Seawater, >2500 ppm)	$/Kgal
1. Desalting equipment capital recovery	0.24
2. Site related capital recovery	0.05
3. Indirect capital recovery	0.07
4. Operating and maintenance labor	0.03
5. G&A supplies and maintenance materials	0.02
6. Chemicals	0.05
7. Power (9 kWhr/Kgal)	0.29
8. Membrane replacement	0.23
9. Membrane replacement labor and labor G&A	0.06
Total Water Costs, $/Kgal	1.04

Table 8

Electrodialysis Process (Seawater, >2500 ppm)	$/Kgal
1. Desalinating equipment capital recovery	0.86
2. Site related capital recovery	0.18
3. Indirect capital recovery	0.20
4. Operating and maintenance labor	0.03
5. G&A supplies and maintenance materials	0.05
6. Chemicals	0.24
7. Power (40 kWhr/Kgal)	1.24
8. Membrane replacement	0.24
9. Membrane replacement labor and labor G&A	0.06
Total Water Costs, $/Kgal	3.10

Table 9

Solar Pond Distillation Process	$/Kgal
1. Desalinating equipment capital recovery	0.52
2. Site related capital recovery	0.11
3. Indirect capital recovery	0.12
4. Operating and maintenance labor	0.03
5. G&A supplies and maintenance materials	0.02
6. Chemicals	0.02
7. Power (6.5 kWhr/Kgal)	0.15
8. Steam	0.44
Total Water Costs, $/Kgal	1.41

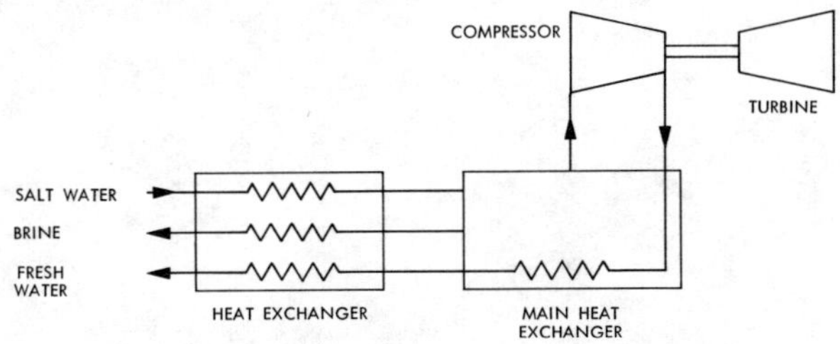

FIGURE 2. VAPOR COMPRESSION DESALINATION PROCESS (REF. 3).

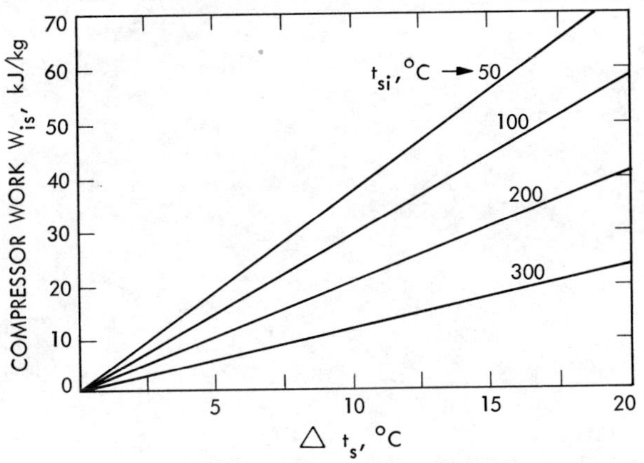

FIGURE 3. EFFECT OF INLET TEMPERATURE ON COMPRESSOR WORK FOR VARIOUS TEMPERATURE DIFFERENCES (ΔT_S, °C) BETWEEN THE STEAM ENTERING AND LEAVING THE COMPRESSOR (REF. 3).

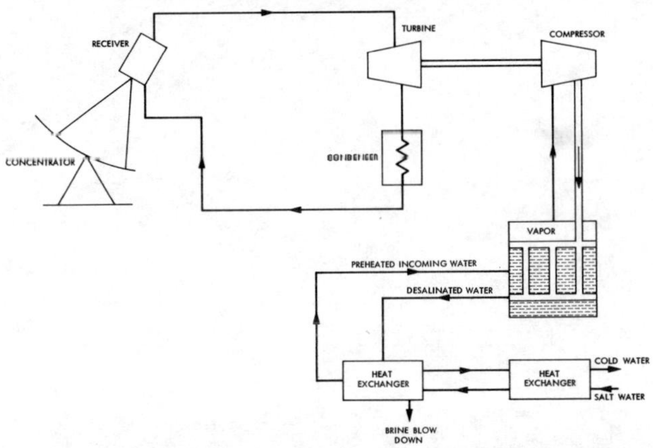

FIGURE 4. SOLAR POWERED/VAPOR COMPRESSION DISTILLATION.

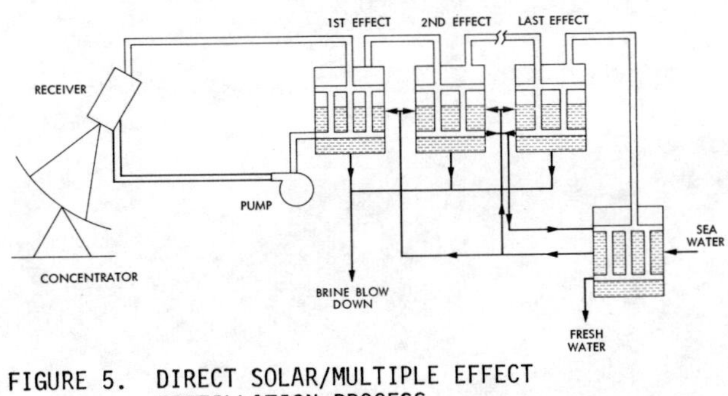

FIGURE 5. DIRECT SOLAR/MULTIPLE EFFECT DISTILLATION PROCESS.

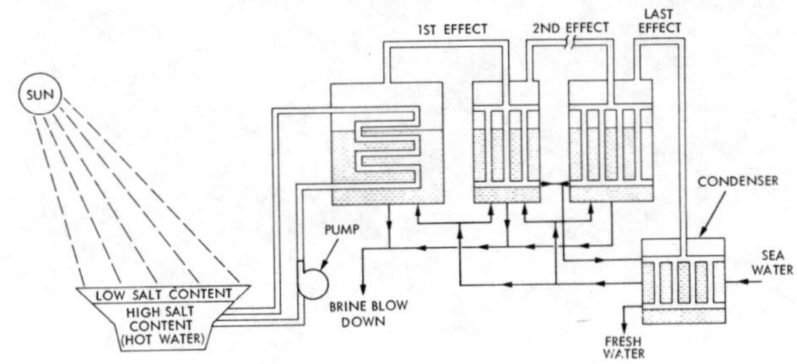

FIGURE 6. SOLAR POND/MULTIPLE EFFECT DISTILLATION PROCESS.

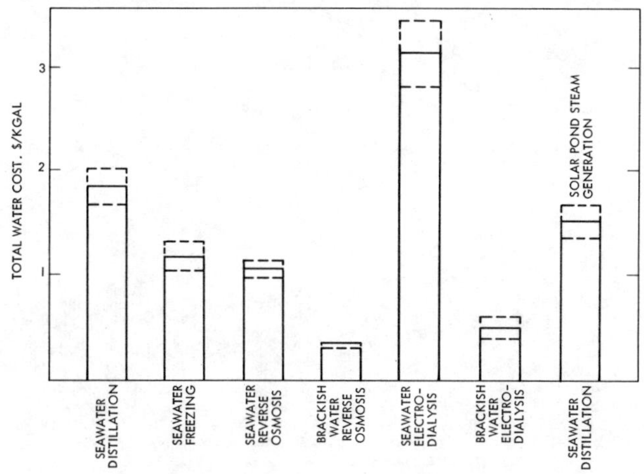

FIGURE 7. COMPARISON OF SOLAR THERMAL/DESALINATION WATER COST FOR VARIOUS DESALINATION PROCESSES. (DASHED LINES INDICATE RANGE OF UNCERTAINTY.)

SECTION 14
ENERGY UTILIZATION: MILITARY FACILITIES

Chapter 56
ENERGY ENGINEERING FOR A TOWN-SIZED MILITARY BASE

K. E. Scheuch, J. P. Gustinis

ABSTRACT

An energy engineering project for the development of energy conservation projects at a large military installation was completed which resulted in economically viable projects which substantially reduced basewide energy consumption. A major milestone was the establishment of a methodology to isolate the major areas of energy use and to logically apply applicable measures to a base of over 2,000 structures comprising ten million square feet constructed over the past 100 years. Energy conserving measures analyzed included packages for family housing, load reducing projects, system and plant modifications, heat recovery devices, a complete utility analysis and a centralized energy monitoring and control system.

INTRODUCTION

The Bickle Group, A Division of The CRS Group, Inc., has completed the first two increments of a comprehensive Energy Engineering Analysis Program (EEAP) for a town-sized military base. The work focused on cost-effective capital improvements to existing buildings, utilities and distribution systems, and the expansion of the existing energy monitoring and control system.

Activities on this project were grouped into three broad categories:

Baseline Assessment: Review of overall US Army energy reduction goals, familiarization with applicable US Army regulations and guidelines, analysis of current and historical energy consumption, evaluation of previous energy conservation studies conducted for the base, and establishment of required guidelines for evaluating specific energy conservation projects.

Development and Application of Systematic Methodology: for identifying and screening Energy Conservation Measures (ECM's). This installation consists of approximately 2,000 buildings comprising almost ten million square feet. Over seventy potentially major capital improvements to buildings, utility systems and energy management and control systems were identified. A methodology for matching and evaluating the effectiveness of each of these possible energy conservation actions was developed and applied.

Project Documentation: To be included among the US Army's funding requests to Congress for energy conservation activities. Specifically, documentation consisted of DOD funding forms for each major energy conservation project reflecting detailed descriptions of engineering analyses used to calculate net energy savings, preliminary cost estimates, and economic analyses.

Sixteen major energy conservation projects were recommended as cost effective for implementation, resulting in an approximately 10% reduction in energy consumption when compared to the specified base year of FY75. This savings is equivalent to over 33,000 barrels of oil per year. The US Army currently ranks energy conservation projects according to the ratio of energy savings to cost of implementation (E/C). The composite of all sixteen projects selected is an E/C of 62, which easily qualifies the projects in view of the FY83 goal of an E/C of 18 or greater.

In addition to analyzing and developing documentation for energy conserving capital improvements, this program included data gathering and preliminary analysis of potential energy savings to be derived from utilization of solar energy and from converting to Total Energy and Selective Energy power plants. Detailed analysis and preparation of documentation for specific projects were beyond the scope of this contract. However, significant additional savings in consumption appeared feasible from implementation of such future projects.

BACKGROUND

The Army Energy Plan dated February 1978, includes Energy Engineering Analysis Programs (EEAP) for specific US Army facilities. The objective of an EEAP is to develop systematically a package of projects that will reduce energy consumption in accordance with the goals of the plan.

Specifically, "The plan sets both short- and long-range goals for the Army consistent with the President's guidance in Executive Order 12003, dated 19 July 1977. The Army Energy Plan directs the Army staff and Major Army Commands (MACOM's) to develop detailed implementation plans and include their cost in appropriate programming and budgeting documents."

The short-range (FY85) goals include at least a 25% reduction in total facility energy consumption compared to FY75 consumption. The Department of Army Advisory Group on Energy provided the following guidance to achieve this 25% reduction:

Energy Conservation Investment Program (ECIP)	12%
Energy Management	8%
New Construction	5%

The long-range (FY2000) goals call for a 50% reduction in total facility energy consumption compared to FY75 consumption. Part of this long-range goal also calls for elimination of natural gas as a fuel source. The current EEAP portion of the Army Energy Plan addresses ECIP goals, specifically the engineering analysis of five types of energy-consuming systems:

Increment A	Existing Buildings
Increment B	Existing Utilities and Energy Distribution Systems; Centralized

	Energy Monitoring and Control System (EMCS)
Increment C	Solar and Renewable Energy Systems
Increment D	Total Energy (TE) and Selective Energy (SE)
Increment E	Central Boiler Plant

Each of these five Increments is completed in three Phases:

Phase I	Data gathering and field inspections,
Phase II	Analysis, project identification, technical feasibility and economic evaluations,
Phase III	Preparation of DD Forms 1391 and PDB's, final documentation of results and recommendations.

Only Increments A and B were completely funded, while Phase I of Increments C and D were also funded.

Baseline Assessment: Our first major activity, The Baseline Assessment, included compilation of current and historical energy consumption patterns, analysis of local climatic parameters, identification of rate structures for purchased energy, review of previous energy conservation work for the base, and assembly and review of all applicable US Army regulations and other relevant documents.

Reviewing historical energy consumption and recent weather data, we concluded that the dominant climatic variable for energy consumption at this specific location was cooling degree-days due to the wide scale use of mechanical refrigeration. We found that the base uses approximately 110 million Btu's (MBtu) per additional cooling degree-day. This highlights an important problem since the Army-wide reference year of FY75 had approximately 29% fewer cooling degree-days than an average or typical year. Reducing energy consumption by 25% below the atypical FY75 is equivalent to an actual 35% reduction when compared to a typical weather year. Thus, the use of FY75 as the reference may make it more difficult for the base to meet the FY85 goals. In accordance with Department of Army "Energy Conservation Investment Program Guidance", all of the calculations and documentation are based on a comparison with FY75 energy consumption.

Review of previous work indicates that several important steps in an EEAP had already been accomplished. A three-volume report by Freese and Nichols, Inc. Consulting Engineers was particularly relevant, and our work drew heavily on their documentation. However, as part of the Baseline Assessment, we conducted a thorough review of the Freese and Nichols work to assure ourselves that their data was sufficiently accurate to serve as our data base.

We further extended the usefulness of the data by organizing it into a computer-accessed data file. Questions such as "How many square feet of South-facing glass are there?" became simple to address. We anticipate that this computerized data base could be valuable in updating the EEAP and in developing actual construction documents for implementing the energy conservation projects.

Methodology: The base consists of approximately 2,000 buildings with a combined square footage of almost 10 million square feet. Developing a list of energy-conserving measures was no problem. Most of the buildings were constructed during a time when energy consumption was not a design constraint; many of the structures require energy conservation improvements. The significant technical problem was to quickly and efficienty compile the list of the most cost-effective capital improvements, when dealing with a large number of different buildings constructed over many decades, characterized by different occupancy schedules, internal loads, facades, etc.

A few simple observations will illustrate the point. The Bickle Group has developed a "Master List" of some seventy potential energy conservation measures. Simply analyzing each measure individually for each building would result in 2,000 buildings x 70 ECM's or 140,000 individual analyses. The problem is significantly more complicated, however. In reality, ECM's are usually undertaken more than one at a time. It is important to recognize that their effects are not simply additive. For example, if one ECM saves 10% and another ECM saves 20%, combining them will not necessarily save 30%, because their functions may conflict. If we assume that as many as five individual ECM's might be taken on a given individual building, the total number of unique configurations or analyses becomes 2,000 buldings x 70 x 69 x 68 x 67 x 66 or 2,900,000,000,000 individual analyses. Many of these calculations are redundant and can be easily automated. Nevertheless, this amount of activity was beyond any reasonable scope of work or funding.

A major breakthrough in our work was the development of a simple, systematic methodology that would allow us to quickly screen and identify the most effective ECM's for each building. A number of potential approaches were identified and analyzed; a method which grouped buildings into fourteen different categories was adopted. A typical building was defined for each category. Then, the list of ECM's was screened to determine which might be applicable to this "typical" building. A proprietary screening program was used to calculate preliminary estimates of energy savings, which were combined with preliminary cost estimates to focus on the "best" ECM's.

Subsequently, detailed analyses were conducted on each of the identified ECM's. ECM's with individual benefit to cost ratios (B/C) that fell within applicable guidelines were analyzed in significantly more detail to determine the expected energy savings. More detailed cost estimates were also calculated, and the ECM's that survived this second screening process were than developed into full-blown ECIP's with complete documentation.

The methodology is not perfect. However, it is a flexible methodology that represents a significant advance over investigating large numbers of individual buildings or components. By grouping buildings into functional types and conducting this three-level screening process we were able to quickly and efficiently identify the most cost-effective energy conservation measures. Furthermore, by identifying and incorporating discrete check or decision points into the process, we were able to involve the client in a meaningful review of the results of our on-going activities. Constructive feedback on the results of each screening step was solicited and received, enabling us to proceed to the next step with confidence that we had not inadvertently omitted or overlooked administrative, operational and/or political constraints on ECM's selected that would normally be beyond the expected familiarity of the "outside" professional consultant.

Results and Recommendations: Thirty-three potential major projects survived the first two levels of screening and were carried through to the third level of analysis. After detailed calculations, 16 of these 33 projects were found to meet the Army's energy conservation guidelines of an E/C ratio of 18 and a B/C ratio of 1.0 for FY83 implementation. These 16 projects were further analyzed and grouped into 12 ECIP's. Careful grouping of projects into ECIP packages can be utilized effectively to minimize disruption of occupants in buildings and to develop work packages which, in our opinion, provide the best overall value to the US Army.

Table 1 summarizes the 12 ECIP's, while Figure 1 shows the impact of these projects on the established goals of the base. Implementation of these 12 ECIP's will cost slightly less than $3,000,000, and will reduce the FY83 utility bill by more than $750,000. This utility cost avoidance is a result of three effects: (1) reduced consumption; (2) reduced electrical demand charges; and (3) reduced maintenance and labor costs. Figure 2 shows the cash flow analysis for implementation of all 12 ECIP's in FY82. Note that the breakeven point occurs after a little more than three years.

In addition to the ECIP recommendations indentified above, we recommended that certain maintenanace and operations improvement projects be undertaken.

Also, we believe that future Increments of the EEAP (solar energy, total and selective energy plants, and central boiler plants) will identify projects to reduce energy consumption even further--well beyond the 12% target for FY85 ECIP's.

The Bickle Group continues to be strong advocates of the "software" approach to energy conservation and we have done pioneering work in establishing administrative and policy methods geared specifically toward reducing energy consumption. The proposed Increments F and G of the EEAP basically deal with these areas of administration and policy changes as well as improved operational and maintenance practices. This work is important to the attainment of the US Army's goals in energy conservation, particularly in view of the demonstrated success rate of achieving significant percentage reductions in consumption at no or very low cost to the installation.

TABLE 1 ECIP'S RECOMMENDED FOR IMPLEMENTATION

Project Title	Payback (years)
Conversion to High-Pressure Sodium Street Lighting	1.9
Flue Gas Analyzers with Feedback Trim for Large Heating Season Boilers	7.2
Installation of Turbulators in Boilers Larger Than One Million Btu/Hr	3.3
Flue Gas Analyzers with Feedback Trim for Large Boilers Used All Year	3.1
Energy Conserving Projects for Family Housing	2.7
Roof Insulation	5.8
Solar Reflective Film	7.0
Energy Monitoring and Control System	2.5
Barracks Chiller Heat Recovery	12.5
Energy Conserving Modifications to Hospital #1	8.1
Waste Heat Recovery from Laundry Equipment	7.1
Energy Conserving Modifications to Hospital #2	9.0

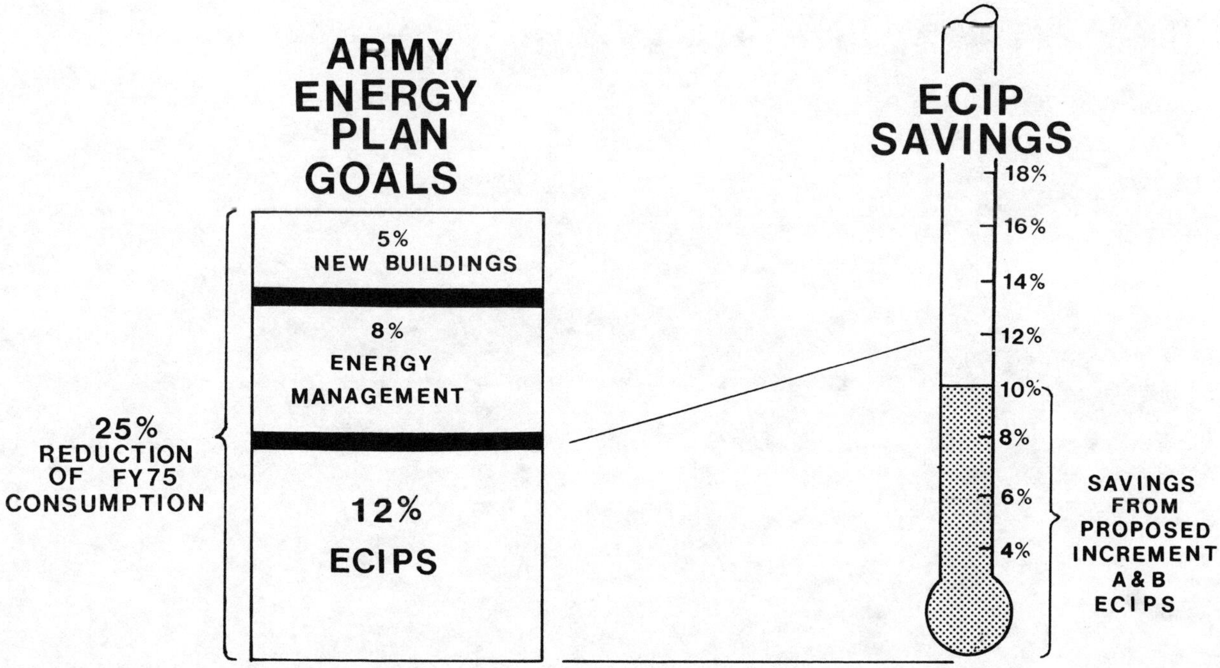

FIGURE 1: ECIP SAVINGS FROM INCREMENTS A AND B

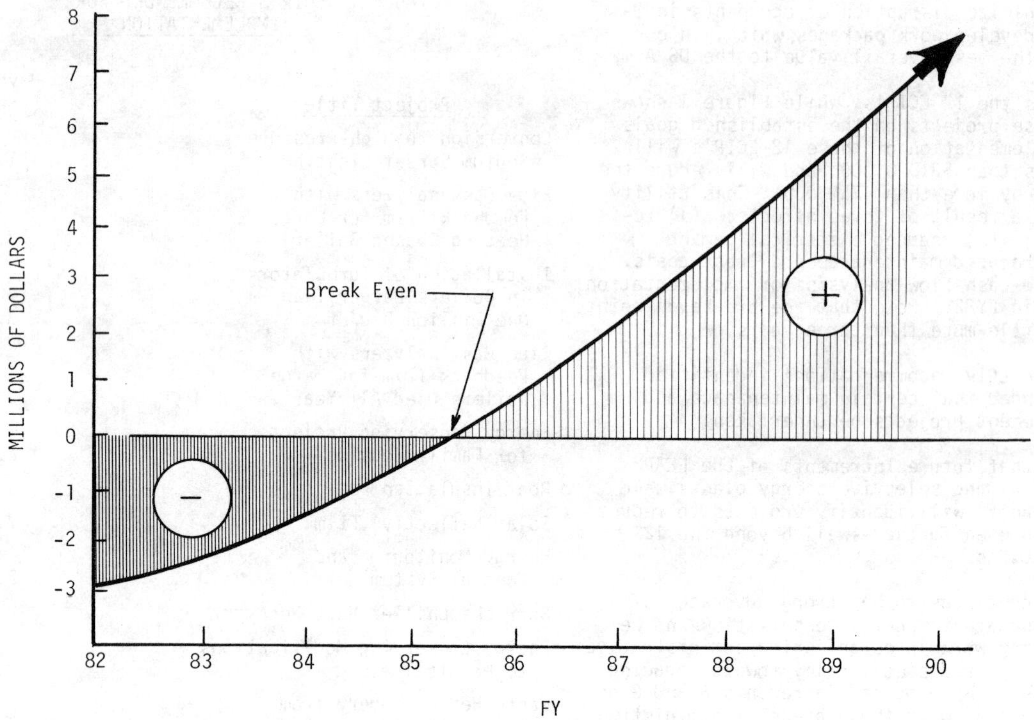

FIGURE 2: CASH FLOW FOR ECIP IMPLEMENTATION

Chapter 57
ENERGY CONSERVATION MANAGEMENT IN ARMY INDUSTRIAL FACILITIES

S. J. Crow

The concept of "Force Readiness" is all encompassing today as it has been in the past whenever we speak of the armed forces. Recently, however, a new consideration has been added to the long list of critical items that determines the readiness of a nation's forces. That consideration is energy, and it is being realized, in many sectors, both in and out of the government, just how many functions are dependent on energy availability.

In DARCOM, the US Army Materiel Development and Readiness Command, readiness is looked at in terms of the Army Chief of Staff's General Meyers, three days of war: Deter the day before war, Fight the day of war, Terminate the conflict the day after war at an acceptable level of security. Translating this to the Army's three top priorities, we have to have an industrial mobilization base that provides near-term readiness and long-term sustainability while, at the same time, conserving our dwindling energy resources. We have to provide the most energy efficient equipment for our troops in our desire to give them the best possible advantage on the battlefield. To have the best equipment in the world will do you absolutely no good if you do not have the energy to power the equipment or the fuel to transport it.

If we do not provide these assurances, our work force, our troops, and the general public will lose confidence in us, our leaders, and our nation. Energy, therefore, has a key role to play in the cycle of events of research, design, engineering, acquisition, maintenance/support, and disposal of our equipment. To accomplish these tasks in an energy efficient manner, we need to be dedicated to energy conservation and involved in comprehensive planning and execution in order to maximize the benefits from our available energy resources.

To set the stage for a discussion of energy conservation management in Army industrial facilities, we need to define the magnitude of the job. DARCOM consists of a network of 66 military installations and 250 separate units occupying over 236 million square feet of building area with a total of more than 110,000 personnel. This includes all peacetime installations, both active and layaway.

Our inventory management responsibility is awesome, in order of magnitude larger than the Defense Logistics Agency (DLA) when $23.5 billion in ammunition and other major items in the hands of troops at overseas depots is included. If DARCOM were ranked by Fortune and Moody's in their 1980 rankings, we would rank about 10th.

The command is strung out across the United States including Alaska. It operates test facilities in Panama and maintenance facilities in Europe. It monitors an annual expenditure in excess of $15 billion.

Being such a large organization with very diversified missions, functions and products, the task of energy management becomes very complex. There are two prime intents of our energy program; 1) reduce energy consumption and 2) reduce dependency on non-renewable energy resources without degrading mission readiness.

The Department of Defense consumes less than 2 percent of our nation's energy, and the Army consumes approximately 18 percent of this. Within the Army, industrial installations encompassing R&D, Production, and storage facilities account for about 19 percent of the Army's total energy consumption. Our consumption of over 8M equivalent barrels of oil in FY 80 equated to an energy bill of approximately $150M. Coupling the strategic and fiscal impacts of today's energy use, it quickly becomes very obvious why we have aggressively gotten behind energy conservation.

Because the Army's and DARCOM's mobility energy consumption are only 15 percent and 7 percent, respectively, this article will concentrate on actions in support of facility/industrial rather than mobility energy objectives.

The Army's and DARCOM's energy programs essentially were founded in the 1973-1974 time frame. They came about primarily as the result of the 1973 oil embargo and the realization that our energy supplies were not inexhaustable and uninterruptable. While energy prices did increase during this time, the financial aspects of energy conservation were not the driving force in establishing the programs. Thus, it was more of a readiness/capability requirement that motivated the program during the early days. As was said previously, the basic goals and objectives of the energy program are to reduce consumption, reduce dependence on nonrenewable resources, and to obtain a position of leadership in pursuit of national energy objectives. Specific numerical goals are established to provide realistic targets that will meet or exceed long-range objectives.

Energy management resources in personnel were very limited and at best were evidenced by a few part-time personnel at the installation level with only one or two full time coordinators available in the major headquarters. The predominance of effort for the first 4-5 years was in developing an accurate consumption reporting system that would be responsive to management needs and would provide a reliable energy consumption history on which sound decisions could be made. Also accomplished during this period was the development of basic regulatory documents and broad facilities energy plans. Most conservation projects were low or no-cost or those that had fast payback. The majority of the housekeeping and obvious retrofit projects were accomplished during this early period. Thermostats were set back to 65°F in winter and up to 78°F in the summer, insulation and storm window projects were accomplished throughout most of our active installations, neglected steam lines and traps were repaired and insulated; insulated panels replaced glass on many of our large industrial buildings and lighting levels were reduced in areas that did not require high levels of lighting.

The Department of Defense (DOD) and the Department of the Army (DA) established goals for 1985 and the year 2000 based on 1975 consumption. In most cases these goals represented estimates derived from preliminary energy surveys and assumed constant mission requirements. They were also based on building envelop energy requirements with little concern given to the internal functions of the building. For the most part, these goals have worked out to be valid targets for the Army. They have not, however, been found to be an acceptable target for DARCOM with its industrial facilities.

Because of an initial lack of experienced energy managers, goals were often passed directly down from higher headquarters to installations with no attempt made to tailor an installation's goal based on mission requirements or known energy project completions. Within DARCOM this resulted in the command achieving its 1985 goal in 1979 because of a substantial reduction in ammunition production requirements in the 1976-1977 time frame. This inexperience also compounded our energy management problems in that we could not tell how much of our consumption reduction was as the result of conservation or was gained from production cut backs. While we still do not have an absolute handle on it, we do believe we are getting it under control. An almost total lack of metering on any installation also added to the problem.

In early 1979, with the Iranian oil cut-off and the meteoric rise in petroleum prices overnight, the impact of energy management took on new importance in the military community just as it did with the rest of the nation. Within DARCOM, the two man energy office grew to six and energy conservation became a major topic of commanders at all levels. The inadequacies of existing plans with respect to industrial facilities became more pronounced and renewed emphasis was placed on intensive energy management. Basewide detailed energy audits were scheduled over a three year period for all installations at a cost of over $150M. This is a congressionally funded program designed to develop energy conservation investment projects that have a 3-5 year payback rather than 1-2 years as was required previously. Add-on increments to these energy studies will include selective energy and total energy packages at some installations as well as developing installation energy master plans at all bases.

In 1980 DARCOM was tasked by DA to develop an industrial energy plan that incorporates the total energy equation with respect to industrial facilities. This plan has been developed and approved and is becoming the foundation of our energy program. It was written to be a comprehensive planning document incorporating energy considerations into all aspects of the Army's industrial mission. As was stated previously, the earlier plans were limited primarily to building envelope or HVAC and lighting considerations. For most of the Army this is sufficient and using MBTU/sq. ft. as an energy efficiency measurement is adequate. In our production, maintenance, R&D and test facilities however, this is not adequate. We have incorporated these considerations and initiatives for our administrative and housing facilities but we have expanded the scope of the program to include our other mission functions. Programs have been developed and more are being planned to look at the industrial process with respect to waste heat recovery if the process itself cannot be changed and at altering the process if more energy efficient methods can be adopted. As an example, in one production plant the material flow was examined and found to travel 6,000 ft. and cross the bays 17 times. After process modifications this was changed so that the distance was reduced to 2200 ft. and it only crossed the bays 5 times. Since the product was a 105mm gun tube, significant energy savings were realized in this materials handling project.

Industrial equipment, much of which was installed in the early 40's, is being looked at for replacement by energy efficient equipment. With more and more energy conservation aspects being designed into modern industrial motors, pumps and equipment, this program could result in significant savings during the coming years.

A major area where significant study effort is being expanded is in power/steam plant fuel use. In FY 80 more than two thirds of our fuel used for steam generation or heating was oil or natural gas. While supplies are currently ample to satisfy our needs, it is not envisioned that this will necessarily be the case as we approach the year 2000. Consequently, a fuel use strategy is being developed that hopefully will insure that we will have the capability to burn that fuel which is both economically and abundantly available. With price controls coming off oil and gas it appears that synthetic fuels could be available in commercial quantities within 15 to 20 years. For this reason, and because of the inherent environmental and transportation limitations, we are somewhat reluctant to convert all of our large boiler plants to coal right away.

In line with the development of a fuel conversion program we are also looking very closely at installing electrical generation capability in any new steam plant. Co-generation has not caught on widely within the Army to date but it is being given very serious consideration and is scheduled to be included in a new central boiler plant facility to be constructed at Redstone Arsenal, Alabama.

Energy management control systems (EMCS) have gotten off to a slow start throughout our installations. This has not been accidental however, since there are so many types of systems on the market the development of a standard specification has been very difficult. This is being resolved though and I would expect that many of our energy audits will include recommendations for EMCS.

Energy conservation management in Army industrial facilities is a major program that has enjoyed moderate success to date. Because of the national security aspects of our capability to continue defense production in peacetime, mobilization or wartime, we cannot ignore any area that might have an energy impact. As a result, and since DARCOM aligns itself with the private industrial sector with respect to energy conservation and management we try to maintain a close working relationship with industry in order to stay current with the latest developments in industrial energy conservation.

We have made a good start in achieving our goals but we are a long way from finding the right answers to all our energy needs. Advanced technology will provide some of the solutions to reducing our consumption and reliance on non-renewable energy sources. We are investing in solar in a very limited scale at the present time but expect major gains will be achieved as the solar industry expands. Fuel cell technology is beginning to emerge as another possible source for filling smaller and more mobile energy requirements. Large multifuel fired co-generation systems also appear as part of the total energy equation of the future.

In energy management we have found that there is no single answer nor is there a single functional area that can or will solve the problem. Because of the economics involved in both supplying energy or suffered as the result of a loss of reliable energy, energy managers must constantly keep the attention of top level executives on energy actions. Large capital investment is going to be required to achieve our goals and unless we can convince those who control the money of the validity of our needs we will not be able to implement the required energy initiatives.

SECTION 15
ENERGY USE PLANNING

Chapter 58
COMMUNITY ENERGY PLANNING

D. C. Hall

ABSTRACT

In this era of uncertainty of raw materials and explosion of possibilities, comprehensive energy planning for any community requires a careful analysis of the actual energy needs of the total community, a realistic projection of future growth, an investigation of all possible alternatives capable of meeting energy needs, and the development of an implementation program to complete all necessary goals. Such a plan must cover all areas of a community, including, but not limited to, physical layout, economic requirements, regional energy situation, and alternative energy supply options. The basic procedures for developing such a plan are:

1. Perform an energy audit of the community, develop a demand profile, and project future consumption requirements.

2. Investigate all possible energy conservation and alternate energy source use options.

3. Select a strategy combining all possible options to decrease nonrenewable energy use, to increase the efficiency of energy use, and to decrease the adverse environmental effects of energy use.

4. Develop an implementation plan to successfully bring the community to a minimum energy use state.

The methodology outlined in this paper could be used as a model for other communities to develop their own en-energy plans.

INTRODUCTION

Many communities, both large and small, are aware of the potentials of the energy problem and how these potentials should be resolved to continue to provide necessary services to the people. This awareness has been precipitated mainly by economic factors since the cost of energy has increased so rapidly over the past decade.

Traditionally, community-wide problems have been handled by one area of the community (i.e., security by police and fire departments, utilities by water or power departments, and so on); however, energy use and its inherent problems involve all sectors of a community and must be addressed by the community as a whole or some areas will be made to suffer unnecessarily. To increase the complexity of the problem, such a solution must be accomplished with the minimum expenditure of funds.

Officials at all levels of government are faced with the formidable task of developing plans and policies that enable the expenditure of funds in a manner that will maximize the short-term return on investment and still allow a long-term continuous energy supply. To accomplish this, officials must know and understand both the actual requirements for energy at the user level and what options exist to meet these needs. Points to be considered are:

- The actual energy requirements of a community (type and amount)
- The effect of conservation on these requirements
- Options available to a particular community for conservation and/or shifting of energy sources
- Policies/procedures necessary to stimulate implementation of those options
- The effect of implementing those options on the regional environment of the community.

METHOD OF ANALYSIS

To develop an integrated energy plan for any community is a large undertaking that requires an analytical integration of technological, socioeconomic, and institutional factors. Figure 1 shows some key elements necessary for inclusion in such an analysis; the specific methods are:

1. Collect or develop a data base on historical energy consumption patterns, past energy conservation actions, available energy resources, and existing environmental conditions

2. Establish a list of all pertinent political, legal, and institutional factors

3. Identify conservation opportunities available to the community

4. Develop an energy profile for future years, based on historical data and future requirements (See Figure 2)

5. Identify alternative energy supply options and determine amount available to the community

6. Determine the optimum mix of conservation and supply options to provide a maximum return investment (energy strategy)

7. Determine the best possible implementation method for the energy strategy.

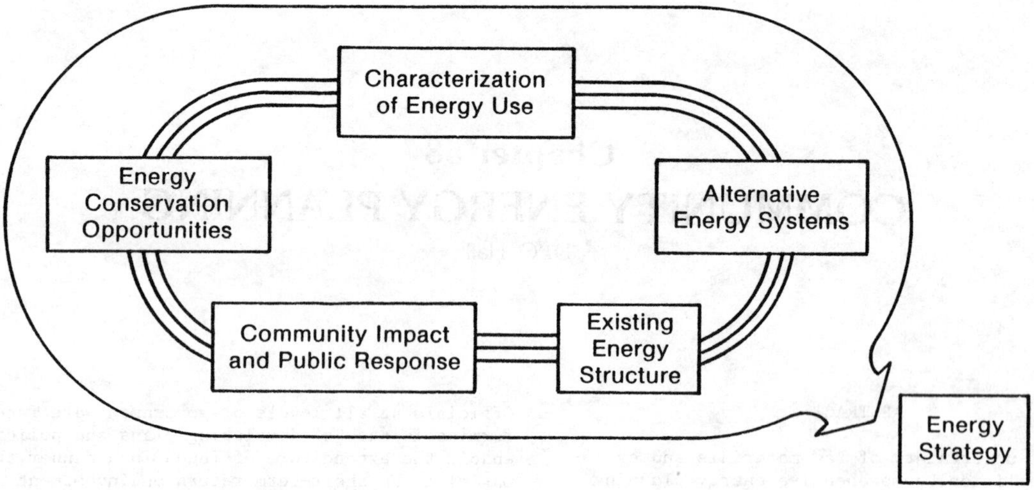

FIGURE 1. KEY ELEMENTS IN DEVELOPING AN INTEGRATED COMMUNITY ENERGY PLAN.

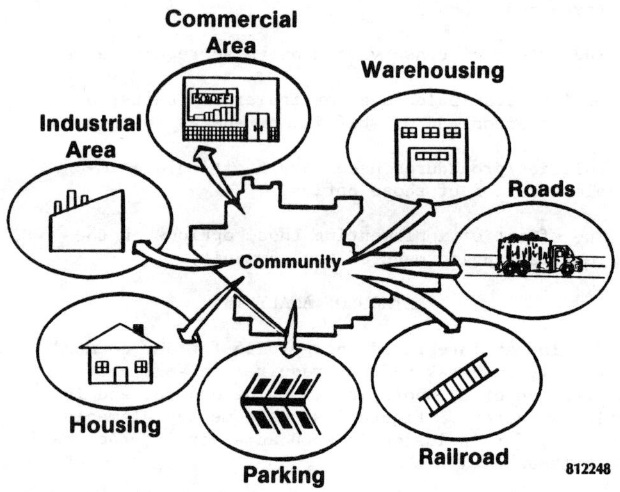

FIGURE 2. COMMUNITY PHYSICAL DATA.

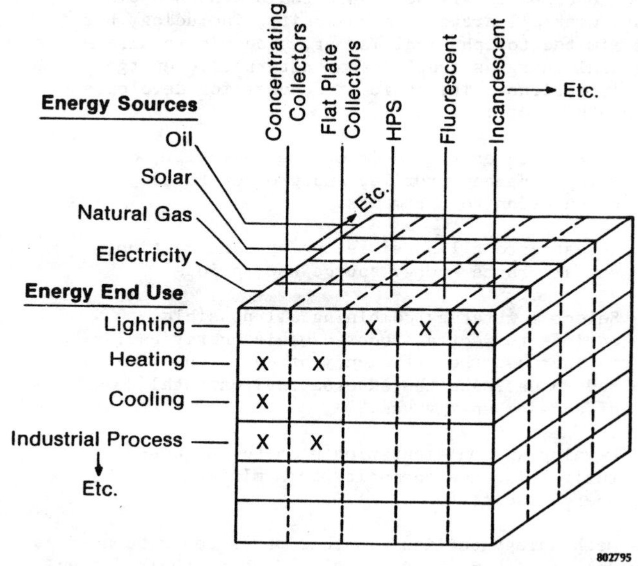

FIGURE 3. ENERGY MATRIX CONCEPT.

The determination of the optimum energy strategy, or mix of technology/implementation methods/management procedures used, is, of necessity, a matrix analysis. See Figure 3. An energy matrix of this type contains the following parameters:

- Facility energy end use and overall consumption
- Potentially useful technologies
- Potentially useful energy sources.

Within each defined parameter are numerous divisions: within the facility energy end use and consumption area are peak demand, base load, and seasonal consumption; within the potential technologies use for lighting are incandescent, fluorescent, high-pressure sodium, polarized lenses, etc.; within the potentially useful energy sources are solar, oil, natural gas, electricity, etc. This type of energy matrix, done for each type of facility within the community, establishes the minimum required design and performance characteristics to supply energy to that particular facility. It also allows prioritization of the various methods to meet those characteristics. After matrices are completed for the major types of facilities, regional matrices are done for established energy regions within the community. Then an overall matrix for the entire community can be accomplished to determine the best central energy generation and distribution schemes.

Obviously, the facility, regional, and overall matrices must be accomplished in an iterative fashion, as the outcome of each is dependent upon the outcome of the other.

The subsequent accomplishment of facility, regional, and overall matrices enables the prioritization of all actions and the development of an implementation plan, as well as appropriate investment levels.

SAMPLE ANALYSIS

A hypothetical community, a composite of several medium-size towns located in the Mohawk Valley region of New York, was used to perform a sample analysis. A composite was used to insure that all types of development procedures would be used in accomplishing the energy plan. The area regional climate, a major driving force in energy consumption, is relatively harsh in the winter and mild in the summer; there are four distinct seasons:

- Winters are three to four months long and are generally cold and damp, with frequent snow, high winds and low clouds prevailing for extended periods.
- Spring is relatively mild with frequent rain and wind.
- Summers are warm, with temperatures generally in the mid 70s to the low 90s.
- Falls are mild and essentially repeat the spring weather patterns.

A typical community in this area normally will contain one or two medium to heavy manufacturing plants, several small commercial areas, and the normal diverse mix of community housing.

RESULTS

Using the analysis method described, several important results have been obtained. Based on assumptions made at the start of the analysis (assumptions such as cost of energy per year, material costs, etc.) this hypothetical community could reduce its annual energy consumption to 25% to 35% below its existing consumption and provide approximately 10% of the remaining requirement from the use of renewable resources. This mix of conservation and energy source change is the optimum for this community in terms of maximum return on investment in the shortest period of time.

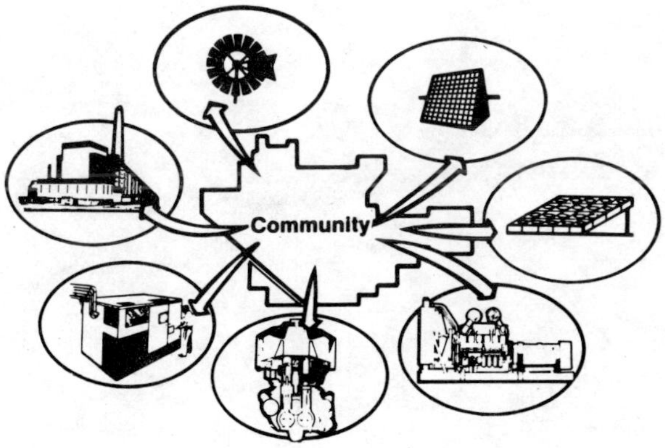

FIGURE 4. ENERGY SOURCES AND TECHNOLOGIES AVAILABLE TO A COMMUNITY.

Examples of energy sources and technologies for possible use are shown in Figure 4. Attendant yearly cost/energy savings expected from implementation of such a plan are shown in Figure 5. The major reduction in energy consumption comes from the use of existing techniques and technologies in a planned and systematic manner on individual facilities, both housing and commercial/industrial types.

CONCLUSIONS

There are several important conclusions to draw from such an analysis. First, there is no firm, fixed, best energy strategy for any particular community. Each analysis depends directly on basic assumptions made at the start of the analysis, and, if a particular assumption changes, then that portion of the analysis must be redone.

Second, in order to achieve anywhere near the maximum energy use reduction that an Integrated Energy Use Plan can allow, the community must be willing to dedicate all the resources necessary to fully implement the plan. (This entails directing a multiyear program through all changes in public opinion, public officials, state and local policy changes, etc.) If the indicated level of investment for a particular year is not met, it will precipitate a "domino effect", and reduce the number of actions possible in the next year; historically, this postponement type of action has always increased the cost of a program, not decreased it. Although most people do not recognize the fact, the availability of energy is fast becoming the economic lifeblood of a community. Availability and reliability of an adequate energy supply has become extremely important not only to the private citizen, but also to industry. For example, if a community can show that it has a good plan to insure energy availability, such a plan could be a deciding factor in industry locating there.

Third, the people of the community must be involved in any conservation plan to make them intimately aware of the requirements for energy conservation. A relatively minor amount of dedicated manhours and resources put to this task have the potential for at least a 10% reduction in the overall energy consumption in any community. (CAUTION: this must be a continuing task; it cannot be done once and then forgotten, for people are only aware of those ideas continually brought before them.)

Lastly, it is much more cost effective to conserve energy than to try to increase or maintain a high level of supply. This energy strategy development emphasizes reducing energy end use requirements first, and then finding the best sources to supply that energy.

This nation must stretch its energy reserves, both renewable and nonrenewable. The more energy used, the more the environment and individuals feel the impact. Using the minimum required, not wasting more than the laws of physics demand, must be a primary goal. See Figure 6. The major energy use reduction is occuring through conservation by the American people on an individual basis, driven by economics. Current national goals (use solar energy more, convert oil users to coal, etc.) are fragmented and directed more toward trying to maintain or increase the needed supply rather than properly evaluating the situation. This paper is an attempt to show what can be accomplished by establishing priorities taht will enable each community to receive the maximum return on its capital investment.

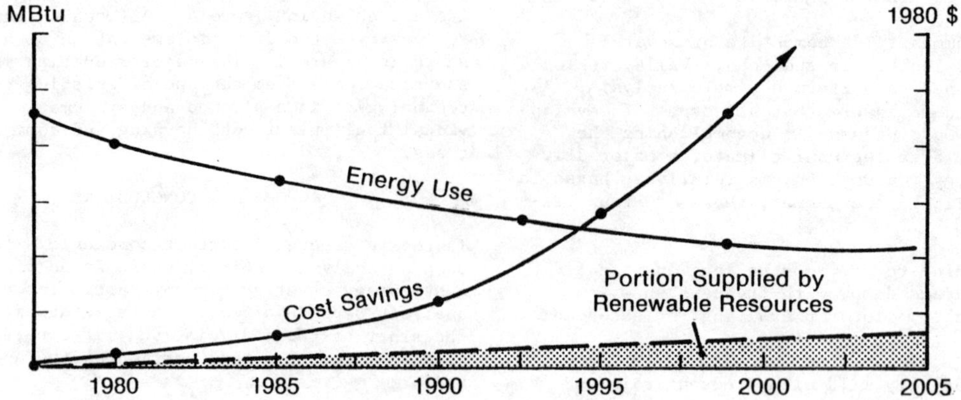

FIGURE 5. ESTIMATED YEARLY ENERGY/COST SAVINGS.

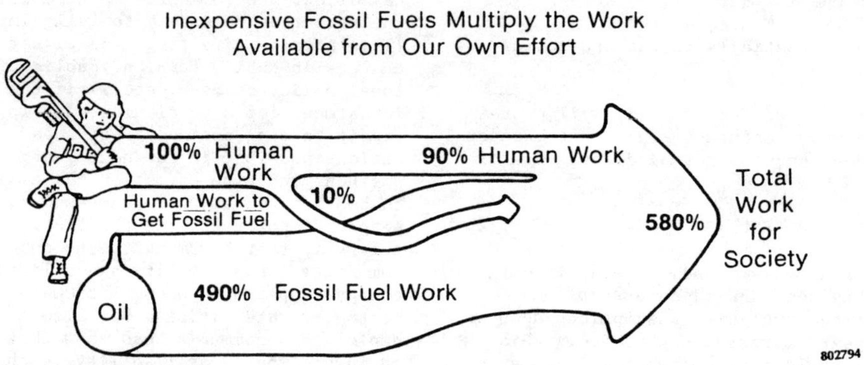

FIGURE 6. STRETCHING OUR RESERVES OF FOSSIL FUELS KEEPS THEM AVAILABLE TO DO DIFFICULT WORK FOR US IN THE FUTURE.

Chapter 59
END USE TECHNOLOGY-TO-MARKET

T. E. Root

INTRODUCTION

This paper addresses the process which has been developed to analyze new end use technologies so that they can be properly integrated into corporate planning strategies. This includes the development of marketing programs, new ventures, system impact strategies, and their effect upon the electrical load which, in turn, impacts the generation and transmission expansion plans.

The technology assessment process at Detroit Edison was formalized with the inception of the Planning Department in the mid-1970's and assigned as a function of the Resources & Facilities Planning group. One of the first products of the group was the Technology Assessment Report which was a screening process for all technologies that might impact the Company, and which provided a basis for a more detailed analysis for those technologies which were identified as high priority. This report has been periodically reissued and updated, and has evolved into a standardized format resulting in a working document for planning purposes, and in a reference and resource document on new technologies. As this report has evolved, it has become increasingly clear that generation and transmission technologies and end use technologies must be treated differently from one another.

In 1979, a decision was made to separate end use technologies from generation and transmission technologies because of the realization of the different impacts and needs of the two types of technologies, and the major shifts in energy prices, economic growth, etc., which were occurring in our service area. The separation was also predicated on the growing need for better interdepartmental communication on end use technology as a result of the formation of the New Technology Marketing group in the Marketing Department and the Energy Utilization Research group in the Engineering Research Department.

DEFINITIONS

New End Use Technologies are those technologies which impact the electrical system on the customer side of the meter, but which have not get gained a significant market penetration in the Detroit Edison service area. These technologies may or may not be commercially available at the present time.

New generation and transmission technologies are those technologies which can be utilized for the production, transmission, or distribution of electricity, but which have not been implemented by the Company, or for which significant ongoing development is occurring.

New ventures represent areas of business outside of the production and transmission of electricity, but which are in some way related to electrical production, transmission, or use.

System impact strategies are plans to accommodate new end use technologies which, if implemented, may impact the reliability of the electrical system, but which offer no potential for the development of marketing programs.

RATIONALE

Generation and transmission technologies and end use technologies differ significantly in their impact. Generation and transmission technologies, as they become commercial, serve as input to the planning process by offering an additional resource which can be utilized in expansion or maintenance of the electrical system. End use technologies, on the other hand, are implemented by customers, and are largely outside the control of the utility except for the influence of marketing programs. Thus, generation and transmission technologies offer a basis for corporate action in reaction to the efforts of manufacturers and customers.

The implementation of new technologies by electrical customers has had a significant effect upon the planning of our generation and transmission systems. Today, the implementation of new end use technologies and their impact upon electric demand and energy use can occur in a much shorter time frame than our planning and construction of major additions to our generation and transmission capacity. Thus, it is becoming increasingly important to assess the development of user technologies and their impact. The information developed in these assessments provides valuable input to the load forecasting area as an input to end use modeling. Since the load forecast serves as the basis for planning generation and transmission expansions, end use technology assessment can ultimately impact the total corporation.

New end use technologies also represent an area of interest to the Marketing Department in developing marketing strategies for the Company. The process which has been developed is designed to provide direction for the Marketing Department by identifying those technologies which have the greatest potential impact upon the use of electrical energy, and for which marketing strategies should be developed. This will allow the most efficient utilization of those generation facilities in existence and under construction.

The combination of factors described above has led to the realization that the methodology used in the past for assessing technologies is not adequate for the assessment of end use technologies. This has led to the development of the following process which takes into account the greater complexity in determining the potential impact upon the Company of technologies which will be implemented by others and produces results directed at load forecasting, system impacts, and marketing instead of system expansion.

METHODOLOGY

The process for developing end use technologies into market programs is divided into four distinct steps: the end use technology assessment report, prioritization of high priority technologies, detailed assessment of those technologies, and corporate action. These steps are shown schematically in Figure I, along with their impact upon the electrical load forecast, and are discussed in detail below.

End Use Technology Report

The first step in the process which ultimately leads to

the development of marketing programs, new ventures, and strategies for accommodating the impacts of new technologies upon the electrical system is the issuance of an end use technology assessment report. This report is intended to consolidate Company expertise and to provide preliminary assessment data on a wide variety of new end use technologies. The preliminary assessments broadly quantify high priority technologies and serve as a basis on which management can intelligently commit manpower and resources toward the coordinated performance of detailed assessments. It is critical that this first step be performed carefully, since it defines the technologies which are to be analyzed in greater detail. This process will insure that the impacts of end use technologies are addressed in the most economic and efficient manner.

A committee comprised of representatives from the Planning, Marketing, and Engineering Research Departments, and led by the Planning Department, is responsible for developing the End Use Technology Assessment Report. If other expertise is needed, it can be acquired on an ad hoc basis. Although input to the report is produced by all three departments, Planning has the overall responsibility for authoring and issuing the report.

Initially, an extensive list of technologies is identified, then ranked according to potential impact and time frame as described below. For those technologies with the highest rankimg and most immediate impact, a preliminary assessment is performed.

Generation of the list of technologies to be considered is performed by the committee. This list comprises all end use technologies that committee members could identify as having a potential impact upon energy use. From this "laundry list" each committee member ranked each technology. The primary criteria used for ranking the technologies are time frame, level of impact, and general interest level as shown in Table I. Time frame subdivisions are based upon the ability of the industry to bring a developed technology to commercialization in eight years or more. In addition, the public acceptance of the technology is considered with commercial availability; thus, even though a technology is commercially available, its impact can be delayed until it is accepted by the general public. Within each time frame, the technology was ranked by its anticipated impact upon the Company, and by the expected interest in the technology by the general public and the government.

Upon completion of individual rankings, a consensus ranking is compiled by the committee. The high priority technologies for preliminary assessments are chosen from the consensus list. Preliminary assessments are then performed. The information required in each assessment was developed to address the technical, economic, marketing and system impact aspect of the technologies. These aspects were translated into a standardized format as shown in Table II. The assessments broadly qualify the impact of each technology upon the Company.

As a result of the preliminary assessments, some of the technologies previously identified as having a potentially significant impact are eliminated. For the remaining technologies, the qualification process provides a sound basis for management review in the market and system impact screenings phase.

Prioritization

The recommendations of the first step are an identification of technologies which warrant detailed assessment. Marketing, Planning, and Engineering Research Departments then establish the priorities for performing the detailed assessments. The Marketing Department prioritizes the technologies for marketing opportunities, and the Planning Department screens for system impacts. Representatives from the three departments, together, meld the two lists into one working plan. For those technologies chosen in the screening process, additional commitments of time and money are required to produce detailed assessments; thus, the involvement of the three departments.

Detailed Assessment

For each technology to be assessed in detail, an ad hoc project team is established consisting of one representative from each of the three key departments. Leadership of the team depends upon the nature of the technology being assessed. Each project team is responsible for developing test plans, schedules, and budgets; overseeing the detailed assessment; and publishing a report on the work. The project organization may have a different composition from the End Use Technology Committee, since the function of the project organization would be designed to oversee the detailed analysis of one specific technology. Actual analysis is performed by, or under the direction of, various departments as appropriate for the technology.

The detailed assessments address individual technologies and include indepth market analysis and technical evaluation. The scope of each assessment is tailored to the technology and could range from a paper study to a multi-year research and development program. The key areas addressed in the detailed assessments are the same as those for the preliminary assessments (Table II); however, the best available expertise is called upon for these analyses, including consultants, manufacturers, and outside research firms.

The detailed assessment report, which combines the market and technical analysis and technical research, provides such information as the expected market penetration rate; area of marketability, e.g., urban, suburban, or rural; basic marketing strategies; the current status of the technology, including outside contact persons. Recommendations of the report include additional development required before initiating or in parallel with, a marketing program, a system impact strategy, or development of a new venture.

The results of each detailed assessment provide a detailed basis for determining the nature of a marketing program, system impact strategy, or new venture, and the appropriate timing for such a program. In addition, recommendations for support facilities to provide additional research or development work are included, along with projected scope and budgets for such facilities. Both written and oral presentations are made to the appropriate department for inclusion in their future plans.

Corporate Action

Marketing programs or new ventures are initiated by the Marketing Department based upon the results of the detailed analyses. System impact strategies are developed by one of several departments depending upon the imminence and degree of impact from the technology.

Support facilities are under the jurisdiction of the Engineering Research Department; however, the budgeting for such facilities could come from either Engineering Research or the initiating department, depending upon the nature and scope of the facility required.

ADDITIONAL PRODUCTS

Each step in the process described above provides increasingly refined information as input to end use load forecasting. The results of the End Use Technology Assessment Report provide input product in the form of a list of technologies and preliminary data on those technologies with potentially significant impacts on electrical energy use and demand, and also a list of technologies which will probably have little or no impact. The detailed assessments provide specific projections of the impact of a technology, over time. Finally, for those technologies for which a marketing program is developed the goals of the marketing program, as well as the ongoing monitoring of the program, then serve as additional input to the load forecast.

Although the goal of the end use technology-to-market process is not to perform load management studies, the technologies of interest and importance are the same in both cases. Since the detailed assessments provide nearly all of the information required on a technology to perform a load management analysis, the process described herein provides additional support to those areas concerned with load management.

TABLE I

CRITERIA FOR RANKING END USE TECHNOLOGIES

Primary Criteria

A) TIME FRAME (Commercially Availably & Accepted)

1) 0 - 3 years
2) 4 - 7 years
3) 8 years and beyond

D) LEVEL OF IMPACT (By Time Frame)

1) High
2) Some
3) Low

C) GENERAL INTERST LEVEL (By Time Frame)

1) High
2) Some
3) Low

TABLE II

ITEMS TO BE ADDRESSED IN PRELIMINARY ASSESSMENTS

1) Size and Characteristics of End-User Market:
   - Seasonality
   - Time of Day
   - By Use -(lighting, heating, etc.)

2) Description of Technology

3) Status

4) R&D Level

5) Projected Commercialization

6) Competitive Technologies

7) Government Interest

8) Barriers to Implementation:
   - Technical
   - Marketing

9) Economics:
   - Real
     - Capital
     - Operating
   - Perceived

10) Socio-Political Impact

11) Rates

12) Recommendations & Conclusions
   - R&D Goals
   - Marketing Goals
   - Load Shape
   - kW-kWh/Unit

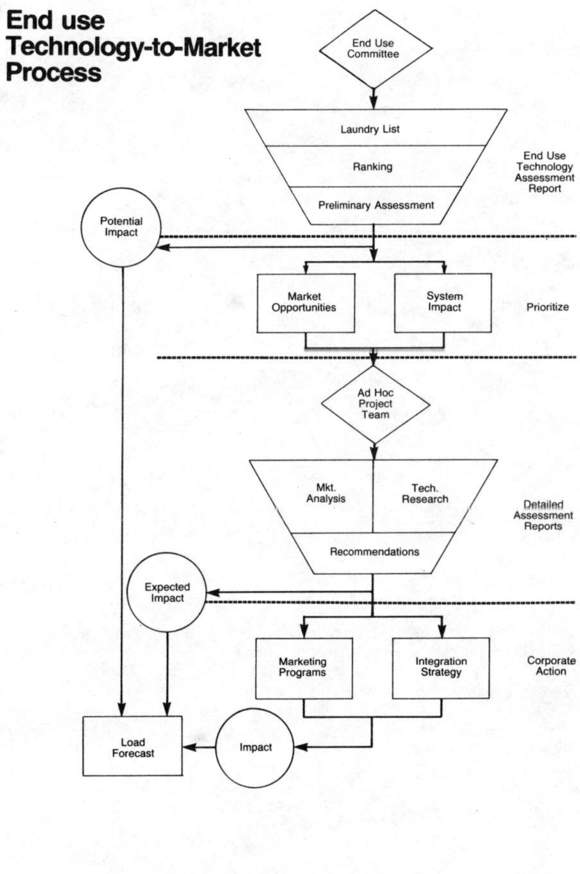

SECTION 16
ELECTRICAL ENERGY MANAGEMENT SYSTEMS

Chapter 60

A CRITICAL ANALYSIS OF ENERGY MANAGEMENT SYSTEMS

C. Bullock

ABSTRACT

The decade of the 1970's has seen many advances in electronics hardware and software and in instrumentation technology in general. The decade of the 1980's will see many more.

These developments of the 1980's have, to a great extent, been assimilated into state-of-the-art energy management system design. Undoubtedly, the assimilation process will continue in the future.

This paper examines the relationship between advances in electronics and instrumentation technology on the one hand, and energy management systems on the other. It speculates about future systems design based on technological advances anticipated in the 1980's.

GENERAL

Energy management systems are important not only because of their potential contribution to energy conservation, but also because of their potential as a significant growth industry in the 1980's. Growth in this sector has been quite impressive. In 1970, there were but a handful of companies making energy management systems. Today, ten years later, there are more than one hundred and fifty. In 1970, there were only a few hundred installed systems. Today, there are several thousand. Today, the market is estimated at approximately 300 million dollars annually, with an annual growth of 15 to 20 percent per year. It is expected that by 1990, this market will exceed $1.8 billion annually. This future important technology in the 1980's should be of great interest to us all as professionals concerned with the efficient use of energy. I would like to share my perspective on where this important technology is going in the 1980's by addressing three central issues. First, I will consider the issue of what is the state-of-the-art in energy management systems today. Second, I will consider the forces motivating changes in current systems. Finally, I will consider how today's system will change in response to these forces.

The state-of-the-art in energy management systems can be seen from several perspectives. I will discuss the functions performed and performance.

FUNCTIONS PERFORMED

There are several classes of control strategies employed in most large systems today. Not all systems, of course, employ all types of strategies.

<u>Demand Limiting Control</u>: Large users pay for electricity based not only on their usage (i.e., kilowatt hours) but also on the basis of demand (kilowatts). The latter refers to the maximum power used in a given period and is computed by measuring usage over a short period of time, such as a 15- or 30-minute interval. Utilities charge dearly for demand and reflect directly the amount of reserve capacity required to serve all users. Typical charges for demand are $3 to $7 per kilowatt hour.

The principal governing demand limiting is very simple. It consists of sensing the instantaneous load in a facility (preferably at the demand meter or meters) and shedding or turning off loads as the demand rises.

Demand limiting is most appropriate when isolated peaks of usage occur. It is especially useful when, in addition, an electric rate has a ratchet clause. (A ratchet clause is a rate method which bases demand charges for the current month on the greatest measured demand in several previous months, for example, the past 11 months. Hence an inadvertent peak in one month costs very dearly over several future months. Ratchet clauses are increasing in popularity.)

Demand limiting is less useful when electrical load doesn't change significantly over the work day.

Duty Cycle Control: Duty cycling is a strategy used to reduce electrical consumption and demand due to the coincidental activation of loads simultaneously. Duty cycling forces a load off for a period of time (e.g., five minutes off, 15 minutes on) to allow the "off time" to be utilized in a scheduled, predictable manner without breaking the thermal inertia of the system the load is servicing.

Time-of-Day Control: Time-of-day control is set to turn loads off and on based on a schedule or schedules. The control strategy is based on a real time clock and calendar which automatically controls loads when the clock and calendar reach a schedule point. Time-of-day control can be used to schedule daily, weekly, weekend, holiday or any other event.

Chiller Reset Control: Chilled water cooling systems use the chilled water in the system as a heat sink to reduce the ambient temperature in desired areas. Many systems maintain a constant chilled water temperature regardless of the system demand. An energy management system can reduce energy consumption by allowing the chilled water temperature to float upward when the cooling demand is reduced due to low outside temperatures.

Optimum Plant Start-Up/Shut-down: Plants without energy management systems generally start-up in a very haphazard manner. Loads turn on either slowly or rapidly depending on how quickly the staff switches on lights and equipment. Because of the unorchestrated manner in which this is done, it can lead to unjustified surges in demand. At the end of the shift, the plant generally turns off in a haphazard manner, which leads to excessive electric usage. Energy management systems can solve both these problems quite easily.

Brownout Control or Directed Load Reduction: Electric grid demand is generally increasing at a faster rate than utility capacity. Because of this, industrial and commercial users must increasingly have the capability to respond quickly to directed load reductions or "brownouts" from the utility. This problem is solved in a manner similar to demand limiting.

Maintenance Forecasting: Energy management systems are frequently used to monitor the status of equipment throughout the facility. They can frequently detect a failing unit long before it actually stops operating. Because of this, equipment is usually maintained better when used in conjunction with an energy management system.

Fire/Security Control: Operationally, fire, security, and energy management are very similar. Because of this, systems which perform energy management functions will often perform fire and security control as well. Indeed, most large systems on the market today offer fire and security control software as a standard software option.

Information: There is an increasing demand for quick, up-to-date information, not only with respect to energy usage and demand, but also for savings projections, use projections, energy-related maintenance reports, to name but a few. Obviously some of this demand has been fueled by the capability of having such reports. Much of this demand, however, has been stimulated by an uncertain energy situation and the realization that better information translates into better decisions as well as dollar savings. Energy management systems, particularly larger systems, can provide this type of information very easily and very quickly.

These are the major functions found in EMS/EMCS systems today. Not all system classes perform all functions. Generally, the smaller systems are confined to demand limiting and simple time related switching functions. Many of the strategies described above are closely related to other strategies described, and sensitivity to this interdependence must be observed in implementing a system for proper operation. In general, smaller systems are less able to handle interdependencies among different strategies. Larger systems, of course, provide more strategies and flexibility and are more able to accommodate interdependencies among strategies.

THE PERFORMANCE PERSPECTIVE

The performance perspective addresses the questions of cost and savings associated with an EMS/EMCS installation. Let's deal with cost first. This varies considerably with the application and installation. Cost, for example, is a function of the number of points and the type of points. There are two point types: analog and digital. The latter represent a simple switch closure and are relatively inexpensive, while the former represent electrical signals which may take on any value within a continuous range. They are more expensive because they require more sophisticated electronics. Obviously a system whose applications needs can be satisfied primarily with digital points (e.g., primarily demand control and duty cycling) will generally be much less expensive than a system whose application demands an equivalent number of points, many of which are analog points (e.g., HVAC control). Having said this, there are some generalizations which can be made. Small systems of less than 30 to 40 points sell in the $300 to $500 range per point installed, with wireless systems often costing much less. Large systems of several hundred points sell in the $750 to $1,200 per point range.

There are several reasons small systems are much less expensive than large systems on a per point basis. First, the information and control requirements of small systems are much less, so the fixed cost portion (i.e., processor hardware and software) is much simpler and therefore less expensive. Second, small systems are predominantly digital which are, of course, less expensive than analog points. Third, small systems usually control points within much smaller geographical areas, so wiring and installation is much easier.

Savings also vary considerably with the application and with the conservation strategies employed. A study of EMS sponsored by Brookhaven National Laboratory and performed by Honeywell in 1978 estimated expected savings to be between 10 and 27 percent of building energy. (See Figure 1.) This is consistent with performance of today's systems as well.

AEMS PRIMARY CONTROL FUNCTION CATEGORY	RANGE OF OVERALL ENERGY SAVINGS (% of Total Building Energy)
ELECTRICAL DEMAND CONTROL	2 to 7
LOAD EFFICIENCY IMPROVEMENT • Supply temperature reset • Boiler optimization • Chiller optimization • Sequence heating and cooling • Combustion efficiency	2 to 10
VENTILATION • Free cooling • Warm-up cycle • Enthalpy	3 to 5
HVAC SCHEDULING • Temperature reset • Start-stop • Optimal start • Duty cycle	3 to 5
TOTALS (DIRECT ADDITION)	10 to 27

FIGURE 1

Small systems pay back very quickly, generally within one to two years. This should not be surprising since the simplest EMS, a clock thermostat, pays back within one year and many small systems are little more than very intelligent timeclocks with demand limiting control.

Paybacks in large systems are quite respectable as well and generally fall in the three to five year range. In fact, it is not uncommon to find large systems realizing two and one-half year paybacks.

THE FUTURE OF EMS/EMCS

What can we expect of EMS/EMCS in the next five to ten years? Let's start by asking what factors are motivating change in system design.

I see four primary forces which influence and will continue to influence EMS/EMCS system utilization as follows:

1. increased energy prices and unstable supply
2. decreasing costs of control technology
3. high inflation, particularly as reflected in increased labor costs for skilled workers
4. increased public awareness and acceptance of computer controls as a way to reduce energy requirements.

<u>Increased Energy Prices and Unstable Supply</u>: Little needs to be said about this issue to this audience. Suffice it to say that the cost of end use energy will continue to increase faster than the general rate of inflation.

<u>Declining Cost of Automated Controls</u>: The cost of automated controls is related to three computation indices: memory cost, secondary storage costs, and processor costs.

The cost of computer memory has declined by approximately 50 percent in actual dollars over three years for the past 20 years. The decline in constant dollars has, of course, been much more rapid.

The cost of secondary storage, which is an index of computer peripherals cost, has declined by more than 60 percent in actual dollars every three years for the past 20 years.

The cost of central processor computing power has declined by 92 percent every five years for the past 20 years.

These trends are slowing; there will, however, continue to be very strong downward price pressure in this technology through the decade.

Continuing Inflation: We will all undoubtedly be very pleased if the Reagan administration succeeds in reducing inflation sharply without severely depressing the economy. Most traditional economic forecasting firms, however, are still projecting annual rates in excess of nine percent for the next five years. Perhaps conventional wisdom will be wrong.

Even if the general rate of inflation abates significantly, however, EMS/EMCS industry will most probably face wage inflation well in excess of the general rate over the next five years for one simple reason: there is a temporary shortage of skilled labor in computer and in plant engineering disciplines, and this shortage is expected to last three to five more years. Since labor content of EMS/EMCS as a percent of sales is relatively high, ranging from 30 to over 50 percent, this is a significant factor.

User Awareness: Potential users of EMS/-EMCS are much more aware and receptive to computer technology than users of even five years ago.

There are two reasons for this. First, computer technology has been widely available and, for the most part, it is a technology which performs well as measured in accuracy and reliability.

Second, young people entering the labor force today are introduced to computers early in high school, and so they readily accept automation. Recall that about a generation has passed since the first minicomputers were introduced (mid-1960s).

Several factors of EMS/EMCS will be affected by these trends. You will notice a close parallel between what can be expected to happen in EMS/EMCS and what has already happened in data processing.

Hardware: Several trends can be anticipated in EMS/EMCS hardware.

First, system networks will tend increasingly to make use of existing, or at least, inexpensive wiring. Wiring installation and check-out now constitute approximately 20-25 percent of typical installation costs, and it is very labor intensive. Using existing or inexpensive wiring could reduce this by more than 50 percent. Of the existing wiring used, we can expect to see use of telephone lines as in the recently released Bell Dimension System or office information systems wiring such as ETHERNET. We can anticipate that interand intra-plant electrical wiring will probably be of lesser consequence in large facilities where electrical wiring is generally an unreliable and uncontrollable communications medium. In small facilities, however, use of wireless systems could become very significant.

With regard to inexpensive wiring, we can expect systems to tend toward common data bus configurations, not unlike most concepts proposed for office information system networks.

Second, at a microscopic level we can anticipate that hardware will become much more complex and smarter, while at a macroscopic level hardware will become more modular and simpler to use, simpler to install, and simpler to repair.

Third, we can anticipate that system monitoring and control will be much more distributed than we see in today's systems. In the short run, the central processor will be devoted much more to information processing user interface requirements, and coordination among distributed units as opposed to specific monitoring and control. In the long run, we can expect to see the demise of the central processor. In the long run, the information processing may very well be distributed among many identical hardware modules or distributed processing units (DPUs) on a common bus while the user interface for accessing that information is just another DPU with a terminal attached.

Software: System software will undergo changes similar to system hardware.

First, at a microscopic level it will be much more complex, while at a macroscopic level it will be simpler and easy to use.

Second, software even on small systems will be configured to teach the user. One revolutionary technique pioneered most successfully by word processing manufacturers like Wang, Vydec, etc., has been the development of machines which teach the user. A consequence of this breakthrough is that very little customer support is required. The technique is used only crudely on EMS/EMCS equipment today and generally only on larger systems.

Third, software will increasingly become firmware as concepts of structured programming become more pervasive in system software design. We will see this trend because it leads to a better return software development and permits adequate support at reduced costs. It also leads a product line in the direction of depot or replacement maintenance, both of which complement the tendency of hardware to be more modular.

Fourth, software will increasingly be implemented only in higher level languages, thus making them less expensive to support.

Prices: Several factors figure into future EMS/EMCS prices. At a superficial level, advances in computer and manufacturing technologies clearly tend to reduce the materials cost of EMS/EMCS in constant dollars. Labor costs, on the other hand, appear to increase the system costs. Given the labor content of a typical job, one would probably expect prices to rise in constant dollars.

However, several trends mitigate against this. The tendency toward common bus systems will tend to reduce labor content in installation and checkout. Increased modularity of hardware will tend to do the same. The trends described in software tend to reduce the system and applications support, both of which are very labor intensive.

In addition, and as is common in computer technologies, many of the trends described in software and hardware are much more advanced in smaller systems. Hence on balance, we should anticipate the price of large systems should continue to fall substantially in constant dollars. Small systems, on the other hand, will probably remain the same price in constant dollars unless significant ways to reduce labor content are proven (e.g., wireless systems).

SECTION 17
FLUIDIZED BED COMBUSTION

Chapter 61
LOW QUALITY FUEL BURNING IN FLUIDIZED-BED COMBUSTORS

J. E. Mesko

ABSTRACT

Technical and economic analysis of the construction and operation of coal washery waste burning steam and by-product electric power generation facility, located at an industrial plant, is being presented. The facility analyzed employs fluidized bed boilers for the generation of steam for process and building heating/cooling purposes, in conjunction with electric power co-generation. Results of the analysis are presented, using life cycle costs and investment payback periods, return on investment, pinpointing the areas, type and magnitude of costs which should be considered in the selection of the combustors, or systems. Capital and operating costs, and recognized economic barriers are also presented and their effects indicated. Life cycle cost of each of the alternatives analyzed are compared and the expected payback periods for different annual fuel quantities, plant sizes and for different annual production levels are discussed.

BACKGROUND

A detailed engineering study was performed, evaluating the application of steam/electric power co-generation systems with fluidized bed combustors (FBC) as a means of burning low quality washery waste efficiently and in an environmentally acceptable manner. The study addresses an existing mid-western steel company, currently furnished with steam from oil/gas fired steam generating plants. Average age of the existing plants is over thirty-five years.

Based on the findings of the economic evaluation, sufficient information was provided to enable corporate management to assess steam and power co-generation, with the use of washery wastes and FBC technology as an alternative thermal energy source, to increasingly scarce and expensive petroleum and natural gas fuels. Technical and economic feasibility, using life cycle cost method, was analyzed and evaluated in detail.

Washery Waste as Fuel

Washery waste burning combustion tests were conducted in the laboratories of an eastern university. Preliminary results of the small scale fluidized bed tests indicated that it is possible to burn washery wastes without supplementary fuel firing. Test results indicate that combustion can be started over a range of combustion temperatures and a stable hot bed of waste ash can also be maintained. The tests also indicate that when a small amount of coke oven gas is co-combusted with washery waste, improved combustion characteristics and increased combustion efficiency are attained. The co-combustion tests were based on using 0.7 cubic feet of coke oven gas per pound of waste and resulted in an increase of overall combustion efficiency from 55% to 70%.

The tests utilized small scale laboratory size equipment, therefore, full scale combustion tests will be required to verify these early results. It is planned that the full scale tests will be conducted by selected boiler manufacturers, as part of the boiler fabrication contract, to permit finalization of the boiler design parameters and expected performance.

It should be noted, that while it is desirable that the boiler units burn only washery wastes, the technical feasibility of using washery wastes alone as a fuel, on a full scale basis, has not been firmly established. The use of washery waste, co-combusted with coke oven gas in the ratio determined by the small scale laboratory tests has been considered, as well as the use of washery wastes as the sole fuel. Also considered was washery wastes co-combusted with coal as a low cost substitute for locally available coke oven gas.

Combustion Characteristics

An eight inch size fluidized bed test unit was used to conduct combustion tests on typical washery waste samples. Laboratory test report indicates that the material had the following characteristics:

	AS RECEIVED	DRY BASIS
Moisture	7.16%	xxx
Volatile Matter	12.72%	13.70%
Fixed Carbon	7.77%	8.37%
Ash	72.35%	77.93%
Sulfur	.25%	.27%
Carbon	13.55%	14.59%
Hydrogen	1.44%	1.55%
Nitrogen	.36%	.38%
Oxygen	4.89%	5.27%
BTU/LB	2,185	2,353

Results of the test indicated the following:

Washery waste consisting of a mixture of jig refuse and underflow can be combusted in a fluidized bed boiler without the use of any supplementary fuel. However, the excess air requirements to sustain combustion was greater than 100%. When the feed rate of washery waste was increased and the excess air quantity lowered, combustion ceased. When propane gas was added this did not occur. A full scale FBC boiler to burn washery waste and coke oven gas, designed for approximately 20% excess air, would have an overall boiler efficiency of approximately 70%.

Because of the high ash content of the washery waste, approximately 72%, it is difficult to burn out the carbon completely. As the fuel burns, an ash layer builds up around the unburned carbon, insulating it, and eventually causing it to extinguish before complete combustion occurs. Co-firing washery waste with coke oven gas provides additional heat and allows more carbon to burn out. Without coke oven gas addition, total combustion efficiency is at 68-69%. With a coke oven gas feed rate of 0.7 cubic feet per pound of washery waste, total combustion efficiency increases to 85%. This increase in combustion efficiency, plus the increase in fuel heating value, due to the addition of coke oven gas, would increase boiler efficiency from approximately 55% to 70%.

STEAM/BY-PRODUCT POWER/CO-GENERATION FACILITY

Basic Plant Concept

The basic plant evaluated would supply thermal requirements of an industrial site. The plant analyzed consists of a fuel preparation building and separate boiler house. Large capacity fuel and ash storage silos and baghouse filters are located outdoors. A conceptual steam plant arrangement is shown in Figures 1 and 2.

The boiler house contains the FBC generators, auxiliaries and the fuel feed systems necessary for proper combustion and load following, while the fuel preparation building houses equipment necessary to produce sized and blended washery wastes, such as: crushers, screens, and conveying equipment.

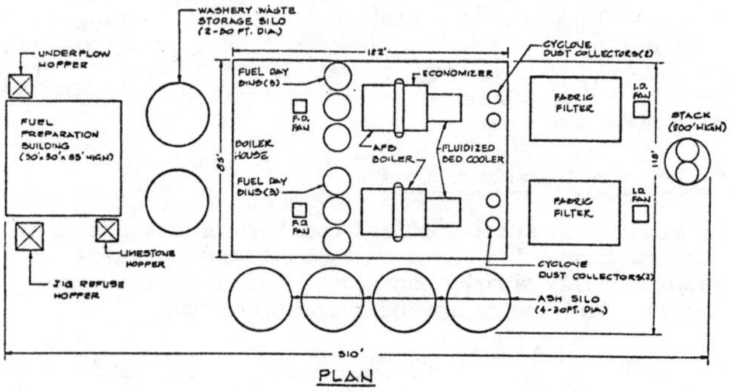

FIGURE 1. BOILER PLANT - 2-100,000 LB/HR STEAM GENERATORS

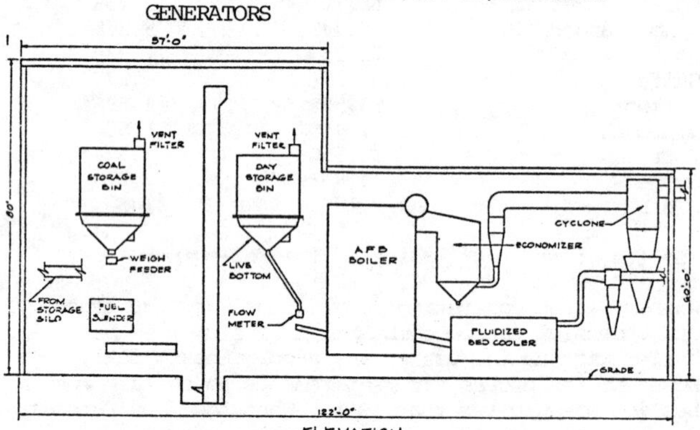

FIGURE 2. BOILER PLANT - 2-100,000 LB/HR STEAM GENERATORS

Plant Capacity

Annual production of washery waste computed for the purpose of this evaluation was about 440,000 tons. Analysis of a sample of washery waste indicated an approximate heating value of 2,200 Btu per pound. An FBC boiler plant using 440,000 tons per year of washery waste and coke oven gas, as the supplementary fuel, in the ratio of 0.7 cubic feet per pound of waste, as suggested by the tests, will produce 200,000 pounds per hour of 300 psig saturated steam. An FBC boiler plant using washery waste as a sole fuel will burn 656,000 tons per year of washery waste and produce 200,000 pounds per hour of steam. The additional waste required to make up for the missing 216,000 tons per year (656,000-440,000) would be reclaimed from local stock pile.

Alternate plant configuration considered was for washery waste co-combusted with coal, which is only slightly different in equipment, overall size and cost than the facility burning washery wastes only.

The new boilers considered were for the generation of 300 psig steam with the lower pressure requirements satisfied through pressure reducing stations. Each boiler was sized to produce 100,000 pounds per hour 300 psig dry and saturated steam of which 90,000 pounds per hour will be exported and the difference used for feedwater heating and in plant usage.

FUEL HANDLING PREPARATION & STORAGE

Fuel handling system was sized to operate one shift, five days a week, equivalent to about 250 tons per hour waste handling capacity. The jig refuse and underflow is to be received in separate hoppers at the fuel preparation building. Fuel preparation system is shown schematically in Figure 3.

Jig refuse is to be transported by bucket elevators to a conveyor which will deliver it to either one or two 60 tons per hour primary crushers which will reduce the jig refuse to one inch top size. The crusher will discharge into a conveyor which will also receive the underflow and will transport the material to either one of the two secondary crushers where the jig refuse will be reduced to the size required by the FBC boilers and blended with the underflow. The blended fuel will then be transported to two 30 diameter silos for storage. Capacity of these silos will provide for three days of full load operation. Fuel will be discharged from these silos from bin activators and transported to the day bins. Each day bin will discharge to the fuel feeder for one cell of the FBC boilers. The fuel blending and feed system is shown schematically in Figure 4.

Pollution Control

The use of washery waste as a fuel will generate three potential pollutants: SO_2, NO_x, and particulates. As a new source of air pollution, the FBC boiler plant must conform to applicable emission standards, and demonstrate that air pollution emissions will not cause significant deterioration of ambient air quality.

In an FBC boiler, combustion air is forced up through a perforated grid into a bed of inert material, such as limestone for sulfur removal, if required, or accumulated ash, into which fuel is injected. The air creates a suspension of the fuel and bed mixture. The fuel is ignited usually by preheated bed material (preheated by the use of auxiliary oil or gas burners) and burns within the bed. Sulfur in the fuel combines with calcium in the bed material, producing calcium sulfate which is continuously drained from the bed. Sulfur capture is generally controlled by the amount of active calcium in the bed, so that the discharge of sulfurous gases to the atmosphere can always be maintained with applicable regulatory limits.

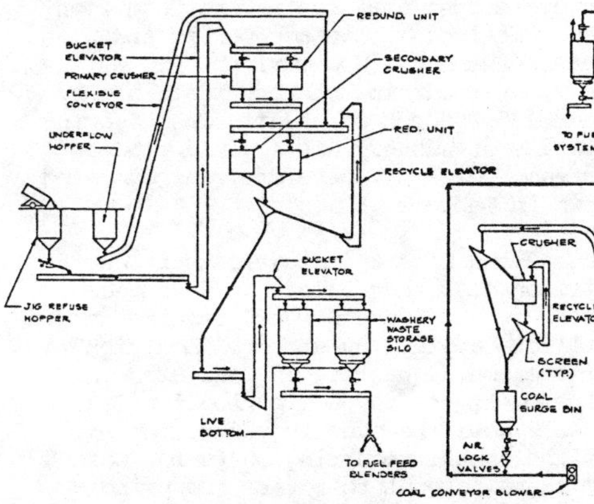

FIGURE 3. FUEL PREPARATION SYSTEM - SCHEMATIC FLOW DIAGRAM

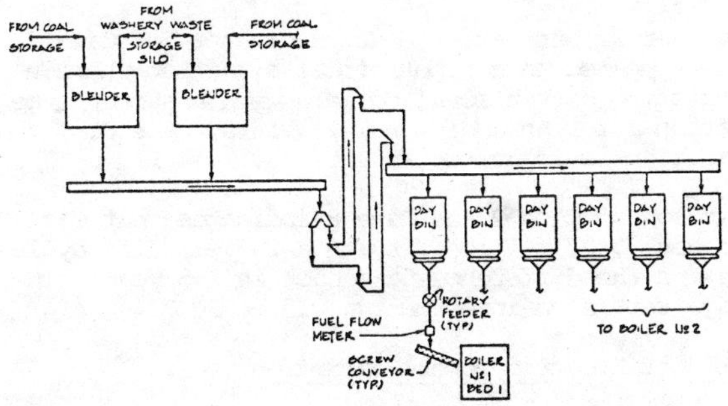

FIGURE 4. FUEL BLENDING & FEED SYSTEM - SCHEMATIC FLOW DIAGRAM

Combustion temperature in an FBC boiler is mantained at 1650°F. At this relatively low combustion temperature, the formation of nitroqen oxides is extremely inhibited, so that discharge of these gaseous compounds to the atmosphere are automatically maintained below allowable limits.

Combustion of washery waste will produce particulate emissions. For the FBC plant, all the flue gases will pass through first mechanical and secondly high efficiency fabric filters, which will reduce the particulate emissions to within allowable limits.

Combustion of washery waste will produce ash as a solid waste product. Bottom ash will be collected and drained from the bed, while the fly ash will be captured by fabric filters. See Figure 5. These wastes will be disposed of on site, or marketed, as applicable.

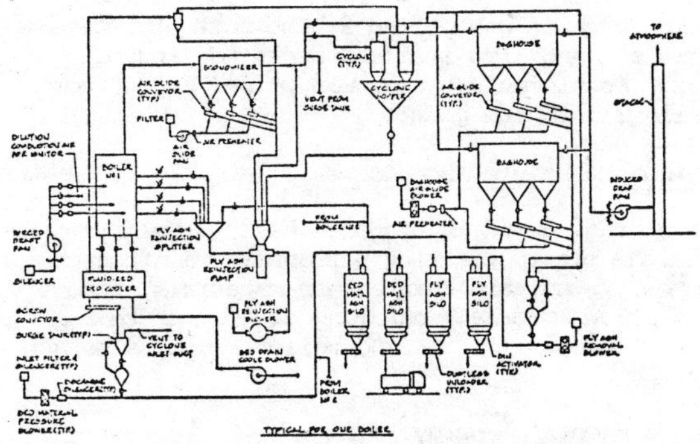

FIGURE 5. FLUE GAS & FLY ASH HANDLING SYSTEM SCHEMATIC FLOW DIAGRAM

ECONOMIC ANALYSIS

Methodology

The economic evaluation of the relative merits of investment was performed using life cycle cost analysis. Life cycle costing techniques are becoming the standard in industry and government for energy investment analysis. The techniques are actively promoted by the United States Department of Energy and other government agencies.

Cost considered in the analysis include:

- One time capital investment required to construct a complete operational generating facility.
- Recurring annual costs for fuel, operation and maintenance of the plant.

The analysis presents the 1983 present value of the twenty-five year life cycle costs for several plant configurations and capacities. It also indicates the difference in life cycle costs and payback period, based on recurring annual costs, as well as the expected average return on the investment.

The analysis considered the plant capital cost, including investment tax credit, and the annual recurring costs for fuel, plant operation and maintenance, and electric power. The annual recurring depreciation tax credit was also considered.

The capital and the recurring costs were developed in 1980 dollars and escalated to scheduled plant operation, assumed to be in 1983. Using the 1983 annual recurring costs, present worth of the twenty-five year life cycle cost was determined, using the general inflation and differential escalation rates, and assuming a 15-1/4% discount rate for capital recovery.

Construction Costs

Capital cost of investment required for construction of a complete steam generating facility including required air pollution control equipment and provisions for fuel handling, preparation and storage, ash handling, storage and disposal were estimated. A one-time investment tax credit was also included.

Fuel Costs

Annual fuel costs were determined for each of the alternate configurations. For comparision purposes fuel costs for conventional boilers burning oil/gas were based on oil at $24 per barrel. The cost of purchased gas was assumed to be equal to the price of oil.

The washery waste was priced at the cost of recovery plus haulage and also included operation and maintenance. Supplementary coal was priced at $40 per ton delivered to the plant.

Operating and Maintenance Costs

Annual operating and maintenance costs were developed and include plant operating personnel, maintenance personnel, repair and maintenance materials, recovery and underflow material, handling of jig refuse and underflow material to the FBC boiler plant, also ash removal.

Purchased Electric Energy Savings

Electric power requirements were assumed for each of the plant configurations based on expenses with similar plants. Costs were determined using the current cost of electricity at 0.026 per kWh. The co-generation of electric power as a by-product of supplying processed steam was also considered as a reduction to the purchased power cost. The savings in purchased electric power was treated as a credit to the plant.

Life Cycle Cost Analysis

The 25 year life cycle cost analysis for the seven boiler plant configurations is presented in Table 1. Category A includes Cases 1, 2 and 3, boiler plant configurations, producing 135,000 pph; and Category C includes Cases 6 and 7, plant configurations, producing 200,000 pph.

In Category A, Case 2 is the least cost configuration. Case 2 represents an FBC boiler plant producing 200,000 pph steam burning washery waste, supplemented with coal. The total 1983 values of the 25 year life cycle cost is $83 million. Compared to replacement in-kind oil/gas fired boiler represented by Case 3, Case 2 would cost $312 million less on a 25 year life cycle basis and the difference paid back in 2.1 years. The average return on investment for Case 2 compared to Case 3 is 30.4%.

In Category B, Case 4 is the least cost configuration. Case 4 is an FBC boiler plant producing 135,000 pph steam, burning washery waste alone. The total 1983 value of the 25 year life cycle cost is $62 million. Compared to replacement in-kind oil/gas fired boiler represented by Case 5, Case 4 would cost $207 million less on a 25 year life cycle basis and the difference paid back in 3.1 years. The average return on investment for Case 4 compared to Case 5 is 17.7%.

In Category C, Case 6 is the least cost configuration. Case 6 is an FBC boiler plant producing 200,000 pph steam, burning washery waste alone. The total 1983 value of the 25 year life cycle cost is $77 million. Compared to replacement in-kind oil/gas fired boilers represented by Case 7, Case 6 would cost $318 million less on a 25 year life cycle basis and the difference paid back in 2.7 years. The average return on investment for Case 6 compared to Case 7 is 22.3%.

Additional Comparisons

Additional comparisons were made to confirm the economic feasibility of Case 2 and variations of Case 2. These included investigation of additional steam generating capacity, continued operation of existing conventional units and of electric power.

Increased Steam Production

The 25 year life cycle cost analysis favors Case 2 on the basis of the capital cost, payback period and average return on investment. However, the FBC boilers for Case 2, when burning washery waste alone, would only produce 135,000 pph of steam. Case 6 burning larger quantities of washery waste would produce 200,000 pph of steam. Case 2, therefore, was compared to Case 6 as shown in Table 2.

Comparison of Case 2 and Case 6 indicates that Case 6 would cost $6 million less than Case 2 on a 25 year life cycle basis. However, Case 6 would require an additional $8 million capital investment. Additionally, because of the small quantity of coal used in Case 2 and its low cost, the payback period for Case 6 compared to Case 2 would be over 10 years. The small savings in the annual recurring costs for Case 6 is less than the annual amount to retire the principal (capital cost) and interest so that the return on investment is a negative quantity.

Continued Present Operation

To further demonstrate the attractiveness of Case 2, it was compared to a hypothetical situation wherein the existing conventional boilers were assumed to be operational for an additional 25 years, Case 8. The comparision is indicated in Table 3.

Comparison of Case 2 and Case 8 indicates that Case 2 would cost $307 million less on a 25 year life cycle basis and the difference paid back in 2.5 years. The average return on investment would be 30.7%.

By-Product Electric Power Generation

The proposed FBC boiler plant can readily incorporate provision for co-generation of inexpensive electric power as a by-product of supplying plant process steam. The additional cost of the plant can be recovered by reduction in purchased electric power costs.

As FBC boiler plant for electric power co-generation would generate high pressure, high temperature superheated steam. This steam would be passed through a back pressure (non-condensing) steam turbine. The turbine would exhaust steam at the pressure required for process purposes. Steam passing through the back pressure turbine gives up some of its thermal energy to electric generation. The remaining energy in the turbine exhaust is used for plant needs. The electric power produced is much less expensive than purchased power due to the lower heat rate. Typically, steam would be generated at 1350 psig and 925°F and exhausted from the turbine at 300 psig for process use. The turbine steam flow through rate for these conditions would be approximately 27 lb. per kWh. For a 200,000 pph boiler plant output, 7500 kW of electricity would be produced.

An FBC boiler plant of 200,000 pph capacity burning washery waste and coal and producing 7500 kW of co-generated electric power, Case 2A, was compared to comparable oil/gas fired conventional boilers, Case 3. The comparison is shown in Table 4. A credit of $1.5 million annually was applied to Case 2A to represent the savings in purchased power.

Comparison of Case 2A and Case 3 indicates that Case 2A would cost $370 million less on a 25 year life cycle basis and the difference paid back in 2.1 years. The average return on investment would be 31.4%.

It should be noted that if the turbine exhaust steam pressure was reduced from 300 psig to 150 psig, additional energy could be extracted from the steam and additional electric power generated. At a 150 psig exhaust condition, the steam rate would be approximately 19.5 lb. per kWh. The boiler plant would produce 10,000 kW of co-generation electricity which would reduce purchased power costs by $2.0 million annually.

CONCLUSIONS

The thermal energy remaining in washery waste which results from refining operations and which over the years has been discarded and stockpiled, can be recovered and utilized by FBC combustion. The technical feasibility of using washery waste as a fuel in an FBC boiler has been supported by results of laboratory combustion tests. The tests confirm that washery waste can be used alone or with a supplementary fuel which would improve overall boiler efficiency from 55 to 70 percent. Supplementary fuels could include coke oven gas and coal.

The laboratory combustion tests of washery waste must be supplemented by full scale combustion tests to confirm laboratory results and establish boiler design parameters. This should be done by the boiler designer and fabricator under a separate contract.

The 25 year life cycle cost analysis indicates that an FBC boiler plant burning washery waste alone, or washery waste with a supplementary fuel, is more economical to own and operate than equivalent capacity oil/gas fired replacement (conventional) boilers. The 25 year life cycle savings in owning and operation costs for an FBC boiler plant compared to oil/gas fired replacement (conventional) boilers varies from $207 million for an FBC plant burning washery waste alone and producing 135,000 pph steam to $318 million for an FBC plant burning washery waste alone and producing 200,000 pph steam. The additional capital cost required to construct a washery waste burning facility would be recovered from the net annual fuel cost savings in less than 2.1 to 3.1 years, depending on the quantity of washery waste burned.

The 25 year life cycle cost analysis indicates that a smaller capacity FBC boiler plant, burning washery waste and coal to produce 200,000 pph steam, or burning washery wastes alone to produce 135,000 pph steam, is more economically feasible due to its shorter capital recovery, than an FBC plant burning washery waste only to produce 200,000 pph steam.

The 25 year life cycle cost analysis indicates that an FBC boiler plant burning washery waste alone, or washery waste with a supplementary fuel and generating by-product electric power is the most economical to own and operate. The 25 year life cycle savings in owning and operating costs for such a plant compared to oil/gas fired replacement conventional boilers is $370 million and the additional capital construction cost can be recovered in 2.1 years.

TABLE 1. ALTERNATE BOILER INSTALLATIONS - 25 YEAR LIFE CYCLE COST ANALYSIS ($ x 1000)

		CATEGORY A		CATEGORY B		CATEGORY C	
PLANT DESCRIPTION	CASE 1	CASE 2	CASE 3	CASE 4	CASE 5	CASE 6	CASE 7
BOILER TYPE	FBC	FBC	CONVENTIONAL	FBC	CONVENTIONAL	FBC	CONVENTIONAL
NUMBER & CAPACITY	2-100,000 PPH	2-100,000 PPH	2-100,000 PPH	2-100,000 PPH*	1-135,000 PPH	2-100,000 PPH	2-100,000 PPH
FUEL TYPE	W.W. + COG	W.W.+ COAL	OIL OR GAS	W.W.	OIL OR GAS	W.W.	OIL OR GAS
FUEL QUANTITY-PER HOUR	55.8 TON, 78,200 CU. FT.	55.8 + 1.9 TONS	1736 GALS.	55.8 TONS	1172 GALS.	83.2 TONS	1736 GALS.
I. PRESENT COSTS (APRIL 1980 DOLLARS)							
1. CAPITAL COSTS (NET)	21,330	22,320	5,000	21,270	3,380	28,237	5,000
2. ANNUAL FUEL COSTS	1,340	539	7,800	0	5,280	0	7,800
3. ANNUAL O&M COSTS	894	929	450	894	409	1,141	450
4. ANNUAL ELECTRIC COSTS	610	615	50	605	27	707	50
5. DEPRECIATION TAX CREDIT	(392)	(411)	(92)	(391)	(62)	(520)	(92)
6. TOTAL ANNUAL RECURRING COST (2+3+4-5)	2,457	1,732	8,208	1,118	5,654	1,328	8,208
II. 1983 COSTS							
7. CAPITAL COSTS (NET)	27,729	29,016	6,500	27,651	4,394	36,708	6,500
8. ANNUAL FUEL COSTS	2,140	885	12,480	0	8,450	0	12,480
9. ANNUAL O&M COSTS	1,162	1,208	585	1,162	532	1,483	585
10. ANNUAL ELECTRIC COSTS	952	959	78	944	42	1,103	78
11. DEPRECIATION TAX CREDIT	(510)	(534)	(120)	(508)	(81)	(676)	(120)
12. TOTAL RECURRING COSTS (8+9+10-11)	3,744	2,519	13,023	1,598	8,943	1,910	13,023
III. 25 YEAR LIFE CYCLE COSTS							
13. CAPITAL COSTS	27,729	29,016	6,500	27,651	4,394	36,708	6,500
14. FUEL COSTS	65,300	19,300	380,800	0	257,800	0	380,800
15. O&M COSTS	15,245	15,849	7,675	15,245	6,980	19,457	7,675
16. ELECTRIC COSTS	25,923	26,104	2,123	25,705	1,143	30,024	2,123
17. DEPRECIATION TAX CREDIT	(6,691)	(7,006)	(1,574)	(6,665)	(1,063)	(8,869)	(1,574)
18. TOTAL LIFE CYCLE COST	127,506	83,263	395,524	61,936	269,254	77,320	395,524
19. LIFE CYCLE COST DIFFERENCE	-268,018	-312,261	BASE	-207,318	BASE	-318,204	BASE
20. PAYBACK (YRS.)	2.2	2.1	BASE	3.1	BASE	2.7	BASE
21. AVERAGE RETURN ON INVESTMENT	26.0%	30.4%	BASE	17.7%	BASE	22.3%	BASE
22. UNIT COST OF STEAM 1983 PRESENT VALUE OF LIFE CYCLE COST ($/1000 LB)	3.24	2.11	10.04	2.33	10.12	1.96	10.44

*TWO FBC UNITS RATED AT 100,000 PPH EACH, BURNING 55.8 TPH OF WASHERY WASTE ALONE WILL PRODUCE 135,000 PPH STEAM.

TABLE 2. ALTERNATE BOILER INSTALLATIONS - 25 YEAR LIFE CYCLE COST ANALYSIS ($ x 1000)

PLANT DESCRIPTION	CASE 2	CASE 6
BOILER TYPE	FBC	FBC
NUMBER & CAPACITY	2-100,000 PPH	2-100,000 PPH
FUEL TYPE	W.W. + COAL	W.W.
FUEL QUANTITY - PER HOUR	55.8 TON + 1.9 TON	83.2 TON
I. PRESENT COSTS (APRIL 1980 DOLLARS)		
1. CAPITAL COSTS	22,320	28,237
2. ANNUAL FUEL COSTS	599	0
3. ANNUAL O&M COSTS	929	1,141
4. ANNUAL ELECTRIC COSTS	615	707
5. DEPRECIATION TAX CREDIT	-411	- 520
6. TOTAL ANNUAL RECURRING COST (2+3+4-5)	1,732	1,328
II. 1983 COSTS		
7. CAPITAL COSTS	29,016	36,708
8. ANNUAL FUEL COSTS	886	0
9. ANNUAL O&M COSTS	1,208	1,483
10. ANNUAL ELECTRIC COSTS	959	1,103
11. DEPRECIATION TAX CREDIT	-534	-676
12. TOTAL RECURRING COSTS (8+9+10-11)	2,519	1,910
III. 25 YEAR LIFE CYCLE COSTS		
13. CAPITAL COSTS	29,016	36,708
14. FUEL COSTS	19,300	0
15. O&M COSTS	15,849	19,457
16. ELECTRIC COSTS	26,104	30,024
17. DEPRECIATION TAX CREDIT	-7,006	-8,869
18. TOTAL LIFE CYCLE COST	83,263	77,320
19. LIFE CYCLE COST DIFFERENCE	BASE	-5,943
20. PAYBACK (YRS.)	BASE	10+
21. AVERAGE RETURN ON INVESTMENT	BASE	NONE
22. UNIT COST OF STEAM, 1983 PRESENT VALUE OF LIFE CYCLE COST ($/1000 LB)	2.11	1.96

TABLE 3. ALTERNATE BOILER INSTALLATIONS - 25 YEAR LIFE CYCLE COST ANALYSIS ($ x 1000)

PLANT DESCRIPTION	CASE 2	CASE 8
BOILER TYPE	FBC	CONVEN.-EXISTING
NUMBER & CAPACITY	2-100,000 PPH	2-100,000 PPH
FUEL TYPE	W.W. + COAL	OIL/GAS
FUEL QUANTITY-PER HOUR	55.8 TON + 1.9 TON	1,736 GALS.
I. PRESENT COSTS (APRIL 1980 DOLLARS)		
1. CAPITAL COSTS	22,320	0
2. ANNUAL FUEL COSTS	599	7,800
3. ANNUAL O&M COSTS	929	450
4. ANNUAL ELECTRIC COSTS	615	50
5. DEPRECIATION TAX CREDIT	-411	0
6. TOTAL ANNUAL RECURRING COST (2+3+4-5)	1,732	8,300
II. 1983 COSTS		
7. CAPITAL COSTS	29,016	0
8. ANNUAL FUEL COSTS	886	12,480
9. ANNUAL O&M COSTS	1,208	585
10. ANNUAL ELECTRIC COSTS	959	78
11. DEPRECIATION TAX CREDIT	-534	0
12. TOTAL RECURRING COSTS (8+9+10-11)	2,519	13,143
III. 25 YEAR LIFE CYCLE COSTS		
13. CAPITAL COSTS	29,016	0
14. FUEL COSTS	19,300	380,800
15. O&M COSTS	15,849	7,675
16. ELECTRIC COSTS	26,104	2,123
17. DEPRECIATION TAX CREDIT	-7,006	0
18. TOTAL LIFE CYCLE COST	83,263	390,598
19. LIFE CYCLE COST DIFFERENCE	-307,335	BASE
20. PAYBACK (YRS.)	2.5	BASE
21. AVERAGE RETURN ON INVESTMENT	30.7%	BASE
22. UNIT COST OF STEAM, 1983 PRESENT VALUE OF LIFE CYCLE COST ($/1000 LB)	2.11	9.91

TABLE 4. ALTERNATE BOILER INSTALLATIONS - 25 YEAR LIFE CYCLE COST ANALYSIS ($ x 1000)

PLANT DESCRIPTION	CASE 2A	CASE 3
BOILER TYPE	FBC-COGENERATION	CONVENTIONAL
NUMBER & CAPACITY	2-100,000 PPH 7,500 kW	2-100,000 PPH
FUEL TYPE	W.W. & COAL	OIL OR GAS
FUEL QUANTITY - PER HOUR	55.8 TON + 1.9 TON	1736 GALS
I. PRESENT COSTS (APRIL 1980 DOLLARS)		
1. CAPITAL COSTS	26,000	5,000
2. ANNUAL FUEL COSTS	599	7,800
3. ANNUAL O&M COSTS	1,091	450
4. ANNUAL ELECTRIC COSTS	(885)	50
5. DEPRECIATION TAX CREDIT	(478)	(92)
6. TOTAL ANNUAL RECURRING COST (2+3+4-5)	327	8,208
II. 1983 COSTS		
7. CAPITAL COSTS	33,670	6,500
8. ANNUAL FUEL COSTS	886	12,480
9. ANNUAL O&M COSTS	1,418	585
10. ANNUAL ELECTRIC COSTS	(1,381)	78
11. DEPRECIATION TAX CREDIT	(621)	(120)
12. TOTAL RECURRING COSTS (8+9+10-11)	302	13,023
III. 25 YEAR LIFE CYCLE COSTS		
13. CAPITAL COSTS	33,670	6,500
14. FUEL COSTS	19,300	380,800
15. O&M COSTS	18,600	7,675
16. ELECTRIC COSTS	(37,600)	2,123
17. DEPRECIATION TAX CREDIT	(8,150)	(1,574)
18. TOTAL LIFE CYCLE COST	25,820	395,524
19. LIFE CYCLE COST DIFFERENCE	-369,704	BASE
20. PAYBACK (YRS.)	2.1	BASE
21. AVERAGE RETURN ON INVESTMENT	31.4%	BASE
22. UNIT COST OF STEAM, 1983 PRESENT VALUE OF LIFE CYCLE COST ($/1000 LB)	0.66	10.04

SECTION 18
ENERGY CONSIDERATION IN INDUSTRIAL SITE SELECTION AND RELOCATION

Chapter 62

PLANNING FOR ADEQUATE REGIONAL ENERGY CAPACITY: A STRATEGIC APPROACH

G. R. Hill

INTRODUCTION

This past year the Southern States Energy Board participated in a multistate strategic planning program for the Commission on the Future of the South. The Future of the South report is published every six years by the Southern Growth Policies Board which represents 13 states.

The 1980 Commission has examined trends, issues and concerns facing the region in the decade of the eighties. Four areas were addressed: The Children of the South, The Economy, The Cities and Energy.

The Task Force on Energy, staffed by the Southern States Energy Board, elected to present the Commission with an action plan for implementing a future strategy. The findings of the Task Force and the policy recommendations formulated for the Commission will be discussed briefly.

The Energy Task Force has attempted to make recommendations that they believed would shape the future of the region beneficially. It is the opinion of the Energy Task Force members that implementation of the recommendations summarized in this report will result in the continued availability of an adequate supply of energy for all sectors, while providing for a clean and protected environment.

The long-range implication of decisions made in the decade of the eighties causes them to be pivotal. Decisions made in the next several years will have a profound impact upon energy availability for many years to come, partly because of the long lead times necessary for development and construction of energy technologies and facilities.

DEVELOPING A REGIONAL STRATEGY

Long-range planning in the private sector has existed for some time. Formal strategic planning in business is, however, a relatively new activity which emerged on a wide scale in the last 15 years. The Committee for Economic Development has attempted to define the process by noting that:

> "Strategic planning focuses on the broad policy questions facing an organization, such as its basic mission and purpose and alternative courses for action or strategies to achieve those missions and purposes. Strategic planning is conducted at higher levels of management, includes a larger range of alternatives, covers a longer period of time, and includes a higher degree of uncertainty and more structured problems (than are encountered in operational planning.)
>
> The process of strategic planning forces management to identify and address its major decisions in an analytical and systematic manner."[1]

The application of the strategic planning process to the public sector is difficult and corporate models have not proved to be adequate. Planning needs between business and government differ because of intrinsic differences in functions, responsibilities and purpose. Government does not have an overriding and narrowly focused single goal similar to the corporate "profit." Instead, government has broader concerns such as providing "for the common defense and general welfare."[2]

Strategic plans can be developed for government and can serve as useful analytical tools. The purpose of strategic planning in government is to identify major problems in advance in order to make better decisions, to understand more clearly the implications for the future of current legislative proposals and actions, and to improve coordination of organizational plans and activities.[3]

Encouraging increased planning and evaluation in government is a difficult task. The political environment tends to discourage long-range analysis because the pay-off often comes at a time when the initiator is no longer in a position to receive the credit.[4]

The development of a strategic plan by representatives from 13 political subdivisions should be viewed as a major accomplishment. The recommendations made in the process by the Energy Task Force will serve as valuable guidelines to the individual decision-makers who will implement policies and plans having an impact upon energy production and consumption in the South.

Objectives

The major objectives of the Energy Task Force were:

1. To assist the Commission on the Future of

the South in the preparation of its report;

2. To review and analyze emerging alternative energy technologies and strategies that can best assist the Southern States in maintaining a solid energy future;
3. To review and analyze federal energy policy initiatives to determine the impacts on the region; and
4. To present to the Commission on the Future of the South a statement of findings and recommendations which, if implemented, will ensure energy self-sufficiency, a stable environment and balanced growth.[5]

Assumptions

The Energy Task Force formulated a number of basic assumptions to guide the study. The assumptions explain much of the current perceptions about our present and future energy situation.

1. The Southern Region is a vital part of the nation and has a key role to play in the nation's energy future.

In 1980 the South contributed 56 percent of the nation's petroleum production, 82 percent of natural gas, 489 percent of total coal production and 37 percent of the electricity generated by nuclear power. While these percentages are projected to decline slightly over the coming decade, mostly due to the maturing of southern oil and natural gas fields and the expanded production of western coal, the South will still be the primary energy producing region of the United States.

2. The South will continue to grow in population and will continue to develop economically.

Present policies must provide for future demand resulting from both a growing population and an expanding economy.

3. Conservation is an essential part of any energy future.

Recent years have demonstrated that conservation can have a significant effect on patterns of energy use and growth in demand. Conservation should be defined as more efficient use of energy rather than non use. For Task Force purposes conservation can be treated as an energy "source."

4. There is no energy shortage.

The region and the nation possess vast, untapped energy resources. While spot shortages of certain kinds of energy occur and some energy sources are indeed finite, energy alternatives are available to take their place. Development of these resources is a matter of economics and environmental and political constraints.

5. The cost of energy will continue to rise.

Higher energy costs reflect several realities. First, energy costs in this country have been kept unnaturally low through governmental controls. As these costs are decontrolled, the prices paid by end users rise to reflect the replacement value of the energy purchased. Second, energy is more difficult to find and more costly to produce since most of the readily recoverable sources are gone. Third, our dependence on imported oil has become vastly more expensive as the producing countries continue to increase the price of their oil.

6. Increased domestic energy production can pose serious environmental problems.

The threat to the environment and human health posed by the increased production and use of energy is a serious matter that must be considered in any discussion of energy policy. All these potential effects add to the economic and social costs of developing a given energy source.

7. The nation will continue to import a significant portion of its petroleum needs for the foreseeable future.

While a continued dependencc on foreign imports presents great risks, both politically and economically, there is no present alternative but to continue such imports. It has been shown, however, that conservation and the increased development of domestic energy sources can reduce imports significantly.

8. Overdependence on imported petroleum leaves the country highly vulnerable to both rapid price increases and supply interruptions.

Imported petroleum is the nation's most critical energy issue at present. The current level of imports, approximately 40 percent of total petroleum demand, is unacceptably high. Although total imports have been reduced slightly in recent years, greater efforts to reduce imports further are needed. The continuing Middle East conflict demonstrates clearly the danger in remaining dependent on an unstable area of the world for such a major portion of energy needs.

9. Unconventional sources of energy will make only a minor contribution to total energy supplies through the 1980s but should be developed as rapidly as possible.

Energy from unconventional sources, such as renewable energy sources, unconventional natural gas and synthetic fuels, will not be a significant contributor to energy supply but can begin to make noticeable contributions during the 1980s. This is basically a matter of economics and regulatory policy.

10. The demand for energy will continue to grow but at a slower annual rate than the growth in demand experienced during the past several decades.

The growth in total energy demand averaged almost 5 percent a year during the period 1965-70. From 1975-80 the average annual increase had fallen to less than 2 percent. Projections indicate that the national average annual increase during the 1980s will be approximately 1.2 percent. This slowdown in the annual increase in energy demand can be attributed to several factors, chief among them high energy costs and conservation. Demand growth in the South can be expected to be somewhat higher than the national average, but it will be much lower than the 5 percent of the late 1960s and early seventies.

11. Over the short term, traditional sources, including nuclear power, are the only energy sources available for base-load electric generating capacity.

Federal energy policy mandates a shift away from oil and natural gas for electric generation. This leaves only coal and nuclear power for base-load generating capacity. Other sources, such as increased use of hydroelectric and renewable resources, could make some contributions. Both coal and nuclear power have severe environmental and human health implications though they appear to be manageable, but the current regulatory climate causes much uncertainty for both coal and nuclear power.

12. Government regulation of the energy industry is a major obstacle to domestic production and energy independence.

During the 1970s the Congress enacted more than 40 major laws dealing with energy and environmental protection. While the intent of most of these legislative initiatives encompassed worthwhile goals, the result can only be described as bizarre. Each act resulted in the promulgation of regulations by a federal agency often with total disregard for the actions of another. All too frequently little apparent regard was shown for the intent of Congress.

13. The Southern Region has unique energy concerns that are not always recognized in the development of national energy policy.

Even though the South has evolved to resemble more closely the rest of the nation, it still has unique characteristics that manifest themselves in energy concerns. For example, transportation is a vital issue in the South with its vast land area, large rural population, underdeveloped public transportation system and heavy dependence on agriculture and tourism. The availability of liquid fuels for transportation will continue to be of major importance to the region.

14. Energy availability should not be a constraining factor in regional economic development.

The South has enjoyed several decades of impressive economic development. This growth has, in part, been fueled by cheap and abundant energy. In the future, energy will no longer be cheap or abundant. A primary goal of regional policy-makers should be to ensure adequate supplies of energy to support economic development.[6]

FINDINGS AND RECOMMENDATIONS

Prior to the 1970s the nation had no real energy policy and no apparent need for one. Historically the economy had been fueled by cheap and abundant supplies of energy. The nation had become heavily dependent on petroleum, most of which was produced domestically. During the 1970s, several events shocked Americans and brought home some political and economic realities in dramatic fashion. These occurrences forced us to realize that we had become dependent upon a highly unstable and often unfriendly portion of the world for a significant proportion of our energy needs and, also, that energy would henceforth cost far more.

The first event, the overthrow of King Idris of Libya by the Khaddafi military coup, brought about a price increase for OPEC oil of some 20 percent. The 1973 Yom Kippur War and subsequent Arab oil embargo resulted in a quadrupling of imported oil prices and in spot shortages of fuels in the United States. The embargo also demonstrated that the Arab nations were willing to use oil as a political and economic weapon. The Iranian revolution and the abdication of the Shah brought about another round of price increases totaling in excess of 140 percent between December 1978 and August 1980. The continuing conflict in the Middle East has resulted in still further price increases and a reduction in the supplies of oil available to the industrial nations.

The Energy Task Force found imported oil and refined products to be the central issue in our current energy situation.[7] The United States is dependent on petroleum for 45 percent of its energy, and more than 40 percent of the petroleum and refined products consumed is imported, most of it from OPEC. We now import about $100 billion worth of petroleum and refined products each year, and our region and nation are highly vulnerable to supply interruptions and price increases. The past two years have shown that conservation can slow the growth in demand for petroleum dramatically over a short period. Even so, the country's total oil bill is still much too high, and much of the oil we import comes from a highly unstable area of the world. Even with conservation, we will still import a large percentage of the oil we will consume during the 1980s, the 1990s and beyond. We must still make every effort to reduce dependence on foreign oil imports. The actions taken during the 1980s will decide whether or not we achieve that goal.

Transition from fossil fuels to other energy sources will not take place during the coming decade, but the decisions and policies put in place now will decide our energy posture 20 years in the future. Owing to the long lead times necessary to build energy facilities or to develop alternative energy technologies, delays in making critical decisions today could mean energy shortages in the 1990s or in the 21st century.

Conservation must be at the heart of any comprehensive energy program. While there is no overall shortage of energy resources and the region and the nation have vast stores of untapped energy, conservation is vital for prevention of spot shortages of finite energy resources before alternatives can be developed to take their place. Conservation and renewable energy resources will play a major role in the energy future of the region. Conservation has already reduced the level of oil imports far below what was projected, and the growth in demand for all energy sources has slowed dramatically. Far more can be accomplished.

If our goal is both to reduce the region's dependence on imported energy and to ensure adequate energy supplies for all sectors of the economy, we must develop those fuels that can replace imported oil in the short term. If we are to accomplish these goals we must also encourage maximum production of domestic petroleum and those fuels that have the potential for displacing petroleum in many applications. The Task Force on Energy has made specific recommendations that will help to achieve these goals in the shortest time possible.

Increased production of domestic petroleum can be achieved by more widespread application of in-fill drilling and enhanced oil recovery. In-fill drilling is the practice of drilling additional wells in existing fields and reducing the well spacing. In-fill drilling and enhanced oil recovery can quickly, and with a minimum of environmental damage, make the most significant contribution to increased domestic supplies of crude oil. For example, a total of 450 billion barrels of oil has been discovered domestically, of which only about 118 billion barrels have been produced. About 28 billion barrels remain to be produced by primary and secondary recovery methods. The U.S. Department of Energy estimates that between 18 and 52 billion barrels of oil can be produced by present day enhanced recovery techniques. Fifty-two billion barrels of oil is the equivalent of 70 synthetic fuels plants, each producing 100,000 barrels per day for 20 years.

Enhanced recovery is not cheap, however. The Windfall Profits Tax diverts capital from such activities to the more speculative ventures of drilling in frontier areas. This capital diversion occurs because "new" oil is taxed at a more favorable rate than "old" oil. The Task Force has recommended that the Crude Oil Windfall Profits Tax Act be amended to exempt from the tax oil produced from stripper wells and enhanced recovery. It was further recommended that a tax credit, to be applied against the Windfall Profits Tax, be allowed for investments in domestic production.

In addition to increased production of domestic petroleum, natural gas and coal can replace petroleum products in many applications. The South has a great abundance of both these fuels. In 1980 the region produced 82 percent of the nation's natural gas and 48 percent of its coal.

For some years the assumption was made that natural gas was in critically short supply and only a limited amount of additional resources could be produced. These assumptions have been shown to be inaccurate, and now the most conservation estimate indicates a 35 to 60-year supply at current consumption levels. This estimate does not include unconventional natural gas which is rapidly becoming competitive with other fuels. The Task Force has recommended that current restrictions on natural gas use be removed since these restrictions have a severe impact on the South. In addition, the Task Force has recommended that Congress immediately repeal the incremental pricing provisions of the Natural Gas Policy Act. There would appear to be no need, though, to hasten decontrol of the wellhead price of natural gas. The current pricing levels provide adequate incentive for exploration and production, and immediate decontrol could have severe impacts on the economics of a number of southern states.

The South has vast coal resources, and the coal industry can easily meet an increase in demand. Coal production and use are constrained by current regulations, even though the expansion of coal as a fuel is a foundation of national energy policy. Coal can serve as the primary boiler fuel in new power plants and large industrial facilities. It is cost competitive with other fuels, but environmental regulations drive up the cost of burning coal, especially high sulfur eastern coal. The Task Force has noted that, while clean air standards need not necessarily be relaxed, industry should be allowed maximum flexibility in meeting these standards. The "bubble concept" is an example. In addition, mining reclamation standards should be altered to allow mined lands to be returned to productivity or some useful state rather than to the approximate original contour.

Coal exports hold great potential for the southern economy. Estimates indicate that unconstrained demand for American coal will continue to rise rapidly through this century. Unfortunately, exports are severely limited by a lack of adequate deepwater port facilities capable of handling the new supercolliers. The Task Force has recommended that a major program of port expansion and development be undertaken, with some federal assistance to augment private development where short-term economic factors are inhibiting.

Finally, the Task Force has recommended the completion of a coal slurry pipeline system serving the South. If the use of coal rises to projected levels by 1990 and beyond, it is doubtful that the railroads can handle the increased tonnage. Slurry pipelines have been shown to be the most economical means of coal transportation over medium distances of 300 to 400 miles. The Task Force noted that state or federal legislation will be needed to grant the right of eminent domain.

The Task Force found that while synthetic fuels from coal hold great promise for the future, they are only competitive in limited circumstances today. The development of a major production capability now will lock us into what is essentially World War II technology and may retard the industry's development potential over the next decade. What is needed is support for research and development of new conversion technology and demonstration or pilot plants using the best available current technology. There appears to be no need for a major effort to develop a large-scale coal-to-pipeline synthetic gas production capability. Conventional and unconventional natural gas from several sources can be produced more cheaply. Liquid fuels from coal do have the potential for displacing petroleum, and the production of cleaner burning liquid fuels from high sulfur eastern coal could be a boon to the region's economy. The most plentiful source of liquid fuel is shale oil. New technologies may permit the production of oil from lower grade eastern shale, but shale oil production also holds the potential for serious environmental damage, and great care should be taken in developing this energy source.

The South is heavily dependent upon nuclear power for electric generation and will be for several decades. The Task Force has recommended that the use of nuclear power be continued and has urged more research into advanced reactors, including completion of the Clinch River Liquid Metal Fast Breeder Reactor, a recommendation that failed to receive unanimous support. The Task Force has also recommended that the Nuclear Regulatory Commission be reorganized and strengthened with a focus on reactor operating safety and that stringent requirements be set and progress toward meeting those requirements be closely monitored.

The Task Force has also called for the establishment of a policy that sets a rational time schedule in which radioactive waste storage sites can be developed and put into operation. In addition, the time required for construction and operating license proceedings for a nuclear power plant should be reduced, consistent with safety and environmental concerns. The Task Force further recommended that off-site nuclear emergency response planning should be improved, with a strong role for state and local governments.

The most serious problem facing the energy industry is government regulation. The need for regulatory review and reform is readily apparent. While the goals of most of the energy and environmental laws passed during the past decade are worthwhile, and great strides have been made in cleaning up the environment, energy production has been severely hampered in the process and consumer energy costs have risen. Fault is not generally found with regulatory goals and objectives but with procedures. The Task Force has stated that they have no intention of surrendering gains made in environmental protection, but such protection can be achieved at far less cost and delay. Llewellyn King, publisher of <u>The Energy Daily</u>, calls this the "bonzai garden club syndrome," where any group of six or more self-styled protectors of the public interest can effectively block construction of a facility which really is in the public interest.[8]

The Task Force has called for a more balanced approach to energy production and environmental protection on the part of the federal government.[9] For example, the government either owns outright or controls on-shore lands equal in size to all the land east of the Mississippi River, plus Arkansas, Oklahoma and Texas. Less than 1 percent of this land is currently producing energy. The federal government also has jurisdiction over 559 million acres of land beneath the waters of the Outer Continental Shelf, an area equal to all the land east of the Mississippi River. Less than 3 percent of this land has been leased and only 5 percent is scheduled to be leased by 1990. Approximately 60 percent of our total petroleum resources are located in the Outer Continental Shelf. We have recommended that these public lands be opened to exploration and production more rapidly. Vast wilderness areas can and should be protected while still allowing exploration in the areas most promising for energy production.

The availability of energy plays a key role in industrial location decisions. In order to continue to attract new and expanded industry to the South we must have adequate supplies of energy. In order to ensure this and still reduce our vulnerability to imported oil, all energy sources must be developed. Each southern state has the potential to produce some of its energy needs from indigenous resources, and it is this energy independence which should be encouraged to the greatest extent possible.

REFERENCES

[1]Michael H. Moshow, Strategic Planning in Business and Government, (New York: Committee for Economic Development, 1978), p. 4.

[2]Ibid., p. 38.

[3]George A. Steiner, Strategic Planning, (New York: Macmillan, 1979), p. 319.

[4]Moshow, Strategic Planning in Business and Government, p. 39.

[5]February 2, 1981 Report of the Task Force on Energy to the Commission on the Future of the South, (Atlanta, Georgia: Southern States Energy Board, 1981), p. 1-2.

[6]Ibid., p. 1-6.

[7]Ibid., p. 2-1.

[8]Llewellyn King, "The Institutional Crisis in Energy Decision-Making," Selected Papers, Second Annual Meeting, New Orleans, Louisiana, (Atlanta, Georgia: Southern States Energy Board, 1979).

[9]Report of the Task Force on Energy, p. 4-20.

Chapter 63
ENERGY FOR INDUSTRY: A CALIFORNIA PERSPECTIVE

D. Greene

Over the last decade, the quest for an energy policy has consumed more paper, more political oratory, more bureaucratic time – and ultimately, more energy – than we care to think. For years we lamented the fact that Congress wouldn't give us an energy policy. And just as the Carter programs were beginning to attain a reasonable degree of coherence, along comes Ronald Reagan to tell us that we don't need an energy policy.

Under the Brown Administration, California does have an energy policy. We are proud of it, and it works. Those may seem brash statements, but the record bears them out. California's energy policy has three themes: efficiency, cost-reduction, and environmental compatibility. Around these themes we have built an array of programs that target particular market sectors and classes of consumers.

Our basic goal is to divert dollars currently flowing to OPEC into energy resources that yield the greatest overall economic benefit, that provide maximum return in terms of energy produced or saved, that are indigenous to California, and that present the least risk to our citizens and our environment. Based on these criteria, investments in energy efficiency have the highest preference, followed by cogeneration and, where cost-effective, renewable energy supplies. Last come conventional power sources, because of their high cost, environmental risk and long lead time. This ranking of priorities provides a guide for investment decisions by utilities, state and local governments, and business, as well as rate and licensing decisions.

ENERGY DEMAND LOWERED

California's recent experience has shattered the conventional wisdom that economic growth cannot be achieved without a proportional growth in energy demand. For the last two years, California's total energy consumption has dropped, while Gross State Product has continued to grow.

In 1974, utility energy planners forecast that by 1995 California would need numerous new central station power plants, including 28 large nuclear plants, to satisfy projected growth in demand for electricity. Today, utilities and the California Energy Commission agree that growth in electricity demand has dropped to less than one-third of the 1974 rate, and that further reductions will occur as the benefits of increased energy efficiency continue to accrue. As a consequence, few of the central station power plants believed necessary in the mid-1970's remain in utility construction plans today. These changes reduce the utilities' projected capital expenditures by over $60 billion, avoiding what would have been a large and inflationary burden on California's economy. Reduction in electricity demand growth coincides with a period of remarkable economic growth for California, and underscores the fact that energy efficiency is compatible with growth in economic activity and jobs, and improvement in the quality of life.

Californians have also reduced their use of natural gas over the last five years. Although there are nearly 10% more customers, overall use of natural gas remains constant, and per capita residential use has fallen by over 10%. This decrease has been possible because Californians have turned to such simple efficiency measures as ceiling insulation. We are meeting our growth requirements through increased energy efficiency.

The profound economic benefits of these trends have been documented by several major recent studies of energy policy choices, including those by the Harvard Business School and the Carnegie-Mellon Institute. As we use energy more efficiently, we export fewer dollars to foreign countries and other states, make capital available for other investments, provide more jobs for Californians, and hold down the cost of energy. By building energy efficiency into California's housing, businesses, and industry, we make our economy stronger, less subject to inflationary pressures, and less vulnerable to foreign events.

RELIANCE ON RENEWABLE RESOURCES, INCREASED PRODUCTIVITY

Our policies have also fostered the generation of energy from renewable resources. California leads the nation in geothermal development. California is on the threshold of a cogeneration boom that will benefit utilities, industries and the state as a whole. California conducts a major loan program to encourage biomass energy projects, and leads the nation in the use of agricultural wastes for energy. Other renewable energy sources – wind, solar, small hydro – have in the last five years moved from experimental status to commercial reality. A strong and growing new industry is devoted to developing and marketing renewable energy technologies.

California has continued to increase production of its fossil fuel resources while protecting its environment. Today, we produce more of the oil we need than we did five years ago. Since 1975, production has risen by 10% - up by 33 million barrels per year. In 1979 and 1980, oil and gas drilling activity in California set new records. Development now underway in California's heavy oil producing region could help increase statewide production by 60 million barrels a year by 1985. This increase, if coupled with conservation efforts in transportation, can enable us to end foreign oil imports by 1990. Achieving this goal restrains rising energy costs and means more jobs for Californians. To make this possible, we support refinery modifications that will enable refiners to process a greater proportion of heavy oil.

CALIFORNIA'S PROGRAMS

What has California done to foster improved energy efficiency and reduce reliance on traditional fuels? Here are our programs.

- o California's regulations require new homes and new appliances to be energy-efficient and affordable. High monthly energy costs strain household budgets and make it more difficult for families to meet basic needs for food and shelter. Our energy efficiency standards, first adopted in 1978, require new buildings and appliances to meet a high level of energy efficiency. Recently revised residential standards will reduce household energy use by 50% and produce annual savings to consumers of $1.5 billion (in today's dollars) annually by the year 2000.

- o California provides substantial tax credits for conservation and solar investments. Our 40% conservation tax credit applies to insulation, weather-stripping, double-glazing, and other conservation measures. Our 55% solar tax credit encourages homeowners and businesses to install solar water-heating systems, and helps homebuilders offset the costs of building "passive solar" houses. As a measure of this program's success, California has almost 100,000 solar installations, or roughly 1/4 of all solar units in the country.

- o Low cost financing will soon be available for cost-effective conservation measures. Thanks to the California Public Utilities Commission's leadership, Pacific Gas and Electric has developed its innovative "Zero Interest Plan" (ZIP) for conservation loans. Under this plan, PG & E customers can receive an interest-free loan for conservation measures. "ZIP" will spread the benefits of conservation to all PG & E customers and help hold rates down.

- o The use of geothermal power, one of our most cost-effective energy sources, is being expanded in California. In 1975, California's geothermal plants produced 405 MW; today that figure is 908 MW. Production is expected to double again by 1985 when capacity to generate 1850 MW should be in place. Through the cooperative efforts of state and local agencies and utilities, geothermal electricity output will continue to grow. Direct uses of geothermal heat have a high potential for reducing fossil fuel use.

- o Wind energy is a practical ally in the fight to reduce reliance on oil. California utilities expect to develop over 400 MW of wind energy by 1990. Small businesses, farms, and investors are increasingly interested in wind development. This change reflects the technical maturity and potential future economic advantage of both large and small wind turbines. Demonstration wind turbines, sponsored by our Office of Appropriate Technology, Energy Commission, and the Department of Water Resources and now in operation, have helped identify and resolve technical problems.

- o California is a leader in generating energy from biomass and agricultural wastes. The State technically and financially supports private sector biomass and alcohol fuels projects. Under this program, California is rapidly expanding its production of pure alcohol and electricity from organic wastes. In addition, we are stimulating the market for pure alcohol by encouraging conversions of car, bus, and truck fleets from gasoline to alcohol.

COGENERATION KEEPS INDUSTRY COSTS DOWN

California emphasizes the use of cogeneration at industrial facilities. Cogeneration is an important near- to mid-term energy resource which offers many benefits to industry. It is a conservation technology which increases the efficiency with which fuel is used through the sequential production of steam and mechanical or electrical energy from the same fuel source. By installing equipment to produce electricity from waste heat, industries reduce overall fuel use and drastically decrease energy costs. Industrial, agricultural and commercial firms are turning to cogeneration in increasing numbers because is makes economic sense, presents fewer environmental problems than new power plants, and avoids the conflict between agricultural and energy uses of water. Cogeneration also makes economic the use of wastes from two major California industries – agriculture and forestry.

Governor Brown has set a goal of 6,000 MW of cogenerated energy in California by 1990 – a goal that is both technolgically and economically feasible. He has appointed a Cogeneration Task Force to identify and remove barriers to cogeneration development, and to clearly articulate California's position before Congress and federal agencies. California's utilities are a key element in the program, and are cooperating to help reach this goal.

Industries initially expressed reluctance to consider cogeneration for two major reasons: unfavorable pricing policies, and regulatory requirements. Developing a pricing standard or guidelines for utility purchase of cogenerated electricity is crucial to promoting the use of this energy source. California's Public Utilities Commission laid the groundwork for guidelines in a series of orders to utilities, giving them the clear message that cogeneration must be part of electric resource plans. These orders further provided industry with a clear economic incentive to invest in cogeneration.

Cogeneration depends on the cooperation of private businesses which are not primarily in the business of producing energy, and it is particularly important that these persons perceive the actions of federal and state government as consistent in encouraging cogeneration development. Industry's second big concern, however, was seemingly excessive regulatory requirements of the state and federal government. To remove this barrier, the California Legislature acted to expedite cogeneration facility siting and make it easier for industries to utilize cogeneration technology within the limits of existing air quality standards. To coordinate the permit process, our Office of Permit Assistance has set up a Cogeneration Desk to help private applicants understand applicable regulations and secure state and local approvals.

California state government also recognizes its responsibility to practice what it's preaching. We have prepared and are implementing the Cogeneration Blueprint for State Facilities, a plan to generate 400 MW of electricity through cogeneration at state facilities. By 1990, this program will make state operations self-sufficient in electricty and save taxpayers $735 million.

These efforts to promote cogeneration are beginning to pay off. There are over a thousand megawatts of cogeneration currently in existence or under construction. An additional 2,000 megawatts have been proposed over the last 18 months. The range of firms involved is great: from refineries and heavy manufacturers to hospitals, wineries, and banks. Cogeneration offers them significant cost reductions and greater security of energy supplies.

While serious challenges remain, public and private sector actions over the last five years have reduced California's energy use and put in place a broad range of programs to improve energy efficiency and develop cost-effective alternative supplies. These actions benefit California's economy by slowing the flow of dollars out of state, freeing capital for other investments, creating more jobs for Californians, and holding down the cost of energy.

Many of California's state and local energy programs are now models for other states. With continued effort in these directions, we can reduce our political and economic vulnerability and pass on to future Americans an energy system that is safe and secure.

SECTION 19
INDUSTRIAL ENVIRONMENT: ENERGY AND HVAC ADVANCES

Chapter 64

RECENT ADVANCES IN FACTORY HVAC DESIGN

D. O. Smith

The use of traditional commercial engineering practices in the design of an air conditioning system for a factory will most always result in overdesign and unnecessary first cost and energy usage.

It has been proven many times during the past several years; there is a better way to design factory air conditioning systems. The better way is to capitalize on several dynamic factors unique to factories, particularly single story factories. This methodology can often result in air conditioned factories with less first cost and less energy than ventilation alone. The principles can be applied both to new and add-on or retrofit situations.

Purpose of This Article

1. Develop an awareness of the importance and benefits of factory air conditioning.

2. Show the limitations and problems associated with conventional factory air conditioning design today.

3. Discuss the dynamic concepts unique to factories and how these concepts can be used to advantage in the optimum design of the air conditioning system.

Why Factory Air Conditioning

Market research indicates the number of new factories that are air conditioned will double between 1981 and 1985.[1]

Studies reveal that air conditioned manufacturing facilities experience summer productivity rates 20 to 25% higher than plants that are not air conditioned.[2] This improvement results from the reduction of production errors and accidents, reductions in absenteeism and attrition -- particularly among high caliber, skilled employees.

Also of major importance are government and union demands for comfortable and healthy work environments. Demand for skilled employees has led to the air conditioning of manufacturing facilities -- an amenity sought by many high caliber craftsmen.

Of growing importance is the use of electronic process control equipment, much of which requires a stable temperature environment.

Industrials (and industrial engineers) are concerned almost entirely with the manufacture of their own products. They do not want problems with the air conditioning to detract from their primary concern. They need well designed systems that work - systems that take into account the uniqueness of their facility, production requirements, and expansion or rearrangement plans. Flexible system designs are a must to accommodate rapidly changing production processes. As mentioned before, industrial engineers are responsible for getting a quality product out the door.

Traditional Design Techniques Result In High Cost and Wasted Energy

The utilization of conventional design techniques for industrial facilities leads to high system and operating cost in several ways:

* Cooling the entire plant volume
* Air distribution system design shortcomings
* Lack of consideration to dynamic aspect of process loads, ventilation requirements, infiltration, the effects of stratification, and thermal storage possiblities
* Overdesign

Complete Cooling of the Entire Space and Poor Air Distribution System Design

A conventional air distribution system design is shown in Figure 1a. In this design, the supply air is introduced at the full height of the plant and return air taken out at the same level. In addition, the entire facility is air conditioned in spite of the fact that the workers, in most cases the original reason to air condition the facility, are only located in the lower level and in certain isolated spots. A better system brings the air supplies and returns down low to minimize conditioning the upper portion of the plant facility. The best approach not only gets the conditioned air down to the level of the workers, but also confines the conditioned air to that area which is occupied (Figure 1b).

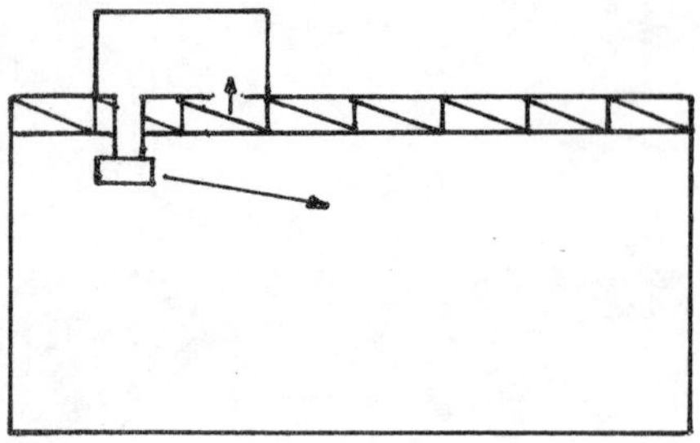

Figure 1a

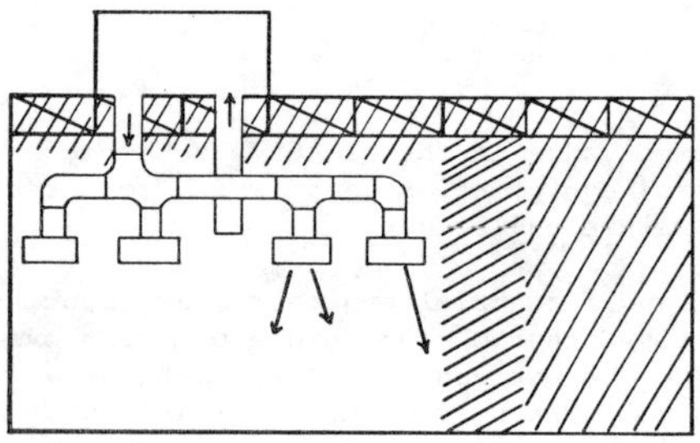

Figure 1b

Details of Process Loads, Ventilation Analysis, and Infiltration Analysis Can Make A Big Difference

Process Loads: The location and concentration of process loads, particularly relative to the space to be conditioned, can be a big factor in the total air conditioning load. Isolation of a portion of the process load from the conditioned area can provide a tremendous savings in air conditioning tonnage.

Of particular importance in the analysis of the process loads is an evaluation of how these loads might be redistributed in the future. The process load diversity is one of the largest single determinants of factory air conditioning design capacity and of present and future costs.

Ventilation Requirements: Another major factor in determining overall capacity required is ventilation and the resultant makeup air requirements. Each industrial process must be analyzed on its own merits to determine the amount of exhaust that is really necessary for that process and to determine if untempered outside air can be brought in as makeup without imposing this load on the refrigeration or heating system.

Infiltration: Few people consider the real impact of an open door on the heating and air conditioning load. Consider for a moment a 12' high by 12' wide door with a ten mile per hour wind blowing against it. Ten miles per hour does not seem like much, because wind velocities often exceed this figure. On a 12' by 12' door, a ten mile per hour wind would result in almost 130,000 cubic feet per minute of infiltration air which must be heated or cooled by the air conditioning system. Obviously, whatever can be done, should be done to limit this factor.

A word about "negative" buildings. How often have you been told that a factory is negative? Regardless of building pressurization, wind will cause air to flow through the facility due to the difference in wind pressure on the windward versus the leeward exposures. During a recent study of an existing large (over 2,000,000 sq. ft.) facility, we were advised that the plant was so negative, they were receiving many complaints. We vented this plant on a 45-50° day when the economizer was bringing in almost 2,000,000 CFM of fresh air - there was no way the plant was negative that day! But, the industrial engineering manager said, "Come with me, I'll prove it to you". We walked along the 1000' north wall; he opened the first large door we came to and, sure enough, the cold air instantly cooled about 20 bays around the door and workers started complaining. We came to the second door and opened it - same result.

Turning the corner and going down the 2000' east wall, we walked along an assembly line coming to a 12' wide by 16' high door that was standing open within a few feet of the line workers. We walked out this door, and walked, and walked - for about 50' before we sensed cool outside air - the exfiltration was that strong, but nobody was complaining. The point is, most plants are not "negative". Wind blowing at open doors causes infiltration and complaints.

Overdesign

More often than not, when using conventional commercial design practices on an industrial facility, the resultant system is overdesigned. Overdesign does not always assure that comfort conditions will be maintained. However, overdesign does always:

Increase first cost: More tonnage and air handling capacity than necessary invariably increases first cost, often beyond budget limitations.

Lower part load efficiency: Refrigeration and air handling equipment selected for maximum efficiency at full load, will operate to consume more power per CFM or per BTU at part load conditions.

Increase demand charges: If 1,000 tons of refrigeration capacity is installed when only 500 or 650 tons are required, you can bet there will be a point where all 1,000 tons are energized - for example, on the initial cool down. Moreover, utilities often have "ratchet" charges which means the 1,000 ton demand peak will be paid for up to twelve months.

Computerized Design

Viron became actively involved in factory air conditioning in the mid 70's when Dr. Ray Dean was appointed as "chairman of the board" of the Viron Corporation. Dr. Dean did considerable research and developed equations to quantify the effects of stratification, thermal storage and spot cooling, particularly as applied to factories. He put these effects, along with an energy optimization system, into computer programs which are used by Viron today.

The first actual usage of the computer program was for the design of an air conditioning system in 1976 for a 34,000 square foot factory addition in Kansas City. Since that time, many factory air conditioning systems have been designed utilizing these concepts.

The concepts and a case study of the first Kansas City factory design were written up in the ASHRAE JOURNAL and presented at the 1976 ASHRAE meeting in Dallas. The most recent of a series of articles appeared in Plant Engineering in July, 1981.

Stratification

Let's discuss how hot air stratification can be utilized in a factory to achieve substantial reductions in air conditioning requirements.

In a single story plant with a high ceiling, natural temperature stratification builds up a layer of hot air under the roof. Under certain conditions, this causes heat to flow out rather than in. The depth and temperature of this static thermal blanket is affected by:

* Ceiling height - it must be high enough to avoid disturbing air currents.
* Ceiling construction - mass, insulation and radiation affect heat storage and heat transfer characteristics.
* Outdoor temperature, wind, and solar load - representing the total effect of outdoor conditions on the plant.
* Internal heat gains - lights, people, and equipment - representing the total effect of internal loads on temperature stratification.
* Type and location of the air distribution system - representing a major effect and one that is not well understood.

Stratification dynamics: The factors just listed all work together in a dynamic fashion to impact on the total air conditioning load in a single story manufacturing facility.

* Solar load and high outdoor temperature cause heat to flow in through the roof membrane heating the roof mass and the air underneath the roof. Internal loads, including wall and glass transmission, glass solar, infiltration, people, lights, and equipment, all place heat into the occupied space which then collects in this stratified layer. If the stratified layer is left calm and undisturbed, much of this heat will flow out through the roof.

Understanding air distribution . . . is the key!: It is essential that the stratified layer be left undisturbed if it is going to be allowed to work. If the air distribution system is not properly designed or if the ceiling is ventilated, the stratification effect will be lost, markedly increasing the load on the system.

Stratification effects on system design: The stratification profile at four different times in a plant operating on a two shift basis in the winter is shown in Figure 2. It can be seen that the stratification effect is not pronounced. The conclusion, therefore, is that for heating, stratification can be essentially ignored.

FIGURE 2

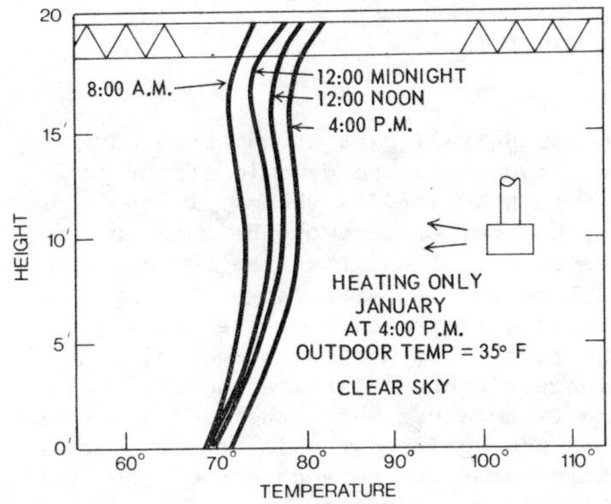

On a summer design day, (Figure 3), it can be seen that the temperature is nearly the same from floor to roof at the beginning of the first shift in the morning. By 4:00 p.m., when the design outdoor temperature is reached, the stratification layer has developed and the temperature just under the roof is approximately 20° higher than the occupied space. As the sun goes down and the outdoor air temperature drops, this stratified layer gradually decays as heat flows out of the building.

FIGURE 3

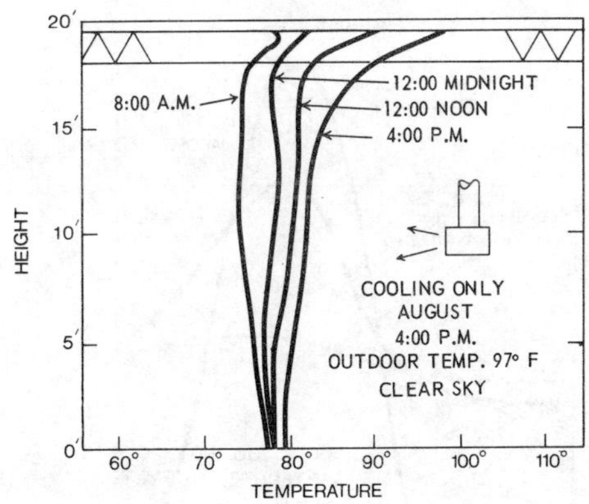

The most dramatic effect of stratification is shown on a spring/fall day. At the beginning of the first shift the temperature is nearly the same at the floor and roof but by 4:00 p.m. the temperature just underneath the roof has risen by 25° to 30° (Figure 4).

FIGURE 4

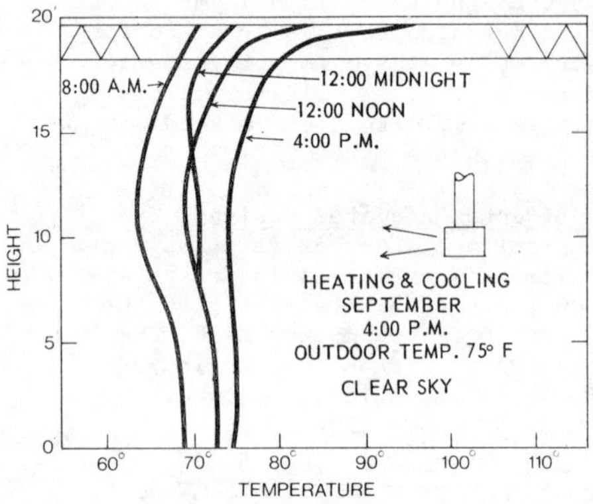

By taking the spring/fall data shown above and replotting it (Figure 5), the dynamic effect of stratification on HVAC load is shown. In the morning hours, the temperature on the outside surface of the roof is lower than the temperature on the inside surface causing heat flow out. During day time hours, the temperature on the outside surface of the roof rises due to outdoor air temperature increases and solar load while, at the same time, the temperature under the roof increases due to internal heat gains. The important feature is that the temperature difference across the roof membrane is minimized. For example, using conventional design at a temperature of 110°F just above the roof and an indoor temperature of 75°F, the resulting temperature difference is 35°. Using the stratified layer concept the temperature just below the roof would be 95°F and the temperature difference would only be 15°F. In both cases, worker comfort is essentially the same. In the evening hours when the sun goes down, the outside roof surface temperature drops rapidly to a level below the stratified layer temperature, allowing heat to flow out. This effect assists the air conditioning system by reducing the load.

FIGURE 5

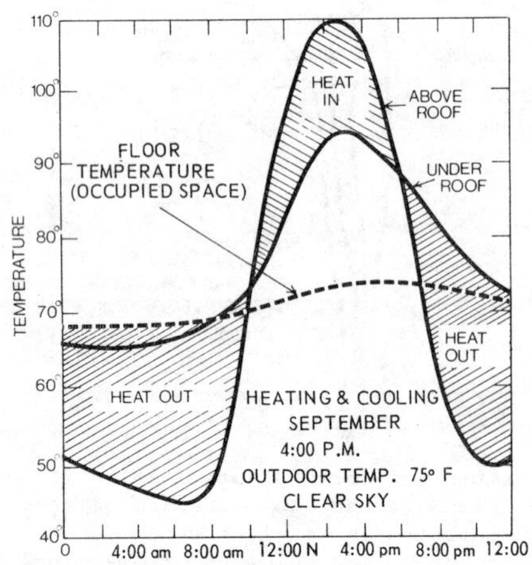

It is obvious, therefore, that stratification can actually reduce the air conditioning load; consequently, conventional calculation methods usually result in substantial overdesign of the required cooling system(s).

Thermal Storage

If the conditioned space temperature is allowed to climb during peak load periods, heat will be absorbed in the plant structure, equipment, and stored material, reducing coincident cooling requirements on the air conditioning system. This thermal storage effect is influenced by:

* Ceiling or roof mass - with a heavy roof, heat normally transferred into the stratified layer will be stored for later release. This effect is more pronounced if the manufacturing facility is operated on a single shift rather than a two or three shift basis.

* Stratified layer temperature swing - obviously, the greater the temperature swing obtained in the stratified layer, the more heat will be stored in the roof mass.

* Floor/equipment/stored goods mass

* Allowable occupied space temperature swing.

  If a significant swing is allowed in occupied space temperature, the mass of the floor, equipment, and stored goods will all store additional heat and attenuate peak loads.

As with stratification, the thermal storage effect is dynamic, not static; therefore, its analysis cannot be accomplished with conventional static design methods. In the daytime, heat from solar loads and outdoor temperature loads, as well as from internal heat sources, is stored to different extents in the roof, floor and mass within the structure. At night during unoccupied periods, part of this stored heat is transferred to the atmosphere and not picked up by the air conditioning system.

An example of this storage effect and stratification is a manufacturing plant in the midwest where conventional static design methods indicated a requirement for 250 tons of installed refrigeration capacity and 102,000 CFM of installed air handling capacity. Utilizing thermal storage in the concrete roof, and a 5° temperature swing, dynamic air conditioning design reduced the tonnage to 120 and the installed air capacity to 54,000 CFM, approximately a 50% reduction of installed capacities required.

Stratification and thermal storage concepts have been applied recently in an energy conservation retrofit of a 90,000 square foot Hackensack, New Jersey facility. This structure is 16' floor to roof which is a fairly low-bay application of these concepts. As part of an energy conservation study, the entire facility was evaluated to determine the extent these concepts applied. The results:

1. The existing installed 200 tons of air conditioning was inadequate to handle the building load which was substantiated by the building operators who had trouble maintaining any degree of summer comfort. This was caused by the addition of production equipment subsequent to the installation of the HVAC system. Computer modeling indicated the installed tonnage to be 50 tons short.

2. Stratification calculations indicated that creating a thermal barrier about 9' above the floor and allowing for a 5°F temperature swing to capitalize on thermal storage would reduce the tonnage required to less than 200 tons and would also reduce the need for ventilation fan energy. This stratified layer effect was achieved by a redesign of the air distribution system. This will result in annual savings of over $4,500; the cost of implementing this modification was $19,800, a 4.37 year payback at 1980 energy cost. Had the initial installed tonnage been adequate and the system operating under control, the savings would have been substantially more.

3. This plan also provided better air distribution and control and reduced worker complaints.

Spot/Area Cooling

One of the most interesting possibilities for capacity reduction in an industrial air conditioning design is the use of spot cooling. Unfortunately, this concept is often misused with bad results.

* Use caution - one of the major problems with spot cooling is the reliance of many designers upon drawings with "trained" arrows (a preconceived notion of cool air flow rather than considering air flow dynamics).

* Pay careful attention to:

  Air distribution.

  Loads and adjacent spaces.

  Overlap of zones.

Whenever air is introduced or discharged into a space, it sets up secondary air motion patterns. These air motion patterns will cause loads from equipment and lights in adjacent spaces to be "induced" into the occupied spot. In order to properly account for these loads, our experience shows that a "buffer" zone must be extended around the space approximately equal in width to the mounting height of the diffusers. In most cases, in our experience, when these buffer zones are included, the spot cooled zones will overlap meaning that total area cooling should be utilized, rather than spot cooling. A common exception to this overlapping consideration is when the area to be conditioned is long and narrow, such as an assembly line and where the buffer zone can extend into the storage area.

Many Factors Influence Proper Design

There are many factors within control of the designer which can have great influence on total system capacity and energy consumption:

* Diffuser height - generally the lower, the better.

* Throw distance - optimize to condition area with minimal effect on the stratified layer (direct or induced air flow).

* Air volume control - often, the savings in fan energy offered with variable air volume design in an industrial environment will pay for refrigeration energy.

* Ceiling reflectivity - generally, the more reflective a ceiling is, the poorer a radiator it is. Therefore, a reflective ceiling will reduce plant air conditioning load by impeding the transfer of heat from the stratified layer to the floor.

* Design temperature/temperature swing - generally, the lower the design temperature in the occupied space, the more cooling capacity and energy is required. However, a low temperature, coupled with a wide allowable temperature swing and a high mass building and stored material, will often require less installed capacity and less energy than a higher temperature and a more restrictive swing due to storage effects.

* Equipment efficiency - use of larger equipment will generally result in improved operating efficiency and reduced maintenance cost.

* System control - the air system should be controlled to minimize conditioned air volume whenever possible and to utilize "free" cooling whenever possible as with an economizer control system.

* Refrigeration equipment should be controlled to minimize compression ratio - to allow head pressures to reduce at low outdoor ambient conditions and allow suction pressures to increase at reduced load.

Equipment Selection

In the selection of equipment, some general rules should be followed:

* Use large rather than small air handling units - considerable first cost savings can result from the use of fewer large rather than many small air handling units. Usage diversity can save on initial installation costs. Over the long run, the use of fewer large air handling units rather than many small ones will save substantially when high heat source areas are relocated in the manufacturing environment.

* Closely evaluate the prospects of utilizing refrigerated air conditioning instead of evaporative spray cooling - the use of refrigerated air conditioning will often result in much lower CFM requirements than evaporative spray cooling. In many industrial applications, the KW and horsepower required to handle evaporative cooled air is more significant than first cost and energy costs associated with the use of refrigeration cooled air.

* Air cooled versus evaporative cooled versus water cooled condensing. The operating efficiencies (and longer equipment life) to be obtained with evaporative or water cooled condensing will often outweigh the first cost advantage of air cooled condensing. An energy comparison should be made in the design stages to evaluate this decision.

* Air versus radiant heating - in most large manufacturing environments, the major heat loss is through the roof. In interior zones, the effects of walls and infiltration are negligible. With an air system, it is often less costly to provide a minimal volume of heated air to offset this roof load rather than use independent air or radiant heaters. In areas of high infiltration, or in warehouse areas where an air system would not be used, radiant "spot heating" is often the most efficient system to use.

System Control

The most energy efficient control system for an industrial plant is a variation on the standard commercial control system utilized in variable air volume systems. Well designed variable air volume systems have many advantages, the most important of which is the ability to track the building load. Only enough air is supplied to satisfy the instantaneous (block) load, thus saving fan horsepower. These systems also tend to be self balancing.

The need for reheat or mixing of hot and cold air for temperature control purposes is minimized or eliminated because the space temperature control is provided through the variable air volume itself.

During refrigeration operation, discharge air temperature controls should be used to maintain a constant supply air temperature. Air volume regulating equipment is then used to control the amount of air going into different zones. Static pressure control should be utilized on the variable air volume air handling unit to modulate the system air quantity to match the block load requirement.

Below the enthalpy changeover point where the total heat (enthalpy) of the return air is more than the total heat (enthalpy) content of the outside air, 100% outside air should be used and the refrigeration used to augment the outside air cooling. At this point, it costs less to refrigerate outside air than to refrigerate return air.

When the outdoor temperature is below the design cooling supply air temperature (refrigeration lockout point), the refrigeration can be shut off and 100% outside air supplied at varying air temperatures down to a no-load point. As the outside air temperature falls, lower volumes are supplied, saving fan energy.

After a no-load (dead) band, heating should be brought on at minimal outside air and supply air volume and the temperature ramped up as heating load increases. If the maximum heating supply temperature is reached, and more heat is required, the air volume should then be increased.

Energy Tradeoff . . . . CFM For Tons

It is obvious from this discussion that air handling energy and refrigeration energy can be optimized to minimize the total energy required.

* Higher supply air temperatures - at higher supply air temperatures, the changeover point from refrigeration to 100% free cooling is also higher. Refrigeration required to augment free cooling above the changeover point will be reduced. However, the fan horsepower required will increase as a result of the increased air requirement at the higher supply air temperature.

* Supply air temperature optimization can best be handled by computer analysis - for each system there is a crossover point on supply air temperature where minimum total energy is required. This is a tedious and time consuming trial and error calculation by hand but is easily and quickly handled by a computer.

A pharmaceutical facility in Kansas utilized this variable air volume concept in the energy conservation retrofit of part of their 156,000 square foot plant. The energy conservation measure entailed conversion of major air handling systems from terminal reheat to variable air volume control. Fan speed was also reduced commensurate with the differences between peak and block CFM requirements. Actual annual savings, determined by monitoring usage for the first 12 months after the change, exceeded $71,000. Cost of implementation was approximately $64,000 and has been paid back completely from first year savings.

One example of all the above mentioned concepts is a large automobile assembly plant in the Southwest.

Following is a comparison of design results and predicted energy usage for this plant utilizing traditional design techniques versus the factory air conditioning design techniques discussed herein.

Concluding Comments

In the final analysis, capitalization of several dynamic factors, unique to factories, can significantly reduce:

* first cost
* operating costs
* maintenance
* involvement of operating personnel

These savings can be achieved while simultaneously providing as good, or better comfort to plant employees.

A working knowledge of physics combined with detailed information on the dynamics of processes and operation for a specific industrial facility yields the optimum functional and energy efficient design.

DESIGN PARAMETERS & ENERGY USE

	Conventional Design	Dynamic Factory Air Conditioning Design	% Savings
HVAC System Design			
Ton of Refrigeration	11,400	7,466	35%
Total CFM	3,300,000	2,094,000	37%
CFM Outside Air - summer	822,700	475,000	42%
- winter	1,095,000	475,000	57%
Heating BTUH	166,240,500	85,036,000	49%
HVAC Energy Requirements			
Refrigeration KWH/YR	10,876,000	9,557,000	12%
Supply Fan KWH/YR	15,419,000	6,299,000	59%
Heating BTU/YR	$127,424 \times 10^6$	$68,776 \times 10^6$	46%

Based on installed cost estimates of $600 per ton for refrigeration, $2.25 per CFM per nominal 30,000 CFM air handling units, and $2.10 per CFM per nominal 60,000 and larger air handling units, the following installed cost comparisons were made:

APPROXIMATE INSTALLED COST

	Conventional Design		Dynamic Factory Air Conditioning Design	% Savings
Refrigeration @ $600/Ton (11,400 Tons)	6,840,000	(7,500 Tons)	4,500,000	34%
Air Handling @ $2.25/CFM (3,300,000 CFM)	7,425,000	(2,100,000 CFM)	4,410,000	41%
TOTAL	14,265,000		8,910,000	38%

Chapter 65

HOW TO REDUCE ENERGY WASTE DUE TO EXCESSIVE EXHAUST/MAKEUP AIR

M. M. Roberts

BACKGROUND

Three general types of ventilation requirements are found in industry today: (A)

* General Ventilation
* Dilution Ventilation
* Localized Exhaust

General Ventilation is used for heat stress control. Ventilation for this purpose in a manufacturing plant is often more costly from a first cost and energy standpoint than refrigerated air conditioning. The reasons for this surprising conclusion and design techniques to accomplish this result are covered in the literature.

Dilution Ventilation utilizes general building exhaust with associated infiltration to dilute high concentrations of undesirable contaminants. The writer's experience has been that, more often than not, dilution ventilation occurs by accident. As the manufacturing process grows and changes, more and more sources of contamination exist in the building and more and more exhaust fans are added. Often times, the amount of exhaust far outstrips the make up air capability in the plant causing negative pressure problems, localized discomfort around doors and other building openings, and excessively high energy costs for heating and/or cooling.

As a result, the use of dilution ventilation is generally not recommended. An excellent discussion on dilution ventilation and its use and misuse is presented in "Industrial Ventilation", a publication of the American Conference of Governmental Industrial Hygienists. (A)

Localized Exhaust more often offers the best combination of effective manufacturing environment control along with minimum air flow and energy comsumption. Use of (and design of) exhaust hoods is covered thoroughly in Section IV of the "Industrial Ventilation Manual".

IMPORTANCE OF EXHAUST/MAKE UP AIR CONTROL

It is well documented that the number of totally air conditioned manufacturing facilities is increasing. In the design of an HVAC system the impact of exhaust and make up air is often critical. This impact is shown on a large southwestern manufacturing facility in the following table:

3,100,000 Sq. Ft. Southwestern Manufacturing Facility

	TOTAL	DUE TO MAKE UP AIR	%
Refrig. Load (Tons)	7,681	2,615	34%
Heating Load (1,000,000 BTUH)	118.6	75.1	63%
Annual Refrig. Electrical (1000 KWH)	8,416	1,400	17%
Annual Heating Fuel (1,000,000 MBTU)	103,436	70,384	68%

Note that the capacity required as a result of make up air alone is 2,615 tons cooling and 75,100,000 BTUH heating. Further, annual energy consumption due to make up air is 1,400,000 kilowatt hours and 70,384 million BTU's, worth $239,000 at 4.5¢ per kilowatt hour and $2.50 per million BTUs. It is obvious that reduction of exhaust and associated make up air will pay big benefits in first cost of capacity installed and operating cost due to energy consumption.

INFILTRATION OR EXFILTRATION - IT'S ALL THE SAME

Sometimes, designers will overlook the fact that any air that is exhausted from a facility will be replaced, one way or the other, by air from the outside. The required outside air will either come through cracks or holes in the building as infiltration, often creating comfort problems, or through make up air equipment which will heat or cool the air before it enters the occupied space. Whether the air is heated or cooled in a make up air unit before it gets into the building or is allowed to get into the building by infiltration, the same amount of energy will be expended to heat and cool it or the space will be too cold or too hot. In fact, if makeup air equipment is not used and air is allowed to infiltrate, more energy will often be expended because larger air quantities will be required to achieve comfortable working conditions.

The designer must consider the building envelope as a box into which air flows and out of which air flows. Infiltration coming in at several points will be balanced by exfiltration leaving at several points. Air cannot be pumped out of or into a building without offsetting infiltration or exhaust.

ARE MANUFACTURING PLANTS REALLY NEGATIVE?

Designers are often involved in factories that are considered "negative", many of which are not. During a recent study of an existing large (over 2,000,000 sq. ft.) air conditioned facility, we were advised that the plant was so negative that the industrial engineers were receiving many employee complaints. We visited this plant on a 45-50° day when the economizer was bringing in almost 2,000,000 cfm of fresh air - there was no way this plant was negative that day! But, the industrial engineering manager said, "Come with me, I'll prove it to you." We walked along the 1000' north wall; he opened the first large door we came to and, sure enough, the cold air instantly cooled about 20 bays around the door and workers started yelling. We came to the second door and opened it - same results.

Turning the corner and going down the 2000' east wall, we walked along an assembly line coming to a 12' wide by 16' high door that was standing open within a few feet of the assembly line workers. We walked out this door, and walked and walked - for about 50' before we sensed cool outside air - the exfiltration was that strong, but nobody was complaining. The point is, most plants are not "negative". Wind blowing at open doors causes infiltration and complaints; exfiltration is not noticeable and causes no complaints, but is a severe energy problem.

HOW TO REDUCE EXHAUST/MAKE UP AIR

With the foregoing in mind, it is obvious that reductions in exhaust will pay big dividends in tonnage, heating, and cfm required in the design of an air conditioned manufacturing facility and even bigger dividends in the total environmental control energy required to run the manufacturing facility. The question is, "How"?

The purpose of this paper is to answer that question by presenting the conclusions of a study made originally as a part of an air conditioning design study for the previously mentioned manufacturer's southwestern plant. The ventilation portion of this study developed design recommendations which would allow reduction of over 40% in exhaust with associated reductions in installed capacity and HVAC energy while, at the same time, maintaining adequate and safe plant environments.

HOOD DESIGN GUIDELINES

As mentioned above, it is not the purpose of this paper to deal at length with the details of exhaust hood design as these design parameters and philosophy are discussed in great detail elsewhere. Rather, its purpose is to suggest modifications to existing hoods (or new hood designs) to reduce conditioned air exhaust.

To provide a background in hood design so the reader may understand and appreciate the changes being suggested, it is well to review the following points:

*An exhaust hood is designed to create air motion (capture velocity) across the source of the contaminant sufficient to entrain the contaminant and contain it within the air stream being exhausted.

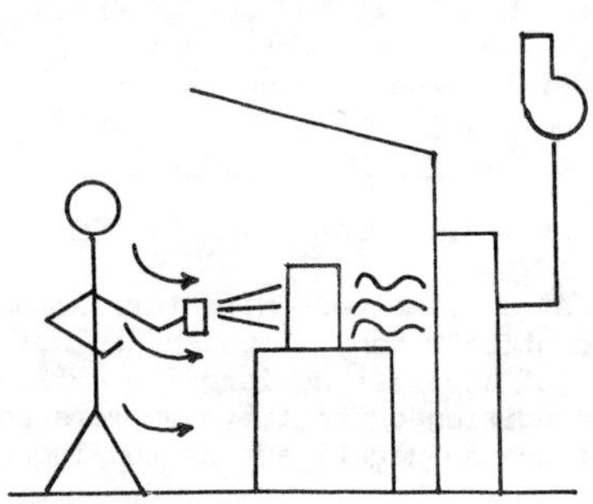

Fig. 1: Paint Spray Booth Showing Good Capture Velocity

*The type of hood to be used can generally be determined by the direction in which the contaminant naturally flows. For example, exhaust from hot processes should be taken <u>over</u> the process as hot gas tends to rise; and, particles from grinding should be collected from the trajectory which they follow leaving the grinding wheel.

Fig. 2: Hood Using Natural Convection Currents

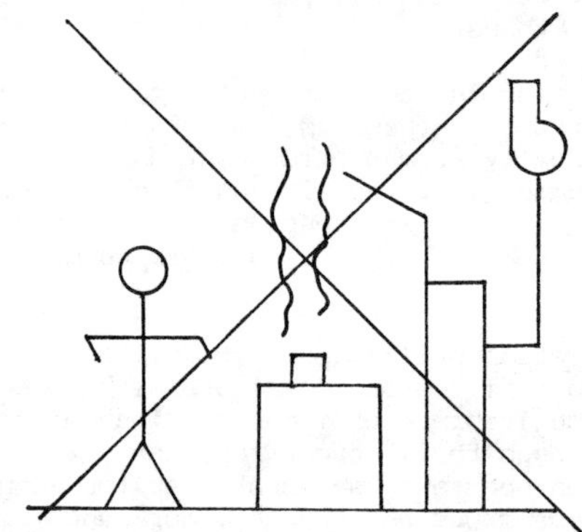

Fig. 3: Poor Design Not Using Natural Convection Currents

*Exhaust should be taken in such a way that the contaminants are not forced to flow past the workers. Rather, fresh air should be drawn over the workers then across the contaminant source and into the exhaust hood.

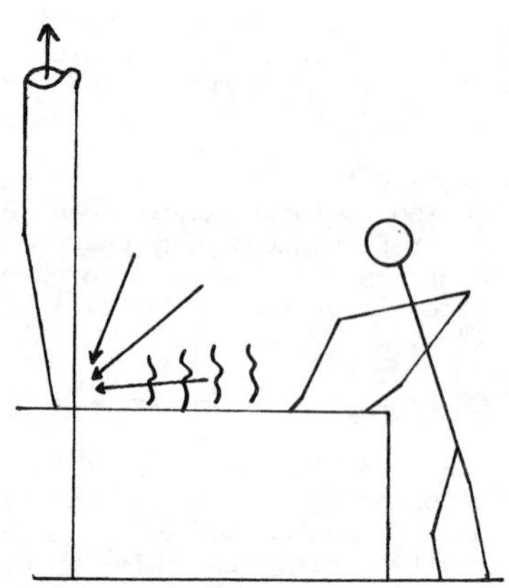

Fig. 4: Contaminants Drawn Away From Worker

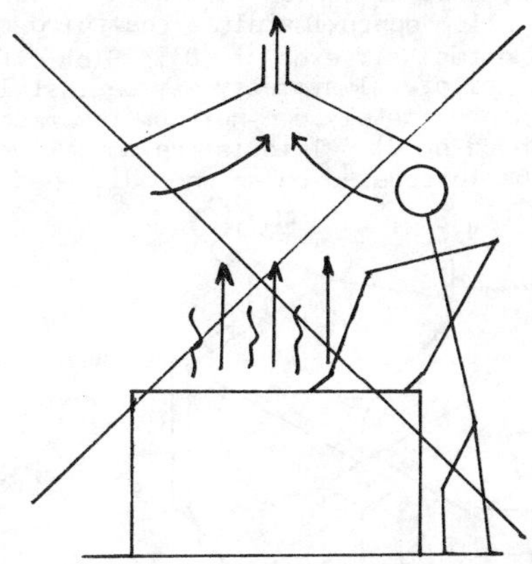

Fig. 5: Poor Design -
Contaminants Drawn By Worker

*Minimized exhaust volumes are accomplished by building the most complete process enclosure possible with the smallest number of and size of holes.

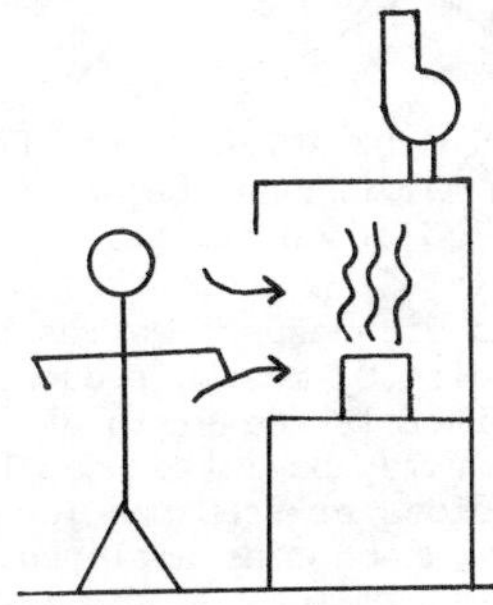

Fig 6: Closely Contained Process,
Small Opening Reduces Exhaust

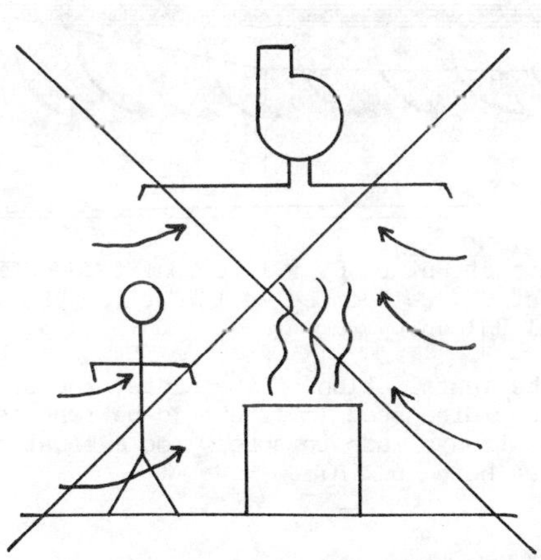

Fig. 7: Poor Design - Large Opening
Requires High Volume Exhaust

*The capture velocity is the velocity required at the point where the contaminant is emitted to pick up the contaminant and draw it into the exhaust hood, overcoming any opposing forces or air currents. The effectiveness of capture velocity can be increased by a "push - pull" system.

*Conditioned air exhaust volume is determined by:

1) Capture velocity.
2) Exhaust volume.
3) Schedule of operation of exhaust (and make up air) system - should operate only when contaminant is being produced.

NOTE: Avoid the commonly used rule of thumb for the design of an exhaust hood - 100 feet per minute across the face of the hood. The "real" volume requirement is based on the velocity required at the contaminant introduction point to overcome opposing air currents or to overcome (or enhance) the natural tendancy of the contaminants to flow in a certain direction. 100 feet per minute was a good rule of thumb several years ago but, as is shown in the following table extracted from "Industrial Ventilation", it is really not applicable in most situations.

Condition of Dispersion of Contaminant	Examples	Capture Velocity, fpm
Released with practically no velocity into quiet air.	Evaporation from tanks; degreasing, etc.	50-100
Released at low velocity into moderately still air.	Spray booths; intermittent container filling; low speed conveyor transfers; welding; plating; pickling	100-200
Active generation into zone of rapid air motion	Spray painting in shallow booths; barrel filling; conveyor loading; crushers	200-500
Released at high initial velocity into zone of very rapid air motion.	Grinding; abrasive blasting, tumbling	500-2000

In each category above, a range of capture velocity is shown. The proper choice of values depends on several factors:

Lower End of Range	Upper End of Range
1. Room air currents minimal or favorable to capture.	1. Disturbing room air currents.
2. Contaminants of low toxicity or of nuisance value only.	2. Contaminants of high toxicity.
3. Intermittent, low production.	3. High production, heavy use.
4. Large hood—large air mass in motion.	4. Small hood—local control only.

Fig. 8: Range of Capture Velocities

THE PROBLEM: REDUCE EXHAUST/MAKE UP AIR QUANTITY -
THE SOLUTION: SLOT DIFFUSER DESIGN TECHNIQUES

In order to evaluate the effectivness of a given hood design and to accomplish a reduction in the total amount of air being exhausted, each specific hood and process exhaust must be reviewed on its own merits. Experiements conducted by the writer and others have found that the use of slot diffuser technology can aid in exhaust reduction while achieving adequate contaminant control. The following design suggestions have proven helpful in the analysis of exhaust reduction opportunities:

1) Evaluate the type and volume of contaminant, location of the contaminant introduction point, and capture velocity required.

2) Utilize natural velocity and direction vectors of the contaminant particles to aid in their capture.

3) Reduce, if possible, hood entrance size.

4) Reduce exhaust air flow to minimum which will achieve adequate capture velocity recognizing points 1, 2 and 3.

5) Introduce air through well designed slot diffuser around edge of the exhaust hood to "protect" edges from contaminant spill over.

6) If practical, introduce unconditioned outside air through the edge protection slots to minimize exhaust of room air.

In almost every case use of these suggestions in conjunction with some "fine tuning" of air flows during process operation will reduce total exhaust dramatically. Specific examples discussed later show reductions of conditioned air exhaust by 50% to 70% with corresponding reductions in make up air, infiltration, energy costs, and personnel discomfort.

THE SLOT DIFFUSER - HOW IT HELPS

The development of slot diffuser technology was a major step in the development of variable air volume systems for commercial buildings. Slot diffusers introduce air at high velocities through a long, thin slot. This air, when blown across a surface, has a characteristically higher velocity pressure and lower static pressure than surrounding, almost still, room air. The difference in static pressures causes this stream of air to hug the surface under widely varying flow rates avoiding cold air "dumping" common at reduced flow rates with conventional diffusers.

The second characteristic of importance is that the wide surface area of air presented to the room tends to pull in large volumes of air from the surrounding space causing efficient room air circulation and efficient mixing or induction of room air into this supply air stream.

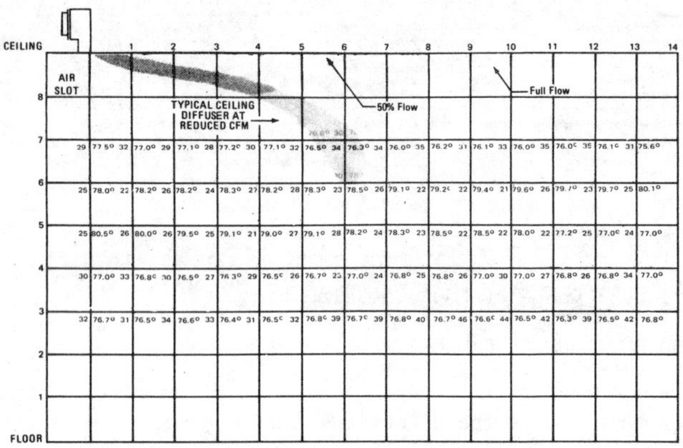

Fig. 9: Air Slot Performance Tests Demonstrates The "Coanda" Effect, and Temperature Profiles Demonstrate Air Mixing Of Slot Diffusers.

The hugging of adjacent surfaces coupled with high secondary induction makes the slot diffuser ideal for introduction of untempered outside air at the edge of an exhaust hood.

APPLICATION OF SLOT DIFFUSERS TO EXHAUST HOODS

Goals in the application of slot diffuser technology to both existing and new exhaust hoods are as follows:

*Provide contaminant reductions to satisfactory levels for worker health and safety.

*Provide for reductions in exhaust of conditioned plant air at the hood entrances.

*Maintain worker comfort in the worker occupied zones.

EXPERIMENTAL STUDIES

As was mentioned before, appliations of slot diffusers with unconditioned air introduction must be made on a case-to-case basis. The experimental process and "rules of thumb" suggested are as follows:

1) Analyze each process exhaust to determine how the hood is used and the nature of and volume of the contaminants released in order to establish minimum hood opening size and capture velocity required.

2) Utilizing acceptable air flow measuring equipment, determine actual exhaust volume being taken. At the same time, analyze the adequacy of the exhaust to reduce contaminants by observing the process in operation.

3) Within the limits of aspect ratio up to approximately 1.2, slot diffusers to protect hood edges will allow approximately a one-third reduction in normal air exhaust (Q). Slot diffusers with bilateral symmetry may be installed to supply approximately one-half of the reduced air (one-third of Q). This is the amount of air which can be introduced on an unconditioned basis. (B)

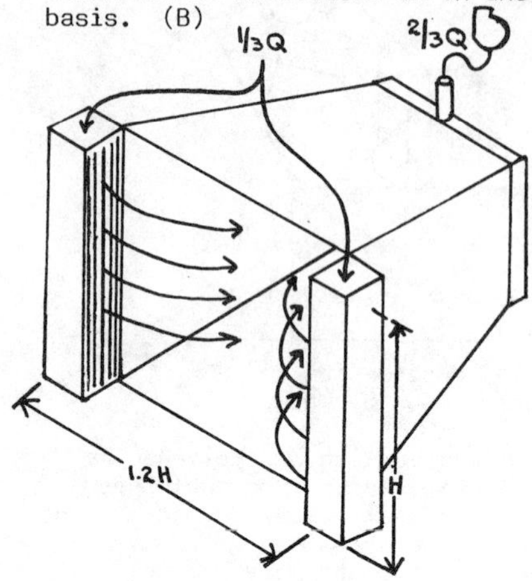

Fig. 10: For Aspect Ratios Of 1.2 Or Less, Conventional Conditioned Air Exhuast Can Be Reduced To Approximately 1/3 of Normal Values.

For exhaust hoods with aspect ratios greater than 1.2, total air (Q) may be reduced by one-third and approximately one-sixth of the original amount of air may be introduced from each of the two long sides or from one long side if bilateral symmetry cannot be employed.

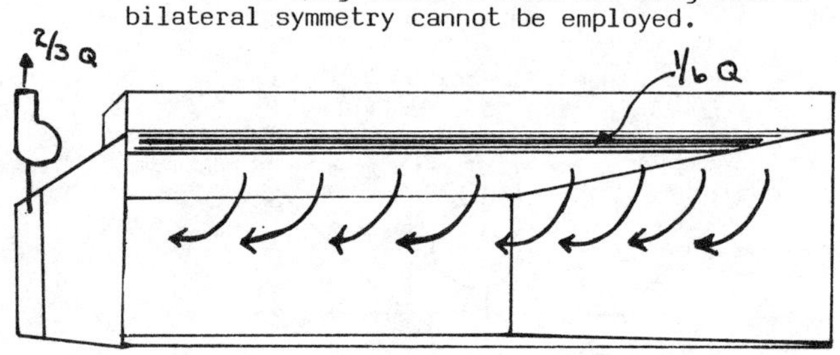

Fig. 11: For Aspect Ratios Over 1.2, Conditioned Air Exhaust Can Be Reduced To 1/2 Or 1/3 Of Normal Values.

4) After the installation of the slots, the observations discussed in #2 should be repeated and adjustments made to supply and exhaust air to reduce both to minimum.

EXAMPLES

Three example cases were analyzed using the above steps:

*Small paint spray booth.
*Auto roll test.
*Large paint booth.

*SMALL PARTS PAINT BOOTH - MANUFACTURING PLANT
(Fig. 12, 13, 14)

Steps (1 & 2): Process Analysis: Adequate Environment With 8000 cfm Of Conditioned Air Exhaust.

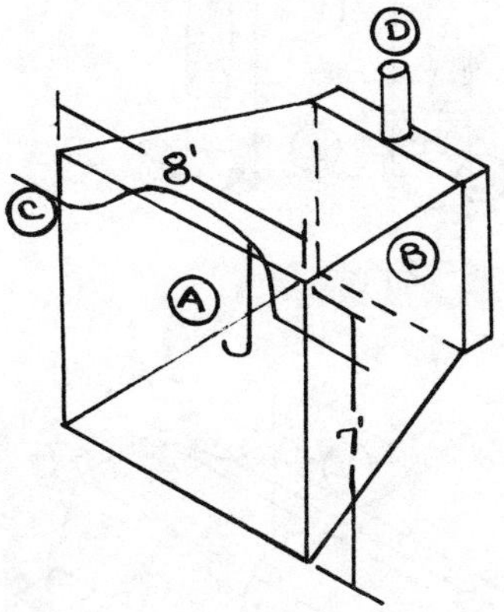

Fig. 12

A) Exhaust booth face area paint sprayed to rear and sides of booth.

B) Paint spray filter.

C) Parts conveyor.

D) Measured conditioned air exhaust: 8000 cfm

Steps (3 & 4): Add Unconditioned Air Supply Slots And Adjust Flow.

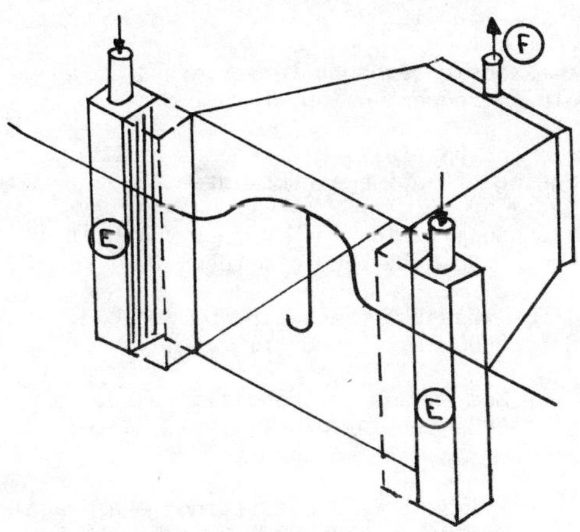

Fig. 13

E) Install supply slots - each side: 1400 cfm each unconditioned air.

F) Readjust exhaust to 5600 cfm - 2800 cfm unconditioned air, 2800 cfm plant air - a reduction of 65%!

NOTE: On a winter day (0-10°) air temperature measurements showed the following profile:

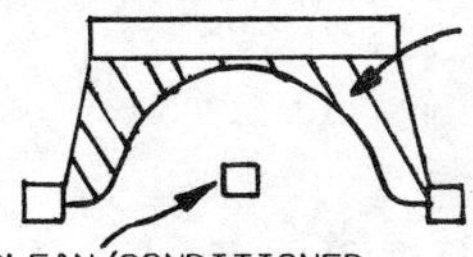

Fig. 14

*AUTO ROLL TEST - AUTO ASSEMBLY PLANT
(Fig. 15, 16, 17, 18)

Steps (1 & 2): Process Analysis: Automotive exhaust at roll test. Existing design gives adequate smoke containment with 7000 cfm of conditioned air exhaust.

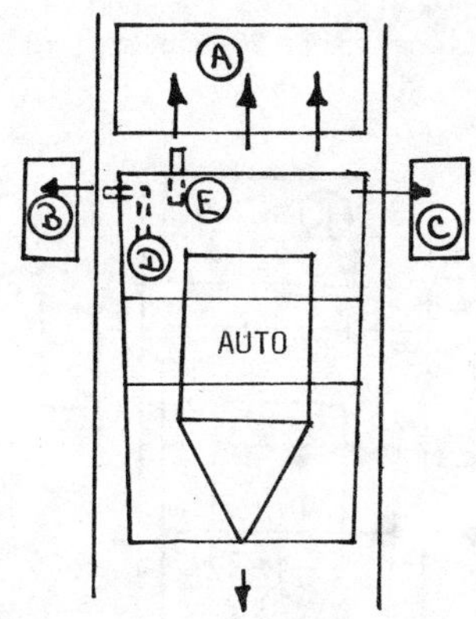

Fig. 15

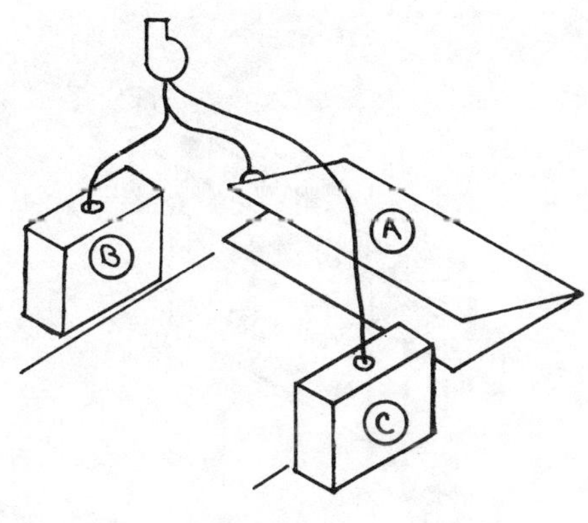

Fig. 16

A)	Raised flap exhaust:	4,400 cfm
B)	Right side exhaust:	800 cfm
C)	Left side exhaust:	1,800 cfm
	TOTAL	7,000 cfm

D) Station wagon exhaust directed to the right.

E) Sedan exhaust directed to the rear.

Steps (3 & 4): Add Unconditioned Air Supply Slots and Adjust Flows. (Fig. 17, 18)

F) Top supply slot, right side: 750 cfm unconditioned air

G) Rear directed supply slot, right side: . . . . 400 cfm unconditioned air

H) Rear directed supply slot, under car: . . . . 450 cfm unconditioned air

I) Cap left exhaust to seal.

J & K) Increase right exhaust: 1150 cfm
Decrease rear exhaust: 2450 cfm

Total exhaust: 3600 cfm - 1600 cfm unconditioned air, 2000 cfm plant air - a reduction of 71%!

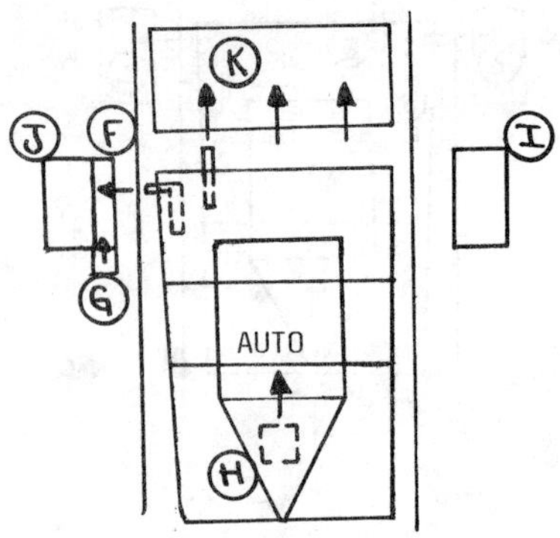

Fig. 17

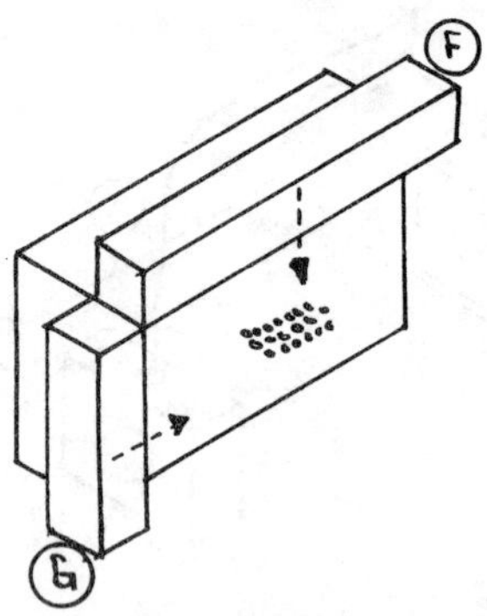

Fig. 18

*LARGE PAINT BOOTH - HEAVY EQUIPMENT MANUFACTURING PLANT (Fig. 19, 20)

Steps (1 & 2): Final Painting Required of Large Items (10'W x 12'H x 23'L) in Booth to be Added to Existing Plant. Lack of Space Required Double Ended Tunnel Type Booth. Existing Plant Air Conditioning Capacity Required Minimum Exhaust.

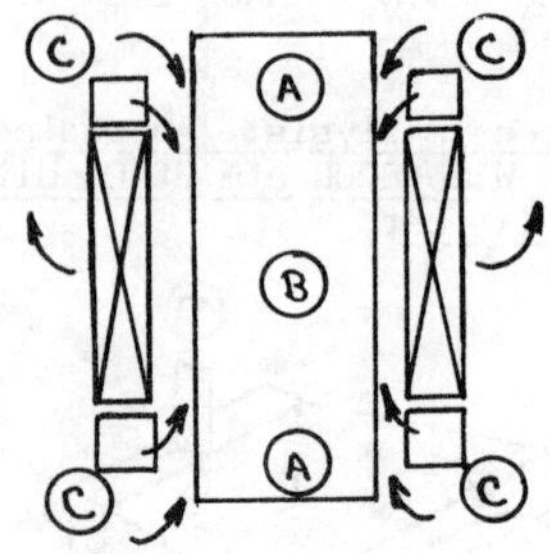

Fig. 19

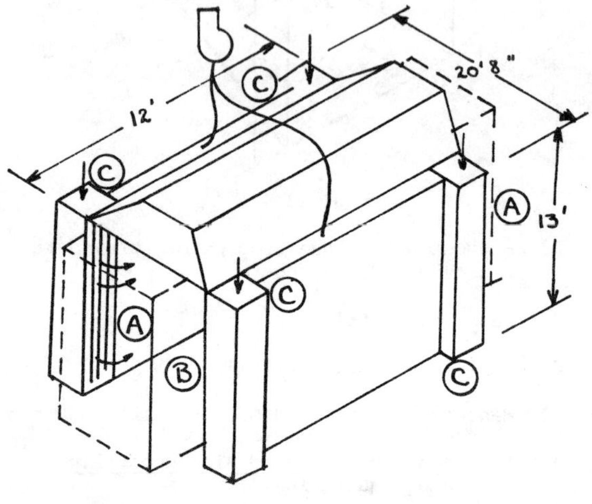

Fig. 20

A) Normal design exhaust to achieve 100 fpm at booth entrance: 41600 cfm

B) Equipment partially blocks ends of booth reducing effective exhaust area 42%.

Steps (3 & 4): Supply slot each corner balanced at 2200 cfm, 8800 cfm total.

Exhaust on each side balanced at 8800 cfm, 17600 cfm total.

Net exhaust of conditioned air 8800 cfm, a reduction of 79%!

REFERENCES: (A) "Industrial Ventilation" - A Manual of Recommended Practice - 16th Edition

(B) Experience has shown that with aspect ratios of 1.2 or less, the amount of exhaust may be reduced to two-thirds of the original acceptable level with one-half of that air (one-third of the original) being introduced on an unconditioned basis from the slot diffusers. The net result is a two-thirds reduction in the exhaust of conditioned plant air.

SECTION 20
ARCHITECTURAL CONSIDERATION IN BUILDING ENERGY CONSUMPTION

Chapter 66
OPTIMUM DESIGN OF UNDERGROUND SERVICE FACILITY

W. Steigelmann, R. Crane, F. McKibban

A new Technical Support Facility (TSF) complex has been designed for the FAA by The Vitetta Group, architect/engineer. Synergic Resources Corporation provided technical/analytical support to Vitetta involving analyses performed to help to determine the optimum use of thermal insulation, solar energy concepts and systems (both passive and active), and other energy conservation features at this facility.

The Technical Support Facility will be constructed at the National Aviation Facilities Engineering Center (NAFEC), located near Atlantic City in Southern New Jersey. The TSF will consist of two one-story buildings. One building will house offices, a number of shops, a large warehous/storage area, and receiving/shipping docks. This "OSW building" (offices, shops, warehouse) will be mostly buried below earth. Only the front wall, facing the access road, and part of the roof (approximately 17,000 square feet compared to a total roof area of 64,200 square feet) will be exposed, with the remainder earth-covered. The second building, the Service Area Compound Building, is located above ground, and contains shop and storage areas and a 5,000 square foot vehicle service bay area. This paper will discuss the various analyses performed to support optimization of the design of the OSW building. The design alternatives which have been evaluated include:

- Insulation R-values for the exposed roof
- Insulation requirements for buried or earth-covered walls and roofs
- Skylight orientation and size
- A direct-gain passive solar system
- Tempering ventilation air by means of underground ventilating ducts.

It must be recognized that there are no rigid criteria to indicate when a particular design feature should be included in the design, or to determine optimum values. Also, the energy savings expected to be produced by solar and other techniques and technologies are difficult to determine accurately. Because of the many interrelationships between design variables, the results from this analysis should be viewed as being indicative of the directions that are most likely to be beneficial in reducing energy requirements at a reasonable cost, rather than being completely definitive.

REFERENCE DESIGN

The starting point for the analyses is termed the Reference Design. The basic decision had already been made to construct the building below grade, as described above. The building is comprised of three discrete zones, as follows:

Zone	Floor Area (Square Feet)
Offices	16,800
Shops	23,600
Storage	23,800
Total	64,200

The heating and cooling season temperature set points are:

Zone	Heating: Occupied Hours	Heating: Unoccupied Hours
Offices	65	55
Shops	65	55
Storage Areas	55	55

Zone	Cooling: Occupied Hours	Cooling: Unoccupied Hours
Offices	78	-
Shops	Ventilation Only	-
Storage Areas	Ventilation Only	-

The exposed section of roof is over the administrative office areas, and includes approximately 2,400 square feet of north-facing skylights and 1,000 square feet of south-facing windows. The shops have approximately 970 square feet of south-facing skylights. The skylights and windows are double glazed. The R-values of the building structure and added thermal insulation for the different segments of the building structure are:

South Wall:	12
Buried Walls:	10.5
Exposed Roof:	19
Buried Roof:	10

and the ventilation/infiltration rates for the various portions of this building, expressed as air changes per hour, are:

Zone	Ventilation (Occupied Hours)	Infiltration (Unoccupied Hours)
Offices	2.1	0.1
Shops	2.1	0.1
Storage Areas	0.3	0.0

The heating and cooling requirements of the building were determined by use of the BEEP (Building Energy Evaluation Procedure) computer program. This program utilizes the average day and average night ambient temperatures (wet-bulb air, dry-bulb air, and soil at various depths), average solar insulation values for each month, building design parameters (dimensions, materials, insulation values, ventilation flow rates, and passive solar design), and building operational parameters (internal heating produced by lights, equipment and occupants; thermostat settings -- winter and summer, occupied and unoccupied) to determine the amount of heat that must be added to the conditioned space during the heating season by the HVAC system, and removed from the space during the cooling season. Storage of heat in the building's structural elements is also represented.

Because portions of the building are to be maintained at significantly different internal temperatures, some portions are buried and others are exposed, some have passive solar heating and others do not, and the ventilation and internal heating rates are also markedly different it is necessary to represent the building in the computer model by a number of zones.

The calculated net flows of heat through the various elements of the building, and those resulting from internal gains and restoration of structural temperatures, are as follows during the heating and cooling seasons:

	Heat Loads (million Btu/Yr.)	
Element	Heating	Cooling
Roof	-502	1
Walls	-31	0
Windows/Doors*	334-195	32+4
Foundation/Base Slab	-106	0
Ventilation/Infiltration	-382	33
Internal Gains	1,157	216
Structural Storage	-593	0
Arithmetic Total:	-318	286
Actual Net Total:	-889	249

Negative signs denote heat losses (heat flow from the building); positive numbers are heat flows to the building's interior spaces.

It should be noted that there is a significant difference between the arithmetic totals and the actual totals. This is because the timing differences between the heat flows. For example, solar gain entering through a window at 11:00 AM cannot heat the building structure at 7:00 AM, and the overheating of the space resulting from solar and internal gains during April and May cannot heat the building in January. The "Actual Net Total" values are derived directly from the computer model, in which the timing of the heat flows are correctly represented. The model also represents transfers of heat between the different zones of the building, so that some of the excessive heat in one zone is picked-up by the HVAC system and carried to a zone that requires heating. In addition, the heat flows through the thermal envelope (including ventilation/infiltration) are only tabulated when they result in interior temperatures that would activate mechanical heating or cooling.

*Solar Gain ± Transmission

The individual heat flow values tabulated above are useful because they show where the designer's attention should be focused to reduce the thermal loads: Roof insulation, ventilation/infiltration, solar gains, internal heat gains from lights and equipment had already been minimized consistent with other design objectives, and the structural design is fixed.

THERMAL INSULATION VARIATIONS

Exposed Roof

Three alternatives were evaluated: the amount of insulation for exposed roofs; the amount and location of insulation for buried roofs and walls; and the benefits of insulating the skylights at night.

The effect of roof insulation R-value on annual heating and cooling loads is shown in the following table.

Insulation	Change in Annual Loads (Million Btu)	
R-Value	Heating	Cooling
10	+51.0	+2.0
19 (Base)	0	0
30	-21.0	-1.0

It can be seen that the insulation R-value of the exposed roof in the main building has a noticable effect on annual heat loss, but has a minimal effect on heat gain. Decreasing the R-value from 19 to 10 increases the heat loss by more than double the reduction achieved by increasing the R-value to 30. This clearly shows the diminishing returns which are obtained from increasing the R-value of insulation beyond a given point. It was concluded that increasing the R-value of the roof insulation beyond a value of approximately 19 will not provide sufficient savings to offset the added costs.

Buried Roof and Walls

A separate computer program was used to analyze the heat flow through the buried roof and walls of the main building. For these analyses, it was assumed that the roof and walls were modeled as one-foot thick concrete slabs covered by earth. Three cases were identified: no insulation, insulation between the earth layer and concrete slab, and insulation on the inner surface of the concrete slab. It was found that adding insulation between the earth layer and concrete slab reduced heat loss. However, an R-value greater than 8 had no appreciable effect in decreasing heat loss. Insulating the inner surface of the concrete slab is more effective in reducing heat loss than locating the insulation between the earth layer and concrete slab. This occurs because concrete is a good storage medium. If the inner surface of the concrete is not insulated, at night, during the heating season, the concrete cools. During the morning, a significant fraction of the energy used for space heating is used to reheat the concrete. By insulating the inner surface of the concrete, the concrete is separated from the conditioned space and consequently does not have an impact on the heating load. It should be noted that while concrete roofs and walls can cause a greater heating load, they can also moderate cooling loads by storing daytime internal gains from people, lights, applicances, etc. If the inner surface of concrete components of the structure are insulated, the concrete is not able to store internal heat gains, and the moderating effect on the cooling load is lost.

Since most of the concrete roof and wall area in the main building is located in zones which are not mechanically cooled, it was recommended that the inner surface of the concrete roofs be insulated with an insulation R-value in the range of 5 to 10. The reduction in heat loss is approximately 375 million Btu per year. Insulation of the walls is much less important because there is a greater thickness of earth insulating these areas.

Night Insulation of Skylights

The effect of night insulation for the skylights is to reduce the annual heating load by 43 million Btu. Night insulation for the skylights should seriously be considered, as well as night insulation for the direct-gain passive solar windows, if it can be done inexpensively. Another consideration is that, for night insulation to be effective, it must be used. Therefore, the insulation system must be reliable and convenient to use.

PASSIVE SOLAR OPTIMIZATION

The amount of solar energy that enters the office zone during the heating season is approximately 90% of the net annual heating load (QTOT) for the zone. The excess solar gains and internal heat generation are greater than the net annual heating load. Consequently, this space will overheat at times during the winter unless the excess heat is vented or transferred to other zones of the building is most desirable, but care must be taken to adequately design the air distribution system so that overheating can be controlled.

During the heating season, the building structure and contents become cooler at night. Conversely, the building structure and contents become warmer at night during the cooling season. In the morning, the building structure and contents must be reheated during the heating season and cooled during the cooling season. This morning load on the HVAC system is appreciable. Unless specific steps are taken, excess solar and internal gains cannot be used to offset this morning load during the heating season because it occurs when the magnitude of these gains is small. Notice that QST represents the major fraction of the load seen by the HVAC system.

Solar and internal gains can be used to decrease the value of QST by increasing the amount of thermal mass in zones where these gains are significant. The optimum locations for the thermal mass are the floor area directly behind the solar glazing, and the east, north and west walls. It is desirable to use floor and wall areas for thermal storage whenever possible. The thermal mass should be spread over as large an area as possible to insure that the space conditions are uniform. A thermal mass thickness greater than 4 inches has no beneficial effect. The thermal mass should have a flat finish, and be reasonably dark in color. Surfaces of furnishings, partitions, etc., which are not used for thermal storage should be light in color.

The analysis of the interaction of thermal mass with the conditioned space is very complex, particularly in areas such as Zone 1 of the main building, which have large internal heat gains. Based on the results obtained with the BEEP computer program, Zone 1 could benefit from more thermal mass than would be contained in the floor and walls of the present design. Thermal mass can be increased by using concrete or filled concrete blocks for partitions rather than light weight materials. It would also be desirable to limit the amount of carpeting, particularly in front of the solar windows.

The use of vertical south facing skylights in the office zone, as compared to the north facing skylights in the "base design" was evaluated. Vertical, south-facing skylights were investigated and found to reduce the annual heating load somewhat (19 million Btu), but would contribute significantly to the annual cooling load (34 million Btu). Consequently, it is recommended that the skylights face north, as per the "base design" to minimize overheating, while still providing natural lighting to the office space.

VENTILATION SYSTEM OPTIMIZATION

The heat losses and gains resulting from the need to heat and cool (respectively) the ventilation make-up air and air that infiltrates into the building are larger than the "building envelope" thermal loads. The heat loads contributing by infiltration is small because of inherent considerations (i.e., underground construction), and good design practices (e.g., weatherstripped doors, well-sealed windows, entrance vestibules, internal doors or other methods of isolating the receiving/shipping area from other conditioned spaces). The heating and cooling required for the ventilation make-up air can be minimized by either of two techniques: (i) heat exchange; or, (ii) earth tempering. In the first, special air-to-air heat exchangers are used to heat and cool the supply air stream using building exhaust air. Because the exhaust air stream using building exhaust air. Because the exhaust air stream is at approximately 65°F during the heating season and 78°F during the cooling season, and because these temperatures can only be approached but not attained, this method works well during the winters but is much less effective during the summer. Also, during the summer months, much of the heat load is due to humidity, and heat exchangers usually do not transfer moisture in the air; only temperature. Because the outdoor air often has high humidity in the summer, even special heat exchangers that can exchange water vapor are likely to be of small value.

Earth tempering is the term given to the concept of passing the ventilation supply air through ducts buried 12-15 feet in the ground. At this depth the temperature of the earth surrounding the ducts is nearly constant all year around and near the annual average air temperature. The actual temperature is affected by the presence or absence of ground water. At the FAA NAFEC site, the surrounding earth temperature is approximately 35°F. This temperature is excellent for cooling and dehumidifying the make-up air during the summer, and is also quite effective for reducing the heating load in the winter.

Earth tempering of the ventilation make-up is more effective in reducing the buildings heating and cooling requirements than a heat recovery heat exchanger installation that is 50% effective. Also, as was the case with the roof insulation, the amounts of savings is somewhat contingent upon the HVAC systems being able to effectively distribute the excess heat. The earth-tempering technique can reduce the "base case" energy requirements by 15% during the heating season and 13% during the cooling season. The system can be installed relatively easily in the main building by routing the buried ducts around the sides and rear of the buried portions.

Chapter 67

THE EFFECT OF SKYLIGHTING ON ANNUAL BUILDING ENERGY USE

D. J. Crise

INTRODUCTION

The rising costs and uncertain availability of fossil fuels for the lighting, heating, and cooling of buildings are leading to ever-increasing numbers of strategies for energy utilization and management. The use of skylights in buildings is a design and construction strategy that shows great promise for energy savings in addition to pleasant architectural and psychological amenities. Building skylights are passive solar collectors making direct use of the sun's light and heat. Skylights admit solar light to reduce the energy demand for artificial lighting while at the same time contributing the sun's heat to the building's heating requirements. During the summer months, undesirable heat gained through the skylights is offset by the reduction in heat given off by the lighting fixtures and the corresponding reduction in air-conditioning loads.

The effectiveness of skylights in harnessing the sun's energy with human comfort is dependent on balancing their daylighting effect and thermal impact with the building's lighting, heating, and cooling energy demands. A procedure for computing the annual energy balance of skylighted buildings is described in this paper with a parametric analysis to demonstrate the procedure and to evaluate the potential for energy-saving benefits of skylights.

METHODOLOGY

Energy losses and gains that are attributable to the skylights in a building* are calculated using hourly local solar and meteorological data together with illumination and thermal parameters and skylight properties. The net annual energy effect of skylights is determined from the calculated balance of savings and losses during the heating and cooling seasons. The impact of skylights on a building's energy usage includes the computed energy demand, the purchased energy required to meet that demand and factored values that account for types of depletable energy resources.

To evaluate the cost-effectiveness of skylights in buildings, equivalent benefits or costs are developed from their corresponding energy savings and losses. These economic values are calculated using local heating fuel and electricity rates. In addition, a payback period is determined that indicates the time required to recoup the skylight investment. Energy-balance and attendant benefit-cost results are computed for buildings with and without air conditioning for comparison.

The analytical model was formulated pursuant to established conventions and published methods of the design and engineering professions.[1,2,3] Structured as a series of conjugate algorithms, the model develops a succession of climatic, thermal, lighting, and economic solutions that culminate in a net annual energy balance for the particular building and set of conditions being examined. The schematic flow of the computations is shown in Figure 1.

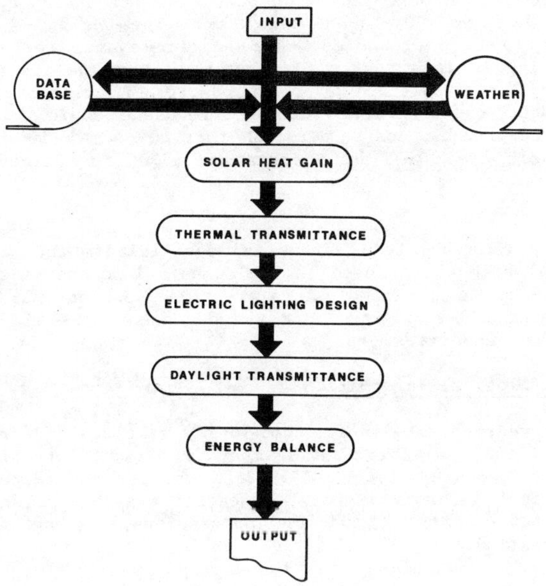

FIGURE 1. SCHEMATIC FLOW OF COMPUTATIONS.

To illustrate the effect of skylighting on annual building energy use, a parametric evaluation of a 10,000 square-foot hypothetical building with 40 acrylic-domed skylights is presented. The generic construction of a domed skylight is indicated by Figure 2.

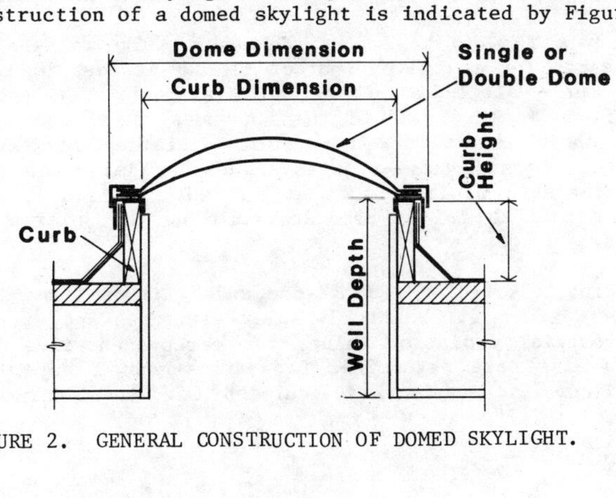

FIGURE 2. GENERAL CONSTRUCTION OF DOMED SKYLIGHT.

*Upper floor of multistory buildings

Climatological and cost data for Atlanta, Georgia are used in the analysis. For geographical comparison, the results of an analysis for the same building located in Chicago, Illinois are included in a tabulated summary. Salient aspects of the analysis are described as follows:

Input Information

An analysis require 54 input variables that define local parameters for the particular building design undergoing analysis. The input data includes information of both numeric and alphabetic character format describing illumination and thermal parameters, local solar and meteorological conditions, and skylight and construction properties.

The hypothetical one-story building being examined in this analysis is 80 feet wide by 125 feet long with a ceiling height of 16 feet. Forty double dome skylights measuring 49 inches square are evenly distributed over the building's flat roof. Based on an approximate in-place cost of $18 per square foot, the total skylight cost would be $12,005. The building is considered to be occupied during the hours from 7:45 a.m. to 5:15 p.m. five days per week. The design light level is 50 footcandles. The heating system for the building in this anaylsis is assumed to be fueled with natural gas costing $3.55 and $3.60 in Atlanta and Chicago, respectively, for each thousand cubic feet. The electrical costs per KWH that are used in the analysis are $0.05 for Atlanta and $0.06 for Chicago.

Data Base

Arrays of tabulated factors and coefficients that are to be subsequently used in the computations are stored in a data base. These values are time dependent due to changing solar positions or are dependent on building or room geometries.

Meteorological Information

Local weather data is attached with a file created from a National Oceanic and Atmospheric Administration (NOAA) computer type. The file consists of meteorological information for 248 weather stations through the United States and its possessions. Among the information on the file for each of the stations is geographic position, mean temperatures, normal degree days, and solar radiation data.

Solar-Heat Gain

The solar heat transmitted through a rooftop skylight is dependent on the suns position relative to the building site, the physical properties of the skylight glazing, and general sky conditions. A quantitative value, expressed in Btu/sq.ft. of horizontal surface of roof, will be developed to represent solar-heat gain.

The position of the sun in the sky relative to a point on earth for any given hour of the day is described by the sun's altitude and the solar azimuth. The solar altitude is the angle of the sun above the horizon and the solar azimuth is its angular deviation from true south. Because dome-type skylights on flat roofs are considered omnidirectional at any solar altitude, the solar azimuth is not used in computing the solar-heat gain.

Knowing a building's longitude and latitude (from the NOAA weather file), hourly solar altitudes and extraterrestrial radiation values for a representative day (the 21st) are determined for each month. The solar altitude in degrees is calculated from the following formula:

$$\theta = \sin^{-1} (\cos L \cos \delta \cos H + \sin L \sin \delta) \quad (1)$$

The reader is directed to the Nomenclature at the end of the paper for definition of terms used in all equations.

Direct normal-solar intensity and diffuse-solar intensity are then computed for each hour that the solar altitude is positive (above the horizon). For computing direct intensity, the following expression is used:

$$I_{DN} = \frac{A}{\exp(B/\sin \theta)} \quad (2)$$

The values of A and B vary during the year because of seasonal changes in the dust and water-vapor content of the atmosphere and due to the changing distance between the earth and sun. These values are stored in the data base.

The diffuse-solar intensity is computed using the formula as follows:

$$I_{dH} = C \times I_{DN}/CN^2 \quad (3)$$

To account for solar transmittance properties and other characteristics of acrylic skylight domes, the hourly direct- and diffuse-solar intensities are tempered by solar-heat gain factors developed by others.[4]) These factors are stored in the data base. Their retrieval and utilization are dependent on the computed hourly solar altitude, and on input options pertinent to the clearness of the dome, the shape of the skylight, and skylight construction details.

The sky conditions must also be considered in determining the solar heat gain through skylights. Three types of general sky conditions are considered: (1) clear skies; (2) overcast skies; and (3) partly cloudy skies. Clear-day solar heat gain is computed as the sum of the factored direct-normal and diffuse solar intensities. For overcast and cloudy days, the solar heat gain is represented by the factored diffuse intensity only. Subsequent calculations are based on weighted-average heat gains that are developed by apportioning the solar heat gains by their respective percentages of sunshine. Monthly solar heat gain for days with clear skies and overcast conditions are shown in Table I together with the daily and monthly weighted-average heat gains.

Thermal Transmittance

The impact on building-energy usage by heat transferred by convection and/or conduction through the skylight domes is significant. Following the Second Law of Thermodynamics, the thermal transfer is assumed to occur in the direction of decreasing temperature. Thermal transfer during the heating season is therefore a loss from the building. In the summer months heat is transmitted through domes and curbs into the building as a heat gain. Daily heat transfer rates through the domes and curbs during both heating and cooling seasons are calculated using the following expression:

$$Q_{h,c} = \frac{DD}{DM} \times U_{w,s} \times 24 \text{ hr/day} \quad (4)$$

A given month is considered to be in the heating or cooling season, depending on the dominance of the heating - or cooling-degree days. If the heating-degree days dominate, winter U values for domes and curbs are used in eq. 4. Similarily, summer U values are used in the months when cooling-degree days are dominant.

In an analysis, heat flowing through skylight domes and curbs is offset by the heat that would be transferred through an equivalent area of roof. By including the equivalent roof-area heat-transfer calculation as a credit in the analysis, the net effect of heat transferred through the skylight is determined.

Air leakage or infiltration is another thermal effect accounted in the computations. The infiltration coefficient (IC) in Btu/hr per foot of curb perimeter and the flow of infiltration (IF) in Btu/day per square foot of skylight is developed as follows:

$$IC = \frac{0.011 \times \Delta t}{\text{curb perimeter, ft}} \qquad (5)$$

and

$$IF = \frac{IC \times DD \times SP \times 24\ \text{hr/day}}{DM \times \text{Skylight area, ft}^2} \qquad (6)$$

Results of thermal transmittance computation for the Atlanta sample building are shown in Table II.

Electric Lighting Design

The Lumen-Zonal Cavity (LZC) method is a classical and well-documented procedure for calculating lighting levels in buildings.[3] This method is used herein to determine the energy required to maintain a design-light level produced at a specific height above the floor as follows:

$$\lambda = \frac{EA \times Ar}{CU \times LLF} \qquad (7)$$

The design-light level (EA) in footcandles (lumens per square foot) is an input variable and is dependent on the intended function of the building space. The illuminated area is signified as A_r. The coefficient of utilization (CU) is retrieved from the data base as a function of the room-cavity ratio and surface reflectances of 50-, 30-, and 20-percent for walls, ceilings, and floors, respectively. The room-cavity ratio is calculated with a relationship of room size and shape. The light loss factor (LLF) accounts for the loss of light due to the aging of the lamp and the accumulation of dirt on a fixture's reflective surfaces and lens. A light-loss factor can be input or, if unknown, a default value of 0.69 will be used in the LZC equation.

Because of their widespread usage in commercial, industrial, and institutional buildings, fluorescent luminaire fixtures with two 4-foot-long 40-watt lamps and prismatic wraparound lenses are used in the lighting design calculations. Each of the luminaires is rated with a flux of 6300 lumens. Therefore, by dividing by 6300, the number of light fixtures required to maintain the design light level is determined. The total electric power consumed by all the fixtures and ballasts is computed by multiplying the number of fixtures by 92 watts. Because each watt represents 3.413 Btu per hour, the equivalent thermal energy consumed by the electric lighting can be easily determined.

Daylight Transmittance

The daylight illumination transmitted through the skylights is time dependent with constant changes in the sun's position and conditions of the sky. The visible-light transmittance property of the skylight glazing and geometries of the interior building space and skylight wells also significantly influence the illumination performance of the skylights.

Skylight illumination performance is measured by the number of footcandles that are projected on the horizontal working plane at a specific height above the floor. To calculate skylight illumination in footcandles, the Toplighting Method described in the IES Lighting Handbook is employed as follows:

$$\mathcal{E} = E_h \times A_s / A_r \times KU \times KM \times WE \qquad (8)$$

Values for illumination values E_h incident on the horizontal plane of skylights are developed in Ref. 3 as a function of solar altitude and are stored in the data base. When determining hourly daylight illumination for days with cloudy or overcast skies, only values for incident illumination from overcast skies are used for the E_h term. For days with clear skies, E_h is the sum of values for incident light from a clear sky during the appropriate season and direct sunlight radiation.

Tabulated data for daylight transmittance values of domed skylights and toplight transmittance ratio factors are retrieved from the data base for determining the toplight coefficient of utilization (KU). The maintenance factor (KM) accounts for illumination depreciation caused by the accumulation of dirt on room surfaces and on the exterior of the skylight domes. A skylight analysis is based on a KM factor of 0.70, which is appropriate for an annual cleaning of glazing materials installed horizontally over clean areas such as office spaces.

Decreases in daylight illumination on the work plane result from the interflections of daylight in the skylight wells shown in Figure 2. Skylight well-efficiency factors (WE) based on 90-percent well-wall reflectance are correlated with well indexes and are stored in the data base as a relationship of well depth, width, and length. Hourly weighted-average illumination with local percentage of possible sunshine are computed from the daylight illumination effects on both clear days and overcast days, and are shown in Table III.

Energy Balance

The skylight-influenced energy factors that have been identified and computed to this point in the analysis are balanced to determine their seasonal and annual impact on building-energy comsumption. The energy balance for Atlanta is indicated by Table IV. The reader is advised that the cooling-season balance is for an air-conditioned building. For buildings without air conditioning, the conductive heat gain, the solar heat gain, and the elimination of heat emanating from the lighting fixtures are disregarded in cooling-season balancing because their presence does not contribute to a purchased energy-intensive cooling load. The seasonal results are combined to determine the net annual result per square foot of skylight.

The winter season is assumed in calculations when the heating degree days outnumber the cooling degree days, and conversely, for the summer season. To compute the net building-energy demand resulting from the skylights for each of the two major seasons, the energy savings and losses are algebraically summed. The seasonal results are combined to determine the net annual results per square foot of skylight.

The quantity of purchased and delivered energy attributable to the skylights that is necessary to satisfy the computed energy demands (savings or losses) is determined by factoring these demands (excepting lighting savings) with the fuel-utilization efficiencies of the heating and cooling systems. The energy savings resulting from the reduction of electric lighting are not factored by fuel-efficiency ratings because the lighting system is independent and not related to the performance of the environmental control system. The light savings could be achieved regardless of the building's interior temperature or comfort level.

Fuel-efficiency ratings of 0.67, 0.60, and 0.95 are used in the calculations for natural gas, heating oil, and electricity, respectively, depending on the type of heating energy provided as input. The efficiency of an air-conditioning system is expressed as the coefficient of performance which is the ratio of tonnage of produced refrigeration to the energy required to operate the equipment. The analyst can input a COP value or use the default value of 2.2. In the present analysis, a COP value of 2.4 was used.

The energy embodied in the various types of delivered energy sources is calculated using weighting factors suggested in the federal building energy standard proposed during the Carter administration.[5] The weighting factors for buildings in the commercial sector that are used in the present analysis are 1.00, 3.08, and 1.20 for natural gas, electricity, and fuel oil, respectively.

Economics

A simplified economic analysis is performed to assist decision makers in evaluating the cost effectiveness of skylights in buildings. Using local heating fuel and electricity rates, benefits and costs that correspond to the skylight influenced energy savings and losses, respectively, are computed. The benefits and costs are equivalently expressed in dollars per year per square foot of skylight.

For the sample building in Atlanta, the use of the skylights will yield a net benefit of $2.46 and $2.83 per square foot of skylight with and without air conditioning, respectively.

RESULTS AND CONCLUSIONS

Comparative results of the annual energy balance for the skylighted building located in Atlanta and Chicago are summarized in Table V.

The simple payback periods in the table represent the number of years required to equate the cumulative energy cost savings with the initial skylight investment exclusive of the time value of money or energy cost fluctuations. The discounted payback periods reflect future values of money and fuel-cost escalation. These latter values were computed using a discount rate of 12 percent and a fuel escalation rate of 15 percent.

With their varying climates, both Atlanta and Chicago suggest that attractive energy savings are possible in different regions of the country. Reported analysis results for Pittsburgh, Boston, and San Diego validate this conclusion.[6] The discounted payback periods for both Atlanta and Chicago would be less than 7 years indicating a great opportunity for cost-effective energy savings in buildings with skylights, particularly those with expansive flat roof areas.

ACKNOWLEDGEMENT

The computational model described in this paper was developed by the author while associated with the United States Steel Corporation Research Laboratory.

NOMENCLATURE

A_r	Illuminated room area, ft^2
A_s	Total gross area of skylights, ft^2
CU	Coefficient of utilization
DD	Heating or cooling degree days
DM	Days in month
EA	Design illumination level, lm/ft^2
E_h	Incident illumination on skylights, lm/ft^2
H	Local hour angle
I_{dn}	Direct-normal intensity, Btu/hr ft^2
I_{dh}	Diffuse-solar intensity, Btu/hr ft^2
KM	Maintenance factor
KU	Toplighting coefficient of utilization
L	Local latitude, degrees
LLF	Light loss factor
SP	Skylight perimeter, ft
$U_{w,s}$	Seasonal coefficient of thermal transmittance, Btu/hr ft^2 °F
WE	Skylight well-efficiency factor
Δt	Outside-Inside temperature difference, F
δ	Solar declination
ε	Daylight illumination, lm/ft^2
λ	Total initial luminous flux, lumens
θ	Solar altitude, degrees

REFERENCES

(1) Architectural Aluminum Manufacturers Association, Voluntary Standard Procedure For Calculating Skylight Annual Energy Balance, Chicago, Illinois, 1977.

(2) American Society of Heating, Refrigerating, and Air-Conditioning Engineers, Handbook of Fundamentals, New York, 1977.

(3) Illuminating Engineering Society, IES Lighting Handbook, New York, N.Y. 1972.

(4) L.F. Schutrum and N. Ozisik, "Solar Heat Gains Through Domed Skylights", ASHRAE Journal, Vol. 3, No. 8, Aug., 1961.

(5) The Bureau of National Affairs, Proposed Energy Performance Standards for New Buildings, Dept. of Energy, Washington, D. C., December 1979.

(6) D. J. Crise, "Daylighting And Thermal Effects Of Skylights On Annual Building Energy Use", Proceedings of 1981 Annual Meeting of AS/ISES, Philadelphia, Pa., May 26-30, 1981.

TABLE I

SOLAR RADIATION DATA (BTU/SQ.FT.)

MONTH	CLEAR DAY INSOLATION	OVERCAST DAY INSOLATION	DAILY WEIGHTED AVERAGE GAIN	MONTHLY WEIGHTED AVERAGE GAIN
JAN	409.36	56.56	225.90	7002.91
FEB	652.08	75.14	386.69	10827.23
MAR	979.27	107.60	604.45	18738.05
APR	1298.64	163.94	912.84	27385.22
MAY	1493.89	216.84	1085.23	33642.14
JUN	1557.84	242.64	1123.82	33714.70
JUL	1492.56	238.16	1015.89	31492.60
AUG	1299.08	198.94	914.03	28335.06
SEP	961.97	132.80	655.17	19655.24
OCT	651.75	88.58	472.90	14659.77
NOV	413.60	61.18	269.10	8073.14
DEC	332.04	49.62	190.83	5915.73

TABLE II

SKYLIGHT THERMAL TRANSFER DATA

(BTU/DAY/SQ.FT.)

	CONVECTION HEAT TRANSFER						INFILTRATION	
	HEATING SEASON			COOLING SEASON				
MONTH	DOME	CURB	EQ.ROOF	DOME	CURB	EQ.ROOF	HEATING SEASON	COOLING SEASON
JAN	380.	3.	54.	0.	0.	0.	9.	0.
FEB	336.	3.	48.	0.	0.	0.	7.	0.
MAR	240.	2.	34.	0.	0.	0.	3.	0.
APR	81.	1.	11.	0.	0.	0.	0.	0.
MAY	0.	0.	0.	60.	1.	12.	0.	0.
JUN	0.	0.	0.	128.	2.	25.	0.	2.
JUL	0.	0.	0.	156.	2.	31.	0.	3.
AUG	0.	0.	0.	150.	2.	30.	0.	3.
SEP	0.	0.	0.	91.	1.	18.	0.	1.
OCT	74.	1.	11.	0.	0.	0.	0.	0.
NOV	228.	2.	32.	0.	0.	0.	3.	0.
DEC	361.	3.	51.	0.	0.	0.	8.	0.

TABLE III

DAYLIGHTING EFFECTS FROM SKYLIGHTS

(LUMINANCE IN FOOTCANDLES)

WEIGHTED ILLUMINATION LEVELS

HOUR	JAN	FEB	MAR	APR	MAY	JUN	JUL	AUG	SEP	OCT	NOV	DEC
1												
2												
3												
4												
5												
6												
7	0.	0.	9.	29.	42.	43.	36.	27.	15.	7.	0.	0.
8	8.	19.	36.	61.	75.	75.	66.	58.	45.	36.	19.	8.
9	29.	41.	61.	90.	101.	100.	92.	88.	76.	61.	40.	29.
10	45.	63.	81.	109.	118.	117.	108.	107.	96.	84.	60.	44.
11	60.	77.	81.	122.	133.	133.	123.	121.	108.	82.	73.	58.
12	67.	73.	88.	132.	125.	128.	119.	131.	114.	87.	79.	65.
13	69.	75.	90.	133.	126.	131.	123.	133.	114.	86.	78.	65.
15	55.	74.	75.	112.	119.	121.	115.	112.	96.	79.	56.	47.
16	38.	57.	72.	95.	103.	105.	99.	96.	76.	53.	36.	32.
17	21.	35.	46.	67.	79.	83.	78.	70.	45.	28.	13.	12.
18												
19												
20												
21												
22												
24												
NO. OF HOURS AT OR ABOVE DESIGN LEVEL	5	7	8	10	10	11	10	10	8	8	6	4

TABLE IV

SUMMARY OF ANNUAL ENERGY BALANCE ANALYSIS *

ENERGY IMPACT
(BTU/YEAR PER SQ. FT. OF SKYLIGHT)

NOTE THAT NEGATIVE NUMBERS REPRESENT COSTS AND POSITIVE NUMBERS REPRESENT SAVINGS

	COMPUTED ENERGY DEMAND	REQUIRED DELIVERED ENERGY	EMBODIED RESOURCE ENERGY	BENEFIT/COST OF SKYLIGHT ($/YR./SQ.FT.)
HEATING SEASON				
WINTER HEAT LOSS				
DOME COVER	-51408.			
CURB	- 473.			
EXFILTRATION	- 942.			
EQ. ROOF AREA	7289.			
NET LOSS SUBTOTAL	-45534.	- 67962.	- 67962.	-.24
SOLAR GAIN	92602.	138212.	138212.	.49
ELEC. LT. SAVINGS	100517.	100517.	309593.	1.53
ELEC. LT. HEAT LOSS	-100517.	-150026.	-150026.	-.53
NET WINTER ENERGY EFFECT OF SKYLIGHT	47068.	20742.	229818.	$ 1.25
COOLING SEASON				
SUMMER HEAT GAIN				
DOME COVER	-17916.			
CURB	- 231.			
INFILTRATION	- 275.			
EQ. ROOF AREA	3556.			
NET GAIN SUBTOTAL	-14866.	- 6194.	-19078.	-.09
SOLAR GAIN	-146840.	-61183.	-188444.	-.93
ELEC. LT. SAVINGS	103912.	103912.	320048.	1.58
LT. HEAT ELIMINATE	103912.	43297.	133353.	.66
NET SUMMER ENERGY EFFECT OF SKYLIGHT	46118.	79831.	245880.	$ 1.22
NET RESULTS				
BUILDING WITHOUT AIR CONDITIONING	150979.	124653.	549866.	$ 2.83
BUILDING WITH AIR CONDITIONING	93186.	100573.	475697.	$ 2.46

*FOR ATLANTA

TABLE V

COMPARATIVE TOTAL ENERGY BALANCE FOR 40 SKYLIGHTS, MBtu

With Air Conditoning	Energy Demand Savings	Purchased Energy Savings	Embodied Energy	Simple Payback (Years)	Discounted* Payback (Years)
Atlanta	62150	67076	317264	7.30	6.60
Chicago	32905	12036	238504	7.67	6.89
Without Air Conditioning					
Atlanta	100695	83137	366730	6.36	5.80
Chicago	51915	19957	262899	7.04	6.37

* Based on 12% discount rate and 15% fuel escalation rate.

SECTION 21
THE ENERGY MANAGER'S FORUM

Chapter 68

THE GROWING ROLE OF THE CORPORATE ENERGY MANAGER

C. F. Feledy

The scope of this paper is essentially limited to conservation and its impact on the corporate energy manager. The purpose is to enlighten top management on the issues, trends, constraints and challenges facing the energy manager.

INTRODUCTION

Every energy manager faces two basic problems today: the high price of energy and the uncertain availability of energy supplies. A once cheap and abundant commodity is becoming increasingly more expensive and scarce. In the future, this scenario will continue to deteriorate at an accelerating rate. Though increasing the challenge, this will result in greater opportunities, recognition and rewards for the energy manager.

CONSERVATION IN PERSPECTIVE

Why conservation? According to the Harvard study, Energy Future, edited by Daniel Yergin and Robert Stobaugh, there are five basic reasons:

- it is a largely untapped incremental energy source
- it is a major alternative to imported oil
- conservation requires no technological breakthroughs
- it could supply up to 40-percent of current U.S. usage
- it is cheaper than discovering new oil supplies

Conservation also has five other advantages worth considering:

- it uses proven, effective technologies
- it is a low risk alternative
- it is cost effective
- it is environmentally acceptable
- it can be implemented in a hurry

Now, let's look at industry's track record. It has significantly improved its energy efficiency, with a 14-percent improvement in 1979 compared to 1973. In 1979 alone, this represented a savings of over 850 million barrels of oil. Industry's record is much better than the commercial, residential, and transportation sectors and has accounted for 61-percent of the nation's total improvement. However, there is a large potential for further savings, which have been estimated at about 40-percent by some sources.

It is important to look at conservation efforts in perspective. Many companies have already "skimmed off the cream" for their conservation projects and are starting to feel the impact of the law of diminishing returns. Therefore, achieving additional savings will be increasingly difficult and will require larger capital investments.

Although there has already been an impressive improvement in energy efficiency in industry, there is still much that can be accomplished. Only limited changes have been made in "process energy" -- in the manufacturing process itself. And most process conservation projects today have not been very energy intensive. In the future, I feel there will be significant changes made in process energy. But limited capital funds will mean that only pure conservation projects will be implemented. And the law of diminishing returns will mean that we will get less conservation "bang" from each conservation capital "buck".

In summary, industry will continue to improve its overall efficiency, but at a diminishing annual rate of return. Nevertheless, industry will continue to exceed the other three sectors.

THE ROLE OF ENERGY PRODUCTIVITY

Now, let's look at the evolution of energy terminology. First, we started with "energy conservation." Then, we switched to "energy management" and then to "energy efficiency." Today, the preferred terminology is "energy productivity."

Energy productivity can be defined as reducing the amount of energy needed to produce one unit of output. It is really a tool for improving profits. Other ways to view it are as negative energy demand or as alternative form of energy supply. Also, from the standpoint of semantics, I feel "productivity" has a more positive connotation than "conservation," which to some people implies deprivation.

The "reindustrialization of America" is a current key issue, or, if you will, a "buzz word." The major thrust of industry in this decade will be to regain its competitive position relative to Japan and other countries. The way to do this is primarily through increased productivity. Energy is a vital element in the overall productivity mix, and I feel it should demand the same attention now being given to labor, material, overhead and so on. Almost every business decision hinges to some degree on energy. Also, energy costs are escalating much faster than the general rate of inflation. The energy manager will play a vital role in reindustrialization, helping to improve his firm's "bottom line" and keeping it in operation by guaranteeing that it will have adequate energy supplies.

I think you can call the 1970's the "Decade of Environmental Protection," and I predict the 1980's will be known as the "Decade of Energy Productivity."

THE EXPANDING ROLE OF THE ENERGY MANAGER

A front-page story in the Wall Street Journal on October 22, 1980 was titled "Energy Managers Proliferate and Grow in Stature at Many Companies." It said:

"When energy prices jumped in 1973, most companies gave division managers the added job of watching energy costs. But in the last two years, most big firms have hired full-time energy managers who earn between $35,000 and $60,000 and work for a vice president."

This reflects a growing trend. As a matter of fact, there are now several corporate recruiting firms concentrating on energy managers alone.

Energy productivity is an emerging speciality in corporate management, and I feel it is a powerful new management tool. This new profession combines the disciplines of engineering, economics, manufacturing, environmental sciences and business management. Thus, the manager must be sophisticated, interdisciplinarian, results-oriented, and, above all, a "salesman" (in a positive sense).

This is my job specification for the ideal energy manager:

- B.S. in Engineering, preferably E.E., M.E., or I.E.
- Master of Business Administration degree
- sound plant experience in plant engineering and/or manufacturing management, and
- consulting experience (optional).

I'm convinced of the validity of the following axiom, which I call Feledy's Law: "As the price of energy continues to escalate rapidly and supplies become even more uncertain, the energy manager's importance and visibility to top management increases proportionately." Here are some organizational predictions I'd make about the energy manager in the future. He will:

- report to a higher level in the corporate structure
- get more attention from top management
- have a larger staff
- command a larger salary, and
- be in great demand

His rapid rise in the corporate structure will be similar to the rise of management information systems executives in the past.

The basic role of the energy manager varies greatly from company to company. In large part it is a function of how energy intensive the firm is. A composite picture of the role an energy manager plays today shows he performs these four core functions:

- planning
- productivity
- policy, and
- governmental and regulatory affairs.

Optional functions performed are:

- supply and purchasing of energy (especially in energy intensive companies)
- environmental affairs
- water conservation, and
- resource recovery.

My prediction is that in the next ten years the energy manager will be responsible for all eight of these functions. His probable new corporate job title will be either director or vice president of resources.

I feel that it is critical that the manager should possess these seven financial management skills:

- industrial cost accounting
- life-cycle costing
- discounted-cash-flow analysis
- return-on-investment analysis
- forecasting models
- capital planning and budgeting, and
- cash-flow analysis

Why does he need these skills? Unfortunately, today many managers have limited skills in financial areas. Consequently, they have limited success in selling management on capital appropriation requests for productivity projects and getting approval for increased departmental budgets. In the future, many managers will have MBA degrees and will acquire the skills to be successful in getting the necessary approvals for appropriations and budgets.

ORGANIZING FOR ENERGY PRODUCTIVITY

Today, in many companies, the initial momentum of the oil conservation program is gone. The ad-hoc conservation committees formed in 1974-75 have been dissolved and have become extensions of cost reduction programs with only part-time staffing. They have lost their identity and momentum.

In the future, I feel energy programs will regain top management support. A corporate energy manager and an adequate staff will be established. Also, it will be necessary for the energy manager to go "back to basics" to ensure success.

The proven elements for a successful ongoing corporate program are:

- establishing specific goals, objectives and policies
- developing achievement strategies and tactics
- preparing an annual action plan
- continuing energy audits of facilities
- improving the energy reporting system
- tracking progress monthly
- initiating corrective action when necessary
- updating energy contingency plans
- preparing five-year energy plans
- implementing productivity projects
- developing alternative energy sources
- motivating employees and communicating with them
- monitoring and forecasting supply and demand
- keeping abreast of new technologies, and
- participating in governmental affairs and rate cases.

More involvement in the corporate planning process is needed for the energy manager. Today, he has little involvement in the strategic planning process in most companies. The manager is somewhat more involved in planning for capital, facilities and manufacturing, but the energy plan, if it exists at all, is not usually integrated into overall corporate plans. In the future, the energy manager will routinely provide inputs into strategic, capital, manufacturing, and facilities planning and will develop close relationships with these four functions. The energy plan will be routinely integrated into overall corporate plans. In addition, energy managers will be confronted with energy issues that represent truly new horizons for them. The energy manager will regularly participate in these eleven energy issues and his ultimate success will hinge on how much he gets involved with them:

- analysis of his product's energy costs
- analysis of the competitor's product energy costs
- development of manufacturing processes
- finding less energy intensive substitute materials
- development of alternative energy supply sources
- analysis of facility closings and relocations
- design of new facilities
- planning of energy capital projects
- review of corporate energy research and development projects
- leasing and buying of captive energy sources, and
- identification of new corporate energy products.

CONDUCTING ENERGY PRODUCTIVITY PROGRAMS

There are two major phases of an energy productivity program, patterned after General Motors' very successful program. Phase 1 involves three elements:

- better administrative and supervisory controls
- improved operations and maintenance methods, and
- improved motivation of employees.

These are basically low capital intensive items.

In Phase 2, we get into more efficient use of energy through engineering improvements, namely:

- changing product specifications
- changing process specifications
- using more waste heat recovery methods
- improved equipment controls
- conserving materials
- upgrading facilities.

Where do we now stand? Most companies have implemented Phase I but few companies have achieved all of Phase 2, where the greatest savings result. My prediction is that many companies will achieve all of Phase 2 in the next ten years. Special emphasis will be placed on changing product and process specifications and using more waste heat recovery methods.

This curve developed by Arthur D. Little Inc. dramatically illustrates the cost effectiveness of productivity modifications:

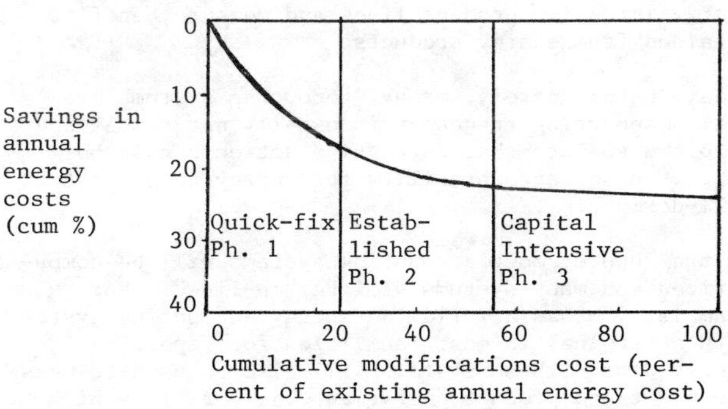

The curve shows how the law of diminishing returns affects energy conservation projects. There are three major phases:

1) Quick-fix: involves little or no capital investment, uses basic changes in maintenance and operating procedures such as lowering temperatures and turning off lights. Has greatest return-on-investment.

2) Established: uses established technologies and medium capital investments.

3) Capital intensive: uses advanced technologies or capital intensive established technologies.

Today, productivity is achieved primarily by the quick-fix approach. Established projects are generally used by larger companies. But few capital-intensive projects are operational.

The following chart shows the mix of four major types of productivity modifications in each phase. The four technologies are:

- changes in external and physical properties
- adjustments in internal load and comfort conditions
- changes in HVAC systems and controls, and
- building operation and maintenance improvements.

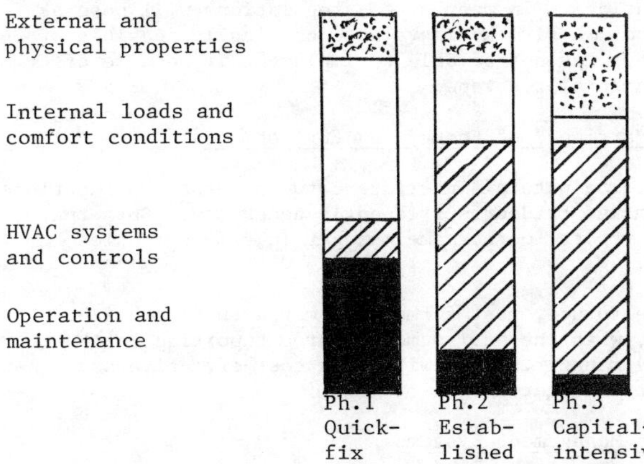

Source: Arthur D. Little Inc.

Today, emphasis has been placed on operations and maintenance improvements and adjustments in internal load and/or comfort conditions. Moderate gains have been made in HVAC improvements, but there are relatively few energy management systems operational.

In the future, there will be increased emphasis on Phase 2 and 3 projects. Energy management systems will be commonplace and inefficient HVAC systems will be replaced. Energy inefficient buildings will also be retrofitted.

SELECTED FUTURE ENERGY TRENDS

Cogeneration

Cogeneration greatly increases overall energy efficiency for power generation from about 32-percent to 70 to 80-percent. PURPA and tax credits make it more attractive now, but capital costs are still very high.

In 1900, almost 40-percent of all U.S. electricity was cogenerated. Today, only about 5-percent is cogenerated. It is limited primarily to process industries. In the future, there will be much greater use of cogeneration in a wide variety of industries. The energy manager must be knowledgeable in feasibility analysis in order to decide whether or not to cogenerate.

Fuel Switching

Today, considerable emphasis is being placed on dual-fired capability, that is, the firing of boilers or furnaces with either oil and/or natural gas. There is low priority on coal conversion and very few companies control or own their own fuel sources.

In the future, there will be a high priority on coal conversion, using either coal-oil mixtures, or other direct-firing methods, including conventional and fluidized-bed combustion. Another possibility is the use of low and medium-Btu gas produced from coal. Numerous large companies will lease or own their own fuel sources, especially in energy-intensive firms. Companies may develop their own oil and gas wells or have it done by developers. Some larger firms will buy coal reserves or operating companies.

The manager must understand these three fuel-switching factors: technology, project economics and governmental regulations. In many cases, regulations will pose a problem. A firm may have an economically feasible project but may not be able to implement it because of governmental red tape.

The "Marriage" of Energy and Environment

There is a natural interface between these two functions that makes tradeoffs frequently necessary. Both functions report to the same manager in some companies today.

In the future, most firms will merge these two functions, with the environmental area reporting to the energy manager. This will be a cost-effective and synergistic merger.

Energy Management Systems

These systems provide a major opportunity for improved energy productivity. EMS technologies have matured and there are numerous excellent systems available commercially. Energy management systems control HVAC systems, lighting, and processes. A typical return on investment after taxes is 25 to 30-percent.

Today, market penetration for energy management systems is moderate. Large systems are very capital intensive and smaller companies lack software capability and trained personnel to run them.

In the future, the manager will prepare system conceptual designs. Direct digital control technology will prevail in energy management systems. Marginal EMS suppliers will be eliminated from the market; and we feel that because of improvements in technology and the escalation of the energy crisis, return on investment for energy management systems will approach 40 to 50-percent.

Water Conservation

Water will become our next major crisis and water conservation will become our next national ethic. There is a natural interface between water conservation and energy productivity; water conservation will evolve in a similar method. This water conservation ethic will represent a big opportunity for the energy manager to ensure his firm's water supplies, thereby further increasing his corporate role.

Today, some energy managers are already responsible for water conservation. However, the focus is on recycling. In the future, managers will be responsible for conservation and the development of alternative sources. The emphasis will be on sophisticated water treatment as opposed to recycling.

Process Energy Productivity

Today, most energy managers have limited involvement in process energy productivity. Most process engineers have a "hands off" attitude and hold processes sacrosanct, preventing energy managers from having much impact. In the future, there will be extensive involvement by the energy manager to optimize Btu savings, review and revise process specifications, and develop alternative processes. However, the lack of capital dollars will impede his progress, since most changes tend to be extremely capital-intensive, especially in the steel and petrochemical industries.

Waste Heat

Waste heat can best be described as America's largest and least-appreciated industrial by-product. In developing waste heat recovery projects, the energy manager must evaluate these four waste heat parameters: use, quality, quantity and transferability.

Today, energy quality is often overlooked. For example, high pressure steam in a boiler is frequently reduced to low-pressure steam for plant use. This represents a large overlooked opportunity for savings. The steam could be used to drive a topping turbine to cogenerate electricity. There is a trend toward better use of thermodynamic analysis to take advantage of high-quality energy through the use of direct drive turbines for cogeneration. The energy manager will have to understand much better the thermodynamic impact of energy quality.

Energy Accounting Systems

These are a proven, valuable tool for measuring energy productivity. They create a data bank for energy consumption, facilitate future energy planning, identify energy-intensive product lines and measure benefits obtained from energy products.

Today, unfortunately, manual reporting systems prevail. Reporting frequency is usually not timely and is often so late that corrective action cannot be taken. Also, accountability for energy use is not well-defined.

In the future, most accounting systems will be computerized and many systems will be "on-line". For example, IBM has a worldwide energy accounting system with a terminal in most facilities for reporting energy data. These systems will also be used to measure plant and product performance. There is a high probability that energy indexes will be included in corporate management reports. In fact, at UTC we put these numbers in management reports today. The energy manager will become responsible for preparing the energy accounting system's conceptual design.

Electrical Cooperatives

Electrical cooperatives represent an organizational means whereby several companies in the same geographical region can join together to share power and shed loads as directed by their electric utility. For the utility, this kind of cooperative effort reduces the demand for electric peaking power. For cooperative members, it reduces electricity costs by up to 25-percent and forces conservation. To be effective, however, electric cooperatives require friendly relations among members. Only a few co-ops are operational now, but the results are favorable so far. Electrical cooperatives will continue to grow significantly, aided by energy management systems that will facilitate their formation. Energy managers must understand the ramifications of these cooperatives.

Solar Energy

There are few significant industrial solar installations because most proposed solar projects are not cost-effective. In addition, few managers understand solar technology. Solar energy will not be the great panacea of the 1980's. There will be limited solar industrial installations, primarily for domestic and process hot water heating and some cooling. There will be very limited use of photovoltaics unless there is a major manufacturing and cost reduction breakthrough soon. However, there will be extensive use of passive solar technology in new buildings. Consequently, energy managers must become more knowledgeable about solar technology.

Computer Simulation for Building Design

Today, the simplistic analytical tools used for building design are not accurate. Computer simulation provides the ability to examine the combined effects of numerous inter-related design parameters. Probably the biggest benefit of computer simulation is that life-cycle costs of buildings can be optimized. Simulation studies have produced energy savings of 30 to 40-percent.

Today, unfortunately, energy efficiency is a low-priority factor in new designs. Annual energy operating costs in Btu per square foot per year are usually not calculated and the future costs or availability of fuel sources are not considered.

In the future, the energy manager will direct simulation studies, life-cycle energy cost analysis will be commonplace and simulation studies will be required for most buildings.

Energy Service Companies

These are an offshoot of the "Least Cost Energy Strategy" study released a few years ago by the Mellon Institute. A service company delivers heat, lighting and cooling for a fixed monthly charge, supplying energy at the lowest cost. The service company sells a complete package of energy efficient equipment, building alterations and energy itself.

Energy service companies are still in the formative stage and very few are operational. Possible service companies include IBM, Honeywell, General Electric and many smaller contractors. Service companies will probably be most attractive to smaller users. They will probably become commonplace in the future and customers will realize significant savings in energy and overhead costs. The energy manager will have to know how to select the right company.

Summary of Other Selected Trends

The energy manager must be knowledgeable of several other trends. First, there will be greater use of:

- microprocessors to control boilers and manufacturing processes
- industrial heat pumps
- energy consultants
- high and low temperature heat exchangers
- organic Rankine bottoming cycles
- incinerators for factory trash
- HID lighting systems, and
- synthetic fuels.

We will also see significantly improved energy efficiencies in all of the following types of equipment:

- HVAC equipment
- electric motors
- gasoline and diesel engines
- advanced cogeneration systems
- pumps
- furnaces, ovens, dryers
- insulation and refractories
- machine tools
- metal finishing processes
- HID lighting systems, and
- turbines

Summary

In summary, there are four key points concerning conservation and its effects on the corporate energy manager:

1) Energy productivity will play a vital role in "reindustrialization of America."

2) The role of the energy manager will expand greatly.

3) We will need better-qualified energy managers.

4) The energy manager must be able to deal with numerous emerging energy trends.

SECTION 22
ENERGY CONSCIOUS DESIGN

Chapter 69

ENERGY CONSCIOUS BUILDING: A CONTRACTOR'S PERSPECTIVE

F. M. Reaves, J. Reaves

A comprehensive energy-conscious building strategy should consider all aspects of energy use connected with building utilization, as well as projected trends in energy sources. Such a strategy should be innovative, combining new equipment, systems, and technologies in complementary and interactive combinations. It should be objective, without bias towards a single technology or manufacturer. And it should take into account the critical importance of proper installation and maintenance.

R.M. Thornton, Inc. is a mechanical contracting firm which is concerned about energy efficiency. We have developed considerable expertise with new energy technologies as well as with conventional mechanical systems. The pace of change in the building industry means the mechanical contractor must work as hard as the engineer or architect to keep pace with new concepts. We feel that the role of mechanical contractors in the implementation of energy-conscious construction is crucial and generally overlooked.

The Thornton Research and Demonstration Building was designed to evaluate a number of innovative systems. The building comprises office, warehouse and shop spaces. It is designed to combine active and passive solar, thermal storage, heat pumps, water and waste recycling into an interactive ecology which is efficient and comfortable. Our goal is a building which is as self-sufficient as possible, independent of oil, gas, water, or sanitary and storm sewer connections. The building is also a test facility for energy-conservation devices and hardware.

In planning the building, we tried to develop a comprehensive energy-conscious strategy. We decided we could reduce energy use for heating and cooling, lighting, and transportation. We also realized that cutting water and sewer use, which is worthwhile in itself, would also save energy locally for pumping and treatment. We tried to look into the future, and decided that electricity (as a medium for the transfer of energy) is likely to be the most secure and practical source for our purposes. We took a careful look at the products and technologies available. And, of course, we followed through on the building from the design phase through installation and maintenance.

The site was carefully chosen. In an industrial park near the Washington Beltway, we found an accessible site with an unshaded southern exposure. We agreed to build the structure in cooperation with United Sheet Metal, another construction firm; our common wall and stairwell reduces energy loss, and first costs. Thornton has been active in promoting a construction park on the site, where other construction-related firms maybe located, in order to cut down on transportation costs.

In designing the building, we consulted two pioneers in energy efficiency, Mr. A.I. McFarlan and Dr. Harry Thomason. The space conditioning system is uniquely flexible and controllable. It incorporates many ideas developed by McFarlan during the 1950's. These include heat recovery from people, lights, office machines and south rooms. By using chillers, and two air handling systems (one for all rooms and one for perimeter rooms) with coils offering large heat exchange surfaces, he was able to transfer heat from the interior to the perimeter of the building to offset losses through the building envelope.

The variable air volume approach to air distribution, also incorporated into the building, insures more comfort with less energy. Temperature is regulated in each room by thermostats and dampers which control the amount of air entering the diffuser (as opposed to conventional systems which regulate the temperature of the distributed air). This allows individual room control as well as waste heat utilization.

Dr. Harry Thomason, also during the 1950's, developed a system to heat homes efficiently and economically with simple flat plate collectors using a trickle design. The system included a steel tank for thermal (water) storage, with rocks around the tank to increase the heat exchange surface. In hot weather, the tank was cooled by running the air conditioning unit at night. This permitted more efficient compressor operation, allowing use of a smaller compressor and off-peak electricity.

Both McFarlan's and Thomason's innovations were practical and economical, developed without aerospace technology and without government support. Wide adoption of the concepts was hampered by designers who refused to pay reasonable royalties or give credit to the inventors. Of course, they were ahead of their time; it was not until 1973 that energy efficiency became a general concern.

What McFarlan and Thomason have in common is a

FIGURE 1. SOUTH FACADE OF BUILDING INCLUDING SUN SHADES.

concern with both low first costs and low temperature differential (low delta-t). They recognized that efficient energy use in space conditioning as well as efficient utilization of solar energy required taking maximum advantage of small differences in temperature. They designed systems which were characterized by efficient air distribution and large heat exchange surfaces.

The Thornton building combines and extends these concepts. Thermal storage comprises four separate water reservoirs. One is 6,000 and two are 8,000 gallon recycled gasoline tanks. One of these tanks can be used to store water from the solar collectors, one for condensor water, and one for chilled water, or they can serve in any combination. A separate moat and cistern made of concrete holds approximately 50,000 gallons. The separate thermal storage tanks allow us to switch quickly from heating to cooling as the need arises.

A fifteen-ton chiller is connected to twelve-row coils in the air handling units. Because of the large surface area the maximum necessary water temperature for heating is 82°, and the minimum necessary water temperature for cooling is 55°. Water can be circulated through the coil from the condenser side, from the chiller side, from any one of the four tanks, or from the solar collectors directly. This allows enormous flexibility in choosing the most efficient operating mode.

Higher heat exchange surfaces permit more favorable return water temperatures, yielding higher efficiency in distribution and heat pumps. The combination of substantial thermal storage with the heat pump and large heat exchange surfaces allows the system to perform several unique tricks.

One mode we have dubbed "night pushback". Many buildings are capable of "night setback": in cold weather the thermostats are turned back to 55 degrees after working hours or on weekends, allowing the building to gradually lose heat through the envelope until the next morning, when heat is added to bring the building up to a comfortable temperature. The Thornton building "pushes" the building temperature back to 55 degrees by using the chiller to warm stored water a few degrees. In the morning, this water is cooled by running the chiller. The building is thereby brought back to 70° and heated for two hours before the water temperature is back to where it was the night before.

Another mode we call "flywheel" heating and cooling. It eliminates the chiller and uses stored water and the heat exchanger to achieve comfortable temperatures. In mild weather, for up to twelve weeks in the spring and twelve weeks in the fall in the Washington area, buildings tend to be too cold in the morning and too warm in the afternoon. By circulating water from storage through the coils we achieve comfortable temperatures at no cost other than running one pump and two fans. In this mode, a few degrees is added to storage in the afternoon and taken away in the morning. Occasionally when the temperatures go beyond the comfort range, the chiller is run.

In the summer and winter, we run the chiller at night to store water at the appropriate temperature for the next day. This means greater efficiency, a smaller required chiller size, lower demand charges, and off-peak electricity utilization. Water is stored at somewhere between the optimum return temperature and the expected supply temperature. Normal storage is 71°.

On the roof are approximately 2,000 square feet of solar collectors, or about one square foot for each five square foot of office space. (The warehouse receives some heat lost from the piping and walls). Most are the "Solaris" type designed by Dr. Thomason, and are manufactured by Robert Porter of Newark, Delaware. They are low cost trickle collectors, constructed of corrugation from a header at the top to a trough at the bottom where it is collected and piped into storage. The design is made more efficient by coupling the collectors with the chiller; supply temperature to the collector is as low as 35°.

FIGURE 2. SOLARIS COLLECTORS AND CLIVUS MULTRUM VENTS.

Delta-t (in this case the difference between the supply water and outside air temperature) is kept at a practical minimum. We are presently testing two Westinghouse collectors and have two other brands on hand and ready to install.

In addition to the active solar collectors, the building is designed to take advantage of passive solar principles. Most of the window area faces south. Over the windows are "mansard"-style sunshades. These consist of redwood slats set at the appropriate angle (29° for the Washington area) so that during winter days the maximum amount of sunlight enters, but during summer months the south wall is completely shaded. The sunshades were designed by architects George Brunati and Bob Kieckhefer from suggestions of architect Stuart Liss and Thornton Co.

During the summer months, we use the roof for cooling (there is no cooling tower). Storm water from the condenser is sprayed on the roof and returned to storage. Evaporation cools the office roof as well as the condenser water.

Stormwater from the roof drains is saved and stored in the moat reservoir. Not only is this resource an addition to the building's thermal storage capacity, it helps to eliminate an increasingly troublesome urban-suburban problem: erosion and flooding from storm run-off in developed areas. The stored water is also used for lawn sprinkling. We plan to experiment with treating and filtering the water from the cistern for drinking and washwater. Fire sprinkler water may also be taken from the storage tanks.

Two Clivus Multrum waterless composting toilets have been installed and are in use under an experimental health permit. These units treat human wastes, kitchen garbage, paper, and other biodegradable wastes through biological digestion, yielding small amounts of dry and liquid fertilizer, safe enough to apply on the lawn out front. Moisture is evaporated and released through a vent on the roof. The Clivus Multrum cuts water use in the building by half. A graywater treatment system operating on the same principles renders water from sinks safe for application to the lawns.

We believe that it is practical to construct a building which is completely independent of water and sewer connections, thereby conserving water, energy, and the costs to society of treatment facilities and piping.

The combination of energy-related systems in the building is effective in reducing both the size of the chiller (half that ordinarily prescribed for this size building) as well as BTU per square foot (one third of the average for the Washington area) and total utility costs (half of the local average). Although payback calculations are very sensitive to varying price projections and the particular fuels for comparison, we expect the additional cost of the building over a conventional design to be paid back in less than ten years.

We believe the principles involved in the Research and Demonstration Building are sound and can be extended and adapted to a wide variety of building systems. In Europe, there has been great success with district heating. On the basis of our experience, we foresee the possibility of a system which connects buildings together with a ductile-iron pipe water loop. Ductile-iron pipe is considerably cheaper and more durable than the insulated conduit currently used for district heating and cooling systems. Each building would have a chiller which would either heat or cool the building by drawing heat from or adding heat to the water loop. Water would be circulated through the uninsulated pipe from building to building, serving as a heat storage and heat transfer medium and subject to the moderating temperature of the surrounding earth. The loop could connect heat sources, heat sinks, and large scale storage, as well as connecting buildings which have complimentary heating and cooling requirements, i.e. one produces heat at the same time another needs heat. One system might join up offices, residences, factories, shopping centers, auditoriums, solar collectors, solar ponds, incinerators, lakes, fountains, waterfalls, groundwater wells, and power plants. Such a system would provide efficient heating and cooling by utilizing low temperature differentials, including solar, all kinds of waste heat, and inexpensive cooling and storage. It would be more economical and more versatile than conventional district heating and cooling systems.

The Research and Demonstration Building has given us a wealth of experience in designing, building, and living with innovative mechanical systems. Being involved as owner, contractor, and tenant allows us to test new equipment and technologies, and gain experience with installation and operation. It has also enabled us to promote and encourage the spread of new ideas in the field, through articles about the building and monthly tours. Partly as a result, we have become one of the leading mechanical contractors in the Washington area in the field of solar and energy conservation installations.

Our experience with the building and as contractors has led us to the conclusion that the key to the creation of energy-conscious buildings is the process by which construction or retrofit projects are designed and built, rather than one particular technology, piece of equipment, or design approach. No single system or approach is going to fit the multiplicity of heating, cooling, and mechanical system needs that exist in the commercial-industrial building industry. No single type of retrofit is going to fit all existing buildings and operating precedures. It is the decision-making process by which a broad variety of options are selected from and implemented which is crucial. If we can build energy-consciousness into this process, we can be assured of energy-efficient and cost-effective buildings.

The traditional approach to construction stresses a linear separation of functions over time: the owner states the requirements that the proposed building will have to meet; then the architect draws up plans; then the engineer designs a mechanical system; then specifications are drawn up; then the job is put out for bid; then a contractor is chosen on the basis of low bid; then the building is constructed; then it is turned over to the owner for operation. The sequence is logical and facilitates the delineation of responsibility, but it also encourages passing the buck. It is a process more suitable for business-as-usual and for conventional technologies which are well understood, pre-packaged, and non-interactive.

Future requirements for energy and resource efficiency may put more pressure on the decision-making process. There are hundreds of items which can be performed in a building to save energy. Many of these are low-cost or no-cost. However, they will not save the owner money if no one identifies and promotes them. New technologies, the wide variety of new equipment and systems, life-cycle costing, and other considerations suggest that extra time and attention spent on analysis will pay back in substantial savings later. Many of these systems will be complementary and interactive, so that the traditional compartmentalization of disciplines will no longer be efficient.

We feel that there are a number of advantages to a design-build team approach, particularly where cost and resource efficiency are important considerations. The design-build team concept is one in which the owner, architect, consulting engineers, and major contractors all participate in the major phases of both design and construction. The design process benefits from the experience and wide variety of expertise of both contractors and designers.

The design-build team goes through a number of steps to achieve the desired result, a functional, cost-effective, resource-efficient building. The steps are:

1. Define the problem.

   Type of building, functions, location, patterns of use, etc. Establish critieria for decision making.

2. Analyze the data.

   Site, climate, load and load patterns. If an existing building, an energy audit and analysis of existing system.

3. List alternative solutions.

   Alternative sites, structural designs, electrical and mechanical systems, energy sources, environmental systems, etc.

4. Estimate costs.

   Each alternative should be analyzed in terms of first costs, energy and resource costs, maintenance costs, and payback period.

5. Select best combination of options.

6. Complete design and schedule work.

7. Arrange financing.

8. Oversee construction.

9. Track operation.

   Does the building live up to expectations and/or what can we do to improve it?

The above list does not constitute a sequence in time; the steps may very well overlap. For instance, the approach facilitates pre-purchase of equipment (i.e. during the design phase), which offers advantages in terms of design efficiency, energy efficiency, pricing, and scheduling. The design-build team concept allows better supervision of the entire process from start to finish.

One of the features of this approach is a closer collaboration between disciplines. Many questions of energy efficiency involve an intimate connection between architectural and engineering considerations. Proper design decisions require collaboration and feedback.

Another feature is the participation of contractors. This can have several advantages. A major part of the contractor's expertise is in cost estimation, scheduling, and budgeting. With a contractor on board the team gets more accurate feedback about the relative costs of design alternatives. This can mean a big savings in first costs as well as accurate appraisal of life-cycle costs.

A good contractor will have experience with a variety of systems and can also provide valuable feedback concerning installation problems and considerations, maintenance, service, and reliability. This is particularly important with systems which depend for their economic advantage on longevity; a higher initial investment should be matched by additional years of efficient, trouble-free operation. A contractor may have more intimate experience with the problems that a design incurs than the designers themselves.

More participation by the owner in the decision-making process means that the person who has to make the bottom-line decisions about costs and functions will be more familiar with the various alternatives and considerations. There has been a tendency on the part of owners to be blind to life-cycle costs, even when first costs are not higher. Sometimes an ornamental lobby or thicker carpets may seem better investments than energy efficiency. Participation in the design team can educate the owner to long-range planning.

The approach works most effectively when all participants are experienced, honest, motivated, and concerned with both resource and cost-efficiency. The selection of participants, on the basis of reputation and credentials, is therefore crucial. In addition, the participants should be compensated for their time in such a way that their decisions are objective and not based on a particular technology, system, or manufacturer. They should not have incentive to choose high-cost options. Many systems of compensation, particularly those based on a percentage of the total cost or the cost of a sub-system, encourage the choice of expensive options, promote compartmentalization to the detriment of efficiency, and do not reward for savings. Also, since most jobs are put out for bid after the design has been completed, contractors have no incentive for opportunity to use their experience to save the owner money, either in first cost or energy.

Present standard architect and engineering fees are not sufficient to allow designers to take into account energy and resource efficiency, pay back, etc. This is undeniably extra work for the designers, and should be compensated. Low percentage-linked fees encourage quick-and-dirty design, and give no incentives to save money either on first costs or long term costs. Designers under time constraints may be biased toward systems which are simple to design. Since some equipment manufacturers offer design assistance, designers are sometimes tempted to select such a manufacturer's equipment whether or not they are the most efficient.

Low fees give the designer no opportunity or incentive to spend the time to keep up with advancing technology. Energy policy studies have identified this situation as a bottleneck in the implementation of energy-conscious building strategies. The continuing education of architects, engineers, and contractors is a investment well worth making.

One approach is simply an hourly fee for time spent for all participants, including the contractors. This could be extended through the construction phase by compensating the contractors for work on a cost-plus-materials basis. Of course, all participants must be honest and motivated. Another approach might be to separate the design and construction phases, compensate a contractor on an hourly basis during design, and then write a separate contract for construction or put the job out for bids.

Another possibility might be a combined formula for compensation based on a number of factors: size of the job; time spent within certain limits; savings (in first cost and energy), as compared to a standard conventional building. Example: designers on the team might be paid a standard fee based on the size of the building or the first cost-estimate, plus a percentage of the savings achieved on the building in terms of all costs compared to a conventional building. Contractors might receive an hourly fee for time spent on the team effort, including the budgets.

The construction management approach might also be appropriate. Instead of a general contractor, there is one firm which is designated the construction manager (e.g. the mechanical contractor in a case where the mechanical work was the largest component, or a general, or a separate CM firm). The construction manager oversees all phases of design and construction from the first meeting to completion, working with architects, engineers, and contractors to give the owner the best possible building for the money.

Some companies, especially in the retrofit field, have experimented with agreements in which they contract to provide energy (or a given comfort level) for a building for a period of time. They make energy efficiency investments out of their own pockets, and pay those investments back out of savings. (Although the retrofit field is different in many ways from the construction of new buildings, many of the same considerations apply. Owners of existing buildings are besieged by peddlers of one energy "fix" or another. An experienced engineer working with an experienced and conscientious contractor can find a wide variety of energy-saving modifications, from no-cost to low-cost, in most buildings. A thoughtful and knowledgeable approach to the problem is more practical than selecting a single product or technology).

Our experience as contractor and building owner has led us to conclude that good planning is crucial to both cost and energy efficiency. On occasion, we have installed high technology energy equipment that, with the best of intentions, has not been practical. We prefer not to make money in correcting mistakes in mechanical systems which could have been avoided with a little forethought. Relatively inexpensive design or operating procedures can save a lot of energy. The energy-conscious building is the result of the combined efforts of owner, architect, engineer, and contractor. We all have something to offer.

SECTION 23
ENERGY EMERGENCY AND CONTINGENCY PLANNING

Chapter 70

CONTINGENCY PLANNING AT THE FEDERAL LEVEL - CAN THE GOVERNMENT LET THE MARKET WORK?

D. M. Di Paolo

The recent change in administrations has brought about a major shift in energy policy. Emphasis has switched from end-user oriented programs developed and run by the Federal Government to free market philosophies pushing economic decisions out to the private sector. Although this change will promote economically efficient development of energy resources, there is some question whether a "hands off" policy, especially in the area of energy contingency planning, is either politically realistic or entirely adequate.

Despite the recent publication of the Reagan Administration's National Energy Policy Plan (NEPP), and the transmittal to Congress of the Department of Energy's (DOE) long-awaited report on "Domestic and International Energy Emergency Preparedness," it is still unclear whether the Administration intends to do more in energy contingency planning than commend the workings of the free market.1/

The Administration has not supported any legislation to provide standby authority to deal with oil supply disruptions, preferring instead to rely on existing authorities. However, in the absence of new, market-oriented legislation, there is a real possibility that Congress will act, in the face of rising oil prices during the next crisis, to fill a perceived policy vacuum by once again imposing some form of federal regulation on the oil industry.

The following discussion outlines the likelihood of oil supply disruptions in this coming decade, the probable federal response, and some measures that might be taken to avoid government controls during the next crisis.

Likelihood of Future Disruptions

There is virtual unanimity among participants in and observers and analysts of the petroleum industry that at least one major oil supply disruption will occur sometime in this decade. The only real disagreements are among the political analysts concerning the immediate cause and type of disruption to expect, and among the industry as to what should be the Federal Government's response.

U.S. dependence on OPEC oil has increased dramatically since the 1973-1974 Arab oil embargo period. In 1973, the U.S. received 46 percent of its imports from eight OPEC countries.2/ U.S. dependence on imported oil as a percentage of consumption peaked in 1978, bringing us just short of the 50 percent mark. Government price controls helped increase rather than lessen that dependence; but even with the end of controls in February 1981, changing production patterns in exporting countries will mean that our relative dependence on OPEC oil will continue well into and perhaps throughout the 1980's. In January 1981, although we imported only 36 percent of our oil consumption, roughly 60 percent of those imports came from only seven OPEC nations. The other 40 percent of our imports was supplied by an equally small number of non-OPEC nations.

This increasing concentration in our sources of supply of imported oil is especially troubling, given the ever-present tension in the Middle East. Total OPEC production figures tend to obscure the fact that one nation, Saudia Arabia, is no longer merely a swing producer, but, in fact, has attained a position of unquestioned dominance within OPEC. While Saudi production constituted approximately 25 percent of total OPEC production in 1973, by January 1981, the Saudi share had risen to over 40 percent of OPEC production.3/ Increasing U.S. dependence, in relative terms, on OPEC oil is thus accompanied by increasing concentration of OPEC production in one country, making the potential for oil supply disruptions even greater.

Political analysts have long warned of the repercussions from a major Middle East war escalating to the point of direct superpower confrontation. The 1978 revolution in Iran and the 1980 war between Iran and Iraq are striking examples of internalized Middle East conflicts that have directly affected the flow of oil coming from the region and thrown world oil markets into panic. Although the disruptions in themselves were relatively minor, Iran's threatened blockage of the vital Straits of Hormuz and attacks against overland pipeline facilities during the height of the Iran/Iraq war vividly illustrate the devastatingly large and long-term disruption of oil supplies that could result from even highly localized conflict.4/

Economic Vulnerability to Disruptions

It is just within the past year or two, since the 1978 Iranian revolution and the attendant cutback in world production, that the nature of our vulnerability to disruptions has been well understood. Prior to that market upheaval, the Nation's energy policy was focused on oil import reduction as a means of achieving energy independence.5/ The major price shocks resulting from the relatively minor shortage in 1979 prompted the realization that insulation from the effects of oil cutoffs via a zero-oil import policy was unattainable.6/

During the first half of 1981, the amount of oil imported into the U.S. dropped to just over 5 million barrels per day, down from a high of 8 1/2 million barrels per day in 1977. Imported oil currently constitutes about 36 percent of total oil use in the United States, and about 20 percent of all energy used in the U.S. It is this latter figure, and the implications that it carries for the costs of quickly substituting other fuels for imported oil lost through a disruption, that the DOE's Office of Policy and Evaluation characterized as "the crux of our problem with 'oil import vulnerability.'"7/

The damaging effects of disruptions would be felt by the U.S., even if we did not import a single barrel of oil. Energy prices would soar domestically as U.S.-produced oil was priced at market levels, and the impacts of the disruption on other oil-consuming nations would reverberate back to the U.S. The economic structure of the Western world is too closely interwoven to enable one country to effectively insulate itself from major upheavals in the economies of other nations. In a severe disruption, common defense imperatives would also require U.S. sharing of the hardships wrought by the cutback in oil availability and higher prices. "Vulnerability" to a supply reduction is clearly not tied only to the level of imports.

The sudden and dramatic increase in oil and oil product prices due to a disruption have been calculated by many analysts both inside and outside of the government. The anticipated price increase is extremely difficult to forecast, since economic modeling tools rely on a set of assumptions of future behavior that are based on simplified measures of past behavior. The best estimates available are, however, frightening. According to various DOE calculations, a three million barrel per day disruption lasting for one year would result in direct losses to the U.S. economy on the order of: over $20 billion lost in direct transfers to oil-producing countries; almost $2 billion in efficiency or dead-weight losses; and about a $60 billion drop in GNP.8/ The increase in the world oil price of over $10 per barrel would also result in an almost 3 percent increase in the consumer price index and a .8 percent increase in the rate of unemployment. A ten million barrel per day disruption would result in a world crude price of well over $50 per barrel, while a complete shutdown of the Persian Gulf (18 million barrel per day loss) would push world crude costs to anywhere from $100 to $200 per barrel, a fourfold to sixfold increase in current prices. Frank Zarb, the former Administrator of the Federal Energy Administration, has testified before a Senate subcommittee that the price of a gallon of gasoline in a future shortfall to the U.S. could be as much as $10, in 1980 dollars.9/

Price increases of this magnitude will cause great economic stress in the U.S. and other industrialized countries (inflation, loss in productivity and GNP, unemployment) and instability in world monetary and financial markets. Reduced economic activity in those countries will lead to a loss in U.S. exports. Less developed countries, increasingly dependent on oil and oil imports, will place greater demands on industrialized countries both for refinancing of loans and for outright aid at a time when the industrialized nations can ill afford it. At the same time, the oil price increases will result in large transfers of wealth from all oil-consuming to oil-producing nations. The personal hardship resulting from uncertain supply and suddenly higher prices will, for some segments of the population, be particularly great -- those people on fixed or low incomes, those who rely heavily on oil products for their own use, and those who are employed in an industry particularly dependent on oil usage. And the economic stress of making the sudden transition from oil to other fuels will quickly be felt throughout the population, as the prices of goods directly or indirectly related to fuel prices also increase.

Current Federal Government Policies

The Reagan Administration from the outset established energy contingency planning as one of its chief priorities. One of the first actions taken by President Reagan after taking office was to decontrol U.S. oil production and distribution. This action was proclaimed as a "positive first step toward a balanced energy program" -- a program designed to make us less dependant on the OPEC nations and to remove the artificial constraints on the market that had exacerbated the oil shortages in the 1970's.10/

The import of that announcement seemed clear. The Federal Government would no longer intrude in oil price and supply decisions, even during oil supply disruptions of the magnitude already experienced. Unless a clear and compelling need for intervention was seen, competitive forces would be allowed to work and the success or failure of individual business enterprises would rest in their own, not the government's hands.

But the Administration has yet to go beyond that first step in completing its energy program. No clear policy has been announced regarding possible speedup of natural gas deregulation, funding of synthetic fuels projects has been delayed again and again without real explanation, and, to a large degree, conservation programs and alternate fuels development have been ignored.11/ And the industry and public have been left to speculate whether the Administration's preference for market-oriented policies will carry over to strategies for responding to oil supply disruptions of greater size or impact than the three relatively small disruptions seen in the past decade.

Congressional authorization for federal price and allocation controls, embodied in the Emergency Petroleum Allocation Act of 1973 (EPAA), expires on September 30, 1981.12/ Although there is very little likelihood that the EPAA itself will be extended, there is substantial support in Congress for some form of narrowly constructed standby regulatory authority to be given to the President to be used during severe supply disruptions.

Legislation has been drafted in both the House of Representatives and in the Senate to give the President discretionary standby authority. Senator McClure (R-Idaho) introduced one such measure on July 20, 1981, and Representative Sharp (D-Ind.) had circulated a discussion draft of a similar bill by the end of July. Neither measure attempts to flesh out the substance of the standby controls authorized by the legislation, and both measures include strict sunset provisions to ensure quick expiration of the controls. Senators Durenberger (R-Mn.) and Andrews (R-N.D.) have also introduced standby control legislation. Their measure, unlike the others, would only provide authority for allocation, but not price, controls.

Support for controls stems largely from the perceived need for the Federal Government to be prepared to do something in response to a disruption. Allowing price increases to take their course and the market to distribute available supplies will inevitably result in many cases of extreme personal hardship and spot shortages in isolated areas. Real questions exist about the ability of the international and domestic market structure to handle a very large disruption without some form of government intervention to at least ensure continuation of essential services while market adjustments take place.13/

Long gasoline lines in both 1973/1974 and 1979, home heating oil shortages during the earlier disruption and surpluses during the latter, trucker demonstrations and excess supplies being brokered by farmers, gas lines in cities and overflowing tanks in resort areas, and misallocations too numerous to mention, have all been generally accepted as evidence that government regulations tend to make shortages worse.14/ Nonetheless, there seems to be a national psychological ambivalence about controls. Government regulations are bad, but can we leave it up to the oil companies to ensure that everyone is treated fairly in a crisis? The oil company conspiracy theories we have all heard during shortage periods -- full crude oil tankers sitting off both the east and west coasts, gasoline being burned in the desert -- have contributed to pressure on Congress for the government to take on a central role in managing a shortage.15/

One alternative to controls has been introduced by bipartisan sponsors, Senators Bradley (D-N.J.) and Percy (R-Ill.). Their bill, the Emergency Preparedness Act of 1981, S. 1354, would allow market forces to distribute oil supplies during disruptions and offset the effects of higher oil prices by quickly recycling increased windfall profits tax revenues through block grants to states and through tax cuts and transfer payments to individuals. The bill also calls for development of a plan for drawdown of the Strategic Petroleum Reserve to minimize the overall impact of disruptions, tax incentives to encourage increased privately-held stockpiles, and coordinated demand restraint and stockpile policies among U.S. allies and trading partners.

The provisions of S. 1354, although not yet endorsed by the Administration, are both in philosophical approach and in specific recommendations very close to the conclusions of staff analysis performed by DOE in cooperation with the Office of Management and Budget (OMB), the Treasury Department, the Council of Economic Advisers, and other executive offices.16/ The ongoing staff analysis, to culminate in a comprehensive contingency planning strategy, recommends against imposition of price and allocation controls and favors the use of fiscal policy measures to minimize the economic damage of an oil supply disruption. Both the Bradley-Percy bill and the draft DOE analysis recognize that the market approach will be highly imperfect, but conclude that the ultimate costs of even a very severe disruption will be higher with controls in place than without.17/

If neither the Bradley-Percy bill nor one of the more narrow standby control bills is passed as a replacement for the EPAA, many observers agree that the Federal Government will have a hard time avoiding a full scale reimposition of EPAA-type controls in the next crisis. As local governments, small businesses, individuals and various interest groups feel the economic bite stemming from a disruption, political pressure will quickly mount for the Federal Government to take precipitous action to put a lid on oil price increases.

The strongest pressure for Federal Government involvement can be expected from two quarters: those individuals and representatives of consumers hardest hit by oil price increases; and oil firms themselves which are unable or unwilling to pay higher prices to obtain scarce oil supplies. During the many debates on oil price decontrol over the past few years, consumer representatives have argued strongly that a price control system is warranted even in non-disrupted periods because of the price-setting role played by the OPEC cartel and the dominance of the major multinational oil companies in the international oil market.

In a disruption, the immediate increase in the price of fuel for transportation, heating and air conditioning, followed by price increases of food and other vital commodities, will create severe hardship for individuals without the means to absorb the cost increases or access to alternative fuels. If assistance is not readily available, especially from state and local governments or community groups, pressure will quickly build in Congress for direct federal assistance or, more likely, full-scale reimposition of federal price and allocation controls.

Industry groups, such as crude-short refiners and independent marketers cut off from their normal source of supplies, can also be expected to press for a government allocation program. In the same way, large oil consumers, such as utilities and groups that were designated "priority" users under government allocation regulations, may also exert pressure for reimposition of controls. And political pressure for federal action can also be expected within government channels themselves, as high prices threaten to force cutbacks in essential services such as police and fire protection at the local level.

The basic pricing policy underlying the Federal Government's response to the next severe oil supply disruption is thus still very unclear. While the Administration would prefer to allow market forces to distribute the shortfall, in the absence of a preauthorized fiscal policy that quickly recycles tax revenues back to states and individuals and thus builds its own constituency, pressures will be strong for reimposition of price and allocation controls.

One major factor which could minimize the possibility that Congress will respond to the next disruption with EPAA-type controls is the Strategic Petroleum Reserve (SPR). Currently approaching 200 million barrels, the SPR will soon be of sufficient size to greatly reduce the magnitude and thus the costs of a short-term supply interruption.

Aside from oil price decontrol, the most significant contribution the Reagan Administration has made toward reducing our vulnerability to disruptions is its open and strong commitment to the SPR. Analysts in the Department of Energy have demonstrated that drawdown of public and/or private oil stockpiles in the early stages of a disruption is the single most effective means of lessening the world price increase due to the disruption and the ultimate economic impact of the disruption.18/

The need for building up SPR size and capacity to even greater levels thus is of the highest priority. But the debate over on- versus off-budget financing for the SPR has not helped advance the program. A great deal of time and effort has been spent by Administration officials and members of Congress and their staffs to develop "smoke-and-mirror" funding schemes for the SPR which would assure resources to purchase oil, but which would not appear on the pared-down federal budget. Bouncing SPR funding from one column to the next has not changed the amount of money needed, nor the speed with which the SPR should be filled. Because of the important psychological aspect of the SPR in providing a buffer against disruptions and an assurance to crude-short refiners that supplies will be available in a disruption outside the spot market, complicated debates over the form of funding should not be allowed to delay SPR fill or cast uncertainty over the program itself.

Instead of focusing on the question of off-budget versus on-budget financing, efforts should be centered on developing a financing scheme that quickly maximizes the size of the SPR and gives assurance that it will be distributed quickly and efficiently during a disruption. One such plan being developed by DOE staff is a SPR certificate plan to provide private individuals with limited ownership rights to SPR oil.19/ Under this plan, as opposed to a straight auction of oil actually contained in the SPR at the time a disruption occurred, certificates would be sold now. Holders of certificates could redeem them for the world price of oil at a 5-year maturation date; return the certificate at any time for 90 percent of the world price; or submit a bid to resell the certificate to the government during a SPR drawdown. Handling of the certificates and drawdown procedures could be done by a SPR corporation provided with borrowing authority.

Acquisition of SPR oil would be financed under the plan by revenues from three sources: sale of certificates; sale of SPR oil during disruptions; and borrowing from the Treasury. Revenues from sale of oil during disruptions would also be used to repurchase certificates and retire debt to the Treasury. This type of low-risk plan reduces the direct budget costs of the SPR, but at the same time helps accelerate SPR fill. It also provides a mechanism for self-protection by individuals and firms that can directly reduce their vulnerability to income loss in a disruption, and indirectly reduce the nation's vulnerability. This insurance aspect is of particular importance to those refiners whose supplies might be particularly affected during a disruption. Finally, it would help to create a climate encouraging increased private-sector stockpiling by helping to create a constituency against reimposition of price controls.

Especially now that the SPR is reaching a significant size, it is imperative that a policy for its use be announced. But the Federal Government's SPR drawdown policy is, as yet, unformulated. Congress has approved a distribution plan that establishes basic options for distribution, either allocation or sale, but does not specify either recipients of SPR oil or how price is to be determined. 20/ Mechanisms were developed by the DOE to distribute SPR crude within the framework of the EPAA price and allocation control structure, but procedures have yet to be developed for a competitive auction or any other variation of straight sale of SPR oil. Judging from the draft analyses done by DOE staff, the competitive auction is the preferred alternative of the Administration. As demonstrated in the Bradley-Percy bill, however, strong pressure will be felt to develop a drawdown plan that also contains some non-competitive access provisions for oil companies in particularly bad circumstances, or refiners with some claim to "priority" needs for such uses as agricultural activities.

DOE analysis also supports federal actions to create a climate favorable to increased private-sector stockpiling and to encourage holding of inventories through use of tax policies. Tax incentives can be used in two ways: 1) to encourage construction of storage facilities and buildup of private reserves prior to a disruption; and 2) to encourage the use of those reserves during disruptions. The Iranian shortage manifested in 1979 showed clearly that privately held inventories will not always be drawn down at the time and rate that the government would deem optimal for the purposes of minimizing the overall effects of the disruption. Instead, private firms may hold onto or even accumulate more stocks in the early stages of a disruption before the extent and duration of the disruption, and the expected price increase, can be judged.21/ Oil storage tax incentives should be constructed to encourage use of private reserves in the early stages of a disruption to maximize available surge capacity.

The government's coordination of SPR and privately-held oil use with stocks held by other large oil-consuming nations is also unclear. Within the framework of the International Energy Agency (IEA), the member nations have put a great deal of effort into developing a coordinated stockpile policy.22/ While the IEA partners have reached general agreement on the importance of coordinating stock drawdown in a disruption, and target-level stocks of 90 days of imports, relatively little progress has been made toward a cohesive strategy. One of the major obstacles is the variation among the partners in relative dependence on imports and in actual stock levels. For example, Japanese stocks, while greater than those in the Federal Republic of Germany in April 1981 by approximately 80 million barrels, would replace Japanese imports for only 90 days. German stocks would last about one and one-half times the length of Japanese stocks.

Another obstacle appears to be the Reagan Administration's own reluctance to take part in a "formal system that would commit the U.S. in advance to a specific course of action" in responding to shortfalls too small to trigger the IEA sharing system.23/ The U.S. would thus seem to prefer an ad hoc approach to stock management, but an internal IEA review is still underway to establish a joint policy on when to use government-controlled stocks and when to encourage drawdown or buildup of company-owned stocks.

The Reagan Administration has announced its complete support of the International Energy Agency and the IEA long-term cooperation and emergency sharing agreements.24/ The IEA has served extremely well as a forum for exchange of information and ideas; its major contribution in facilitating the development of a coordinated stock strategy hopefully may be made soon. The sharing agreement itself, however, contains some fundamental flaws that render it useful more as a backup to less formal, ad hoc resolution of temporary supply imbalances among oil-importing countries.

State and Local Involvement

Under the philosophies of the new Administration, which favor a lessening of direct federal involvement and greater responsibility at the State and local levels, the main burden for developing contingency plans to respond to an oil crisis will fall to State and local governments and community organizations. The manifestations of a sudden oil supply disruption will be seen first at the State and local levels. Prices will rise and shortages will be experienced in differing degrees dependent upon the region of the country, proximity to product distribution facilities, and the supply situation of the oil firm supplying the community. The immediate energy needs and resources of each community may therefore differ widely, and individual communities may be best able to develop contingency plans that are specifically tailored to fit their particular needs.

However, State and local planning efforts in most communities are still in the very early stages. State energy conservation plans initially drafted in accordance with the requirements of the Energy Policy and Conservation Act of 1975 (EPCA) have in many cases not been completed due to withdrawal of federal funds. Many State and local emergency preparedness programs and a number of State energy offices will also be eliminated as a result of the budget cuts proposed by the Administration. Another factor that has contributed to lack of community planning is the reversal of market conditions from shortages or imminently threatened shortages through the fall of 1980 to a relative "glut" in mid-1981. A General Accounting Office (GAO) study has found that contingency transportation planning initiated by the Department of Transportation has come to a virtual standstill because of funding problems, lack of information about federal and State measures, and lack of guidance and perception of a lack of urgency.25/ The GAO also found that those plans that had been drafted by some cities would have limited effectiveness in resolving transportation bottlenecks, since they focused mainly on mass transit systems and did not develop innovative measures. There are several cities across the country, notably Seattle, Miami and several smaller towns in New England, that have devised detailed contingency plans. In addition, several states, particularly California, New York and Illinois, have developed comprehensive state-wide plans. The majority of communities, however, have given little thought to planning for an energy emergency, and a sudden supply cut-off would find these communities badly prepared.26/

The consequences of this lack of energy preparedness on the part of the Nation's cities and towns could be extremely serious. Preplanned community efforts to conserve use of energy are particularly important in the face of quickly rising prices both to reduce uncertainty or panic and to lessen the possibly devastating impact of a severe disruption on individuals, continuation of essential services, local industries, and the local economy in general.

As discussed earlier, DOE and other Administration staff are studying the possibility of preplanning and preauthorizing the use of increased windfall profits tax revenues during a disruption to reduce some of the economic damage expected from high oil prices. This concept is the cornerstone of the Bradley-Percy bill introduced in July 1981. If such a plan is adopted, state governments and state energy offices will find themselves with a new role and complex new responsibilities.

Exact amounts of government revenue to expect from the windfall profits tax in any given disruption are difficult to estimate, since they are dependent on the actual oil price rise resulting from the disruption. Expected revenues in a significant disruption, however, can be estimated to be on the order of tens of billions of dollars per year.27/ An increase in the windfall profits tax rate, especially when combined with expected increased revenues from the corporate income tax, provide a large pool of resources which, under the Bradley-Percy approach, would be used to increase transfer payments to low-income individuals, decrease income and payroll taxes for all taxpayers, and provide emergency block grants to state governments to assure the continuation of essential services and to address other priority concerns in individual states. These measures are intended to provide individuals with incentives to take self-protection measures prior to disruptions, but, at the same time, ensure that those who need help can receive it.

The Federal Government's role probably will not stop with determination of a revenue distribution formula. States and local governments are likely to receive, in advance of a disruption, some form of encouragement or assistance to enable them to develop adequate and non-conflicting contingency programs. Without some form of federal oversight or coordination, it is possible that policies adopted by individual states would differ in ways that work against the attainment of maximum efficiency in the overall emergency program.

Short-Term Availability of Alternate Fuels

There are a number of other actions currently on the Federal Government's contingency planning list that are designed to enhance supply during a disruption or assist energy distribution. These measures include industrial fuel switching, power wheeling, removal of restrictions on use of gas, relaxed environmental standards for use of low sulphur petroleum and increased coal use, enhanced domestic oil production, stepped-up nuclear plant licensing, and increased use of alternate and renewable fuels such as solar power, wood and gasohol. Utility of some of these measures is limited by technological or safety constraints. But the overriding reason why these measures in the aggregate currently can contribute little in the way of surge capacity available in an oil disruption is that, despite oil decontrol, the energy market is still restricted by government regulation.

Natural gas regulation is the key barrier to truly efficient use of fuels, in normal as well as disrupted times. A recent DOE study shows that replacement cost pricing of both oil and gas would result in lower levels of oil imports and a more attractive long-term economic prospect for alternative fuels.28/ Because natural gas is priced under federal regulations below market levels even in normal times, the potential for switching from oil to gas during an oil disruption is limited severely. Substantial portions of industrial fuel-switching capacity have already been utilized. More importantly, price regulations have limited incentives for natural gas suppliers to increase production during a disruption.

Government regulations also inhibit full utilization of available fuel-switching and power-wheeling during a disruption, by limiting the pass-through of cost increases to consumers. For example, in the electric utility sector, ability to recover increases in operating costs is not only limited by Federal Energy Regulatory Commission (FERC) regulation of transfer of power between systems, but state public service commission regulation of ultimate consumer prices. Provision for increased costs by including fuel cost adjustment clauses in contracts is not a complete answer, since in many states adjustments are not automatic even when such clauses are present.

Long-Term Measures

Despite the realization that past energy policies have increased our dependence on oil and delayed development of alternative fuel resources, and that a severe oil supply disruption could result in an economic disaster, Federal Government support for new technologies has not been directed toward the most productive areas. To date, government monies have been spent for research and development in a wide range of areas in scattergun fashion. DOE-sponsored projects have included everything from coal gasification and oil shale recovery to electric automobiles and wood stoves. While some of the projects funded have produced valuable results, or at least pushed development in an area further ahead, it is not at all clear that the same progress would not have been made if funding had come solely from the private sector. In fact, in the absence of price controls, private industry would have had significant incentive to make even greater progress in these areas over the past decade than has been accomplished with government encouragement.

The synthetic fuels program is the most visible and costly example of inefficient use of government resources in the energy area. As originally established, the Synthetic Fuels Corporation (SFC) was to commit up to $88 billion over 10 years in order to stimulate creation of an industry capable of replacing 2 million barrels per day of oil imports by 1992. Funding by the semi-private corporation is to be in such forms as loans, loan guarantees, and price guarantees. In addition, both the DOE and the Department of Agriculture have ongoing programs to fund synthetic fuels demonstration projects, as well as separate alcohol fuels and biomass programs. This programmatic support is in addition to favorable tax treatment designed to assist the development of alternate fuels, in particular, alcohol fuels. Reservations about the utility of this massive amount of federal support, taken together, stem not so much from the amount of federal support as from the extensive federal intrusion into a developing industry already skewed by years of federal price controls on oil and gas.

The careful scrutiny being given to individual synthetic fuels projects by the Reagan Administration at first blush seems encouraging. At this state of development, it is vitally important to send the proper signals to industry -- not to expect subsidies of projects that the private sector should and would undertake under the right economic conditions. But the Federal Government to date has been sending mixed signals -- extremely liberal leasing policies pursued by Interior Secretary James Watt leave the impression that environmental concerns run a very poor second -- or even third or fourth -- to energy production. Yet investors in extremely expensive coal gasification and other synthetic fuels projects already underway have been left hanging as funding decisions have been delayed time and time again. Appointment of directors to the board of the Synthetic Fuels Corporation has also been delayed, but statements from the Administration still indicate federal support for at least limited synfuels development.

Both voluntary as well as mandatory conservation measures have been given extremely short shrift by the current Administration. Except for a tax credit structure put in place under the Carter Administration, there is currently little federal support for development of renewable sources of energy or conservation activities in general.

Aside from tax subsidies for alcohol fuels, existing tax credits generally pertain to sources of energy for heating and cooling. The three most important tax credits provided under the Energy Tax Act of 1978 and the Windfall Profit Tax Act of 1980 are a residential credit for renewable energy resources, a business/investment credit and a conservation credit. Briefly, the residential credit provides a 40 percent credit on the first $10,000 invested in solar, wind, and certain geothermal energy. The business credit is at a lower rate, 15 percent for investment in certain energy properties, plus an additional 10 percent credit on qualifying energy property expenditures. The energy conservation credit, for home insulation and weatherization expenditures, is set at a 15 percent rate with an individual $400 ceiling.

Although the statutory language authorizing these credits is fairly specific, the Internal Revenue Service (IRS) is currently taking a very restrictive view in interpreting the statutes through their regulations. In fact, the IRS has in some areas further narrowed the scope of technologies eligible for the credits. One example of this is the geothermal credit. The statute included no temperature restrictions in its definition of geothermal energy qualifying for the credit, but the IRS placed temperature restrictions in their regulations which in effect excluded from the credit geothermal heat pumps using ground water.29/ The IRS regulations also limit use of the credit for energy investments that serve a dual function, as in solar collectors which also serve as part of a roof structure.30/ There are currently attempts underway in Congress to amend the statutory language so as to broaden the scope of the tax credit applications, but the Administration has not yet taken a position on this issue.

Conclusion

The fundamental step toward reducing our vulnerability to disruptions and higher prices has already been taken -- oil price decontrol. The new Administration's change in oil policies was taken in recognition of the fact that market forces are a more efficient and less costly means of allocating supplies, not just during periods of adequate supplies, but also during shortages. But the Administration has yet to announce a comprehensive contingency planning policy that carefully defines the government's role in oil supply disruptions; removes remaining disincentives to self-protection by individuals, industries, communities, firms, and consortia of firms; creates new incentives where justified for increased private-sector stockpiling and other insurance schemes and development of alternative energy resources; and most importantly, communicates through straightforward, unambiguous legislation, the Federal Government's intention to let market mechanisms handle the allocation functions during severe, as well as moderate, disruptions.

1/ Both documents were transmitted to Congress in July 1981.

2/ Data in this paragraph are calculated from DOE data published in the Monthly Energy Review.

3/ Id.

4/ See generally The Library of Congress, Congressional Research Service, "The Oil Market Impact of the Iran-Iraq Conflict and its Potential Expansion" (Washington, D.C., October 13, 1980).

5/ See generally Federal Energy Administration, Project Independence (Washington, D.C., November 1974); and Executive Office of the President, Energy Policy Planning, the National Energy Plan (Washington, D.C.,April 1977).

6/ See T. Neville,J. Blankenship, M. Barron, J. Eschbach, and W. Lane, The Energy Problem: Costs and Policy Options, a staff working paper of the Office of Oil and Gas Policy, Policy and Evaluation, U.S. Department of Energy (May 23, 1980).

7/ Assistant Secretary for Policy and Evaluation, U.S. Department of Energy, Reducing U.S. Oil Vulnerability (Washington, D.C., November 10, 1980) (hereinafter cited as OVS Study), p. 5.

8/ Data in this paragraph are taken from various unpublished DOE staff analyses, and rounded to give approximations of the magnitude of expected economic losses.

9/ Hearings before the Subcommittee on Energy, Nuclear Proliferation and Government Processes of the Senate Committee on Government Affairs, May 1, 1981.

10/ Statement by the President, Office of the Press Secretary, the White House (January 27, 1981).

11/ U.S. Department of Energy, The National Energy Policy Plan (Washington, D.C., July 1981).

12/ Pub. L. No. 93-159 (November 27, 1973).

13/ See, for example, R. Corrigan, "The Next Energy Crisis -- A Job for the Government or the Free Market?" National Journal (June 20, 1980), pp. 1106-1109.

14/ See W. Lane, Jr., The Mandatory Petroleum Price and Allocation Regulations: A History and Analysis (Report to the American Petroleum Institute, Washington, D.C., May 1981).

15/ Congressional hearings convened to investigate the real cause of the various oil shortages during the past decade are too numerous to mention here. See, for example, Current Energy Shortages Oversight Series, The Major Oil Companies, Hearings before the Permanent Subcommittee on Investigations of the Senate Committee on Government Operations, 93d Cong., 2d Sess., January 21-23, 1974.

16/ Initial drafts of the unpublished staff analysis directed by DOE and begun in the fall of 1980 were entitled "Petroleum Emergency Preparedness Strategy Study" and incorporated into an ongoing staff analysis currently entitled "Comprehensive Emergency Preparedness Strategy."

17/ Both also recognize, however, that some controls might be necessary in a wartime situation. Authority for wartime controls already exists under the Defense Production Act of 1950, as amended (64 Stat. 798).

18/ OVS Study, supra note 7, pp. II-A-6 to II-A-8.

19/ Unpublished draft staff analysis, Office of Policy, Planning and Analysis, U.S. Department of Energy (June 1981).

20/ Strategic Petroleum Reserve Distribution Plan, Amendment No. 3 (1979).

21/ OVS Study, supra note 7, p. II-A-6.

22/ The International Energy Agreement was signed in November 1974. The IEA currently has 20 voting members and one non-voting member (Norway). France is not a member by choice.

23/ Robert D. Hormats, Assistant Secretary of State for Economic and Business Affairs, Statement before the Senate Subcommittee on Energy, Nuclear Proliferation and Governmental Processes, July 14, 1981.

24/ Id.

25/ General Accounting Office, "Transportation Contingency Plans for Future Gas Shortages Will Not Meet Commuter Needs" (July 1, 1981).

26/ Hearings before the Senate Subcommittee on Energy, Nuclear Proliferation, and Government Processes on Community Efforts to Cope With Energy Shortages (May 1, 1981).

27/ Sen. Bill Bradley, 127 Congressional Record 88 (June 11, 1981), p. S. 6050.

28/ U.S. Department of Energy, Office of Policy, Planning and Analysis, draft report on Natural Gas Deregulation (June 11, 1981).

29/ Title 26 U.S.C. Section 1. 44C-2(h), 45 Fed. Reg. 57, 716 (August 29, 1980).

30/ Title 26 U.S.C. Section 1. 44C-2(f)(4).

Chapter 71
WHAT TO EXPECT FROM STATE ENERGY EMERGENCY PLANS

G. D. Simon

Since the 1973 embargo caught the nation by surprise, government at all levels has maintained a working effort to do shortage contingency planning. Under the Carter Administration this effort was quite extensive and involved local agencies and states as well as the federal government. Under the Reagan Administration, this effort has been pared back considerably. Even so, the threat of a disruption remains high given the political and social unrest in the small corner of the world that provides 40 percent of all the oil consumed. The current surplus of oil on world markets should not be interpreted as the end of the energy crisis. There is a long way to go before relative calm is restored to the production and use of energy. The need for preparations to create a "shock absorber" for potential shortages still exists.

This paper focuses on the status of state-level energy shortage contingency planning and its implications for emergency preparedness in the business community. The paper concludes that while states will now bear the burden falling on government for coping with future energy shortages, they have made few advances in their emergency responses. This lack of true preparedness has become much more serious as a result of fuel price decontrols. Most states assume the presence of price controls during an energy emergency, and have constructed their emergency plans accordingly. These plans may now be obsolete given the new free market environment that currently exists. Of greatest importance to the business community is government's apparent move away from intervention as a shortage response. Responsibility for future shortages is being shifted to the private sector, leaving individuals and firms to fend for themselves.

KINDS OF SHORTAGES TO EXPECT

Clearly the most likely shortages in the next ten years tend to be disruptions in world oil supplies. The prospects for an oil shortage are closely tied to the instabilities in the Middle East. As the Iran-Iraq war has demonstrated, the tensions in the area are not only a function of the Arab-Israeli confrontation but are also related to long-standing religious, ethnic, and political disputes among the Islamic nations. Future disruptions may be less in the form of politically-inspired embargoes and more like the interruption of deliveries experienced when Iran and Iraq went to war. A one million barrels per day (b/d) shortfall to the United States (about a 6 percent shortage) could easily happen, even several times. A 2 million b/d shortfall (12 percent) is less likely, but analysts still rate its chances as better than 50 percent in the next ten years. A 3 to 4 million b/d shortfall (20 to 25 percent) begins to look like an extreme case, but still possible.

The chances of a natural gas shortage in the next ten years appear more slim now than they looked just a few years ago. Higher prices have stimulated interest in conservation and have encouraged more exploration. Even delays in large gas projects like the Alaska pipeline are not generating substantial anxiety over shortages. So far, the natural gas picture is only getting brighter.

Chances for an electricity shortage in many cases depend on the chances for oil or gas shortages. As a result, the outlook for electricity shortages tends to be an intermediate case. The majority of any electricity shortage problems tend to be a lack of fuel supplies to maintain the output of electricity, not a problem of meeting peak loads. Fuel shortages could spell serious problems for cutting back use over many weeks. Fortunately, utility storage tanks are reasonably full right now, and could avoid a serious crunch for many months after a disruption. Some parts of the country, however, also face potential shortages of peak generating capacity. This is a problem independent of oil and gas supplies. An additional 3 to 5 percent of peak shaving or supplemental supplies may be needed in these areas to ensure adequate reliability.

One must recognize that during an oil shortage, consumers may be asked to cut back on electricity and gas use both to save oil and to make more fuel available for substitution for oil elsewhere. While the chances of a gas shortage by itself or a peak generating capacity problem in certain parts of the country do not look high, an energy manager in a firm, for example, may still be confronted with mandates to cut back electricity and gas use temporarily.

STATE RESPONSIBILITIES

The U.S. style of federalism creates overlapping responsibilities among local, state, and national governments. This situation is no less true for energy emergency responses. Which responsibilities end up at what level is really a function of the public pressures for action at each level, the ability and willingness of any level of government to intervene, and the level of government best suited to handle the problem. The current temperament in the nation's capital results in more responsibility focused on the states to respond in a crisis. The Reagan Administration apparently will avoid or ignore pressures to intervene and allow the free market to handle most of the shortage response. This approach deflects much of the public pressure to "do something" onto state and local governments, especially for the smaller (less than 12 percent) shortages which are the most likely.

---

*Former Director of Shortage Contingency Planning, California Energy Commission

Putting states on the front line of the emergency response is not actually a major change from the Carter Administration policy. In 1979 Congress passed the Emergency Energy Conservation Act or EECA (P.L. 96-102) which established the process to be followed in a future energy shortage. Once the President declares an energy emergency, states are mandated to submit response plans for approval, and to act according to those plans unless the shortage worsens beyond 20 percent. A federal "back-up" plan was to be prepared if states did not come up with effective steps to handle the crisis and ultimately in an oil crisis when the threshold of 20 percent was crossed, coupon rationing of gasoline would start. These rather weighty responsibilities were not thrust on unwilling states. In fact, the National Governor's Association lobbied vigorously for placing the states in charge.

EECA is still in effect, but will expire in July 1983. The Reagan Administration is still bound by it, although much has changed since it was enacted. The President halted or eliminated the "back-ups" to state action by rescinding five of seven adopted elements of the stand-by federal conservation plan (leaving only public information campaigns and odd-even gasoline purchase plans) and dropping all funding for setting up coupon rationing in advance. This federal pull-out leaves the states even more alone than EECA envisioned.

Of greater significance, the President also dropped all price and allocation controls on oil and oil products. These controls were the foundation for the oil emergency response through the Nixon, Ford and Carter years. The authority to reimpose these controls expires September 30, 1981, and the Administration's lack of interest in extending this authority or devising a replacement for it is certainly a strong signal that Washington intends to stay out of the private sector's response to another interruption. The only major federal preparations for shortages appears to be continued filling of the Strategic Petroleum Reserve, but even this is now jeopardized in disputes over how best to finance the purchases for the stockpile. The jury is still out whether states will step into the oil crisis vacuum in any substantial way, or merely focus on maintaining order and appealing for cooperation.

For electricity shortages, states have traditionally played a large role and likely will continue to do so. Electric utility regulation has been predominantly a state function and this has been applied in most states to include contingency planning. Natural gas shortage contingency planning has been nearly equally shared between the states and the federal government, owing to federal regulation of the interstate pipeline companies. With the federal government's phased decontrol of natural gas, the focus on natural gas contingencies has shifted to the states, and is now similar to the electricity emergency response.

The states' roles and responsibilities are then relatively large. Where will the public pressures for action face? Due to the views of the current administration and existing federal law, the states will probably be seen as having the more sympathetic ear. To what extent can and will any government intervene? The federal government apparently has chosen to avoid intervention. Many states may be more activist. What level of government can best handle the problem? For electricity and to a growing extent for natural gas, states have the predominant regulatory role. For oil, states need to deal directly with affected groups and businesses, promoting emergency conservation, resolving hardships and providing credible information. Unilateral state attempts to control prices would not be beneficial.

STATUS OF STATE PREPARATIONS

After the enactment of EECA and the Iranian revolution in 1979, states were very enthusiastic about preparing for future shortages. DOE's promise of federal funds fanned this enthusiasm to an even higher level. In return DOE asked that state emergency plans be submitted before the President declared an energy emergency. DOE felt an early submittal was preferable to the 45-day time limit that EECA placed on states once an emergency was declared. With initial DOE grants in-hand, and more grants promised for the future, state emergency planning began in earnest during 1980. By 1981 the retraction of federal EECA funds in conjunction with the oil glut had virtually eliminated state emergency planning activities. To date, only Nebraska and Missouri have submitted a draft emergency plan to DOE. However, all states, with the exception of New Jersey, have new contingency plans in some phase of development. Twenty-eight states have produced some draft available for review, but progress in finalizing their work is quite slow.

In general, state plans propose a campaign for voluntary conservation in early crisis stages, with mandatory prohibitions and fuel redistribution as a second line of defense. Characteristically, the voluntary conservation campaigns would encourage large employers to reduce energy use by a certain percentage target, increase the number of employee car pools and adopt staggered work hours to prevent overloading public transit. The campaign asks the general public to change thermostat settings, eliminate unnecessary driving and to car pool.

Mandatory restrictions generally involve:

- Strict 55 mph enforcement,
- Lighting use restrictions,
- Dropping speed limit to 50 mph,
- Parking bans,
- One or two day a week car use restrictions,
- Building temperature limits, and
- Conservation quotas.

Thus far 40 states have passed laws giving the Governor specific authority to impose these types of mandatory measures. Several of these laws also establish a <u>state set-aside of fuel</u> for resolving hardships for certain "high priority" users, grant state authority to <u>control the distribution and allocation of fuel</u>, <u>allow suspension of state transportation or environmental quality regulations</u>, and authorize the imposition of <u>marketing restrictions</u> such as odd-even sales and minimum purchase.

Electricity and natural gas utilities have their own plans for shortages, but these are generally approved by a state's utility regulatory agency. These plans also contain mandatory restrictions (usually banning decorative uses, nighttime lighting, advertising) beyond voluntary appeals. Responses to severe shortages involve rotating outages (electricity) or curtailing service in a pre-established order of priority (gas).

Overall, current state plans present several problems. First, the restrictions to be used during a serious shortage are harsh, very difficult to enforce, and unpopular. Since their effectiveness is in doubt, voluntary conservation may be the only real "tool" states have at their disposal during an energy emergency.

Second, oil emergency plans tend to address the wrong problem. Under price decontrols, the next oil shortage will quickly become an economic crisis. Most studies see gasoline at $3 per gallon in a 10 to 15 percent shortage. As prices rise sharply at the pump, personal income to be spent elsewhere will decline. The business community will find the demand for its products dropping as inflation and unemployment rapidly move upward. Interest rates will jump as businesses and families try to replace lost income with credit. Nevertheless, most state plans implicitly assume the problem will be lines at the pump and people without fuel. In reality, gasoline lines may be replaced with unemployment lines. A state's emergency role may then shift to encouraging moderate price hikes, cushioning the resulting economic stress, and shifting budgets to cover large fuel costs for essential government services.

Third, state plans which envision extensive allocation or price controls, or which involve a Governor taking control of in-state supplies, are likely to raise the ire of energy suppliers. Also, neighboring states whose supplies are adversely affected by this action may challenge its legality in the courts.

Finally, federal and state plans for natural gas emergencies assume that large end-users have plenty of fuel (oil) stored in the event of a curtailment, and as a result, cut service to these industries first. Given the higher probability for an oil shortage and the subsequent price run-up that would ultimately threaten these industries, shifting residential and commercial natural gas supplies to industry in order to save oil may realize greater long-term benefits during an oil crisis.

IMPLICATIONS FOR BUSINESS

In future energy shortages the business community is likely to deal primarily with state agencies. These agencies will determine the emergency actions to be taken, encourage cut-backs in energy use, and continually request real-time data. Some steps can be taken now to make this experience less difficult for everyone involved.

First, the innovative and attractive ideas scattered throughout the state plans should be consolidated and adopted nationwide. These include:

- Load Curtailment Cooperatives - Avoid a disruptive rotating outage by allowing commercial/industrial customers to tailor their own electricity use cut-back. Utilities have offered to guarantee service to groups of customers willing to pledge a substantial cut-back in energy use when requested by the utility. This arrangement allows businesses to protect their essential needs while temporarily curtailing less important uses. In comparison, an indiscriminate rotating outage could spell big financial losses for a company. California utilities are now signing contracts to pay companies for load reductions once the companies are called on.

- Auxiliary Power Contracts give reduced rates to customers who have stand-by generators and agree to allow utility use of their equipment.

- Emergency Surcharges on natural gas or electricity usage above a specified "conservation quota" will allow flexibility in reducing consumption. From an operational standpoint, this is a much better approach than banning certain energy uses or curtailing operation hours. Experience in Los Angeles in 1973 with this technique was impressive.

- Emergency Park and Ride Lots set up in shopping center parking lots can provide an important reduced cost option for employees to get to work. Considerable advanced planning is required of local governments to make this option work.

Second, businesses should work toward a goal of "self sufficiency." Larger fuel stockpiles can be valuable shortage insurance, since government is unlikely to locate emergency fuel supplies for anyone other than health and law enforcement services. The feasibility of holding a fuel reserve now should be checked by comparing the high cost of fuel storage against that fuel's decontrolled price during a crisis. Emergency conservation plans should be worked out in advance to facilitate good cooperation with government early in a crisis, and potentially forestall more drastic governmental measures. Pains should be taken to ensure that any state "conservation quotas" calculated will be based on historical energy consumption data that does not penalize investments in "non-emergency" conservation. Emergency conservation plans should include not only facility energy use, but energy savings in employee commuting, business travel and freight shipping. An easily understood emergency process (e.g., yellow, orange and red stickers on equipment switches) should also be incorporated into conservation plans. Larger inventories of critical parts or raw materials should be maintained to avoid disruptions in freight transport due to short fuel supplies. Finally, the capability for fuel switching can prevent a business from being backed into a corner by either the price or availability of fuel. Companies that have been disconnected from natural gas service when curtailments were severe in the mid-1970s should consider reconnection.

Unless a future energy shortage is extremely severe, most businesses will find themselves predominantly dealing with states. Innovative state responses to energy emergencies are too few and too scattered, and generally consist of voluntary appeals followed by mandatory restrictions. Since the latter response may be unworkable, hurried retrenchment at the time of an emergency should be expected. Energy managers in the business community can help avoid the resulting confusion by recognizing state roles, promoting the more flexible approaches, and preparing company contingency plans that ensure greater self sufficiency. Above all, dependence on outside help during a crisis could be a mistake. The trends toward "free market" and "decentralized" solutions are strong.

SECTION 24
PROCESS INNOVATION AND FACILITIES MODERNIZATION

Chapter 72
THE IMPACT OF ENERGY COSTS ON FACILITIES MODERNIZATION

W. P. Smith, Jr.

A recent Gallup survey found energy costs a major concern of U.S. industry's chief executive officers looking at 1981. And if your company happens to depend on petrochemical feedstocks, there's very little doubt you would put energy at the head of the list. Badische Corporation, a manufacturer of chemicals, fibers and yarns, is one of those. In 1981, for the first time in Badische's history, conversion energy(all energy other than feedstock) costs, which depend directly on oil prices, are surging upward as well. What can a company like Badische do to contain these costs?

Like many other U.S. companies, Badische has tackled hard since 1973 the quick-fix housekeeping area and has even gotten into some of the costlier conservation technologies. Up to the latter part of 1978, Badische was partly owned by Dow Chemical Company, and under that arrangement was included in Dow's highly acclaimed energy management program.

Now a wholly owned subsidiary of BASF American Corporation, Badische is intensifying its energy efforts and benefitting greatly from certain high-technology processes developed by BASF in West Germany. "We have a full exchange of technology between our European operations and here," explains Hans Kopper, president and chief executive officer of Badische.

KEEPING COMPETITIVE

Though up against some pretty stiff competition in this country - such as Celanese, American Cyanamid, E.I. du Pont de Nemours, and Monsanto - Badische has captured a significant part of the market. Its Sylvania, Georgia plant, for example, is one of the world's largest producers of colored spun carpet yarn.

"To be competitive, we always try to have the lowest production cost," says Hanns Pohl, director of corporate planning. "In a situation of low energy prices and high labor costs, it makes economic sense to use more energy to reduce labor costs. And that's what we and industry in general did wherever labor costs were high. Unfortunately, past energy prices did not represent the true costs of producing and transporting the energy itself, so we ended up with an imbalance in the system. Any artificial pricing that induces the impression energy is cheap really encourages consumption of energy. As a nation, we have only been delaying the resolution of the problem. With increasing energy costs, energy-related goals will step up to the highest priority."

Process innovations and energy management, indicates Kopper, are significantly increasing Badische's conversion efficiency and yield from oil-based raw materials. "We are working on improving our chemical processes to get more caprolactam, for example, out of every gallon of cyclohexane," says Kopper. "Sometimes we even sacrifice a little conversion energy to get a lot more out of the basic raw material. But most of the time this extra use of energy isn't necessary to improve the process. From year to year, our processes have become far better than they were before."

Badische finds by agressively working on saving energy, it is gaining a competitive edge in some product areas.

FACING FEEDSTOCK SHORTAGES

Kopper points out another reason Badische is working so hard on process energy efficiency. "Energy - not just as means for conversion but as petrochemical raw materials - was the main problem we had in 1979," he says. "We were short of raw materials, and our costs almost doubled. We could not, of course, absorb these costs, and we had difficulty passing them on to our customers as price increases."

Most U.S. companies are just beginning to take a closer look at the potential for process energy improvements. Up to now, industrial processes have been taboo territory to energy-efficiency hunters in many plants.

At Badische, on the other hand, the German experience has fine tuned the energy sensibilities of top management. "Energy prices in Europe have always been much higher and we've always been much more aware of energy," says Kopper. "Energy was a cheap item here, at least up to 1974, when it got a little more costly. Now is the second push, and it won't be the last. Gasoline prices at the pump have doubled in a year's time.

"In Ludwigshafen, where I come from, we operate the same processes as we do at our Freeport, Texas plant. Though the plants in Germany were built before those in the U.S., they were much more energy efficient, with all the small conservation measures, than we were here. It just paid off earlier to emphasize energy efficiency in Europe.

"We have now converted any of our older processes, built in the sixties, which made sense to convert. We did that particularly when it came time to replace a piece of equipment. And we are continually looking into the chemistry of our processes to see if there are any steps we can eliminate or make more energy efficient."

NOURISHED BY PARENT

BASF has committed about $1 billion between 1979 and 1984 for capital improvement programs at its North American operations. The recipient of a substantial portion of this, Badische is agressively investing in new plants, plant expansions and modern equipment and processes.

The company completed construction this past year of a new neopentyl glycol plant at its Freeport, Texas chemicals site. This plant incorporates an advanced process developed by BASF and already successfully used in Germany. The new facility has already become a major U.S. supplier and is capable of expanding. It has a present capacity of 25 million pounds per year of neopentyl glycol, which is used in the production of polyesters, alkyd resins, urethanes, plasticizers and synthetic lubricants.

Project engineers paid careful attention to energy efficiency in the design and construction of this new plant. They included in their plans an environmentally sound system of burning streams of gas and liquid waste to generate steam, thereby reducing the company's dependence on purchased energy.

FRUSTRATION WITH FUA

The idea of using this heat rather than letting it go into the air seemed like a good one. But under the law, because the heat recovery potential exceeded 50 million Btu, Badische had to get the waste stream classified as alternate fuel. "That's where we ran into a problem," says Kopper. "More than a year went by and we couldn't get a decision from the Department of Energy (DOE), under the Fuel Use Act (FUA), on whether or not the waste streams could properly be classified as alternate fuels within the meaning of FUA and DOE regulations.

"DOE applied its own restrictive definition on the issue of the marketability of waste streams; that is, if an organic waste stream could be sold for any price that would exceed the cost of storage and handling only, then the waste stream must be classified as petroleum. Therefore, it could not be burned in a boiler that exceeded a rating of 50 million Btu. This definition effectively ruled out the possibility that the market price might have no relation to the actual fuel value of the waste systems. We thus had to modify the incinerator-boiler and incur substantial additional cost."

Badische recently broke ground for a new $120 million acrylic monomer complex to be completed by mid-1982 at its Freeport site. The complex will include giant new acrylic acid and acrylate plants, as well as the expansion of existing acrylate facilities. Again, Badische applied for the necessary fuel classification determination to burn the new plant's waste streams.

The acrylic monomers project was similarly affected by the DOE interpretation. Badische waited to order the necessary equipment, but never did get the acrylic monomers' waste streams classified as alternate fuels.

"We endured very costly delays until we were forced to move on," explains Ravi Singhania, engineering and maintenance service manager at Freeport. "It took more than a year and cost more than a million extra dollars, because the design of the incinerator-boiler system had to be changed."

Kopper points out that it's not the law he objects to but the lack of judgment and long-range vision in governmental regulatory agencies. "The agencies that are established to administer these laws don't have an overview of the whole picture. And there's so much struggling between agencies you are left wallowing in uncertainty. That's really the big problem industry has to face," Kopper says.

Badische feels that by spending capital to reuse its waste streams it would not only be saving energy but would be supporting government policies. This is why the company finds the whole situation so difficult to understand. But other efficiency avenues are open.

INCREASING YIELD, SAVING ENERGY IN PROCESS

"We have an acrylic acid plant in Freeport now," says Kopper, "operating on a totally different chemical process, using acetylene as the basic feedstock. The new, larger plant will use a new BASF-developed propylene oxidation process, which will place us at the technological forefront of this industry. There is much more energy efficiency planned into this process than in the one at the existing Freeport plant. We will really squeeze out every last Btu, giving us a big cost advantage."

Another new process Badische hopes to bring into operation would eliminate several energy-consuming steps in the production of nylon staple. What Badische and many other fiber producers presently do, according to Kopper, is take their monomer (caprolactam), polymerize it, extract the monomers out of this product, cool and dry the product, make strands out of it and cut it up into what are called polycaprolactam chips. From storage, these chips are then loaded onto hopper cars and transported to the fiber plant (Badische's Anderson, South Carolina plant), where they are remelted and put through an extruder. This process of going back and forth twice between a liquid and solid state, along with the necessary material handling and transportation, consumes a great deal of energy.

Badische and others in the fiber industry have been carrying on extensive research and development in this area, trying to eliminate some of these steps. Ideally, once they have polymerized the caprolactam, they would like to spin the polycaprolactam immediately in the warm stage.

"We are considering a pilot plant for this R&D project," says Kopper. "It will take several years to be installed and become operational."

"It would save energy in several ways," explains Roger Zoccolillo, vice president, Fibers and Yarns Group, "by avoiding the need for cooling, chopping, blowing the chips into storage, transporting, unloading and reheating. It would cost tens of millions of dollars to implement this project, but it would save a tremendous amount of energy."

Yarn dyeing is another process area in which Badische has moved out front and is saving a substantial amount of energy. According to a recent study by Georgia Institute of Technology, the two primary methods Badische uses to color yarn - solution and stock dyeing - are the most energy efficient in the industry.

"Our solution dyeing process in Anderson, South Carolina," explains Zoccolillo, "allows the colorant to be added directly to the polymer in one continuous stream so that the resulting yarn comes out already dyed."

The Georgia Tech study indicates less than 100 Btu are required in solution dyeing a pound of yarn, while 100 or more times as much energy is used in dyeing carpet yarn by conventional methods.

The alternative to precoloring the yarn is to take baled uncolored fiber, spin it into yarn, put the yarn on a package, place it in an autoclave and then dye it. "The amount of energy consumed by this process is substantial," says Zoccolillo. "Our predyeing processes at Anderson and Williamsburg save energy for both the buyer and Badische, because we are also a consumer of our own Fiber."

Sylvania's stock dyeing operation is also highly efficient. When the company took over full-ownership of the plant in 1977, it quickly initiated a complete modernization and expansion program, incorporating new processes and technology to improve output and energy efficiency. With the addition of nearly 100,000 square feet, the Sylvania complex now covers 400,000 square feet.

A heat recovery system in this plant saves a considerable amount of energy in the dyeing operation by using heat from waste hot water to raise the temperature of incoming water.

Badische's stock dyeing process, employing sophisticated heat recovery systems, requires an average of about 4,000 Btu per pound of yarn, according to the Georgia Tech study. By contrast, the study indicates the Kusters continuous dyeing method, a standard industry process, requires about 10,000 Btu to color a pound of carpet yarn, and the atmospheric Beck process, another common dyeing method, requires about 13,000 Btu per pound.

Putting these figures in perspective, the equivalent of a gallon of oil will provide enough energy to dye more than 1,000 square yards of carpet by solution dyeing and more than 19 square yards by stock dyeing, but less than 6 or 7 square yards by the other conventional methods.

One more noteworthy process development, in the production of nylon fibers for carpets and home furnishings, is the basis of a $20 million expansion currently underway in Anderson.

"We are now installing high technology machinery at our Anderson plant to put us at the forefront of BCF (bulked continuous filament) nylon production," says Kopper. "This project was not implemented primarily to save energy, but we eliminate one or two process steps with it and save some energy as a result. Its main advantages are increased yields, labor savings and quality improvement."

MULTIPLYING STEAM SAVINGS

At Williamsburg, Virginia, site of Badische's corporate headquarters and acrylic fibers and yarns operations, a capital project has just been completed, which according to James R. Simonsen, consulting engineer and chairman of the plant's energy committee, was definitely undertaken for the sole purpose of saving energy. "We went from single-effect evaporation in our acrylic fiber plant in 1958 to a double-effect in 1967 and converted to a triple-effect evaporator in 1980 - all because of the price of oil," says Simonsen. "The economics of this project were figured at a 36 percent payback in October of 1977. Now it's over 50 percent because of oil cost increases."

Badische requires the use of this evaporating equipment to remove water from the solvent used in its acrylic fiber polymerization process. The addition of a third evaporator involved the costly construction of a tower to accommodate it, but the installation is significantly reducing the plant's steam requirements.

"Essentially, single-effect evaporation means you need a pound of steam to evaporate a pound of water," explains Simonsen. "Two effects mean you need half of a pound of steam. And three effects mean you need one-third of a pound of steam. Or instead of 1,000 Btu, only about 333 Btu are needed to evaporate a pound of water. So at Williamsburg we're saving about 7,000 pounds of steam an hour."

This project has proved so successful and has such a fast payback, Simonsen is considering installing a quadruple-effect evaporator this year.

Simonsen also reports that the installation at Williamsburg of a polymer stripper, costing over $1/2 million, is increasing the yield from the raw material (acrylonitrile) by stripping residual monomer from polymer solutions. Ironically, this installation was required by OSHA, and an initial payback of five years was calculated. That payback is now down to about three years.

"In the acrylic fibers area, we're up about 35 percent in production since 1972, and we're down about 10 percent in total energy use," says Simonsen. "We've achieved a 30 percent reduction in energy per pound of product produced. Greater capital expenditures will be necessary to achieve further increases in energy efficiency, but with engineering and research, we can still find projects with very attractive paybacks."

Excluding raw material costs, Badische spent over $40 million in 1980 for purchased energy. According to the company's energy accounting system, that's $5.5 million less than it would have had to pay out if it were still operating at its 1972 efficiency level.

In some areas, however, Badische lacks an earlier platform for determining its energy efficiency improvements. "We have changed our operations so much in the last few years, comparisons are difficult to make," says Kopper. He indicates further that the company's energy savings might not sound as good as those of some others in the chemical industry, because Badische does not have the same energy-saving opportunities. For example, Badische buys most of its steam for its chemical operations at Freeport rather than producing it, and this is an area in which some of the big chemical companies are now saving large amounts of energy.

"We have turned down no reasonable energy-saving project in this company," says Kopper. "Energy savings in my point of view are directly related to cost savings, very reliable cost savings. You can pocket them almost immediately. So energy-saving projects almost always end up with very high priority."

In addition to looking favorably at most energy-saving projects, Badische has just instituted a procedure requiring every new capital investment to undergo a detailed study of its energy impact.

NEW TREND IN FIBERS INDUSTRY

Walter Smith, energy coordinator of the Anderson, Sylvania and Allendale plants, points out that Badische has been carrying on an aggressive energy management program in many areas besides processes, and with good reason. "While our overall labor costs are still higher than energy," Smith says, "one of our products at Anderson now costs more in energy than labor. And other products will cross over the line this year if we can't do something to control our energy use and costs. This trend of energy surpassing labor costs is new in the fibers industry, and that makes it significant.

Over the past eight years, the Anderson plant has implemented full-scale programs of steam and electrical metering, steam trap maintenance, reinsulation, lighting retrofit, heat recovery and numerous other energy-saving projects. "The result," says Smith, "is that we are saving the equivalent of more than three million gallons of oil a year. This annual savings amounts to more energy than all Badische Anderson employees use in about three years in their personal cars."

Several energy audits of the Anderson plant have already been conducted, and a total Btu accounting system has been in effect for years. This past summer Smith coordinated the first intensive 4-day audit of the Sylvania and Allendale plants, two smaller, more recent Badische acquisitions. Although energy consciousness already runs high at these plants, the audits should provide the basis for further efficiency-improving measures.

Meanwhile, conservation efforts continue mounting at the Anderson plant. Currently underway there is the installation of a demand control and energy management system, which is computer-based with smart terminals. "An interesting feature," explains Smith, "is we're going to move the system from a control room in the middle of the plant out to the utilities building, where we'll consolidate our utilities functions and multiplex all of the signals between the two buildings. This system will shed 32 process loads on a priority basis. To determine potential savings, we went through shedding loads manually before trying to do it automatically, and I recommend others do the same."

CURBING KILOWATT CONSUMPTION

Because electricity accounts for 75 percent of total conversion energy use at Anderson, 80 percent at Sylvania and 95 percent at Allendale, a strong electrical energy management program is paying off. "We're expanding our compressed air operations but trying to hold electricity usage constant by cutting back on lighting and increasing efficiency in other areas," says Smith.

Main features of the Anderson's plant's lighting program include retrofitting the parking lot with high pressure sodium (hps) fixtures and lamps, installing dummy tubes to cut out unnecessary fluorescent lighting and using hps in the new plant expansion. "The payback on the retrofit installation of sodium vapor lighting in the parking wasn't great," says Smith. "We calculated 3.8 years on a life cycle cost basis. But we did it anyway because the risk was so low."

After determining there were certain lighting areas where a wide application of dummy tubes would make sense, Smith evaluated several alternative products and found one, containing a capacitor in the tube, much more effective than the others. The use of this dummy tube throughout the plant, wherever appropriate, had an average payback of four and a half months. It produced cost savings, according to Smith, in three ways: reduction in direct lighting electricity costs, reduction in airconditioning load, and extension in the life of the remaining tubes.

Fifty percent of the Anderson plants's electrical

energy goes for airconditioning required to maintain process air conditions. "Temperature conditions and moisture levels are critical concerns during all fiber production stages," says Kopper. "Our fiber operations are expressly excluded from the government's temperature restriction regulations, and there's not much opportunity for us to lower our energy use in that area. But we are determined to do all the little things we can."

At Anderson, for example, 108 meters make supervisors accountable for the electrical energy consumption in each work area by providing a clear indication of precisely how many kilowatt hours (kwh) are being consumed by the plant's warping, drawtwist and spinning operations.

It would be possible to detail an impressive list of conservation projects going on at all of Badische's operations. The most important feature of the company's energy program, however, is top management's determination to save energy every way possible while complying with national energy policy.

"The percentage of energy consumption in our total cost has not increased in the last several years," says Kopper. "That means we have been able to increase our energy efficiency so much that energy price increases have been offset."

Simonsen and Smith are currently evaluating programs to install coal burners in place of the gas and oil-fired boilers at the Williamsburg and Anderson plants. They have determined they cannot simply convert their present boilers, so they are planning to maintain these boilers as back-up systems. Trying to install coal capability, they find, is fraught with huge costs and regulatory problems.

THE PETROCHEMICALS PROBLEM

As a member of the Chemical Manufacturers Association (CMA) and the Man-Made Fiber Producers Association (MMFPA). Badische has another concern related to future shortages and the possible government allocation of energy. "In my opinion," says Kopper, "first priority should be given to feedstock. Any oil product is much better used when converted into a usable product rather than burned up in an automobile or elsewhere."

In 1979, the government regulation on lead-free gas cut sharply into the supply of cyclohexane, creating a supply shortage for Badische. "I'm not opposed to lead-free gas," says Kopper. "We need to go in this direction. But such sudden legislative changes cut unnecessarily into the supply of valuable chemicals, like aromatics. The oil industry is now developing alternatives in making lead-free gas. When the government reacts quickly without being aware of the effects of its actions, it creates new problems."

According to Kopper, Badische can use no substitute for petrochemical feedstocks. So a strong energy management program - including good raw material purchasing contracts, an emphasis on improving process efficiency and continuing energy conservation strides - is essential to the company's future survival and growth.

Chapter 73

FACILITIES MODERNIZATION IN THE MANUFACTURING WORK PLACE

J. E. Van Schelt

1. ABSTRACT

This paper deals with a case study of how Hamilton Standard conceived and installed an energy management program. An overview of the program is presented. Specific actions are outlined, including the conception and installation of a computer controlled plant operation system and the development of a comprehensive 5 year plan.

The paper shows how we reduced our usage of oil by over 50% and the energy used per unit of output by 67%. Our energy plan became the model for all divisions of United Technologies and our efforts earned an award for the "Best Energy Conservation Performance of all Divisions" of United Technologies. These efforts contributed to the award given to Hamilton Standard by the American Institute for Plant Engineering for the "Most Outstanding Plant Engineering Program in the United States."

2. BACKGROUND

Hamilton Standard is one of the several divisions of United Technologies Corporation. We manufacture a variety of products including Environmental Control Systems for aircraft and spacecraft, engine controls, propellors, life support systems for space vehicles, including the Lunar Landing Module and the space suit and backpack worn on the moon, as well as a variety of electronic devices for auto emissions and safety testing, engine monitoring and Inertial Guidance Systems for space probes.

Our Plant Engineering Department was involved in energy conservation long before the Arab Oil Embargo made it fashionable. This paper will show how we used two basic management tools, planning and technology, to make our energy management program what it is.

3. LOGICAL STEPS

We like to think that there are three logical steps to energy conservation. First, take out the fat, then bring engineering talent to bear on all energy uses in the plant, and finally optimize the use of energy with an Energy Management System.

a. The first step consists of the things with which most firms are familiar by now; organizing an Energy Conservation Committee, lowering heating temperatures, turning things off when not in use, and generally eliminating waste.

b. Next, all uses of energy must be examined to ensure that the most efficient use is being made. This involves getting back to the basics of re-engineering systems which may be twenty years old or more and using the latest materials and techniques to increase efficiency.

c. Finally, when the fat is gone, and energy use is as efficient as it can be made, optimize the operation of the plant with an Energy Management System.

4. PLANNING

Our guide at Hamilton Standard is our Five Year Energy Plan, which is revised each year. The Plan contains all the things we always wanted to know about energy use in our plant and a few things we didn't think we wanted to know until we started developing the plan.

In order to attack energy use, we had to know where the energy came from. #6 fuel oil and electricity make up about 90% of the BTU's entering our plant. Since this is the case, most of the gains were made in these areas. We had already cut the use of propane, auto fuel, diesel and other fuels by taking out the fat with our Energy Conservation Team. (See Figure 1) Of course, patriotic as energy conservation may be, our plant has a controller, just as other plants do, so we have to look at costs, too. When we do this, the #6 fuel and electricity change places in importance, but they still make up over 90% of the cost of energy, and remain the important targets. (See Figure 2)

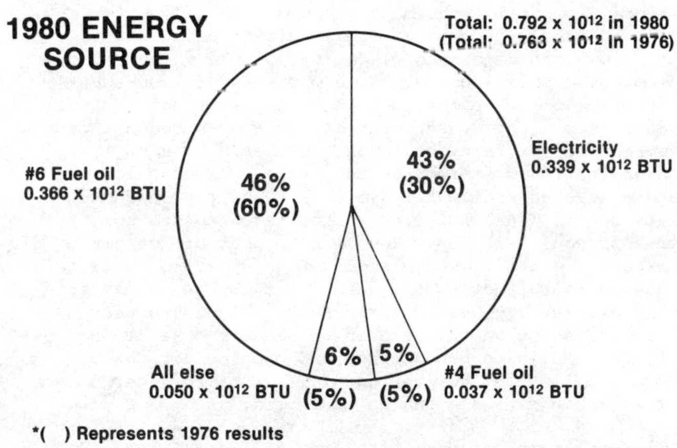

FIGURE 1

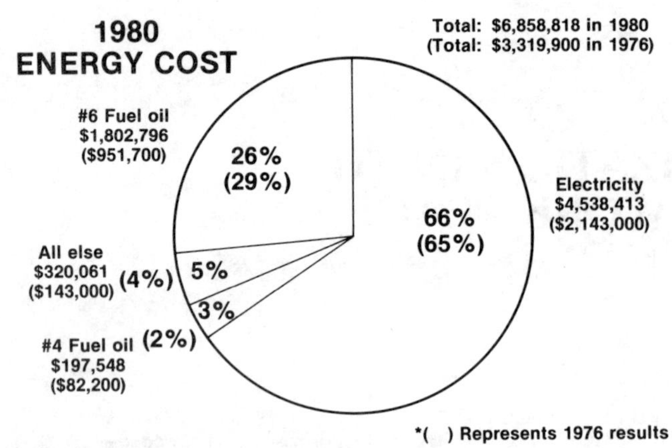

FIGURE 2

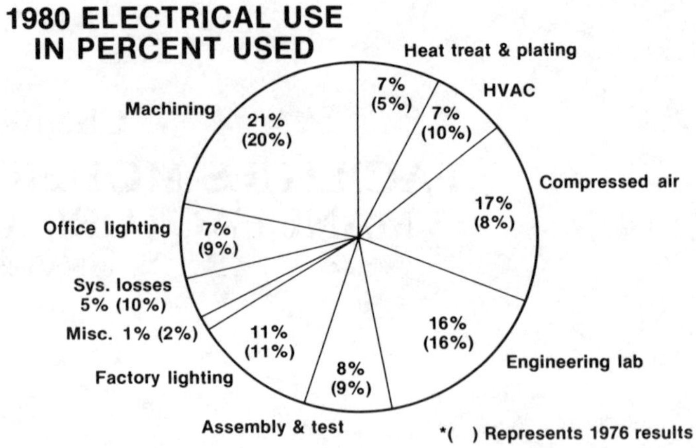

FIGURE 4

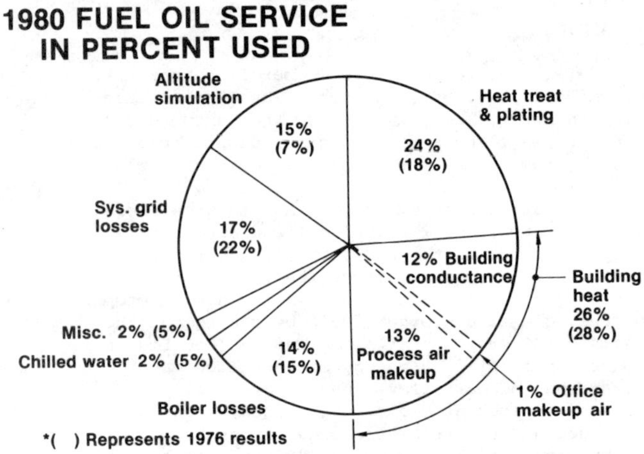

FIGURE 3

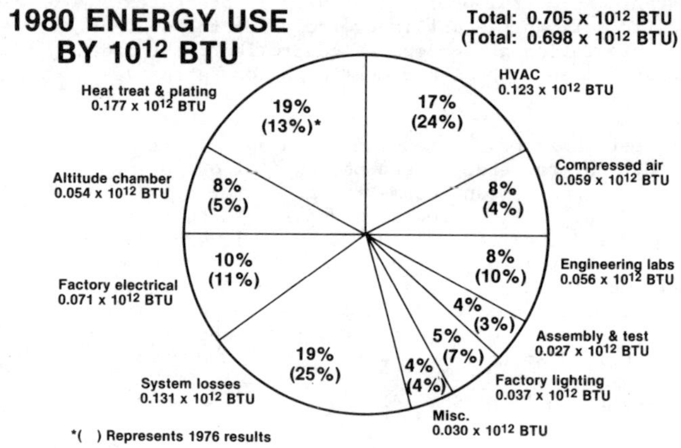

FIGURE 5

The next step is to find out where these two commodities are used. We chose the categories shown in Figure 3 to depict the major uses of #6 fuel, but this choice isn't magic. The important thing was to breakdown the uses into pieces of the pie which could be studied individually. This pie sort of shocked a lot of people when they saw that the major use of fuel was losses. The measurement of these uses doesn't require a lot of fancy equipment, either. Some of this determination was from records, some from calculations, and some even from putting a bucket under a condensate return line and using a stop watch to see how fast it would fill up. You'll note that office air makeup is a pretty small piece of the pie, so cutting back any more there won't gain too much. As a matter of fact, the aggravation felt by the Plant Engineer per BTU is just about as high as we dare make it in the office. Building conductance and process air makeup are fair targets, however and I'll describe later some of the things being done in that area.

The same thing was done with uses of electricity (Figure 4) and then both fuel and electrical energy combined gave us a pie which clearly showed what areas would probably be most likely to prove rewarding (Figure 5). We then set out to create projects in each of these categories and projected the savings we expected from each project.

We chose to relate energy use to sales dollars. Since sales are reported publicly on a corporate level, rather than by division, the dollars are omitted from Figure 6; however, we do deflate the dollars to 1972, and the figures show the results of efforts through 1980.

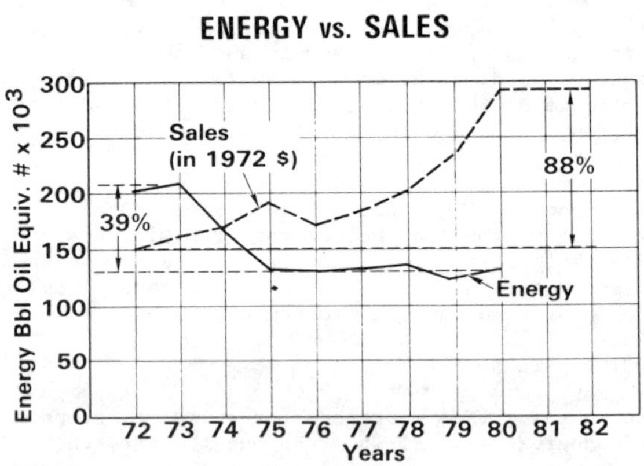

FIGURE 6

Integrating the two in Figure 7, we see that energy use per unit of output, with inflation taken out of the picture, has decreased dramatically. By the end of 1980, the energy use per sales dollar (in constant 1972 dollars) had been reduced by a few tenths more than 67%.

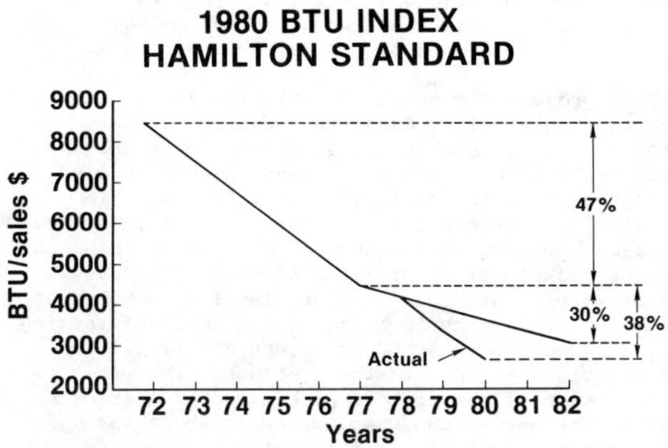

FIGURE 7

4. TECHNOLOGY

This section will describe several of the projects which are contained in the plan. In the area of system losses, one of the culprits turned out to be the common steam trap. First we outlawed the disc trap which, although it is neat looking, just doesn't last in service. We developed a training manual for our designers and maintenance people and provided them with an ultrasonic detector. By using this device, which allows the user to actually hear the trap operating or not operating, as the case may be, our steam trap maintenance program became much more effective.

We were than able to check the traps more quickly and assure that live steam wasn't finding its way to the atmosphere through flash tank vents. We calculate the annual savings to be about 1,500 barrels of oil. A visual check through a test cock is even better than the gun, but in most systems these test cocks are not available. We have installed visual test cocks on all of our steam traps.

One of the major actions taken by our Plant Engineer in the early seventies was shutting down the Power House on weekends. Many of his staff argued against this action, predicting leaks and other problems. As a former Marine Engineer, I must admit that I was skeptical, but we set about installing small electric steam generators in those places which absolutely needed steam on weekends, for instance to provide the necessary humidity for the environment in our space systems department clean room. In spite of all our skepticism, it has not created the serious problems we predicted. Even the Power House crew are now fully behind the shut-downs. We calculate the savings at approximately 2,000 barrels of oil per year.

We also found that we could eliminate some portions of the steam distribution system. A reducing station was installed in one of our buildings. It provides steam for the heating system, formerly provided by a 12" low pressure line which ran some 1,400 feet from the Power House. In addition to saving 500 barrels of oil annually, the old low pressure line is used to provide compressed air to our production test facility in this building. By using this line, we avoided the cost of a new line which was estimated at $257,000.

Also in the area of system losses, we began to question why our low pressure steam line had to run at 25 psi. With the cooperation of production personnel from the Conservation Team, we gradually lowered the pressure to 17 psi. The savings just in the line are about 1,000 barrels, not even taking into account the increased efficiency of Power House auxiliaries which provide this low pressure steam.

One of the pieces of pie in Figure 1 is labeled boiler losses, and naturally we took a hard look at our Power House. In addition to replacing the original boiler controls, we automated our boiler blowdown. Before we installed this system, we tested for dissolved solids and chemical residual in the conventional way, and adjusted our blowdown accordingly. This device continuously monitors the conductivity of the blowdown and adjusts the blowdown valve. We still test the water manually, of course, but adjustment of the valve is done continuously now, so we have more precise control. Now we don't have to blowdown an excess, which wastes boiler water and heat, nor do we run the risk of not blowing down enough, which, of course, is bad for the boiler. Annual savings are about 200 barrels, which is a small part of our fuel bill, but the device is not expensive.

We also installed viscosity measurement devices to keep the fuel at the correct atomization point and oxygen measurement to keep our excess air to a minimum. These devices cut our annual fuel use by about 1,500 barrels.

When we looked at building heating, we found about half of it was conductance, and since we have a fairly well constructed plant, I doubted that we could justify any changes. I was very skeptical about the fans people wanted to sell me - but they really work! Stand under them and you'd swear there was a heater overhead. We even got complaints that people were too warm. The destratification fans worked so well in bringing the heat from the truss level down to the floor that we got an outrageous idea. We installed a pipe about two thirds the length of a football field and a couple of feet in diameter in the trusses over our Heat Treat Department. It can draw in up to 17000 CFM of outside air. Holes on top of the duct distribute the air through the trusses and force the warm air down to the floor. The system has a set of dampers to allow it to modulate to recirculate inside air if it gets too cold in the trusses. Savings are in the neighborhood of 600 barrels per year. Since then we have installed several other such systems in areas where we had high enough ceilings to make it worthwhile.

Still in the heating area, our jet fuel controls are tested in an Engineering Test Lab. Because of the hazardous nature of this operation, we require 100% fresh air for this lab 24 hours per day, seven days a week. Right next to this lab, located in the basement are a number of compressors and transformer vaults. A duct now takes the exhaust air which is heated by the compressors and transformers and feeds it to the intake of the fuel lab's ventilation system. About 90% of the lab's air requirement is satisfied this way. The savings are about 1,000 barrels per year.

In our high temperature fuel lab we also had some success with modification of the ventilation system. Because of the nature of the testing, the control room is located in the center of this building, with the test cells along the outside. The same requirement for 100% fresh air exists here for the test cells where fuel is located, but not in the control room. We modified the system, with the approval of our Safety Engineering Department, and we now take in only 10% makeup air. Since this is a 24 hour per day, seven day a week requirement, the savings are about 300 barrels even though the system is not large.

The final project in the area of heating and ventilating is a paint spray booth used to paint propeller blades. Designed in 1947, this booth had three painting stalls for the three colors used, and two steam heated ovens to dry the paint between painting stations. The unit exhausted 24,000 CFM. Of this 6,000 was for the ovens, so this had to be heated even in the summer.

A newly designed booth sprays all three colors in one station, and makes use of infra red lamps for drying, which eliminates the need for ventilation. The exhaust requirement has dropped from 24,000 CFM to 5,000 CFM and the savings are close to 2,500 barrels per year. A major change like this in a process where there are customer specifications, Quality Control requirements and the like doesn't happen overnight, but persistence paid off, and I'm told that the quality acceptance rate is higher now than it was originally, which does not help our energy budget too much, but makes it easier to sell the next process change.

Our method of providing propane for process and test furnaces gave another opportunity for change. We purchase propane in liquid form and store it on our property. The facility holds 60,000 gallons of propane. The farm is designed to vaporize enough propane to satisfy our requirements in the summer simply by absorbing heat from the atmosphere. In the wintertime, however, steam coils had to be used to keep the system pressurized. We've changed the system now to a liquid pump to get the pressure required and vaporizers which will heat only the propane going to the factory. We'll save about 300 barrels of oil annually by getting rid of these big radiators.

And speaking of radiators, before we installed viscosity measurement devices, our two 75,000 gallon #6 fuel tanks were held at 120°F for proper flow of the oil to the Power House. These tanks got insulated very early in the game with a savings of about $18,000 the first year for a $12,000 investment. Probably the easiest appropriation we ever sold. Since we've installed the viscosity measurement system, we've dropped the storage temperature to about 90°F, also.

Our plating area was singled out as a significant user of energy. As such, it received special attention by the Energy Conservation Committee. For years we've been afraid to return condensate from plating for fear of boiler contamination. A few years ago we installed a condensate recovery system in one of our plating areas. This system continuously monitors the condensate's conductivity and sends it back to the boilers as long as it stays within set limits. Results to date are better than we expected, and it has saved about 500 barrels of oil annually. When we expand this to our main plating area, we'll multiply this figure by 5 or 10 times and also considerably reduce our flow to the Waste Treatment Plant, which won't hurt our relationship with the Department of Environmental Protection.

Our engineering studies,verified with the bucket of condensate I mentioned earlier, showed that up to 75% of the heat loss in our plating tanks was from the surface. We tried what the platers referred to as ping-pong balls first, but these were light and got sucked into vents or they eventually leaked and sank to the bottom of the tank. Our next attempt was with solid polypropylene balls which didn't sink.

A more permanent solution to the problem is now being fashioned by the use of fairly lightweight hinged covers formed of polypropylene. These covers will save up to $3,000 per tank annually when reduced exhaust is also allowed by our Safety Department.

The last project I'm going to deal with is our steam ejectors. When Walter Cronkite shows animated cartoons of a space ship course correction those little rockets that go "phfft" have to work precisely as programmed or the correction doesn't work. In order to test these attitude control systems, as they are called, we use an altitude chamber in which we can fire the rockets at simulated altitudes of 100,000 feet or more. This is a pressure level of about one millimeter of mercury absolute. 3500 hours of testing per year uses about 10,000 or 12,000 BBL of oil to make the steam to operate the ejectors. The products of combustion of the rockets present severe corrosion problems for any vacuum pump, which is why steam ejectors have been used in the past. With increased cost of fuel, however, a number of companies have developed vacuum pumps which can withstand the severe conditions, and we have made a number of proposals to replace some of the stages of this ejector. The mechanical pumps are, of sourse, expensive, but we see a potential savings of more than $200,000 per year in this project.

Finally, when all of these energy use areas are made efficient, we believe they should be optimized with a system which uses the latest electronic technology. We call our system an Energy Management System.

Back in the early seventies, before the oil embargo, I had just joined the Plant Engineering Department. The Plant Engineer put me to work going around the plant determining which systems might be connected to a computer which could be programmed to start, stop, report and otherwise control these systems. Several people wondered at the time about this effort, since computerization looked so expensive, and fuel costs were relatively low at the time. When energy costs shot up so rapidly, we were ready, and by drawing on the expertise of our Electronics Department, we were able to create what is probably the most glamorous of our projects, and certainly the one which has had the most influence over the plant.

The objective of our Energy Management System is to provide the most efficient management of utilities to best utilize manpower, to conserve energy and to provide incipient failure recognition. In other words, its objective is to save money. In addition to "fine tuning" our operations, we can anticipate maintenance problems in a more timely fashion and avoid the cost of breakdown or replacing parts too soon. The method chosen to reach our objective was to centralize monitoring control to the area of our Power House where our main energy sources are controlled. Our plan was to optimize the heating, ventilating and air conditioning systems, steam generation, compressed air generation, and electrical energy demand. We also monitor the distribution and usage of steam, electricity, compressed air, and the effluent of our Industrial Waste Treatment Plant.

The key to the savings of the Hamilton Standard Energy Management System is in the savings of utility cost, rather than in the elimination of people, although the system certainly allows people to be better utilized. Our experience at Hamilton Standard indicates that a 2-3 year payback is typical of this type of system.

Hamilton Standard personnel designed and built the necessary hardware, and wrote the software application modules necessary to do the job. The more significant of these modules are:

- Electrical Load Shedding
- Electrical Load Cycling
- Optimized Equipment Start/Stop
- Building Precooling
- Zone Temperature Night Setback
- Enthalpy Damper Control
- Hot/Cold Deck Temperature Optimization
- Chilled Water Temperature Optimization
- Inventory Control
- Equipment Trending

5. THE BOTTOM LINE

As a result of the planning and technology described above, our present utility costs, while higher than in 1972, have increased at a much lower rate than the unit costs of all forms of energy. Our use of fuel oil has decreased by 77,000 barrels annually (over 50%) and our electrical use has remained approximately the same. During this time our sales have increased by almost 90% with inflation removed. The annual savings to our division is between nine and ten million dollars.

Our Five Year Energy Plan has been used as a model for all other Divisions of United Technologies, and our efforts earned an award for the "Best Energy Conservation Performance of all Divisions" of United Technologies. Our Electronics Department is now designing and installing Energy Management Systems modeled after ours in other Divisions of the Corporation. These efforts contributed to the award given to Hamilton Standard by the American Institute for Plant Engineering for the "Most Outstanding Plant Engineering Program in the United States."

We are convinced that the use of the two management tools - planning and technology - has rewarded our Division and Corporation and can be of great value to other enterprises who are willing to undertake a comprehensive Energy Management Program.

SECTION 25
POWER GENERATION AND FUELS UPDATE

Chapter 74

ELECTRIC POWER FROM HOT DRY GRANITE - FIRST COMMERCIAL PROGRAM

J. W. Foster, Sr.

BACKGROUND INTRODUCTION

The largest heat energy resource available to mankind is that of the earth itself. Temperatures increase with depth and are about 7000° C. at the earth's core. Fortunately for us all even these temperatures did not sustain a thermo-nuclear reaction at the earth's beginning, as did the sun. Even at a mere five to ten kilometers beneath our feet the rock temperatures are 480° F. The shallower half of these depth-ranges is well within modern drilling access.

A tiny fraction of the total geothermal resource is extractable via naturally ocurring deep ground water, a classic example being the Geysers area in California where more than 900 MW is presently generated from steam. However, the great mass of geothermal heat is trapped within dry crystalline rock containing no natural extractable fluid. This is HDR energy, i.e., hot dry rock.

There is in our time, an urgent and strategic need to prove that under today's technology heat energy in quantity can be produced from deep dry rock for conversion to electric power at commercially competetive costs. The Atomic Energy Commission, ERDA and DOE, recognizing the enormity of this resource base, have spent more than $60 million over the past eight years toward academically oriented research in HDR technology. Dry rock heat has been proven to be extractable at The Los Alamos National Laboratory. We intend to show that the Government's results can be enhanced on a scale of 16 and perhaps 30 to 1 and thereby attain economic viability at a level attractive to industry.

Nearly all American states and most countries have areas of suitable geology for HDR energy production. Under the Corporate design and patented procedures the ideal medium for the construction of the deep-earth heat exchange "oven" is considered to be granite. The "oven" serves for the transfer of heat from in-situ rock to extractable fluid. Fortunately, granite plutons and batholiths (younger than pre-Cambrian) are not only hotter than most crystalline rocks but they occur in abundance in wide geographic distribution.

HDR Energy Development Corporation is instigating the world's first prototype project for the commercial production of HDR energy, currently targeted for construction in the massive granite of the White Mountain volcanic area of New Hampshire. If the design is successfully executed, electricity at the prototype site can be generated at a busbar cost under four cents/kwh based upon a three earth-loop complex. If levels of energy production even approach the calculated expectations, a new billion dollar industry will have been born. We think that we have the design which unlocks dry rock heat to commercial requirements.

PHYSICAL SYSTEM OF CLOSED LOOPS

The Earth Loop

Because massive granite has extremely low effective permeability, particularly under the compressive stresses of great depth, the principal components of a heat extraction system must include an injection well, production well, a man-induced fluid and a very large transmissible oven in the geothermal zone. The Corporation's patented procedures relate solely to the construction of the oven, specifically, unique fracturing techniques which are thought by ourselves and many third-party scientists to be the key to the building of a commercial-size rock to fluid heat exchange complex. These procedures are described in some detail later in this paper.

Our particular design for the closed earth loop specifies that the injection well, drilled vertically to a rock temperature of about 200° C., be deviated to 45° and thereafter drilled through a mid-oven temperature of 250° C. The production-well is an image of the injection well except that the deviated length of the borehole lies approximately 457 m. shallower than the deviated length of the injection well. Both deviated boreholes lie in the same general vertical plane, with generous tolerances for imprecision in the execution of deviated drilling. Each well is lined only in the vertical section with 11 3/4 in. API 60 lb. casing. The deviated boreholes are turbo-drilled with 9 7/8 in. bits and left unlined.

The oven, or the body of granite serving for heat transfer, lies between the two parallel deviated boreholes and consists of a huge complex of man-induced in echelon fractures.

If constructed to design, the total fracture area amounts to about 7.5 million m^2, which equals 15 million m^2 of exposed rock face for rock to fluid heat transfer. This is the size oven necessary for long-life delivery of the heat energy we intend to extract, 356 million Btu/hr.

The heat exchanger at ground surface closes the hydraulic circuit of an earth loop, and a battery of heat exchangers can serve several earth loops energizing one power plant.

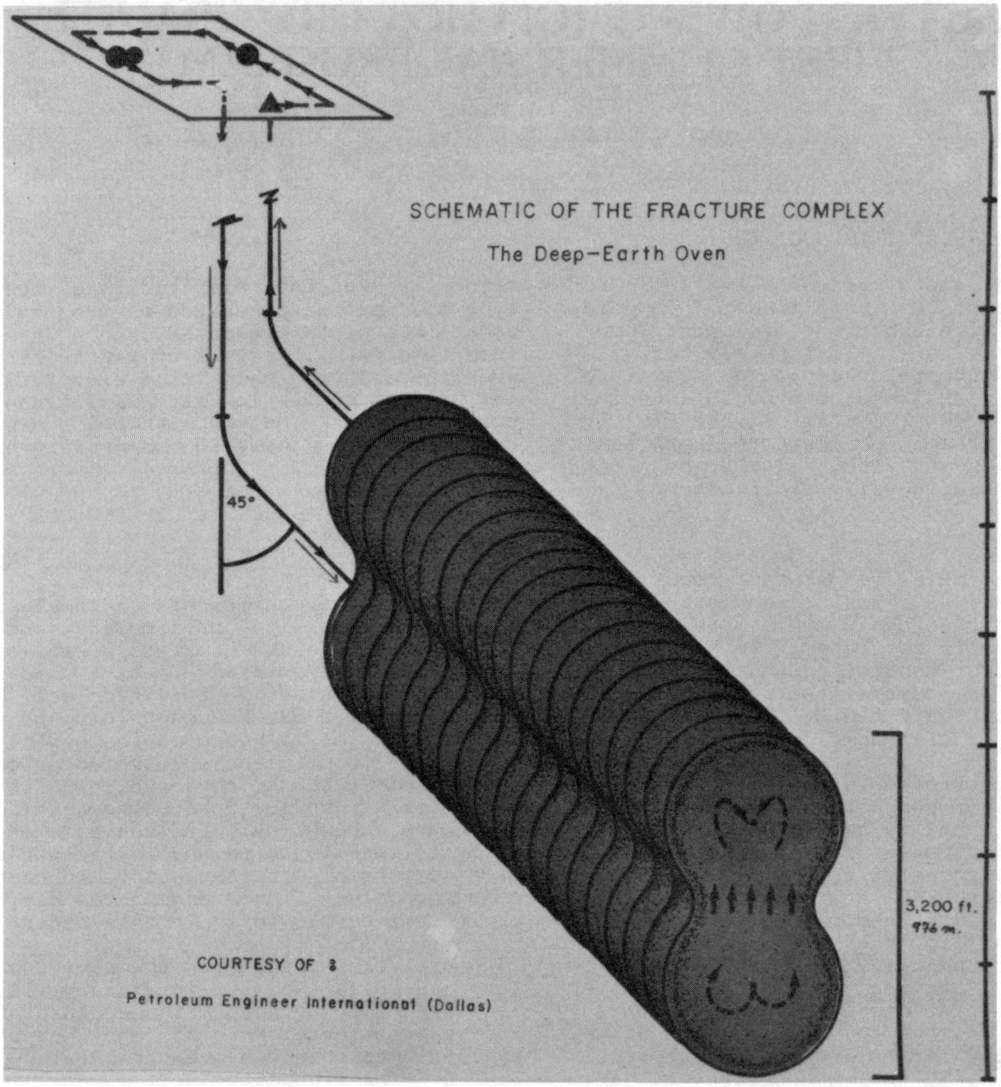

FIGURE 1. A RENDERING OF AN IDEALIZED DEEP-EARTH HEAT EXCHANGE COMPLEX. THE 18 REPETITIVE FRACTURES CONSTITUTE A ROCK-TO-FLUID CONTACT AREA OF APPROXIMATELY 15 MILLION SQUARE METERS.

Energy Conversion Loops

Under a binary-fluid configuration the principal top-side loop runs an organic fluid from the secondary of the heat exchanger to vapor-driven turbo-generators to a water-cooled freon condenser and back to the heat exchanger, a closed circuit (1). A tertiary fresh-water loop passes cold water through the condenser, dissipates its Btu's to the maximum beneficial use at a given site area and returns to the condenser. In middle and higher latitudes this by-product heat can be dissipated effectively at least six to eight months of the year through large greenhouse facilities, industrial heating, fish farms and even bathing spas. These options are site specific and need not be dwelled upon in this paper.

The possibility exists that low pressure reciprocating steam engines or steam turbines could be shown cost-effective in converting HDR energy to mechanical torque. The closed earth loop could be maintained through a water-to-water heat exchanger yielding steam for these less sophisticated units.

THERMO-DYNAMICS OF THE EARTH LOOP

Design of the earth loop targets a circulation rate of 150 kg/sec. of pressurized fresh water at a delivered temperature of about 240° C. This translates to 1.188 million lbs./hr. With a heat exchanger designed for a pinch-point of 10° C. the net yield is 356 million Btu/hr. or 104 MWt.

If one could assume zero hydraulic impedence in the deep oven, which one cannot, calculations indicate that the loop's free-flow production would exceed 150 kg./sec. simply from the head difference between the cold injection well and the hot production well. However, we must presume the necessity of applying substantial hydraulic pressure at the well head on the injection side of the loop. This injection pressure serves three functions, (1) to partially inflate the oven fractures in accordance with the laws of elasticity, (2) to negate pressure loss associated with hydraulic impedence in the oven and (3) to permit back-pressure on the production well to prevent flashing. We anticipate an injection pressure of about 1500-2000 p.s.i. but will be prepared to test the system at higher injection pressures if required. Every earth loop will have its own signature in respect to optimum injection pressures because oven transmissivity will be peculiar to any single earth loop.

The size of the fracture complex at depth required for a minimum 20-30 year life with less than 10% temperature decay is a calculation involving the volume-rate of heat extraction and the efficiency of the buoyancy sweep of fluid across any one fracture. The Los Alamos National Laboratory's published conclusions indicate that about 50,000 m^2 of fracture in granite rock should support one kilogram per second of fluid extraction at our intended energy levels. Our design yield of 150 kg/sec. therefore requires about 7.5 million m^2 of fracture (15 million m^2 of transfer surface). The oven as designed and illustrated here meets those size requirements, partly as a result of the proprietary procedures on fracture propagation.

PROPRIETARY PROCEDURES - ROCK MECHANICS

Granite can be fractured under the controlled application of confined hydraulic pressure. Rock fails at that pressure which overcomes the horizontal component of the in-situ compressive stress plus the tensile strength of the rock. If our mid-oven depth is, say, 22,000 ft. (6.7 km), a gross hydraulic pressure of about 15,000 p.s.i. will approach the level of rock failure, about 65% of this pressure deriving from simple hydrostatic under clear water conditions. With weighted mud the hydrostatic p.s.i. at the fracturing depth can be readily elevated to about 12,500 pounds, if necessary, reducing the net applied pressure to only 2500 p.s.i. The required pressures are well within the capacity of hydraulic pumps used routinely by the oil industry for fracturing operations.

Straddle-packers will be used. These devices, now manufactured and tested for use in high temperature environments, enable the application of gross hydraulic pressure at a specific confined zone, perhaps only three meters of borehole length.

Finally, as introduction to the pertinent rock mechanics of fracturing operations, the plane of induced fracture lies normal to the plane of least principal in-situ stress. At depth the plane of least principal stress is horizontal, and, therefore, the induced fractures are vertical. Natural preference for verticality, in fact, increases with depth. This phenomenon is the prime reason for our selection of a site, for the prototype construction, where the geothermal gradient is moderate, where rock in the 250° C. zone is relatively deep, at least 18-20,000 ft (5.5-6.1 km).

Prime Patent

U.S. Patent 4,223,729, by the author of this paper, issued in 1980, is simple in format but complex in terms of rock mechanics and the laws of elasticity. In a nutshell the procedures are as follows:
(a) Construct both dry boreholes in the deviated configuration previously described, the deviated portion of the production well lying about 450 m above and parallel to its brother. (The configuration itself is not proprietary).
(b) Insert two straddle-packers, one in each borehole, positioned so that the packers lie in the same vertical plane.
(c) Induce two small fractures, each approximately vertical, by the hydraulic pressurization of the two packers. These fractures need not occur at the same instant, and they are not expected to interconnect. Induced pressure is lost when the fracture events take place.
(d) Re-pressurize, by independent hydraulic systems, both fractures simultaneously and thereby cause both fractures to propagate. Growth of the penny-shaped fractures will, of course, occur in all directions in the fracture plane.
(e) Continue independent propagation of both fractures. As the tips of the fractures reach a certain proximity, one to the other, there shall occur a coalescence of the stress fields and, in a sense, one fracture begins to "feel" the presence of the other.
(f) Upon significant coalescence of the stress fields off the tips of the two fractures, the stresses actually embrace, and the intervening rock shall promptly fail. At that instant the two fractures become one, and hydraulic communication borehole to borehole is accomplished.
(g) Repeat the procedures at 17 up-hole positions at pre-selected intervals, thus achieving an enormous fracture complex. (3).

The heart of patent's validity is the nature of the stress-field (5) off the tips (in full circle) of a fracture undergoing propagation, as in Figure 2. Also, the unique change in a stress-field's geometry once it senses its counterpart in coalescence is highly significant. It may be analogous to metal fatigue where, for example, cracks from adjacent boltholes as in an engine mounting can grow and ultimately interconnect with catastrophic results.

A secondary U.S. Patent, co-authored by J.W. Foster and Clarence R. Jones, P.E., relates to the application of a sudden pressure surge simultaneously into the packers for the control of the timing of the final fracture event.

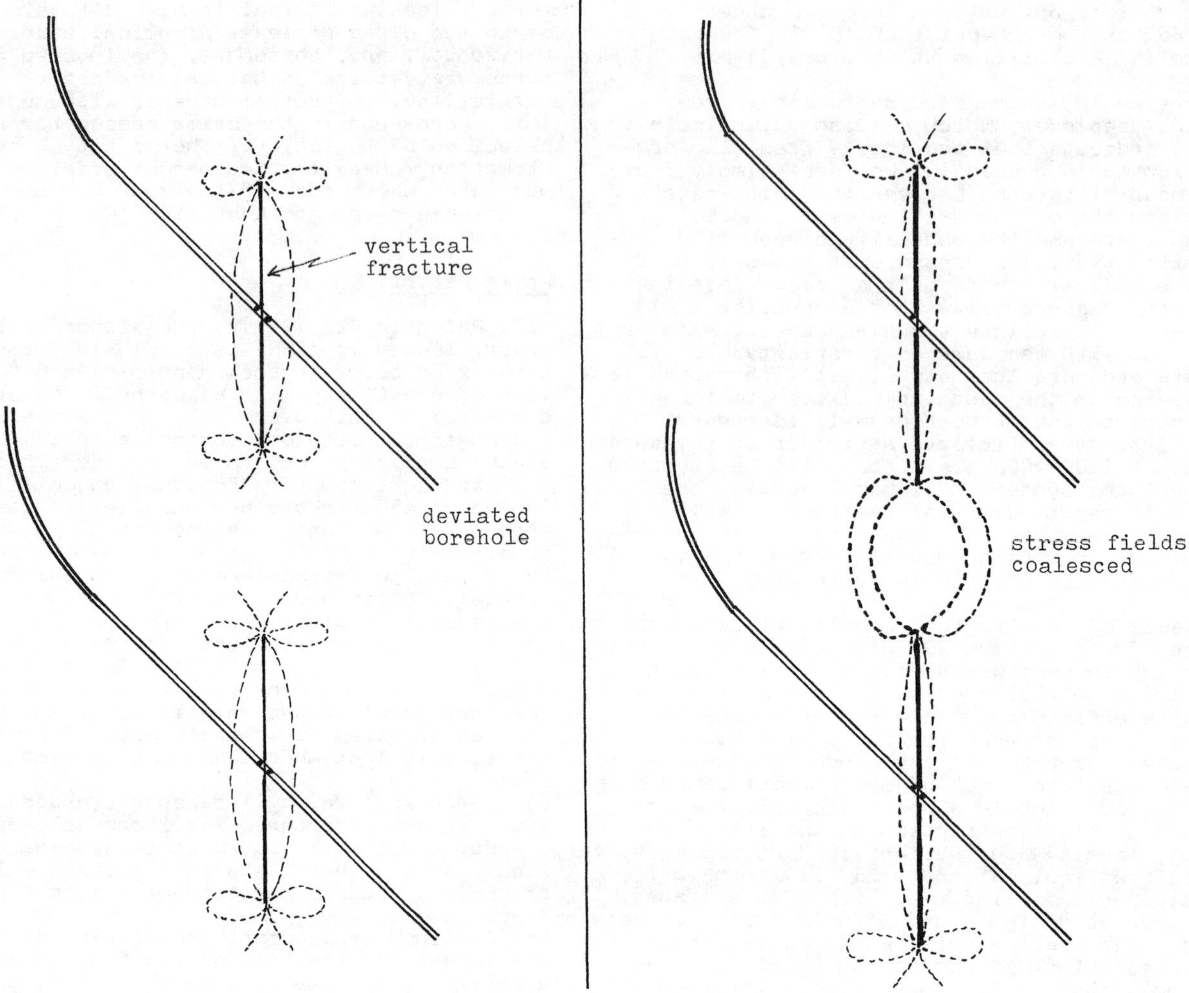

FIGURE 2. SCHEMATIC SHOWING STRESS-FIELD PATTERNS OFF THE TIP FRINGE OF TWO PENNY-SHAPED FRACTURES IN ROCK DURING THEIR SIMULTANEOUS PROPAGATION BY INTERNAL HYDRAULIC PRESSURE. THE EMBRACE OF TWO STRESS FIELDS (RIGHT) SHOWS THE PATTERN AN INSTANT BEFORE FINAL ROCK FAILURE WHEN THE TWO FRACTURES BECOME ONE, THUS PROVIDING POSITIVE HYDRAULIC COMMUNICATION BETWEEN BOREHOLES.

ADVANCED DRILLING TECHNOLOGY

Since the drilling of the two boreholes represents the largest part of the construction budget there is great interest in new technology which portends very significant savings. Future prospects for economic HDR energy is therefore considerably enhanced.

Even in the first prototype operation we anticipate cost-effective use of new hard-rock bits, the turbo-drill, and possible the Terra-Drill.

Modern dragbits, some with angled poly-crystalline diamond cutters, are likely to prove the more cost-effective in the coarse-grained Conway granite of New Hampshire. These bits avoid the waste of energy associated with tricone bits which grind rather than rip. Dragbits maximize the shearing weakness of coarse rock and permit relatively large cuttings to be recovered.

Maurer Engineering Inc. of Houston has developed a high-temperature turbo-drill specially suited to geothermal operations. All deviation drilling in the prototype construction is expected to be executed at 9 7/8 in. with this turbo, operating at 200-450 r.p.m. We also intend to test turbo-drill performance in the 14 3/4 in. vertical drilling, weighing the higher rates of penetration against the shortened bit life of high speed bit rotation. This trade-off is applied most critically in the greater drilling depths where round-tripping for bit changes becomes very costly and where bit life becomes more important.

The Terra-Drill is cited favorably in Dr. William Maurer's book "Advanced Drilling Techniques". (2). This patented system, owned by Tround International, is predicated on the impact of three or more self-destructing ceramic projectiles at micro-second intervals, fired through the bit and into undisturbed rock at bottom hole. Tests of Terra-

Drill hardware and performance conducted by Sandia Laboratories (4) indicate that hard marble is shattered to a depth of 12-15 cm. below the bit, and further tests by Tround International in Barre granite confirm similar success in this volcanic rock. The potential advantage of this advanced hardware lies largely in longer bit life in that, by firings every 30 seconds, the bit is continuously cutting rubblized rock rather than the fresh undisturbed face at bottom-hole.

While the Terra-Drill technology and hardware remain to be tested in a deep, rigorous rock environment, there is clear possibility that the costs of making hole can be substantially reduced over conventional budget.

The hope and, perhaps, the expectation is that in a matter of a few years a 30,000 ft. well in granite can be drilled for the same cost as a 22,000 ft. well in 1981, particularly likely if there is strong industry incentive to develop the technology which is on the brink of availability.

GEOLOGICAL CONSIDERATIONS

Until HDR energy is a more established industry very special criteria should be applied for site acceptability. Eventually these will be relaxed as confidence grows from experience. The first prototype operation is targeted for a site which appears to meet the following conditions - (1) occurrence of massive granite at a temperature-depth of 250° C., (2) a benign tectonic history in modern time, (3) geology which permits open-hole drilling below shallow surface pipe, thus saving the cost of a first casing string, (4) heat-flow values of 2.5 to 3.5 calories/cm^2/sec. (equating to a geothermal gradient in granite of about 35-50° C/1000 m at shallow to intermediate depths), (5) a supply of fresh water of at least one million U.S. gallons/day (3,785 m^3), and (6) a heat flow value which derives entirely from primal conductive heat or from radiogenic sources or a combination thereof, not, in other words, from convection.

A principal concern is the avoidance of geological conditions which can escalate drilling costs by exposing the operation to potentially serious problems down-hole. This concern develops the criteria for benign tectonics, thus avoiding excessive horizontal stresses at depth, and no convective heat which risks lost circulation problems and other difficulties.

We intend to undertake a small test fracture at the bottom of the first vertical borehole and do so primarily to establish the azimuth of the preferred fracture plane. This plane will be parallel to the plane of least principal horizontal tectonic stress. Deviation is then executed 90° to the azimuth of the test fracture.

THE SITING OF THE PROTOTYPE PLANT

Our Corporation has studied potential site areas in more than 30 American states. At a glance one might conclude that Western regions would be attractive and indeed they are. But when all special criteria for Operation No. One are considered we believe that the Conway granite of the White Mountain batholith in New Hampshire best fits. A more ideal site may exist, but we have not identified it.

The Conway granite is believed to offer sites where its massive roots extend to at least 8000 m., has a long history of benign tectonics, is exposed at ground surface, and has a measured heat-flow value of about 2.7 at shallow depths. Large capacity water wells can be installed in valley alluvium without excessive State restrictions and, not least, the Project is warmly received in prospect by area residents and political dignitaries.

About half of the moderately high heat flow in the Conway granite derives from above-normal conductive, primal sources, perhaps as a result of an anomolous magma plume at great depth. Heat flow values are further enhanced by radiogenic sources, the Conway containing widely disseminated uranium (± 16 ppm), thorium (± 60 ppm) and the isotope of pottassium. No significant heat in New Hampshire is known to result from deep convective ground water.

RESOURCE BASE - WORLDWIDE GLANCE

If, as many scientists believe, HDR energy becomes a booming new industry, increasing attention will be given by researchers toward improved knowledge of the plutons and batholiths of the world. Since the oil industry has historically avoided crystalline rock, at lease until recent interest in overthust belts, the granites of the world remain virtually unpenetrated in depths of significance to HDR energy.

Every major continent displays plutons and batholiths, some being very old pre-Cambrian and others much younger. Since the half-life of uranium is a thousand million years, youthful age is not a prime consideration, although the very old granites seem not as hot as some younger ones. For example, the Conway granite of N.H. is Jurassic-Cretaceous, about 100 m.y.o. Devonian granite in the same region yields very little radiogenic heat.

European geologists particularly have long had interest in deep magma plumes and regions of subduction where above normal heat flow values are common. In those many regions of the world with conductive heat-flow values of ~ 2.5 in granitic rock, radiogenic heat is not a needed supplement.

The fact remains that HDR energy sources worldwide, within accessible drilling depths by 1981 technology, constitute a heat resource base of staggering dimensions. Any area of a size 10 km x 10 km, which meets the essential criteria, can in theory support energy production equivalent to several thousand megawatts of power, yet actually occupy only a few tens of hectares of precious land surface. However, we envision that electric plants based upon HDR energy will tend to be of modest size, 40-200 MW, and widely scattered across a power network, thereby reducing the transmission costs associated with very large nuclear or fossil-fuel power plants.

CONSTRUCTION COSTS - THREE EARTH LOOPS NEW HAMPSHIRE SITE

This budget is based upon the best advice available in America and presumes no spectacular successes with very advanced drilling techniques. Penetration rates and bit life are derived in part from the records of the Los Alamos National Laboratory's 1980 drilling in crystalline metamorphic rock, Fenton Hill, N.M. No drilling budget can allow for catastrophic disasters such as unrecoverable lost hardware at great depths. We allow funds for preventive measures, e.g., frequent sophisticated examination of drill pipe for metal fatigue.

Earth Loop Alpha	1981 $ x 1000
Geophysics, gravity survey--------$	200
Site acquisition, preparation and mobilization----------	1,820
Long-lead orders------------------	5,720
Injection well drilling-----------	2,928
Production well drilling----------	2,184
Miscellaneous direct costs (fuel)-	1,611
Fracturing operations-------------	2,058
Circulation-production testing----	275
Engineering management and Operators functions-------------	4,000
Weighted contingencies------------	2,694
	$23,490

Several substantial items in this first-loop budget will serve any subsequent loops at the same site area. For this reason and allowing for additional savings stemming from experience, Earth Loops Beta and Gamma are each budgeted at about 35% less than Alpha Loop. We believe, furthermore, that once the results of the first project encourage the launching of a second generation of projects the cost of these will be on the order of $40 million per three-loop installation. Proof of effectiveness of ultra-advanced drilling techniques, such as the Terra-Drill, could eventually reduce costs by several million dollars.

EQUIVALENT ENERGY VALUES

These calculations on comparative energy demands presume, for the HDR system, a net yield of 104.4 MWt/earth loop, 21% conversion efficiency, 22 MWe/earth loop, and an installation where three earth loops energize one power plant producing 66 MWe.

Oil Fuel Equivalents

1 Kwh requires 10,500 Btu (high efficiency plant)
1 bbl. oil = 6,000,000 Btu
therefore, one megawatt hour requires 1.75 bbls/hr.
66 MW requires 1,011,780 bbls/year.

Gas Fuel Equivalents

1 Kwh requires 10,000 Btu (high efficiency plant)
1 MCF gas = 1,000,000 Btu
therefore, one megawatt hour requires 10 MCF/hr.
66 MW requires 5,781,600 MCF/year

Coal Fuel Equivalents

1 Kwh requires 10,500 Btu (high efficiency plant)
1 ton coal - 25,000,000 Btu
therefore, one megawatt hour requires 0.42 ton/hr.
66 MW requires 242,827 ton/year.

If an HDR energy production system with a gross output of 66 MWe costs $50 million and under an operating design which is highly automated, it is readily seen that the capital cost of installation is amortized in less than two years if compared with, say, oil fuel at $30 per barrel.

ECONOMICS

Cost Per Annum	Total System Cost	
	$80,000,000	$90,000,000
Internal power (4 MW)	1,401,600	1,401,600
Operations and maintenance	975,400	975,400
Depreciation on power plant 10 year straight line at $40 million	4,000,000	4,000,000
Return to total capital at 15%	12,000,000	13,500,000
Total	$18,377,000	$19,877,000

Kilowatt Hours of Production

549,252,000

Busbar Cost

	3.35¢	3.62¢

Gross Annual Revenue

Installed capacity-----------------66,000 KW
Utilization factor------------------95 percent
Kilowatt hours--------------------549,252,000
Sale price--------------------------$0.077/kwh
Gross revenue---------------------$42,292,404

ENVIRONMENTAL ASPECTS

The Los Alamos National Laboratory has declared that HDR energy is considered to be the most "innocuous" of all the alternate energy sources. That description derives from the Laboratory's empirical findings and theoretical analyses. Heat extraction at the Phase I earth loop, Fenton Hill, has revealed no seismic events related to the fracturing or rock contraction. (6).

At ground surface the configuration of closed loops assures no contamination of streams or soil, and well construction under the Corporate design precludes effect on shallow ground waters by the man-induced geothermal fluid, which is itself fresh water. It would appear that, in summer, cooling towers may be needed for supplemental dissipation of byproduct heat, which would result only in a discharge of clean water vapor to the atmosphere.

POTENTIAL IMPACT ON ENERGY STRATEGY

The significance of the introduction of HDR energy onto the world scene can be measured only by the rate of growth of this new industry and by the type power plant which it could make redundant. In America's New England, for example, where imported oil is still a major source of energy for electric power plants, a vigorous development of HDR energy would have marked effect on imports and in time would negate the necessity.

We think the nuclear power industry would be effected only in that starts of new plant construction could eventually be reduced. Given the cost of nuclear plants in 1981, it could hardly be expected that any would be declared redundant once built.

Regions and countries which should have the keenest interest in the evolution of HDR energy are surely those which are remote from oil, gas and coal resources, which is one consideration favoring the Corporate endeavor in New Hampshire as site-area number one.

The drilling industry in our time is ill-prepared, in respect to available equipment, to respond immediately to a logarithmic increase in demand for big rigs - but respond they would in due course.

If Earth Loop Alpha meets basic expectations, the first project will spawn five new programs and each of those will spawn yet another generation of facilities. It is not precocious to envision the reality of 70,000 megawatts of power from HDR-based electric plants in America by the end of this century - or approximately today's output of the Country's nuclear facilities.

REFERENCES

(1) S.L. Milora, J.W. Tester; Geothermal Energy as a Source of Electric Power, The MIT Press, Cambridge, 1976.

(2) W.C. Maurer; Advanced Drilling Techniques, Petroleum Publishing Co., Tulsa, 1980.

(3) J.W. Foster; Geothermal Projects to Use Modern Drilling, Frac Techniques, Petroleum Engineer Int., Dallas, Sept. 1980.

(4) M.M. Newsome, R.L. Alvis, D, Dardick; The Terra-Drill Program, Sandia Laboratories, SAND 76-0228, Albuquerque, 1976.

(5) I.N. Sneddon, M. Lowengrub; Crack Problems in the Classical Theory of Elasticity, John Wiley and Sons, Inc., New York.

(6) G.M. Cremer et al; Hot Dry Rock Geothermal Energy Development Program, Annual Report FY 1979, Los Alamos Scientific Laboratory, 1980.

Chapter 75

SYNTHETIC FUELS FROM COAL - ENERGY IN TRANSITION

V. K. Gupta

ABSTRACT

The world's supply of oil and natural gas is being depleted at an alarming rate, while the demand for these fuels remains constant or is increasing. Thus alternate ways of obtaining fuels must be secured if our energy-consuming and highly mobilised society is to continue to exist at its present level. Since coal makes up majority of the fossil fuel reserves in this country, far outweighing the known reserves of petroleum and natural gas. Unfortunately the most abundant fossil fuel the coal is of the form which makes it difficult to distribute cleanly and efficiently. In addition the coal often contains high level of pollutants which make its direct use as environmentally damaging. Coal conversion is the process of making clean burning and easily handled gaseous, liquid, and solid fuels from coal. While coal conversion goes back many years, only in the past few years has coal conversion become a technologically viable alternate to the use of petroleum and natural gas. The basic concept of coal conversion chemistry is to enrich the raw coal with some form of hydrogen to produce hydrocarbon products with relatively large hydrogen content as compared to coal. This can be accomplished by reaction of coal with hydrogen or steam at high temperatures and high pressures in the presence of catalyst in some cases or by removing excess carbon from coal to give hydrogen rich carbonaceous byproduct. Coal conversion processes can be divided into two categories, gasification and liquefaction. These two types of processes produce much overlapping, and many systems produce both gaseous and liquid products. These processes can be further classified on the basis of type of reactor, residence time, operating pressures, and the nature of the product like low Btu, medium Btu, and high Btu pipeline gas. Several such processes have been commercially developed or are being developed and many more are in the pilot plant or demonstration stage of the development. This paper will describe the status of coal conversion technology into synthetic fuels from technical, commercial, and environmental point of view. Future prospects of this technology being commercialised will also be discussed.

INTRODUCTION

The energy crisis in U.S. was recognised about ten years ago, yet the crisis is felt only at times of occasional symptoms (the oil embargo of 1973, the natural gas shortage of 1976-77, the long gasoline lines of the summer of 1979, and the steeply rising prices of gasoline during 1979-80). As the symptom is relieved, the public sense of crisis fades away, and the future crisis however remains to be solved. The diagonisis of the U.S. energy crisis is simple: the energy demand is increasing while the supplies of oil and natural gas are diminishing. Total energy production in U.S. in 1978 was about 3 percent less than in 1972, the year of peak production, whereas in the same period the energy consumption rose by 9 percent. The difference is being made up by increasing the oil imports at continually rising prices. The imports now provide about half of all the oil consumed in U.S. at a total cost of about 45 billion dollars. In 1978 oil and gas accounted for 74 percent of U.S. energy needs, whereas coal contributed only 18 percent as pointed out in the energy consumption data (1) given in table 1.

TABLE 1. U.S. ENERGY CONSUMPTION IN 1978

Energy Source	Consumption Standard Units	Consumption Quads	Conversion Factor[a]
Coal	623.5 million short tons	14.09	44.3 million short tons
Natural gas	19.41 trillion cubic feet	19.82	0.79 trillion cubic feet
Petroleum	6838 million barrels	37.79	181 million barrels
Hydropower	301.6 billion kilowatt-hours	3.15	95.7 billion killowatt-hours
Nuclear power	276.4 billion kilowatt-hours	2.98	92.9 billion kilowatt-hours
Geothermal and other[b]	3.3 billion kilowatt-hours	0.07	46.3 billion kilowatt hours
Net imports of coke	5.0 million short tons	0.13	38.5 million short tons
Total[c]		78.01	

a. Values are equivqlent to one quad.
b. Includes wood, refuse, and other organci matter burned to generate electricity.
c. Details do not add to total due to rounding.

Recent trends of energy demand and the dependence of U.S. economy on imported oil have caused serious concerns. Long term interruption of oil and gas supplies could cause serious damage to our industrial output, employment, and even endanger national security. For these reasons, industrialised nations have been looking for alternatives to oil and gas in order to make them energy independent. In addition the data on proven reserves from F.E.A. Project Independence Report do not indicate long term availability of oil and natural gas. It is clear from the data given in table 2 that U.S has abundant coal reserves that would last a long period of time under the current levels of usage. When the same data is displayed in the form of a pie as shown in figure 1, it is clear that 90 percent of the energy that is available in the form of coal provides about 18 percent of the needs. Thus coal offers a source of energy to be used in transition period while alternate technologies such as fusion or solar are fully developed and realised.

TABLE 2. PROVEN RESERVES FROM F.E.A. PROJECT INDEPENDENCE REPORT

Source	Fuel Units	Quads	Years left at 1972 consumption levels
Coal			
high sulfur (more than 1 percent)	273 billion tons	6908	
low sulfur (less than 1 percent)	160 billion tons	3838	
Total	433 billion tons	10746	823
Oil			
lower 48 (crude)	30 billion barrels	176	
natural gas liquids	6 billion	37	
Alaska	10 billion barrels	59	
Total	46 billion barrels	272	8
Gas			
lower 48	218 TCF	225	
Alaska	32 TCF	32	
Total	250 TCF	257	11
Shale	20-170 billion barrels	116-986	3-28
Tar Sands	29 billion barrels	168	28

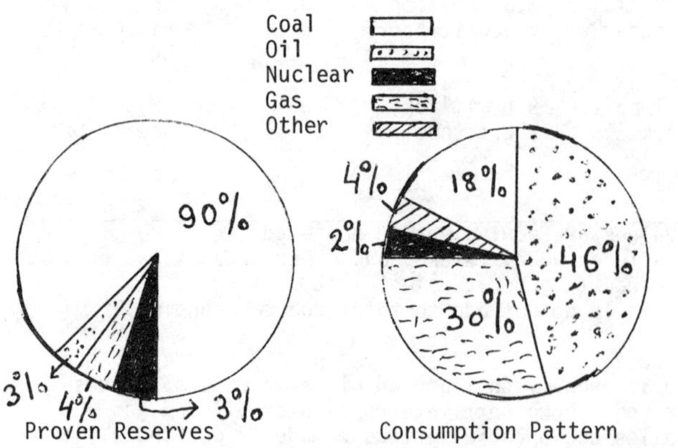

FIGURE 1. PROVED RESERVES (ECONOMICALLY RECOVERABLE) WITH EXISTING TECHNOLOGIES(2)

PRINCIPLES OF COAL CONVERSION

Coal has experienced some problems in utilization for variety of reasons. These include 1) it is a solid, 2) it contains inorganic material that is not useful as a fuel, 3) it contains some elements, the oxidation of which produces environmentally objectionable compounds(sulfur, mercury, nitrogen etc.,), 4) it is not of uniform composition. Conversion of coal to liquids and gases essentially eliminates these four disadvantages, although some separation and chemical reactions are required. Such conversion can be accomplished by the addition of hydrogen. The amount of addition of hydrogen determines the properties of the synthetic fuel and the cost of conversion. A comparison of some representative materials illustrates this situation and is given in table 3.

TABLE 3. HYDROGEN/CARBON RATIO AND HYDROGEN CONTENT OF FUELS(3)

Fuel	Typical Atomic Ratio H/C	Hydrogen Content Weight %
Methane	4.0	25
Iso-octane(C_8H_{18})	2.25	15.8
Pennsylvania crude oil	1.98	15
Shale Oil	1.60	11.8
Texas crude oil	1.55	11.5
Tar Sand Bitumen	1.50	11.1
Coal Tar	1.33	10.0
Benzene(C_6H_6)	1.00	7.7
Bituminous Coal	0.8-1.0	5-6.6

A very large number of processes have been attempted for coal conversion to synthetic fuels, all of which were pioneered by German scientists before World War II. The processes for the production of synthetic fuels and chemicals may be categorised in several ways, but the most simple is by the type of products as follows:

1. Low Btu fuel gas - produced from coal by reaction with steam and air.
2. Medium Btu fuel gas (synthesis gas) produced from coal by reaction with steam and oxygen.
3. High Btu fuel gas (substitute natural gas)- this can be produced by a three step process.
   a. Coal when reacted with steam and oxygen produces medium Btu fuel gas consisting mainly of hydrogen, carbon monoxide and carbon dioxide.
   b. The synthesis gas in step a is then shifted, so that it is altered to produce a proper ratio of the hydrogen and carbon monoxide mixture.
   c. Methanation of the H_2/CO mixture to produce pipeline gas.
4. Heavy liquid fuels- produced from coal by pyrolysis, solvent refining, hydropyrolysis, liquid phase hydrogenation, or by synthesis gas production followed by catalytic reduction of the synthesis gas to higher hydrocarbons.

5. Light liquid fuels and chemicals- produced from coal by one of the above processes, in some cases additional hydrogenation (vapor phase) or other catalytic or high temperature reaction. Such processing could lead to mixture of liquids with no predominant compounds produced or to high yields of methanol or other chemicals.

Gasification Chemistry: Coal is chemically a complex solid substance that primarily consists of carbon (C) with relatively small amounts of hydrogen (H), oxygen (O), nitrogen (N), sulfur (S), and ash. Major reactions by which carbon is gasified are shown in table 4. Reaction with oxygen to form CO provides

TABLE 4. CARBON GASIFICATION REACTIONS

$C + 1/2\ O_2 \longrightarrow CO + 47{,}600$ Btu

$C + O_2 \longrightarrow CO_2 + 169{,}300$ Btu

$C + H_2O(g) \longrightarrow H_2 + CO - 56{,}500$ Btu

$C + 2H_2O(g) \longrightarrow 2H_2 + CO_2 - 38{,}800$ Btu

$C + 2H_2 \longrightarrow CH_4 + 32{,}200$ Btu

$C + CO_2 \longrightarrow 2CO - 74{,}200$ Btu

47,600 Btu for each pound atom of carbon gasified. When carbon reacted completely to form CO_2, larger quantity of heat is produced, whereas if carbon is reacted with steam to form H_2 and CO or H_2 and CO_2 the reactions are endothermic. Carbon may also be gasified by direct addition of hydrogen (hydrogasification) to form methane an exothermic reaction or by CO_2 to form CO an endothermic reaction. Gasification processes are so adjusted that the exothermic and endothermic reactions essentially balance the opposite heat effects. For example the following reactions for the production of the mixture of CO and H_2 or CO and CH_4 have been thermally balanced. When CO and H_2 are the desired products

$$2.2\ C + H_2O(g) + 0.60\ O_2 \longrightarrow H_2 + 2.2\ CO$$

$$O_2/C = 0.27 \qquad H_2O/C = 0.45$$

$$2.35\ C + H_2O(g) + 0.425\ O_2 \longrightarrow 1.85\ CO + 0.5\ CH_4$$

$$O_2/C = 0.18 \qquad H_2O/C = 0.43$$

the amount of oxygen required is 0.27 moles of oxygen per atom of carbon, and when methane is the desired product the oxygen requirement is appreciably reduced. Relative proportions of CO and H_2 are adjusted by reacting CO with steam (water gas shift reaction). This reaction may be used alongwith CO_2 removal to produce hydrogen as the final product. Similarly when methane is the desired product adjustment is made by reacting CO with H_2 or CO_2 with H_2 (methanation reactions) to form methane. The reactions are given below:

$$CO + H_2O(g) \longrightarrow H_2 + CO_2 + 17.700\ \text{Btu}$$

$$CO + 3\ H_2 \longrightarrow CH_4 + H_2O + 88{,}700\ \text{Btu}$$

$$CO_2 + 4\ H_2 \longrightarrow CH_4 + 2\ H_2O + 71{,}000\ \text{Btu}$$

Figure 2 shows the two principal routes for converting coal to gas. They differ in how heat necessary for gasification is supplied. In the first case the heat is supplied by combustion of carbon with air directly in the process so that nitrogen from the air is mixed into the product gas with low Btu(90 to 150 Btu/SCF). Alternately, the heat may be supplied by other means so that nitrogen does not become part of the product gas. This is done by using oxygen instead of air, or by supply of heat through indirect heat transfer from hot gases or solids that may or may not take part in gasification reactions, or hydrogasification where gasification energy is supplied by the exothermic reactions of hydrogen with carbon and carbon oxides to form methane. These are the routes for the production of medium Btu gas that may be used for fuel, or converted by methanation or hydrogen removal to SNG or by methane removal to hydrogen or Syngas. In addition to air or oxygen steam is introduced into the gasifier either as supplementary gasification agent as moderators to exothermic reactions, or to supply the needed hydrogen not available in coal.

FIGURE 2. COAL - TO-GAS PROCESS STEPS

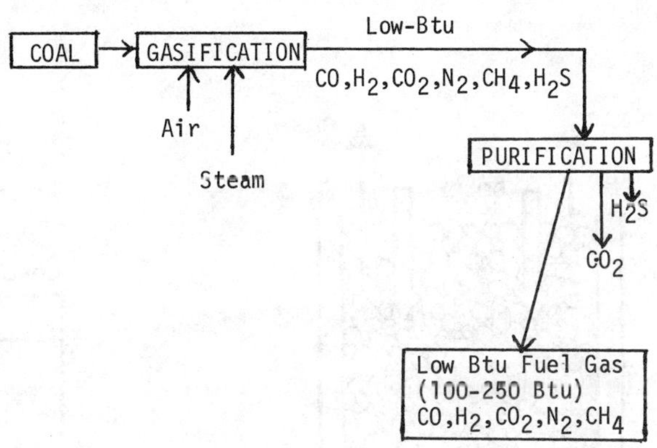

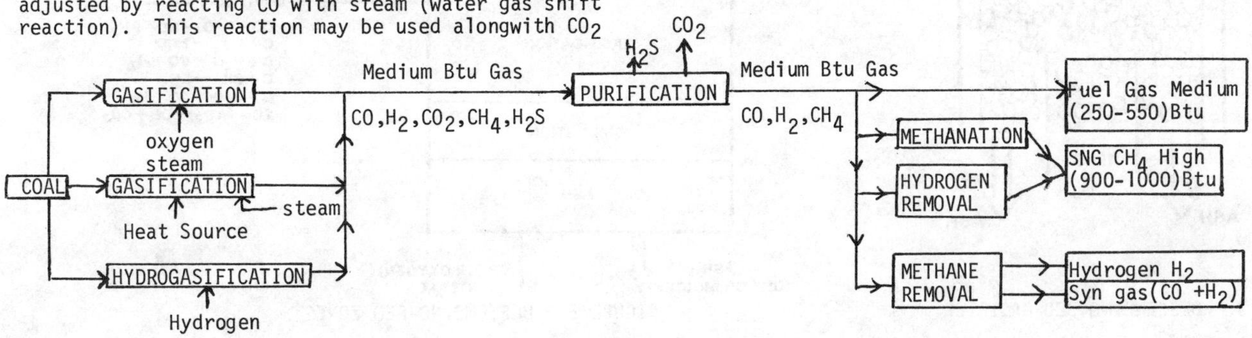

Gasification Reactor Types: Methods of contacting the solid coal with reacting gases to accomplish the required gasification is another major mechanical consideration. There are three principal types of contacting methods (4):

A descending bed of solids with upflowing gas.
Fluidized bed of solids.
Solids entrained in a gas.

Additionally, contacting is practiced less extensively in these reactors:

Molten baths.
Tumbling beds.
Stationary beds (underground gasification).

Descending bed gasifier is often referred to as slow moving or fixed bed shown in figure 3. Lumps of coal 1/8 inch to 1 inch diameter in size are laid down at the top of the vessel full of such solids. The reactant gases such as oxygen and steam are fed at the bottom and flow upward slowly through the interstices between coal lumps. While the coal is coming down it is heated by the ascending gases from the combustion zone at the bottom of the reactor. The temperature distribution through the height of the reactor is shown in figure 4. The amounts of steam and air (oxygen) control the temperature in the reactor and the nature of the ash whether molten or dry. If dry ash is desired then enough steam is added so that the endothermic reaction of coal with steam offsets the combustion reactions to the extent that the ash-fusion temperature is not attained. On the other hand if slagging operation is desired then the intake of steam is reduced accordingly. The temperature changes of coal as it descends down the gasifier cause changes in coal as shown in figure 5. The temperature zones are not as distinct as shown but will blend more or less gradually into each other. In the drying zone water is removed and when the temperature rises above 600° F, devolatilization of coal takes place. This is followed by gasification region where gasification of char resulting from devolatilization takes place by reaction with gases and steam moving up from the combustion zone. In the combustion zone remaining char is burned and reacted with steam and the ash is cooled by the entering reactant gases, and collected at the bottom. The product gas exits at top in the temperature range of 500° to 1000° F. This temperature is not high enough for the decomposition of tars and oils, thus the product gas contains appreciable quantities of tars and oils that have to removed. In order to facilitate this cleanup some descending-bed gasifiers are operated in a two stage mode where a portion of the gas is removed directly from the gasification zone (Primary gas of figure 5) and only the remainder sufficient to heat the incoming coal moves through the devolatilization and drying zones. This provides a gas stream free of tars and oils. The residence times in this type of reactor are fairly large.

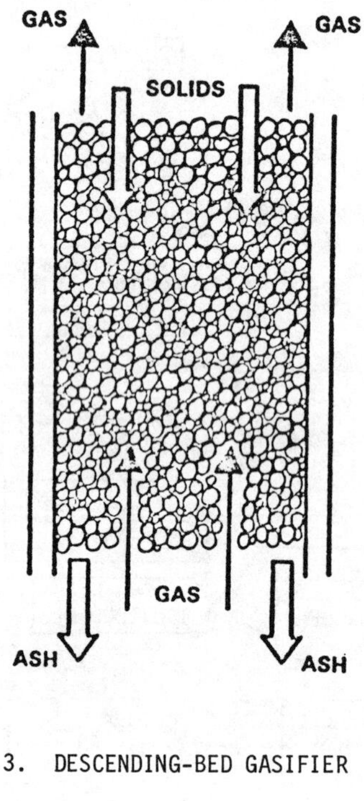

FIGURE 3. DESCENDING-BED GASIFIER

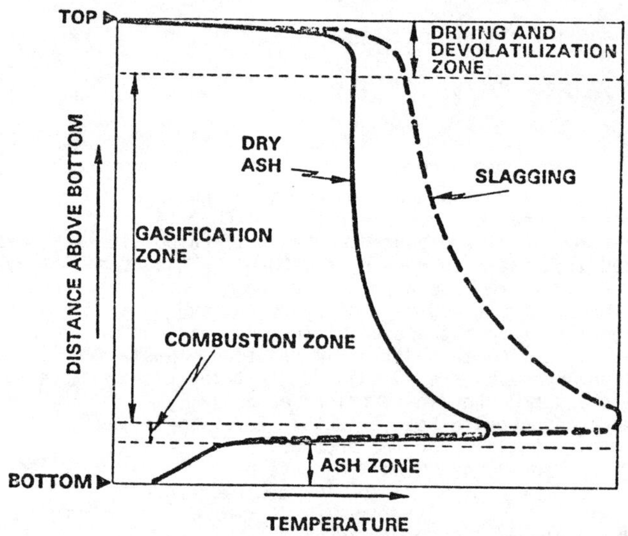

FIGURE 4. A DESCENDING-BED TEMPERATURE DISTRIBUTION

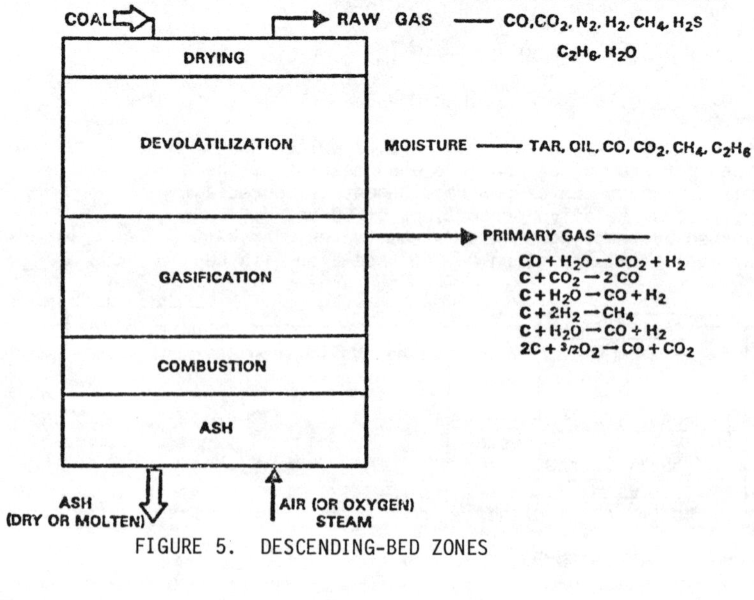

FIGURE 5. DESCENDING-BED ZONES

In the fluidized-bed system shown in figure 6, coal is ground to about 8 mesh or less and introduced into the reactor. The reactant gases are introduced through a perforated deck near the bottom of the vessel. The volume rate of gas flow is such that its velocity is high enough to suspend the solids but not enough to blow them out with the gas. This makes a violently boiling bed of solids having very intimate contact with the upflowing gases and results in uniform temperature through out. Reaction rates in the fluidized-bed system are faster than the moving-bed system because of the intimate contact between solids and gases and the increased surface area due to small particle size.

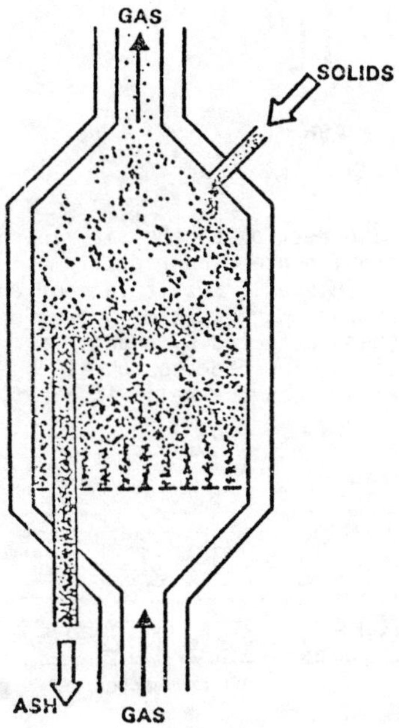

FIGURE 6. FLUIDIZED-BED GASIFIER

The entrained flow gasifier shown in figure 7 uses a finer grade of coal (80% through 200 mesh). It is fine enough that it can be readily conveyed pneumatically by the reactant gases. Velocity of the mixture must be 20 ft/s or higher depending upon the fineness of the coal. There is little or no mixing between the solids and the gases, except where the gas initially meets the solid and the reactions occur in a completly cocurrent fashion. This type of reactor is used only for very rapid reactions like combustion of coal with oxygen or the initial reaction of fresh coal with H_2. There are two types of entrained-bed reactors: single stage and double stage. In single-stage operation, the coal is essentially completely gasified in the atmosphere of air(or oxygen) and steam. In the double stage operation the initial portion of the feed coal is gasified in an atmosphere of hydrogen or synthesis gas, the remaining char is then gasified in a second stage that may or may not be entrained flow. The residence time in such reactors is very small of the order of few seconds. All of the single stage entrained flow gasifiers operate at high temperatures (2500-3000 F). In the two stage process the first stage can be carried out at lower temperature because of the higher reactivity of coal in the initial gasification stage. Pressure operation then results in appreciable formation of methane gas with savings in heat and coal feed requirements due to the exothermic heat effect of methane formation.

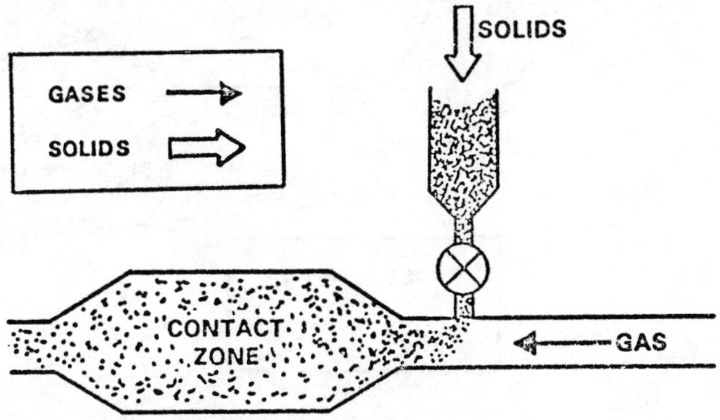

FIGURE 7. ENTRAINED FLOW GASIFIER

Molten bath gasifiers (figure 8) depend upon feeding coal and gasification medium (oxygen, air, steam) below the surface of a hot liquid bath. Reaction takes place in the bath. Gas leaves from the surface of the bath and the unreacted carbon and ash floats on the surface from which it is withdrawn. There are three types of molten baths, molten salt, molten iron, and molten coal ash. The reaction temperature is 1700-1800 F in the salt bath and up to 3000 F in the coal ash bath. Very little methane is formed due to high temperature. These gasifiers have the capability to accept any kind of coal regardless of fines content, caking properties, or crushing strength.

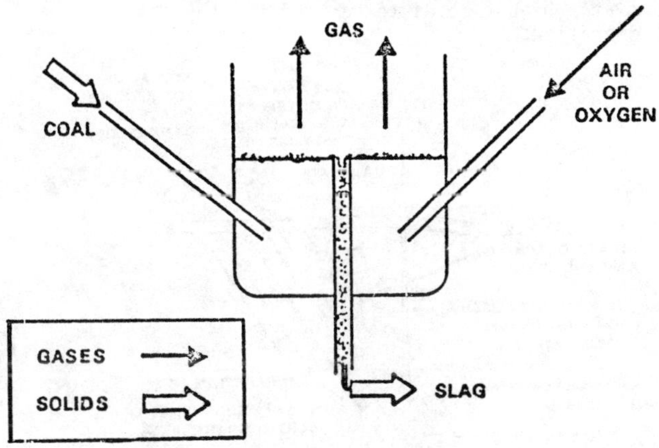

FIGURE 8. MOLTEN BATH GASIFIER

Allis Chalmers Corp. has developed a coal gasifier employing a rotary ported kiln that has been used industrially since 1973 for direct reduction of iron ore. The gasifier is shown in figure 9. The ports, which are openings in the kiln wall allow air and steam to be injected under the tumbling bed of solids. Coal is fed to the high end of the kiln and moves counter current to the gases. Most of the ash is entrained with the product gases but some overflows at the lower end of the kiln. A 44-ton/day pilot gasifier has been operated and a 600-ton/day demonstration plant is contemplated.

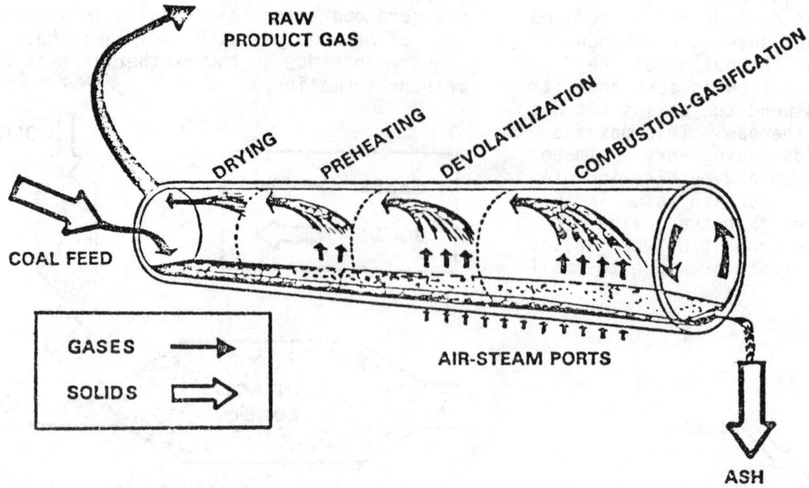

FIGURE 9. TUMBLING BED GASIFIER (5)

Underground Coal Gasification: There are two configurations of wells employed in underground (in situ) coal gasification(UCG) depending upon the nature of the coal seams. For steeply dipping seems the stream method is employed. For horizontal seams the 'linked vertical wall' percolation method is used. In the stream method a primary injection well is drilled through surrounding strata to the base of the coal seam. A production well is drilled through the length of the coal bed (see figure 10). As gasification proceeds from the base of the seam ash falls down and will eventually clog the injection tube, so a second injection tube is drilled into the seam at a higher level to be used for the later phase of operation.

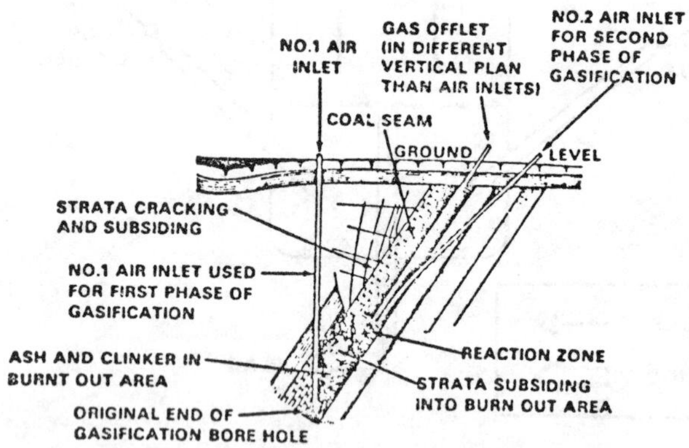

FIGURE 10. A STEEPLY DIPPING BED CONCEPT (6)

The progressive steps of linked vertical wall operation are shown in figure 11. In figure 11B, that air is introduced into and ignition is initiated at the bottom of the well to the right. After ignition is initiated the gas flow is reversed by injecting air into the well at left (figure 11C). From that point onward combustion proceeds from right to left where as air moves counter current from left to right. This produces a relatively narrow passage between the bottoms of injection and the production points. If the flame front were to move in the same direction as the air flow a wide volume will be opened, proceeding radially outward from the air inlet. This will result in poor coal utilization and unstable gasification operation. After linkage has been effected by counter current combustion the gasification proceeds until the complete volume of coal between the wells is gasified

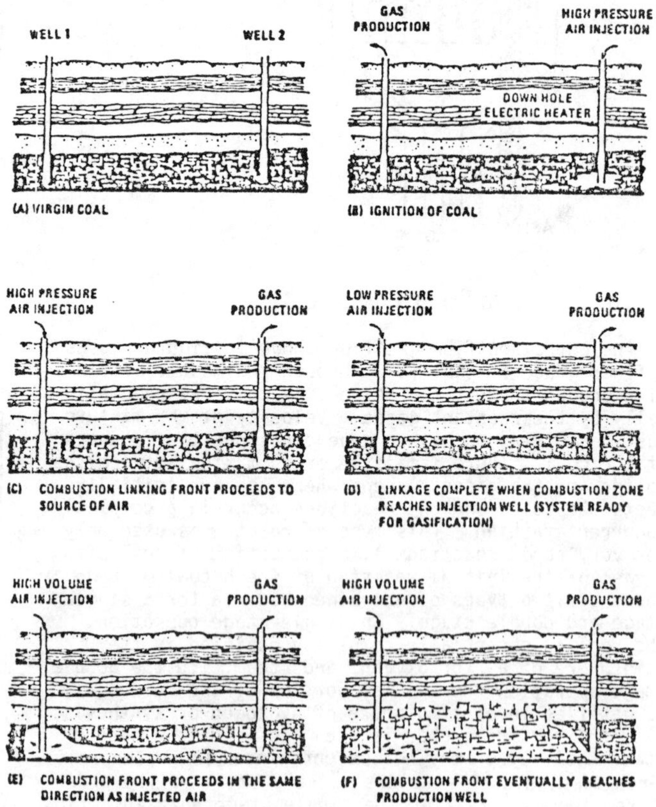

FIGURE 11. LINKED VERTICAL WALL OPERATION (7) WITH COUNTER CURRENT COMBUSTION FOR LINKING

(figure 11F). As shown in table 5 there are a number of UCG projects presently being undertaken or planned in the United States.

TABLE 5. UNDERGROUND GASIFICATION PROJECTS

U.S. DEPARTMENT OF ENERGY OPERATIONAL PROJECTS

Laramie Energy Technology Center, Western Low-Btu.

Lawrence Livermore Laboratory, Western Med-Btu.

Morgantown Energy Technology Center, Eastern Coal.

Gulf Research and Development Co with TRW (subcontractor), Steeply Dipping Coal Beds.

OTHER OPERATIONAL PROJECTS

Texas Utilities Services, Inc., Texas Lignite.

Texas A & M University, Texas Lignite.

Alberta Research Council, Alberta Sub-bituminous Coal.

PLANNED PROJECTS

Atlantic Richfield, Wyoming Sub-bituminous Coal.

New Mexico Consortium, New Mexico Sub-bituminous Coal.

Various Gasification Processes: At present gasification processes fall into three categories as shown in table 6. The processes have been summarized in

Commercial processes available on the market

Developing processes, and

Processes where gas is a major by-product to oils and char.

terms of the heating value of the product gas, the operating pressures, and the types of reactors. The details of the processes can be found in several publications (3-5, 8-16).

COAL LIQUEFACTION

Coal can be used as a raw material to provide a non-polluting liquid fuels suitable for firing power plant boilers, or transportation purposes. The general reaction for coal liquefaction is

$$C_nH_{.7n} + .5n/2\ H_2 \longrightarrow C_nH_{1.2n}$$

coal hydrogen synthetic crude oil

The requirement for producing synthetic crude oil is to increase the H/C ratio from .7 in bituminous coal to 1.2 in liquid hydrocarbons. This may be accomplished in several ways as depicted in figure 12. The processes for liquefaction are not as far along toward commercialisation as processes for gasification. Two general approaches are used, one is to force hydrogen on to the carbon in the solid structure and produce a liquid, and the other is to break down the coal into hydrogen and carbon monoxide and shift the CO/H_2 mixture and catalytically convert this gas to liquid product.

	Route	Processes
COAL →	Fischer-Tropsch Synthesis	SASOL Plant
	Pyrolysis and Hydrocarbonization	COED COALCON's Hydrocarbonization Garret's Coal Pyrolysis
	Solvent Extraction	Consol Synthetic Fuel Exxon Donor Solvent Process SRC
	Direct Catalytic Hydrogenation	H-Coal Synthoil Gulf Catalytic Coal Liquids

FIGURE 12. CLASSIFICATION OF LIQUEFACTION PROCESSES

Using the methods listed in figure 12, many processes have been used for liquefaction of coal. Not many of these have been carried out at full scale production levels. The exceptions are Germany during World War II and South Africa presently. For Germans economics was secondary consideration relative to national security. A similar situation led to synthetic liquid fuels production in South Africa called SASOL Plant. Because of diminishing supplies of oil, the prospects of coal liquefaction by pyrolysis, direct hydrogenation, solvent extraction (and refining), and by gasification to synthesis gas with subsequent liquid production.

Pyrolysis causes thermal decomposition of coal by heating it in the absence of oxygen to drive of volatile matter producing a liquid product as well as by-product gas and char. The products are separated into gaseous and liquid products, the gases are freed from unwanted substances (sulfur compounds, water, etc.), and a usable gas is obtained in addition to the liquid fuel. The gas and char may be used to produce hydrogen for hydrotreating the liquid reaction product to lower its sulfur content and to produce a high-Btu synthetic crude oil. Char may also be burned with air or oxygen to provide the necessary thermal energy for pyrolysis. Since pyrolysis processes yield low volumes of liquids and require, char disposal means, pyrolysis methods will not be a major source of synthetic petroleum. Some current pyrolysis processes are COED (Char-Oil-Energy Development of FMC Corporation), TOSCOAL (The Oil Shale Company), and Garrett processes (Garrett Research & Development Company, subsdiary of Occidental Petroleum Company).

Solvent extraction (dissolution) is a more severe method of liquefying coal as compared to pyrolysis. The crushed coal is dissolved in selective solvent, the ash is filtered out, and the oil is treated by hydrocracking. Pyritic sulfur alongwith insoluble ash is removed from the solvent, however, organic sulfur in the coal is carried along with the liquid fuels. After separation from the solids, the liquids can be catalytically upgraded to produce synthetic petroleum. The solids may be used as fuel or used for process heat or for generation of hydrogen. The hydrogen treatment with or without a catalyst improves

TABLE 6. COAL GASIFICATION PROCESSES

Commercial Processes	Heating value Btu/Scf			Operating pressure psia			Type of reactor		
	100-250	250-550	950-1000	15-30	90-500	1000-up	Fluid bed	Entrained bed	Moving bed
Lurgi	x	x	x		x				x
Koppers-Totzek		x	x	x				x	
Welleman-Galusha	x	x		x					x
Winkler	x	x	x	x			x		
WD/GI	x	x		x					x
Wilputte	x			x				x	
Otto Rummel	x			x				x	
Developing Processes									
ATGAS	x	x	x	x			Molten iron bath		
CO_2 Acceptor		x	x		x		x		
Molten salt		x	x			x	Molten salt bath		
Synthane	x	x	x			x	x		
U-Gas	x				x		x		
Agglomerating ash		x	x		x		x		
Westinghouse	x				x		x		
Riley Morgan	x			x				x	
General Electric	x				x				x
Combustion Engineering	x	x		x				x	
Bi-Gas	x	x	x			x		x	
Texaco	x	x	x			x		x	
Hydrane		x	x		x		x		
Hygas		x	x			x	x		
Stone & Webster/General Atomic		x	x					x	
Multiple Catalyst Process			x						
Deco Process	x	x	x	x					
Exxon		x							
Solar Gasifiers (17)									
Moving-bed coal gasifier		x		x					x
Fluidized-bed gasifier		x		x			x		
Molten-salt solar coal		x		x			Molten salt bath		
Coal gasifier with Helium heat-exchanger		x		x				x	
CO_2 Acceptor Process (18)		x	x		x		x		

Processes where gas is a major by product of coal to oil/char	comments
COED-COGAS	Multistage pyrolysis followed by char gasification.
Garrett	One stage pyrolysis: heat transfer provided by char.
Coalcon	Hydrocarbonisation process.
Toscoal	Single-stage pyrolysis; heat transfer by circulating ceramic balls.
CSP	Hydrogenation is carried out non-catalytically with hydrogen donor solvent.
SRC	Hydrogenation carried out non-catalytically with hydrogen and solvent.
H-Coal	Catalytic hydrogenation with hydrogen and solvent in ebullating bed.

TABLE 7. CHARACTERISTICS OF COAL LIQUEFACTION TECHNOLOGIES (11)

Process	Coal Feedstock	Reactor temperature (°F)	Reactor pressure (psi)	Product oil grade (°API)	Heating value of oil (Btu/lb)	Yield (bbl/ton coal)
			HYDROGENATION			
Synthoil	Pulverized, dried caking or noncaking	850	2000-4000	Not applicable	17,700	3.0
H-Coal	Pulverized, dried caking or noncaking	850	2700	5-17	a	4.4
SRC	Pulverized, dried caking or noncaking	800	1000	b	b	b
CSF	Pulverized, dried caking or noncaking	800	1000	b	b	b
			PYROLYSIS			
COED	Pulverized, dried caking or noncaking[c]	600-1600 (in four reactors)	6-10	25	b	1.0-1.5
TOSCOAL	Pulverized, dried caking or noncaking	970	atmospheric	6-13	16,000	0.5
			CATALYTIC CONVERSION			
Fischer-Tropsch	Depends on gasification process	Two reactors at different temperatures	330-360	various	various	b

a Different coal feedstocks may require a greater number of pyrolysis feedstocks.
b Unknown.
c Lower temperature gives lower yield of liquids.

the product quality and removes the organic sulfur. There are three types of dissolution processes, 1) those that use no catalyst and no hydrogen, 2) those that use hydrogen but no catalyst, and 3) those that use both catalyst and hydrogen. A method using neither catalyst nor hydrogen is the Consol Synthetic Fuel (CSF) process of Consolidation Coal Company. The Solvent Refined Coal (SRC) process of Pittsburgh & Midway Coal Company uses hydrogen but no catalyst. Three processes that use both catalyst and hydrogen are the Synthoil (U.S. Bureau of Mines), H-Coal (Hydrocarbon Research, Inc.), and Catalytic Coal Liquids (CCL)(Gulf Research & Development) processes. Table 7 lists the operating conditions for some of the coal liquefaction technologies. This list by no means is thorough and complete.

THE ENVIRONMENTAL COSTS OF SYNTHETIC FUELS

One reason to adopt synthetic fuel technology is the environmental protection provided by synthetic fuels. Coal conversion is substituted for stack gas clean up in the combustion process. Combustion of medium Btu gas results in 40 ppm NO_2 and 48 ppm CO (dry volumetric basis), where as combustion of natural gas gives 60 and 70 ppm respectively. Direct combustion of coal causes higher and additional emissions (17). Conversion is not without its effluent control issues, thus it is important to consider the production of effluents from selected technologies for conversion processes, as well as combustion systems, must meet environmental standards. Effluents causing both water and air pollutants, plus solid wastes are considered here as a result.

Synthetic Fuel Plant Emissions: Effluents that can cause water pollution include dissolved solids, suspended solids, and organic matter (e.g., phenols). Thermal discharges can also be considered, however, these vary to the extent water is employed for cooling purposes. Table 8 presents representative values for water pollutants for various conversion processes now available or emerging.

TABLE 8. WATER POLLUTANTS FROM COAL CONVERSION PROCESSES (18)

Process	Effluent[a] Total Dissolved Solids	Total Suspended Solids	Total Organics
Low Btu[b]	-[d]	-	-
Medium Btu[b]	-	-	-
High Btu[c]	43.1	0.9	0.426
Solvent Refined Coal	52.3	0.017	0.003
Light Liquids	63.6	0.008	0.002

a in tons/10^{12} Btu produced as fuel.
b Atmospheric low Btu gasification; Koppers-Totzek medium Btu gasification.

TABLE 9. ESTIMATED CONTROLLED AIR EMISSIONS FROM COAL CONVERSION FACILITIES (in lbs/10^6 Btu of fuel produced)*

Process/Coal	Products					
	Particulates	SO_x	NO_x	Hydrocarbons	CO	Aldehydes
Low Btu/eastern	0.008	0.115	0.084	0.002	0.006	3.1×10^{-5}
Low Btu/western	0.009	0.046	0.135	0.003	0.010	5.0×10^{-5}
High Btu/eastern	0.011	0.149	0.108	0.002	0.008	4.0×10^{-5}
High Btu/western	0.036	0.059	0.173	0.004	0.013	6.5×10^{-5}
Solvent Refined/eastern	0.008	0.044	0.066	0.001	0.006	6.0×10^{-6}
Solvent Refined/western	0.014	0.014	0.075	0.001	0.007	1.0×10^{-5}
Light Liquids/eastern[a]	0.029	0.322	0.270	0.006	0.020	1.0×10^{-4}
Light Liquids/western[a]	0.030	0.127	0.415	0.009	0.031	1.5×10^{-4}

* Source (19).
a Assumes the Fischer-Tropsch Process.

c Lurgi Process.
d Indicates no appreciable amount.

The emissions which must be controlled by air pollution technology include particulates, sulfur oxides, nitrogen oxides, hydrocarbons, carbon monoxide, and aldehydes. Table 9 summarizes the generation of such effluents by the various conversion processes. From the table one can see that for producing 1 Quad Btu of energy will need to control some 4000 tons of particulates, 20000 to 80000 tons of sulfur oxides, 35000 to 60000 tons of nitrogen oxides, 800 to 1300 tons of hydrocarbons, 2500 to 4500 tons of carbon monoxide, and 10 to 20 tons of aldehydes if the process employed is low Btu gasification. If the Fischer-Tropsch process is used the amount of pollutants will be much higher. Air emissions are a function of both process configuration and process efficiency.

Coal conversion generates solid wastes, principally ash, that must be disposed of. Table 10 enumerates representative amounts of solid wastes from various processes. The generation of solid wastes depends, in large part on the specific coal being processed.

TABLE 10. SOLID WASTES BY COAL CONVERSION PROCESSES (in tons/10^{12} Btu of energy produced) (18)

Process	Tons of Solid Waste Generated
Low Btu Gasification	7060
Medium Btu Gasification	8430
High Btu Gasification	5270
Solvent Refined Coal	3460
Light Liquids Production	3260

Gas Clean Up And Processing: Use of coal conversion technology does not eliminate the environmental problems, rather it makes them somewhat easier to deal with. For example in the conversion process sulfur from the coal reacts with hydrogen from steam to form H_2S in contrast to SO_2 in direct Combustion. H_2S is more easily removed from the gas stream than sulfur dioxide. Further the volume of fuel gas produced by any process is much less than the volume of stack gas resulting from combustion. The level of oxides of nitrogen is also low due to the lower temperatures in the gasifiers and controlled amount of air in the reactor. The raw gas from any reactor will be in the temperature range of 250 to 3000 F and at pressures from atmospheric to 1500 psi. It will contain significant amounts of H_2, CO, CH_4, and higher hydrocarbons, and will also contain impurities such as H_2O, CO_2, coal fines, tars, oils, phenols, variety of sulfur and nitrogen compounds, and a number of trace metals in the ash. Removal of undesirable components and adjustment of the final gas composition is generally as important in both technical and economical sense as the basic gasification process.

Due to the number of pollutants and different operating conditions in different processes, a "universal" purification system cannot be designed to be applicable in all cases. However, one possible sequence of steps is shown in figure 13. Various steps include a cyclone or a precipitator to removeparticles of coal and other solid materials, followed by cooling,quenching,or washing, to condense tars and oils and remove dust and water soluble materials - phenols, chlorides, ammonia, hydrogen cyanide, thiocyanate, and perhaps some sulfur compounds from the gas stream.

The next step in the gas processing is to adjust H_2/CO ratio, if necessary, in a water-gas shift reactor, so that the down stream methanation reaction can take place properly. The sulfur will be present as H_2S and is removed as elemental sulfur using Claus technology. CO_2 is removed by the number of techniques. If high

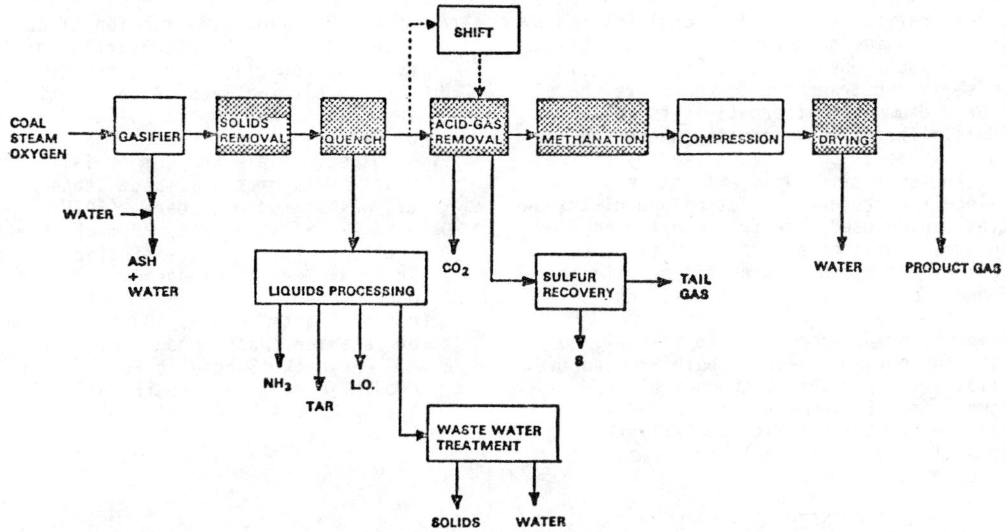

FIGURE 13. GAS CLEAN UP AND PROCESSING (20)

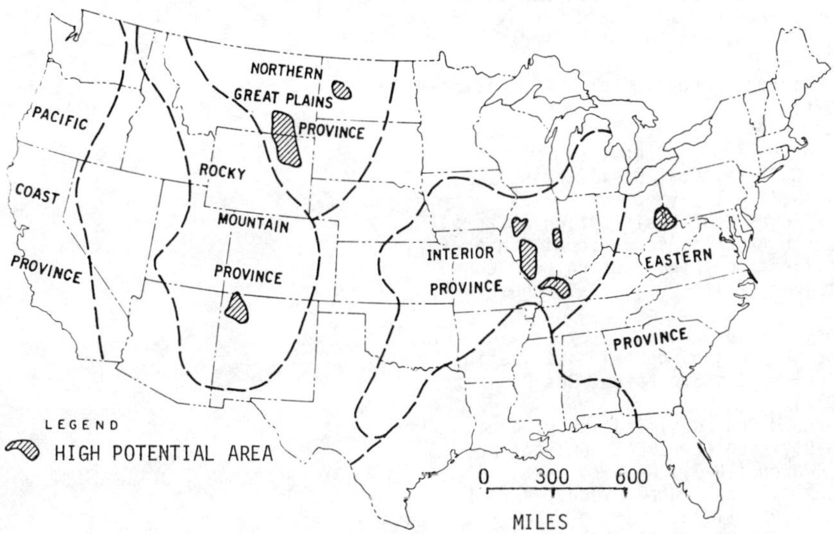

FIGURE 14. AREAS OF HIGH POTENTIAL FOR COAL CONVERSION PROCESSES (21)

Btu gas is to be produced it is necessary to remove CO_2 and H_2S. Final processing steps are methanation, compression, and drying if SNG is the final product, or other suitable operation such as methane removal to produce pure hydrogen.

COMMERCIALISATION OF SYNTHETIC FUELS

Future commercialisation of synthetic fuels will depend upon a number of factors like the price and availability of oil and natural gas, the ability of conversion plants to produce fuels at a competitive cost without damaging the environment. It will also depend upon the availability of coal and water. The matching of coal and water availability has been performed by the U.S. Bureau of Mines as shown in figure 14. The map shows where coal and water resources can both be found. The areas of potential are particularly appropriate for large SNG or coal liquefaction plants.

A recent report by Fryback (22) has pointed out some of the problems and promises of synthetic fuels. The main problems in getting the synthetic fuels to the market place are: 1) economic, 2) environmental, 3) regulatory. On the economic side synthetic fuel plants require large capital investments to build the plants. After the plants are built large investments are needed to produce raw materials and transport them to the conversion site. In addition huge capital is involved in storing enough raw material in hand for continous operation of the facility. On the environmental and regulatory side any synthetic fuels plant of significant size will disturb the land, add air and water emissions to the environment, and effect the water availability and possibly the quality. Reasonable environmental laws and regulations will be needed to ensure the quality of environment and the production of synthetic fuels at affordable and competitive cost. In the final analysis, if the industry and

government agree that it is in the national interests to develope synthetic fuels industry, and the public agrees, it will happen. Synthetic fuels hold their greatest promise in the transport sector where there are no short or medium term alternatives to replace the liquid fuels. The utility industry where the use of oil and natural gas is being restricted by choice or regulation, the synthetic fuels will be very appropriate since the present production and distribution facilities can be used. The trend in the energy consuming industries will be similar to utilities. Synthetic fuels will contribute significantly to the increasing demand for petrochemical feedstocks.

Future research and development effort should be directed toward the optimal balance, both from an over all energy utilization and cost between fuel production and fuel consumption. In summary, it can safely be concluded that the synthetic fuels specially the liquid fuels can and will make significant contribution to U.S. future energy needs.

REFERENCES

1. 'Energy in Transition 1985-2010,' Final Report of the committee on Nuclear and Alternate Energy Systems, N.R.C., National Academy of Sciences, Washington, D.C. (1980).
2. National Energy Outlook, Federal Energy Administration Report, (1976).
3. L.L. Anderson, and D.A. Tillman,'Synthetic Fuels From Coal', John Wiley & Sons, Inc., (1979).
4. W.W. Bodie, and F.C.Shora,'Coal Gasification Technology Overview,'I.G.T. Symposium (1979).
5. R.W. Woods,'The Kilngas Coal Gasification Systems,' Presented at the 5th Annual International Conference on Coal Gasification, Liquefaction, and Conversion to Electricity, Pittsburgh, Pa, August 1-3, (1978).
6. P.R. Wieber,'Prospects for Producing Synthetic Natural Gas Through Underground Coal Gasification,' 9th Synthetic Pipeline Gas Symposium, Des Plaines Ill., Sept., (1977).
7. B. Allen,'Underground Coal Gasification Program,' Applied Research Division for ERDA, Contract No. Ex-76-C-01-2343, March (1977).
8. A. Verma,'From Coal to Gas,' Chem. Tech., Part I 372 (1978); Part II 626 (1978).
9. Coal Processing, Gasification, Desulfurization-A Bibliography 1930-1974, Atomic Energy Commision, National Technical Information Service, U.S. Dept. of Commerce, Washington, D.C.(1976).
10. A. McRae, J.L. Dudas, and H. Rowland,'The Energy Source Book,' Aspen Systems Corporation (1977).
11. Environmental, Health, and Control Aspects of Coal Conversion, An Information Overview. Vol. I ORNL/E 1S-94UC-11-41-48-90, April (1977).
12. H. Perry,'Clean Fuels From Coal,' Advances in Energy Systems and Technology, Vol. I, Academic Press, New York p. 307 (1978).
13. R.C. Bailie,'Energy Conversion Engineering,'Energy Science & Technology Series, Addison Wesley Publishing Co., Reading Massachussetts (1978).
14. Proceedings of the 13th Intersociety Energy Conversion Engineering Conference, San Diego Ca., August 20-25 (1978).
15. 'Energy to the 21st Century,' Proceedings of the 15th Intersociety Energy Conversion Engineering Conference, Seattle Washington, August 18-22 (1980).
16. N. Berkowitz,'An Introduction to Coal Technology,' Academic Press, New York (1979).
17. Ad Hoc Panel of Liquefaction of Coal,Assesment of Technology for the Liquefaction of Coal,National Research Council, Washington, D.C. (1976).
18. 'Energy Alternatives: A Comparative Analysis,' The Science and Public Policy Program, University of Oklahoma, May (1975).
19. 'Synthetic Fuels For Commercialisation Program,' Draft Environmental Impact Statement, ERDA, Vol. IV, Washington D.C. Dec., (1975).
20. D.K. Fleming, and H.S. Primack,'Purification Processes for Coal Gasification,' A.I.Ch.E. 81st National Meeting Kansas City Mo., April (1976).
21. A.E. Lindquist,'Siting Potential for Coal Gasification Plants in U.S.,'Bureau of Mines, U.S. Dept. of Interior, Washington D.C., I.C. 8735 (1977).
22. M.G. Fryback,'Synthetic Fuels - Their Problems and Their Promises,' 88th National meeting A.I.Ch.E. Vol.1, 947. (1980).

Chapter 76

FEASIBILITY OF USING WOOD FUEL FOR ELECTRICITY GENERATION IN A SMALL CITY

Y. Omurtag

I. INTRODUCTION

The energy problems facing the nation have led to indepth reviews of renewable resources, such as biomass, for power generation. This report will deal specifically with wood as an energy source for possible power generation.

Wood as an energy source on an industrial scale, especially in developed countries, has almost been completely replaced by the more convenient fossil fuels. Now that the real costs of fossil fuels in terms of unit costs for energy, reliability of supply, and future depletion have become apparent, there has been a resurgence of interest in wood as an energy source, because of its relatively low cost, abundance, and renewability.

In order for wood to be considered a feasible energy source, it must prove to be economical, and meet environmental pollution criteria. Therefore, wood as an energy source will be critiqued in terms of availability, potential as fuel source, and different modes of energy conversion.

II. POTENTIAL WOOD RESOURCES

Fuel choices of the future will be governed by two fundamental considerations, cost and availability. Therefore, these are the areas in which present emphasis will be placed. The primary industrial fuels since the 1950's have been fuel oil and natural gas. The major concern facing the world now is how long will these finite resources last.

At the present time, using known consumption rates, these reserves would last approximately 50 and 90 years respectively. (1) Today with oil and natural gas providing approximately 75 percent of our nation's energy, it becomes evident that alternative forms of energy must be developed and then utilized more heavily in order to avert the impending crisis. This energy need has brought about the re-examination of wood as a fuel.

A. Material Availability

The supply picture for fuel wood is difficult to quantify since there are many competing end uses for wood. Using the Aerospace Research Corporation's approach to estimating the availability of these forests after lumber and pulpwood needs are met, there is an apparent yield of 308 million excess acres of biomass for fuel. (2) Approximately 41 percent of the forest land or 308 million acres would be available for fuel use, because of ongoing needs for lumber and pulpwood. The average electrical energy yield per acre of rough forest can be determined as being 2516 kw hours using the following data: (2)

Average higher heating value of wood, Btu/lb	7680
Average power plant thermal efficiency, %	32.5
Average biomass yield per acre, tons	1.72

Since not every region throughout the United States possesses the same potential for biomass production, this paper shall now pertain more specifically with southern Missouri, the area of interest.

The initial step was to compile Table I, which has wood residues by county, noting that 47 percent of this region is wooded. This unsold wood residue amounts to 884.35 billion Btu's or 86.7 million kw hours of energy that goes unutilized each year. (3) The following facts should give a better indication as to the amount of energy available from wood residues. In order to supply the city of Lebanon's approximate 9000 people with a monthly amount of 1000 kw hours per person of power, it would require about 190 millions pounds (OD) of wood or equivalent 114 million kw hours of power. With adequate forest management, this potential power supply could indefinitely serve Lebanon and its residents.

All comparisons made in this text will hold true for a 50 MW unit, since scaling up to it presents the worse case while lowering incremental costs.

B. Cost Analysis

At present, the biggest drawback to widespread use of wood waste for power generation is the lack of a developed supply line. As of late, tremendous advances have been made in both harvesting and packaging of wood fuel.

One way to increase feasible transportation distances is achieved by enriching the fuel wood through pelletization. Pelletizing removes the moisture and air from the bulk of wood residue, thereby compacting the fuel and increasing the energy yield per pound of fuel. (4) The manufacturing of wood pellets eases fuel handling and could be the key to commercialization of wood-fired plants, since existing facilities can be easily converted to burn wood pellets.

TABLE I

SOUTH CENTRAL MISSOURI WOOD RESIDUES BY COUNTY

COUNTY	AREA (thousand acres) Total	Forest	ANNUAL PRODUCTION OF WOOD RESIDUES OVEN DRY BASIS (millions of lbs)	VOLUME OF EXISTING WOOD RESIDUES--OD (millions of lbs)	USABLE HEAT FROM ANNUAL PRODUCTION (billion Btu)	USABLE HEAT FROM EXISTING RESIDUES (billion Btu)
CAMDEN	409.6	230.9	15.54--total 6.90--unsold	121.43	95.55--total 42.42--unsold	746.66
DALLAS	343.7	131.2	5.58--total 2.75--unsold	0.10	34.03--total 16.78--unsold	0.61
LACLEDE	492.8	204.2	25.92--total 14.06--unsold	4.66	159.24--total 86.38--unsold	28.63
PULASKI	352.3	183.5	21.81--total 12.60--unsold	54.78	134.01--total 77.42--unsold	336.60
TEXAS	757.1	384.3	159.96--total 86.14--unsold	106.52	972.61--total 523.76--unsold	647.68
WRIGHT	437.7	166.9	44.30--total 22.39--unsold	43.86	272.23--total 137.59--unsold	269.53
TOTALS	2793.2	1301.0	273.11--total 144.79--unsold	331.35	1667.67--total 884.35--unsold	2029.71

Transportation is the key to pelletizing, since within the 50-mile radius of a wood-fired plant it results in an increased cost. But for distances greater than 50 miles, pelletizing becomes necessary for wood to compete economically with other conventional fuels. Table II shows the comparative costs of fuels within a 50-mile radius, while Table III emphasizes how pelletization increases effective transport radii. (1)

TABLE II

AVERAGE FUEL COST PER MILLION BTU's

No. 2 Fuel Oil	$2.95
No. 6 Fuel Oil	$2.30
Natural Gas	$1.75
Coal	$1.65
Wood Chips	$1.55
Pelletized Wood	$1.80

Reference (1)

These prices were estimated as average for Southeastern United States, and represent costs to a typical industrial customer, delivered within a 50-mile radius. While these figures are simply rough averages that are dependent on location and time, it still becomes apparent that wood can compete with fossil fuels on a cost basis.

TABLE III

EFFECTIVE TRANSPORTATION RADII FOR WOOD FUELS (1977)

Fuel	Transport Method	Energy Available for a Transport (as a % of total energy content)	Effective Transport Radii (in miles)
Softwood (45% moisture)	Truck	1.3	20
		2.2	30
		4.0	40
	Rail	1.3	70
		2.2	110
		4.0	150
Hardwood (35% moisture)	Truck	1.3	25
		2.2	40
		4.0	50
	Rail	1.3	90
		2.2	150
		4.0	200
Softwood Pellets (15% moisture)	Truck	1.3	30
		2.2	60
		4.0	90
	Rail	1.3	120
		2.2	200
		4.0	260
Charcoal	Truck	1.3	60
		2.2	100
		4.0	170
	Rail	1.3	220
		2.2	370
		4.0	490

Reference (1)

C. Conversion Methods of Wood For Energy

Wood can be used directly for generation of energy by direct combustion or gasification.

1. Direct Combustion. Direct combustion for process steam or electricity is an old technology that is commercially proven.

In direct combustion, all organic wood components are combined with oxygen to form carbon dioxide and water byproducts. It has been roughly estimated that one oven-dry ton of wood is required for each 12,000 pounds of steam.

Three basic approaches exist for complete combustion of wood: (1) the Dutch oven, (2) the spreader-stoker, and (3) the fluidized bed combuster. Each system seeks to maximize and achieve complete oxidation of the volatiles and the fixed carbon of the wood fuels, while providing: (1) the flexibility of furnace operation, (2) wet fuels handling, and (3) pollution control from stack gas emissions.

In the first technique, the Dutch oven system, a heavy bed of wood is placed on the combuster grate. Under fire air flows through the grate to the fuel on the bed, where partial combustion occurs. Volatiles are given off by heating the bed and are combusted with over-fire air. Either natural or forced circulation can be used with ashes dropping through the grate into a cleanup area.

The second approach, the spreader-stoker system, is the most popular wood-burning system presently employed. The fuel is fed into the fire box above the grate and begins to burn while falling to the grate. Moisture contained in the fuel is partially driven off while in suspension. The depth of the fuel bed is proportional to the fuel's moisture content.

Fluidized beds (Figure 1) have a large heat flux rate and can burn wet, dirty and non-uniform fuels. Air is forced by a blower through flow nozzles into the air plenum and up through the fluid bed which contains the fuel in suspension while combustion and thermal conversion takes place.

2. Gasification. Gasification is a major modification of fundamental pyrolysis. This process produces fuel gases to meet fuel demands and burns the remaining char to provide heat for gasification.

Some of the potential advantages for gasifiers are:

1. Energy conversion efficiency is quite high, maximum about 80%.

2. High water content wastes, including wet wood and even sewage sludge, can be completely converted to a low Btu gas and ash.

3. Char combustion eliminates the need for a treatment plant for the produced fuel gases.

The major disadvantage of a gasifier is the low Btu rating of the gas, only about 10 to 50 percent of that of natural gas. Another potential disadvantage for wood gasification is its limited experience with large-scale systems.

III. EVALUATION

It can be said that wood fuels can be efficient and highly desirable sources of energy if the correct conditions exist, such as: (1) extraction costs are minimized, (2) transportation distances are held down (usually a 50-mile radius with a vehicle, or possibly a 200-mile radius with rail), and (3) thermally efficient conversion processes are employed. As an added note, the radius boundaries can be kept small by utilizing the idea of co-conversion. The basis behind this principle is that wood and another fuel substance such as coal, would be burned together, thus decreasing the overall dependence on one fuel or another.

A. Direct Combustion

All existing wood-based electric plants use direct combustion and most of them were built in the early 1930's. Therefore, the level of investment associated with them is not representative of the current cost experience. Thus, the cost of produced electricity at these older plants is extremely low. In one plant (1931 vintage) the annual operating cost is expected to be approximately 2.7 mills/kwh. (7)

B. Gasification

The largest wood gasification research unit in use in this country is the University of Missouri-Rolla's GROW (Gasification Research On Wood) unit. GROW, which started operation in the fall of 1979, has already achieved a low Btu gas production of approximately 200 Btu's per cubic foot using sawdust as feedstock. The GROW project uses a fluidized bed gasifier. One major advantage of this unit is its suitability of gasifying all types of biomass including wood chips, sawdust, animal manure, corn cobs, and other agricultural by-products.

C. Environmental Considerations

The increased use of wood for energy by the utility sector can affect the environment in two general areas. The first being the impact on the forest resources with increased withdrawal of wood, specifically if whole tree chipping is utilized. The other would be the environmental impact on both air and water quality. Since trees recycle most of their nutrients through leaf fall, harvesting on a 20-year cycle for continuous yield coupled with leaving the foliage behind should minimize nutrient depletion. As wood usage increases, forestry techniques concerning soils, watersheds, wildlife, and other natural phenomena, will have to improve so yield can stay abreast of demand. (6)

Emission control for wood-fired systems is normally limited to particulate control. Particulate emissions can be controlled by either preventing the formation of fire particulates or by actually collecting the particulates to meet environmental constraints. Meanwhile, it appears as though gasification of wood is probably the most satisfactory process from an environmental point of view. When the product gas is cleansed by cyclone separators the tar and char content of the produced gas is reduced by re-feeding into the fluidized bed. Therefore, when the gas is burned it has very low emissions, well within EPA standards.

D. Economics

Despite the escalation in fossil fuel prices which make the life-cycle economics of wood-fueled electricity production increasingly attractive, a major financial constraint is the higher capital investment per kilowatt required for wood plants compared to that of fossil plants. This is basically caused by wood's higher transportation costs combined with its lower heating value limits, which in turn economically limits the feasible wood plant size to the 50 megawatt range.

In addition to the aforementioned problems, the wood fuel supply line is still in the drawing board stages. Assurance of longterm wood supplies at reasonably stable prices is a critical issue for commercialization. This barrier can be overcome by issuing long-term supply contracts and by working jointly with local entrepreneurs in establishing this new enterprise.

In all probability, the decisions whether or not to use wood for energy for power generation will not be based entirely on technical feasibility or classical marketplace economics. Government incentives and disincentives will influence these decisions. An incentive for wood energy would be the Energy Tax Act of 1978 as it provides an additional 10 percent "energy percentage" business investment tax credit. This credit, which applies during the period of October 1, 1978 through December 31, 1982, may be claimed for investment in "alternative energy property". (5) This is defined to be property such as boilers whose primary fuel will be an "alternate substance", non-boiler burners for alternative fuels, and equipment used for the unloading, transfer, storage, reclaiming from storage, and preparation of an alternative fuel for burning.

IV. CONCLUSION

It has become apparent that fossil fuels will soon be depleted at present consumption rates. Therefore, mankind will have to learn to utilize renewable resources such as solar energy or fusion energy for their future energy needs.

Wood is a form of solar energy that, with proper forest management, is both renewable and abundant. Another major advantage for wood is its energy has already been stored and does not require expensive capital investment for storage facilities.

When looking at Missouri's six-county region centered around Lebanon, it is noted that annual unused wood residues total 145 million pounds, and annual growth totals 4.5 billion pounds. This figure combined with the fact that 47 percent of this region is wooded insures a secure supply for power generation for years to come.

Systems using direct combustion or gasification of wood to generate electricity are technically ready, although further research and development is warranted to advance fluidized-bed technology. Wood technology would be most compatible with municipal and rural electric utilities' smaller power requirements.

There are no real inherent barriers to the use of wood for energy, since all so-called constraints are only relative in comparison to the cost or properties of fossil fuels such as wood's low-density and high-moisture content, making it more expensive to transport than fossil fuels. For economic reasons, the cost of transportation combined with not exceeding the regional wood resources available place limits on the maximum size of individual wood processing installations, which in turn prevents attaining the economics of scale available from fossil fuels. This barrier can be partially offset with the use of pelletization technology, which increases transportation distances while stabilizing costs.

Looking more specifically at the city of Lebanon, as of 1980 a 10 MW wood unit could be 99.2 percent powered by unutilized wood residue produced in this region. Using the minimum biomass yield of 1.72 tons per acre, the study region could support a 50 MW unit in addition to meeting the wood industry's demand. A 50 MW unit would totally satisfy Lebanon's power needs and would lower their incremental costs, since Lebanon receives power from Laclede Electric, who receives power from Show-Me Electric, and who purchases power from Associated Electric Cooperative for 2.5¢/kwh.

BIBLIOGRAPHY

1. Tillman, David A. Wood as an Energy Resource. Academic Press, Inc., 111 Fifth Avenue, New York, N.Y. 1978.
2. Hamrick, Joseph T. Development of Wood as an Alternative Fuel for Large Power Generating Systems. Aerospace Research Corporation, Roanoke, VA, October 1978.
3. Spencer, John S. and Burton L. Essex. Timber In Missouri. Forest Service, U.S. Department of Agriculture, 1976.
4. Gallagher, Alissa M. "The Feasibility of Using Wood Wastes as Fuel for Steam Generators." Paper prepared for English 160, UMR, May 5, 1980.
5. Branch. "Solid Fuels Technology Workshop." Paper prepared for Georgia Office of Energy Resources, May 15, 1979, Georgia Institute of Technology.
6. Pierce, Bill. "Economic, Social, and Environmental Advantages of a Wood-Fired Electrical Cogeneration Plant for the Laclede Electric Cooperative." Paper prepared for Laclede Electric Cooperative, October, 1979, Elridge, Missouri.
7. Gordon, Judith and others. Commercialization of Wood Combustion Systems. MITRE Corporation, 1820 Dolly Madison Blvd., McLean, VA, August 1979.